Ecological Studies, Vol. 177

Analysis and Synthesis

Ecological Studies

Volumes published since 1998 are listed at the end of this book.

James R. Ehleringer Thure E. Cerling
M. Denise Dearing
Editors

A History of Atmospheric CO_2 and Its Effects on Plants, Animals, and Ecosystems

With 151 Illustrations

Springer

James R. Ehleringer
M. Denise Dearing
Department of Biology
University of Utah
Salt Lake City, UT 84112
USA

Thure E. Cerling
Department of Geology and Geophysics
and
Department of Biology
University of Utah
Salt Lake City, UT 84112
USA

Cover illustration: Illustrated are the changes in the atmospheric carbon dioxide concentrations over three time periods. The left plate shows long-term decreases in atmospheric carbon dioxide levels over the last 550 million years and the role of the biota in significantly decreasing carbon dioxide levels when plants invaded land. The middle plate shows the variations in carbon dioxide levels over the last 400,000 years. The right plate shows the imprint of humans over the last half century, increasing carbon dioxide levels significantly well above interglacial levels. Data are based on graphics in Chapters 1, 2, 4, and 5.

ISSN 0070-8356
ISBN 0-387-22069-0 Printed on acid-free paper.

Printed in the United States of America. (WBG/MVY)

9 8 7 6 5 4 3 2 SPIN 10940464

springer.com

Preface

Our planet's atmosphere is thought to have changed gradually and over a very wide range of CO_2 concentrations throughout history. From ancient atmospheric gases trapped in ice bubbles, we have strong evidence indicating that atmospheric CO_2 values reached minimum concentrations of approximately 180 parts per million during the Last Glacial Maximum, which was only 15,000 years ago. At the other extreme, calculations suggest that some 500+ million years ago the atmospheric CO_2 concentrations may have been about 4000 to 5000 parts per million. The available evidence suggests that the decline in atmospheric CO_2 over time has been neither steady nor constant, but rather that there have been periods in Earth's history when CO_2 concentrations have decreased and other periods in which CO_2 levels were elevated. The changes in atmospheric CO_2 concentrations over geological time periods are the result of biological, chemical, and geological processes. Biogeochemical processes play a significant role in removing organic matter from the active carbon cycle at Earth's surface and in forming carbonates that are ultimately transported to the continental plates, where they become subducted away from Earth's surface layers.

Atmospheric CO_2 concentrations rose rapidly as Earth transitioned out of the last Ice Age, and the atmosphere has changed dramatically since the dawn of the Industrial Age with especially large increases over the past five decades. Both terrestrial and aquatic plant life significantly influence the sequestration of atmospheric CO_2 into organic matter. There is now substantial evidence to show that these terrestrial and marine photosynthetic organisms currently play a major

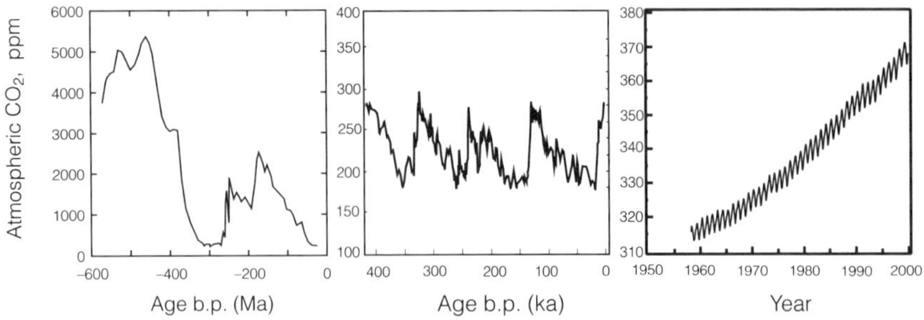

Illustration. Illustrated are the changes in the atmospheric carbon dioxide concentrations over three time periods. The left plate shows long-term decreases in atmospheric carbon dioxide levels over the last 550 million years and the role of the biota in significantly decreasing carbon dioxide levels when plants invaded land. The middle plate shows the variations in carbon dioxide levels over the last 400,000 years. The right plate shows the imprint of humans over the last half century, increasing carbon dioxide levels significantly well above interglacial levels. Data are based on graphics in Chapters 1, 2, 4, and 5.

role in reducing the rate of atmospheric CO_2 increase. It is thought that, in turn, and over much longer time periods, atmospheric CO_2 concentrations greatly affected both the evolution and the functioning of biota.

In this volume, we explore interactions among atmospheric CO_2, ecosystem processes, and the evolution and functioning of biological organisms. The motivation for this volume originated from a Packard Foundation award to the University of Utah to promote basic interdisciplinary research involving geochemistry, paleontology, and biology and interactions among atmospheric CO_2, plants, and animals. More specifically, this volume is the result of a symposium that expanded on the original interdisciplinary research by bringing together a diverse scientific audience to address these interactions from an even broader perspective. Too often by working in separate disciplines, we fail to realize our common ground and overlapping interests. In this volume, we try to capture some of the excitement revolving around the known changes in atmospheric CO_2, how these changes in our atmosphere have influenced biological processes across our planet, and in turn how biological processes have produced feedback to the atmosphere and thereby have influenced the rate of change in atmospheric CO_2.

The first section of this volume focuses both on understanding processes that have influenced atmospheric CO_2 and on quantitatively documenting the known variations in atmospheric CO_2 over the past several hundred million years. The chapters include coarse-scale calculations of the possible ranges of atmospheric CO_2 based on geological proxies as well as fine-scale, high-precision measurements based on ice cores over the past four hundred thousand years. We complete this documentation with direct observations over the past five decades from the longest continuously running atmospheric measurement program to date.

In the second section, we examine how changes in atmospheric CO_2 have influenced the evolution and expansion of not only terrestrial plants but also animals, including humans. Here we see that there is strong physiological, ecological, and evolutionary evidence suggesting that changes in atmospheric CO_2 have had both direct and indirect effects on the kinds of plant taxa that dominate Earth's surface. Certainly, atmospheric CO_2 concentrations influence primary productivity, but apparently they also influence the abundance of different types of plants. In turn, these changes in the food supply are likely to have influenced the evolution of herbivorous animal systems. Paleontological studies provide convincing evidence of the changes in mammalian taxa, with stable isotope analyses providing information about the dietary preferences of different mammalian herbivore lineages.

The third and fourth sections of this volume explore the functioning of historical and modern ecosystems and, in particular, how the structure and functioning of current ecosystems are expected to change under future elevated atmospheric CO_2 concentrations. In these sections, we focus on understanding carbon sequestration patterns of terrestrial ecosystems and how they are influenced by interactions with other aspects of the climate and physical environment. Terrestrial ecosystems cannot be fully grasped without an understanding of the animal herbivores that selectively graze on different vegetation components. Elevated atmospheric CO_2 concentrations are expected to impact herbivorous animal systems indirectly through the effect of changes in food quality on herbivory. The work presented in these sections points to the need for bringing together plant, animal, and climate studies if we are to try, ultimately, to predict the consequences of changes in atmospheric CO_2 on the functioning of future terrestrial landscapes.

This volume is an effort to bridge disciplines, bringing together the different interests in atmospheric science, geochemistry, biology, paleontology, and ecology. It is difficult to understand historical changes in atmospheric CO_2 without appreciating the role of plants in modifying soil chemistry and photosynthesis as a carbon-sequestering process. Our understanding of the future atmosphere, and therefore our future climate system, will hinge on having knowledge of the carbon-sequestering capacities of future vegetation, especially in light of changes in the thermal environment and the nutrients required for sustained plant growth.

We thank the Packard Foundation for its generous support and for promoting interdisciplinary science.

James R. Ehleringer
Thure E. Cerling
M. Denise Dearing

Contents

Contributors

Bacastow, Robert B.	Scripps Institution of Oceanography, University of California, San Diego, CA 92093-0244, USA
Barnola, Jean-Marc	CNRS Laboratoire de Glaciologie et Géophysique de l'Environnement (LGGE), Grenoble, France
Beerling, David J.	Department of Animal and Plant Sciences, University of Sheffield, Sheffield S10 2TN, UK
Berner, Robert A.	Department of Geology and Geophysics, Yale University, New Haven, CT 06520, USA
Bhattacharya, S.K.	Physical Research Laboratory, Navarangpura, Ahmedabad 380 009, India
Blunier, Thomas	Climate and Environmental Physics, Physics Institute, University of Bern, Switzerland

Braswell, Bobby H.	Morse Hall, University of New Hampshire, 39 College Road, Durham, NH 03824-3525, USA
Cerling, Thure E.	Department of Geology and Geophysics and Department of Biology, University of Utah, 257 South 1400 East, Salt Lake City, UT 84112, USA
Chamberlain, C. Page	Department of Geological and Environmental Sciences, Stanford University, Stanford, CA 94305, USA
Chapman, David S.	Room 719, Department of Geology and Geophysics, University of Utah, 135 S 1460 E, Salt Lake City, UT 84112-0111, USA
Churkina, Galina	Max Planck Institute for Biogeochemistry, Carl-Zeiss-Promenade 10, 07745 Jena, Germany
Dearing, M. Denise	Department of Biology, University of Utah, 257 South 1400 East, Salt Lake City, UT 84112, USA
Ehleringer, James R.	Department of Biology, University of Utah, 257 South 1400 East, Salt Lake City, UT 84112, USA
Enting, Ian	CSIRO Atmospheric Research, Private Bag 1, Aspendale, Victoria 3195, Australia
Etheridge, David	CSIRO Atmospheric Research, Private Bag 1, Aspendale, Victoria 3195, Australia
Francey, Roger	CSIRO Atmospheric Research, Private Bag 1, Aspendale, Victoria 3195, Australia
Freeman, Katherine H.	Department of Geosciences, Pennsylvania State University, University Park, PA 16802, USA

Ghosh, Parthasarathi	Geological Studies Unit, Indian Statistical Institute, 203, B.T. Road, Calcutta 700 035, India
Ghosh, Prosenjit	Physical Research Laboratory, Navarangpura, Ahmedabad 380 009, India
Harris, John M.	George C. Page Museum of La Brea Discoveries, 5801 Wilshire Boulevard, Los Angeles, CA 90036, USA
Harris, Robert N.	Room 719, Department of Geology and Geophysics, University of Utah, 135 South 1460 East, Salt Lake City, UT 84112, USA
Heimann, Martin	Max Planck Institute for Biogeochemistry, Jena, Germany
Huxman, Travis E.	Ecology and Evolutionary Biology, University of Arizona, Tucson, AZ 85721-0088, USA
Joyce, Linda A.	Rocky Mountain Research Station, USDA Forest Service, Fort Collins, CO 80526, USA
Keeling, Charles D.	Scripps Institution of Oceanography, University of California, San Diego, CA 92093-0244, USA
Leafloor, James O.	Canadian Wildlife Service, Winnipeg R3C 4W2, Canada
Leakey, Meave G.	The National Museums of Kenya, P.O. Box 40658, Nairobi, Kenya
Lindroth, Richard L.	237 Russell Laboratories, Department of Entomology, University of Wisconsin, Madison, WI 53706, USA
Lund, Christopher P.	Carnegie Institution of Washington, Institute for Global Ecology, 260 Panama Street, Stanford, CA 94305, USA

Lupia, R.	Sam Noble Oklahoma Museum of Natural History and School of Geology and Geophysics, University of Oklahoma, 2401 Chautauqua Avenue, Norman, OK 73072, USA
MacFadden, Bruce J.	Florida Museum of Natural History, P.O. Box 112710, Gainesville, FL 32611, USA
McElwain, Jennifer C.	Department of Geology, Field Museum of Natural History, 1400 South Lake Shore Drive, Chicago, IL 60605, USA
McMurtrie, Ross E.	School of Biological, Earth and Environmental Sciences, University of New South Wales, Sydney 2052, Australia
McWilliams, Scott R.	Department Natural Resources Science, University of Rhode Island, 105 Coastal Institute in Kingston, Kingston, RI 02881, USA
Medlyn, Belinda E.	School of Biological Earth and Environmental Sciences, University of New South Wales, Sydney, New South Wales 2052, Australia
Meijer, Harro A.	Centre for Isotope Research, University of Groningen, Groningen, Netherlands
Monnin, Eric	Climate and Environmental Physics, Physics Institute, University of Bern, Switzerland
Norby, Richard J.	Environmental Sciences Division, Oak Ridge National Laboratory, Oak Ridge, TN 37831, USA
Pagani, Mark	Department of Earth and Planetary Sciences, Yale University, New Haven, CT 06520, USA

Piper, Stephen C.	Scripps Institution of Oceanography, University of California, San Diego, CA 92093-0244, USA
Rayner, Peter	CSIRO Atmospheric Research, Private Bag 1, Aspendale, Victoria 3195, Australia
Sage, Rowan F.	Department of Botany, University of Toronto, 25 Willcocks Street, Toronto, Ontario M5S 3B2, Canada
Schimel, David	National Center for Atmospheric Research, 1850 Table Mesa Drive, Boulder, CO 80305, USA
Shaw, M. Rebecca	Carnegie Institution of Washington, Institute for Global Ecology, 260 Panama Street, Stanford, CA 94305, USA
Trenbath, James	Max Planck Institute for Biogeochemistry, Carl-Zeiss-Promenade 10, 07745 Jena, Germany
Trudinger, Cathy	CSIRO Atmospheric Research, Private Bag 1, Aspendale, Victoria 3195, Australia
van der Merwe, Nikolaas J.	Archaeology Department, University of Cape Town, Rondebosch 7701, South Africa
Wahlen, Martin	Scripps Institution of Oceanography, University of California, San Diego, CA 92093-0244, USA
Waldbauer, Jacob R.	Department of Geological and Environmental Sciences, Stanford University, Stanford, CA 94305, USA
Ward, Joy K.	Department of Biology, University of Utah, 257 South 1400 East, Salt Lake City, UT 84112, USA

Whorf, Timothy P. — Scripps Institution of Oceanography, University of California, San Diego, CA 92093-0244, USA

Willis, K.J. — School of Geography and the Environment, University of Oxford, Mansfield Road, Oxford OX1 3PS, UK

Wullschleger, Stan D. — Environmental Sciences Division, Oak Ridge National Laboratory, Oak Ridge, TN 37831-6422, USA

1. The Rise of Trees and How They Changed Paleozoic Atmospheric CO_2, Climate, and Geology

Robert A. Berner

1.1 Introduction

Large vascular plants with deep, extensive root systems arose and spread over the continents starting about 380 million years ago during the Devonian Period. Previously there were only bryophytes, algae, and small vascular plants restricted to the edges of water courses (Gensel and Edwards 2001). Large plants are important because their vast root systems produce a larger interface between the geosphere and the biosphere than do the more primitive species, where the plant/mineral interface is greatly reduced. This large interface allows plants to take up nutrients more rapidly, to grow bigger and faster (Algeo and Scheckler 1998), and to accelerate mineral weathering (Berner 1998). In addition, the larger plants, upon death, supply a much greater mass of organic matter for burial in sediments. Because of these effects, the rise of large vascular plants brought about a dramatic change in the level of atmospheric CO_2, the climate, and the formation of carbon-rich deposits (coal) during the late Paleozoic.

1.2 Plants, Weathering, and CO_2

The level of CO_2 is controlled, on a long-term, multimillion-year timescale, by two carbon cycles: the silicate-carbonate cycle and the organic matter cycle. The silicate-carbonate cycle can be represented succinctly by the following reactions (stated in words by Ebelmen 1845 but symbolized by Urey 1952):

$$CO_2 + CaSiO_3 \rightarrow CaCO_3 + SiO_2 \quad (1.1)$$

$$CO_2 + Mg\ SiO_3 \rightarrow MgCO_3 + SiO_2 \quad (1.2)$$

The arrows, from left to right, refer to all Ca-Mg silicate weathering (the silicate formulae are generalized) plus sedimentation of marine carbonates. These two weathering reactions summarize many intermediate steps, including photosynthetic fixation of CO_2, root/mycorrhizal respiration, organic litter decomposition in soils, the reaction of carbonic and organic acids with primary silicate minerals thereby liberating cations to solution, the conversion of CO_2 to HCO_3^- in soil and ground water, the flow of riverine Ca^{++}, Mg^{++}, and HCO_3^- to the sea, and the precipitation of Ca-Mg carbonates in bottom sediments. (The reactions, from right–to–left, represent thermal decomposition of carbonates at depth resulting in degassing of CO_2 to the atmosphere and oceans by diagenesis, metamorphism, and volcanism.)

Plants accelerate the rate of weathering and liberation of Ca^{++}, Mg^{++}, and HCO_3^- to solution in the following ways:

1. Rootlets (+ symbiotic microflora) with high surface area secrete organic acids/chelates, which attack primary minerals in order to gain nutrients (in this case Ca and Mg).
2. Organic litter decomposes to carbonic and organic acids providing additional acid for mineral dissolution.
3. Plants recirculate water via transpiration and thereby increase water/mineral contact time.
4. Plants anchor clay-rich soil against erosion allowing retention of water and continued dissolution of primary minerals between rainfall events.

Based on present-day field studies of the quantitative effects of plants on weathering rate (Drever and Zobrist 1992; Arthur and Fahey 1993; Bormann et al. 1998; Moulton, West, and Berner 2000), these effects combine to accelerate silicate weathering rates by a factor of approximately 2 to 10. When the field results are applied to global carbon cycle modeling, the calculated effect on atmospheric CO_2, due to the rise of trees during the Paleozoic Era, turns out to be dramatic (Fig. 1.1).

1.3 Plants, the Organic Cycle, and CO_2

The organic cycle can be represented succinctly by the following reaction (stated in words by Ebelmen 1845):

$$CO_2 + H_2O \rightarrow CH_2O + O_2 \quad (1.3)$$

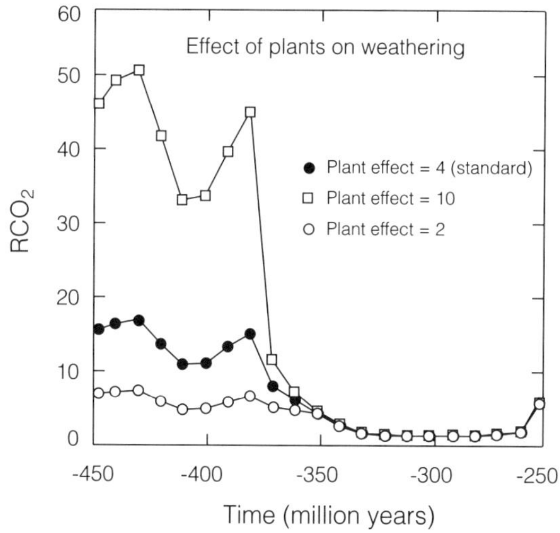

Figure 1.1. Plot of RCO_2 versus time based on GEOCARB III modeling (Berner and Kothavala 2001). The three curves illustrate sensitivity to the quantitative effect of the rise of large land plants on weathering rate. RCO_2 is the ratio of mass of carbon dioxide in the atmosphere at a past time divided by the present preindustrial mass.

The reaction, from left to right, refers to the burial of organic matter in sediments. This represents a net excess of global photosynthesis over respiration and is a major sink for atmospheric CO_2. (From right to left, the reaction refers to weathering of old organic matter (kerogen) on the continents or thermal decomposition of organic matter upon deep burial combined with the oxidation of reduced gases emitted to the atmosphere and oceans.)

Enhanced burial of organic matter occurred after the rise of large land plants. This is because of the production of a new compound, lignin, which is relatively resistant to biodecomposition. The burial of lignin-derived humic material, and other plant-derived microbiallyresistant substances, in terrestrial and coastal swamps and in the oceans (after transport there via rivers) resulted in not only large increases in the global burial of organic matter (Holland 1978; Berner and Canfield 1989) but also the formation of vast coal deposits of the Carboniferous and Permian periods (Bestougeff 1980). In fact, production and preservation of terrestriallyderived organic debris was so large that it may have dominated over the burial of marine-derived organic matter at this time (Berner and Raiswell 1983; Broecker and Peacock 1999).

1.4 Carbon Cycle Modeling

A carbon cycle model, GEOCARB, has been constructed for calculating weathering rates, carbon burial rates, degassing rates, and levels of atmospheric CO_2

over Phanerozoic time (Berner 1994; Berner and Kothavala 2001). The model quantifies the effects of changes in climate, tectonics, paleogeography, paleohydrology, solar evolution, and plant evolution on the rates of silicate weathering. Sensitivity analysis indicates that for the Paleozoic Era, the most important factor affecting CO_2 was the rise of land plants. Feasible variations in tectonic, paleogeographic, and other factors result in CO_2 variations that are far less divergent than those brought about by plant evolution. Sensitivity of atmospheric CO_2 to the value used for the acceleration of Ca-Mg silicate weathering by plants, based on the results of modern plant-weathering studies mentioned above, is shown in Fig. 1.1. Use of a factor of 4, based on our own modern plant studies in Iceland (Moulton, West, and Berner 2000) results in the plot of CO_2 versus time shown in Fig. 1.2.

The theoretical values of Fig. 1.2 agree well with those obtained for paleo-CO_2 by independent methods (Mora, Driese, and Colarusso 1996; McElwain and Chaloner 1995; Mora and Driese 1999; Cox, Railsback, and Gordon 2001). The figures show that tree-accelerated weathering brought about a large drop in atmospheric CO_2. However, the cause of this drop in CO_2 is often misrepresented as resulting from simply an increased weathering carbon flux. Instead, the acceleration of weathering by plants was balanced by greenhouse-induced decel-

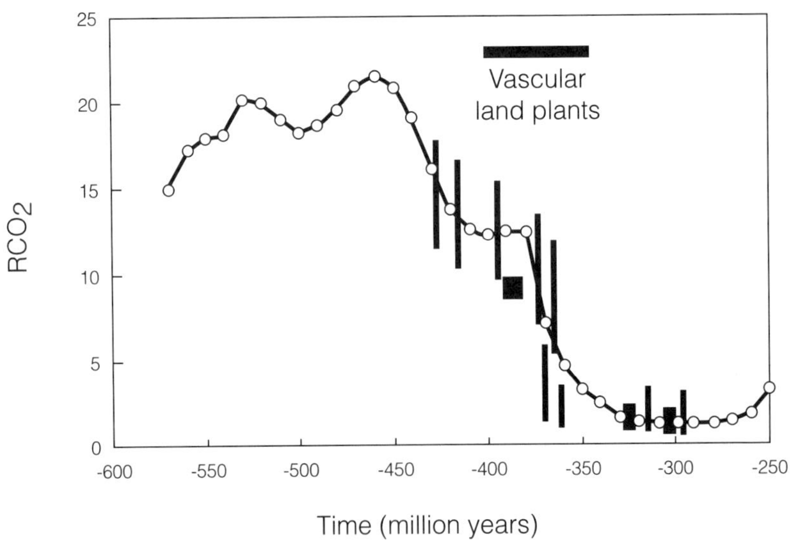

Figure 1.2. Plot of RCO_2 versus time based on GEOCARB III modeling for a fourfold effect (Moulton, West, and Berner 2000) of plants on weathering rate (line connecting dots). The superimposed vertical bars and larger squares represent independent estimates of paleo-CO_2 via the carbonate paleosol method (Mora, Driese, and Colarusso 1996; Mora and Driese 1999; Cox, Railsback, and Gordon 2001), and the stomatal ratio method (McElwain and Chaloner 1995), respectively.

eration of weathering due to falling CO_2, and this resulted in the stabilization of CO_2 at a series of lower levels.

Further drop of CO_2 into the Carboniferous and Permian periods is due to the increased burial of organic matter accompanying the production of bioresistant organic matter by large woody land plants. This is shown in Fig. 1.3 for the result of maintaining the carbon isotopic composition of the oceans and atmosphere constant with time and comparing the result to that based on recorded carbon isotopic data (Berner 1994).

Organic burial rate in GEOCARB modeling is calculated mainly from the carbon isotopic record, with elevated oceanic $^{13}C/^{12}C$ representing faster removal of ^{13}C-impoverished carbon from seawater and the atmosphere due to greater photosynthesis and burial. The additional drop in CO_2 is due to a rise in oceanic $^{13}C/^{12}C$ to high values between 400 and 250 million years ago. (Rapid isotopic equilibration of carbon isotopes between the oceans and the atmosphere is assumed so that burial of plant-derived organic matter on land can affect the $^{13}C/^{12}C$ of the oceans via atmospheric and riverine transport.)

1.5 Climatic and Geological Consequences

The large decrease in atmospheric CO_2 beginning in the Devonian and continuing into the Carboniferous (see Fig. 1.2) correlates with the initiation of continental glaciation. The Parmo-Carboniferous glaciation was the longest and most extensive glaciation of the entire Phanerozoic (Crowley and North 1991),

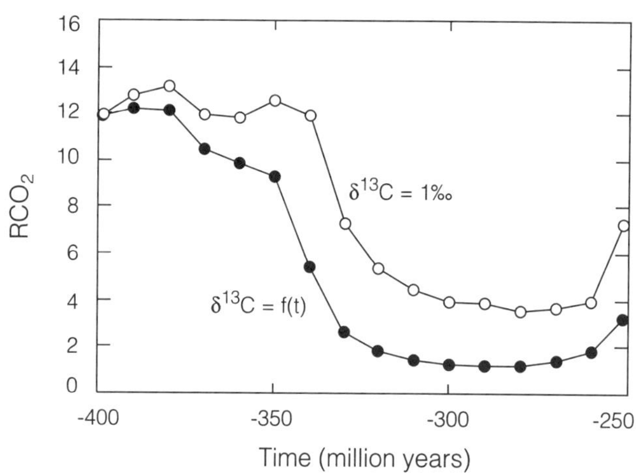

Figure 1.3. Plot of RCO_2 versus time showing the effect of the variation of $^{13}C/^{12}C$ during the Permian and Carboniferous. The lower curve is derived from GEOCARB II modeling (Berner 1994) based on published carbon isotope data. The upper curve represents the effect of holding the value of $\delta^{13}C = 1‰$ for the entire period.

lasting about 80 million years and extending at times from the South Pole to as far north as 30°S. Coincidence of this glaciation with a drop in CO_2 (Crowley and Berner 2001) strongly suggests that CO_2, by way of the atmospheric greenhouse effect, was a major factor in bringing about the glaciation. Although the waxing and waning of the glaciers during this period could have been caused by, for example, variations in Earth's orbit (Crowley and North 1991), the most reasonable explanation is that the lowering of CO_2 to a level sufficient for the year-round accumulation of snow and ice at high latitudes allowed glaciation to be initiated and to occur on a continental scale. This lowering of CO_2 resulted primarily from the rise of large vascular land plants.

The massive burial of plant-derived organic matter also led to the formation of vast coal deposits during the late Paleozoic Era. Permian and Carboniferous coals are much more abundant than coals from any other period (Bestougeff 1980), in spite of the fact that these coals are much older and have been subjected to loss by erosion for a much longer time. This must mean that original production and/or preservation for burial was unusually large. Perhaps preservation was enhanced by a lag in the evolution of lignin-decomposing microorganisms (Robinson 1991). However, the coal abundance also owes something to paleogeography. During the Permian and Carboniferous periods there was one large continent (Pangaea) with vast lowlands under wet climates that were topographically and geomorphically suitable for the growth of swamp plants and the preservation of their debris. Thus, the Permo-Carboniferous increase in organic burial probably was due to both biological and geological factors.

1.6 Summary

The effect on atmospheric CO_2 of the spread of large vascular plants beginning in the Devonian was twofold. First, the uptake of nutrients from rocks resulted in the enhanced weathering of Ca-Mg silicate minerals resulting in the transfer of CO_2 from the atmosphere to marine Ca-Mg carbonates. Second, the rise of trees caused the production of large amounts of microbially resistant organic matter, in the form of lignin, which resulted in increased sedimentary organic burial and further CO_2 removal. These changes in the carbon cycle led to a large drop in atmospheric CO_2, massive long-term glaciation, and the formation of vast coal deposits. Computer models of the long-term carbon cycle, based partly on field studies of the effects of plants on modern weathering, have been employed to calculate atmospheric CO_2 levels over this time period; the resulting values agree with independent estimates.

Acknowledgments. This research was supported by DOE Grant DE-FG02-01ER15173 and NSF Grant EAR 0104797.

References

Algeo, T.J., and S.E. Scheckler. 1998. Terrestrial-marine teleconnections in the Devonian: Links between the evolution of land plants, weathering processes, and marine anoxic events. *Philosophical Transactions of the Royal Society,* Series B, 353:113–28.

Arthur, M.A., and T.J. Fahey. 1993. Controls on soil solution chemistry in a subalpine forest in north-central Colorado. *Soil Science Society of America Journal* 57:1123–30.

Berner, R.A. 1994. GEOCARB II: A revised model of atmospheric CO_2 over Phanerozoic time. *American Journal of Science* 294:56–91.

———. 1998. The carbon cycle and CO_2 over Phanerozoic time: The role of land plants. *Philosophical Transactions of the Royal Society,* Series B, 353:75–82.

Berner, R.A., and D.E. Canfield. 1989. A new model for atmospheric oxygen over Phanerozoic time. *American Journal of Science* 289:59–91.

Berner, R.A., and Z. Kothavala. 2001. GEOCARB III: A revised model of atmospheric CO_2 over Phanerozoic time. *American Journal of Science* 301:182–204.

Berner, R.A., and R. Raiswell. 1983. Burial of organic carbon and pyrite sulfur in sediments over Phanerozoic time: A new theory. *Geochimica et Cosmochimica Acta* 47: 855–62.

Bestougeff, M.A. 1980. Summary of world coal resources. Twenty-Sixth International Geological Congress of Paris, Colloquiem C-2, 35:353–66.

Bormann, B.T., D. Wang, F.H. Bormann, G. Benoit, R. April, and R. Snyder. 1998. Rapid plant-induced weathering in an aggrading experimental ecosystem. *Biogeochemistry* 43:129–55.

Broecker, W.S., and S. Peacock. 1999. An ecologic explanation for the Permo-Triassic carbon and sulfur isotope shifts. *Global Biogeochemical Cycles* 13:1167–72.

Cox, J.E., L.B. Railsback, and E.A. Gordon. 2001. Evidence from Catskill pedogenic carbonates for a rapid large Devonian decrease in atmospheric carbon dioxide concentrations. *Northeastern Geology and Environmental Science* 23:91–102.

Crowley, T.J., and R.A. Berner. 2001. CO_2 and climate change. *Science* 292:870–72.

Crowley, T.J., and G.R. North. 1991. *Paleoclimatology.* New York: Oxford University Press.

Drever, J.I., and J. Zobrist. 1992. Chemical weathering of silicate rocks as a function of elevation in the southern Swiss Alps. *Geochimica et Cosmochimica Acta* 56:3209–16.

Ebelmen, J.J. 1845. Sur les produits de la decomposition des especes minerales de la famille des silicates. *Annales des Mines* 7:3–66.

Gensel, P.G., and D. Edwards. 2001. *Plants invade the land: evolutionary and environmental perspectives.* New York: Columbia University Press.

Holland, H.D. 1978. *The chemistry of the atmosphere and the oceans.* New York: Wiley.

McElwain, J.C., and W.G. Chaloner. 1995. Stomatal density and index of fossil plants track atmospheric carbon dioxide in the Palaeozoic. *Annals of Botany* 76:389–95.

Mora, C.I., and S.G. Driese. 1999. Palaeoenvironment, palaeoclimate, and stable carbon isotopes of Palaeozoic red-bed palaeosols, Appalachian Basin, USA, and Canada. Special Publications of the International Association of Sedimentologists 27:61–84.

Mora, C.I., S.G. Driese, and L.A. Colarusso. 1996. Middle and late Paleozoic atmospheric CO_2 levels from soil carbonate and organic matter. *Science* 271:1105–107.

Moulton, K.L., J.A. West, and R.A. Berner. 2000. Solute flux and mineral mass balance approaches to the quantification of plant effects on silicate weathering. *American Journal of Science* 300:539–70.

Robinson, J.M. 1991. Land plants and weathering. *Science* 252:860.

Urey, H.C. 1952. *The planets: Their origin and development.* New Haven: Yale University Press.

2. Atmospheric CO_2 During the Late Paleozoic and Mesozoic: Estimates from Indian Soils

Prosenjit Ghosh, S.K. Bhattacharya, and Parthasarathi Ghosh

2.1 Introduction

Carbon dioxide (CO_2) and water vapor are major greenhouse gases in Earth's atmosphere that control the planet's surface temperature (Houghton and Wood 1989). Variations in CO_2 levels can lead to major changes in climate, surface processes, and biota. For example, over the past two centuries the combustion of fossil fuels has raised the atmospheric CO_2 level from about 275 ppmV (Barnola et al. 1987) to the current level of 365 ppmV (Keeling 1994); this has caused already discernible increases in Earth's near-surface temperature (IPCC 1990). Over a longer timescale, analysis of trapped gases in the polar icecores (Barnola et al. 1987) showed that during the past 150,000 years CO_2 levels have oscillated between approximately 200 and 300 ppmV. Knowledge of past carbon dioxide levels and associated paleoenvironmental and paleoecological changes is useful for predicting future consequences of the current increase in atmospheric CO_2.

While it is possible to directly measure the CO_2 content of the late Pleistocene and Holocene atmosphere, direct estimation of CO_2 levels in the atmosphere is not possible for the pre-Pleistocene epoch.. Several indirect methods (proxy) to estimate the paleo-CO_2 concentration in the atmosphere have been proposed: for example, carbon isotopic analysis of pedogenic carbonates, stomatal index count of fossil leaves, and isotopic composition of marine sedimentary carbon and boron from carbonate fossils (Cerling 1991; McElwain and Chaloner 1996; Ekart et al. 1999; Ghosh, Ghosh, and Bhattacharya 2001; Crowley and Berner 2001). The $\delta^{13}C$ of pedogenic carbonates provide the best $_pCO_2$ estimates for the pre-Tertiary (Royer, Berner, and Beerling 2001). The experimental proxies play an important role in putting constraints on the theoretical models of carbon cycle based on mantle evolution (Tajika and Matsui 1992) and biological and tectonic changes in the past (Berner 1994; Berner and Kothavala 2001), which provide estimates of the atmospheric [CO_2] during the Phanerozoic.

How do paleosol carbonates record the past CO_2 level in the atmosphere? These carbonates are precipitated in the root zone of plants when groundwater supersaturated with carbonate ions can release CO_2 by some process. They are common in regions receiving an annual rainfall of less than 800 mm. The addition of CO_2 in the groundwater during plant respiration and subsequent evaporation and transpiration of water from a plant can induce supersaturation and carbonate precipitation. Paleosol carbonates record the isotopic composition of local soil CO_2, which primarily reflects the type of vegetation (fraction of C_3 and C_4 plants) in the ecosystem (Cerling 1984). The soil CO_2 is a mixture of two components: plant respired CO_2 and atmospheric CO_2. It is important to note that $\delta^{13}C$ of atmospheric CO_2 (about-7‰) is very different from that of soil CO_2 (about-25‰). Atmospheric CO_2 penetrates inside the soil by diffusion and mixes with the soil CO_2 leading to its isotopic change. Today, except for ecosystems with very low productivity, such as deserts, the atmospheric contribution to total soil CO_2 is very small because of very low concentration of CO_2 in the modern atmosphere. However, high atmospheric CO_2 can make a significant contribution to total soil CO_2; in times when few or no C_4 plants were present, this contribution could result in significant isotopic shifts in the $\delta^{13}C$ of soil carbonate precipitated in isotopic equilibrium with soil CO_2. Therefore, $\delta^{13}C$ of pedogenic carbonates can act as a proxy indicator for [CO_2] variations in geologic past (Cerling 1991; Ghosh, Bhattacharya, and Jani 1995; Mora, Driese, and Colarusso 1996; Ekart et al. 1999).

To understand the past [CO_2] variations quantitatively, Berner (1994) proposed the GEOCARB II model based on equations governing the CO_2 outgassing and CO_2 consumption through weathering; he then improved it further in GEOCARB III (Berner and Kothavala 2001). Both of these models predict that in the early Phanerozoic (550 Ma) the [CO_2] was 20 times the present atmospheric level (PAL). Subsequently, the $_pCO_2$ declined in the middle and late Paleozoic (450–280 Ma) to reach a minimum value (approximately similar to the PAL) at about 300 million years ago. The period from 300 to 200 million years ago was again characterized by a rapid rise in the [CO_2] when it increased to 5 times the PAL. Next came a gradual decline, down to the PAL, with a small peak in the early Tertiary. Such large changes in [CO_2] during the geologic past must have had significant influences on the climate, biota, and surface processes of Earth (Berner 1991 and 1997; Mora, Driese, and Colarusso 1996).

The motivation for the present study came from the discovery of well-developed and well-preserved paleosols in the Gondwana sediments of central India; these paleosols cover the period of significant CO_2 change mentioned above in well-spaced intervals. The stable isotopic composition of pedogenic carbonates formed in these paleosols was investigated to decipher the CO_2 concentrations and to compare them with those predicted by the Berner-model.

2.2 Description of Paleosols from Central India

The Gondwana sediments and the overlying Deccan trap of the Satpura basin (Fig. 2.1) of central India range in age from the Permo-Carboniferous to the uppermost Cretaceous.

The thickness of the whole sedimentary succession is about 5 km (Fig. 2.2). The sediments comprise alternate layers of coarse clastics (sandstones along with extrabasinal conglomerates) and fine clastics (red mudstone/carbonaceous shale/white mudstone). The basal unit of this succession is a Permo-Carboniferous glaciolacustrine deposit called the Talchir Formation. The formations overlying the Talchir represent several episodes of fluvial, lacustrine, and alluvial deposition (Robinson 1967; Casshyap and Tewari 1988; Casshyap and Qidwai 1971; Casshyap, Tewari, and Khan 1993; Veevers and Tewari 1995; Ghosh 1997). Occurrences of fossil vertebrates (Chatterjee and Roychowdhury 1974; Mukherjee and Sengupta 1998; Bandyopadhyay and Sengupta 1998) and freshwater

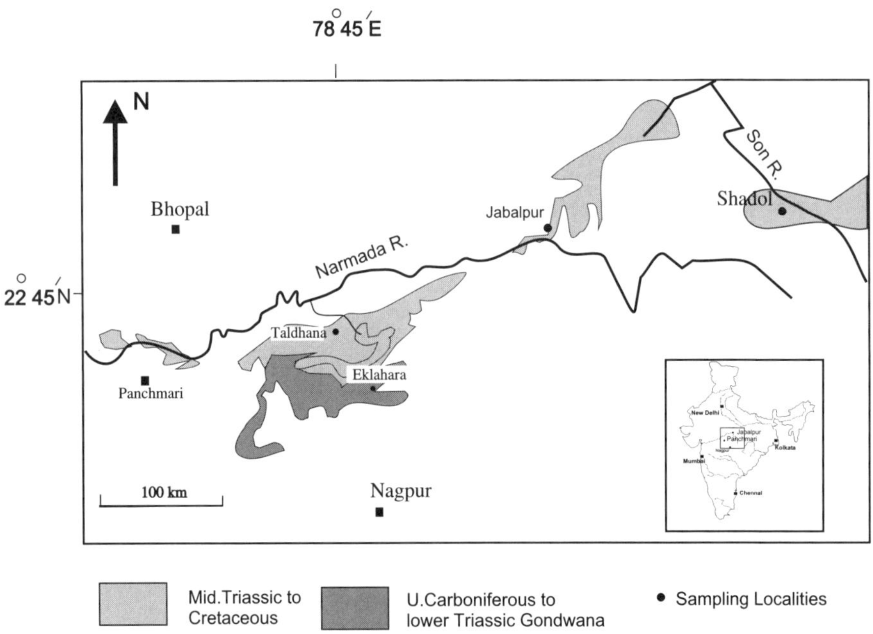

Figure 2.1. Sketch of paleosol locations in the geological map of the Satpura basin and Son Valley basin of the Gondwana supergroup in central India.

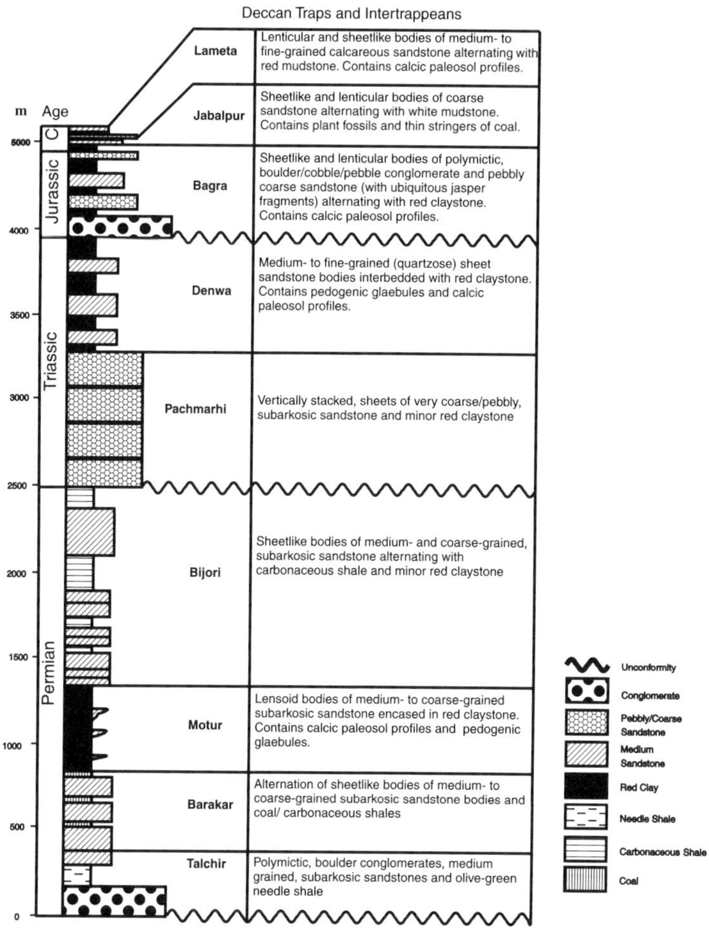

Figure 2.2. Generalized lithostratigraphy of the Gondwana succession of the Satpura basin. (From Ghosh, Ghosh, and Bhattacharrya 2001, with permission from Elsevier Science.)

bivalves, coupled with such evidences of pedogenesis as the presence of rootlet horizons, paleosol profiles, and so forth (Ghosh, Bhattacharya, and Jani 1995, Ghosh, Rudra, and Maulik 1998; Ghosh 1997; Tandon et al. 1995, 1998) give credence to the alluvial origin of these sediments.

In four litho-formations (the Motur, Denwa, Bagra, and Lameta) of the Satpura basin and one ormation (the Tiki) of the Son Valley basin, preserved calcic paleosols have been identified and characterized (Ghosh, Bhattacharya, and Jani 1995; Ghosh 1997, Ghosh, Rudra, and Maulik 1998; Tandon et al. 1995, 1998; Andrews, Tandon, and Dennis 1995). Stable isotopic compositions of the pedogenic carbonates and associated organic matters from the Motur, Bagra, Denwa, and Lameta formations were investigated in an earlier study (Ghosh,

Ghosh, and Bhattacharya 2001). This chapter describes a more refined analysis of the earlier results and presents an additional analysis of soils from the Tiki Formation. A brief description of the five formations is given below.

2.2.1 Motur Formation

The Motur Formation is a 700 m thick succession of fluvial channel sandstone bodies alternating with floodplain complexes made up of red claystone and thin sandy splay deposits. In the middle part of the Motur succession a few calcic paleosols occur within the floodplain deposits. These paleosols are characterised by three to four vertically superposed distinct pedo-horizons forming paleosol profiles (Fig. 2.3A).

Two types of profiles can be recognized. One type is around 50 cm thick, whereas the other is thicker (3–4 m). The thinner variety comprises an uppermost horizon (3–5 cm thick) of coalesced platy globules overlying a 10 to 30 cm thick horizon of closely spaced vertically oriented rhizocretions (Fig. 2.3B).

The rhizocretion horizon grades downward to a horizon with profuse subspherical globules that overlies a gleyed horizon. The uppermost horizon of fused platy globules is similar to the K horizon of modern aridisols whereas the zones of rhizocretions and globules can be compared with the Bk soil horizons (Soil Survey Staff 1975).

The thicker paleosol profiles show a meter-thick upper zone with a number of curved, mutually intersecting inclined surfaces (see Fig. 2.3A), coated with centimeter-thick carbonate layers, within a red claystone host. The underlying horizon is characterized by small, subvertical, dispersed, calcareous rhizocretions. Underneath this horizon is a thin (3–5 cm thick) horizon of subspherical globules overlying a gleyed horizon (~20 cm thick).

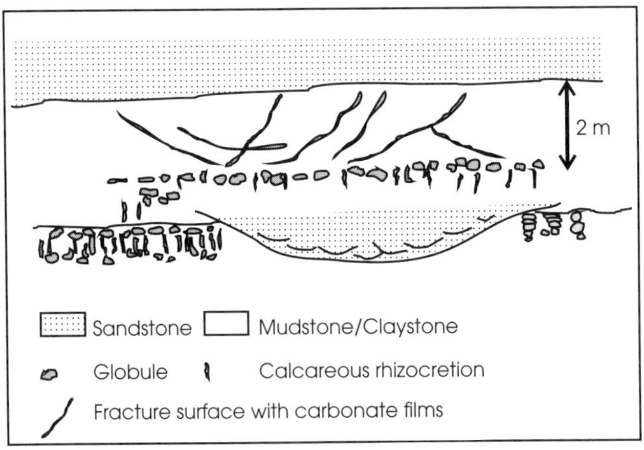

Figure 2.3A. Field sketch of the two vertically superposed calcic paleosol profiles of the Motur Formation, exposed near Eklahara colliery.

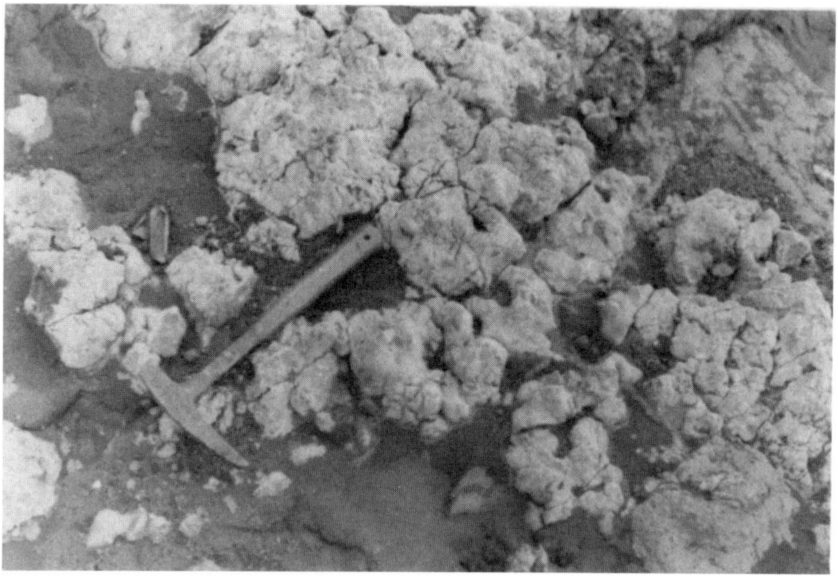

Figure 2.3B. Close-up of globular horizon (Motur Formation, near Eklahara colliery).

The curved inclined surfaces, within the clayey host, in the upper part of these paleosols resemble the zone of pedogenic slickensides (ss horizon) that develop in modern vertisols in response to shrinking and swelling of the soil clay matrix. The underlying horizons of rhizocretions and globules can be compared with modern Bk horizons. The field features of these paleosol profiles are similar to the Appalachian Paleozoic vertic paleosol profiles described by Driese and Mora (1993) and Mora et al. (1998).

A total of 21 samples were collected from the Motur Formation near Eklahara colliery (22°12'N, 78°41'E). Studies on the vertic paleosols have demonstrated that the globules occurring within the vertic horizons are enriched in ^{13}C compared to those occurring below the vertic horizons and compared to the rhizocretions in general (Driese and Mora 1993; Mora, Fastovsky, and Driese 1993). The cracks that develop in the vertic soils in response to the shrinking and swelling of the soil clay matrix possibly allow direct and nondiffusive penetration of the heavier atmospheric CO_2 deeper down the soil; hence, samples from this horizon may provide an incorrectly high estimate of the atmospheric CO_2. The vertic paleosols, therefore, were sampled for rhizocretions and globules from the horizons that occur considerably below the horizon of slickensides and above the gleyed zone. For the thinner paleosols, samples of rhizocretions and globules were collected from the basal part of the horizon of rhizocretions and the underlying horizon of globules. The horizons of platy globules were not sampled; thus, the studies avoided a possible anomalous large contribution of atmospheric CO_2 near the soil-to-atmosphere contact (Cerling 1984, 1991).

2.2.2 Denwa Formation

The Denwa Formation is about 600 m thick at its maximum. The lower half of the formation is characterized by a regular alternation of medium- to fine-grained, thick (3–15 m) fluvial channel sandstone bodies and floodplain deposits. The floodplain deposits comprise red mudstones intercalated with centimeter-to-decimeter thick, fine-grained sandstone layers. In contrast to the lower part of the formation, the upper half is fines-dominated. The red mudstones encase

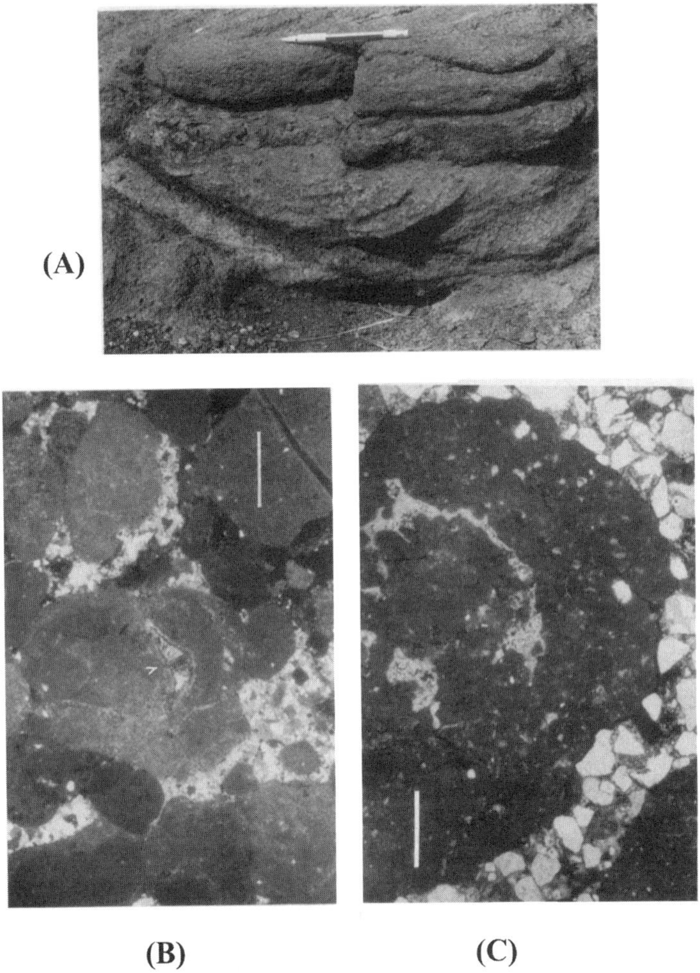

Figure 2.4. (**A**) Close-up of a trough cross-stratified channel-fill deposit, made up of pedogenic globules within the upper part of the Denwa Formation. (**B**) Photo-micrograph of pedogenic globules; note the spar-filled cracks within the calcareous matrix of the globules. Scale bar = 0.8 mm. (**C**) Photomicrograph of a globule; note the sharp and rounded outline of the globule and the spar-filled (lighter) circum-granular crack. Scale bar = 1 mm. (**C** from Ghosh, Ghosh, and Bhattacharrya 2001, with permission from Elsevier Science.)

fine-grained (fine sandstones and siltstones) point bar deposits that are 2 to 4 m thick and lenticular channel-fill bodies that are about 1 m thick. Detrital subspherical globules that are the size of pebbles to coarse sand constitute the bulk of these channel-fill bodies. These bodies are internally trough cross-stratified (Fig. 2.4A).

Microscopic observations of the detrital globules reveal that they are composed mostly of micrite and minor microspar with one or two sand-sized detrital quartz grains floating in the carbonate groundmass (Fig. 2.4B).

A number of globules show radial fractures filled with either blocky spars or barite. Spar-filled, circum-granular cracks also have been noted in some of the globules along with such pedogenic features as clotted micrite and corroded detrital quartz grains (Fig. 2.4B,C).

The field occurrence of paleosols in the Denwa Formation is limited, and only the widespread occurrence of detrital globules with pedogenic microfabrics provides indirect evidence of pedogenesis during the later part of the Denwa sedimentation. However, exposed paleosols with two distinct pedohorizons can be studied in a single exposure. The profile is more than 4 m thick. Its upper 2 to 3 m is characterized by a number of inclined and mutually intersecting nearly planer surfaces and dispersed globules the size of small pebbles to coarse sand and pale yellow blotches. Larger inclined features are about 5 m long and 5 to 7 cm thick, whereas the length of the smaller ones is 1 to 2 m and these are less than a centimeter thick. Large inclined features are filled with calcareous very fine sandstone. The surfaces of the smaller inclined surfaces have a polished appearance and are, at places, coated by 1 to 5 cm thick carbonate layers. This horizon passes gradually to an underlying horizon of numerous isolated pebble-sized globules, small (2–7 cm long) calcareous rhizocretions and pale yellow-gray mottles. The larger inclined features of the overlying horizon, however, cut across the lower horizon. The discordant relationship between the horizons and the large inclined features along with their sandy fills suggest that these are possibly desiccation cracks. The smaller inclined features can possibly be equated with the pedogenic slickensides noted in the upper part of the modern day vertisols (Soil Survey Staff 1975). The lower horizon represents the zone of carbonate accumulation at a deeper part of the soil. The macroscopic characters of the Denwa paleosol profiles are also similar to the Appalachian vertic paleosols (Driese and Mora 1993; Mora et al. 1998).

Nineteen samples of soil carbonates were collected from the formation (see Fig. 2.4D). The bulk of the samples comprise detrital pedogenic globules of the channel-fill bodies (between 22°38'N, 78°20'E and 22°35'N, 78°38'E). The rhizocretions and globules from the basal horizon of the paleosol near Taldhana village (22°37'N, 78°32'30"E) also were sampled.

2.2.3 Tiki Formation

The Tiki Formation is a 1200 m thick succession of fluvial channel sandstone bodies alternating with floodplain complexes made up of shale and thin sandy floodplain deposits. Floodplain deposits consist essentially of bright red colored

Figure 2.4D. A vertic paleosol profile of the Denwa Formation, exposed near Taldhana village showing a zone of well-developed vertical rhizocretions (R) overlying a zone of subspherical calcareous globules (G). Each division of scale bar is 50 cm.

clays, interbedded with soft, light colored sandstone. There are three different types of paleosols occurring in the area. Type I shows calcareous rhizocretions in a host of medium- to fine-grained, cross-stratified fluvial channel sandstone deposit (Fig. 2.5A).

Most of the concretions are rhizocretions with a few subspherical globules. The majority of the rhizocretions are subvertical, whereas the rhizocretions occurring near the top are typically subhorizontally aligned. Type II shows the development of calcareous globules and a few rhizocretions in the red mudstone/siltstone of the floodplain deposits associated with the fluvial channels (Fig. 2.5B).

The majority of the concretions are subspherical and platy globules, and, in contrast to type I, rhizocretions are rare. Type III paleosol (Fig. 2.5C) shows cross-bedded detrital calcareous concretions at the basal part of a sandy fluvial channel deposit. These are immature aridisols or calcisols with only a partially modified host at the base (C horizon) and a weakly developed Bk on top of it.

Nine samples of soil carbonates were collected from the fFormation. The bulk of the samples comprise detrital pedogenic globules of the channel-fill bodies, the rhizocretions, and globules from the basal horizon of the paleosol near Beohari village, 12 km north of Shadol district, Madhya Pradesh (21°N, 81°30 E; see Fig. 2.1).

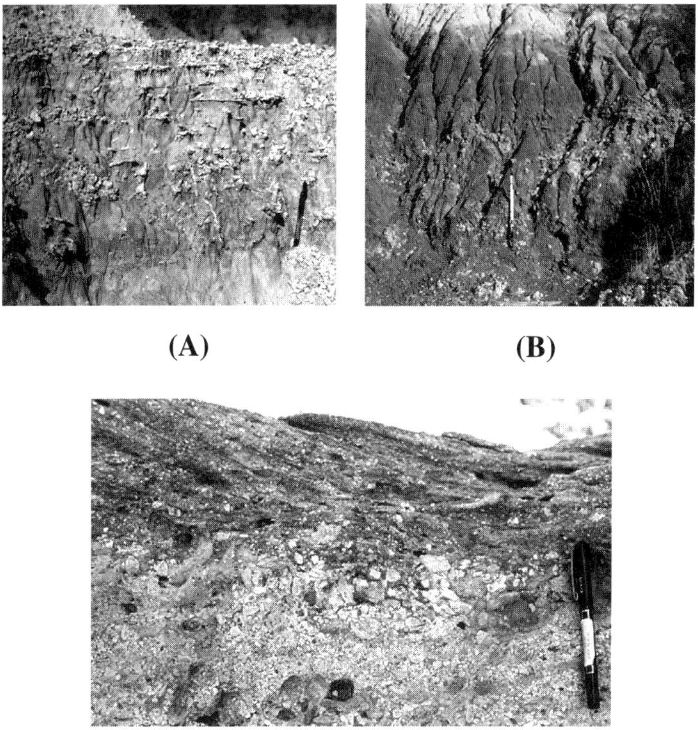

Figure 2.5. (**A**) Calcareous rhizocretions in a host of medium- to fine-grained, cross-stratified fluvial channel sandstone deposit. Most of the concretions are rhizocretions with a few subspherical globules. Majority of the rhizocretions are subvertical, whereas the rhizocretions occurring near the top are typically subhorizontally aligned. (**B**) Development of calcareous globules and a few rhizocretions in the red mudstone/siltstone of the floodplain deposits associated with the fluvial channels. Majority of the concretions are subspherical and platy globules (scale is 1.5 m in length). (**C**) Cross-bedded detrital calcareous concretions at the basal part of a sandy fluvial channel deposit.

2.2.4 Bagra Formation

The Bagra Formation is 250 to 500 m thick. Thick units of polymictic conglomerates and pebbly sandstones, interbedded with red claystone units, characterize this formation. The internal architecture of the clastic bodies indicates that most of them were deposited within the channels of braided streams, whereas the claystones were formed in the associated floodplains. The paleosol profiles are associated mainly with these floodplain deposits and are found in the lower half of the formation. At places, pedogenic modification extended up to the coarse-grained abandoned channels and the wings of the main channel

(A)

(B)

(C)

Figure 2.6 **(A)** Calcic paleosol profile of the Bagra Formation; note a horizon of coalesced globules in the upper part of the profile and a zone of well-developed, vertical rhizocretions in the lower part (hammer included for scale). **(B)** Bedding plane view of a rhizocretion associated with paleosol profile of the Bagra Formation. The outline of the rhizocretion is marked by arrows; note the coaxial zonation within the rhizocretion. **(C)** Transverse section through a rhizocretion; note the slightly tortuous shape of the central zone of the rhizocretion (pen included for scale). (**A**, **B**, and **C** from Ghosh, Ghosh, and Bhattacharyya 2001, with permission from Elsevier Science.)

bodies. The paleosol profiles are 70 cm to 2 m thick. They are characterized by a 10 to 50 cm thick well-developed horizon of fused calcareous globules at the top (Fig. 2.6A).

A thick to very thick (50 cm to 1.5 m) horizon of closely spaced, vertically oriented, large cylindrical rhizocretions occurs below the top globule horizon. The cross-sectional diameter of the rhizocretions ranges from 3 to 5 cm; the length ranges from 30 to 100 cm. The rhizocretions internally show two distinct coaxial cylindrical regions with a sharp contact in between (Fig. 2.6B,C).

The inner region (1–2.5 cm in cross-sectional diameter) comprises large spars and minor carbonaceous clays (Fig. 2.6C). The outer region is made up of reddish gray micritic limestone. The field features of the paleosol profiles of the Bagra Formation are comparable to the present-day aridisols with a well-developed K-horizon at the top and a thick Bk horizon below it (Soil Survey Staff 1975).

A total of 29 samples from the rhizocretions and pedogenic globules were collected from different paleosol profiles of the Bagra Formation between 22°35'N, 78°18'E and 22°40'N, 78°37'30"E (see Fig. 2.1).

2.2.5 Lameta Formation

Lameta beds represent an extensive lacustrine-palustrine deposit occurring as discontinuous patches along the Narmada lineament in central India below the Deccan basalts. Diverse freshwater and terrestrial fossils are reported from the Lameta beds; the most important are dinosaur skeletal remains and eggshells (Sahni et al. 1994; Sarkar, Bhattacharya, and Mohabey 1991). The present section is located at Lameta Ghat on the Narmada River, 15 km southwest of Jabalpur. Here, fluviatile sediments that are about 35 m thick transitionally overlie fluvial sandstone and mudstone facies of Cretaceous Jabalpur Formation. A number of calcic paleosols, stacked vertically, have been recorded in the Lameta Formation. Ghosh (1997) suggested that development of the calcic paleosols took place in alternation with fluvial depositional events. The paleosol profile shows vertical root traces, and globules occur where the water table was low. More details of Lameta Formation are given in Ghosh, Bhattacharya, and Jani (1995) and Tandon and Andrews (2001).

2.3 Reappraisal of Paleosol Ages

The importance of uncertainty regarding the ages of these strata became apparent when we compared our pCO_2 estimates of paleosols from the four stratigraphic horizons of the Satpura Valley basin (Ghosh, Ghosh, and Bhattacharya 2001) with the revised model curve (GEOCARB III) of Berner and Kothavala (2001). Ghosh, Ghosh, and Bhattacharya (2001) found a major discrepancy between predicted values and pCO_2 estimates in the case of Jurassic (Bagra) and Cretaceous (Lameta) paleosols, which prompted us to do a reappraisal of the ages.

The age estimates of the Satpura soils are based mainly on the paleontological records and allow a great deal of uncertainty. It is crucial to ascertain the nature and extent of uncertainty of the soil ages to compare our observations with the predictions. We discuss below our earlier consideration and present understanding based on available literature.

The ages of the soils studied here are derived from faunal remains of vertebrates and microfossils, such as diatoms. Additional clues are obtained by considering the stratigraphic position of soil horizons with respect to strata of known ages. A literature survey indicates the following ages (with uncertainty) for Motur, Denwa, and Lameta formations: early middle Permian (275 ± 15 million years ago), middle Triassic (235 ± 5 million years ago), and late Cretaceous (80 ± 15 million years ago), respectively (Raja Rao 1983; Bandyopadhyay and Sengupta 1998; Casshyap, Tewari, and Khan 1993; Satsangi 1988; Chatterjee and Roychowdhury 1974). In the case of the Bagra Formation, the absence of fossil records was confounding. Several workers contended that it might be considered either equivalent to or younger than Denwa (Sastry et al. 1977). Since the age of the Denwa beds is thought to be between late Lower Triassic and middle Triassic (Chatterjee and Roychowdhury 1974), the age of Bagra was thought to be around late Triassic (Singh and Ghosh 1977). However, Casshyap, Tewari, and Khan (1993) studied the field relationship of Bagra Formation with the underlying Denwa sequence in central India and based on various lines of stratigraphic, tectonic, and sedimentologic evidence suggested that the Bagra Formation should be younger than late Triassic but older than or equivalent to late Jurassic to early Cretaceous. We now consider the Bagra Age as 175 ± 30 million years ago instead of the 200 million years ago (Jurassic) adopted earlier (Ghosh, Ghosh, and Bhattacharya 2001). The Tiki shale bed yielded a rich microfloral assemblage equivalent to upper Triassic microfloral elements (Sundaram, Maiti, and Singh 1979). In addition, it contains reptile fossils of Carnian Age (Anderson 1981; Kutty, Jain, and Roychowdhury 1998; Sengupta 1992). The upper part of the sedimentary sequence is dated as Carnian-Norian (210 million years ago) on the basis of vertebrate fossil evidence. Therefore the age of Tiki Formation is considered to be middle to upper Triassic, 225 ± 15 million years ago (Dutta and Das 1996).

2.4 Analytical Procedures

The thin sections of the samples were carefully studied to identify portions rich in pedogenic micrite so that sparite-rich portions could be avoided. Micritic-rich parts of the samples were thoroughly cleaned with deionized water in an ultrasonic bath. XRD analysis of fine-grained powders showed dominance of calcite with subordinate clays and detrital quartz grains. A few milligrams of powdered samples were reacted with 100% orthophosphoric acid at 50°C in a vacuum using an online extraction system (Sarkar, Ramesh, and Bhattacharya 1990). The evolved CO_2 gas was thoroughly purified and analyzed in a VG 903 mass spectrometer. Some of the samples were analyzed in a GEO 20-20 dual inlet mass spectrometer at 80°C using an online CAPS extraction system. Calibration and checkup of the system were done using NBS-19.

In order to determine the $\delta^{13}C$ value of organic matter associated with the soil carbonates, powdered samples were treated with 20% HCl for 24 hours to remove the carbonates. Dried residue was loaded in a 10cm long quartz break seal tube along with CuO (wire form) and Ag strip. The quartz tube was evacuated, sealed, and combusted at 700°C for 6 hours. Evolved CO_2 was purified and analyzed using GEO 20-20 IRMS. To check the reproducibility of measurements, UCLA glucose standard was analyzed along with each set of samples. Many of the samples were re-analyzed to check the homogeneity of the samples. Isotopic ratios of carbon and oxygen are presented in the usual d notation in units of per mil (‰) with respect to international standard V-PDB and V-SMOW, respectively, and are reproducible within ± 0.1‰ at 1σ level.

2.5 Observations

The detailed results of isotopic analysis of soil carbonates from Motur, Denwa, and Bagra formations are given in Ghosh, Ghosh, and Bhattacharya (2001), and the mean $\delta^{18}O$ values (excluding a few outliers that deviate by more than 2σ) are 18.2‰ (σ = 0.9‰), 24.2‰ (σ = 0.7‰), and 24.1‰ (σ = 0.8‰), respectively. The corresponding mean $\delta^{13}C$ values are as follows: -6.5‰ (σ = 0.8‰), -6.7‰ (σ = 0.5‰), and -6.1‰ (σ = 0.8‰), respectively. The $\delta^{18}O$ and $\delta^{13}C$ values of all the samples are distributed (Fig. 2.7) such that except for a few outliers the values are closely clustered around their respective means.

The mean values for pedogenic carbonate samples from the Lameta Formation (Ghosh, Bhattacharya, and Jani 1995) are: 24.7‰ for $\delta^{18}O$ and -9.1‰ for $\delta^{13}C$. The mean $\delta^{13}C$ values of organic matter associated with the paleosols are: -23.5‰ (σ = 0.5‰) for the Motur Formation, -24.6‰ (σ = 0.5‰) for the Denwa Formation and -26.1‰ (σ = 1.2‰) for the Bagra Formation. Organic matter from a single sample of the Lameta soil yielded a value of -27.3‰. Andrews, Tandon, and Dennis (1995) reported $\delta^{13}C$ values of -22.1‰ and -17.1‰ for two samples of organic matter from the Lameta Formation. These latter values are somewhat inconsistent with the normal range of C_3 plants and, therefore, not considered here. The results of isotopic analysis of soil carbonates from the Tiki Formation are given in Table 2.1. The mean $\delta^{18}O$ value for the Tiki samples (excluding the outliers that deviate by more than 2σ) is 26.1‰ (σ = 1.2‰). The corresponding mean $\delta^{13}C$ value is -6.9‰ (σ = 0.8‰) for carbonates and -24.7‰ (σ = 1.2‰) for associated organic matter.

In order to derive paleoclimatic information from the stable isotopic composition of paleosol carbonate and organic matter, it is necessary to establish that the samples have not undergone any postdepositional alteration. The fine-grained fabric in the samples and the dominance of pedogenic micrite along with general absence of sparry calcite rule out major recrystallization subsequent

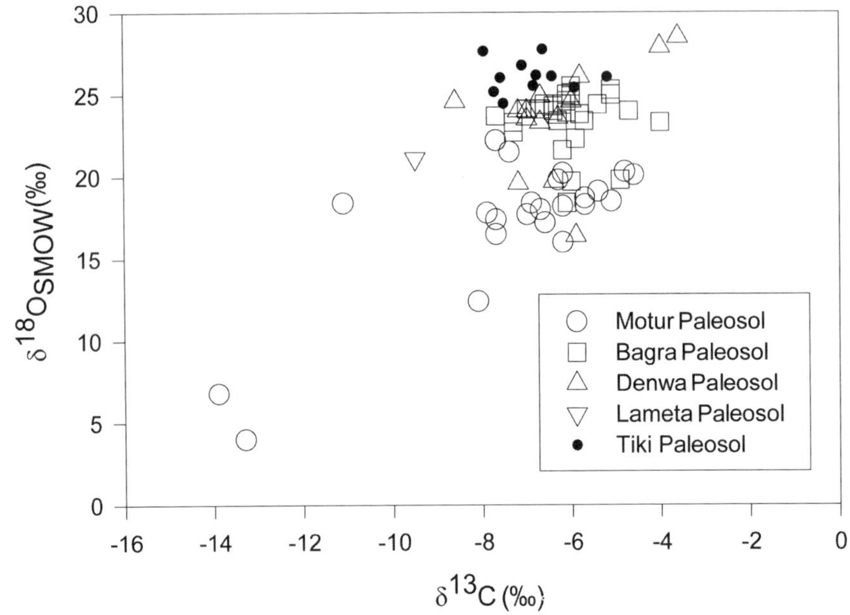

Figure 2.7. Plot of $\delta^{18}O$ versus $\delta^{13}C$ values of the pedogenic carbonates from the Motur, Denwa, Tiki, Bagra, and Lameta formations.

Table 2.1. Carbon and oxygen isotopic composition of pedogenic carbonate and $\delta^{13}C$ of soil organic matter from the Tiki Formation (middle Triassic)

Sample	$\delta^{13}C$, ‰	$\delta^{18}C$, ‰	$\delta^{13}C_{org}$, ‰
TK-21	−6.6	−3.1	−24.2
TK-37	−7.1	−4.1	−25.0
TK-2	−6.4	−4.7	
TK-22	−7.5	−6.3	−22.1
TK-15/4	−7.6	−4.8	
TK-16	−6.8	−5.3	−25.2
TK-28	−7.7	−5.6	
TK-20	−5.9	−5.4	−25.3
TK-18	−5.2	−4.7	
TK-3	−6.8	−4.7	−25.6
TK-14	−8.0	−3.2	−25.2
Mean ± standard deviation	−6.9 ± 0.8	−4.7 ± 1.0	−24.7 ± 1.2

to the precipitation of original carbonate. The isotopic data also show only minor spread in carbon and oxygen isotopic composition (see Fig. 2.7) among various samples from the same formation (except in a few outliers which are probably due to diagenetic alteration). The consistency in isotopic composition within 1‰ would be unlikely if alteration affected majority of the samples. Additionally, our results are comparable with the published data of other workers on soil carbonates of similar ages and environments (Cerling 1991; Yapp and Poths 1992; Mora, Driese, and Colarusso 1996).

The observed $\delta^{13}C$ values of the organic matter associated with the pedogenic carbonates lie within a range that closely matches that of the C_3 vegetative biomass. Because all the samples are from horizons older than Miocene (when C_4 plants were absent), these values provide additional evidence against any late diagenetic modifications. Moreover, the differences in $\delta^{13}C$ between coexisting organic matter and carbonate have a range (14‰–20‰) similar to that obtained for other paleosol samples of similar age (Mora, Driese, and Colarusso 1996).

2.6 Model for Estimation of Atmospheric pCO_2

Cerling's (1991) paleobarometer model is basically an isotopic mixing model where the soil-CO_2 is made up of atmospheric CO_2 (through diffusion) and plant-respired CO_2 from vegetative sources. Since this plant-respired CO_2 flux dominates, soil CO_2 in well-aerated soils can be modeled as a standard diffusion-production equation (Baver, Gardner, and Gardner 1972; Kirkham and Power 1972; Cerling 1984):

$$\frac{\partial C_s}{\partial t} = D_s \frac{\partial^2 C_s}{\partial z^2} + \phi_s(z) \tag{2.1}$$

where C_s = soil CO_2 (molecule cm^{-3}), t = time, z = depth in the soil profile (cm), and $\phi_s(z)$ = CO_2 production rate as a function of depth (molecule sec^{-1} cm^{-3}), D_s is the diffusion coefficient, and ϕ_s is the production function with respect to the depth z (Cerling 1984). Solving this equation with a no-flux boundary at depth and Cs (at z =0) set equal to atmospheric carbon dioxide yields:

$$C_s(z) = S(z) + C_a \tag{2.2}$$

where $S(z)$ is CO_2 contributed by soil respiration, and C_a is atmospheric CO_2. Cerling (1984, 1991) solved this equation for ^{12}C and ^{13}C and showed that the isotopic composition of soil CO_2 is controlled by diffusion process (diffusion of CO_2 from soil to atmosphere) and CO_2 production. In a paleosol, $CaCO_3$ precipitates in equilibrium with soil CO_2, which is modified by atmospheric contribution, thus allowing the $\delta^{13}C$ value of carbonate to be used for atmospheric [CO_2] calculation. The solution to the soil CO_2 equation of Cerling (1984, 1991, 1999) can be recast in terms of the isotopic composition of atmospheric CO_2:

$$C_a = S(z)\,\frac{\delta^{13}C_s - 1.0044\delta^{13}C_\phi - 4.4}{\delta^{13}C_a - \delta^{13}C_s} \tag{2.3}$$

under the assumption that $^{12}C_s/^{13}C_a \approx C_s/C_a$ where $\delta^{13}C_s$, $\delta^{13}C_\phi$ and $\delta^{13}C_a$ are the isotopic compositions of soil CO_2, soil respired CO_2 and atmospheric CO_2 respectively. The term *S(z)* is a function of depth but approaches a constant value below about 20 to 30 cm in depth (Cerling 1984; Cerling and Quade 1993). Soil CO_2 is enriched in ^{13}C relative to respired CO_2 by 4.4‰ due to mass dependent rate of diffusion (Cerling 1984), independent of CO_2 concentration.

The $\delta^{13}C$ of soil carbonate is governed by the $\delta^{13}C$ value of the soil CO_2 and fractionation during precipitation, which is temperature dependent (Deines, Langmuir, and Harmon 1974). The $\delta^{13}C$ values of the two sources of soilCO_2 (e.g., respiration and atmosphere) are quite different. Whereas the recent pre-industrial atmospheric CO_2 is −6.5‰, the CO_2 respired by C_3 type vegetation can have value between −20‰ and −35‰, with a mean around −27‰ (Ehleringer 1989). The degree of infiltration of the atmospheric CO_2 in the soil matrix (which is dependent on the atmospheric $[CO_2]$) thus influences the $\delta^{13}C$ value of the soilCO_2 and consequently the $\delta^{13}C$ value of the pedogenic carbonate. Therefore, this model allows us to estimate the $[CO_2]$ values of the ancient atmosphere from the $\delta^{13}C$ value of the pedogenic carbonates if the following parameters are known:

1. The temperature of calcite precipitation in the soil
2. The $\delta^{13}C$ value of the plant-respired CO_2
3. The $\delta^{13}C$ value of the atmospheric CO_2
4. The difference between the concentration of the soil pCO_2 and the atmospheric pCO_2

We discuss below the procedures for estimating each of these parameters.

2.6.1 Estimation of Soil Temperature

Estimation Based on Paleolatitude. The temperature of the calcite precipitation in the soil is close to the mean soil surface temperature. Scotese (1998) surmised that the Permian was a period during which large global temperature changes took place. In the early Permian, global temperature was low with a mean around 10°C; at the end of the Permian, the temperature was extremely high, near about 35°C (Scotese 1998). The global temperature was maintained at a slightly reduced value of 30°C till the middle Jurassic. The widespread glacial deposits at the Permo-Carboniferous boundary (Talchir Formation) followed by thick coal deposits of the overlying Barakar Formation (see Fig. 2.2); the occurrence of red beds and calcic paleosols in the Motur, Denwa, Tiki, and Bagra Formations; and the presence of thick deposits of carbonaceous claystones in the Bijori Formation (late Permian), all comprise constitute supporting evidence for the suggested trend of the global temperature. However, the surface temperature of a locality is also influenced by its altitude and latitudinal position. The paleo-

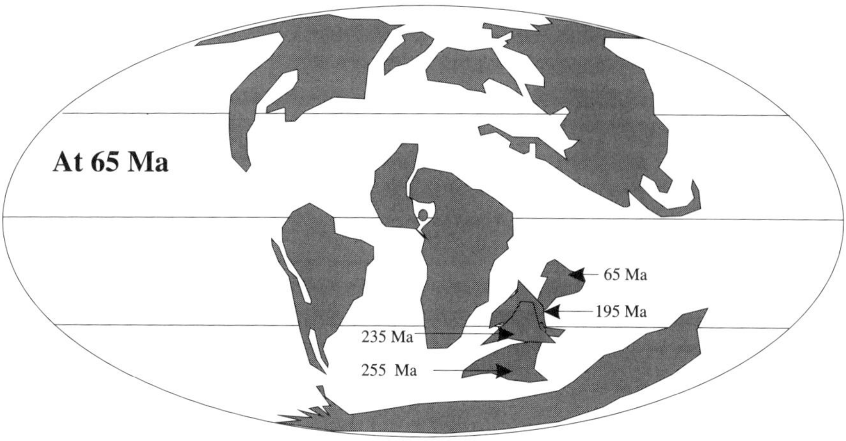

Figure 2.8. Reconstruction of palaeolatitude of India between 255 and 65 Ma as proposed by Scotese (1997). The figure also shows the distributions of the ocean basins and continents at 65 Ma.

geographic reconstructions of the Gondwana landmasses indicate that the study area was situated at 60°S during the Permian (275 ± 15 million years ago), at 50 to 40°S during the middle to upper Triassic (240–205 million years ago), and at 25°S during the late Jurassic to early Cretaceous (175 ± 30 million years ago), as shown in Fig. 2.8. During this period the altitude seems to have remained consistently low and probably close to the sea level (Scotese 1997).

The temperature of soil carbonate precipitation in the present case can be inferred based on the latitudinal position of the Indian landmass during Paleozoic and Mesozoic periods and by using the global data set relating latitude and average temperature (IAEA/WMO 1998). We assume that the major part of the soil carbonate precipitates at the maximum possible temperature at a given latitude. The paleosols described in this study were formed in paleolatitude range of 60°S to 25°S. Modern growing season temperature close to sea level in these latitudes typically ranges from 10° to 30°C. A scatter plot of maximum monthly temperature for the 60°S to 25°S latitudes reveals that 20°C is a typical temperature at 60°S while for 45°S to 25°S one can consider 25°C as representative (see Fig. 5 of Ekart et al. 1999). We have, therefore, chosen 20°C as the temperature of soil carbonate precipitation for the Motur Formation and 25°C for the Denwa, Tiki, and Bagra formations.

Estimation Based on Oxygen Isotope Ratios. An independent estimate of surface temperature can be obtained based on $\delta^{18}O$ values of the carbonates and composition of the ancient soil water in these locations. Soil water is derived from precipitation the composition of which depends on the latitude and the composition of the ocean water. The average $\delta^{18}O$ values of the present-day rainfall at 60°S, 45°S, and 25°S are available from the compilation of Rozanski, Araguas, and Gonfiantini (1993). The $\delta^{18}O$ of the ocean water during the Permian, Triassic, and Jurassic periods was slightly less and commonly assumed to be −1‰ (Veizer, Fritz, and Jones 1986). Assuming that the global moisture circulation was similar to that of today, one can estimate the composition of the precipitation at those locations and times (Table 2.2).

However, a correction is needed for the Permian Period (Motur soil) because, according to Kutzbach and Gallimore (1989), a megamonsoonal circulation operating at that time gave intense seasonal precipitation. We estimate a 2.5‰ depletion in the $\delta^{18}O$ value due to attendant amount effect in the rainfall based on difference in rainfall during the Permian (Kutzbach and Gallimore 1989) and today (Rozanski, Araguas, and Gonfiantini 1993). Similarly, a depletion of 3‰ was postulated for the Lameta Period due to continental effect (Ghosh, Bhattacharya, and Jani 1995). Using these values and an enrichment of 2‰ due to evapotranspiration (Ghosh, Bhattacharya, and Jani 1995) during conversion of rainwater to soil water, one can estimate the soil temperatures during these four periods (see Table 2.2). The values are 17°, 17°, 23°, and 25°C, respectively, and are in reasonable agreement with the estimates derived in the preceding paragraph.

The above method of estimating the soil temperature is critically based on the assumption that no significant diagenetic alteration took place that could have changed the original $\delta^{18}O$ of the carbonate. For example, an increase of approx 1‰ in $\delta^{18}O$ due to postdepositional diagenesis would imply a approx 4°C decrease in temperature estimate leading to reduction in the [CO_2] estimate by about 20% (Cerling 1999; Ekart et al. 1999).

Table 2.2. Estimation of CO_2 concentration and its isotopic composition in the atmosphere at different times in the past

Formation Age (My)	$\delta^{13}C$ (carbonate)	$\delta^{13}C$ (OM)	Temperature (°C)	$\delta^{13}C$ of atmospheric CO_2, ‰@	pCO_2† A	pCO_2† B	Atmospheric CO_2 (ppmV)
MOTUR (275 ± 15)	−6.5	−23.5	20	−2.5	540	890	715
DENWA (235 ± 5)	−6.7	−24.6	25	−3.6	910	1510	1210
TIKI (225 ± 15)	−6.9	−24.7	25	−3.7	880	1460	1170
BAGRA (175 ± 30)	−6.1	−26.1	25	−5.1	1675	2775	2225
LAMETA (80 ± 15)	−9.1	−27.6	25	−6.6	1110	1850	1480

@ $\delta^{13}C$ of atmospheric CO_2 obtained by subtracting 21‰ from $\delta^{13}C$ (OM) (see text).
† Column A and B refer to $S_{(z)}$ values of 3000 and 5000 ppm in Cerling's model, respectively (see text).

2.6.2 $\delta^{13}C$ of the Plant-Respired CO_2

The $\delta^{13}C$ value of the organic matter associated with each soil was determined in this study to act as a suitable proxy for the $\delta^{13}C$ of plant-respired CO_2 in that soil environment. Prior to the Miocene, C_4 plants were not present in the ecosystem; C_3 plants were the dominant vegetative biomass. The $\delta^{13}C$ values of the plant-respired CO_2 of the modern C_3 plants range from −20‰ to −35‰ with an average of −27‰ (Ehleringer 1989). The mean $\delta^{13}C$ values of −23.5‰, −24.6‰, −24.7‰, −26.1‰, and −27.6‰ for the Motur, Denwa, Tiki, Bagra, and Lameta Formations (see Table 1.1) lie close to the mean $\delta^{13}C$ of the modern C_3 plants and provide robust estimates for the isotopic composition of the plant-respired CO_2 at those times.

2.6.3 $\delta^{13}C$ of the Atmospheric CO_2

The $\delta^{13}C$ of atmospheric CO_2 is an important parameter in Cerling's model, but it is also one of the least sensitive (Cerling, Wright, and Vanstone 1992). In earlier studies (Cerling 1991; Andrews, Tandon, and Dennis 1995; Ghosh, Bhattacharya, and Jani 1995), the $\delta^{13}C$ was assumed to be constant and its value was thought to be −6.5‰, equal to the preindustrial atmospheric CO_2 value. However, since then several studies (Mora, Driese, and Colarusso 1996; Thackeray et al. 1990) have demonstrated variation in the $\delta^{13}C$ of atmospheric CO_2 in the past. There are two ways of estimating this $\delta^{13}C$: one based on $\delta^{13}C$ of marine brachiopods (Mora, Driese, and Colarusso 1996), the other based on $\delta^{13}C$ of organic matter (Farquhar, Ehleringer, and Hubick 1989; Thackeray et al. 1990). Since the brachiopod data (Veizer, Fritz, and Jones 1986) show a large scatter, we adopt the second alternative here. Mora, Driese, and Colarusso (1996) observed that the fractionation between atmospheric CO_2 and soil organic matter remained relatively constant between approx 21‰ to 22.5‰ through middle to late Palaeozoic. Ghosh, Ghosh, and Bhattacharya (2001) used this information for estimating the $\delta^{13}C$ of the atmospheric CO_2 from the mean $\delta^{13}C$ of the C_3 organic matter assuming approximately 21‰ fractionation. We note that some workers (Ekart et al. 1999) have assumed a fractionation of 18‰ between $\delta^{13}C$ of atmospheric carbon dioxide and soil organic carbon. However, the magnitude of this fractionation varies significantly with pCO_2 concentration in the atmosphere, which determines the stomatal pore size. Carbon isotope studies of modern plants indicate that increased $[CO_2]$ lowers the plant $\delta^{13}C$ (Gröcke 1998). van de Water, Levitt, and Betancourt (1994) showed that a 30% increase in $[CO_2]$ results in a plant $\delta^{13}C$ shift of −1.5‰. Therefore, in a regime of $[CO_2]$ (Phanerozoic) the atmosphere-plant fractionation is expected to be more than 19‰ (corresponding to modern-day values when average atmospheric $\delta^{13}C$ is −7‰ and average C_3 plant $\delta^{13}C$ is −26‰) as observed in recent atmosphere. The assumed $\delta^{13}C$ shift of 21‰ for our $[CO_2]$ estimation is consistent with this argument. The mean $\delta^{13}C$ of the organic matter from the Motur, Denwa, Tiki, Bagra, and Lameta formations are −23.5‰, −24.6‰, −24.7‰, −26.1‰, and −27.6‰, respectively (see Table 1.1). Therefore, the $\delta^{13}C$ values of the atmospheric CO_2 in those periods are estimated to be −2.5‰, −3.6‰, −3.7‰, −5.1‰, and −6.6‰, respectively. It is interesting to note that there is a systematic decrease in the $\delta^{13}C$ of atmospheric CO_2 from 260 to 65 million years.

2.6.4 Difference Between the Soil [CO_2] and the Atmospheric [CO_2]: S_Z Parameter

The difference in concentration between the soil CO_2 and the atmospheric CO_2 depends on soil temperature, pressure, porosity, and tortuosity; soil respiration rate; and depth of CO_2 production. This difference (S_z) plays an important role in Cerling's mixing model and needs to be assessed independently because of its strong influence on pCO_2 estimate. An approximate value of the soil porosity can be obtained from observations made on recent soils of the same type. However, the mean CO_2 production depth and the soil respiration rate for geologically old soils cannot be estimated with certainty. The rooting depth, the dimensions of the rhizoliths and their density of occurrence in the paleosol profiles may probably be used for such estimation, but no such model is available at present. Observations made on the modern soils show that for the desert soils S_z is less than 3000 ppmV and for well-drained temperate and tropical soils S_z ranges between 5000 and 10,000 ppmV. Mora, Driese, and Colarusso (1996) assumed a value between 3000 and 7000 ppmV for the middle to late Paleozoic semi-arid tropical to temperate vertic paleosols. Some of these paleosols (of the Silurian Age) indicate a very shallow rooting depth. In contrast, Ghosh, Bhattacharya, and Jani (1995) adopted a range between 3000 and 5000 ppmV for the well-drained, semi-arid late Cretaceous paleosols with higher rooting depths than the Silurian soils (Ghosh, Bhattacharya, and Jani 1995; Tandon et al. 1995, 1998; Ghosh, Rudra, and Maulik 1998). We note that the gleyed pedohorizons (signature of clay rich component with less permeability) are, in general, absent in the profiles of the Denwa, Tiki, Bagra, and Lameta formations. Some amount of gleying is present close to the Bk pedohorizons of the Motur and the Denwa formations that may suggest a higher value of S_z. However, the high latitudinal position (60°S) of the study area along with a low mean annual temperature during the Motur time implies reduced plant productivity and consequently low S_z. All the paleosols analyzed in this work show characteristics of well-drained, semi-arid soils. In view of the above considerations, a range of S_z values between 3000 and 5000 ppmV is adopted for these paleosols. However, given the sensitivity of the model to this variable, a critical assessment of its estimation is required.

2.6.5 Calculation of pCO_2 in the Ancient Atmosphere

Using the mean $\delta^{13}C$ values of the paleosol carbonates along with the above-mentioned values of the required parameters and estimated $\delta^{13}C$ of the atmospheric CO_2 based on soil organic matter in the model of Cerling (1991), we have calculated [CO_2] values for the five time windows mentioned above (Ghosh, Ghosh, and Bhattacharya 2001). The values are (in ppmV): 540 to 890 for Motur, 910 to 1510 for Denwa, 880 to 1460 for Tiki, 1675 to 2775 for Bagra, and 1110 to 1850 for Lameta (Table 2.2). The uncertainties in average $\delta^{13}C(OM)$ (see Table 2.1) is not included in the quoted uncertainty of the $[CO_2]$. However, variations in $\delta^{13}C(OM)$ (0.5‰ for Motur and Denwa and 1.2‰ for Tiki and Bagra formations) would change the calculated mean $[CO_2]$ by 20% for Motur and Denwa and 40% for Tiki and Bagra formations, respectively. It is to be noted that the soil samples studied here belong to the same geochemical milieu because they all formed either in the same basin or in a neighboring basin (Tiki) at different times. This implies that the relative CO_2 variation deduced here is expected to be well constrained.

These values represent the first independent estimates of atmospheric CO_2 level during 260 to 65 million years based on soils formed in a Gondwana continent located in the Southern Hemisphere. This is considered an important period in the evolution model (GEOCARB III) since it predicts an increase in the CO_2 level after the low in the early Permian (310–285 millin years ago) and ascribes it to the enhanced rate of degassing. This prediction is verified by our results, but the detailed nature of variation is found to be slightly different. Fig. 2.9 shows a plot of CO_2 concentration in ppm against age, in million years ago. It is seen that there is excellent agreement of derived concentrations with

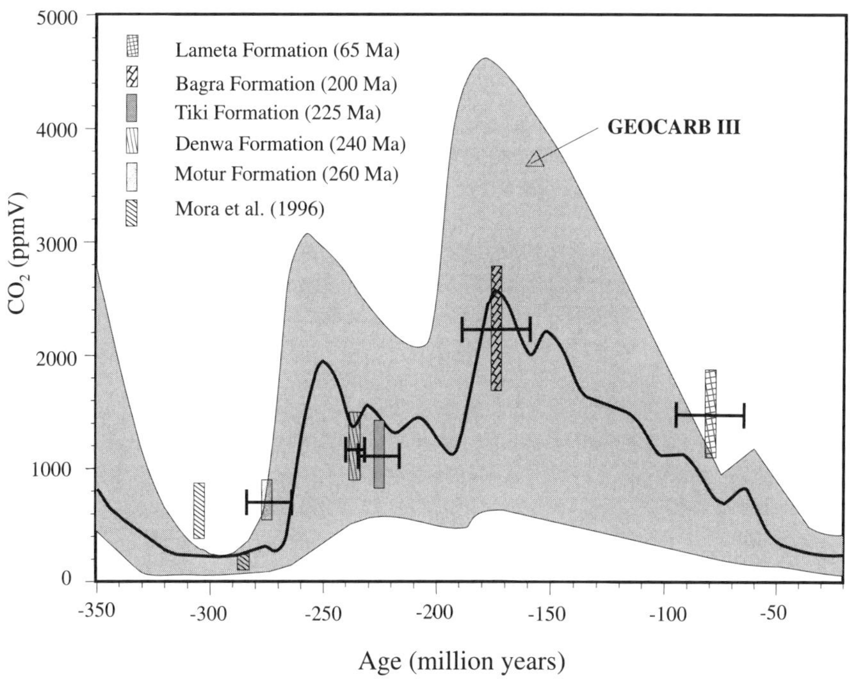

Figure 2.9. Estimated concentration of atmospheric CO_2 (expressed relative to present-day concentration) as a function of time. The points are based on isotopic analysis of pedogenic carbonates. The dashed curve along with envelope (error range) is obtained from GEOCARB III (Berner and Kothavala 2001).

the prediction of Berner and Kothavala (2001) prediction, which was between 275 and 160 million years ago (corresponding to ages of Motur, Denwa, Tiki, and Bagra formations). The abundance of carbon dioxide for 80 million years (Lameta), however, is 1480 ppm in contrast to the prediction of Berner and Kothaval (2001) estimate of about 1000 ppm. The disagreement persists even at the lower limit of our estimate. In this context, it is to be noted that all available evidence indicates very high CO_2 concentrations in the middle Cretaceous to late Cretaceous atmosphere. Emiliani (1995) points out that middle Cretaceous was characterized by widespread anoxia, huge marine ingression indicating active mid-ocean ridges, and warm oceans. All these factors indicate a large increase in the CO_2 flux to the atmosphere and consequent exacerbation of the greenhouse effect and vigorous expansion of biotic productivity. Based on these considerations, an upward revision of model CO_2 input flux at late Cretaceous can, therefore, be considered, and this revision lessens the disagreement.

2.7 Summary

The fluvial deposits of Motur (middle Permian), Denwa (late Triassic), Tiki (early Triassic), Bagra (middle Jurassic), and Lameta (late Cretaceous) formations of the Gondwana supergroup in central India contain a number of calcic paleosols. All of these paleosols are characterized by pedogenic carbonates occurring as rhizocretions and glebules. Soil carbonates from several horizons of these formations were analyzed for oxygen and carbon isotopic ratios. Oxygen isotopic ratio is used to infer about the environment of precipitation and paleotemperature. The carbon isotopic composition of these carbonates and the coexisting soil organic matters provided indirect means to determine the isotopic composition and the partial pressure of atmospheric CO_2 using the CO_2 paleobarometer model of Cerling (1991). The estimate shows that the concentration of CO_2 was low (~710 ppmV) during the early Permian (275 ± 15 million years ago) followed by continuous rise during the middle Triassic (~1215 ppmV at 235 ± 5 million years ago), late Triassic (~1170 ppmV at 225 ± 15 million years ago), and the Jurassic (~2240 ppmV at 175 ± 30 million years ago) (see Table 2.2 and Fig. 2.9). It is seen that the CO_2 level increased by a factor of 8 from the Permian to the Jurassic and declined again during the Cretaceous. The nature of the change agrees well with the result of the CO_2 evolution model of Berner (GEOCARB III) except that the magnitude of the CO_2 increase in the late Cretaceous was slightly higher than the predicted value. The late Cretaceous [CO_2] estimate is approx 1480 ppmV, higher than the predicted concentration (1000 ppmV). Degassing of Earth's interior caused by the rapid breaking up of the Gondwana landmass during the Triassic and Jurassic periods could have caused the rapid evolution of CO_2; subsequent biotic proliferation in the Cretaceous probably lead to its slight decline.

References

Anderson, J.M. 1981. World Permo-Triassic correlations: Their biostratigraphic basis. In *Gondwana Five,* ed. M.M. Cresswell and P. Vella, 3–10. Rotterdam: A.A. Balkema.

Andrews, J.E., S.K. Tandon, and P.F. Dennis. 1995. Concentration of carbon dioxide in the late Cretaceous atmosphere. *Journal of Geological Society London* 152:1–3.

Bandyopadhyay, S., and D.P. Sengupta.1998. Middle Triassic vertebrate from India. *Gondwana 10, Cape Town. Journal of African Earth Sciences* 27:16–17.

Barnola, J.M., D. Raynaud, Y.S. Korotkevitch, and C. Lorius.1987. Vostok ice core: A 160,000 years' record of atmospheric CO_2. *Nature* 329:408–14.

Baver, L.D., W.H. Gardner, and W.R. Gardner. 1972. *Soil physics*. New York: John Wiley.

Berner, R.A. 1991. A model for atmospheric CO_2 over Phanerozoic time. *American Journal of Science* 291:339–76.

———. 1994. GEOCARBII: A revised model of atmospheric CO_2 over Phanerozoic time. *American Journal of Science* 294:56–91.

———. 1997. The rise of plants and their effect on weathering and atmospheric CO_2. *Science* 276:544–45.

Berner, R.A., and Z. Kothavala. 2001. GEOCARB III: A revised model of atmospheric CO_2 over Phanerozoic time. *American Journal of Science* 301:182–204.

Casshyap, S.M., and H.A. Qidwai. 1971. Paleocurrent analysis of Lower Gondwana sedimentary rocks, Pench Valley Coalfield, Madhya Pradesh [India]. *Journal Geological Society of India* 5:135–45.

Casshyap, S.M., and R.C. Tewari. 1988. Depositional model and tectonic evolution of Gondwana basins: Lucknow. *Palaeobotanist* 36:59–66.

Casshyap, S.M., R.C. Tewari, and A. Khan. 1993. Alluvial fan origin of the Bagra Formation (Mesozoic Gondwana) and tectono-stratigraphic implications. *Journal Geological Society of India* 42:269–79.

Cerling, T.E. 1984. The stable isotopic composition of modern soil carbonate and its relationship with climate. *Earth Planetary Science Letters* 71:229–40.

———. 1991. Carbon dioxide in the atmosphere: Evidence from Cenozoic and Mesozoic palaeosols. *American Journal of Science* 291:377–400.

———. 1999. Stable carbon isotopes in palaeosol carbonates. In *Palaeoweathering, palaeosurfaces, and related continental deposits,* ed. M. Thirty and Simm-Coincon, 43–60. Oxford: Blackwells.

Cerling, T.E., and J. Quade. 1993. Stable carbon and oxygen isotopes in soil carbonates. In *Climate change in continental isotopic records,* ed. P.K. Swart, C. Lohmann, C. Kyger, J.A. McKenzie, and S. Savin, 217–31. Washington, D.C.: American Geophysical Union.

Cerling, T.E., J. Quade, Y. Wang, and J.R. Bowman. 1989. Carbon isotopes in soils and palaeosols as ecology and palaeoecology indicators. *Nature* 341:138–39.

Cerling T.E, V.P. Wright, and S.D. Vanstone. 1992. Further comments on using carbon isotopes in paleosols to estimate the CO_2 content of the palaeo-atmosphere. *Journal of the Geological Society of London* 149: 673–76.

Chatterjee, S. and T. Roychowdhury. 1974. Triassic Gondwana vertebrates from India. *Indian Journal of Earth Sciences* 11:96–112.

Crowley, T.J., and R.A. Berner. 2001. CO_2 and climate change. *Science* 292:870–72.

Deines, P., D. Langmuir, and R.S. Harmon. 1974. Stable carbon isotope ratios and existence of a gas phase in the evolution of carbonate ground waters. *Geochimica Cosmochimica Acta* 38:1147–164.

Driese, S.G., and C.I. Mora. 1993. Physico-chemical environment of pedogenic carbonate formation in Devonian vertic palaeosols, central appalachians, USA. *Sedimentology* 40:199–216.

Dutta, P.M., and D.P. Das. 1996. Discovery of the oldest fossil mammal from India. *Indian Minerals* 50:217–22.

Ehleringer, J.R. 1989. Carbon isotope ratios and physiological processes in arid land plants. In *Stable isotopes in ecological research,* ed. P.W. Rundel, J.R. Ehleringer, and K.A. Nagy. New York: Springer-Verlag.

Ekart, D.D., T.E. Cerling, I.P. Montanez, and N.J. Tabor. 1999. A 400 million year carbon isotope record of pedogenic carbonate: Implications for paleoatmospheric carbon dioxide. *American Journal of Science* 299:805–27.

Emiliani, C. 1995. Planet Earth: *cosmology, geology, and the evolution of life and environment,* ed. C. Emiliani, 718. Cambridge University Press.

Farquhar, G.D., J.R. Ehleringer, and K.T. Hubick. 1989. Carbon isotope discrimination and photosynthesis. *Annual Review of Plant Physiology and Plant Molecular Biology* 40:503–37.

Ghosh, P. 1997. Geomorphology and palaeoclimatology of some Upper Cretaceous palaeosols in central India. *Sedimentary Geology* 110:25–49.

Ghosh, P., S.K. Bhattacharya, and R.A. Jani. 1995. Palaeoclimate and palaeovegetation in central India during the Upper Cretaceous based on stable isotope composition of the palaeosol carbonates. *Palaeogeography, Palaeoclimatology, Palaeoecology* 114: 285–96.

Ghosh, P., P. Ghosh, and S.K. Bhattacharya. 2001. CO_2 levels in the late Paleozoic and Mesozoic atmosphere from soil carbonate and organic matter, Satpura basin, Central India. *Palaeogeography, Palaeoclimatology, Palaeoecology* 170:219–36.

Ghosh P., D.K. Rudra, and P. Maulik. 1998. Palaeoclimatic indicators in the Satpura Gondwana succession, India, Gondwana 10. *Journal of African Earth Sciences* 27: 1A.

Gröcke, D.R. 1998. Carbon-isotope analyses of fossil plants as a chemostratigraphic and palaeo-environmental tool. *Lethaia* 31:1–13.

Houghton, R.A., and G.M. Wood. 1989. Global climate change. *Scientific American* 260: 36–44.

IAEA/WMO. 1998. *The GNIP [Global network for isotopes in precipitation] database, Release 2.* International Atomic Energy Agency.

IPCC. 1990. *Climate change: The IPCC scientific assessment,* ed. J.T. Houghton, B.A. Callander, and S.K. Varney, 200. Cambridge, U.K.: Cambridge University Press.

Keeling, C.D. 1994. Global historical CO_2 emissions. In *Trends '93: A compendium of data on global change,* ed. T.A. Boden, D.P. Kaiser, R.J. Sepanski, F.W. Stoss, and F.W. Logsdon, 501–505. Springfield, Va.: American Geological Institute.

Kirkham, D., and W.L. Powers. 1972. *Advanced soil physics.* New York: Wiley-Interscience.

Kutty, T.S., S.L. Jain, and T. Roychowdhury. 1998. Gondwana sequence of the northern Pranhita-Godavari Valley: Its stratigraphy and vertebrate faunas. *Palaeobotanist* 36: 214–29.

Kutzbach, J.E., and R.G. Gallimore. 1989. Pangean climates: Megamonsoons of the megacontinent. *Journal of Geophysical Research* 94:3341–357.

McElwain, J.C., and W.G. Chaloner. 1996. The fossil cuticle as a skeletal record of environmental change. *Palaios* 11:376–88.

Mora, C.I., S.G. Driese, and L.A. Colarusso. 1996. Middle to late Paleozoic atmospheric CO_2 levels from soil carbonate and organic matter. *Science* 271:1105–107.

Mora, C.I., D.E. Fastovsky, and S.G. Driese. 1993. Geochemistry and stable isotopes of palaeosols, 66. A Geological Society of America Continuing Education Short Course.

Mora, C.I., B.T. Sheldon, C.W. Elliott, and S.G. Driese. 1998. An oxygen isotope study of illite and calcite in three Appalachian Paleozoic vertic paleosols. *Journal Sedimentary Research* 68:456–64.

Mukherjee, R.N., and D.P. Sengupta. 1998. New capitosaurid amphibians from the middle Triassic Denwa Formation of Satpura Gondwanas, central India. *Alcheringa* 22: 317–27.

Raja Rao, C.S., ed. 1983. Coalfields of India: Coal resouces of Madhya Pradesh, Jammu, and Kashmir. Geological Society of India Bulletin 3:195–98.

Robinson, P.L. 1967. The Indian Gondwana formations: A review. In *Gondwana stratigraphy, International Union of Geological Sciences Symposium, Buenos Aires, 1–15 October 1967,* ed. A.J. Amos, 201–68. Paris: United Nations Educational, Scientific, and Cultural Organizations.

Royer, D.L., R.A. Berner, and D.J. Beerling. 2001. Phanerozoic atmospheric CO_2 change: Evaluating geochemical and paleobiological approaches. *Earth Science Reviews* 54: 349–92.

Rozanski, K., L.A. Araguas, and R. Gonfiantini. 1993. Isotopic patterns in modern global precipitation. In: *Continental isotope indicators of climate,* geophysical monograph 78:1–35. Washington, D.C.: American Geophysical Union.

Sahni, A., S.K. Tandon, A. Jolly, S. Bajpai, A. Sood, and S. Srinivasan. 1994. Cretaceous dinosaur eggs and nesting sites from the Deccan volcano-sedimentary province of peninsular India. In *Dinosaur eggs and babies,* ed. K. Carpenter, K. Hirsch, and J. Horner. New York: Cambridge University Press.

Sarkar, A., S.K. Bhattacharya, and D.M. Mohabey. 1991. Stable isotope analyses of dinosaurs egg shells: Paleoenvironmental implications. *Geology* 19:1068–71.

Sarkar, A., R. Ramesh, and S.K. Bhattacharya. 1990. Effect of sample pretreatment and size fraction on the $\delta^{18}O$ and $\delta^{13}C$ values of foraminifera in Arabian Sea sediments. *Terra Nova* 2:489–93.

Sastry, M.V.A., S.K. Acharya, S.C. Shah, P.P. Satsangi, P.K. Raha, G. Singh, and R.M. Ghosh, eds. 1977. Stratigraphic lexicon of Gondwana formations of India. *Geological Survey of India* (miscellaneous publication) 36:13–15.

Satsangi, P.P. 1988. Vertebrate fauna from the Indian Gondwana Sequence. *Paleobotanist* 36:245–53.

Scotese, C.R. 1997. *Paleogeographic atlas.* PALEOMAP progress report 90-0497, 45. Arlington, Tex.: Department of Geology, University of Texas at Arlington.

———. 1998. Gondwana's climate changes. Abstract in *Journal of African Earth Sciences* 27:172–73.

Sengupta, D.P. 1992. Metoposaurus maleriensis Roychowdhury from the Tiki formation of the Son-Mahanadi valley of central India. *Indian Journal of Geology* 64:300–305.

Singh, G., and R.M. Ghosh. 1977. Bagra conglomerate. In Stratigraphic lexicon of Gondwana formations of India, ed. Sastry et al. *Geological Survey India* (miscellaneous publication) 36:13–15.

Soil Survey Staff. 1975. Soil Taxonomy. *Soil Conservation Survey.* Handbook 436. U.S. Department of Agriculture.

Sundaram, D., A. Maiti, A., and G. Singh. 1979. Upper Triassic microflora from Tiki formation of South Rewa Gondwana basin, Madhya Pradesh, India. In *Fourth international Gondwana symposium: Delhi, India* 1:511–14, ed. B. Laskar and C.S. Raja. Delhi, India: Hindustan Publishing Corp.

Tajika, E., and T. Matsui. 1992. Evolution of terrestrial proto-CO_2 atmosphere coupled with thermal history of the earth. *Earth and Planetary Science Letters* 113:251–66.

Tandon, S.K., and J.E. Andrews. 2001. Lithofacies associations and stable isotopes of palustrine and calcrete carbonates: Examples from an Indian Maastrichtian regolith. *Sedimentology* 48(2):339–55.

Tandon, S.K., J.E. Andrews, A. Sood, and S. Mittal. 1998. Shrinkage and sediment supply control on multiple calcrete profile development: A case study from the Maastrichtian of central India. *Sedimentary Geology* 119:25–45.

Tandon, S.K., A. Sood, J.E. Andrews, and P.F. Dennis. 1995. Palaeoenvironments of the dinosaur-bearing lameta beds (Maastrichtian), Narmada valley, central India. *Palaeogeography, Palaeoclimatology, Palaeoecology* 117:123–54.

Thackeray, J.F., N.J. van der Merwe, J.A. Thorp-Lee, A. Sillen, J.L. Lanham, R. Smith,

A. Keyser, and P.M.S. Monteiro. 1990. Changes in carbon isotope ratios in the late Permian recorded in the therapsid tooth apatite. *Nature* 347:751–53.

van de Water, P.K., S.W. Leavitt, and J.L. Betancourt. 1994. Trends in stomatal density and $^{13}C/^{12}C$ ratios of *Pinus flexilis* needles during last glacial-interglacial cycles. *Science* 264:239–43.

Veevers, J.J, and R.C. Tewari. 1995. Gondwana master basin of peninsular India between Tethys and the interior of the Gondwanaland province of Pangea. *Geological Society of America* (memoir) 187:1–72.

Veizer, J., P. Fritz, and B. Jones. 1986. Geochemistry of brachiopods: Oxygen and carbon isotopic records of Paleozoic oceans. *Geochimica et Cosmochimica Acta* 50:1679–96.

Yapp, C.J., and H. Poths. 1992. Ancient atmospheric CO_2 pressures inferred from natural goethites. *Nature* 355:342–44.

3. Alkenone-Based Estimates of Past CO_2 Levels: A Consideration of Their Utility Based on an Analysis of Uncertainties

Katherine H. Freeman and Mark Pagani

3.1 Introduction

Several molecular-based isotopic methods are employed for the reconstruction of past CO_2 levels. The focus in this chapter is on the use of organic materials derived from marine phytoplankton, in particular the lipid biomarkers called "alkenones" derived from haptophyte algae. These compounds are known to be useful for $[CO_2]$ and temperature reconstructions; they have the added advantages of a high potential for preservation in marine sediments and ubiquity in the modern ocean. The latter property has enabled their study under a variety of marine conditions.

A number of comprehensive reviews give in-depth coverage of isotope fractionation by marine algae: in particular, Laws et al. (2001), who discuss carbon isotope fractionation by algae in general and provide a detailed consideration of fractionation by *Emiliania huxleyi*, which is the most cosmopolitan alkenone-producing species in the modern ocean. Pagani (2002) advances the Laws et al. review with more recent data and addresses in particular alkenone-based CO_2 estimates from ancient records. Hayes (2001) provides a detailed review of the fractionation of the isotopes of carbon during metabolic reactions. Included in his work is a discussion of carbon isotope fractionation during carbon fixation by organisms that employ the Calvin cycle, such as most marine algae. Freeman (2001) provides a general review of the isotopic biogeochemistry of organic matter in the modern ocean, which includes modern controls on the ^{13}C content of inorganic carbon, and the influence of heterotrophic and photosynthetic effects on the ^{13}C abundance of organic carbon.

The theory and observations that are invoked in the calculations of ancient CO_2 levels are reviewed below; in particular, the uncertainties in the observed properties employed and the limitations imposed by the assumptions inherent in the method. To guide the reader in a practical application of these topics, also included is an analytical treatment of the propagation of errors for calculations related to CO_2 estimates, along with an example of these calculations based on previous works and showing that relative uncertainties are about 20% for Miocene CO_2 reconstruction based on alkenone $\delta^{13}C$ data. Also discussed below is the interpretation of isotopic data in the study of climate on longer timescales, including two examples from the Paleozoic.

3.2 Fractionation of Carbon Isotopes During Photosynthesis

The fixation of CO_2 by marine algae can be represented (Francois et al. 1993; Hayes 2001) by the flow of carbon into and out of the cell, and the flow of carbon taken up by enzyme-catalyzed fixation reactions. In Fig. 3.1, the flux of carbon that flows into and out of the cell is represented by ϕ_{in} and ϕ_{out}, respectively, while the flux of carbon fixed into carbohydrate is represented by ϕ_{fix}.

From mass balance,

$$\phi_{in} = \phi_{fix} + \phi_{out} \tag{3.1}$$

and the fraction of entering carbon that is fixed is represented by f, where

$$f = \phi_{fix}/(\phi_{fix} + \phi_{out}) = \phi_{fix}/\phi_{in} \tag{3.2}$$

Fractionation resulting from the combined effects of transport and fixation is represented as ε_P:

$$\begin{aligned} \varepsilon_P &= 10^3\,[(\delta_{CO2aq} + 1000)/(\delta_{org} + 1000) - 1] \\ &\sim \delta_{CO2aq} - \delta_{org} \end{aligned} \tag{3.3}$$

where δ_{CO2aq} is the carbon isotopic composition of dissolved CO_2, and δ_{org} represents the ^{13}C content of organic carbon in the whole cell (Popp et al. 1989; Freeman and Hayes 1992). Based on mass balance considerations, ε_P can be expressed as a function of the relative amount of carbon fixed, and the isotope effects associated with transport (ε_t) and fixation (ε_f):

$$\varepsilon_P = \varepsilon_f - f(\varepsilon_f - \varepsilon_t) \tag{3.4}$$

Thus, total isotopic discrimination (ε_P) is greatest and approaches fractionation during the enzymatic fixation reaction (ε_f) when a small portion of incoming

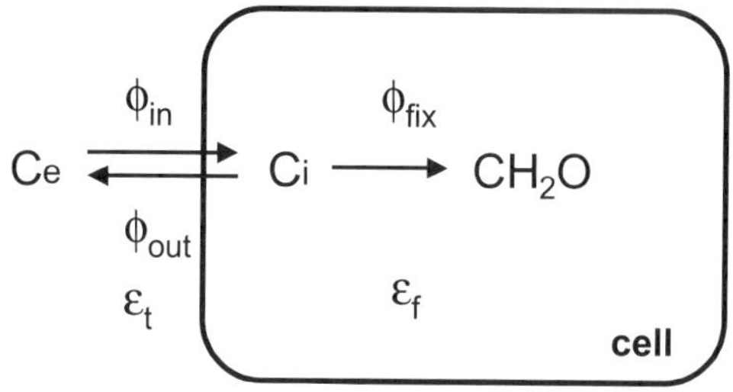

Figure 3.1. Schematic representation of carbon flow during marine photosynthetic carbon fixation. See text for definition of terms.

carbon is fixed. Total fractionation is small and approaches fractionation during transport (ε_t) when most of the carbon entering the cell is fixed into organic carbon (i.e., $f \rightarrow 1$). The ratio of carbon fixed relative to incoming carbon (f) is influenced by a multitude of factors, and it is the primary focus of many field and laboratory studies (e.g., Francois et al. 1993; Laws et al. 1995; Rau, Riebesell, and Wolf-Gladrow 1996; Popp et al. 1998; Burkhardt, Riebesell, and Zondervan 1999; Laws et al. 2001).

3.2.1 Controls on the Relative Amount of Carbon Fixed

In general, the rate of inorganic carbon fixed is a function of the concentration of available carbon and cellular growth rate. The cell is separated from its external environment by the cell membrane, or plasmalemma. The permeability of the membrane with respect to CO_2 is, to a large extent, proportional to its surface area (Laws et al. 1995; Laws, Bidigare, and Popp 1997). If the concentration of ambient dissolved carbon dioxide exceeds demand and/or the organism lacks the ability to actively increase the availability of carbon, then the supply of dissolved carbon dioxide is controlled by diffusion, with a flux that is proportional to the concentration gradient across the membrane (Farquhar et al. 1989). Leaking of carbon back to the external environment, as well as the demand for fixed carbon by the growing cell, will deplete the intracellular pool of dissolved carbon (c_i). Growth rates can be limited by a variety of factors, especially nutrient concentrations and light availability. Growth rates are typically reported in units of per day (day^{-1}) and must be scaled by the amount of organic carbon within the cell (which is proportional to its volume; Verity et al. 1992; Popp et al. 1998) in order to reflect the change in mass per unit time. Thus the demand for carbon is proportional to both growth conditions and the size and shape of the cell.

In summary, if carbon is diffusively transported across the cell membrane, then the fraction of carbon fixed relative to the incoming flux (f) is proportional to the concentration of CO_2 outside the cell (c_e), the growth rate of the cell (μ), the quantity of organic carbon of the cell (C), and the permeability of the membrane with respect to CO_2 (P):

$$f = \phi_{fix}/\phi_{in} = \mu C/Pc_e \tag{3.5}$$

This equation (modified from Laws et al. 2001; Hayes 2001) assumes that carbon enters and exits by diffusion across the cell membrane and that the relative fluxes are driven by the concentration gradient across the membrane. In addition, it assumes that the carbon isotopic discrimination related to transport (as well as the resistance to transport across the cell membrane) is similar for the flow of CO_2 into and out of the cell.

The parameters that influence f can be substituted such that the cell volume (V) serves as a proxy for the amount of carbon in the cell (C), and the cell surface area (A) is used to approximate the permeability of the cell membrane (P). Thus the equation for fractionation during photosynthesis can be recast in terms of parameters that are more readily measurable in laboratory studies:

$$\varepsilon_P = \varepsilon_f - (\varepsilon_f - \varepsilon_t)(V/S)(\mu/c_e) \tag{3.6}$$

with the term V/S representing the ratio of cell volume to cell surface area (Popp et al. 1998).

This formulation emphasizes the influence of three fundamental controls on carbon isotopic fractionation by marine algae: (1) the concentration of carbon dioxide outside the cell, (2) the availability of a growth-limiting property, and (3) specific cell qualities, especially volume and surface area. Thus in field or culture studies, one might expect (and indeed observe) differences among species that reflect both the size and the shape of algal cells (Popp et al. 1998; Pancost et al. 1997). Further, one would expect significant control for a given species by both carbon concentrations and the availability of nutrients and/or light in the growth environment. The influence of carbon dioxide concentrations is, of course, of particular concern to those interested in paleoclimate reconstructions. However, growth conditions and cell geometry cannot be ignored in studies of ancient climate, and this continues to provide the biggest challenge to such applications.

In field studies of marine algae, growth rates and cell geometries can be difficult to characterize, and, further, older data sets rarely include these parameters. Thus Rau et al. (1992) and Bidigare et al. (1997) proposed the folding in of growth properties into a single term (b):

$$\varepsilon_p = \varepsilon_f - b/c_e \tag{3.7}$$

The term b represents all influences on ε_p other than CO_2(aq) concentrations and fractionation during enzymatic fixation.

3.2.2 The Problem of Active Transport

In the preceding discussion, it is assumed that carbon moves across the cell membrane by diffusion only. From a geological perspective, CO_2 concentrations in the modern ocean are low and possibly have been low for the past 25 million years or more (Pagani, Arthur, and Freeman 1999a) relative to past warm periods (Berner and Kothavala 2001). Thus it is not surprising that many modern algae can actively enhance the concentration of intracellular carbon dioxide as CO_2 concentrations become limiting to growth. Such carbon concentrating mechanisms (CCM) are inducible (Sharkey and Berry 1985; Laws, Bidigare, and Popp 1997) under low CO_2(aq) concentrations and involve active transport and/or CO_2 enhancement by way of enzymatic conversion of HCO_3^- into CO_2(aq). Laws et al. (1997) suggested that cells adjust the mechanism of transport (active vs. diffusion) in order to minimize energy costs associated with the demands of active transport. Accordingly, they propose a nonlinear model, where fractionation (ε_P) is a function of the relative amount of carbon delivered to the cell by active means. Keller and Morel (1999) propose a similar mathematical model, which can account for the uptake of either bicarbonate or aqueous CO_2. Based on available isotopic data, Laws et al. (2001) suggest aqueous CO_2 is the likely substrate for active transport.

Although active transport has been associated with lower ε_P values, Laws et al. (2001) emphasize that they are not diagnostic of its presence. In particular, a decrease in ε_p due to the use of ^{13}C-enriched bicarbonate can be offset by an increase in the flux of carbon into the cell such that the ratio of fixed to incoming carbon (f) decreases. For the purposes of ancient climate reconstruction, active transport poses a difficult problem, since it can decouple isotopic fractionation from carbon availability. In studies of past CO_2, the authors seek biomarkers for a taxonomic group for which active transport is unlikely. For example, laboratory evidence suggests that neither active transport, nor a carbon concentrating mechanism, nor calcification are significant influences on ε_p of certain haptophyte algae and that the diffusion model is appropriate (Bidigare et al. 1997; Nimer, Iglesias-Rodriguez, and Merrett 1997; Sikes, Roer, and Wilbur 1980), at least under conditions similar to the modern and near-recent ocean. However, when one is working on very long timescales, modern analogs of ancient species are less common; thus an assumption of biological uniformitarianism becomes increasingly risky. Under these circumstances, paleo-CO_2 research requires the assumption that past CO_2 levels were sufficiently high such that a carbon concentrating mechanism was not required, or that the ratio of active to diffusive carbon transport was relatively invariant.

3.2.3 Isotopic Biogeochemistry of Alkenone-Producing Organisms

Alkenones are long-chain ketones (Fig. 3.2) derived from a species of haptophyte algae (Marlowe et al. 1984).

Most notably, they are produced by *Emiliania huxleyi* (Volkman et al. 1980), which is found throughout the modern ocean (Westbroek et al. 1993) and *Geo-*

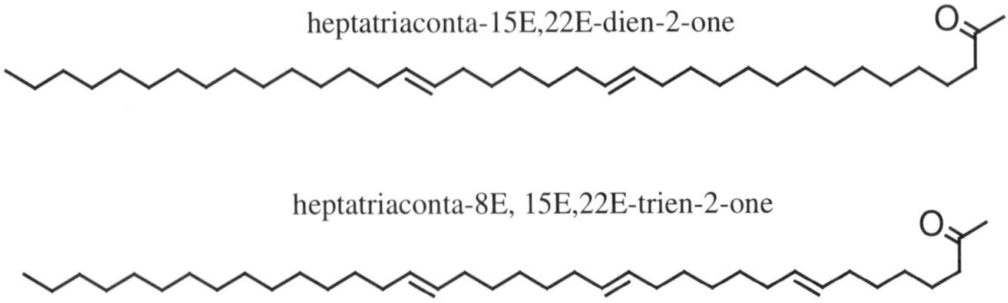

Figure 3.2. The molecular structures of $C_{37:2}$ and $C_{37:3}$ alkenones.

phyrocapsa oceanica (Volkman et al. 1995). The most commonly observed structures are the C_{37}, C_{38}, and C_{39} methyl and ethyl ketones (de Leeuw et al. 1980), with heptatriaconta-15E,22E-dien-2-one ("$C_{37:2}$ alkenone") and heptatriaconta-8E, 15E,22E-trien-2-one ("$C_{37:3}$ alkenone") most often employed in sediment studies because of their utility in paleotemperature reconstructions (e.g., Brassell et al. 1986; Reviewed by Brassell 1993; Sachs and Lehman 1999).

Rechka and Maxwell (1988a, b) confirmed the *trans* configuration of the double bonds and suggested this property makes them more resistant to biological degradation. Indeed, they are more refractory than other lipids in marine waters and sediments (Wakeham et al., 1997a; 1997b, 2002); there is, however, evidence that under certain conditions, the more unsaturated forms may be lost preferentially (Freeman and Wakeham 1992; Gong and Hollander 1999).

Alkenones have attracted a great deal of attention from paleoceanographers. There are many field and laboratory studies of the temperature dependence of the relative abundance of $C_{37:2}$ and $C_{37:3}$ alkenones (Brassell 1993; Herbert 2001), as well as their isotopic properties. For purposes of reconstructing ancient CO_2 levels, alkenones are perhaps the most popular choice because of their ubiquity, preservation, and taxonomic specificity. In addition, as noted above, alkenone-producing haptophytes apparently do not employ active transport mechanisms at the CO_2 concentrations found in the oceans today, allowing the application of equations 4 and 6.

3.2.4 Laboratory Observations

The isotopic properties of alkenones and their source organisms are documented in numerous laboratory experiments that use a nitrate-limited continuous culture system called a chemostat. The chemostat system permits investigators to control growth rates by regulating the supply of a limiting nutrient (e.g., nitrate) and to perform experiments under steady-state conditions. The method has been successfully applied to evaluate the relationship between ε_p, c_e, μ, and cell geometry (reviewed by Laws et al. 2001), for haptophytes *Emiliania huxleyi* (Bidigare et al. 1997) and *Isochrysis galbana* (Laws et al. 2001), as well as the diatoms *Phaeodactylum tricornutum* (Laws, Bidigare, and Popp 1997) and *Porosira gla-*

cialis (Popp et al. 1998) among others. Light- and temperature-limited growth of *E. huxleyi* using a turbidostat (Rosenthal et al. 1999) gives similar response to the nitrate-limited growth found in the chemostat experiments, with a linear relationship between ε_p and μ/c_e.

The more common and traditional methodology involves batch cultures grown in a medium containing nutrients in excess required for growth. Although batch cultures are sensitive to initial conditions, they can be effective in isotopic studies, provided carbon is harvested while cell densities are low to prevent self-shading and to prevent substantial changes in the inorganic carbon pools (Hinga et al. 1994). Unfortunately, there is an unresolved difference between the results from chemostat experiments and some dilute, light-limited but nitrate-replete batch cultures. Batch experiments results record overall lower values for ε_p, and nonlinear relationships between ε_p and μ/c_e, and this may indicate that the regulation of carbon uptake is influenced by the abundance of nitrate or possibly other nutrients (Burkhardt, Riebesell, and Zondervan 1999; Riebesell et al. 2000a), or it may reflect differences in isotopic systematics between stationary (chemostat) and exponential (batch cultures) growth phases. However, the specific cause for the differences among various batch experiments, and between continuous culture and batch cultures, is unresolved and remains an area for further investigation (Laws et al. 2001).

Culture studies are useful to evaluate intermolecular isotopic variations. For studies of continuous culture, alkenones are, on average, 3.9‰ ($\sigma = 0.8$) depleted relative to whole cells (Laws et al. 2001). Values reported for batch cultures (Riebesell et al. 2000b) are slightly more depleted and average -5.2 ($\sigma = 0.3$), with the combined data set (continuous culture + batch) giving an average $\Delta\delta$ value of -4.5 ($\sigma = 0.8$). This value is close to the value used by Jasper et al. (1994), -3.8‰; and it is generally consistent with observed depletion of linear lipids from whole biomass of 3 to 4‰ (Hayes 1993; Schouten et al. 1998). In general, field-based studies of alkenones assume an invariant $\Delta\delta$ value of 4‰; (Bidigare et al. 1997; Jasper et al. 1994). As reviewed by Hayes (2001), the isotopic composition of an individual cellular component can be influenced by changes in the flux of carbon to various fates within the cell, and thus there is a prospect for $\Delta\delta$ values to be sensitive to conditions in the growth environment. Subtle variations in growth conditions could potentially explain the range in observed values of $\Delta\delta$ as well as the approximate 1‰ difference observed between continuous and batch culture experiments.

Laboratory results have helped to determine the fractionation of carbon isotopes associated with enzymatic fixation (ε_f). In vitro results with ribulose-1,5-bisphosphate caroxylase/oxygenase (Rubisco) document values of 30–29‰; for eukaryotes, and lower values (22–18‰) for the bacterial form of Rubisco (Guy, Fogel, and Berry 1993; Roeske and O'Leary, 1985). Microaglae have, in addition to Rubisco, a small activity of β-carboxylase enzymes that attenuate the net fractionation. Chemostat data suggest the combined enzymatic value is around 25‰ for haptophytes, and this value is generally consistent with field observations (Bidigare et al. 1997; Popp et al. 1998). As discussed by Goericke and

Fry (1994) et al. (1994), a range of values is possible, depending on relative enzyme activities, with 25‰ corresponding to approximately 10% β-carboxylase activity. A lower activity would lead to a greater value for ε_f, and Goericke and Fry (1994) et al. (1994) calculate a possible range of 25–28‰. A range of values was observed in the chemostat experiments of Bidigare et al. (1997) for *E. huxleyi,* and calculations incorporating the 95% confidence limits of these data yeild 25–27‰; for ε_f (Pagani et al. 1999a).

3.2.5 Field Observations

Bidigare et al. (1997; 1999a) compiled alkenone-based measurements of ε_p to evaluate relationships between ε_p, growth conditions, and observed concentrations of CO_2(aq). Laws et al. (2001) and Pagani (2002) have enlarged the data set with the inclusion of more recent studies (Popp et al. 1999; Eek et al., 1999; Laws et al., 2001); in total, the observations are drawn from environments that span the ocean and the spectrum of nutrient availability (Fig. 3.3). Values for b are calculated from equation 7, using ε_p values determined from $\delta^{13}C$ values of alkenones and CO_2(aq). We present the b values calculated using $\varepsilon_f = 26‰$ in order to represent the median in the range of values estimated from the chemostat results noted in the preceding discussion. The calculations assume an isotopic difference between biomass and alkenones of 4‰.

The range of values for b reflect the influence of factors other than the concentration of carbon dioxide, principally cell geometry (V/S) and growth rate (μ), on the expression of ε_p. Importantly, this data set represents isotopic characteristics of only alkenone-producing haptophyte algae, thus most probably minimizing the influence of cell geometry. It is proposed, therefore, that the parameter b represents largely the influence of growth rate. Bidigare et al. (1997) showed that this approach is consistent with growth rates measured in the field and that it is generally consistent with continuous-culture laboratory experiments employing both calcifying and noncalcifying strains of *E. huxleyi*. Unfortunately, most of the field studies that constitute this compilation did not include direct measurements of growth rates because this procedure is difficult to do in the field. Thus, the link between growth rates and values for b is based on the correlation between b and the measured concentration of phosphate in the same samples (Fig. 3.3).

The strong correlation between b and PO_4^{3-} is less than compelling, however, when one considers that phosphate concentrations are generally too high to be growth limiting in most of the samples (Bidigare et al. 1997). Nevertheless, the strong correlation suggests that some growth-limiting property is proxied by the phosphate measurements, such as a trace element that serves as a micronutrient (Bidigare et al. 1997).

Both Laws et al. (2001) and Pagani (2002) discuss the relationship between phosphate and b in detail, with consideration of additional controls. For example, as shown in Fig. 3.4A, there is some correlation between phosphate and CO_2 concentrations, which influences the relationship with b values. Correlations

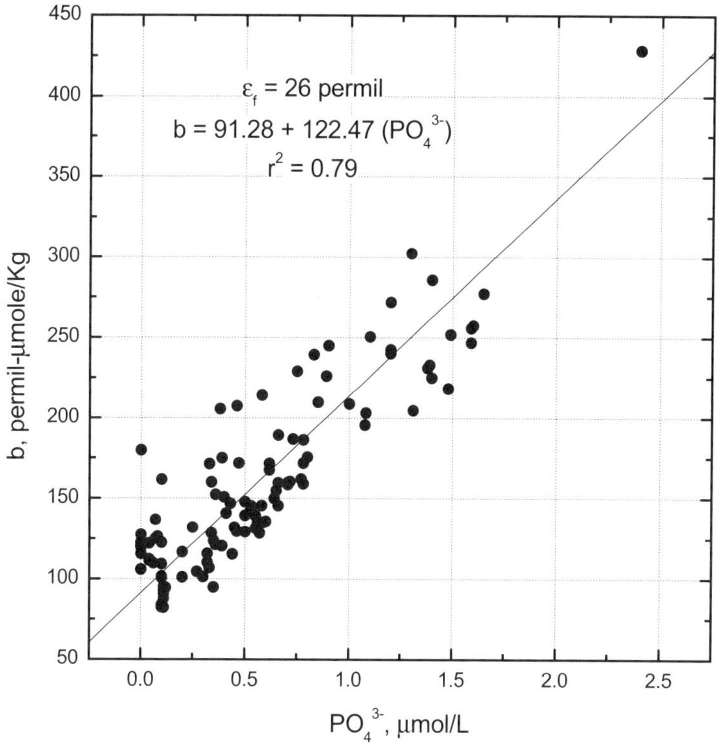

Figure 3.3. Alkenone-based estimates of b plotted as a function of observed PO_4^{3-} concentrations. Data are taken from values compiled by Pagani (2002), with ε_f = 26‰.

between ε_p values and either phosphate ($r^2 = 0.23$) and $CO_2(aq)$ ($r^2 = 0.1$) are obviously not as strong as the correlation between ε_p and $PO_4^{3-}/CO_2(aq)$, where $r^2 = 0.31$ (Fig. 3.4B,C). Although there is a visible trend in Fig. 3.4C, there remains significant variation in ε_p that is not accounted for by the $PO_4^{3-}/CO_2(aq)$ ratio and it is possible that other factors in the growth environment (temperature, light quality, etc.) or cell geometry are important in the open-ocean photic zone. Exact controls on isotopic fractionation continue to be debated, and future field and laboratory studies will likely shed further light on the chemical and biological controls on ε_p in modern waters.

3.2.6 A Sediment Test of the Alkenone Method

In the modern ocean, phosphate concentrations are more highly correlated to alkenone-derived ε_p values than other factors, including $CO_2(aq)$. Although a range of CO_2 concentrations exists in surface waters today, it is relatively narrow, which reflects the fact that ocean surface waters approach (but do not reach) chemical and isotopic equilibrium with the atmosphere (reviewed by Freeman,

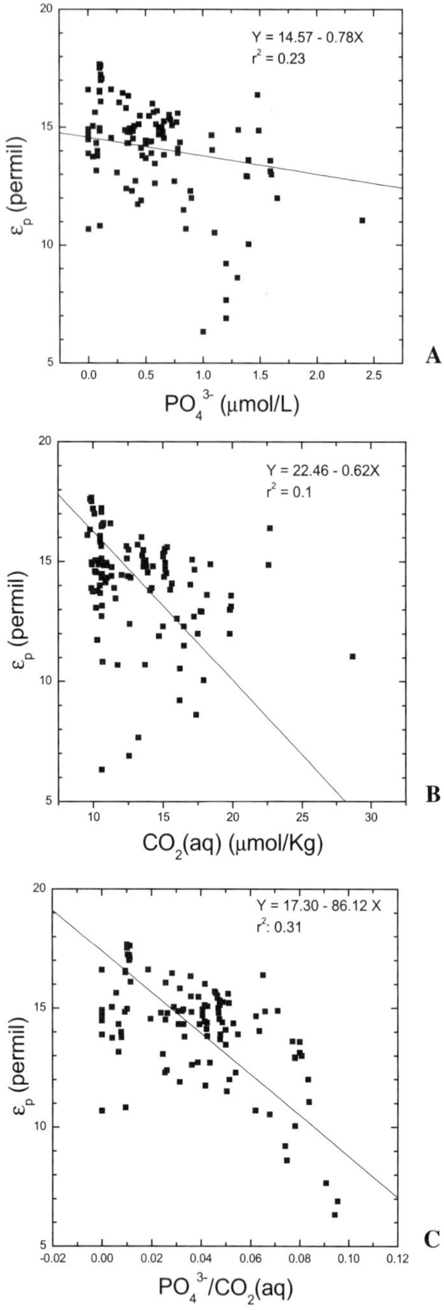

Figure 3.4. Cross plots of ε_p values with phospate concentrations **(A)**, aqueous carbon dioxide **(B)**, and the ratio of phosphate to carbon dioxide **(C)**. All lines represent geometric mean regressions (Sokal and Rohlf 1995).

2001). In order to test the influence of CO_2 on ε_p values, one is forced to exploit laboratory studies, which may or may not reconstruct natural conditions adequately (Laws et al. 2001). An alternative approach is to look to the sedimentary record of recent atmospheric changes.

We recently (Pagani et al. 2002) tested the relationship proposed from alkenones in modern ocean waters by using samples collected from sediments along a transect in the central Pacific Ocean. The depth of alkenone production was inferred from temperatures estimated based on the unsaturation ratio of alkenone abundances (Uk_{37}' values; Ohkouchi et al. 1999), and we employ these depth estimates to determine phosphate values along with both the concentration and $\delta^{13}C$ values of dissolved CO_2. Based on measured inorganic carbon and alkenone $\delta^{13}C$ values, water-column phosphate concentrations at the depth of alkenone synthesis, and the relationship expressed in Fig. 3.3, we calculated $CO_2(aq)$ concentrations at the time of alkenone production.

Alkenone-based $CO_2(aq)$ values were consistently lower than concentrations observed in the modern ocean. However, this difference can be largely accounted for by the removal of anthropogenic contributions from modern values. The resulting match is strong: 84% of the alkenone-based estimates fall within 0–20% of the preindustrial water-column values. A subset of these results is shown in Fig. 3.5, which represents estimates based on water-column data collected in the spring.

The strength of this agreement suggests that the empirical relationship between b and phosphate derived from a wide variety of oceanographic environments (including high- and low-phosphate concentrations) can be used to reconstruct ancient CO_2 levels from alkenone $\delta^{13}C$ values, provided an estimate of paleophosphate concentration or a reasonable range of values are available. It further supports earlier findings that the isotopic composition of alkenone-producing haptophytes is not substantially affected by the active transport of inorganic carbon or calcification. This is promising news for efforts to reconstruct greenhouse gas variations in the past.

3.3 Alkenone-Based Estimates of Ancient CO_2 Levels

3.3.1 Measured Properties in the Calculation of Ancient CO_2 Levels

When employing alkenones for ancient CO_2 reconstructions, the rearranged form of Equation 3.7 is employed:

$$c_e = (\varepsilon_f - \varepsilon_p)/b \tag{3.8}$$

Although this equation contains only three variables, they represent a large number of observations and underlying calculations. The observed and derived properties are listed in Table 3.1, along with our estimate of the approximate uncertainty in the value and the primary basis for each error estimate.

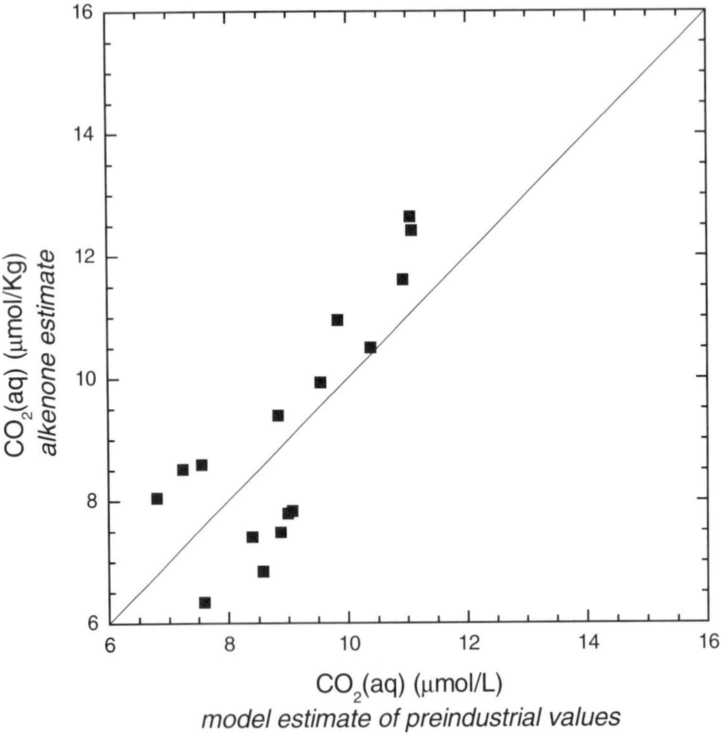

Figure 3.5. Cross plot of the estimated preindustrial CO_2 in western Pacific waters (from Pagani et al. 2002). The ordinate represents values based on alkenone measurements; the abscissa represents values determined from measurements in modern waters minus an estimated contribution of anthropogenic carbon. The estimate of anthropogenic CO_2 is based on the Geophysical Fluid Dynamics Laboratory's Modular Ocean Model (Pacanowski et al. 1991). Values plotted are based on spring production data. The line represents a 1:1 correlation.

Values for ε_p are based on the isotopic difference between an estimate of the $\delta^{13}C$ values of the organic matter produced by ancient algae ($\delta^{13}C_{org}$) and the coexisting dissolved CO_2 in the photic zone ($\delta^{13}C_{carb}$) as noted earlier and defined in Equation 3.3. The value for $\delta^{13}C_{org}$ is determined by adding 4‰ to the measured value of the $C_{37:2}$ alkenone extracted from the ancient marine sediment. This correction reflects lipid-biomass isotopic differences ($\Delta\delta$) discussed previously and is consistent with values employed by other researchers and in our earlier CO_2 reconstructions (Pagani et al. 1999a, b). We have selected the median of the estimate for ε_f (= 26‰) derived from chemostat culture studies as discussed previously, giving an uncertainty of 1‰.

The isotopic composition of carbon dioxide in the ancient growth environment is determined from measurements of a selected carbonate phase known to derive

Table 3.1. Properties involved in estimating ancient CO_2 concentrations and the nature and sources of their respective error estimates

Property	Source	Approx. Error	Basis for Error Estimate
$\delta^{13}C_{carb}$	Foraminifera or some other planktonic carbonate phase	0.2‰	Analytical
$\delta^{13}C_{org}$	$\delta^{13}C$ of $C_{37:2}$ alkenones + 0.4‰	1.0‰	Analytical; culture studies
Temperature	$\delta^{18}O$ of planktonic carbonate source	3°C	Analytical (0.1‰), estimated uncertainty in SMOW
Phosphate	Estimated from ocean or sediment properties	0.1 μMol/kg	Best guess
b values	Field data	20%	Regression statistics
ε_f	Lab studies	1.0‰	Available data

from the photic zone in the upper water column, such as single species or genera of planktonic foraminifera (Pagani et al. 1999a). The carbonate values are used to calculate $\delta^{13}C$ values of CO_2(aq) in equilibrium with the carbonate phase, based on isotopic relationships determined by Romanek, Grossman, and Morse (1992) and Mook, Bommerson, and Staberman (1974) and as detailed in the papers by Freeman and Hayes (1992) and Pagani et al. (1999a). For example, combining the equations of Romanek et al. (1992) and Mook et al. (1974), $\delta^{13}C$ values for dissolved CO_2 are calculated by the following equation:

$$\delta^{13}C_{co2aq} = 373/T + 0.19 + \delta^{13}C_{carb} - 11.98 + 0.12T \tag{3.9}$$

where T is temperature in kelvin.

This calculation requires an estimate of ancient sea surface temperatures, which are typically evaluated using the measured $\delta^{18}O$ values of the planktonic carbonate phase. The temperature estimates require the $\delta^{18}O$ value of coexisting seawater (i.e., ancient SMOW), which can be difficult to determine and is the principle source of uncertainty in this aspect of the calculation. Analytical uncertainties in both carbon and oxygen isotopic measurements are small, approximately 0.2 and 0.1‰, respectively.

A detailed discussion for the strategy and methods for calculating ε_p from isotopic analyses of sedimentary alkenones is provided by Pagani et al. 1999a. In that work, we employed alkenones extracted from sediment samples, in concert with 250–354 and 354–420 μm size-fractionated planktonic foraminifera. The selection of larger foraminifera specimens is routine in isotopic studies in order to avoid isotopic effects associated with juvenile organisms. In addition, there are documented metabolic effects, for example, due to interactions with symbiont phototrophs, which can cause foraminiferal $\delta^{13}C$ values to deviate from equilibrium with the water column (Spero and DeNiro 1987; Spero and Lea

1993). These influences are typically less than 1.5‰, but if not accounted for, they will generally lead to an overestimation of CO_2 concentrations.

A value for b is the final variable required to determine ancient CO_2 levels. The term b is calculated from estimates of ancient phosphate concentrations and the regression shown earlier, in Fig. 3.3. Phosphate values can be estimated based on modern ocean properties in the near recent (i.e., Pagani et al. 2002) or from properties of the accumulated sediments. In this regard, it is recommended that researchers use samples from oceanographic sites that have experienced relatively stable nutrient conditions in the photic zone. Such sites include open gyre localities, where nutrient concentrations are low and stable. In the ancient record, linear accumulation of biogenic materials (e.g., Pagani et al. 1999a) provides strong evidence for nutrient stability over time. Multiple sites and records should be studied in order to dissect global signals from the influence of local conditions on an isotopic record from an individual site (Freeman 2001).

3.3.2 Propagation of Uncertainties

The propagation of errors in the calculation of ancient CO_2 levels can be carried out using basic equations for the handling of random errors, with the assumption that the errors are independent and Gaussian in distribution. In general, for a property, x, that is a function of more than one variable (u, v):

$$\sigma_x^2 = \sigma_u^2 \left(\frac{\partial x}{\partial u}\right)^2 + \sigma_v^2 \left(\frac{\partial x}{\partial v}\right)^2 + 2\sigma_{uv}^2 \left(\frac{\partial x}{\partial u}\right)\left(\frac{\partial x}{\partial v}\right) \tag{3.10}$$

The first two terms represent the sum of squares of individual uncertainties weighted by the squares of the partial derivatives of the function. The third term, when the errors of u and v are independent, becomes negligible and is typically ignored.

Taylor (1982) and Bevington and Robinson (1992) provide excellent discussions and examples of the application of error propagation. Readers with experience in analytical chemistry or physics will be most familiar with such an approach and the various error relationships derived from Equation 3.10. For example, the combined error for added or subtracted properties is simply the square root of the sum of the squares of the errors for the properties.

Equation 3.10 is used for the propagation of errors in Equation 3.8. Because errors are assumed to be independent, the final cross term is ignored. In order to simplify this presentation, we have combined the difference ($\varepsilon_f - \varepsilon_p$) into a single term $\Delta\varepsilon$:

$$\sigma_{ce}^2 = \sigma_{\Delta\varepsilon}^2 \left(\frac{\partial(b/\Delta\varepsilon)}{\partial\Delta\varepsilon}\right)^2 + \sigma_b^2 \left(\frac{\partial(b/\Delta\varepsilon)}{\partial b}\right)^2 \tag{3.11}$$

Differentiating the terms in Equation 3.11 yields:

$$\sigma_{ce}^2 = \sigma_{\Delta\varepsilon}^2 \left(\frac{b}{\Delta\varepsilon^2}\right)^2 + \sigma_b^2 \left(\frac{1}{\Delta\varepsilon}\right)^2 \tag{3.12}$$

This equation forms the basis for the calculation of uncertainties discussed below.

3.3.3 Errors in Δε

Δε is calculated from the subtraction of ε_p from ε_f, and the cumulative error is the square root of the sum of the squares of the errors in each property:

$$\sigma_{\Delta\varepsilon}^2 = \sigma_{\varepsilon f}^2 + \sigma_{\varepsilon p}^2 \tag{3.13}$$

As noted above, we assume $\varepsilon_f = 26‰ \pm 1‰$. The uncertainty in values of ε_p is based on two terms (see Equation 3.3): the $\delta^{13}C$ values of dissolved CO_2 and organic carbon derived from phytoplankton. The error estimate for $\delta^{13}C_{org}$ values is based on both the analytical error in the isotopic measurements of alkenones (ca. 0.7‰; Pagani et al. 1999a; 0.4‰ Pagani et al. 2002) and the laboratory studies reporting about 4‰ lipid-biomass differences (0.8‰). Here we employ 0.6‰ for the error in the isotopic analysis of individual alkenones and 0.8‰ for the uncertainty in the intermolecular correction. Thus the error in $\delta^{13}C_{org}$ is the square root of the sum of the squares of these values, or 1‰.

Error determination for inorganic carbon is based on the combined errors for the temperature estimate (ca. 3°C) and the analytical uncertainty in the carbon isotopic measurements (0.2‰). The calculation is based on Equation 3.10, using Equation 3.9 as the differentiated function. Thus:

$$\sigma_{co2aq}^2 = \sigma_T^2 \left(\frac{\partial f}{\partial T}\right)^2 + \sigma_{carb}^2 \left(\frac{\partial f}{\partial \delta_{carb}}\right)^2 \tag{3.14}$$

where the terms σ_{co2aq}, σ_T, and σ_{carb} represent the uncertainties in the values for $\delta^{13}C$ of CO_2(aq), temperature, and $\delta^{13}C_{carb}$, respectively. The term δ_{carb} represents the $\delta^{13}C$ values of planktonic carbonate. The term f is the function used to calculate $\delta^{13}C_{co2aq}$ represented by Equation 3.9. Using this approach, differentiating Equation 3.9 in terms of both T and δ_{carb}, employing the error estimates listed in Table 3.1, and assuming T = 298K, we estimate an uncertainty of 0.4‰ for the $\delta^{13}C$ value of aqueous CO_2.

Based on the above calculations, we now calculate the uncertainty in the estimate of ε_p values:

$$\begin{aligned} \sigma_{\varepsilon p}{}^2 &= \sigma_{co2aq}{}^2 + \sigma_{org}{}^2 \\ &= (0.4)^2 + (1.0)^2 \\ \sigma_{\varepsilon p} &= 1.1‰ \end{aligned} \tag{3.15}$$

This value can be applied to Equation 3.13, and the error for Δε determined:

$$\begin{aligned}\sigma_{\Delta\varepsilon}{}^2 &= \sigma_{\varepsilon f}^2 + \sigma_{\varepsilon p}{}^2 \\ &= (1.0)^2 + (1.1)^2 \\ \sigma_{\Delta\varepsilon} &= 1.5‰\end{aligned} \quad (3.16)$$

The uncertainty for b is based on the standard error determined from regression statistics for the alkenone data set discussed earlier and presented in Fig. 3.3. This data set includes 109 pairs of values for b and phosphate determinations from a wide variety of localities in the modern ocean. The standard error is calculated as discussed by Sokal and Rohlf (1995):

$$\sigma_b = \sqrt{s_{yx}^2\left[1 + \frac{1}{n} + \frac{\left(x_i - \bar{x}\right)^2}{\Sigma x^2}\right]} \quad (3.17)$$

where

$$s_{yx}^2 = \left(\Sigma y^2 - \frac{\Sigma xy^2}{\Sigma x^2}\right)$$

and the properties Σy^2 and Σx^2 represent the sum of the squares of the deviations for each variable, with Σxy^2 representing the covariance (Sokal and Rolf, 1995). In the regression (shown in Fig. 3.3), x is represented by phosphate values, while y represents values for b. For values of phosphate ranging from 0.2 to 1 μmol/kg (using $\varepsilon_f = 26‰$), the error for b is 27.2‰-μmol/kg. Uncertainties based on the regression statistics increase for values for phosphate concentrations that are higher or lower than those in the calibration data set.

This represents an approximately 20% uncertainty derived from the regression of b versus phosphate concentrations from all available modern field observations. If ancient samples are from a locality for which a specific phosphate-b calibration is possible, it may reduce the magnitude of this uncertainty. For example, Popp et al. (1999) used a subset of the above data set for the Pacific Ocean and estimated approximately 3% uncertainty for the relationship derived from the regression of b and PO_4^{3-}. This resulted in an overall lower estimate of the uncertainty in the alkenone method for CO_2 reconstruction (ca. 11%).

By employing $\varepsilon_f = 26‰$, $\sigma_b = 27.2‰$ μmol/kg, and $\sigma_{\Delta\varepsilon} = 1.6‰$, we can use Equation 3.12 to calculate the uncertainty in estimated CO_2 concentrations for given values of b and ε_p. These values are presented as absolute uncertainty (in μmol/kg) in Fig. 3.6 and as a relative uncertainty (%) of calculated c_e estimates in Fig. 3.7.

The shaded regions represent values of b that correspond to a range of phosphate concentrations between 0.1 and 1.0 μmol/kg. As shown in these figures, the uncertainties in estimated ancient CO_2(aq) concentrations range between 20 and 30% for ε_p values below 20‰, and they increase dramatically with greater isotopic fractionation (ε_p). This increase represents the decreased sensitivity of

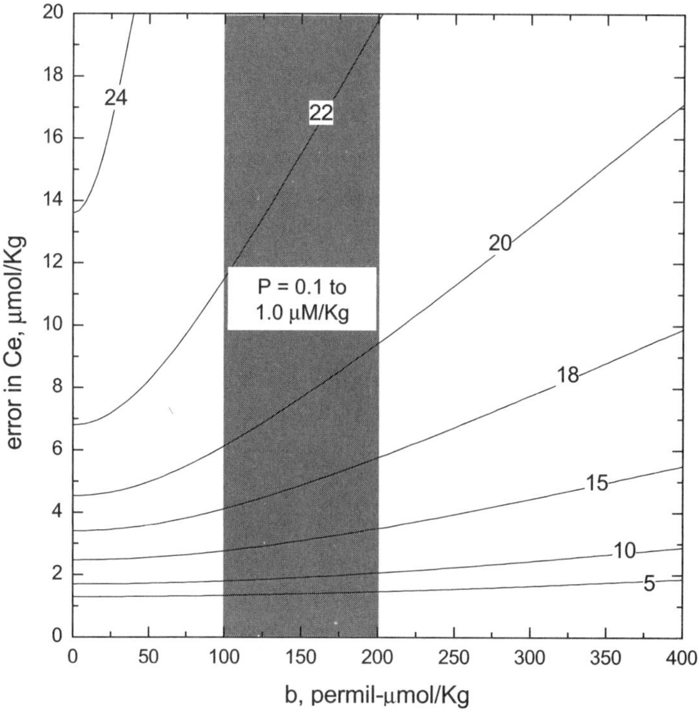

Figure 3.6. Calculated values of absolute uncertainties in estimates of dissolved CO_2 concentrations based on alkenone analyses. Shaded region represents the range of values of b corresponding to phosphate concentrations in the modern ocean. Contours represent ε_p values in units of ‰.

the technique at high CO_2 levels. Once CO_2 reaches higher than 8 to 10 times modern levels, carbon saturates fixation by algae, and, as noted in the discussion of Equation 3.4, $f << 1$, and $\varepsilon_p \rightarrow \varepsilon_f$.

When CO_2 levels are relatively constrained, isotopic analyses of alkenones and inorganic carbon can be used to evaluate changes in growth rates and, by association, changes in nutrient concentrations (Pagani, Arthur, and Freeman 2000; Bidigare et al. 1999b). A similar approach to error propagation can be invoked to estimate uncertainties in the reconstructed values. Although in this case, the uncertainty in the estimate will rely on how well one knows values for CO_2 concentrations in addition to the quality of the isotopic results.

3.3.4 The Problem with Phosphate

The observed relationship between b values and phosphate concentrations plays an important part in the proxy method for CO_2 reconstruction, while also serving as a major source of uncertainty in the results. Constraining phosphate concen-

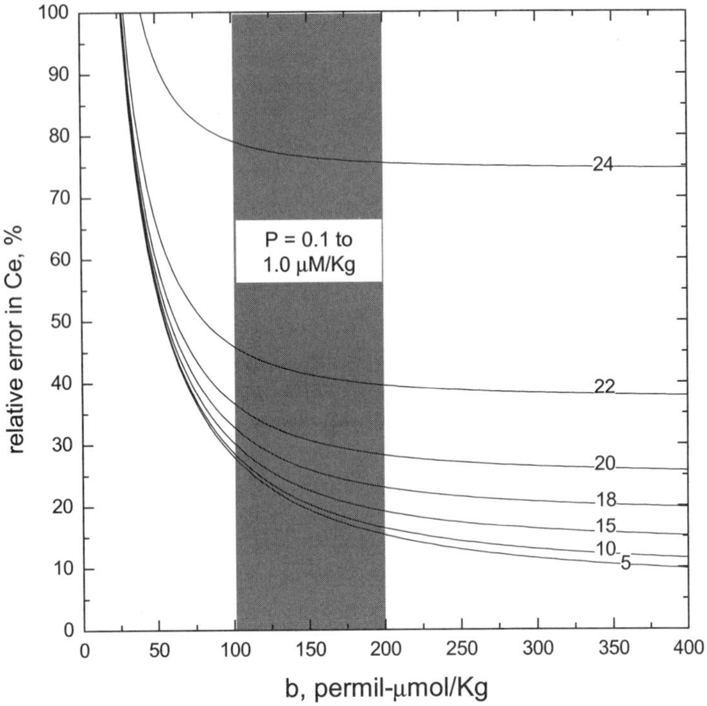

Figure 3.7. Calculated values of the relative uncertainties in estimates of dissolved CO_2 concentrations based on alkenone analyses. Shaded region represents the range of values of b corresponding to phosphate concentrations in the modern ocean. Contours represent ε_p values in units of ‰.

trations in the past is an additional challenge to this approach. In Fig. 3.8, we present a contoured plot of the value for dissolved CO_2 estimated from b for measured ε_p values.

The shaded regions represent the range in the concentrations of CO_2(aq) (horizontal region) and phosphate concentrations (vertical region) observed in the modern ocean. We note, however, that the range of phosphate concentrations could be significantly higher in upwelling regions (see Fig. 3.3), which can lead to values of b that are 300‰-µmol/kg or even higher.

The error in the estimated value of b (ca. 20%) from phosphate concentrations is relatively constant when concentrations of phosphate are within the range observed throughout much of the modern ocean (ca. 0.1–ca. 1 µmol/kg). However, this describes the uncertainty in the *relationship* between these properties based on the variance of phosphate and b measured in the modern ocean. Thus, for ancient studies, it is assumed that phosphate uncertainties are similar to that for modern phosphate values in the data set presented in Fig. 3.3. With good sample site selection, the variation in phosphate can be minimized, and the

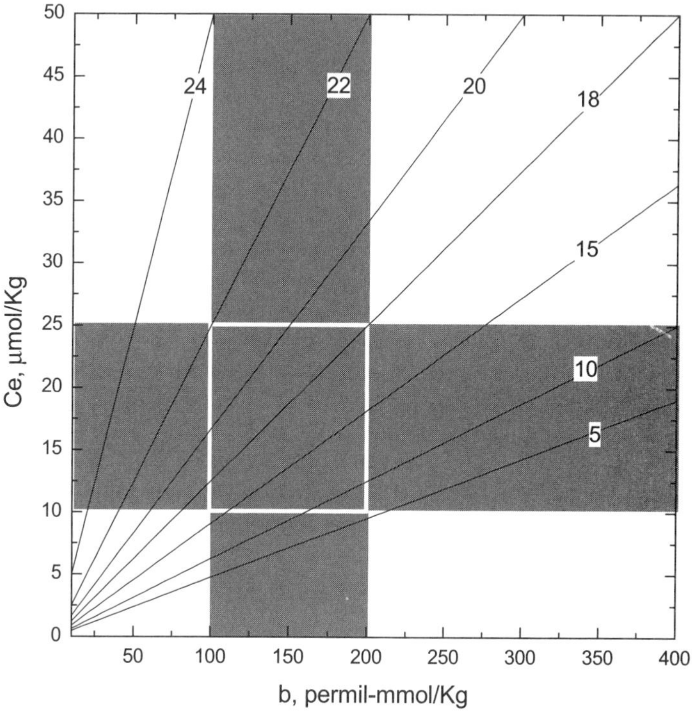

Figure 3.8. Contours of εp values represented estimates of dissolved CO_2 concentrations as a function of b values. The vertical shaded region respresents the range of values of b corresponding to phosphate concentrations in the modern ocean. The horizontal shaded region represents the general range in dissolved CO_2 concentrations observed in the modern ocean.

uncertainty in a paleo-PO_4^{3-} estimation may well be similar to that in the modern data set. In particular, it is essential that the researcher have some estimation of whether the location was characterized by enhanced nutrient concentrations through upwelling, runoff, or other mechanisms as opposed to a low nutrient environment, such as the subtropical gyres today. As noted above (and by Pagani et al. 1999a), the use of sediment accumulation characteristics is essential to sort out these possibilities. These two environmental extremes result in the range in phosphate concentrations shown in Fig. 3.3 (Popp et al. 1999). ε_p values in the modern ocean are relatively more sensitive to variations in phosphate concentrations than CO_2(aq) concentrations, since the latter are partially reset by equilibration with the atmosphere. The dominance of the isotopic signature by phosphate as observed in today's ocean is not likely true through time, since significant changes in atmospheric CO_2 levels in the past will ultimately dominate the record of ε_p variations over Earth's history (Hayes, Strauss, and Kaufman 1999).

It clearly would be helpful to have a better tool for estimating phosphate values than relying on the averaged depositional condition for a given site. Popp et al. (1999) recommend the use of elemental ratios of Cd and Ca in calcite tests from planktonic foraminifera. This ratio is documented to vary with the abundance of phosphate in the modern ocean and may be a useful proxy for reconstructing phosphate concentrations in the ancient ocean (e.g., Boyle 1992; Oppo and Rosenthal 1994; Mashiotta, Lea, and Spero 1997). However, this relationship is not documented for timescales longer than the last glacial era, and there are observations that document a temperature effect (Rickaby and Elderfield 1999). Popp et al. (1999) estimate this approach could potentially provide a phosphate estimate with an uncertainty of about 10%. However, this estimate predated the discovery of the temperature effect, which has discouraged its use in such applications. Currently estimates of paleophosphate concentrations take into account past productivity based on sediment properties and accumulation along with reconstructed circulation and wind patterns. As noted below, these efforts can provide reasonable constraints on phosphate values, giving a range of estimated values.

3.3.5 Miocene Reconstructions

In our previously published CO_2 reconstructions for the Miocene based on alkenones (Pagani et al. 1999a, b), we estimated uncertainties by bracketing our estimates with minimum and maximum values for Equation 3.7. For example, we employed two phosphate values (0.2–0.3 μmol/kg), two values for ε_f (25 and 27‰), the maximum 95% confidence band, along with the geometric mean regression Equation in Fig. 3.3, to yield a range of CO_2 estimates. We considered DSDP site 588 in the southwestern Pacific to represent the sample locality with the greatest stability in terms of nutrient dynamics, although the range of CO_2 estimates was similar for all the sites. At DSDP site 588, the median CO_2 estimates ranged from about 280 to 200 ppmv, with the calculated maximum and minimum values ranging approximately ± 40 ppmv. By employing the error propagation approach presented here, we would raise the error estimate slightly, with uncertainty ranging between 40 and 56 ppmv (ca. $\pm$ 20%). This error is slightly higher than that calculated in an earlier publication using Monte Carlo procedure (Pagani et al. 1999a), which estimated 15% uncertainty for pCO_2 estimates.

Our Miocene CO_2 estimates agree with the results from alternative methods for estimating past CO_2 levels. These include fossil leaf stomatal densities (Royer, Berner, and Beerling 2001), boron isotopes (Pearson and Palmer 2000), and geochemical models (Berner and Kothavala 2001), which all suggest that the Miocene atmosphere contained relatively low and stable CO_2 concentrations.

3.3.6 Paleozoic Studies

Significant problems emerge with this method when applied to the study of climate in more ancient time periods. In particular, the preservation of open-ocean sediments decreases back in time, such that Jurassic and older sediments are rare or nonexistent. For Paleozoic reconstructions, workers are required to use sediment samples from ancient epicontinental seas. Constraining nutrient concentrations in the ancient shallow seas is difficult, due to regional variations in the influences from land runoff, restrictions in circulation, and the sensitivity of the shallow waters to relative changes in sea level. Thus it is a significant understatement to claim that precise reconstruction of very ancient CO_2 using phytoplankton organic materials is challenging. Researchers must employ as many localities as possible in order to document and account for the influence of environmental dynamics of individual localities. Interpretation of these records should be done with caution because of the environmental challenges noted above. In addition, our knowledge of phytoplankton carbon assimilation properties is limited for these ancient times. There are few, if any, studies of the controls on isotopic fractionation by phytoplankton groups that dominated the pre-Mesozoic seas, such at the green algae and cyanobacteria. On top of these challenges, the error curves shown in Fig. 3.6 and Fig. 3.7 demonstrate a rapid increase in uncertainty at higher values of ε_p. For times when CO_2 was presumably much higher than the current level (Berner and Kothavala 2001; Rothman 2002), uncertainties in this approach will likely make it unreasonable for reconstruction of absolute CO_2 values, although they can be used to demonstrate very high levels of CO_2 at any particular time.

Nevertheless, records of organic materials isotopic composition can be used to constrain CO_2 levels associated with climate change recorded in sediments, even though precise estimates of CO_2 levels are impossible. For example, the pattern of sedimentary inorganic and organic carbon isotopic excursions can reveal sensitivity of phytoplankton isotopic fraction to CO_2 changes and therefore identify if CO_2 was above a given threshold for sensitivity. For example, if $\delta^{13}C$ values for organic and inorganic carbon track each other, such that ε_p values do not vary across an interval of climatic shift, it suggests that CO_2 levels were high and that phytoplankton were not carbon limited (Kump and Arthur 1999). For example, in the late Frasnian, despite strong evidence for significant carbon burial events and resulting climate change, $\delta^{13}C$ values of phytoplankton biomarkers tracked the inorganic carbon record precisely, suggesting high CO_2 levels during the middle to upper Devonian (Joachimski et al. 2001). This observation does not address the role of CO_2 changes in climate during the isotopic excursion. However, it does help constrain our understanding of CO_2 levels for this time period, and it suggests they were relatively high.

In contrast, for the late middle Ordovician, isotopic values for organic carbon change by several ‰ more than inorganic carbon during a interval of pronounced oceanic cooling. This produced a shift in ε_p values and Patzkowsky et al. (1997) and Pancost, Freeman, and Patzkowsky (1999) have argued this represents a time when CO_2 was at a low enough level that phytoplankton isotopic fractionation was sensitive to changing environmental conditions. If modern algae (i.e., those studied by Popp et al. 1998) are accurate analogs for Paleozoic marine phytoplankton, the CO_2 threshold is approximately 8 to 10 times current atmospheric level. Thus isotopic data suggest atmospheric CO_2 was above this level in the late Devonian, and below this level in the late Ordovician.

Hayes et al. (1999) compiled inorganic and organic isotopic data in order to constrain CO_2 variations over the past 800 million years. Unlike Pancost et al. (1999) or Joachimski et al. (2001), their reconstructions are based on a compilation of bulk organic carbon analyses as a proxy for phytoplankton isotopic values. Additional caution is needed if bulk organic materials are used because of the potential for mixed signatures from phytoplankton, microorganisms, or other sources of organic carbon (Hayes et al. 1989). Nonetheless, this work extends previous studies (i.e., Popp et al. 1989) and affirms that ε_p has varied significantly over Earth's history.

3.4 Summary

The reconstruction of CO_2 values based on biomarker and isotopic analyses provides many challenges and as many opportunities. The method is best applied when CO_2 levels were relatively low, when phosphate concentrations can be constrained, and with multiple sample localities representing stable oceanographic regimes. With our currently available understanding of alkenone isotopic biogeochemistry and under the best of circumstances, this will lead to uncertainties of approximately 20%. The relative utility of this information depends entirely on the question being addressed. It is inappropriate to expect precise reconstructions on very long timescales because the quality of our understanding of ancient oceanic and biological processes diminishes back in time, while the uncertainty increases at times of high CO_2. However, isotopic variations do record information about very ancient CO_2. Observed variations in e_p can be used to suggest relative changes in CO_2 and whether levels were above or below a threshold level of sensitivity for isotopic fractionation during carbon fixation. Elevated ε_p values approaching the maximum values are associated with significant uncertainty, although when constrained by phosphate estimates, they nonetheless indicate elevated CO_2 levels sufficient to result in the expression of maximum fractionation by enzymatic fixation. In such situations, it is especially necessary to establish multiple records from both high and low phosphate environments to reduce the uncertainty and constrain the range of surface water CO_2 concentrations.

For Cenozoic and potentially late Mesozoic time periods, this approach has valid utility in reconstructing relative changes in CO_2. However, reconstruction of absolute values requires caution, along with the right sample set where the relative influence of nutrient dynamics can be estimated. If a reliable proxy for either nutrient abundances or growth rates becomes available, then researchers will be able to address a fuller range of paleoceanographic and paleoclimate questions than is currently possible. For this approach to be applied with greater confidence on longer timescales, much more work is needed with modern organisms that can serve as analogs for Paleozoic phytoplankton.

Acknowledgments. We thank the National Science Foundation for support of research discussed in this work. K.H. Freeman thanks T. Eglinton, J. Hayes, and the Woods Hole Oceanographic Institution for hosting her sabbatical visit and providing a stimulating environment for the writing of this manuscript.

References

Berner, R.A., and Z. Kothavala. 2001. GEOCARB III: A revised model of atmospheric CO_2 over phanerozoic time. *American Journal of Science* 301:182–204.

Bevington, P.R., and D.K. Robinson. 1992. *Data Reduction and Error Analysis*. New York: McGraw-Hill.

Bidigare, R.R., A. Fluegge, K.H. Freeman, K.L. Hanson, J.M. Hayes, D. Hollander, J.P. Jasper, L.L. King, E.A. Laws, J. Milder, F.J. Millero, R. Pancost, B.N. Popp, P.A. Steinberg, and S.G. Wakeham. 1997. Consistent fractionation of C-13 in nature and in the laboratory: Growth-rate effects in some haptophyte algae. *Global Biogeochemical Cycles* 11:279–92.

———. 1999. Consistent fractionation of C-13 in nature and in the laboratory: Growth-rate effects in some haptophyte algae (1997, 11:279). *Global Biogeochemical Cycles* 13:251–52.

Bidigare, R.R., K.L. Hanson, K.O. Buesseler, S.G. Wakeham, K.H. Freeman, R.D. Pancost, F.J. Millero, P. Steinberg, B.N. Popp, M. Latasa, M.R. Landry, and E.A. Laws. 1999b. Iron-stimulated changes in C-13 fractionation and export by equatorial Pacific phytoplankton: Toward a paleogrowth rate proxy. *Paleoceanography* 14:589–95.

Boyle, E.A. 1992. Cadmium and $\delta^{13}C$ paleochemical ocean distributions during the stage-2 glacial maximum. *Annual Review of Earth and Planetary Science* 20:245–87.

Brassell, S.C. 1993. Applications of biomarkers for delineating marine paleoclimatic fluctuations during the Pleistocene. In *Organic geochemistry: Principles and applications,* ed. M.H. Engel and S.A. Macko, 699–738. New York: Plenum Press.

Brassell, S.C., G. Eglinton, I.T. Marlowe, U. Pflaumann, and M. Sarnthein. 1986. Molecular stratigraphy: A new tool for climatic assessment. *Nature* 320:129–33.

Burkhardt, S., U. Riebesell, and I. Zondervan. 1999. Effects of growth rate, CO_2 concentration, and cell size on the stable carbon isotope fractionation in marine phytoplankton. *Geochimica Et Cosmochimica Acta* 63:3729–41.

Eek, M.K., M.J. Whiticar, J.K.B. Bishop, and C.S. Wong. 1999. Influence of nutrients on carbon isotope fractionation by natural populations of Prymnesiophyte algae in NE Pacific. *Deep-sea research part Ii-topical studies in oceanography* 46:2863–76.

Farquhar, G.D., J.R. Ehleringer, and K.T. Hubick. 1989. Carbon isotope discrimination and photosynthesis. Annual Review of Plant Physiology and Molecular Biology 40: 503-537.

Francois, R., M.A. Altabet, R. Goericke, D.C. McCorkle, C. Brunet, and A. Poisson. 1993. Changes in the $\delta^{13}C$ of surface-water particulate organic-matter across the subtropical convergence in the SW Indian Ocean. *Global Biogeochemical Cycles* 7:627–44.

Freeman, K.H. 2001. Isotopic biogeochemistry of marine carbon. In *Stable isotope geochemistry,* ed. J.W. Valley and D.R. Cole. Reviewed in *Mineralogy and Geochemistry* 43:579–605.

Freeman, K.H., and J.M. Hayes. 1992. Fractionation of carbon isotopes by phytoplankton and estimates of ancient CO_2 levels. *Global Biogeochem Cycles* 6:185–98.

Freeman, K.H., and S.G. Wakeham. 1992. Variations in the distributions and isotopic compositions of alkenones in Black Sea particles and sediments. *Organic Geochemistry* 19:277–85.

Gong, C., and D.J. Hollander. 1999. Evidence for differential degradation of alkenones under contrasting bottom water oxygen conditions: Implication for paleotemperature reconstruction. *Geochimica et Cosmochimica Acta* 63:405–11.

Guy, R.D., M.L. Fogel, and J.A. Berry. 1993. Photosynthetic fractionation of the stable isotopes of oxygen and carbon. *Plant Physiology* 101:37–47.

Hayes, J.M. 1993. Factors controlling ^{13}C contents of sedimentary organic compounds-principles and evidence. *Marine Geology* 113:111–25.

———. 2001. Fractionation of carbon and hydrogen isotopes in biosynthetic processes. In *Stable Isotope Geochemistry,* ed. J.W. Valley and D.R. Cole. Reviewed in *Mineralogy and Geochemistry* 43:225–77.

Hayes, J.M., B.N. Popp, R. Takigiku, and M.W. Johnson. 1989. An isotopic study of biogeochemical relationships between carbonates and organic carbon in the Greenhorn Formation. *Geochimica et Cosmochimica Acta* 53:2961–72.

Hayes, J.M., H. Strauss, and A.J. Kaufman. 1999. The abundance of ^{13}C in marine organic matter and isotopic fractionation in the global biogeochemical cycle of carbon during the past 800 Ma. *Chemical Geology* 161:103–25.

Herbert, T.D. 2001. Review of alkenone calibrations (culture, water column, and sediments). *Geochemistry Geophysics Geosystems* 2:2000GC000055.

Hinga, K.R., M.A. Arthur, M.E.Q. Pilson, and D. Whitaker. 1994. Carbon-isotope fractionation by marine phytoplankton in culture: The effects of CO_2 concentration, pH, temperature, and species. *Global Biogeochemical Cycles* 8:91–102.

Jasper, J.P., J.M. Hayes, A.C. Mix, and F.G. Prahl. 1994. Photosynthetic fractionation of ^{13}C and concentrations of dissolved CO_2 in the Central Equatorial Pacific during the last 255,000 years. *Paleoceanography* 9:781–98.

Joachimski, M.M., C. Ostertag-Henning, R.D. Pancost, H. Strauss, K.H. Freeman, R. Littke, J.S.S. Damste, G. and Racki. 2001. Water column anoxia, enhanced productivity, and concomitant changes in $\delta^{13}C$ and $\delta^{34}S$ across the Frasnian-Famennian boundary (Kowala Holy Cross Mountains/Poland). *Chemical Geology* 175:109–31.

Keller, K., and F.M.M. Morel. 1999. A model of carbon isotopic fractionation and active carbon uptake in phytoplankton. *Marine Ecology-Progress Series* 182:295–98.

Kump, L.R., and M.A. Arthur. 1999. Interpreting carbon-isotope excursions: Carbonates and organic matter. *Chemical Geology* 161:181–98.

Laws, E.A., R.R. Bidigare, and B.N. Popp. 1997. Effect of growth rate and CO_2 concentration on carbon isotopic fractionation by the marine diatom *Phaeodactylum tricornutum. Limnology and Oceanography* 42:1552–60.

Laws, E.A., B.N. Popp, R.R. Bidigare, M.C. Kennicutt, and S.A. Macko. 1995. Dependence of phytoplankton carbon isotopic composition on growth-rate and [CO_2 (aq)]: Theoretical considerations and experimental results. *Geochimica et Cosmochimica Acta* 59:1131–38.

Laws, E.A., B.N. Popp, R.R. Bidigare, U. Riebesell, and S. Burkhardt. 2001. Controls on the molecular distribution and carbon isotopic composition of alkenones in certain haptophyte algae. *Geochemistry Geophysics Geosystems* 2:2000GC000057.

de Leeuw, J.W., F.W. van der Meer, W.I.C. Rijpstra, and P.A. Schenck. 1980. Unusual long-chain ketones of algal origin. In *Advances in organic geochemistry 1979, Proceedings of the ninth international meeting on organic geochemistry, University of Newcastle-Upon-Tyne, U.K., September 1979,* ed. A.G. Douglas and J.R. Maxwell, 211–17. New York: Pergamon.

Marlow, I.T., J.C. Green, A.C. Neal, S.C. Brassell, G. Eglinton, and P.A. Course. 1984. Long chain (n-C_{37}-C_{39}) alkenones in the *Prymnesiophyceae*: Distribution of alkenones and other lipids and their taxanomic significance. *British Phycology Journal* 19:203–16.

Mashiotta, T.A., D.W. Lea, and H.J. Spero. 1997. Experimental determination of cadmium uptake in shells of the planktonic foraminifera Orbulina universa and Globigerina bulloides: Implications for surface water paleoreconstructions. *Geochimica et Cosmochimica Acta* 61:4053–65.

Mook, W.G., J.C. Bommerson, and W.H. Staberman. 1974. Carbon isotope fractionation between dissolved bicarbonate and gaseous carbon dioxide. *Earth and Planetary Science Letters* 22:169–76.

Nimer, N.A., M.D. Iglesias Rodriguez, and M.J. Merrett. 1997. Bicarbonate utilization by marine phytoplankton species. *Journal of Phycology* 33:625–31.

Ohkouchi, N., K. Kawamura, H. Kawahata, and H. Okada. 1999. Depth ranges of alkenone production in the central Pacific Ocean. *Global Biogeochemical Cycles* 13: 695–704.

Oppo, D.W. and Y. Rosenthal. 1994. Cd/Ca changes in a deep cape basin core over the past 730,000 years: Response of circumpolar deep-water variability to northern-hemisphere ice-sheet melting. *Paleoceanography* 9:661–75.

Pacanowski, R., K. Dixon, and A. Rosati. 1991. *The G.F.D.L. Modular Ocean Model users guide*. NOAA/Geophysical Fluid Dynamics Laboratory.

Pagani, M. 2002. The alkenone-CO_2 proxy and ancient atmospheric carbon dioxide. *Philosophical Transactions of the Royal Society of London Series a—Mathematical Physical and Engineering Sciences* 360:609–32.

Pagani, M., M.A. Arthur, and K.H. Freeman. 1999. Miocene evolution of atmospheric carbon dioxide. *Paleoceanography* 14:273–92.

———. 2000. Variations in Miocene phytoplankton growth rates in the southwest Atlantic: Evidence for changes in ocean circulation. *Paleoceanography* 15:486–96.

Pagani, M., K.H. Freeman, and M.A. Arthur. 1999. Late miocene atmospheric CO_2 concentrations and the expansion of C_4 grasses. *Science* 285:876–79.

Pagani, M., K.H. Freeman, N. Ohkouchi, and K. Caldeira. 2002. Comparison of water column [CO_2aq] with sedimentary alkenone-based estimate: A test of the alkenone-CO_2 proxy. *Paleoceanography* 17:1069–81.

Pancost, R.D., K.H. Freeman, and M.E. Patzkowsky. 1999. Organic-matter source variation and the expression of a late middle Ordovician carbon isotope excursion. *Geology* 27:1015–18.

Pancost, R.D., K.H. Freeman, S.G. Wakeham, and C.Y. Robertson. 1997. Controls on carbon isotope fractionation by diatoms in the Peru upwelling region. *Geochimica et Cosmochimica Acta* 61:4983–91.

Patzkowsky, M.E., L.M. Slupik, M.A. Arthur, R.D. Pancost, and K.H. Freeman. 1997. Late middle Ordovician environmental change and extinction: Harbinger of the late Ordovician or continuation of Cambrian patterns? *Geology* 25:911–14.

Pearson, P.N., and M.R. Palmer. 2000. Atmospheric carbon dioxide concentrations over the past 60 million years. *Nature* 406:695–99.

Popp, B.N., E.A. Laws, R.R. Bidigare, J.E. Dore, K.L. Hanson, and S.G. Wakeham. 1998. Effect of phytoplankton cell geometry on carbon isotopic fractionation. *Geochimica et Cosmochimica Acta* 62:69–77.

Popp, B.N., R. Takigiku, J.M. Hayes, J.W. Louda, and E.W. Baker. 1989. The post-Paleozoic chronology and mechanism of ^{13}C depletion in primary marine organic matter. *American Journal of Science* 289:436–54.

Popp, B.N., T. Trull, F. Kenig, S.G. Wakeham, T.M. Rust, B. Tilbrook, F.B. Griffiths, S.W. Wright, H.J. Marchant, R.R. Bidigare, and E.A. Laws. 1999. Controls on the carbon isotopic composition of Southern Ocean phytoplankton. *Global Biogeochemical Cycles* 13:827–43.

Rau, G.H., U. Riebesell, and D. Wolf-Gladrow. 1996. A model of photosynthetic ^{13}C fractionation by marine phytoplankton based on diffusive molecular CO_2 uptake. *Marine Ecology-Progress Series* 133:275–85.

Rau, G.H., T. Takahashi, D.J. Desmarais, D.J. Repeta, and J.H. Martin. 1992. The relationship between $\delta^{13}C$ of organic matter and [CO_2(aq)] in ocean surface-water data from a JGOFS site in the Northeast Atlantic Ocean and a model. *Geochimica et Cosmochimica Acta* 56:1413–19.

Rechka, J.A., and J.R. Maxwell. 1988a. Characterization of alkenone temperature indicators in sediments and organisms. *Organic Geochemistry* 13:727–34.

———. 1988b. Unusual long-chain ketones of algal origin. *Tetrahedron Letters* 29:2599–2600.

Rickaby, R.E.M., and H. Elderfield. Planktonic formainiferal Cd/Ca: Paleonutrients or paleotemperature? *Paleoceanography* 14:293–303.

Riebesell, U., S. Burkhardt, A. Dauelsberg, and B. Kroon. 2000a. Carbon isotope fractionation by a marine diatom: Dependence on the growth-rate-limiting resource. *Marine Ecology-Progress Series* 193:295–303.

Riebesell, U., A.T. Revill, D.G. Holdsworth, and J.K. Volkman. 2000b. The effects of varying CO_2 concentration on lipid composition and carbon isotope fractionation in Emiliania huxleyi. *Geochimica et Cosmochimica Acta* 64:4179–92.

Roeske, C.A., and M.H. O'Leary. 1985. Carbon isotope effect on carboxylation of ribulose bisphosphate carboxylase from *Rhodospirillum rubum. Biochemistry* 24:1603–607.

Romanek, C.S., E.L. Grossman, and J.W. Morse. 1992. Carbon isotopic fractionation in synthetic aragonite and calcite—effects of temperature and precipitation rate *Geochimica et Cosmochimica Acta* 56:419,430.

Rosenthal, Y., H. Stoll, K. Wyman, and P.G. Falkowski. 1999. Growth related variations in carbon isotopic fractionation and coccolith chemistry in Emiliania huxleyi. *EOS Trans., Ocean Science Meeting Supplement* 80(90), OS11C-27.

Rothman, D.H. 2002. Atmospheric carbon dioxide levels for the last 500 million years. *Proceedings of the National Academy of Sciences* 99:4167–171.

Royer, D.L., R.A. Berner, and D.J. Beerling. 2001. Phanerozoic atmospheric CO_2 change: Evaluating geochemical and paleobiological approaches. *Earth-Science Reviews* 54: 349–92.

Sachs, J.P., and S.J. Lehman. 1999. Subtropical North Atlantic temperatures 60,000 to 30,000 years ago. *Science* 286:756–59.

Schouten, S., W. Breteler, P. Blokker, N. Schogt, W.I.C. Rijpstra, K. Grice, M. Baas, and J.S.S. Damste. 1998. Biosynthetic effects on the stable carbon isotopic compositions of algal lipids: Implications for deciphering the carbon isotopic biomarker record. *Geochimica et Cosmochimica Acta* 62:1397–1406.

Sharkey, T.D., and J.A. Berry. Carbon isotope fractionation of algae as influenced by an inducible CO_2 concentrating mechanism. In *Inorganic carbon uptake by aquatic photosynthetic organisms,* ed. W.J. Lucas and J.A. Berry, 389–401. Rockville, Md: American Society of Plant Physiologists.

Sikes, C.S., R.D. Roer, and K.M. Wilbur. 1980. Photosynthesis and coccolith formation: Inorganic carbon sources and net inorganic reaction of deposition. *Limnology and Oceanography* 25:248–61.

Spero, H.J., and M.J. Deniro. 1987. The influence of symbiont photosynthesis on the $\delta^{18}O$ and $\delta^{13}C$ values of planktonic foraminiferal shell calcite. *Symbiosis* 4:213–28.

Spero, H.J., and D.W. Lea. 1993. Intraspecific stable-isotope variability in the planktic foraminifera *Globigerinoides sacculifer*: Results from laboratory experiments. *Marine Micropaleontology* 22:221–34.

Sokal, R.R., and F.J. Rohlf. 1995. *Biometry*. New York: W.H. Freeman.

Taylor, J.R. 1982. *An Introduction to Error Analysis*. University Science Books, Mill Valley, CA.

Verity, P.G., C.Y. Robertson, C.R. Tronzo, M.G. Andrews, J.R. Nelson, and M.E. Sieracki. 1992. Relationships between cell-volume and the carbon and nitrogen content of marine photosynthetic nanoplankton. *Limnology and Oceanography* 37:1434–46.

Volkman, J.K., S.M. Barrett, S.I. Blackburn, and E.L. Sikes. 1995. Alkenones in *Gephyrocapsa oceanica*: Implications for studies of paleoclimate. *Geochimica et Cosmochimica Acta* 59:513–20.

Volkman, J.K., G. Eglinton, E.D. Corner, and J.R. Sargent. 1980. Novel unsaturated straight-chain C37–C39 methyl and ethyl ketones in marine sediments and a coccolithophore Emiliania huxleyi. In *Advances in organic geochemistry 1979; Proceedings of the ninth international meeting on organic geochemistry, University of Newcastle-Upon-Tyne, U.K., September 1979,* ed. A.G. Douglas and J.R. Maxwell, 219–27. New York: Pergamon.

Wakeham, S.G., J.I. Hedges, C. Lee, M.L. Peterson, and P.J. Hernes. 1997a. Compositions and transport of lipid biomarkers through the water column and surficial sediments of the equatorial Pacific Ocean. *Deep-Sea Research Part II-Topical Studies in Oceanography* 44:2131–62.

Wakeham, S.G., C. Lee, J.I. Hedges, P.J. Hernes, and M.L. Peterson. 1997b. Molecular indicators of diagenetic status in marine organic matter. *Geochimica et Cosmochimica Acta* 61:5363–69.

Wakeham, S.G., M.L. Peterson, J.I. Hedges, and C. Lee. 2002. Lipid biomarker fluxes in the Arabian Sea, with a comparison to the equatorial Pacific Ocean. *Deep-Sea Research Part Ii-Topical Studies in Oceanography* 49:2265–2301.

Westbroek, P., C.W. Brown, J. Vanbleijswijk, C. Brownlee, G.J. Brummer, M. Conte, J. Egge, E. Fernandez, R. Jordan, M. Knappertsbusch, J. Stefels, M. Veldhuis, P. Vanderwal, and J. Young. 1993. A model system approach to biological climate forcing: The example of *Emiliania huxleyi*. *Global and Planetary Change* 8:27–46.

4. Atmospheric CO_2 Data from Ice Cores: Four Climatic Cycles

Thomas Blunier, Eric Monnin, and Jean-Marc Barnola

4.1 Introduction

Ice core records of CO_2 reach as far back as 420,000 years before present. The Vostok long-term record shows a consistent pattern of glacial-interglacial CO_2 changes over the past four climatic cycles. It is likely that the CO_2 increase started before a large northern hemispheric warming for each termination. The long-term records have a relatively low time resolution, and the propagation of the atmospheric signal is recorded significantly smoothed by the gas occlusion process. Several records with higher resolution reach back into the last glacial. The CO_2 record in the last glacial shows minor variations that parallel temperature changes in Antarctica. It is possible that this characteristic of the CO_2 record is linked to the bipolar sea-saw between northern and southern hemispheric temperature. A model study over the Younger Dryas period supports this hypothesis. Generally the CO_2 record is highly correlated with the Antarctic temperature over glacial-interglacial cycles. Detailed analyses over the last termination reveal that some characteristics of northern hemispheric climate change are imprinted in the succession of the CO_2 increase. It appears that the large CO_2 changes are dominated by southern hemispheric climate change with an overlaid northern influence.

There are still open questions about the reliability of the CO_2 record. Greenland records are evidently altered by physical or chemical processes; moreover, such effects are not completely ruled out for Antarctic records. However, even including this possibility, the uncertainty for recent high resolution CO_2 measurements from Antarctica is about 1%. Only ice core records are able to reconstruct the atmospheric CO_2 concentration to this level of confidence.

4.1.1 Occlusion of Trace Gas Records in the Ice

Polar ice sheets are an important climate archive for atmospheric air. Snow accumulating on the ice sheet sinters to firn and finally ice under the pressure of following precipitations. This process takes place with no melting in polar regions. A layered record of past precipitation builds up. Mass-balance leads to the ice flowing from the center of the ice sheet (Dome) downwards and outwards to the ablation zone. In central Greenland the thickness of the ice sheet is about 3km. The longest reliable climate record spans about one climatic cycle, roughly 100,000 years. The ice thickness found in Antarctica is about equal. However, some sites in Antarctica have accumulation rates about one order of magnitude lower than in Greenland, and thus their climate record extends over several (four at Vostok) climatic cycles. Ice cores from Dome F and Dome C will extend the ice core climate reconstruction by several climatic cycles.

Glaciers and ice sheets are unique as they are the only paleoarchives that record atmospheric composition in the form of trapped air bubbles. The top 50–150 m of ice are comprised of porous firn (Fig. 4.1).

Atmospheric air exchanges with the air in this top layer. Air in the porous firn thus consists of a mixture of air that was last in equilibrium with the atmosphere at different times (Schwander 1996). Because of physical and (in case of reactive gases) chemical processes, at a given time the concentrations of gas species in the firn are not the same as their atmospheric concentrations. Therefore, and due to gradual air occlusion at the bottom of the firn column, air occluded in ice does not have an exact age reflecting its last contact with the atmosphere; instead, it has an age distribution, and the temporal resolution of any trace gas record from ice cores is inherently limited. This age distribution is a function of both temperature and accumulation, which are highly variable from site to site. Generally, it increases with decreasing temperature and accumulation. The width of the age distribution is as low as seven years at high accumulation/high temperature sites but can be several centuries for Antarctic low accumulation/low temperature sites (Schwander and Stauffer 1984).

In the top few meters of firn, air is well mixed by convection and thus has essentially the same composition as the atmosphere. Below this zone, the air in the firn is static and mixes mainly by molecular diffusion. An equilibrium between molecular diffusion and gravitational settling is reached for each gas component (Craig, Horibe, and Sowers 1988; Schwander 1989). As a consequence, two gas components with different molecular weights fractionate with depth relative to their initial relationship in the atmosphere. The magnitude of this fractionation is well known, allowing the construction of an accurate, corrected ice core record. The process is most important for isotope records but can be neglected when dealing with concentration records. For CO_2, the con-

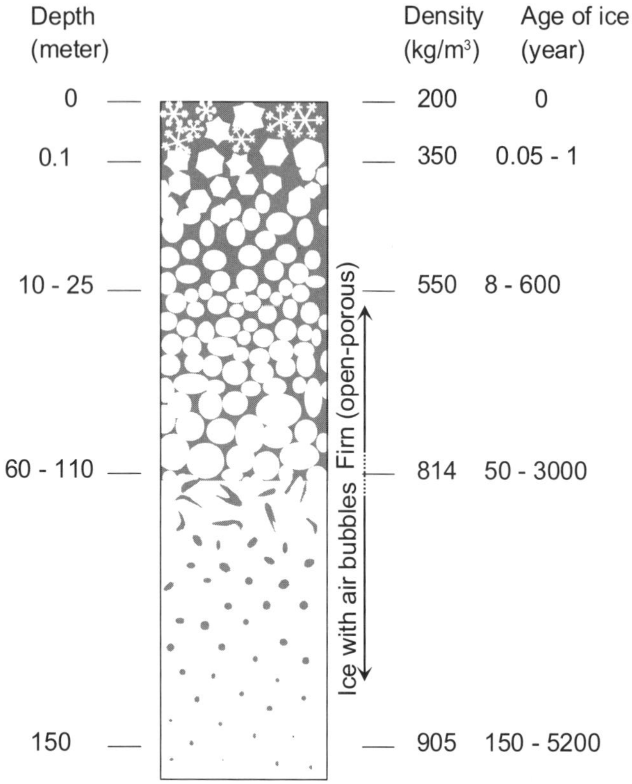

Figure 4.1. Sketch of the firn column. The indicated depth and age ranges are typical for polar ice sheets.

centration measured in ice cores is about 1% elevated compared to the true atmospheric value. This number varies slightly with temperature and the depth of the close off.

Another consequence of the air occlusion at the bottom of the firn column is that the age of the air in an occluded air bubble is less than the age of the surrounding ice. This age difference (Δage hereafter) is the difference between the age of the ice and the mean age of the air at the depth of occlusion. The age of the ice at close off is the dominant term of Δage, and Δage can be calculated using models for the firn densification process (Arnaud, Barnola, and Duval 2000; Barnola et al. 1991; Schwander 1989). Although the process is well understood, the accuracy of the calculation is limited due to uncertainty in past accumulation and temperature. However, Δage calculated with the densification model is in excellent agreement with independent measurements of Δage based on temperature diffusion processes in the firn for central Greenland ice cores (Lang et al. 1999; Leuenberger, Siegenthaler, and Langway 1999; Severinghaus and Brook 1999; Severinghaus et al. 1998). The Δage can be substantial. Under present-day conditions Dage is on the order of decades to a few centuries for high accumulation/high temperature sites: e.g., 210 years for central Greenland (Schwander et al. 1993). At Vostok, where accumulation and temperature are low, Δage is about 3000 years (Barnola et al. 1991; Schwander and Stauffer 1984). Under colder climate conditions that are paralleled by lower accumulation, Δage increases significantly.

Below the close off zone the air bubbles shrink in size as the ice flows to deeper strata under the increasing pressure of overlying layers. When the pressure gets high enough, the gas is transformed from air bubbles into air hydrates (Miller 1969). Air hydrates constitute cage structures made up of water molecules that are fully or partially occupied by air molecules. For example, first hydrate formation has been observed at 500 m depth at Vostok and 1022 m at Summit Greenland, respectively (Shoji et al. 2000). Hydrates decompose into bubbles again once the core is recovered and relaxed. However, a full reformation of air bubbles takes several decades (Uchida et al. 1994). In the zone where bubbles and hydrates coexist, the composition of air in bubbles and hydrates deviate (Anklin et al. 1997; Ikeda et al. 1999). Fractionation takes place during hydrate formation and decomposition.

4.1.2 How Reliable Are CO_2 Data from Ice Cores?

How sure are we that the CO_2 concentration measured in ice cores represents the atmospheric concentration at trapping time? Where melting occurs, gas content and gas composition may be altered by chemical reactions taking place in aquatic systems or by physical gas exchange between the gaseous and the aquatic sections. In Greenland and Antarctica, surface melting is sporadic. At these sites the gas occlusion occurs by dry sintering of the firn described above. However, very slow chemical reactions are able to alter gas concentrations in ice cores due to the long time available for the reaction to proceed.

In specific time intervals, CO_2 from Greenland ice cores has been measured as significantly higher than Antarctic levels, too high given that atmospheric CO_2 is well mixed in the atmosphere and interhemispheric differences of more than a few ppmv are not realistic. In records from Greenland, concentrations fluctuate on the order of 50 ppmv between cold and warm phases of the last glacial (Stauffer et al. 1984), whereas no such variations are seen in Antarctic records (Indermühle et al. 2000; Neftel et al. 1988; Oeschger et al. 1988). It is now established that the elevated CO_2 concentrations in the Greenland ice cores do not represent the atmospheric concentration but are a spurious signal caused by an acid-carbonate reaction or by oxidation of organic material in the ice (Anklin et al. 1995; Delmas 1993; Tschumi and Stauffer 2000).

It has been demonstrated that the Antarctic records provide the most reliable data of changes in global atmospheric CO_2 (Raynaud et al. 1993). Antarctic results are consistent despite the fact that the coring sites have different accumulation, temperature, and impurities. CO_2 measurements made several years apart on the same core show no significant changes. Detailed centimeter-scale CO_2 analyses of several ice cores indicate that also Antarctic ice core records may deviate systematically from the atmospheric CO_2 concentration at the time of air occlusion. However, it is unlikely that this deviation exceeds 1% of the measured concentration (Stauffer et al. 2003).

4.1.3 Extraction of Air for CO_2 Measurements

Principally there are three ways to extract air from the ice: melting, crushing, and sublimation of the ice. It is crucial to exclude alteration of the CO_2 concentration during the extraction or analysis of the gas. In light of the explanations in the previous section, it is obvious that extraction by melting is not suitable for CO_2 measurements. Carbonate reactions in the meltwater can significantly change the CO_2 concentration (Anklin et al. 1995; Delmas 1993). Where air is occluded fully or partly in hydrates, only a fraction of the gas can be extracted by the crushing method. This may cause the composition of the extracted gas to deviate from the true composition (Anklin et al. 1997; Stauffer and Tschumi 2000). The sublimation method has so far not resulted in routine measurements (Güllük, Slemr, and Stauffer 1998). Sorption processes at the apparatus walls (Zumbrunn, Neftel, and Oeschger 1982) lead to contamination of the sample that cannot be excluded so far. Reliable results are produced by the crushing methods that are used routinely in all laboratories measuring CO_2 concentrations from ice cores.

4.2 Glacial-Interglacial CO_2 Variations

Shown in Fig. 4.2 is the long-term perspective of atmospheric CO_2 changes over four glacial-interglacial cycles covering roughly the past 400,000 years. The Vostok ice core shows a remarkable correlation between temperature and greenhouse gas concentrations on glacial-interglacial scales (Petit et al. 1999). CO_2 concentrations oscillate between 180 and 200 ppmv during the coldest glacial periods; they oscillate between 280 and 300 ppmv during full interglacials.

One question regarding glacial-interglacial time scales that is still unanswered is why the atmospheric CO_2 rose by 80 to 100 ppmv during glacial to interglacial warmings. Changes in the atmospheric CO_2 concentration are driven by physical and biogeochemical changes in the ocean or in the terrestrial biosphere, which has grown between the last glacial and the Holocene (Adams and Faure 1998; Crowley 1995). Therefore, going from glacial to interglacial, the terrestrial biosphere is a sink, not a source, to the atmospheric carbon reservoir. The ocean has certainly played the dominant role, since it represents the largest pool of available CO_2 that could have been delivered to the atmosphere.

There is a third player, namely, changing rates in the weathering of silicate and carbonate rocks. Although it certainly played a role, this process is too slow to explain the rapidity of transitions seen in the Vostok record. We can therefore

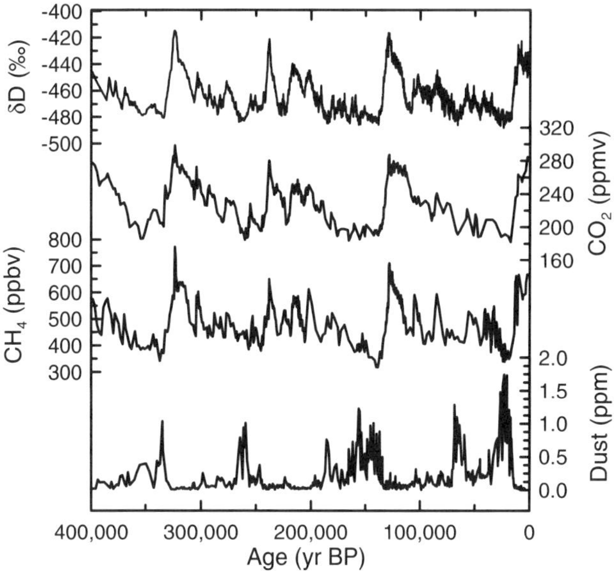

Figure 4.2. Climate records over the past 420,000 years. Deuterium (δD) is a proxy for local temperature, CO_2, CH_4, and dust from the Vostok ice core (Pépin et al. 2001; Petit et al. 1999).

assert that the 170–190 Gt of carbon that accumulated in the atmosphere during the several millennia corresponding to glacial to interglacial transitions must have been the result of a net flux out of the oceans. This is a minimum estimate, since during the same time the uptake of atmospheric CO_2 by the continental biosphere, although not well constrained, was almost certainly positive.

CO_2 is one of the major greenhouse gases. The initial forcing due to the direct radiative effect of increasing greenhouse trace gases (CO_2 + CH_4 + N_2O) during the glacial-interglacial transitions is estimated to have produced a global warming of about 0.95°C (Petit et al. 1999). This initial forcing would have been amplified by rapid feedbacks due to associated water vapor and albedo modifications (e.g., sea ice, clouds, etc.), as is also the case with the increasing load of anthropogenic greenhouse trace gases. Results from different climate simulations make it reasonable to assume that these trace gases and the associated rapid feedbacks have contributed significantly (possibly about half, that is 2–3°C) to the globally averaged glacial-interglacial temperature change (Weaver et al. 1998).

For the interpretation of the CO_2 record it is important to know the timing and the rate of the CO_2 variations with respect to other climatic records. From the long-term Vostok record we can deduce that the decrease of CO_2 to minimum values is slower than the CO_2 increases to interglacial levels. For the last glaciation, the CO_2 decrease lags the southern temperature decrease by several thousand years (Barnola et al. 1991; Cuffey and Vimeux 2001). Generally the CO_2 decrease lags the Antarctic temperature decrease at the entrance of glacial periods (Barnola et al. 1991; Petit et al. 1999).

At the end of glacial periods, the Vostok record suggests that CO_2 and Antarctic temperature begin to increase in phase (Pépin et al. 2001; Petit et al. 1999). However, the low resolution of the Vostok profile prevents a firm statement about the precise timing of the Vostok temperature versus the CO_2 increases. Here we turn to a higher resolution record for the last termination from the Dome C core (Monnin et al. 2001). The Dome C record confirms and details earlier measurements from Antarctica over the last glacial termination (Monnin et al. 2001). We present these records in Fig. 4.3 together with records from Byrd station, Antarctica, and a $\delta^{18}O$ record from Greenland. The $\delta^{18}O$ and δD measured on ice are proxies for the local temperature. In this figure, the Dome C time scale has been adjusted so that the "global" CH_4 signals between the Greenland and the Dome C record are consistent over the Younger Dryas period.

First of all let us note that the temperature increase in Greenland and Antarctica is fundamentally different. Antarctic temperature increases steadily with a dip to colder temperatures toward the end of the transition. This dip is called the Antarctic Cold Reversal (ACR) (Jouzel et al. 1995). In contrast, Greenland temperature increases within decades from almost full glacial to Holocene temperature levels in the Bølling-Allerød period and drops back to near glacial conditions during the Younger Dryas. For the last glacial-interglacial termination, the main and fast temperature increase in Greenland lags the Antarctic temperature increase by several millennia. We now compare the CO_2 increase to the Arctic and the Antarctic temperature increases.

The CO_2 measurements highly correlate with the temperature reconstruction from Dome C. The start of the temperature increase occurs 800 ± 600 years before the increase of the CO_2 concentration confirming earlier estimates from various ice cores (Fischer et al. 1999; Neftel et al. 1988). It is likely that the start of the glacial-interglacial CO_2 increase generally lags the temperature evolution. This assumption was confirmed recently for Termination III ($\sim$240 kyr BP) on the Vostok ice core. Caillon et al. (2003) concluded, based on a temperature proxy in the air bubbles, that the CO_2 increase lagged the Antarctic temperature increase by 800 ± 200 years. However, even a time lag of a few hundred years does not cast doubt on the importance of CO_2 as an amplification factor of the temperature due to the fact that this time lag is much smaller than the 6-kyr duration of the CO_2 and temperature increase.

The rapid glacial-interglacial temperature increase observed in Greenland obviously lags the rise in CO_2 and the Antarctic temperature increase. Greenland temperature and the global CH_4 signal change in concert (at least for the main features over the last termination). Therefore we can regard the CH_4 signal recorded in the Vostok ice core as a proxy for Greenland temperature. If this analogy holds, the CO_2 increase over the last four glacial-interglacial terminations has always occurred before a major temperature increase in the Northern

Hemisphere (Pépin et al. 2001). In summary, the CO_2 increase probably lags the Antarctic temperature increase by a few hundred years but precedes the Greenland temperature increase by a few millennia as does the Antarctic temperature.

The high correlation of the CO_2 record with the Dome C temperature reconstruction (δD record) points to the Southern Hemisphere as the main driver of the CO_2 change over the last termination. Assuming that the Dome C δD record is a qualitative indicator for Southern Ocean surface temperature, one possible cause of the CO_2 increase would be a reduced solubility of the Southern Ocean surface waters owing to increasing temperatures. However, a quantitative estimate shows that the observed CO_2 increase is too large to be explained by this mechanism alone (Broecker and Henderson 1998). An additional mechanism could be an enhanced air-sea exchange in the Antarctic region due to the reduction of Antarctic sea ice that increases the deep sea ventilation (Stephens and Keeling 2000). The decrease of dust also could have influenced the atmospheric CO_2 concentration: e.g., through a decrease of iron fertilization of the Southern Ocean bioproductivity (Broecker and Henderson 1998; Martin 1990). We note, however, that the dust decrease most probably started significantly before the onset of the CO_2 increase and that the dust level already dropped to about Holocene values before 15 kyr BP.

The detailed CO_2 record from Dome C can be subdivided into four intervals over the glacial to interglacial increase. Surprisingly, the steps in CO_2 and CH_4 concentration changes are synchronous. Since the CH_4 signal is a first-order approximation for northern hemispheric temperature (Fig. 4.3), this points to a northern hemispheric influence to the CO_2 increase.

A first CO_2 increase of about 35 ppmv happens over a first moderate increase of the CH_4 concentration from about 17 to 15.5 kyr BP. Over a section of approximately 15.5 to 14 kyr BP where the CH_4 concentration is roughly constant, CO_2 increases slowly. At the sharp CH_4 increase, which parallels the sharp Greenland temperature increase into the Bølling-Allerød period, CO_2 rises almost instantaneously by about 8 ppmv. CO_2 slightly decreases over the Bølling-Allerød period, which corresponds to the ACR to finally increase by 30 ppmv to its Holocene level over the Younger Dryas period.

The slight CO_2 decrease during the Bølling-Allerød period could be attributed to the cooling in the Southern Ocean in conjunction with the ACR, but it will be difficult to find a Southern Hemisphere explanation for the sudden CO_2 increase of 8 ppmv at the beginning of this period. A cause for this sudden increase is more likely connected with the fast temperature increase observed in the Northern Hemisphere parallel with a reorganization of the formation of North Atlantic Deep Water (NADW): an enhanced NADW could increase the deep-sea ventilation and therefore increase the atmospheric CO_2 concentration (Siegenthaler and Wenk 1984; Toggweiler 1999). During the Younger Dryas, the CO_2 increase resumes again and is terminated by a sudden CO_2 rise. Simulations of Younger Dryas-type events with a global ocean circulation-biogeochemical

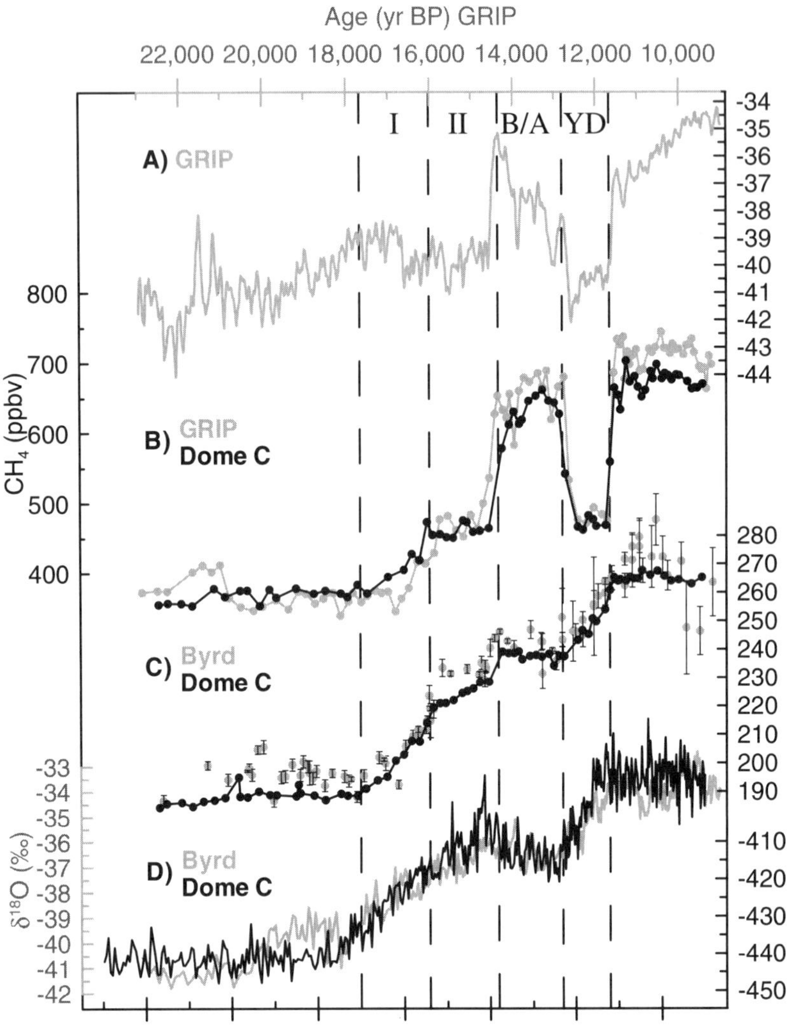

Figure 4.3. Ice core records covering the last glacial-interglacial termination. **(A)** $\delta^{18}O$ from Greenland (GRIP ice core, Dansgaard et al. 1993). **(B)** Methane from Greenland (GRIP ice core, Blunier and Brook 2001) and Antarctica (Dome C, Monnin et al. 2001). The CH_4 concentration in Antarctica is generally slightly lower than the CH_4 concentration in Greenland (Dällenbach et al. 2000). **(C)** CO_2 concentrations from Byrd station (Marchal et al. 1999) and Dome C (Monnin et al. 2001), Antarctica. **(D)** $\delta^{18}O$ from Byrd station (Johnsen et al. 1972) and δD from Dome C (Monnin et al. 2001). The GRIP and Byrd station records are plotted on the GRIP ss09 timescale (top x-axis). The Byrd station records have been synchronized to the GRIP record over the CH_4 signals (Blunier et al. 1998). Dome C records are plotted on the Dome C timescale (bottom x-axis, Schwander et al. 2001). The two timescales are adjusted so that the "global" CH_4 signal is consistent between the GRIP and the Dome C record over the Bølling-Allerød (B/A), Younger Dryas (YD) period.

model suggest that the Younger Dryas increase could have been triggered by a freshwater-induced reduction of the Atlantic thermohaline circulation (Marchal et al. 1999).

The general shape of the glacial-interglacial CO_2 increase corresponding to the Antarctic temperature points to a Southern Ocean explanation for the increase. However, the details of the CO_2 increase have some characteristics of the northern hemispheric climate change pointing to a non-negligible northern hemispheric influence on the global carbon cycle at least over the last termination. The northern hemispheric influence presumably involves changes in the NADW formation (Marchal et al. 1999).

4.2.1 Smoothing in the Vostok CO_2 Record

As pointed out above, the progression of the atmospheric gas composition is recorded smoothed in the ice. This smoothing may be significant and averages over centuries for Vostok station (Barnola et al. 1991; Rommelaere, Arnaud, and Barnola 1997; Schwander and Stauffer 1984). Would a concentration increase like the one we observe over the past few centuries be visible in the Vostok record? In Fig. 4.4 we show how the present anthropogenic CO_2 increase will be recorded at Vostok.

For the propagation of the atmospheric CO_2 concentration to the future we

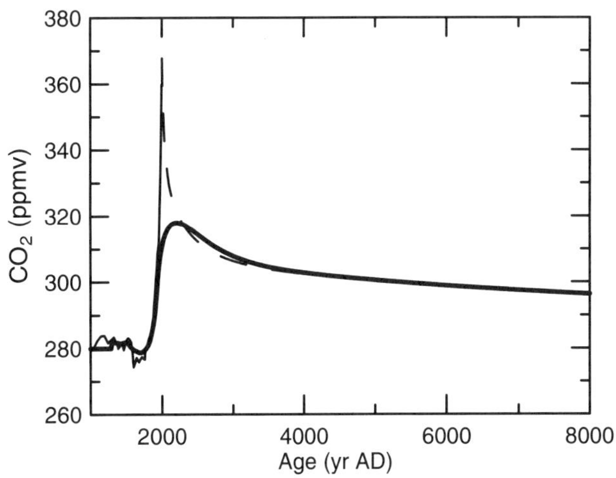

Figure 4.4. Solid line: Atmospheric CO_2 concentration from ice cores and atmospheric measurements up to the year 2000. Broken line: Propagation of the atmospheric CO2 concentration calculated with the Bern model (Joos et al. 1996) setting the hypothetical anthropogenic carbon source to zero after the year 2000. Heavy solid line: prediction of how the atmospheric signal will be recorded in the Vostok ice core, calculated with the Schwander model for gas occlusion (Schwander et al. 1993) with an extension for gradual gas occlusion over the close-off interval (Spahni et al. 2003).

are using the Bern model (Joos et al. 1996). As a lower limit for the future CO_2 concentration we set the anthropogenic CO_2 emission to zero after the year 2000. The smoothed "Vostok record" is obtained using the Schwander model for gas occlusion (Schwander et al. 1993) with an extension for gradual gas occlusion over the close-off interval (Spahni et al. 2003). The resulting concentration propagation with a maximum of about 315 ppmv and a very slow decrease to the preindustrial background clearly stands out. At no place in the Vostok record has such a high concentration with such a concentration trend been measured. Strictly speaking the Vostok record does not exclude a pulselike atmospheric CO_2 signal of few-decades duration with concentrations as high as those found today. However, this would require both a large carbon release (order 200 GtC) and an equally large uptake within only a few decades. Such an oscillation is not compatible with our present view of the global carbon cycle. We conclude that the present CO_2 concentration is unprecedented in the Vostok record.

4.3 Millennial Changes in the Last Glacial

The ice core isotopic records from Greenland (Dansgaard et al. 1993; Grootes et al. 1993) reveal 24 mild periods (interstadials) lasting 1 to 3 kyr during the last glacial, known as Dansgaard-Oeschger (D–O) events (Oeschger et al. 1984), where temperature increased by up to 15°C compared to full glacial values (Johnsen et al. 1995; Schwander et al. 1997). These millennial scale changes are observed in large portions of the world on land and in the ocean (Broecker 1994; Voelker and workshop participants 2002). What is the effect of Dansgaard-Oeschger events on the CO_2 budget?

As outlined above, the CO_2 records must originate from Antarctic records. However, central Antarctica is about the only place where millennial-scale climate change is principally different than in Greenland (Blunier and Brook 2001). Therefore, CO_2 changes recorded in Antarctic ice cores have to be compared to temperature variations recorded in Greenland to identify a potential influence of millennial-scale climate variation on the CO_2 budget.

Millennial-scale variability in Greenland is characterized by abrupt temperature increases, followed by gradual decreases and abrupt returns to baseline glacial conditions. In contrast, Antarctic warmings and coolings are gradual and fewer than in Greenland. Blunier and Brook (2001) and Blunier et al. (1998) showed that the onset of Antarctic warmings preceded by more than 1.5 to 3 kyr the onset of major D–O events, i.e., those that are longer than 2 to 3 kyr, namely, D–O events 8, 12, 14, 16/17, 19, 20, and 21 (Fig. 4.5).

In general, Greenland temperatures were cold or cooling during the gradual warmings in the Antarctic Byrd record. The Antarctic temperature rise was apparently interrupted at or near the time when Greenland temperatures rose abruptly to an interstadial state within only a few decades (Grootes et al. 1993). Subsequently, temperatures decreased in both hemispheres to full glacial level, but cooling in the Byrd record was more rapid.

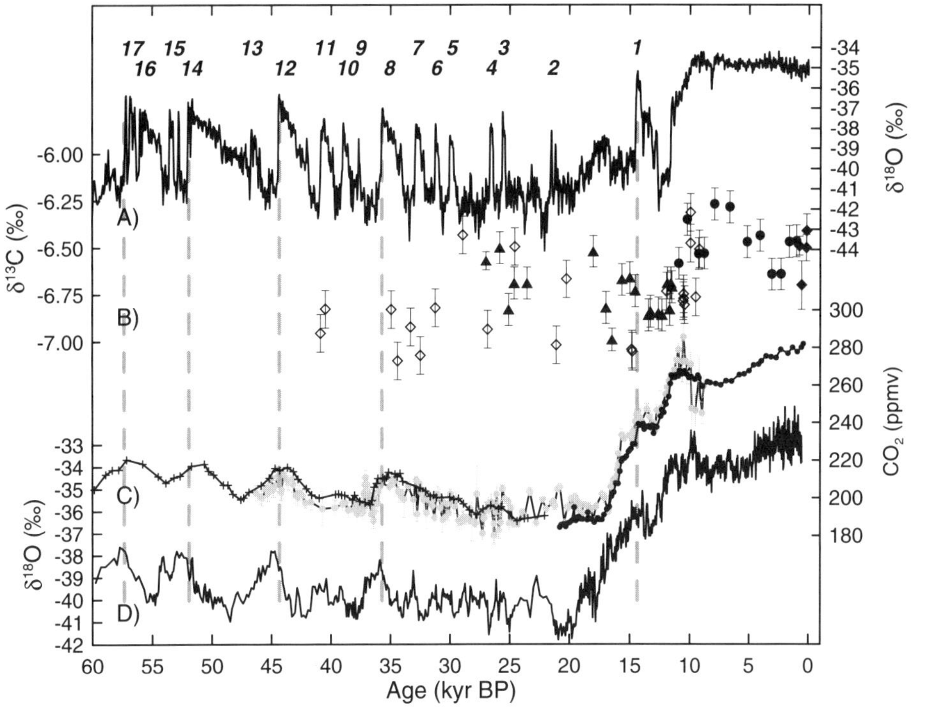

Figure 4.5. Climate records from the last glacial to the preindustrial Holocene: **(A)** $\delta^{18}O$ of the ice, a proxy for local temperature, from the GRIP ice core, Greenland (Dansgaard et al. 1993). **(B)** $\delta^{13}C$ of CO_2; dots (Indermühle et al. 1999b) and filled triangles (Smith et al. 1999) from the Taylor Dome ice core. Open diamonds are from Byrd station and filled diamonds from several ice cores covering the preindustrial Holocene (Leuenberger et al. 1992). **(C)** Composite record of atmospheric CO_2 from Antarctic ice cores. Gray dots with error bars are from the Byrd station ice core (Marchal et al. 1999; Neftel et al. 1988; Staffelbach et al. 1991) on a timescale synchronized to the GRIP temperature record over the global CH_4 record (Blunier et al. 1998; Stauffer et al. 1998). The Taylor Dome record (crosses, 60–20 kyr BP) has been adjusted to the GRIP timescale so that the relative position of CO_2 and Antarctic temperature variations is preserved from the original tie to the Vostok GT4 timescale (Indermühle et al. 2000). The CO_2 record over the last glacial termination and the Holocene (dots) is from the (EPICA) Dome C core (Flückiger et al. 2002; Monnin et al. 2001). **(D)** $\delta^{18}O$, a proxy for local temperature, from the Byrd ice core, Antarctica (Johnsen et al. 1972). $\delta^{18}O$ from Byrd station is plotted on the GRIP timescale (Blunier and Brook 2001).

High resolution CO_2 data from Byrd station (Marchal et al. 1999; Neftel et al. 1988; Staffelbach et al. 1991) span from approximately 47 to 8kyr BP. This record is well synchronized to the Greenland ice core records (Stauffer et al. 1998). The Byrd CO_2 concentration varies between about 180 and 210 ppmv during 47 to 17kyrBP (see Fig. 4.5). Two distinct millennial-scale peaks of 20 ppmv during this period correlate well with the Antarctic temperature reconstruction from Byrd station.

The high-resolution record from the Taylor Dome ice core reaches back to about 60 kyr BP (Indermühle et al. 2000). Indermühle et al. (2000) tied this record to the temperature variations at Vostok and found also a strong covariance between the CO_2 and the temperature record. We present the Taylor Dome record in Fig. 4.5. The time scale has been adjusted so that the relative position of CO_2 and Antarctic temperature variations is preserved from the initial publication (Indermühle et al. 2000).

The Southern Hemisphere warming and, apparently, the atmospheric CO_2 concentration increase occur when Greenland temperatures are lowest (Indermühle et al. 2000). Marchal et al. (1998) simulated the evolution of the atmospheric CO_2 concentration in response to a freshwater-induced collapse of the thermohaline circulation using a coupled ocean circulation-biogeochemical model. In their simulation, a thermohaline-circulation shutdown leads to an increase of the CO_2 concentration due to a warming of the Southern Ocean (the so-called see-saw effect, Broecker 1998; Stocker 1998), which overcompensates the North Atlantic cooling. Changes in alkalinity and dissolved inorganic carbon concentration in the North Atlantic surface, mainly due to dilution, amplify the increase. Sensitivity studies reveal that the response of the temperature in the North Atlantic and the Southern Ocean and of the atmospheric CO_2 concentration depend on the amount, duration, and geographic location of freshwater input (i.e., iceberg discharge, see Fig. 12 in Marchal et al. 1998). That would explain why some D-O events are preceded by an Antarctic warming and an increase of atmospheric CO_2, while others are not (Stocker and Marchal 2000).

4.4 The Holocene

The Antarctic Taylor Dome CO_2 and $\delta^{13}C$ records reveal that the global carbon cycle has not been in steady state during the Holocene (Indermühle et al. 1999b). The CO_2 record was recently confirmed by measurements on the Dome C core and shows an 8 ppmv decrease in the CO_2 mixing ratio (Flückiger et al. 2002). This decrease is paralleled by a 0.3‰ increase in $\delta^{13}C$ between 10.5 and 8.2 kyr BP. Over the following 7 kyr, a fairly linear 25 ppmv CO_2 increase is accompanied by a approximately 0.2 $\delta^{13}C$ decrease (see Fig. 4.5). Inverse methods based on a one-dimensional carbon cycle model applied to the Taylor Dome record (Indermühle et al. 1999b) suggest that changes in terrestrial biomass and sea surface temperature are primarily responsible for the observed CO_2 changes. In particular, the CO_2 increase from 7 to 1 kyr BP could correspond to a cumulative continental biospheric release of about 195 GtC, in connection with a change from a warmer and wetter mid-Holocene climate to the colder and drier preindustrial conditions. This model result is not sensitive to surface-to-deep ocean mixing and air-sea exchange coefficients because of their faster time scales. Changes in the biological carbon isotope fractionation factor, which varies according to the contribution from C_3 and C_4 plants and due to environmental changes, may contribute up to 30 to 50% of the observed and modeled changes. The resulting additional uncertainty in the cumulative biospheric carbon release is ± 30 GtC at 7 kyr BP and ± 70 GtC at 1 kyr BP.

However, this model calculation is very dependent on the $\delta^{13}C$ data and the smoothing line. Better, high resolution δ^{13} data will improve our knowledge of the causes of CO_2 variations not only over the Holocene period but also in the glacial and over the glacial-interglacial transition.

4.5 The Anthropogenic Increase

A continuous record of CO_2 over the past millennium, which overlaps with the period of instrumental data, is contained in the Law Dome ice core (Etheridge et al. 1996), as shown in Fig. 4.6.

There is no doubt that the atmospheric CO_2 concentration has been increasing

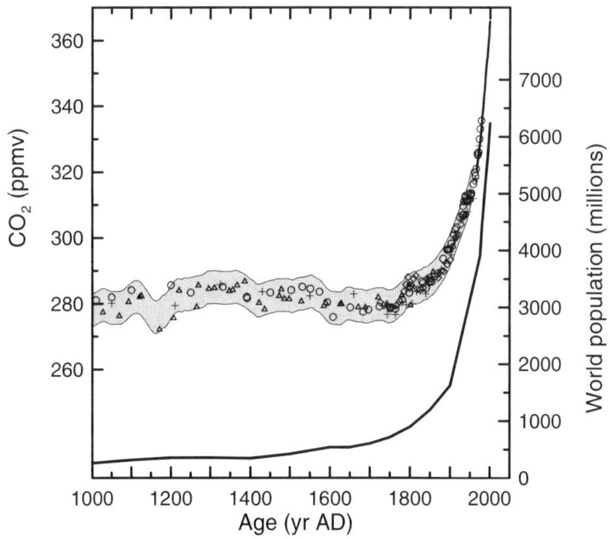

Figure 4.6. The last millennium. CO_2 data from Antarctic ice cores compiled by Barnola (1999). The ice core measurements overlap with direct atmospheric measurements, shown with the solid line (Keeling and Whorf 2000); world population (McEvedy and Jones 1979).

since the industrial revolution and reached a concentration that is unprecedented over more than 400,000 years. The CO_2 record suggests slightly increased concentrations from approximately AD 1200 to 1400, during Medieval times, and slightly reduced concentrations from approximatly AD 1550 to 1800, during the Little Ice Age. However, caution should be used in interpreting these variations as natural. Although all ice core records from different Antarctic locations show the same general picture, we cannot exclude a small contribution from chemical reactions in the ice (Barnola 1999). The pre-anthropogenic level of CO_2 during the last millennium is approximately 280 ± 10 ppmv (Etheridge et al. 1996).

4.6 Comparison of Ice Core CO_2 Data to Other Methods

The smooth increase of atmospheric CO_2 suggested by low-resolution CO_2 records has been questioned for two reasons: (1) the records did not have the resolution to exclude any fast changes (at least over some time intervals) and (2) the smoothed recording of atmospheric signals could potentially mask such fast changes. Indirect CO_2 measurements from $\delta^{13}C$ in peat (Figge and White 1995) and stomatal density and stomatal index measurements (Beerling, Birks, and Woodward 1995; McElwain, Mayle, and Beerling 2002; Wagner et al. 1999) suggest century or even decadal changes of the CO_2 concentration in the range of the Younger Dryas to the early Holocene.

The ice core CO_2 record over the last glacial termination to the Younger Dryas period has been subject to intensive studies in recent years. The results from low accumulation ice cores and high accumulation ice cores are in excellent agreement. From two ice cores we have records with a high time resolution (Byrd station ~160 years; Dome C ~140 years).

Besides the study from Wagner et al. (1999), which has been criticized for its calibration process (Indermühle et al. 1999a), indirect studies of CO_2 concentrations agree fairly well with ice core data on the CO_2 level. However, CO_2 records based on ice core and stomatal findings differ significantly on short-term fluctuations (Fig. 4.7).

It is intriguing that stomatal density data from various sites agree fairly well on a proposed CO_2 variation during the Younger Dryas (Beerling, Birks, and Woodward 1995; McElwain, Mayle, and Beerling 2002). Is it possible that the ice core archive did not record the variations observed in the stomatal records because of its limitations concerning the trapping of atmospheric gas? To investigate this question we treated the Span Pond CO_2 record from McElwain, Mayle, and Beerling (2002) with the smoothing function of Byrd station and Dome C. The width of the age distribution is 20 to 25 years for Byrd station and one order of magnitude larger for Dome C. The effect of the Byrd smoothing function on the stomatal record is minor. Even applying the Dome C smoothing function (see Fig. 4.7) does not remove the short-term fluctuations in the stomatal record. Therefore the disagreement between ice core and stomatal-based CO_2 reconstructions cannot be explained by the smoothing of the ice core record.

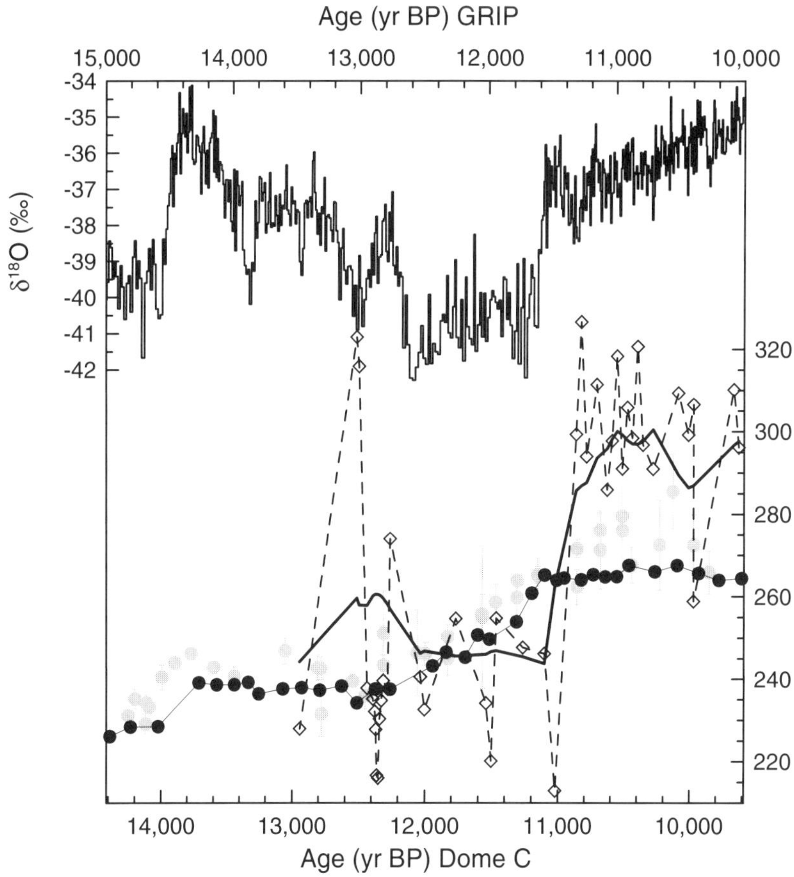

Figure 4.7. Comparison of CO_2 records from Byrd station (gray dots, Marchal et al. 1999) and Dome C (black dots, Monnin et al. 2001) with a reconstruction from stomata frequency measurements (open diamonds McElwain et al. 2002). The Dome C record is plotted on the Dome C timescale (bottom x-axis, Schwander et al. 2001); all the other records are plotted on the GRIP ss09 timescale. The two timescales are adjusted so that the "global" CH_4 signal is consistent between the GRIP and the Dome C record over the Bølling-Allerød (B/A), Younger Dryas (YD) period (see Fig. 4.3). The stomata record shows higher variability than do the ice core records. Ice core reconstructions of atmospheric gas records are smoothed in time. Smoothing the stomata record with the Byrd smoothing function barely affects the record (not shown). The smoothing in the Dome C record is one order of magnitude larger than in the Byrd station record. Still, the two records basically show the same variations. Smoothing the stomatal record with the Dome C smoothing function (heavy solid line) does not remove the suggested steep CO_2 increase around 11.2 kyr BP. The stomatal record is in discrepancy with both ice core records. The origin of this discrepancy is currently not known.

Also we are not aware of any other process during the trapping or measuring of CO_2 from ice cores that would explain the discrepancy. We are convinced that the ice cores are reliable to an uncertainty of about 1% (Stauffer et al. 2003). The advantage of the ice core based method compared to other methods to reconstruct past CO_2 concentrations is that it is the only approach that accesses the atmospheric CO_2 concentration directly as CO_2 in air.

4.7 Summary

The paleo CO_2 record from ice cores spans over four glacial-interglacial cycles. It is highly probable that the CO_2 concentration over this time period has never been as high as it is today. The Vostok CO_2 record and Antarctic temperature are correlated. This points to the Southern Ocean as the main driver for the glacial-interglacial CO_2 variations. The timing of the CO_2 variations is such that the CO_2 decreases lag Antarctic temperature decreases by a few millennia. The CO_2 increase, in contrast, lags the Antarctic temperature increase by a few centuries. This has been demonstrated over the last termination but is probably also the case for terminations II to IV. The CO_2 increase precedes by several millennia the northern glacial-interglacial temperature increase, which dominates over large portions of the world. The detailed CO_2 record over the last transition shows that the CO_2 increase can be divided into several intervals with different increase rates and fast CO_2 increases on decadal timescales. The timing of these intervals shows characteristics of northern hemispheric climate variations. The CO_2 increase over termination I shows a combination of Northern and Southern Hemisphere climate change. CO_2 variations over the Holocene show that the CO_2 budget has not been in steady state over this period. We are certain that the Antarctic CO_2 record represents the atmospheric CO_2 signal and is presently the best reconstruction available.

References

Adams, J.M., and H. Faure. 1998. A new estimate of changing carbon storage on land since the last glacial maximum, based on global land ecosystem reconstruction. *Global and Planetary Change* 17:3–24.

Anklin, M., J.-M. Barnola, J. Schwander, B. Stauffer, and B. Raynaud. 1995. Processes affecting the CO_2 concentrations measured in Greenland ice. *Tellus* 47B:461–70.

Anklin, M., J. Schwander, B. Stauffer, J. Tschumi, A. Fuchs, J.M. Barnola, and D. Raynaud. 1997. CO_2 record between 40 and 8 kyr B.P. from the Greenland ice core project ice core. *Journal of Geophysical Research* 102:26539–46.

Arnaud, L., J.-M. Barnola, and P. Duval. 2000. Physical modeling of the densification of snow/firn and ice in the upper part of polar ice sheets. In *Physics of ice core records,* ed. T. Hondoh, 285–305. Sapporo: Hokkaido University Press.

Barnola, J.-M. 1999. Status of the atmospheric CO_2 reconstruction from ice cores analyses. *Tellus* 51B:151–55.

Barnola, J.-M., P. Pimienta, D. Raynaud, and Y.S. Korotkevich. 1991. CO_2-climate relationship as deduced from the Vostok ice core: A re-examination based on new measurements and on a re-evaluation of the air dating. *Tellus* 43:83–90.

Beerling, D.J., H.H. Birks, and F.I. Woodward. 1995. Rapid late-glacial atmospheric CO_2 changes reconstructed from the stomatal density record of fossil leaves. *Journal of Quaternary Science* 10:379–84.

Blunier, T., and E.J. Brook. 2001. Timing of millennial-scale climate change in Antarctica and Greenland during the last glacial period. *Science* 291:109–112.

Blunier, T., J. Chappellaz, J. Schwander, A. Dällenbach, B. Stauffer, T.F Stocker, D. Raynaud, J. Jouzel, H.B. Clausen, C.U. Hammer, and S.J. Johnsen. 1998. Asynchrony of Antarctic and Greenland climate change during the last glacial period. *Nature* 394: 739–43.

Broecker, W.S. 1994. Massive iceberg discharges as triggers for global climate change. *Nature* 372:421–24.

———. 1998. Paleocean circulation during the last deglaciation: A bipolar seesaw? *Paleoceanography* 13:119–21.

Broecker, W.S., and G.M. Henderson. 1998. The sequence of events surrounding Termination II and their implications for the cause of glacial-interglacial CO_2 changes. *Paleoceanography* 13:352–64.

Caillon, N., J.P. Severinghaus, J. Jouzel, J.-M. Barnola, J. Kang, J., and V.Y. Lipenkov. 2003. Timing of atmospheric CO_2 and Antarctic temperature changes across Termination III. *Science* 299:1728–31.

Craig, H., Y. Horibe, and T. Sowers. 1988. Gravitational separation of gases and isotopes in polar ice caps. *Science* 242:1675–78.

Crowley, T.J. 1995. Ice age terrestrial carbon changes revisited. *Global Biogeochemical Cycles* 9:377–89.

Cuffey, K.M., and F. Vimeux. 2001. Covariation of carbon dioxide and temperature from the Vostok ice core after deuterium-excess correction. *Nature* 412:523–27.

Dällenbach, A., T. Blunier, J. Flückiger, B. Stauffer, J. Chappellaz, and D. Raynaud. 2000. Changes in the atmospheric CH_4 gradient between Greenland and Antarctica during the Last Glacial and the transition to the Holocene. *Geophysical Research Letters* 27:1005–1008.

Dansgaard, W., S.J. Johnsen, H.B. Clausen, D. Dahl-Jensen, N.S. Gundestrup, C.U. Hammer, Hvidberg, J.P. Steffensen, A.E. Sveinbjörnsdottir, J. Jouzel, and G. Bond. 1993. Evidence for general instability of past climate from a 250-kyr ice-core record. *Nature* 364:218–20.

Delmas, R.J. 1993. A natural artefact in Greenland ice-core CO_2 measurements. *Tellus* 45B:391–396.

Etheridge, D.M., L.P. Steele, R.L. Langenfields, R.J. Francey, J.-M. Barnola, and V, I. Morgan. 1996. Natural and anthropogenic changes in atmospheric CO_2 over the last 1000 years from air in Antarctic ice and firn. *Journal of Geophysical Research* 101: 4115–28.

Figge, R.A., and W.C. White. 1995. High-resolution Holocene and late glacial atmospheric CO_2 record: Variability tied to changes in thermohaline circulation. *Global Biogeochemical Cycles* 9:391–403.

Fischer, H., M. Wahlen, J. Smith, D. Mastroianni, and B. Deck. 1999. Ice core records of atmospheric CO_2 around the last three glacial terminations. *Science* 283: 1712–14.

Flückiger, J., E. Monnin, B. Stauffer, J. Schwander, T.F. Stocker, J. Chappellaz, D. Raynaud, and J.M. Barnola. 2002. High resolution Holocene N_2O ice core record and its relationship with CH_4 and CO_2. *Global Biogeochemical Cycles, 16* (1), 1010, doi: 10.1029/2001GB001417.

Grootes, P.M., M. Stuiver, J.W.C. White, S.J. Johnsen, and J. Jouzel. 1993. Comparison of oxygen isotope records from the GISP2 and GRIP Greenland ice cores. *Nature* 366:552–54.

Güllük, T., F. Slemr, and B. Stauffer. 1998. Simultaneous measurements of CO_2, CH_4 and N_2O in air extracted by sublimation from Antarctica ice cores: Confirmation of the data obtained using other extraction techniques. *Journal of Geophysical Research* 103:15971–78.

Ikeda, T., H. Fukazawa, S. Mae, L. Pépin, P. Duval, B. Champagnon, V.Y. Lipenkov, and T. Hondoh. 1999. Extreme fractionation of gases caused by formation of clathrate hydrates in Vostok Antarctic ice. *Geophysical Research Letters* 26:91–94.

Indermühle, A., E. Monnin, B. Stauffer, T.F. Stocker, and M. Wahlen. 2000. Atmospheric CO_2 concentration from 60 to 20 kyr BP from the Taylor Dome ice core, Antarctica. *Geophysical Research Letters* 27:735–38.

Indermühle, A., B. Stauffer, T.F. Stocker, D. Raynaud, J.-M. Barnola, H.H. Birks, W. Eide, H.J.B. Birks, F. Wagner, W.M. Kürschner, H. Visscher, S.J.P. Bohncke, D.L. Dilcher, and B. van Geel. 1999a. Early Holocene atmospheric CO_2 concentrations. *Science* 286:1815a.

Indermühle, A., T.F. Stocker, H. Fischer, H.J. Smith, F. Joos, M. Wahlen, B. Deck, D. Mastroianni, J. Tschumi, T. Blunier, R. Meyer, and B. Stauffer. 1999b. Holocene carbon-cycle dynamics based on CO_2 trapped in ice at Taylor Dome, Antarctica. *Nature* 398:121–26.

Johnsen, S.J., D. Dahl-Jensen, W. Dansgaard, and N. Gundestrup. 1995. Greenland palaeotemperatures derived from GRIP bore hole temperature and ice core isotope profiles. *Tellus* 47B:624–29.

Johnsen, S.J., W. Dansgaard, H.B. Clausen, and C.C. Langway, Jr. 1972. Oxygen isotope profiles through the Antarctic and Greenland ice sheets. *Nature* 235:429–34.

Joos, F., M. Bruno, R. Fink, U. Siegenthaler, T.F. Stocker, C. Le Quéré, and J.L. Sarmiento. 1996. An efficient and accurate representation of complex oceanic and biospheric models of anthropogenic carbon uptake. *Tellus* 48B:397–417.

Jouzel, J., R. Vaikmae, J.R. Petit, M. Martin, Y. Duclos, M. Stievenard, C. Lorius, M. Toots, M.A. Mélières, L.H. Burckle, N.I. Barkov, and V.M. Kotlyakov. 1995. The two-step shape and timing of the last deglaciation in Antarctica. *Climate Dynamics* 11:151–61.

Keeling, C.D., and T.P. Whorf. 2000. Atmospheric CO_2 records from sites in the SIO air sampling network. In *Trends: A compendium of data on global change*. Oak Ridge, Tenn.: Carbon Dioxide Information Analysis Center, Oak Ridge National Laboratory, U.S. Department of Energy.

Lang, C., M. Leuenberger, J. Schwander, and S. Johnsen. 1999. 16°C rapid temperature variation in Central Greenland 70,000 years ago. *Science* 286:934–37.

Leuenberger, M., C. Lang, and J. Schwander. 1999. δ^{15} N measurements as a calibration tool for the paleothermometer and gas-ice age differences: A case study for the 8200 BP event on GRIP ice. *Journal of Geophysical Research* 104:22163–69.

Leuenberger, M., U. Siegenthaler, and C.C. Langway. 1999. Carbon isotope composition of atmospheric CO_2 during the last ice age from an Antarctic ice core. *Nature* 357: 488–90.

Marchal, O., T.F. Stocker, and F. Joos. 1998. Impact of oceanic reorganisations on the ocean carbon cycle and atmospheric carbon dioxide content. *Paleoceanography* 13: 225–44.

Marchal, O., T.F. Stocker, F. Joos, A. Indermühle, T. Blunier, and J. Tschumi. 1999. Modelling the concentration of atmospheric CO_2 during the Younger Dryas climate event. *Climate Dynamics* 15:341–54.

Martin, J.H. 1990. Glacial-interglacial CO_2 change: The iron hypothesis. *Paleoceanography* 5:1–13.

McElwain, J.C., F.E. Mayle, and D.J. Beerling. 2002. Stomatal evidence for a decline in atmospheric CO_2 concentration during the Younger Dryas stadial: A comparison with Antarctic ice core data. *Journal of Quaternary Science* 17:21–29.

McEvedy, C., and R. Jones. 1978. *Atlas of world population history,* 368 pp. New York: Penguin Miller, S.L. 1969. Clathrate hydrates of air in Antarctic ice. *Science* 165: 489–90.

Monnin, E., A. Indermühle, A. Dällenbach, J. Flückiger, B. Stauffer, T.F. Stocker, D. Raynaud, and J.-M. Barnola. 2001. Atmospheric CO_2 concentrations over the last glacial termination. *Science* 291:112–14.

Neftel, A., H. Oeschger, T. Staffelbach, and B. Stauffer. 1988. CO_2 record in the Byrd ice core 50,000–5000 years BP. *Nature* 331:609–11.

Oeschger, H., J. Beer, U. Siegenthaler, B. Stauffer, W. Dansgaard, and C.C. Langway. 1984. Late glacial climate history from ice cores. In *Climate processes and climate sensitivity,* Geophysical Monographs Series, ed. J.E. Hansen and T. Takahashi 29: 299–306. Washington, D.C.: American Geophysical Union.

Oeschger, H., A. Neftel, T. Staffelbach, and B. Stauffer. 1988. The dilemma of the rapid variations in CO_2 in Greenland ice cores. *Annals of Glaciology* 10:215–16.

Pépin, L., D. Raynaud, J.-M. Barnola, and M.F. Loutre. 2001. Hemispheric roles of climate forcings during glacial-interglacial transitions as deduced from the Vostok record and LLN-2D model experiments. *Journal of Geophysical Research* 106:31885–92.

Petit, J.R., J. Jouzel, D. Raynaud, N.L. Barkov, J.M. Barnola, I. Basile, M. Bender, J. Chappellaz, M. Davis, G. Delaygue, M. Delmotte, V.M. Kotlyakov, M. Legrand, V.Y. Lipenkov, C. Lorius, L. Pepin, C. Ritz, E. Saltzman, and M. Stievenard. 1999. Climate and atmospheric history of the past 420,000 years from the Vostok ice core, Antarctica. *Nature* 399:429–36.

Raynaud, D., J. Jouzel, J.M. Barnola, J. Chappellaz, R.J. Delmas, and C. Lorius. 1993. The ice record of greenhouse gases. *Science* 259:926–33.

Rommelaere, V., L. Arnaud, and J.M. Barnola. 1997. Reconstructing recent atmospheric trace gas concentrations from polar firn and bubbly ice data by inverse methods. *Journal of Geophysical Research* 102:30069–83.

Schwander, J. 1989. The transformation of snow to ice and the occlusion of gases. In *The environmental record in glaciers and ice sheets,* ed. H. Oeschger and C.C. Langway Jr., 53–67. New York: John Wiley.

———. 1996. Gas diffusion in firn. In *Chemical exchange between the atmosphere and polar snow,* ed. E.W. Wolff and R.C. Bales, NATO ASI Series. I 43, pp. 527–540, Springer-Verlag, Berlin Heidelberg.

Schwander, J., J.-M. Barnola, C. Andrié, M. Leuenberger, A. Ludin, D. Raynaud, and B. Stauffer. 1993. The age of the air in the firn and the ice at Summit, Greenland. *Journal of Geophysical Research* 98:2831–38.

Schwander, J., J. Jouzel, C.U. Hammer, J.-R. Petit, R. Udisti, and E. Wolff. 2001. A tentative chronology for the EPICA Dome Concordia ice core. *Geophysical Research Letters* 28:4243–46.

Schwander, J., T. Sowers, J.-M. Barnola, T. Blunier, B. Malaizé, and A. Fuchs. 1997. Age scale of the air in the summit ice: Implication for glacial-interglacial temperature change. *Journal of Geophysical Research* 102:19483–94.

Schwander, J., and B. Stauffer. 1984. Age difference between polar ice and the air trapped in its bubbles. *Nature* 311:45–47.

Severinghaus, J.P., and E.J. Brook. 1999. Abrupt climate change at the end of the last glacial period inferred from trapped air in polar ice. *Science* 286:930–34.

Severinghaus, J.P., T. Sowers, E.J. Brook, R.B. Alley, and M.L. Bender. 1998. Timing of abrupt climate change at the end of the Younger Dryas interval from thermally fractionated gases in polar ice. *Nature* 391:141–46.

Shoji, H., A. Miyamoto, J. Kipfstuhl, and C.C. Langway Jr. 2000. Microscopic observations of air hydrate inclusions in deep ice core samples. In *Physics of ice core records,* ed. T. Hondoh, 363–71. Sapporo: Hokkaido University Press.

Siegenthaler, U., and T. Wenk. 1984. Rapid atmospheric CO_2 variations and ocean circulation. *Nature* 308:624–26.

Smith, H.J., H. Fischer, M. Wahlen, D. Mastroianni, and B. Deck. 1999. Dual modes of the carbon cycle since the Last Glacial Maximum. *Nature* 400:248–50.

Spahni, R., J. Schwander, J. Flückiger, B. Stauffer, J. Chappellaz, D. and Raynaud. 2003. The attenuation of fast atmospheric CH_4 variations recorded in polar ice cores. *Journal of Geophysical Research* 30(11):1571, doi:10.1029/2003GLO17093.

Staffelbach, T., B. Stauffer, A. Sigg, and H. Oeschger.1991.CO_2 measurements from polar ice cores: More data from different sites. *Tellus* 43B:91–96.

Stauffer, B., T. Blunier, A. Dällenbach, A. Indermühle, J. Schwander, T.F. Stocker, J. Tschumi, J. Chappellaz, D. Raynaud, C.U. Hammer, and H.B. Clausen. 1998. Atmospheric CO_2 concentration and millennial-scale climate change during the last glacial period. *Nature* 392:59–62.

Stauffer, B., J. Flückiger, E. Monnin, T. Nakazawa and S. Aoki,. 2003. Discussion of the reliability of CO_2, CH_4 and N_2O records from polar ice cores, in: Global scale climate and environment study through polar deep ice cores: Proceeding of the international symposium on Dome Fuji ice core and related topics, 27–28 February 2001, Tokyo, H. Shoji and O. Watanabe, eds., Memoirs of National Institute of Polar Research, Special Issue No. 57, pp. 139–152, National Institute of Polar Research, Tokyo.

Stauffer, B., H. Hofer, H. Oeschger, J. Schwander, and U. Siegenthaler. 1984. Atmospheric CO_2 concentration during the last glaciation. *Annals of Glaciology* 5:160–64.

Stauffer, B., and J. Tschumi. 2000. Reconstruction of past atmospheric CO^2 concentrations by ice core analyses. In *Physics of ice-core records,* ed. T. Hondoh, 217–41. Sapporo: Hokkaido University Press.

Stephens, B.B., and R.F. Keeling. 2000. The influence of Antarctic sea ice on glacial-interglacial CO_2 variations. *Nature* 404:171–74.

Stocker, T.F. 1998. The seesaw effect. *Science* 282:61–62.

Stocker, T.F., and O. Marchal. 2000. Abrupt climate change in the computer: Is it real?, *Proceedings of the National Academy of Science U.S.A.* 97:1362–65.

Toggweiler, J.R. 1999. Variation of atmospheric CO_2 by ventilation of the ocean's deepest water. *Paleoceanography* 14:571–88.

Tschumi, J., and B. Stauffer. 2000. Reconstructing the past atmospheric CO_2 concentration based on ice core analyses: Open questions due to in situ production of CO_2 in the ice. *Journal of Glaciology* 46:45–53.

Uchida, T., T. Hondoh, S. Mae, H. Shoji, and N. Azuma. 1994. Optimized storage condition of deep ice core samples from the viewpoint of air-hydrate analysis. Memoirs. *National Institute for Polar Research* 49:306–13.

Voelker, A.H.L., and workshop participants. 2002. Global distribution of centennial-scale records for marine isotope stage (MIS) 3: A database. *Quaternary Science Reviews* 21:1185–1212.

Wagner, F., S.J.P. Bohncke, D.L. Dilcher, W.M. Kurschner, B. van Geel, and H. Visscher. 1999. Century-scale shifts in early Holocene atmospheric CO_2 concentration. *Science* 284:1971–73.

Weaver, A.J., M. Eby, A.F. Fanning, and E.C. Wiebe. 1998. Simulated influence of carbon dioxide, orbital forcing and ice sheets on the climate of the Last Glacial Maximum. *Nature* 394:847–53.

Zumbrunn, R., A. Neftel, and H. Oeschger. 1982. CO_2 measurements on 1-cm^3 ice samples with an IR laserspectrometer (IRLS) combined with a new dry extraction device. *Earth and Planetary Science Letters* 60:318–24.

5. Atmospheric CO_2 and $^{13}CO_2$ Exchange with the Terrestrial Biosphere and Oceans from 1978 to 2000: Observations and Carbon Cycle Implications

Charles D. Keeling, Stephen C. Piper,
Robert B. Bacastow, Martin Wahlen, Timothy P. Whorf,
Martin Heimann, and Harro A. Meijer

5.1 Introduction

Not only direct observations of CO_2 but also predictions from geophysical and biogeochemical models are needed to establish unequivocally the consequences of human activities on Earth's carbon cycle. During the industrial era, in which the combustion of fossil fuel contributed to a sharp rise in atmospheric CO_2 (Fig. 5.1), measurements of both the concentration and the $^{13}C/^{12}C$ ratio of atmospheric CO_2 are especially relevant. Our ability to predict correctly the time-varying gradients in atmospheric CO_2 that these data establish is an indispensable requirement for trusting models that link the storage of carbon in the atmospheric, terrestrial, and oceanic reservoirs of the carbon cycle to transfers of carbon between these global carbon pools.

These transfers, if thus validated, in turn can be compared with integrations of local flux measurements, carried out on land, for example by the AmeriFlux and EuroFlux eddy-correlation flux networks (Aubinet et al. 2000; Canadell et al. 2000) and in the oceans, for example by the joint Global Ocean Fluxes Study (Karl and Michaels 1996). Thus it is possible to combine top-down and bottom-up approaches to assess human impacts on the global carbon cycle.

To acquire an atmospheric CO_2 database adequate to establish human impacts on continental and global scales, it is necessary to sample air worldwide. This is most directly accomplished by collecting and analyzing flasks of atmospheric air or by continuous measurements of air at field sites (Keeling et al. 1989a).

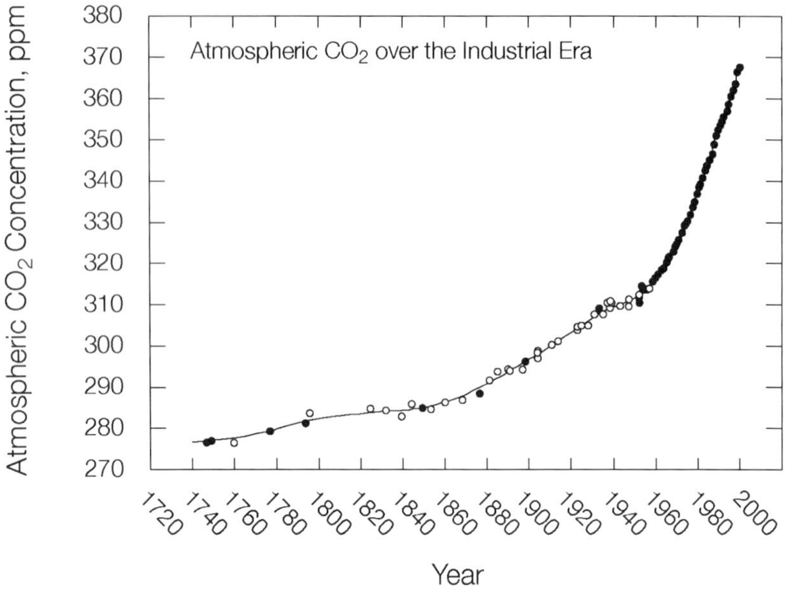

Figure 5.1. Time trend in the concentration of atmospheric CO_2, in ppm, from 1740 to 2000. Data before 1957 are proxies from measurements in air extracted from ice cores at Law Dome, Antarctica (open circles, Francey et al., 1995; closed circles, Etheridge et al., 1996). Data from 1957 to 1978 are averages of measurements from the South Pole and Mauna Loa Observatory. Data from 1978 on are averages of direct measurements of air collected from 6 to 9 locations (closed circles). The curve is a spline function that combines separate fits to the proxy and direct data as described by Keeling et al. (1989a).

Because atmospheric turbulence tends to smooth out the effects of local sources and sinks on the atmospheric CO_2 distribution, only a limited number of sampling locations are required. They must, however, be remote from large local sources and sinks of CO_2. In this chapter we interpret atmospheric CO_2 data, updating an earlier study (Keeling et al. 1989a). This new study again depends on data from fewer stations than do most other recent global studies, but that lack is offset to a considerable degree because it uses a more extensive isotopic data set.

The study has two approaches: first, as reported provisionally by Keeling et al. (2001), to calculate and interpret global average exchanges of atmospheric CO_2 with the terrestrial and oceanic carbon reservoirs that together sum to all globally significant exchanges; second, as reported provisionally by Piper et al. (2001), to calculate a set of regional "source components" that together sum to all globally significant exchanges.

Here we describe results mainly on the global approach. The source components, which also apply to the global scale study, include industrial emissions of CO_2. These components include industrial emissions of CO_2 and long-term human-induced disturbances to terrestrial vegetation and of soils; also included are the perturbations induced by these disturbances, such as the uptake of CO_2 by the oceans and terrestrial vegetation, and the natural CO_2 fluxes. In constructing these components, which portray both sources and sinks of atmospheric CO_2, we make use of *a priori* information as much as possible, including global data sets, on a spatial grid, of observed sea and land surface temperatures and of terrestrial photosynthetic activity derived from remote sensing data observed by satellites.

For the less reliably known components, we prescribe, *a priori,* only their spatial and temporal structures, adjusting their overall strengths simultaneously to predict optimally our observations of CO_2 concentration and $^{13}C/^{12}C$, taking account of the entire set of component fluxes. Here, we compute only global average exchanges of atmospheric CO_2. These exchanges, which transfer CO_2 to and from terrestrial vegetation and soils (henceforth, together called the "terrestrial biosphere") and to and from the world oceans (henceforth, "oceans"), are inferred from global averages of time-varying concentration and $^{13}C/^{12}C$ ratio determined from our observations.

A major challenge is to determine how time-varying sources and sinks of atmospheric CO_2 reflect the interplay of natural processes and human activities, including feedbacks between Earth's carbon cycle and its physical environment. Of paramount interest is that Earth's heat balance is being altered by an enhanced greenhouse effect caused by rising concentrations of CO_2 and other infrared-absorbing gases. Global warming is likely to be occurring as a consequence, altering the carbon cycle globally (Santer et al. 1996; Crowley 2000). The picture is complicated, however, because natural variations in climate also impact Earth's heat balance (Free and Robock 1999; Andronova and Schlesinger 2000) and the carbon cycle (Falkowski et al. 2000).

Although the observations most essential to this study are measurements of CO_2 concentration, valuable information is afforded by its $^{13}C/^{12}C$ ratio, expressed in delta notation as

$$\delta^{13}C = (R_{sample} - R_{standard})/R_{standard} \tag{5.1}$$

where R denotes the measured $^{13}C/^{12}C$ molar ratio of a specific sample of CO_2, and the subscripts refer to the measured sample and the international standard (PDB). Because the rare stable isotope (^{13}C) is strongly fractionated during photosynthesis by land plants, the terrestrial biosphere alters the $\delta^{13}C$ of atmospheric CO_2 far more than do the oceans (Farquhar, Ehleringer, and Hubick 1989; Keeling et al. 1989a). Terrestrial and oceanic CO_2 fluxes can thus be distinguished using $\delta^{13}C$ data, with certain additional information, including variability in the fractionation factor for photosynthesis for different types of plants and under different growing conditions and, on long timescales, evidence of the degree of vertical mixing of the oceans.

5.2 Atmospheric Observations

Observations of atmospheric CO_2 concentration and its $^{13}C/^{12}C$ ratio, expressed by $\delta^{13}C$, were obtained from an array of 10 stations situated along a nearly north-south transect mainly in the Pacific Ocean basin. The stations extend from the Arctic to the South Pole (Fig. 5.2), at sites on land located as far as possible from biological activity and combustion of fossil fuels.

By sampling the air upwind, we minimized the effects of local interferences, so the observations mainly reflect broadscale CO_2 fluxes. The concentration data set begins in 1957, the isotopic data set in 1977. The data provide complete time-series for all stations of the array from 1986 onward. Here we mainly consider data after 1977. The CO_2 concentration C is expressed as a mole fraction in parts per million of dry air (ppm), the isotopic data, $\delta^{13}C$, in per mil (‰) departures from the standard PDB (Craig 1957; Mook and Grootes 1973). Methods of sampling, measurements, and calibrations are described in Keeling et al. (2001).

The CO_2 concentration decreases almost monotonically from north to south such that two Arctic stations, Alert and Point Barrow, register concentrations 3 to 5 ppm higher than the most southerly stations, in New Zealand and at the South Pole (Fig. 5.3A, Alert not shown).

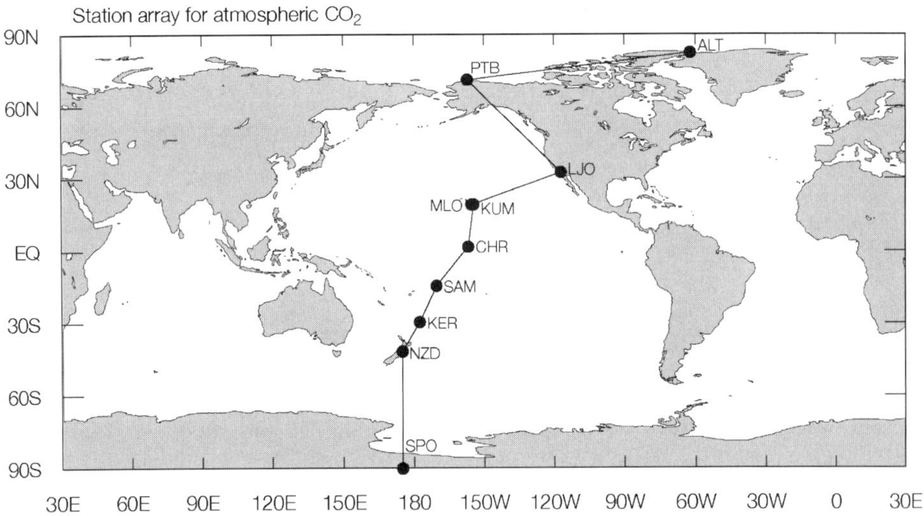

Figure 5.2. Locations of carbon dioxide sampling stations that furnish data for Keeling et al. (2001). Station symbols: ALT = Alert; PTB = Point Barrow; LJO = La Jolla; KUM = Cape Kumukahi; MLO = Mauna Loa Observatory; CHR = Christmas Island; SAM = Samoa; KER = Kermadec; NZD = New Zealand; SPO = South Pole. Further information is listed in Table 5.1. A transect connecting the stations is shown.

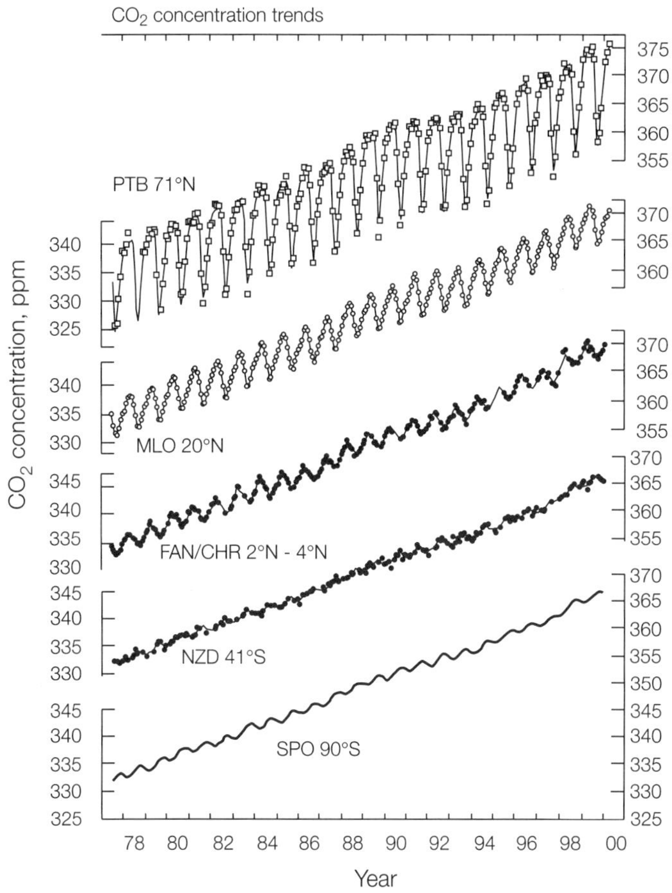

Figure 5.3. Trends in the measured atmospheric CO_2 concentration and its carbon isotope ratio, $\delta^{13}C$. **(A)** Concentration, in ppm, in the Northern and Southern Hemispheres, respectively, shown by monthly averages (dots) and by a smooth curve consisting of the sum of four seasonal harmonics and a spline function (solid lines). The seasonal harmonics include a linear gain factor, to represent increasing amplitude with time. **(B)** Same, respectively, for carbon isotope ratio, $\delta^{13}C$, in ‰. Station code names are as defined in Fig. 5.2. FAN/CHR refers to data for Fanning and Christmas Islands, combined. The scale of $\delta^{13}C$ is inverted so that seasonal patterns of concentration and $\delta^{13}C$ appear with the same phasing. Data are from Keeling et al. (2001).

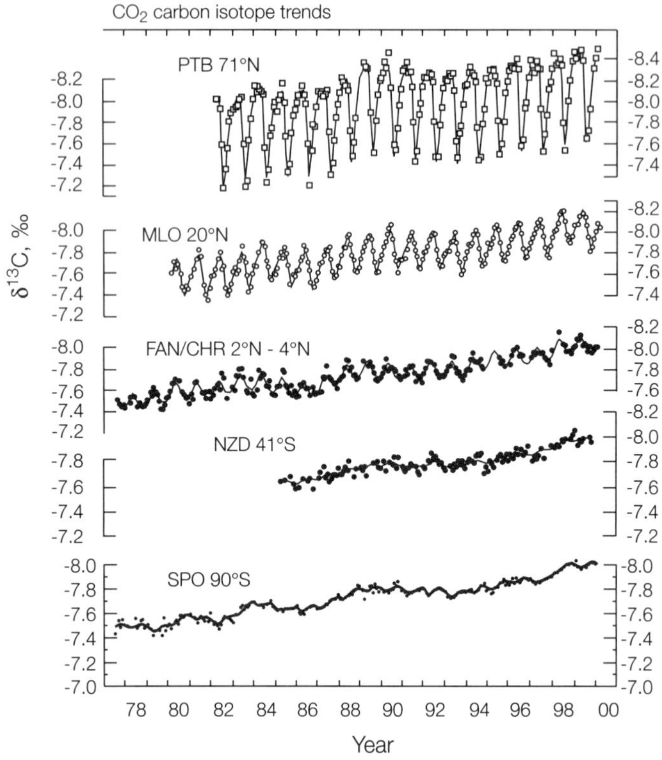

Figure 5.3. *Continued.*

The main cause of the southward decrease is the predominance of industrial emissions of CO_2 in the Northern Hemisphere (Keeling, Piper, and Heimann 1989b). These emissions, in addition, have caused $\delta^{13}C$ of atmospheric CO_2 (Fig. 5.3B) to increase from north to south, because fossil fuels are strongly depleted in the heavy carbon isotope relative to atmospheric CO_2, a reflection of their origins as plant carbon. The explanation for the north-south isotopic gradient in CO_2 is more complicated, however, owing to isotopic fractionation processes not associated with photosynthesis of land plants. Of these processes, the most important is temperature-dependent isotopic fractionation in the air-sea exchange of CO_2 which promotes negative $\delta^{13}C$ at high latitudes relative to low latitudes, and in the Southern Hemisphere relative to the Northern (Keeling et al. 1989a).

Short-term interannual variations in concentration and $\delta^{13}C$ exhibit similar temporal patterns at individual stations and, to a considerable extent, similar patterns from station to station. Emissions of industrial CO_2, because they vary only slightly from year to year, contribute little to these temporal patterns. Therefore, because the $\delta^{13}C$ variations are relatively large, the similarities seen in the patterns imply that terrestrial biospheric fluxes are the dominant cause of short-term variations in CO_2 concentration, as well as in $\delta^{13}C$.

We identified quasi-periodic variability in atmospheric CO_2 distinguished by time-intervals during which the seasonally adjusted CO_2 concentration at Mauna Loa Observatory, Hawaii, rose either more rapidly or less rapidly than a long-term trend line proportional to industrial CO_2 emissions. The Mauna Loa data and the trend line are shown in Fig. 5.4 with vertical gray bars demarking the intervals of rapidly rising CO_2.

Data from Mauna Loa Observatory were chosen for this identification because the measurements there are continuous and thus provide a more precisely determined rate of change than could any other station in our observing program. An association of rapidly rising CO_2 with temperature is seen consistently in the Mauna Loa record since it began, in 1958 (Keeling et al., 1989a; Keeling et al., 1995). As first noted by Bacastow (1976), rapidly rising CO_2 concentrations typically occur during El Niño events, involving almost global scale changes in temperature and identifiable by low Southern Oscillation Index (SOI) values (Rasmusson and Wallace 1983; Meehl 1987). The SOI indeed shows low values at times of some of the gray bars, but not all (see Fig. 5.4).

The SOI tracks the surface barometric pressure difference across the Pacific Ocean in the tropics. This pressure difference weakens during El Niño events and strengthens during the opposite phase in a so-called El Niño–Southern oscillation (ENSO) cycle. On three occasions during the time-period (1983, 1987, and 1998), there were pronounced minima in the SOI followed by sharp increases. Very strong El Niño events occurred at these times, accompanied by both strong warming and prominent increases in the rate of the rise of atmospheric CO_2 (Slingo and Annamalai 2000). The timing of an SOI minimum for these events almost exactly coincides with the commencement of a gray bar, and this suggests a remarkably close phase relation to the rate of atmospheric CO_2 rise. The SOI-CO_2 relation during a prolonged period of low SOI values from 1992 through 1994 is less consistent, indeed, the least consistent for the entire 42 years of the Mauna Loa record. A sharp decrease in rate of rise of CO_2 began immediately after the volcanic eruption of Mt. Pinatubo in 1991 and persisted until 1993. Then CO_2 began to rise in evident association with the dissipation of a volcanic dust veil that had promoted cool temperatures after the eruption (Keeling et al. 2001). The commencement of a gray bar in late 1993 is evidently not related to the ENSO cycle, although the long duration of the gray bar interval probably marks a return to the SOI-CO_2 association seen in the three prominent events already noted. The gray bars near 1980 and 1990 are associated with global warming, not evidently correlated with El Niño events. Also, there is no gray bar at the time of a weak El Niño event in 1992, probably because this event was during the period of cooling associated with the Pinatubo eruption. Thus the gray bars shown in Fig. 5.4 identify times of the ENSO cycle of the past 22 years when it involved strong oscillations in the

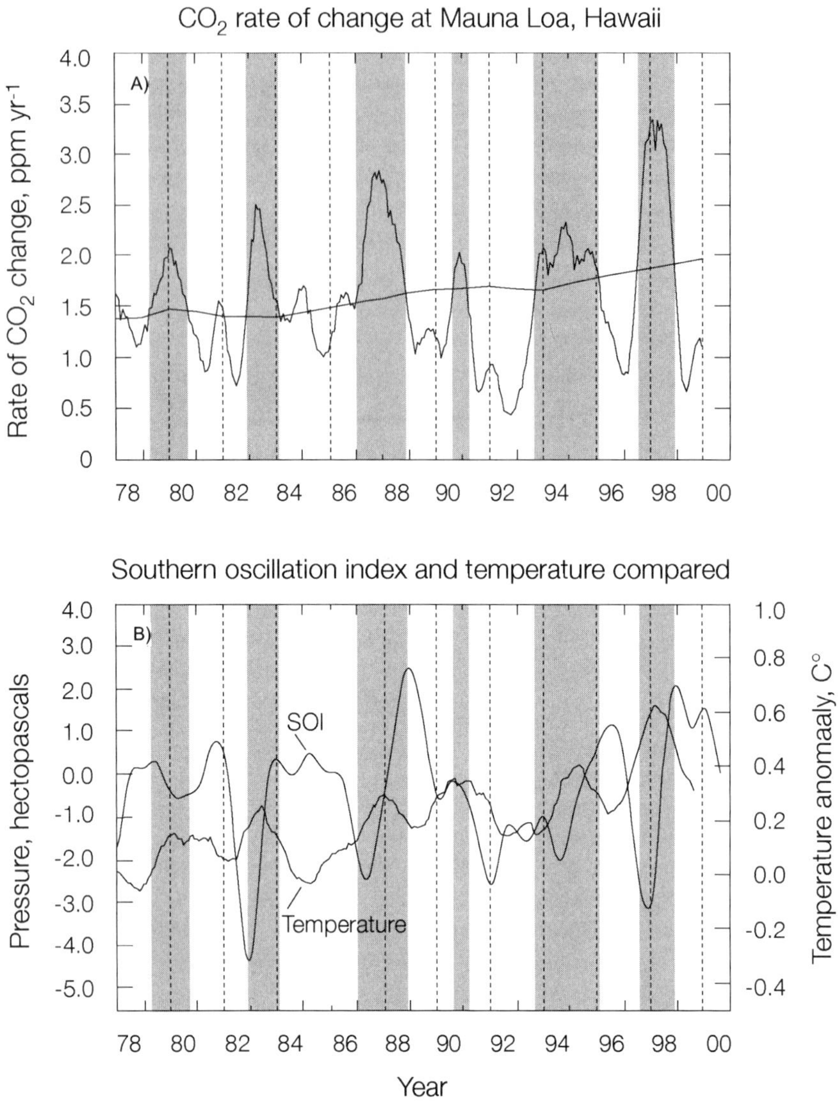

Figure 5.4. Comparison of quantities that vary with the El Niño–Southern Oscillation (ENSO) Cycle. **(A)** Rate of change of atmospheric CO_2 concentration, in ppm yr^{-1}, at Mauna Loa Observatory, Hawaii (strongly fluctuating curve), plotted together with the industrial CO_2 release (slowly rising curve). The latter trend is defined as the average fraction of emissions of industrial CO_2 that remained airborne as determined by Keeling et al. (1995) to be 58.1%. **(B)** The Southern Oscillation Index (SOI, thin line), determined as the barometric pressure difference of Tahiti minus Darwin, Australia, in hPa, and global average temperature anomaly (thicker line). The SOI data were derived from Climate Prediction Center (2000), the temperature data from Jones (1994, and personal communication). Vertical gray bars in both panels indicate timeintervals during which the rate of change of atmospheric CO_2 at Mauna Loa exceeded the industrial CO_2 trend rate.

SOI and was not complicated by a very powerful volcanic eruption; they also, however, identify times of rapid warming not clearly associated with the ENSO cycle.

We estimated the global average concentration and carbon isotope ratio of atmospheric CO_2 from 1978 through 1999 (Keeling et al. 2001, Appendix D). The resulting time-series are shown in Fig. 5.5.

For modeling purposes, these time-series were extended back to A.D. 1740. The extension of the CO_2 concentration series (see Fig. 5.1) consists of data from air trapped in glacial ice collected at Law Dome, Antarctica, assumed to represent the global average concentration, and after 1955 from direct observations. The extension of the $\delta^{13}C$ series was created by a deconvolution procedure, making use of the global CO_2 concentration time-series to establish chemical disequilibria of the global carbon cycle, assuming that the atmospheric, terrestrial biospheric, and oceanic carbon reservoirs were at chemical and isotopic equilibria in 1740. The $^{13}C/^{12}C$ ratio in 1740 was set so that the $\delta^{13}C$ value in 1978 exactly agreed with observations.

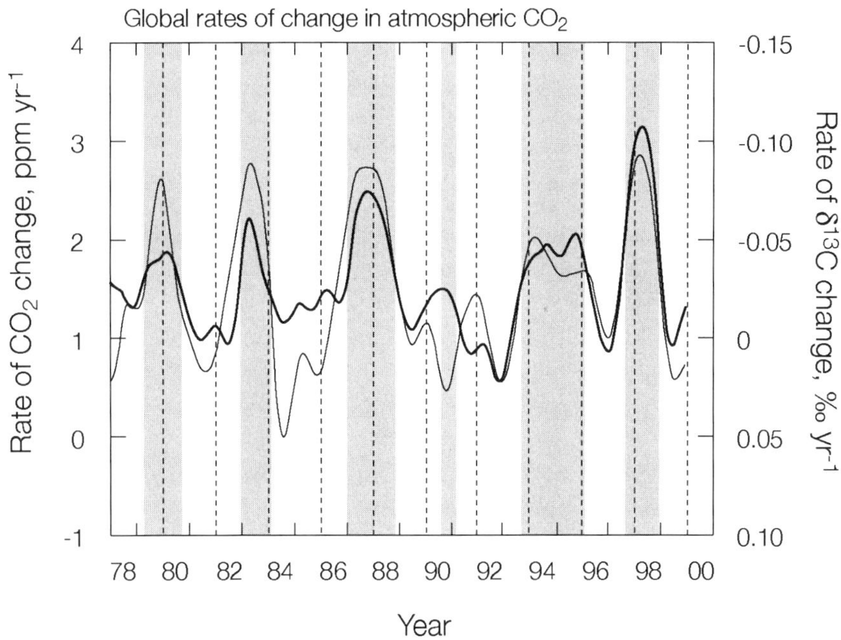

Figure 5.5. Rates of change of globally averaged atmospheric CO_2 concentration, in ppm (thick line), and $\delta^{13}C$, in ‰ (thin line), computed from the time-trends shown in Keeling et al. (2001). Gray bars are as in Fig. 5.4.

5.3 Isotopic Signature of the Seasonal CO_2 Cycle

The concentration of atmospheric CO_2 and its reduced carbon isotope ratio (see Fig. 5.3) covary over their seasonal cycles, because the $\delta^{13}C$ value of the carbon of land plants is distinctly more negative than that of atmospheric CO_2, and the seasonality of plant growth and respiration are the dominant causes of seasonality in atmospheric CO_2. Over the Northern Hemisphere, where this seasonality is a prominent feature of the CO_2 records, the $\delta^{13}C$ values inferred to be exchanged between the terrestrial biosphere and the atmosphere over each annual cycle tend to approach that of CO_2 respired by land plants (Mook et al. 1983; Heimann, Keeling, and Tucker 1989; Pataki et al. 2003). In the Southern Hemisphere, however, plant activity produces only small seasonal cycles in CO_2 concentration and $\delta^{13}C$; consequently, the average $\delta^{13}C$ value that explains the covariance of the seasonal cycle in that hemisphere is not precisely established (Keeling et al. 2001).

In our atmospheric records in the Northern Hemisphere, slight trends can be discerned toward more negative $\delta^{13}C$ values inferred from CO_2 fluxes from decade to decade, in addition to short-term interannual variability (Keeling et al. 2001). The decadal trends do not significantly differ from the average $\delta^{13}C$ trend of atmospheric CO_2, shown on each plot by a straight line. The carbon isotope discrimination term ($^{13}\Delta_{cov}$) that explains the seasonal covariance of $\delta^{13}C$ and CO_2 concentration is

$$^{13}\Delta_{cov} = \delta^{13}C_I - \delta^{13}C_0 \tag{5.2}$$

where the subscript I refers to CO_2 attributed to plant metabolism combined with oceanic CO_2 exchange, and subscript o refers to background atmospheric CO_2. The Arctic stations and La Jolla show little interannual variability in $\delta^{13}C_I$; Mauna Loa and Cape Kumukahi show possibly significant annual variations in the range of 2‰. For these northern stations, where oceanic CO_2 exchange has only a small or a negligible effect on the seasonal cycle of atmospheric CO_2, $^{13}\Delta_{cov}$ is close to the isotopic discrimination for photosynthesis of land plants, suggesting that this discrimination, averaged over large regions, has varied temporally by 1 to 2‰ during the past decade. With respect to latitude, averages of $\delta^{13}C_I$ from 1992 to 1999 show little evidence of variation within either hemisphere, but the hemispheric averages differ sharply: −27.5‰ in the Northern, −17.5‰ in the Southern (Keeling et al. 2001). The lower average for $\delta^{13}C_I$ in the Southern Hemisphere is consistent with a significant contribution to the seasonal cycle of atmospheric CO_2 by oceanic exchange, with its far lower isotopic discrimination than terrestrial exchange. To an indeterminate extent, the lower average may also reflect that a greater part of the vegetation on land in the Southern Hemisphere uses a photosynthetic pathway, C_4, with much lower fractionation than that of the more common C_3 pathway.

5.4 Deconvolution of Global Data

To resolve globally averaged temporal variations in atmospheric CO_2 into terrestrial and oceanic components, we employ an inverse procedure. First, by what we call a "single deconvolution," we deduce the oceanic and terrestrial CO_2 exchange fluxes solely from atmospheric CO_2 concentration data. Then, by a "double deconvolution," we adjust these estimates to be consistent with the $\delta^{13}C$ value of atmospheric CO_2.

In these calculations we identify only the most significant global pools of carbon affecting contemporary atmospheric CO_2: an oceanic pool consisting mainly of bicarbonate and carbonate salts; a terrestrial biospheric pool consisting of organic carbon stored in living plants, detritus, and soils; and a fossil fuel pool representing the available resource of coal, petroleum, and natural gas, combined. The associated global average exchange fluxes of CO_2 we denote as F_{oce}, F_{bio}, and F_{ind}, respectively. The industrial flux (F_{ind}) includes a small contribution from the manufacture of cement (Andres et al., 2000), while F_{oce} and F_{bio} represent net fluxes that are the relatively small differences between large one-way fluxes. All fluxes will be expressed here in petagrams of carbon (Pg C $= 10^{12}$ kg of carbon). To a close approximation, the sum of the three time-dependent fluxes determine the rate of change of atmospheric CO_2 abundance, N_a, expressed by the mass-conserving global atmospheric carbon cycle budget equation

$$dN_a/dt = F_{ind} + F_{oce} + F_{bio} \tag{5.3}$$

Over the past two decades the average strengths of the terms of Equation 5.3 are approximately as follows: The concentration of atmospheric CO_2 has risen at about 3.5 Pg C yr^{-1}, a little over half the rate of industrial emissions, which are about 6 Pg C yr^{-1} (see Chapter 18). The atmospheric CO_2 budget is balanced by an oceanic sink of about 2 Pg C yr^{-1} and by a terrestrial sink of about 0.5 Pg C yr^{-1}.

To specify this budget in greater detail, we distinguish several different contributions to the net exchange fluxes involving the oceanic and terrestrial biospheric carbon pools (Keeling et al., 1989a). We divide the net oceanic exchange flux into two components

$$F_{oce} = F_{ex} + F_{ano,oce} \tag{5.4}$$

where F_{ex} denotes the expected oceanic response to rising atmospheric CO_2 concentrations, and $F_{ano,oce}$ an "anomalous" flux representing a remainder not captured by F_{ex}. We divide the net terrestrial biospheric exchange flux into three components

$$F_{bio} = F_{fer} + F_{des} + F_{ano,bio} \tag{5.5}$$

where F_{fer} represents plant growth stimulated ("fertilized") by elevated atmospheric CO_2 concentration relative to the concentration in 1740, and F_{des} represents the release of CO_2 from the terrestrial biosphere owing to biologically destructive human-caused land-use changes, including deforestation and biomass burning. An "anomalous" flux, $F_{ano,bio}$, denotes a remainder not explained by either of the previously identified processes. In the procedure for single deconvolution, which we begin with atmospheric CO_2 concentration data for 1740 (Keeling et al., 1989a, 2001), we compute the expected oceanic response, F_{ex}, to rising atmospheric CO_2 concentration, using the one-dimensional vertically resolved oceanic box-diffusion model of Siegenthaler and Oeschger (1987). This oceanic submodel is characterized by a vertical diffusion coefficient, K, and an air-sea CO_2 exchange coefficient, k_{am}; it yields an oceanic sink as a function of atmospheric CO_2 concentration that agrees quite closely with estimates from three-dimensional oceanic circulation models (Keeling et al. 1989a). We do not regard its temporal variability as a reliable estimate of the true short-term interannual variability in F_{oce}. A substantial portion of that variability is found to reside in the anomalous oceanic flux, $F_{ano,oce}$, established later, in the double deconvolution calculation.

In the single deconvolution, we also compute the expected terrestrial biospheric response, F_{fer}, to rising atmospheric CO_2, using a terrestrial submodel controlled by a "growth factor." This factor, denoted β_a by Keeling et al. (1989a), expresses the degree to which the increase in atmospheric CO_2 concentration after 1740 caused enhanced plant uptake of CO_2, expressed as a linear growth response of plants to increasing CO_2 concentration resulting from a stimulation of net primary production (NPP) (Keeling et al. 1989a). For β_a, we have adopted the value 0.41, signifying that the increase in uptake by long-lived biospheric carbon has been 41% of the fractional increase in CO_2 concentration, where the NPP of long-lived carbon in 1740 was calculated to be 24.6 Pg C yr^{-1}. (The deconvolution calculations would yield almost the same outputs if, by the inclusion of short-lived biospheric carbon, β_a had been assigned a value of 0.18 with respect to total NPP, the latter calculated to be about 56 Pg C yr^{-1} for 1982 (Heimann and Keeling 1989). The destructive land-use flux, F_{des}, and the anomalous flux, $F_{ano,\ bio}$, are not determined in this calculation.

The calculations of the single deconvolution are repeated via an iterative scheme that adjusts $F_{ano,oce}$ and the sum $F_{ano,bio} + F_{des}$ at each step so that both the $\delta^{13}C$ values and concentration of atmospheric CO_2 agree with observations. This double deconvolution procedure reconciles the global atmospheric CO_2 budget, expressed by Equation 5.3 for the sum of the two isotopes, ^{13}C and ^{12}C, and a similar budget equation for the rare isotope, ^{13}C, alone.

The terrestrial biospheric exchange flux, F_{bio}, of Equation 5.3 represents the difference between two large one-way CO_2 fluxes. These we denote by F_{ab}, an atmospheric CO_2 sink owing to assimilation of CO_2 by plants through photosynthesis, and F_{ba}, an atmospheric CO_2 source owing to respiratory processes involving decay of vegetative matter. Thus,

$$F_{bio} = F_{ba} - F_{ab} \tag{5.6}$$

Both one-way fluxes apply to carbon in general, and thus to the isotopic sum, $^{13}C + {}^{12}C$. Correspondingly, for ^{13}C the net flux is

$$*F_{bio} = *F_{ba} - *F_{ab} \tag{5.7}$$

where the asterisks denote fluxes of ^{13}C alone. The one-way ^{13}C fluxes are related, respectively, to the one-way fluxes for $^{13}C + {}^{12}C$ by the expressions

$$*F_{ab} = \alpha'_{ab} R_a F_{ab} \tag{5.8}$$

$$*F_{ba} = \alpha'_{ba} R_b F_{ba} \tag{5.9}$$

where R_a and R_b denote, respectively, the ratios $^{13}C\,/(^{13}C + {}^{12}C)$ of atmospheric CO_2 and of carbon in the terrestrial biospheric pool, and α'_{ab} and α'_{ba} denote fixed $^{13}C\,/(^{13}C + {}^{12}C)$ isotopic fractionation (discrimination) factors. Corresponding factors with respect to $\delta^{13}C$ values are denoted below by α without the prime sign.

We set α_{ab}, which represents an isotopic discrimination of 15.32‰ against ^{13}C associated with photosynthesis (see below), to 0.98468 (i.e., to 1–15.32/1000), and we set α_{ba}, equal to unity, by assuming, consistent with earlier calculations (Keeling et al., 1989a), that no isotopic fractionation accompanies respiration. We then compute their α' equivalents, making use of the relationship (Heimann and Keeling 1989)

$$R_i = r_i(R_s/r_s) \tag{5.10}$$

where R_s and r_s denote, respectively, the $^{13}C\,/(^{13}C + {}^{12}C\,)$ and $^{13}C/^{12}C$ ratios of the international standard PDB, and the subscript i refers to any given carbon pool. Equations 5.8 and 5.9 are then rewritten

$$*F_{ab} = \alpha'_{ab} r_a (R_s/r_s) F_{ab} \tag{5.11}$$

$$*F_{ba} = r_b (R_s/r_s) F_{ba} \tag{5.12}$$

where r_a and r_b denote $^{13}C\,/^{12}C$ ratios corresponding respectively to R_a and R_b. Similar expressions apply to CO_2 exchange with the oceans, as follows:

$$F_{oce} = F_{ma} - F_{am} \tag{5.13}$$

$$*F_{am} = \alpha'_{am} r_a (R_s/r_s) F_{am} \tag{5.14}$$

$$*F_{ma} = \alpha'_{am} \langle \alpha'_{eq} \rangle r_m (R_s/r_s) F_{ma} \tag{5.15}$$

where F_{ma} denotes the one-way gross flux of CO_2 from the well-mixed surface layer of the oceans to the atmosphere, F_{am} the reverse flux, and R_m and r_m, respectively, the ^{13}C (^{13}C + ^{12}C) and $^{13}C/^{12}C$ ratios of the well mixed surface layer of the ocean. The symbol $<\alpha'_{eq}>$, equal to $\overline{\alpha'_{ma}/\alpha'_{am}}$, denotes the global annual average $^{13}C/(^{13}C + ^{12}C)$ equilibrium fractionation factor for air-sea exchange of CO_2. For convenience, we next express most of the results of our double deconvolution calculations as reduced isotope ratios, defined by Equation 5.1, where δ_a, δ_b, and δ_m denote δ^{13} C of atmospheric CO_2, terrestrial carbon, and carbon in surface seawater, respectively. Also, we express isotopic fractionation between carbon pools by fractionation terms with respect to $^{13}C/^{12}C$ ratios by the expression

$$\varepsilon_{ij} = \alpha_{ij} - 1 \tag{5.16}$$

where i and j denote, respectively, donor and receiver pools.

In most of the calculations, the $^{13}C/^{12}C$ fractionation factor, α_{ab} (equivalently expressed by ε_{ab}), is assumed to be constant (Keeling et al. 1989a). However, below we will discuss the consequences if this factor, which represents the isotopic discrimination of plants during photosynthesis, is allowed to vary. The factor α_{am}, which represents fractionation attending oceanic uptake, is assumed to be constant. The associated factor α_{ma}, however, is assumed to vary with sea surface temperature, and hence with time, because the equilibrium fractionation, α_{eq} (equal to the quotient α_{ma}/α_{am}), is temperature dependent, as discussed below.

The $^{13}C/^{12}C$ isotopic ratio for F_{fer} expressed by annual averages, is set equal to $\alpha_{ab}r_a$, the same ratio as that of CO_2 assimilated by plants during photosynthesis (Heimann and Keeling 1989). The $^{13}C/^{12}C$ ratio associated with the sum, $F_{des} + F_{ano,bio}$, is set equal to r_b, computed by taking account of the average storage time of long-lived carbon in the terrestrial pool (Keeling et al. 1989a).

The magnitude of the global average $^{13}C/^{12}C$ fractionation factor for CO_2 assimilation during plant growth, α_{ab}, affects the computed values of $\delta^{13}C$ for all of the biospheric fluxes in the deconvolution calculations (see Heimann and Keeling 1989). The value that we assign to α_{ab} depends on the relative contributions from C_3 and C_4 plants. These pathways have distinctly different degrees of discrimination against ^{13}C, which may vary with time; in addition, the relative contribution of the two plant types may vary with time. These possible temporal variations are disregarded in our double deconvolution calculations.

The single deconvolution predicts monthly values of all of the exchange CO_2 fluxes defined above and the amounts and $^{13}C/^{12}C$ ratios of carbon in the atmospheric, oceanic, and terrestrial carbon pools. The $^{13}C/^{12}C$ ratios for atmospheric CO_2, so obtained, are then used in initializing the double deconvolution calculation in 1978 by setting the value in 1740 such that the predicted annual value agrees with our isotopic observations in 1978.

The destructive land-use flux, F_{des}, always positive, is computed only for the time-period of the double deconvolution; it is set equal to 2.0 Pg C yr^{-1}, the average for 1980–1989 as given by Houghton (1999). The term, $F_{ano,bio}$, is then calculated by difference to be consistent with Equation 5.5. The relative contributions of F_{fer}, F_{des}, and $F_{ano,bio}$ to the overall net biospheric flux, F_{bio}, though uncertain, only slightly affect the computations of the double deconvolution, because the $^{13}C/^{12}C$ ratios for the three fluxes are nearly the same.

5.5 Isotopic Discrimination by Terrestrial Vegetation

The C_3 photosynthetic pathway of terrestrial plants produces an isotopic discrimination with a typical value of about 18‰ relative to the $^{13}C/^{12}C$ ratio of atmospheric CO_2, and a range for different species of several per mil (Farquhar et al. 1989). Although C_3 plants account for most of net primary production (NPP), plants with the C_4 pathway contribute about 20‰ of global NPP with a discrimination of only about 4‰ (Farquhar al. 1989; Still et al. 2003). Consequently, the global average fractionation factor, α_{ab}, defined earlier, is less than for C_3 plants alone.

To estimate α_{ab}, we adopted a carbon isotope discrimination for C_3 plants of 17.8‰, and for C_4 plants of 3.6‰, as reported by Lloyd and Farquhar (1994). We then computed NPP for C_3 and C_4 plants for selected geographic zones using a vegetation map (Hunt et al. 1996) that prescribes types of biomes at a resolution of 1 degree. For each grid point of the map, NPP was computed by a method that makes use of remotely sensed radiometric data, from 1982–1990, the only period for which reliable radiometric data were available. The resulting global average fraction of NPP contributed by C_4 plants is found to be 17.44‰, leading to a global average discrimination for C_3 and C_4 plants combined of 15.32‰. This calculation of global average carbon isotope discrimination does not take into account subzonal-scale or temporal variability in discrimination of the separate plant types, C_3 and C_4, issues that we discuss next.

For the Northern Hemisphere, discrimination varies in the range from 17.8‰ at high latitudes, to 14.0‰ in the tropics, as the C_4 fraction of NPP varies from 0 to 27%. For the two extratropical zones, where NPP is mainly owing to C_3 plants, discrimination is 1.8‰ less than that inferred from the covariance of atmospheric CO_2 concentration and isotope ratio, $^{13}\Delta_{cov}$, observed for stations in the Northern Hemisphere.

In the tropics and farther south, the likelihood of substantial oceanic influence on $^{13}\Delta_{cov}$ precludes obtaining reliable estimates of plant discrimination from our station data. It is probably reasonable to assume that the discrimination of C_3 plants south of the tropics is nearly the same as north of the tropics, and that the effect of discrimination of C_4 plants is too small for our neglect of its zonal variability to influence our calculations.

Temporal variability in discrimination of C_3 plants is more likely to be a serious neglect in our computations than spatial variability. Such variability is expected because the degree of isotopic discrimination varies with stomatal conductance of plant leaves. When stressed by lack of adequate water, plant stomata tend to close, reducing the carbon isotopic fractionation that can occur at the primary sites of CO_2 fixation within the leaves (Farquhar et al. 1989; Ehleringer, Hall, and Farquhar 1993). Ometto et al. (2002) and Fessenden and Ehleringer (2003) have presented evidence of interannual variability in C_3 plant discrimination. Since variable water stress affects much of the world's vegetation, a global average temporal variation of 2‰ or more seems likely based on these studies.

To examine how variable discrimination against the rare isotope, ^{13}C, by land plants may affect the net biospheric exchange flux, F_{bio}, as calculated in our double deconvolution procedure, we invoke isotopic additivity to establish the shift in $\delta^{13}C$ of atmospheric CO_2 that takes place when atmospheric CO_2 is exchanged with the terrestrial biosphere:

$$(\delta^{13}C)\mu = (\delta^{13}C_o)\mu_o + \delta^{13}C_I(\mu - \mu_o) \tag{5.17}$$

where μ denotes the atmospheric CO_2 concentration, the subscript o represents the reference or background values, and the nonsubscripted concentration and isotopic ratio values represent the observations from any single, individual flask (Keeling et al. 2001).

We first establish this shift (Case 1) assuming, as in our double deconvolution calculations, that isotopic discrimination attending photosynthesis, α_{ab}, is constant. Then, for comparison (Case 2), we compute the shift if the proportions of photosynthesis owing to C_3 and C_4 plants vary, i.e., the forward and reverse fluxes, F_{ab} and F_{ba} of Equation 5.6 are equal. Finally, in Case 3, we compute the shift, if the properties of photosynthesis owing to C_3 and C_4 plants vary.

When we addressed concentration and isotopic ratio data that were obtained only after 1991, we found the latter more precise than obtained hitherto (Keeling et al. 2001). As expressed by Equations 5.6 to 5.9, the net flux, F_{bio}, and its ^{13}C equivalent, $^*F_{bio}$, denote differences between much larger one-way fluxes, F_{ab}, F_{ba} and $^*F_{ab}$, $^*F_{ba}$, respectively. As previously noted, the respiratory flux, F_{ba}, is assumed to have no fractionation (α'_{ba} of Equation 5.8 of unity). Flux F_{ab}, which represents assimilation of CO_2 by plants through photosynthesis, exhibits fractionation, expressed by the factor, α'_{ab}.

We now consider the possibility that α'_{ab} varies temporally, whether directly, or by varying the proportions of F_{ab} contributed by C_3 and C_4 plants that have distinctly different degrees of carbon isotope discrimination. We disregard the small difference between α'_{ab} and α_{ab} (Heimann and Keeling 1989) and adopt the latter in subsequent expressions.

Substituting in Equation 5.7 the relations for $^*F_{ab}$ and $^*F_{ba}$, as given by Equations 5.11 and 5.12, we obtain for the net ^{13}C flux:

$$^*F_{bio} = (r_b F_{ba} - \alpha_{ab} r_a F_{ab})(R_s/r_s) \tag{5.18}$$

To proceed we define a hypothetical chemical steady state for the terrestrial biosphere, in which assimilation and respiration balance globally for both ^{12}C

and ^{13}C. At this steady state F_{ba} equals F_{ab}, and F_{bio} and $*F_{bio}$ are both zero. Hence,

$$r_b = \alpha_{ab} r_a \tag{5.19}$$

and therefore the $^{13}C/^{12}C$ ratios of F_{ab} and F_{ba} are both equal to the $^{13}C/^{12}C$ ratio of the terrestrial biospheric carbon pool, r_b.

Let δ_a and δ_b denote, respectively, the reduced isotopic ratios, $\delta^{13}C$ of atmospheric CO_2 and carbon in the terrestrial biosphere. From the definition of $\delta^{13}C$,

$$r_i = (\delta_i + 1) r_s \tag{5.20}$$

where i stands for either a or b. Replacing r_a and r_b in Equation 5.19 according to Equation 5.20, replacing δ_{ab} by $\varepsilon_{ab} + 1$, according to Equation 5.16, and rearranging, the isotopic steady state condition, expressed in per mil notation, is such that

$$\delta_b - \delta_a = \varepsilon_{ab} \tag{5.21}$$

(disregarding a product of second order terms, $\varepsilon_{ab}\, \delta_a$).

For each of the three cases under consideration, let us assume that an initial steady state is displaced by an abrupt imbalance either in the one-way fluxes, F_{ab} and F_{ba}, which pertain to both isotopes, or only in the one-way ^{13}C fluxes, $*F_{ab}$ and $*F_{ba}$. In each case, for simplicity, we disregard a subsequent approach to a new steady state involving a change in δ_b of the terrestrial biospheric carbon pool. Thus, we compute the maximum possible shift in δ_a in response to a specified displacement from steady state.

Let N_{ao} denote the amount of CO_2 in the atmosphere, and δ_{ao} and δ_{bo} denote δ_a and δ_b, respectively, at the initial steady state. It follows from Equation 5.21 that the reduced isotopic ratios of both F_{ab} and F_{ba} are equal to δ_{bo}. We adopt 747.6 Pg C for N_{ao} based on a CO_2 concentration of 352.2 ppm for 1990, approximately the midpoint of our deconvolution calculations. In accord with our analysis of global average isotopic discrimination, $\varepsilon_{ab} = 15.32‰$.

Let ΔF_{bio} denote an abrupt departure of either of the one-way fluxes, F_{ba} or F_{ab}, or in both combined. Thus, in place of Equation 5.17

$$\Delta F_{bio} = F_{ba} - F_{ab} \tag{5.22}$$

Both F_{ba} and F_{ab}, as noted above, are assumed initially to have reduced isotopic ratios equal to δ_{bo}. In accord with Equation 5.17, and assuming that the departure from steady state has continued for a time-interval, ΔT, too short to cause significant changes in these isotopic ratios

$$N_{ao}\, \delta_{ao} + \Delta F_{bio} \Delta T\, \delta_{bo} = (N_{ao} + \Delta F_{bio} \Delta T)(\delta_{ao} + \Delta\delta_a) \tag{5.23}$$

where $\Delta\delta_a$ denotes the shift in the $^{13}C/^{12}C$ ratio of atmospheric CO_2 owing to the displacement from steady state. Solving for $\Delta\delta_a$ with δ_{bo}–δ_{ao} replaced by ε_{ab}, which is time-invariant according to Equation 5.21, and with $\Delta F_{bio}\Delta T << N_a$

$$\Delta\delta_a = \varepsilon_{ab}\Delta F_{bio}\Delta T/N_{ao} \qquad (5.24)$$

$$= (-0.0205‰)\Delta F_{bio}\Delta T \qquad (5.25)$$

Thus, a 1 Pg C yr^{-1} net flux of CO_2 from the biosphere to the atmosphere, acting for one year ($\Delta F_{bio}\Delta T = 1$), decreases the $\delta^{13}C$ value of atmospheric CO_2 by 0.0205‰; an opposite flux increases $\delta^{13}C$ by the same amount. (The product, $\varepsilon_{ab} \cdot \Delta F_{bio}$ represents an isotopic flux, 'isoflux,' between the terrestrial and atmospheric carbon pools, for this example equal to 15.32‰ Pg C yr^{-1}).

For Case 2, let $\Delta\varepsilon_{ab}$ denote a shift in discrimination for the full one-way flux, F_{ab}, with no change in $\delta^{13}C$ of the return flux, F_{ba}. Invoking Equation 5.17 as in Case 1, but with respect to both addition and subtraction of CO_2 by biospheric exchange:

$$N_{ao}\,\delta_{ao} + F_{ba}\Delta T\delta_{bo} - F_{ab}\Delta T\,(\delta_{bo} + \Delta\varepsilon_{ab}) = N_{ao}\,(\delta_{ao} + \Delta\delta_a) \qquad (5.26)$$

Solving for $\Delta\delta_a$ with $F_{ab} = F_{ba}$,

$$\Delta\delta_a = -\Delta\varepsilon_{ab}\,F_{ab}\Delta T\,/N_{ao} \qquad (5.27)$$

With F_{ab} set equal to our estimate of global NPP, 61.3 Pg C yr^{-1} and $\Delta\varepsilon_{ab}$ set, as an example, to 1‰, we obtain an isotopic shift, $\Delta\delta_a$, of 0.082‰ in one year, four times that evaluated in Case 1. The isoflux for this example, $\Delta\varepsilon_{ab} \cdot F_{ab}$, is 61.3‰ Pg C yr^{-1}. Even if the change in discrimination in Case 2 should occur only in the tropics, where our regional analysis indicates that most of the correlation of biospheric flux with El Niño events occurs, a 1‰ change in carbon isotope discrimination (with respect to an estimated tropical NPP of 32.8 Pg C yr^{-1}) would cause a shift of 0.044‰, more than twice that for a transfer of 1 Pg C of CO_2 in one year to or from the atmosphere according to Case 1.

For Case 3, a change occurs in the proportion of assimilation by C_4 plants, without a change in the fluxes with respect to C_3 and C_4 plants combined. The disequilibrium flux is equivalent to Case 1 except that it is applied separately for C_3 and C_4 plants with the added condition that the changes in net biospheric fluxes are equal and opposite for C_3 and C_4 plants. Denoting the mutual changes by ΔF_b, positive for increasing net C_3 flux to the atmosphere (cf. Equation 5.24)

$$\Delta\delta_a = ({}^3\varepsilon_{ab} - {}^4\varepsilon_{ab})\,\Delta F_b\,\Delta T/N_{ao} \qquad (5.28)$$

$$= (-0.019‰)\,\Delta F_b\Delta T \qquad (5.29)$$

where superscripts 3 and 4 distinguish the fractionation factors for C_3 and C_4 plants, respectively (17.8‰ and 3.6‰). A disequilibrium flux, ΔF_b, of 1 Pg C

yr^{-1} causes almost the same isotopic shift, $\Delta\delta_a$, as a 1 Pg C yr^{-1} change in NPP or respiration in Case 1, a result that could be anticipated because the substitution of a C_4 flux for a C_3 flux has almost the same effect on the $^{13}C/^{12}C$ of atmospheric CO_2 as a diminution in the overall net flux, F_{bio}.

Given the relatively small differences in isotopic discrimination by plants that may cause signals in the $^{13}C/^{12}C$ ratio of atmospheric CO_2 shown by these calculations, it may be unrealistic to assume in a double deconvolution that these signals are caused solely by variations in biospheric flux with isotopic discrimination fixed. Variations in carbon isotope discrimination of the order of 2‰ suggested by recent studies cited above, either from changes in discrimination or the fraction of NPP owing to C_4 plants, can produce signals that are comparable to those from varying net CO_2 exchange.

5.6 Global CO_2 Fluxes

Plots of the terms of the global atmospheric carbon cycle budget, expressed by Equation 5.3, are shown in Fig. 5.6 together with subordinate terms appearing in Equations 5.4 and 5.5. Fig. 5.6A shows the rate of industrial CO_2 emissions, F_{ind} (upper curve), and the rate of change in atmospheric concentration, dN_a/dt (middle curve)—terms that serve as inputs to the single deconvolution procedure. Their difference, $F_{bio} + F_{oce}$ (lower curve), constitutes a combined terrestrial biospheric and oceanic sink of industrial CO_2. In Fig 5.6B this sink is separated into biospheric and oceanic components, F_{bio} (thick curve), F_{oce} (thin curve), respectively, as determined by the double deconvolution procedure. In Panel C the biospheric component is shown divided into three fluxes: F_{fer}, a "CO_2 fertilization sink," representing the response of the terrestrial biosphere to increasing atmospheric CO_2 concentration, an assumed constant "land-use source," F_{des}, of 2.0 PgC yr^{-1} representing human-induced release of terrestrial CO_2, and an "anomalous biospheric flux," $F_{ano,bio}$, that is presumed to result mainly from natural variability in the carbon cycle, although it may also reflect variability in human disturbances not accounted for by F_{fer} and F_{des}. Also shown (dashed curve) is a revision of the anomalous biospheric oceanic flux in which isotopic variations are ignored, equivalent to assuming that they principally reflect changes in discrimination by plants. In Panel D the oceanic component is shown divided into an oceanic "uptake sink," a response to increasing atmospheric CO_2, F_{ex}, and an "anomalous oceanic flux," $F_{ano,oce}$, mainly, or entirely, of natural origin. If isotopic variations are ignored, as in the dashed curve in Fig 5.6C, the uptake sink remains as shown, but the anomalous oceanic flux is reduced to zero for all times.

We first address the average strengths of these terms of the carbon cycle budget equation. As shown in Table 5.1, averages of the terms plotted in Fig. 5.6 for the 1980s agree within 0.2 Pg C yr^{-1} with corresponding fluxes reported by Houghton et al. in the Intergovernmental Panel on Climate Change, Second Assessment Report to the United Nations (IPCC, 1996), altered only slightly in a Third Assessment Report (IPCC, 2001). Agreement is also close for the 1990s

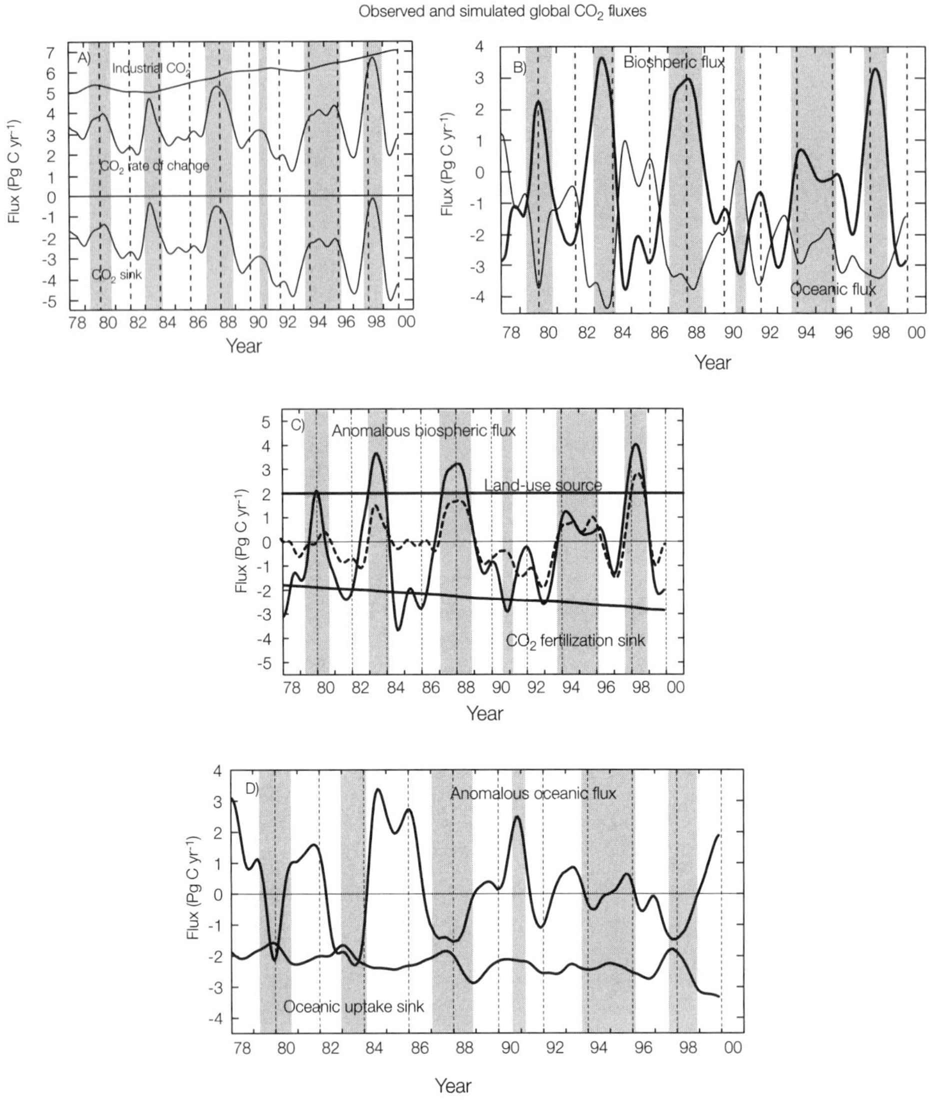

Figure 5.6. Time varying global CO_2 fluxes, in Pg C yr^{-1}, as labeled, positive to the atmosphere, shown as spline fits to monthly data. Gray bars are as in Fig. 5.4. **(A)** Industrial CO_2 emissions, rate of change of CO_2 abundance, and an inferred atmospheric CO_2 sink, defined as the difference between emissions and abundance change. **(B)** Net terrestrial biospheric and oceanic CO_2 exchange fluxes with the atmosphere (thick and thin lines, respectively), computed by double deconvolution as described in the text, assuming constant isotopic discrimination for C_3 and C_4 plants. **(C)** The terrestrial biospheric exchange flux, shown in Panel B, divided into a contribution related to rising atmospheric CO_2 concentration, F_{fer} ("CO_2 fertilization flux," lower curve), a constant "land-use source", F_{des} (straight line), and an anomalous flux, $F_{ano,bio}$, as defined in the text. Also shown (dashed curve) is an alternative estimate of the anomalous biospheric flux in which isotopic variations are ignored. **(D)** Oceanic exchange flux of Panel B, divided into an oceanic uptake flux (F_{ex}, lower curve) and an anomalous flux, $F_{ano,oce}$, as defined in the text.

Table 5.1. Global Net Carbon Dioxide Fluxes, in PgC yr^{-1}, obtained by double deconvolution compared with fluxes of IPCC Reports. Fluxes are positive into the atmosphere. Fluxes reported by IPCC (1996, 2001) have errors quoted here at 90% confidence ($\pm 2\sigma$). Those for 1990 to 1999 are doubled from 67% confidence levels quoted by IPCC (2001).

	1980–1989		1990–1999	
Period covered:	IPCC	Present Analysis	IPCC	Present Analysis
Atmosphere increase:	$+3.3 \pm 0.2$	+3.3	3.2 ± 0.2	+3.3
Emissions from fossil fuel combustion and cement production:	$+5.5 \pm 0.5$	+5.44	6.4 ± 0.8	+6.41
Net oceanic uptake:	-2.0 ± 0.8	−2.0	-1.7 ± 0.5	−2.4
Net emissions from changes in tropical land-use:	$+1.6 \pm 1.0$		insufficient data	
Uptake by Northern Hemisphere forest regrowth:	-0.5 ± 0.5		insufficient data	
Other terrestrial sinks:	-1.3 ± 1.5		insufficient data	
Combined terrestrial biospheric flux:	-0.2 ± 1.9	−0.2	-1.4 ± 1.4	−0.8

with respect to atmospheric CO_2 increase and emissions of fossil fuel CO_2, but we find a substantially larger oceanic uptake of atmospheric CO_2 (by 0.7 Pg C yr^{-1}) than the IPCC and a smaller biospheric uptake (by 0.6 Pg C yr^{-1}). These disagreements are within the range of uncertainty in the absolute magnitude of the IPCC budget estimates when expressed at 90% confidence, but this range is so large that the relative importance of the oceans and terrestrial biosphere in removing industrial CO_2 from the air evidently cannot yet be determined reliably.

The flux estimates just described also reveal variability in the atmospheric CO_2 budget on the decadal time-scale. In agreement with the IPCC we find that atmospheric CO_2 rose at nearly the same rate in the 1980s and 1990s in spite of an increase in CO_2 emissions from fossil fuel combustion of 0.9 Pg C yr^{-1}. These changes are well established, because uncertainty in absolute magnitudes as stated by the IPCC reflect, to a considerable degree, biases that don't change appreciably from one decade to the next.

Estimates of decadal changes in the oceanic and terrestrial biospheric CO_2 exchange fluxes are less certain. We find an increased oceanic sink by relying on a double deconvolution calculation that takes account of changing $\delta^{13}C$, based on data acquired using consistent procedures of both sampling and measurement. The calculations are sensitive to model assumptions regarding how global terrestrial biospheric and oceanic reservoirs respond to perturbations induced by changing atmospheric CO_2. The IPCC in their Third Assessment Report found a reduced oceanic sink by interpreting a shift in the O_2/N_2 ratio in the atmosphere as measured in the 1980s from archived air samples compared to direct measurements in the 1990s. The change in data source from decade to decade contributes uncertainty to some unknown degree, and the results are subject to modeling uncertainty not yet fully established, because of aspects of the oceanic carbon cycle, such as outgassing of oxygen, that are not fully understood. Thus, it is not possible with any great confidence to decide whether a diminishing oceanic sink and an increasing terrestrial biospheric sink, as reported by the IPCC, are more likely to be correct than our finding of increases in both sinks. An important finding on which we agree with the IPCC qualitatively is that the terrestrial biospheric exchange flux increased from the 1980s to the 1990s.

Interannual variations in the fluxes of the budget Equation 5.3 over a shorter term, summarized by the IPCC (2001), are also of interest. Such variations in the rate of change in atmospheric CO_2 concentration, dN_a/dt, and in industrial CO_2 emissions, F_{ind}, shown by the two upper curves of Fig. 5.6A, are both well determined, the former derived from our atmospheric observations, the latter from Andres et al. (2000), the same international statistical data used by IPCC. Variations in their difference, equal to $F_{oce} + F_{bio}$, are therefore also well determined. This combined oceanic and terrestrial sink (lower curve in Fig. 5.6A), indeed, has nearly the same interannual pattern of variability as the change in atmospheric CO_2 concentration, dN_a/dt, because global industrial CO_2 emissions, F_{ind}, vary only slightly from year to year. Most notable are maxima in $F_{oce} + F_{bio}$ close to time-intervals shown by vertical gray bars, suggesting a coupling of both oceanic and terrestrial carbon cycle components with globally average warm periods, often El Niño events, as discussed earlier.

The oscillatory pattern of variations in F_{bio} (Panel B), especially its phasing, is so similar to that of dN_a/dt and $F_{oce} + F_{bio}$ (Panel A) that F_{bio} appears to represent the predominant cause of short-term interannual variability in dN_a/dt. The calculated amplitudes of the prominent fluctuations in F_{bio}, however, substantially exceed those in dN_a/dt, resulting in opposing patterns in the oceanic and biospheric fluxes, as discussed below.

In summary, the decadal averages of F_{bio} and F_{oce}, as listed in Table 5.1, are dependent on the choice of oceanic submodel used in the deconvolution procedure. The short-term interannual variability in the anomalous fluxes, $F_{ano,bio}$ and $F_{ano,oce}$, however, is closely prescribed by the CO_2 observations, provided that isotopic discrimination is correctly specified, as discussed below.

5.7 Reliability of Deduced Fluxes

Before discussing our findings regarding the global average terrestrial and oceanic fluxes, F_{bio} and F_{oce}, we will examine the extent of possible errors in the calculations. We address uncertainty in our results on several time-scales.

Statistical errors in F_{bio} and F_{oce}, on short time-scales, are contributed by scatter in the data for both concentration and $^{13}C/^{12}C$ ratio, expressed by standard errors of the fit of individual observations to the spline curves shown in Fig. 5.3. These errors range from 0.15 ppm and 0.02‰ at the South Pole to 1.1 ppm and 0.07‰ at Point Barrow, Alaska, and result in errors of the order of 0.5 Pg C yr^{-1} with respect to F_{bio} and F_{oce}, as discussed by Piper et al. (2001). Thus, although significant, these errors are not large enough to challenge our finding of large short-term interannual fluctuations in biospheric and oceanic fluxes (of the order of several Pg C yr^{-1}). A more important issue is whether the calculation of these fluxes is affected by systematic errors, which is possible on all time-scales.

We first discuss possible errors associated with the double deconvolution procedure, which deduces fluxes from simultaneous data for the concentration and $^{13}C/^{12}C$ ratio of atmospheric CO_2. We thus add to our earlier discussion of uncertainties on the decadal time-scale. We then discuss errors on the short-term interannual time-scale, approximately that of the ENSO cycle.

To deduce the relative strengths of F_{bio} and F_{oce}, which include responses to the combustion of fossil fuels, we have made use of the same inverse double deconvolution procedure used previously by Keeling et al. (1989a, 1995). Previously, however, only anomalous fluxes were calculated, similar to $F_{ano,bio}$ and $F_{ano,oce}$, plotted in Fig. 5.6C,D. Here, full net fluxes, denoted by F_{bio} and F_{oce}, are also calculated, as plotted in Fig. 5.6B. As pointed out above, the validity of the deduced average magnitude of these fluxes depends on the correctness of the oceanic submodel used in the deconvolution procedure.

Although our study addresses interannual variability of the global carbon cycle over two decades, it does not resolve decadal variability and the so-called missing CO_2 sink issue (Wigley and Schimel 2000) regarding whether the terrestrial biosphere is an important global sink for CO_2 produced by the combustion of fossil fuels as discussed above.

The principal components of the global atmospheric CO_2 budget are known with widely varying uncertainties. As estimated by the IPCC (Table 5.1) at 90% confidence, the independent errors associated with the rate of increase of atmospheric CO_2 (± 0.20 Pg C yr^{-1}) and industrial CO_2 emissions (± 0.50 Pg C yr^{-1}) are so small that the error in the combined oceanic-terrestrial biospheric flux, $F_{oce} + F_{bio}$, is uncertain to only ± 0.54 Pg C yr^{-1} (calculated by quadrature). The errors for the separate oceanic and terrestrial biospheric fluxes alone are substantially larger. When determined from separate errors for three components, the error for F_{bio} is ± 1.9 Pg C yr^{-1}. Such a large error is not surprising since data must be assembled globally for many diverse pools of carbon. A substantially lower error for F_{bio}, ± 0.9 Pg C yr^{-1}, is obtained when determined by subtracting estimates of F_{oce} from the sum $F_{oce} + F_{bio}$, the approach of the IPCC, underscoring the value in obtaining precise estimates of oceanic uptake of atmospheric CO_2 as an indirect approach to determining the global average of the terrestrial biospheric flux F_{bio}, as discussed further below.

Both the terrestrial and oceanic CO_2 fluxes inferred from our atmospheric CO_2 data show a distinct association with fluctuating temperature, most evident during El Niño events when the concentration of atmospheric CO_2 has risen most rapidly. This association with the ENSO cycle is not surprising because these events are attended by pronounced changes in the temperature and circulation of both the atmosphere and the oceans that are likely to affect the global terrestrial and oceanic carbon pools. Thus the phasing of the large fluctuations seen in our atmospheric data are not difficult to explain, but our finding of large opposing terrestrial releases and oceanic sinks of CO_2 phased with the ENSO cycle has been challenged on the grounds that our $^{13}C/^{12}C$ data through 1988 (Keeling et al. 1989a) contain substantial systematic errors, specifically that they do not agree with other measurements showing only small isotopic fluctuations with little association with the strong El Niño events in 1982–3 and 1987–8 (Francey et al. 1995).

Although two succeeding El Niño events, in 1992 and 1994, were not strong enough to substantiate the plausibility of the large fluctuations that our data show for the 1980s, we now possess data for an El Niño event in 1997–98, for which the rate of change in $^{13}C/^{12}C$ ratio (see Fig. 5.5) fluctuated with an amplitude as large as that found for the strong events in 1982–83 and 1987–88. Furthermore, this fluctuation, as were fluctuations for the El Niño events of 1982–83 and 1987–88, is seen separately for every station in our array.

Although calibrating errors cannot be ruled out, a more likely source of systematic error is some shortcoming in either the terrestrial biospheric or oceanic submodel of the double deconvolution procedure. A likely possibility is our assumption that isotopic discrimination attending photosynthesis on land is constant. If we assume that isotopic variability is caused solely by variable isotopic discrimination by plants, fluctuations in the biospheric flux for the strong events of 1982–83, 1987–88, and 1997–98 are reduced by nearly two-thirds, as shown by comparing the solid and dashed anomalous biospheric flux curves of Fig. 5.6B. For these events, fluctuations in the oceanic flux are not just reduced but are reversed in phase to the pattern shown by the oceanic uptake flux in Fig. 5.6C. Even if variability in discrimination is only 1 ‰, fluctuations in the biospheric flux are substantially reduced and the opposing oceanic fluxes eliminated. Tropical drought and heat attending El Niño events are expected to cause stress in plants so that net primary production (NPP) decreases and atmospheric CO_2 rises anomalously, as observed. Perhaps a decrease in discrimination accompanies this presumed decrease in NPP, such that the two effects cannot be distinguished because their timing is too similar.

Evidence, reported by Piper et al. (2001), shows that the association of terrestrial biospheric and oceanic CO_2 fluxes with warming events occurs mostly in the tropics. Correlations of these inferred fluctuations with variations in climatic factors therefore should reflect a relation of climate forcing to the metabolic activity of plants in which the phasing is well determined even if the magnitudes of the fluctuations are poorly established from the atmospheric data. We thus may gain insight into the relationship of the carbon cycle to climate without yet knowing the extent of variable discrimination by tropical plants.

5.8 Short-Term Variability in the Global Carbon Cycle

To compare the magnitude of fluctuations in exchange fluxes with the strength of El Niño events, we again turn to the southern oscillation index (SOI), which tracks the ENSO cycle. For the period of our study, from 1978 to 1999, the most pronounced fluctuations in SOI occurred near the times of strong El Niño events of 1982–83, 1987–88, and 1997–98 (henceforth, events of 1983, 1987, and 1998). We ask, to what degree did extremes seen in the anomalous biospheric and oceanic fluxes, $F_{ano,bio}$ and $F_{ano,oce}$ represent responses to climatic variability at these times?

Because the amplitudes of the fluctuations in SOI are nearly equal for these three El Niño events, it is possible that they reflect approximately equal climatic forcing. Calculated amplitudes of fluctuations in the anomalous biospheric flux, $F_{ano,bio}$ (see Fig. 5.6C, solid curve) for these events are also nearly equal, based on our observing nearly equal amplitudes of corresponding fluctuations in rate of change of $^{13}C/^{12}C$ ratio (see Fig. 5.5). The calculated fluctuation in the anomalous oceanic flux, $F_{ano,oce}$ (see Fig. 5.6D), however, is of lesser amplitude for the 1998 event than for either of the two previous events, reflecting a greater fluctuation in the rate of change of atmospheric CO_2 for this event (see Fig. 5.5). How can this lesser oceanic fluctuation be explained?

Either the fluctuations in the oceanic flux, indeed, were different in spite of similar signals in the SOI for all three events, or the terrestrial biospheric flux has been incorrectly calculated. We first discuss the likelihood of variable oceanic fluctuations. In a previous double deconvolution calculation with data only through 1988 (Keeling et al. 1989a), variations in $F_{ano,oce}$ showed a correlation with tropical sea surface temperature. In the updated analysis, east tropical Pacific sea surface temperature showed greater warming for the 1998 El Niño event than for the events of 1983 and 1987, whereas the oceanic sink was less pronounced. Thus variable fluctuations in tropical sea surface temperature do not explain quantitatively a smaller oceanic fluctuation in 1998.

Alternatively, neglecting to account for temporal variability in isotopic discrimination by plants may have caused errors in the calculated fluxes. Although the nearly equal amplitudes of fluctuations in the first derivative of $\delta^{13}C$ of atmospheric CO_2 that we observe in association with El Niño events (see Fig. 5.5) suggest corresponding nearly equal amplitudes in $F_{ano,bio}$, perhaps this is not a correct conclusion. This flux reflects both anomalous assimilation of carbon by net primary production (NPP) and anomalous release of biospheric carbon by respiration. Although respiration, as well as NPP, generally responds directly to variability in temperature and precipitation, the sensitivities are not necessarily the same. Variations in NPP, respiration, and variable isotopic discrimination could fortuitously lead to our calculating nearly the same amplitudes for fluctuations associated with the three strong El Niño events of 1983, 1987, and 1998 by ignoring variations in discrimination.

5.9 Ways to Reduce Uncertainty in Estimation of Exchange Fluxes

Given the present difficulties in inferring CO_2 fluxes from atmospheric CO_2 data as discussed above, several possible improvements come to mind. One that is obvious is to increase the precision and absolute accuracy of time-series data. Better calibrations would almost surely improve the database for studying fluxes, especially with respect to isotopic data. In particular they would prevent the finding of such large differences in results between laboratories that one laboratory may find a large isotopic El Niño signal while another does not. Most helpful would be the reintroduction of international isotopic standard reference materials, in the range of oceanic and atmospheric CO_2, for systematic long-term intercalibration between laboratories.

Uncertainty in establishing the rate of uptake of atmospheric CO_2 by the oceans presently limits our ability to identify which part of the combined oceanic and terrestrial sink of industrial CO_2 is oceanic, because the terrestrial biospheric sink is, and probably will remain, far less well determined than the oceanic sink. Several indirect means to improve knowledge of the oceanic sink have been pursued: studies of the fundamental behavior of the oceanic carbon cycle through oceanic campaigns, such as the JGOFS program; investigations of the ocean circulation, such as was pursued by the World Ocean Circulation Experiment; and the use of transient tracers, such as chlorofluorocarbons to define better in general the transport of chemical substances in the oceans. The present uncertainty in the oceanic sink (see Table 5.1) takes into account all of these techniques.

Given the dependency of the double deconvolution procedure on the choice of oceanic submodel, as discussed above, there is no means of estimating the correctness of the average magnitudes of net oceanic fluxes from atmospheric CO_2 data alone. The determination of F_{oce} would be better determined if time-series were available adequate to establish the global average of $^{13}C/^{12}C$ ratio of dissolved inorganic carbon (DIC) in surface ocean water, and still better if adequate subsurface isotopic data were also available (Quay, Tilbrook, and Wong 1992). However, no indirect method is likely to be as reliable as direct measurements of DIC, such as were pursued during the World Oceans Circulation Experiment (Needler 1992). Such measurements, however, should be carried out repeatedly to establish the rate of change of the inventory of carbon in the world oceans.

Independently establishing the global exchange of atmospheric CO_2 with the terrestrial biosphere, either by direct flux measurements or by measuring the changing inventory of carbon in plants and sinks, is not likely to produce a precise global inventory for terrestrial biospheric carbon because of the great heterogeneity of the reservoir. Thus the extent to which the terrestrial biosphere is a source or sink of atmospheric CO_2 globally will probably be best determined by subtracting from the well-established combined sink, $F_{bio} + F_{oce}$, the estimates of the oceanic flux, F_{oce}, based on direct oceanic inventory data. This approach has long-term merit because the data gain more and more significance as the time-interval of measurements lengthens.

A third approach is to use precise measurements of atmospheric oxygen concentration concurrently with measurements of CO_2 concentration and $^{13}C/^{12}C$ ratio, further to constrain the relative magnitudes of the oceanic and terrestrial CO_2 exchange fluxes. Atmospheric oxygen is influenced by exchanges with the land and oceans in different proportions than is carbon dioxide, so the combination of atmospheric O_2 and CO_2 measurements can help to resolve the relative proportions of the land and ocean carbon sinks (Keeling and Shertz 1992; Battle et al. 2000). This method, as all other known methods, has complications but is likely to produce valuable evidence of the time-evolution of oceanic uptake on all time-scales from seasonal to long term, complementing direct measurements of increasing carbon in seawater that can be made worldwide only at infrequent intervals.

5.10 Summary

Combined measurements of the concentration and $^{13}C/^{12}C$ isotopic ratio of atmospheric CO_2 have provided us with a basis for distinguishing terrestrial biospheric and oceanic exchanges of CO_2, from 1978 through 1999. We find remarkably similar patterns of fluctuations of variations in concentration and isotopic ratio on the time-scale of the El Niño–southern oscillation (ENSO) cycle, from which we infer increased releases of CO_2 by the terrestrial biosphere, partially offset by greater uptake of CO_2 by the oceans during El Niño events and at other times of above-average global temperature. Not all details of these correlations, however, are readily explained.

The best-established correlation of our atmospheric CO_2 data with the ENSO cycle pertains to the sum of the global biospheric and oceanic fluxes, established unambiguously by the rate of change in CO_2 concentration. The global average rate of change in $^{13}C/^{12}C$ ratio, though less well established, varies closely in phase with the global CO_2 concentration. It follows, therefore, that the global terrestrial biospheric flux, which the isotopic record reflects, almost certainly has correlated with the ENSO cycle. The global oceanic flux probably also correlates because it is determined by subtracting the biospheric flux from the sum of the two fluxes.

The amplitudes of fluctuations in this sum, derived solely from CO_2 concentration data, are also well established, but the amplitudes relating to the separate global biospheric and oceanic fluxes are uncertain. First, miscalibration of the isotopic data, and an inability to interpret these data correctly, may have produced errors in calculating amplitudes. Second, our isotopic data may reflect phenomena, particularly interannual variability in isotopic discrimination by plants, not considered in our calculations of fluxes. The similarity in phasing of variations in concentration and isotopic ratio demands, however, that possible variations in isotopic discrimination by plants, or other neglected phenomena, be phased consistently with fluctuations in the global biospheric flux.

It is, indeed, possible that phasing of variable discrimination is closely tied to the ENSO cycle. The regional analysis of Piper et al. (2001) demonstrates that most of the correlation of biospheric flux with the ENSO cycle occurs in the tropics, where drought may cause both the net primary production of plants and the discrimination to decrease during El Niño events. Also, as discussed above, the existence of small average variations in tropical discrimination may explain a substantial part of the amplitude in biospheric fluctuations that we infer. However, the anticorrelation of oceanic flux with tropical sea surface temperature that we deduce supports the hypothesis that discrimination by plants may have varied only slightly over the ENSO cycle, or not at all. Such an anticorrelation is consistent with observed decreases in upwelling of cold CO_2-laden subsurface water in the tropics during El Niño events, expected to produce increased sea surface temperature together with a release of CO_2 to the air.

Quantitatively it is difficult to decide which mechanism best explains the observed amplitudes of major short-term interannual fluctuations in CO_2 concentration and $^{13}C/^{12}C$ ratio. The southern oscillation index (SOI), a measure of the intensity of El Niño events, shifts by nearly the same amount during three major events in 1983, 1987, and 1998 (the timing of which shows a close relation to the gray bars in Fig. 5.4). Tropical sea surface temperature also shows nearly equal shifts for these events. Nevertheless, the oceanic flux that we infer shows a substantially lesser fluctuation in 1998 than during the two earlier events, a finding not readily explained. Fluctuations in the rate of change of $^{13}C/^{12}C$ ratio for these three events are nearly the same, consistent with nearly equal strengths of the El Niño events as indicated by the SOI and sea surface temperature data. A reduction in amplitude owing to reduced discrimination is therefore expected to diminish, by similar amounts, the biospheric signals for all three events. Thus, if our isotopic data are correct, at least one signal, either oceanic or biospheric, must vary in strength between the events of the 1980s and the 1998 event. At present we cannot reconcile these uncertainties in interpretation.

Because so much of the interannual variability in CO_2 exchange deduced in our study is short term and attributed to a naturally occurring ENSO cycle, we have not dwelt at length on long-term variability in exchange or on anthropogenic factors that cause variability. The double deconvolution procedure that we have used takes account of the unquestionably human-caused rise in atmospheric CO_2 resulting from the combustion of fossil fuel, but this combustion process is sufficiently steady over the short term that it contributes but little to the interannual variability in atmospheric CO_2 on the ENSO timescale. Short-term variability is likely, however, to be partly of human cause, because of recent changes in land-use that may have increased the sensitivity of the terrestrial biosphere to natural variations in climatic factors: for example, indirectly by human-caused changes in albedo and latent heat exchange, or directly by causing more and greater forest and grass fires during periods of drought.

Our principal finding from this analysis of atmospheric CO_2 data is that interannual fluctuations in net exchanges of atmospheric CO_2 are of the order of several Pg C yr^{-1} and correlate with strong El Niño events. The fluctuations clearly involve the terrestrial biosphere and probably the oceans, but their separate amplitudes and phasing cannot yet be precisely determined.

References

Andres, R.J., G. Marland, T. Boden, and S. Bischof. 2000. Carbon dioxide emissions from fossil fuel consumption and cement manufacture, 1751–1991 and an estimate of their isotopic composition and latitudinal distribution. In *The carbon cycle*, ed. T.M.L. Wigley and D.S. Schimel, 53–62. Cambridge: Cambridge University Press.

Andronova, N.G., and M.E. Schlesinger. 2000. Causes of global temperature changes during the 19th and 20th centuries. *Geophysical Research Letters* 27:2137–40.

Aubinet, M., A. Grelle, A. Ibrom, C. Rannik, J. Moncrieff, T. Foken, A.S. Kowalski, P.H. Martin, P. Berbigier, C. Bernhofer, R. Clement, J. Elbers, A. Granier, T. Grunwald, K. Morgenstern, K. Pilegaard, C. Rebmann, W. Snijders, R. Valentini, T. Vesala. 2000. Estimates of the annual net carbon and water exchange of forests: The EUROFLUX methodology. *Advances in Ecological Research* 30:113–75.

Bacastow, R.B. 1976. Modulation of atmospheric carbon dioxide by the southern oscillation. *Nature* 261:116–18.

Battle, M., M.L. Bender, P.P. Tans, J.W.C. White, J.T. Ellis, T. Conway, and R.J. Rancey. 2000. Global carbon sinks and their variability inferred from atmospheric O_2 and $\delta^{13}C$. *Science* 287:2467–70.

Canadell, J., H.A. Mooney, D. Baldocchi, J.A. Berry, J.R. Ehleringer, C.B. Field, S.T. Gower, D. Hollinger, J. Hunt, R. Jackson, S. Running, G. Shaver, S.E. Trumbore, R. Valentini, and B.Y. Bond. 2000. Carbon metabolism of the terrestrial biosphere: a multi-technique approach for improved understanding. *Ecosystems* 3:115–30.

Climate Prediction Center. [date] Monthly Atmospheric and SST Indices. Website: *www.cpc.ncep.noaa.gov/data/indices*.

Craig, H. 1957. Isotopic standards for carbon and oxygen and correction factors for mass-spectrometric analysis of carbon dioxide. *Geochimica et Cosmochimica Acta* 12:133–49.

Crowley, P.J. 2000. Causes of climate change over the past 1000 years. *Science* 289:270–77.

Ehleringer, J.R., A.E. Hall, and G.D. Farquhar, eds. 1993. Stable isotopes and plant carbon/water relations. San Diego: Academic Press.

Etheridge, D.M., L.P. Steele, R.L. Langenfelds, R.J. Francey, J.M. Barnola, and V.I. Morgan. 1996. Natural and anthropogenic changes in atmospheric CO_2 over the last 1000 years from air in Antarctic ice and firn. *Journal Geophys. Research—Atmospheres* 101(D2):4115–28.

Falkowski, P., R.J. Scholes, E. Boyle, J. Canadell, D. Canfield, J. Elser, N. Gruber, K. Hibbard, P. Hogberg, S. Linder, F.T. Mackenzie, B. Moore, T. Pedersen, Y. Rosenthal, S. Seitzinger, V. Smetacek, and W. Steffen. 2000. The global carbon cycle: A test of our knowledge of earth as a system. *Science* 290:291–96.

Farquhar, G.D., J.R. Ehleringer, and K.T. Hubick. 1989. Carbon isotope discrimination and photosynthesis. *Annual Review Plant Physiology and Molecular Biology* 40:503–37.

Fessenden, J.E., and J.R. Ehleringer. 2003. Temporal variation in $\delta^{13}C$ of ecosystem respiration in the Pacific Northwest: Links to moisture stress. *Oecologia* 136:129–36.

Francey, R.J., P.P. Tans, C.E. Allison, I.G. Enting, J.W.C. White, and M. Trolier. 1995. Changes in oceanic and terrestrial carbon uptake since 1982. *Nature* 373:326–30.

Free, M., and A. Robock. 1999. Global warming in the context of the Little Ice Age. *Journal of Geophysical Research* 104:19,057–19,070.

Heimann, M., and C.D. Keeling. 1989. A three-dimensional model of atmospheric CO_2 transport based on observed winds: 2. Model description and simulated tracer experiments. In *Aspects of climate variability in the Pacific and Western Americas*, Geophysical Monograph 55, ed. D.H. Peterson, 237–75. Washington D.C.: American Geophysical Union.

Heimann, M., C.D. Keeling, and C.J. Tucker. 1989. A three-dimensional model of atmospheric CO_2 transport based on observed winds: 3. Seasonal cycle and synoptic time scale variations. In *Aspects of climate variability in the Pacific and the Western Americas,* Geophysical Monograph 55, ed. D.H. Peterson, 277–303. Washington, D.C.: American Geophysical Union.

Houghton, J.T., Y. Ding, D.J. Griggs, M. Noguer, P.J. van der Linden, X. Dai, K. Maskell, and C.A. Johnson. 2001. Climate change 2001, the scientific basis. Contribution of Working Group I to the Third Assessment Report of the Intergovernmental Panel on Climate Change (IPCC). Cambridge: Cambridge University Press.

Houghton, J.T., L.G. Meira Filho, B.A. Callander, N. Harris, A. Kattenberg, and K. Maskell. 1996. Climate change 1995, the science of climate change. Contribution of Working Group I to the Second Assessment Report of the Intergovernmental Panel on Climate Change (IPCC). Cambridge: Cambridge University Press.

Houghton, R.A. 1999. The annual net flux of carbon to the atmosphere from changes in land use 1850–1990. *Tellus* 51B:298–313.

Hunt, E.R., S.C. Piper, R. Nemani, C.D. Keeling, R.D. Otto, and S.W. Running. 1996. Global net carbon exchange and intra-annual atmospheric CO_2 concentrations predicted by an ecosystem process model and three-dimensional atmospheric transport model. *Global Biogeochemical Cycles* 10:431–56.

Jones, P.D. 1994. Hemispheric surface air temperature variations: a reanalysis and an update to 1993. *Journal of Climate* 7:1794–1802.

Karl, D.M., and A.F. Michaels. 1996. Tropical studies in oceanography. *Deep-Sea Research* 43:127–28.

Keeling, C.D., R.B. Bacastow, A.F. Carter, S.C. Piper, T.P. Whorf, M. Heimann, W.G. Mook, and H. Roeloffzen. 1989a. A three-dimensional model of atmospheric CO_2 transport based on observed winds: 1. Analysis of observational data. In *Aspects of climate variability in the Pacific and the Western Americas,* Geophysical Monograph, 55, ed. D.H. Peterson, 165–236. Washington, D.C.: American Geophysical Union.

Keeling C.D., S.C. Piper, R.B. Bacastor, M. Wahlen, T.P. Whorf, M. Heimann, and H.A. Meijer. 2001. Exchanges of atmospheric CO_2 and $^{13}CO_2$ with the terrestrial biosphere and oceans from 1978 to 2000. I. Global aspects. *SIO Reference Series,* No. 01–06. San Diego: Scripps Institution of Oceanography.

Keeling, C.D., S.C. Piper, and M. Heimann. 1989b. A three-dimensional model of atmospheric CO_2 transport based on observed winds: 4. Mean annual gradients and interannual variations. In *Aspects of climate variability in the Pacific and the Western Americas,* Geophysical Monograph, 55, ed. D.H. Peterson, 305–63. Washington, D.C.: American Geophysical Union.

Keeling, C.D., T.P. Whorf, M. Wahlen, and J. Van der Plicht. 1995. Interannual extremes in the rate of rise of atmospheric carbon dioxide since 1980. *Nature* 375:666–70.

Keeling, R.F., and S.R Shertz. 1992. Seasonal and interannual variations in atmospheric oxygen and implications for the global carbon cycle. *Nature* 358:723–27.

Lloyd, J., and G.D. Farquhar. 1994. C-13 discrimination during CO_2 assimilation by the terrestrial biosphere. *Oecologia* 99:201–15.

Meehl, G.A. 1987. The annual cycle and interannual variability in the tropical Pacific and Indian ocean regions. *Monthly Weather Review* 115:27–50.

Mook, W.G., and P.M. Grootes. 1973. The measuring procedure and corrections for the high-precision mass-spectrometric analysis of isotopic abundance ratios, especially referring to carbon, oxygen and nitrogen. *International Journal of Mass Spectrometry and Ion Physics* 12:273–98.

Mook, W.G., M. Koopmans, A.F. Carter, and C.D. Keeling. 1983. Seasonal, latitudinal and secular variations in the abundance and isotopic ratios of atmospheric carbon dioxide: 1. Results from land stations. *Journal of Geophysical Research* 88:10915–33.

Needler, G.T. 1992. The World Ocean Circulation Experiment. *Oceanus* 35:74–77.

Ometto, J.P.H.B., L.B. Flanagan, L.A. Martinelli, M.Z. Moreira, N. Higuchi, and J.R. Ehleringer. 2002. Carbon isotope discrimination in forest and pasture ecosystems of the Amazon Basin, Brazil. *Global Biogechemical Cycles* 16 (December):article 1109.

Pataki, D.E., J.R. Ehleringer, L.B. Flanagan, D. Yakir, D.R. Bowling, C.J. Still, N. Buchmann, J.O. Kaplan, and J.A. Berry. 2003. The application and interpretation of Keeling plots in terrestrial carbon cycle research. *Global Biogeochemical Cycles* 17 (March):article 1022.

Piper, S.C., C.D. Keeling, M. Heimann, and E.F. Stewart. 2001. Exchanges of atmospheric CO_2 and $^{13}CO_2$ with the terrestrial biosphere and oceans from 1978 to 2000. II. A three-dimensional tracer inversion model to deduce regional fluxes. In *SIO Reference Series,* no. 01–07. San Diego: Scripps Institution of Oceanography.

Quay, P.S., B. Tilbrook, C.S. Wong. 1992. Oceanic uptake of fossil fuel CO_2: Carbon-13 evidence. *Science* 256:74–79.

Rasmusson, E.M., and J.M. Wallace. 1983. Meteorological aspects of the El Niño/southern oscillation. *Science* 222:1195–1202.

Santer, B.D. et al. 1996. Detection of climate change and attribution of causes. In *Climate Change 1995,* ed. J.T. Houghton et al., 407–11. Cambridge: Cambridge University Press.

Siegenthaler, U., and H. Oeschger, H. 1987. Biospheric CO_2 sources emissions during the past 200 years reconstructed by deconvolution of ice core data. *Tellus* 39B:140–54.

Slingo, J.M., and H. Annamalai. 2000. 1997: The El Niño of the century and the response of the Indian summer monsoon. *Monthly Weather Review* 12:1178–97.

Still, C.J., J.A. Berry, G.J. Collatz, and R.S. DeFries. 2003. Global distribution of C_3 and C_4 vegetation: Carbon cycle implications. *Global Biogeochemical Cycles* 17 (January):article 1006.

Wigley, T.M.L., and D.S. Schimel. 2000. *The carbon cycle.* New York: Cambridge University Press.

6. Evolutionary Responses of Land Plants to Atmospheric CO_2

David J. Beerling

6.1 Introduction

Fossil spores dating to the Ordovician/early Silurian (440–430 Ma) provide the earliest direct evidence for plants colonizing the land (Edwards 1998). The spores are thought to derive from land plants because their resilient walls are composed of the complex biomolecule, sporopollenin, and are typically a composite of four individual subunits. The ensuing evolution and diversification of plants took place over the next 400 million years, so they encompassed a sufficiently long duration to witness large variations in the geologic processes controlling the rate of CO_2 supply to the atmospheric carbon reservoir by geotectonic events and its removal by silicate rock weathering (see Chapters 1 and 8). It follows that the complex patterns of land-plant evolution occurred against a background of fluctuating atmospheric CO_2 levels. Further, because atmospheric CO_2 is a radiatively active greenhouse gas, any large variations in CO_2 were accompanied by similarly large swings in global climate over the Phanerozoic (Crowley and Berner 2001). The effects of CO_2 and climate on plant evolution are therefore likely to be closely interlinked.

A central question, following from the realization that evolution and the geologic controls on atmospheric CO_2 operate on the same timescales, is whether changes in atmospheric CO_2 actually shaped evolutionary innovations in vascular land plants. Addressing this issue is a new endeavor in ecological inquiry, because CO_2 has tended to be overlooked as a force driving evolution (Knoll and Niklas 1987; Cornette, Lieberman, and Goldstein 2002). The field is a particularly challenging one because the fossil record merely tells what has already happened and provides no direct evidence for the underlying causes. Meeting the challenge requires a history of atmospheric CO_2 variations over the past 400 million years to provide a framework against which to compare evolutionary events seen in the plant fossil record. Phanerozoic patterns of CO_2 change used for this purpose are derived from computer models of the long-term carbon cycle and from a range of proxy (indirect) indicators of ancient CO_2 levels (reviewed in Royer, Berner, and Beerling 2001). The largest shifts in CO_2 shown by models and paleosols over the past 500 million years correspond to geologic evidence for the waxing and waning of the major continental ice sheets (Crowley and Berner 2001). Such correspondence is taken as independent confirmatory evidence for the link between calculated CO_2 changes and climate, via the greenhouse effect (Crowley and Berner 2001).

With the exception of the current exponential rise in atmospheric CO_2 (Prentice et al. 2001), the most dramatic CO_2 excursion of the entire past 400 million years was the substantial (~ 90%) decline between the Devonian and the Carboniferous (Fig. 6.1).

This huge drawdown in CO_2 is supported by a wealth of paleosol-derived CO_2 estimates (Royer et al. 2001), and it occurs largely because of the spread and activity of terrestrial ecosystems, which themselves were undergoing enormous evolutionary development and diversification (Edwards 1998). The late Paleozoic therefore clearly represents a time when we might anticipate identi-

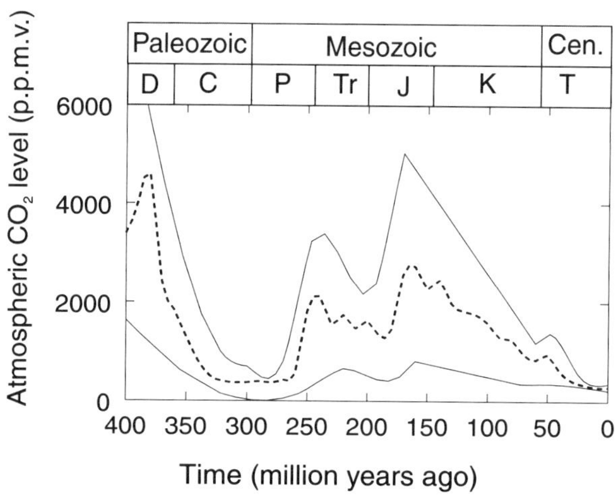

Figure 6.1. Long-term changes in atmospheric CO_2 predicted from a geochemical model of the Carbon cycle (Berner and Kothavala 2001). Broken line indicates the 'best guess' predictions, solid lines the envelope of error. D = Devonian, C = Carboniferous, P = Permian, Tr = Triassic, J = Jurassic, K = Cretacaeous, and T = Tertiary.

fying the action of CO_2 on plant evolutionary processes. Furthermore, because the large drop in atmospheric CO_2 resulted in part from the burial of enormous quantities of photosynthetically derived organic carbon, atmospheric O_2 levels are predicted to have risen to between 26% (Lenton 2001) and 35% (Berner et al. 2000; Berner 2001). A low CO_2, high O_2 atmosphere severely limits the productivity of terrestrial plants with the C_3 photosynthetic metabolism (Beerling and Berner 2000). Such conditions would exert a strong selection pressure for the evolution of a more efficient type of photosynthetic metabolism, possibly involving some kind of carbon concentrating mechanism.

Here, I review the likely effects of falling late Paleozoic atmospheric CO_2 levels on stomatal development (McElwain and Chaloner 1995; Beerling and Woodward 1997) and its implications for the evolution of the broad leaf form, which is termed a megaphyll leaf (Beerling, Osborne, and Chaloner 2001; Osborne, Chaloner, and Beerling 2002; Osborne et al. 2004), and plant stature (Chaloner and Sheerin 1979). I especially consider evidence that the unusual low CO_2/high O_2 atmospheric composition at that time (Beerling and Berner 2000) led to the selection of an alternative photosynthetic pathway (Cerling et al. 1997; Gale et al. 2001). In the final section, I critically discuss the postulated action of CO_2 on the evolution and emergence of the major plant groups and speciation (Woodward 1998; Beerling and Woodward 2001; Barrett and Willis 2001; Willis and McElwain 2002).

6.2 Stomata

Observations on extant land plants with megaphyll leaves, that is leaves with a broad lamina (flat blade) and a branching veination system, indicate that the atmospheric CO_2 concentration at the time of growth influences the development of stomata (Woodward 1987). Leaves operating with few stomata reduce the loss of water vapor by lowering transpiration rates, but without diminishing rates of carbon gain by photosynthesis. By responding in this manner, plants increase their efficiency of water use during growth (Woodward 1987). However, measurements on historical collections of primitive vascular plants with microphyll leaves (characterized by a single strand of vascular tissue supplying water and nutrients to the photosynthesizing mesophyll cells without a leaf gap) show they also respond to CO_2 changes in a similar way (Fig. 6.2).

It is curious that CO_2 appears to influence stomatal production in microphyllous species, because they tend to occupy moist, humid habitats where the need to conserve water during periods of carbon gain is diminished. This suggests that operating with fewer stomata serves a more fundamental metabolic purpose, possibly allowing plants to save energy in the regulation of leaf gas exchange processes (Farquhar and Wong 1984).

Is there evidence that falling Paleozoic atmospheric CO_2 levels influenced stomatal development in land plants so that they could achieve the same ecological and physiological benefits? The question is addressed by comparing pub-

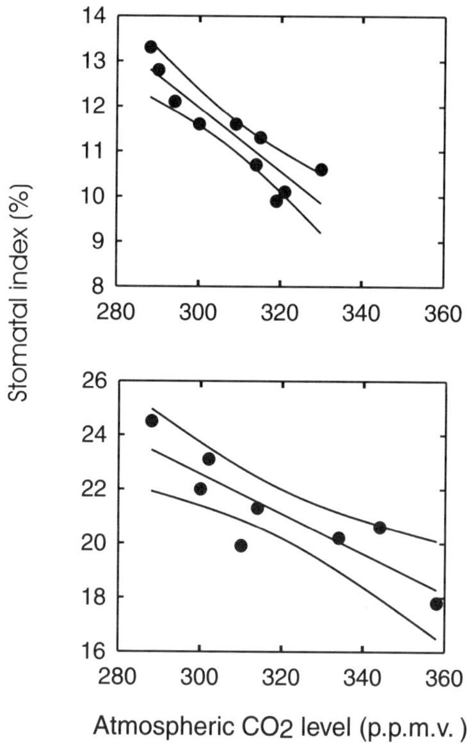

Figure 6.2. Reductions in the stomatal index (fraction of epidermal cells that are stomata) of microphyll leaves of U.K. populations *Selaginella selagenoides (upper panel,* $r^2 = 0.8$, $P < 0.01$) and *S. kraussiana (lower panel,* $r^2 = 0.7$, $P < 0.05$) in response to atmospheric CO_2 increases over the past two centuries (D.J. Beerling, unpublished). Each plot shows the linear regression line with the ±95% confidence limits.

lished measurements of the stomatal density of fossilized leaf cuticles of Devonian, Carboniferous, and Permian plants, which grew under very high and low atmospheric CO_2 levels respectively (see Fig. 6.1). These data, together with the observations of McElwain and Chaloner (1995), indicate that early land plants operating in the very high CO_2 atmosphere of the late Silurian to early Devonian consistently possessed rather few stomata (5–10 mm^{-2}), but these were sufficient to overcome diffusion limitations on rates of photosynthesis (Raven 1993; Beerling, Osborne, and Chaloner 2001). Fossil cuticles of the arborescent lycopsids and seed ferns, which dominated the equatorial coal swamp forests (Phillips and DiMichele 1992), indicate that much higher stomatal densities were the norm by the Carboniferous (Fig. 6.3).

A similar trend is seen in the Permian, with the Glossopterids, which were widespread on the high southern latitude landmasses (Stewart and Rothwell

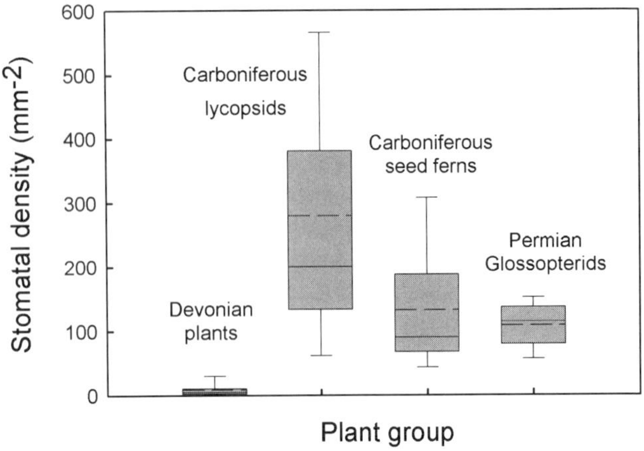

Figure 6.3. Stem and box plots of the variations in stomatal density within various Paleozoic plant groups. Each box delineates the 25th and 75th percentiles; error bars, the 5th and 95th percentiles; the solid horizontal line, the median values; and the dashed horizontal line, the mean values. Data from Beerling and Woodward (1997), Edwards (1998), Pant and Gupta (1968, 1971), Royer (2001), Thomas (1970).

1993), possessing higher stomatal densities than the early Devonian land plants (see Fig. 6.3). Evidence from the fossil record is circumstantial but nevertheless shows that changes in the stomatal characters of successively dominant plant groups declined through the late Paleozoic. Such a change is consistent with the action of CO_2 on stomatal formation seen in contemporary plants (Beerling and Woodward 1997).

A similar conclusion was reached by Edwards (1998), who tracked changes in the stomatal density of the same species of leafy herbaceous Devonian lycopsid, *Drepanophycus spinaeformis*, from North America through the late Devonian. Over this time, the stomatal density of *D. spinaeformis* cuticles increased by 41%, possibly in response to the falling atmospheric CO_2 levels over this time (Berner and Kothavala 2001). Paleobotanical observations, in conjunction with paleo-CO_2 estimates, therefore suggest the stomata-CO_2 response is one of considerable antiquity, although it is likely that stomatal characteristics are intimately linked to the evolution of other structures regulating water balance (Edwards 1998; Raven and Edwards 2001).

6.3 Leaves

The regulation of stomatal development by CO_2 in the late Paleozoic is thought to have had major evolutionary implications for the evolution of leaves in vascular land plants (Beerling, Osborne, and Chaloner 2001; Osborne, Chaloner, and Beerling 2002). A comparison of the timing of the evolutionary origins of megaphyll leaves, implied by fossils, and the major CO_2 drop suggests a casual linkage between the two. Recent work points to a mechanistic explanation for how the linkage operates (Beerling, Osborne, and Chaloner 2001). The explanation is grounded in two key elements. First, it recognizes that early Devonian land plants had few stomata, while the true planated leaves of Carboniferous plants possessed somewhat higher stomatal densities (see Fig. 6.3). Stomata in this context are important because their density, and the degree to which they are open or closed, control the rate of water loss by transpiration and hence the loss of latent heat to the atmosphere. Second, it recognizes that the temperature of a photosynthesizing organ is, to a large extent, determined by the quantity of solar energy intercepted. Cylindrical axes, such as those typical early Devonian plants, intercept less solar energy than do flat leaf blades because an upright stem exposes a substantially smaller surface area to direct-beam solar radiation, particularly at midday in the low latitudes when the sun is directly overhead (Fig. 6.4).

Simulations using a mathematical model of the biophysical principles governing leaf gas exchange and energy budget, which incorporate information on light interception and stomatal anatomy, indicate that the evolution of leaves

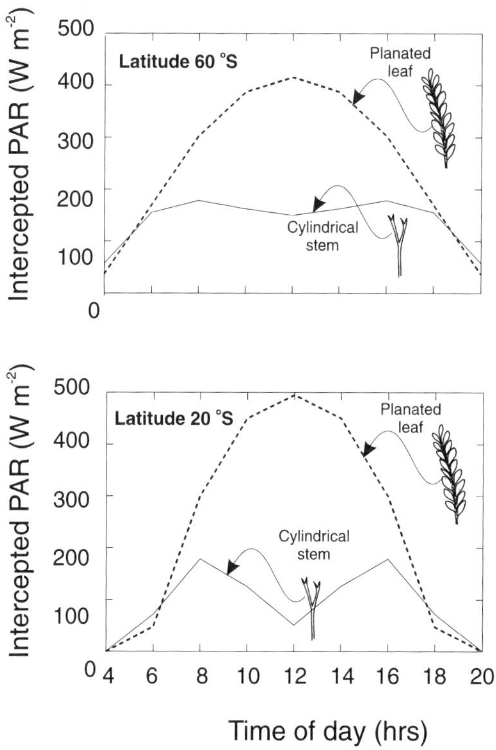

Figure 6.4. Intercepted photosynthetically active radiation (PAR) by a planated (flat) leaf blade and a cylindrical axis in the high *(upper panel)* and low latitudes *(lower panel).* Calculated as described by Beerling, Osborne, and Chaloner (2001).

with a broad lamina in the early Devonian would not have provided a selective advantage, because low stomatal densities limit cooling through latent heat loss and lead to lethal overheating (Beerling, Osborne, and Chaloner 2001). Efficient absorption of solar energy for photosynthesis by laminate leaves in the early Devonian therefore would have carried a fatal penalty. In contrast, the erect axial form of early Devonian plants protected them from overheating, especially at noon when the sun is directly overhead, by virtue of their minimal interception of the sun's energy (see Fig. 6.4). Some arid-land plants adopted a similar strategy to avoid thermal damage by means of evolving secondarily reduced erect leaves (Valladares and Pearcy 1997) or photosynthetic stems (Haase et al. 1999). By the time laminate leaves appeared in the fossil record, during the late Devonian/Carboniferous, atmospheric CO_2 levels had dropped (see Fig. 6.1) and high stomatal densities were evident (see Fig. 6.3). Simulated transpiration rates of these leaves under ambient CO_2 levels were then sufficiently high to maintain leaf temperatures well below those at which damage to the photosynthetic system occurs (Beerling, Osborne, and Chaloner 2001).

Furthermore, it is calculated that as stomatal density rose in response to falling Devonian CO_2 levels, large laminate leaves increasingly gained a selected advantage in terms of carbon gain by photosynthesis (Beerling, Osborne, and Chaloner 2001). The increase in size occurs without the penalty of high temperature damage because of improved transpirational cooling due to the accompanying rise in stomatal density. Quantitative testing of this prediction, by measurement of the terminal widths of photosynthetic appendages for a wide range of Devonian fossil plants, provisionally confirms that laminate leaf evolution has followed the predicted trajectory, as shown in Fig. 6.5 (Osborne, Chaloner, and Beerling 2002; Osborne et al. 2004). The upward trend in leaf width shown by the fossils in this data set continues into the Carboniferous when leaves of pteridiosperms exceed 20 mm.

Later in the fossil record, across the Triassic-Jurassic (Tr-J) boundary (200 Ma), evidence for the converse effect of CO_2 on the leaf size has been reported (McElwain, Beerling, and Woodward 1999). Isotopic analyses of terrestrial and marine organic and carbonate sediments from a worldwide set of locatives revealed the geochemical fingerprint of a major global carbon cycle perturbation marking a substantial, but transient, increase in atmospheric CO_2 (McElwain, Beerling, and Woodward 1999; Tanner et al. 2001; Beerling 2002). The CO_2 rise was probably brought about by a combination of massive volcanic CO_2 outgassing during the emplacement of the Central Atlantic Magmatic flood basalts in conjunction with the release of large quantities of methane from seafloor sedimentary hydrate reservoirs (Beerling and Berner 2002).

Palaeobotanical evidence suggests that canopy-dominant forest trees with large laminate leaves went into decline and were replaced by species with more finely dissected leaves, particularly in the North Atlantic sector (Visscher and Brugman 1981; Fowell, Cornett, and Olsen 1994). The taxonomic replacements seen in the fossil record can be explained by a similar line of reasoning invoked to account for the delayed evolution of megaphyll leaves during close of the Devonian (McElwain, Beerling, and Woodward 1999). It is thought that, as the atmo-

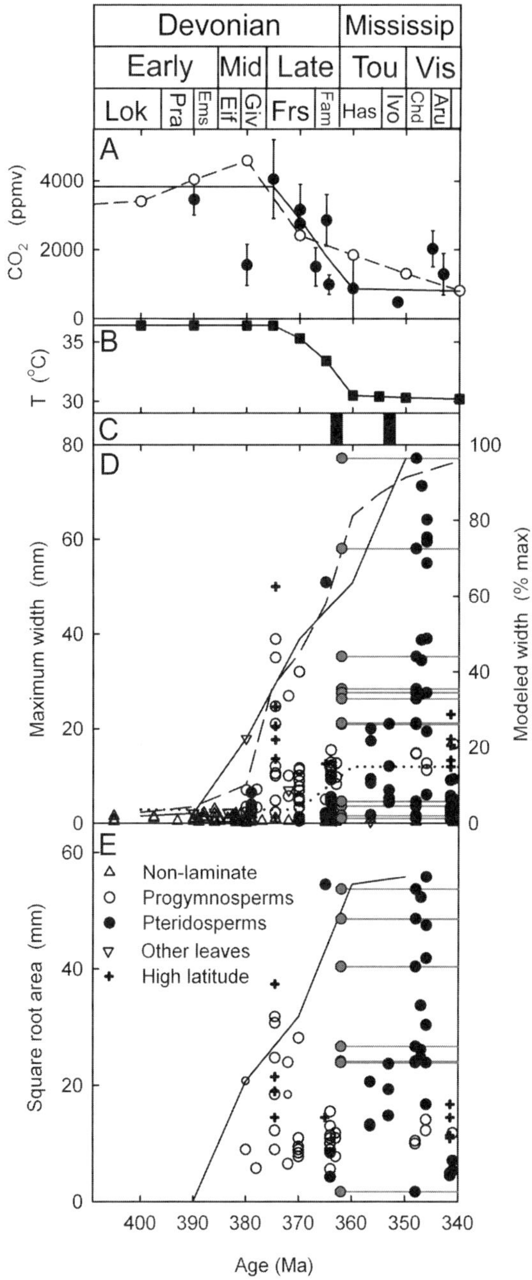

Figure 6.5. Paleozoic changes in atmospheric CO_2 concentration from paleosols (closed symbols) and the geochemical model of Berner and Kothavala (2001) (open symbols) **(A)**, inferred changes in air temperature, based on radiative forcing **(B)**, geological evidence for glaciation **(C)** (after Crowley and Berner 2001), maximum width of the terminal photosynthetic structure **(D)** and square root of the leaf area, taken as a measure of the path length for air flow over the leaf in any direction. Horizontal lines in **(D)** and **(E)** indicate age uncertainty of Missipssippian leaves. Solid lines in **(D)** and **(E)** indicate maximum size for each 10 million year interval.

spheric CO_2 concentration rose and the climate warmed across the Tr-J boundary, those forest trees with large leaves probably became more susceptible to suffering severe temperature damage because (1) CO_2-induced reductions in stomatal densities limited transpirational cooling; and (2) large leaves have low boundary layer conductances and a limited capacity for sensible heat loss. Those species replacing them possessed finely divided or dissected leaves (Harris 1937; Lundblad 1959) and so were better adapted at avoiding thermal damage because such leaf morphologies increase air flow turbulence over the leaf and hence convective cooling.

These quite specific examples from the fossil record suggest that leaf morphology represents one feature of plant anatomy susceptible to being directed by past variations in atmospheric CO_2 concentrations. In both instances the action of CO_2 has overprinted other biotic (e.g., herbivory) and abiotic (e.g., climate) determinants of leaf form. It is interesting to note that some CO_2 emission scenarios for the future suggest a global mean CO_2 concentration approaching 1100 ppmv by the year 2100 (Prentice et al. 2001). If such CO_2 levels are reached, the geologic record suggests our tropical forests will be operating close to a critical threshold CO_2 value, beyond which future CO_2 increases could induce considerable physiological stress on the dominant broad-leaved evergreen taxa and, perhaps, their ecological decline.

6.4 Photosynthetic Pathways

Following the late Paleozoic CO_2 drop, atmospheric CO_2 was maintained at a low level throughout what remained of the Carboniferous and into the Permian (see Fig. 6.1). Low Permo-Carboniferous atmospheric CO_2 levels were largely due to extensive organic carbon burial in the Carboniferous coal swamps; this burial was itself a feature that probably drove the atmospheric O_2 content above the present-day value of 21% (Berner et al. 2000; Berner 2001; Lenton 2001; Beerling et al. 2002). Assessing the potential for these atmospheric conditions to select for a more efficient type of photosynthesis can be done by simulating their effects on the photosynthetic productivity of plants. The simulations were made for plants with the C_3 photosynthetic pathway and with the late evolving, but more efficient, C_4 pathway. The results show that, depending on the climate, the photosynthetic rates of C_3 plants decline by 60 to 80% during the Permo-Carboniferous, 330 to 260 million years ago (Fig. 6.6).

The decline occurs because CO_2 fixation in the primary carboxylating enzyme in C_3 plants, Rubisco, is competitively inhibited at higher-than-present O_2 concentrations. The calculations, supported by experimental evidence with angiosperms (Gale et al. 2001), further show that if plants had evolved the more efficient C_4 photosynthetic pathway, their productivity would have remained unaffected by a rise in O_2 (see Fig. 6.6), because the carbon concentrating mechanism of the C_4 pathway is O_2-insensntive (see Chapter 9).

So, on the basis of model simulations (see Fig. 6.6) and experiments (Gale et al. 2001), the unusual atmospheric CO_2/O_2 mixing ratio of the Permo-Carboniferous could have exerted a strong evolutionary selection pressure for

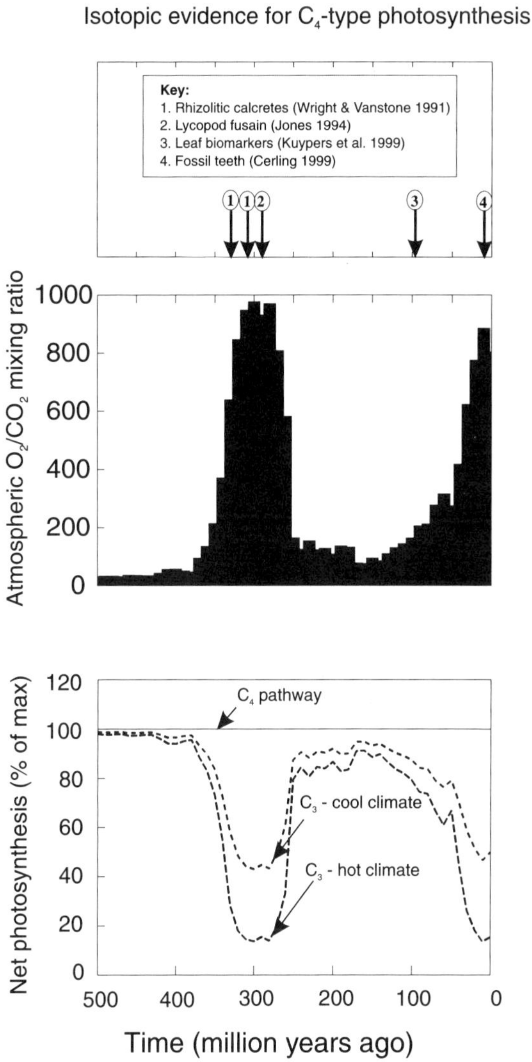

Figure 6.6. Isotopic evidence for the possible occurrence of C_4 type photosynthesis in the geologic record *(upper panel)*, changes in the atmospheric O_2 to CO_2 ratio over the past 500 million years (middle panel), and their simulated impact on the photosynthetic productivity of plants with the C_3 and C_4 photosynthetic pathway *(lower panel)*. Leaf gas exchange simulations were performed using an appropriate biochemical model of photosynthesis (Von Caemmerer 2002) at either 20°C (cool climate) or 30°C (hot climate) and an irradiance of 1000 $\mu mol\ m^{-2}\ s^{-1}$.

some kind of CO_2-concentrating mechanism, such as C_4 photosynthesis (Cerling et al. 1997; Beerling and Woodward 2001). Despite this potential, the oldest accepted C_4 fossils are from North American grasses just 12 million years ago (Cerling 1999). Molecular fossils of leaf waxes suggest an earlier, but unconfirmed, appearance of C_4 plants in North African communities during the late Cretaceous (Kuypers, Pancost, and Damste 1999). But secure evidence for C_4 photosynthesis in Permo-Carboniferous plants is extremely limited (see Fig. 6.6), and equivocal data are restricted to carbon isotope measurements on Carboniferous terrestrial organic matter and soil carbonate concretions (Jones 1994; Wright and Vanstone 1991). These preliminary analyses are potentially of considerable significance since, if substantiated, they would push back the evolutionary origins of C_4 metabolism by over 200 million years.

A detailed survey of the isotopic composition of Permo-Carboniferous fossil plants can be used to search more comprehensively for the isotopic signature of C_4 photosynthetic metabolism because (1) C_3 and C_4 plants fractionate against the heavy isotope of carbon very differently (Farquhar, Ehleringer, and Hubick 1989); and (2) the isotopic signal is generally well preserved in the geologic record. More than 250 measurements were made (Beerling unpublished) on fossil plants of the Carboniferous equatorial coal swamp forests (sphenopsids, seed ferns, tree ferns, horsetails, and arborescent lycopsids) and on high southern latitude Permian Glossopterids from Antarctica (McLoughlin and Drinnan 1997). Carbon isotope discrimination was also calculated by correcting for the isotopic composition of the source CO_2 used during photosynthesis. For the fossil plants this was taken from marine carbonate records (Veizer et al. 1999) with a standard 7 per mil offset to account for the difference between the oceans and the atmosphere.

The results of this survey indicate that none of the fossilized organic carbon samples of Carboniferous or Permian plant specimens analyzed displayed carbon isotope composition or discrimination signatures characteristic of plants operating with the C_4 photosynthetic pathway (Fig. 6.7).

In fact, the spread of values gives a frequency distribution similar to that seen in modern C_3 plants (Farquhar, Ehleringer, and Hubick 1989). It is possible that the arborescent lycopods operated with CAM metabolism and utilized CO_2 respired from the swamps (Phillips and DiMichele 1992), as seen in modern *Styletes* plants in the high Andes (Keeley, Osmond, and Raven 1984). In this mode, the stomata are closed during the day and open to take up soil-respired CO_2 during night. If this were the case, no distinctive carbon isotopic signature would remain.

The absence of C_4 photosynthesis in Permo-Carboniferous can be considered against current ideas relating to those factors thought to have driven the evolution of C_4 pathway (Ehleringer et al. 1991; Cerling et al. 1997; Ehleringer, Cerling, and Helliker 1997; Sage 2001). On the basis of current evidence, summarized by Sage (2001), the evolution of C_4 photosynthesis requires (1) the occurrence of advanced groups of angiosperms, (2) aridity, to promote fire and open habitats, place a premium on water-use efficiency, and establish drought-adapted floras, and (3) low atmospheric CO_2 levels. Obviously, late Paleozoic

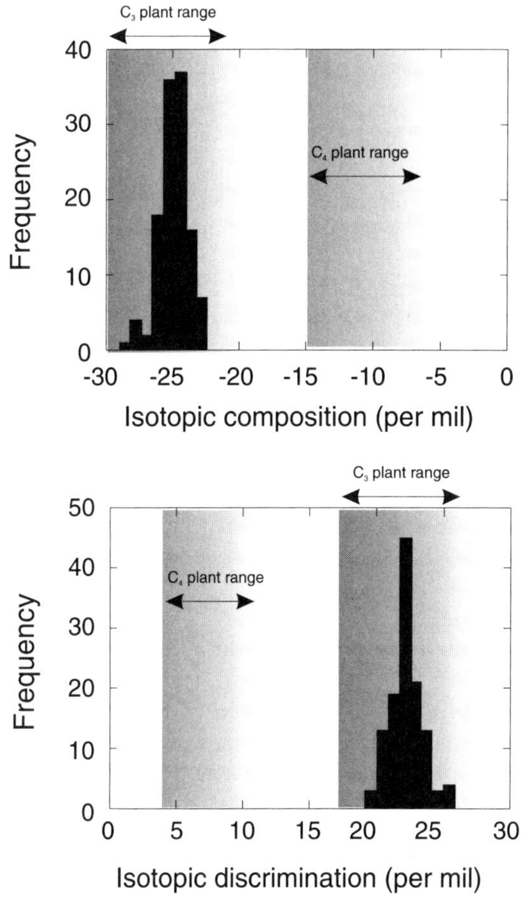

Figure 6.7. Frequency histograms for the carbon isotopic composition *(upper panel)* and discrimination *(lower panel)* of organic carbon samples of Permo-Carboniferous fossil plants (D.J. Beerling, unpublished). The typical ranges for extant C_3 and C_4 plants are as indicated.

ecosystems lacked later-evolving angiosperm representatives, and this might explain the apparent absence of C_4 plants (see Fig. 6.6). But this explanation for the absence of C_4 plants before the Tertiary is unsatisfactory because the reason C_4 photosynthesis is found only in advanced groups of flowering plants is unknown. New discoveries both of C_4 physiology in the green stems of C_3 plants (Hibberd and Quick 2002) and of species without the specialized anatomy of most C_4 plants (Voznesenskaya et al. 2001) raise additional uncertainty in this regard. However, as with the possibility of CAM, if Paleozoic plants had had the capacity to recycle CO_2 from the xylem with enzymes characteristic of C_4 plants (Hibberd and Quick 2002), this type of physiology would not have resulted in a distinctive carbon isotope signature and therefore remains undetected using isotope ratio analyses.

The other preconditions of climate and atmospheric CO_2 are admirably met in the Carboniferous and Permian. Geologic evidence for the climate at the time shows extensive regions of the supercontinent Pangea became progressively more arid, particularly in the Permian, when the equatorial regions were dry and indicators of seasonality and aridity became widespread (Parrish 1993). Furthermore, fossil charcoal from early Carboniferous deposits indicates that the tropical swamp communities were subject to frequent fires, with fire return intervals of 3 to 35 years, providing the disturbance regimes necessary to maintain an open-shade free habitat (Falcon-Lang 2000). Geochemical models and paleosol data indicate the Paleozoic high atmospheric O_2/CO_2 mixing ratio and low atmospheric CO_2 levels matched those of the Tertiary atmosphere when C_4 plants evolved (Ehleringer, Cerling, and Helliker 1997). Moreover, in the Paleozoic these conditions persisted for several times the duration of the CO_2 decline in the Tertiary. All of these considerations suggest that further isotopic investigations of fossil plants from the Permo-Carboniferous with a more global coverage are warranted.

6.5 Whole Plants and Plant Species

I have so far considered CO_2-driven aspects of plant evolution considered at a progressively increasing scale, from cellular to whole leaves. The next level of scale to be examined is that of plants as whole organisms and the emergence of plant groups and species. Earlier considerations indicate that the sensitivity of stomata to CO_2 probably had important evolutionary consequences for leaves. But it is further possible that the same response exerted an influence on the evolution of plant height and, therefore, competitive ability amongst individuals. Studies on modern species show that plants with higher stomatal densities are better able to protect themselves from xylem cavitation (Woodward 1998), a feature especially important for tall plants, which have to transport water between the roots and leaves through a longer length of xylem. By affording this protection, the late Paleozoic rise in stomatal density offered a mechanism allowing taller, more competitive plants to evolve, assuming the development of more efficient rooting systems (Algeo and Scheckler 1998). Such an expectation is borne out by the fossil record, which demonstrates that the maximum plant size of vascular land plants rapidly increased from 1 m to over 30 m during the Devonian (Chaloner and Sheerin 1979; Algeo and Scheckler 1998).

Atmospheric CO_2 probably influenced competitive interactions in late Paleozoic terrestrial ecosystems. But what was its role in driving the evolution of different plant groups, e.g., ferns, horsetails, conifers, and angiosperms, and even different species? By comparing Phanerozoic patterns of vascular plant species richness with CO_2 trends, Barrett and Willis (2001) postulated the major plant groups evolved at times of increasing CO_2. Further, origination rates (that is, the birth of new species) have been positively correlated with atmospheric CO_2

(Willis and McElwain 2002). The mechanism by which high CO_2 might drive these evolutionary events is not clear. However these authors reasoned that the generation times of tropical tree species over the past 40 years has increased, perhaps as a response to the rise in atmospheric CO_2 levels over this time (Phillips and Gentry 1994). They recognized that such a response obviously increases the rate of transmission of genetic adaptations through plant populations, a feature likely to be further enhanced if trees reach maturity faster and are more fecund at high CO_2 (LaDeau and Clark 2001).

Nevertheless, these important claims warrant closer scrutiny. Inspection of the estimated times of first appearance in the fossil record of the major groups of extant vascular plants shows quite clearly that all emerged when atmospheric CO_2 levels calculated by models, and inferred from proxy measurements, were usually high but falling (Fig. 6.8).

This observation implies that periods of falling atmospheric CO_2 were perhaps an effective selective force for evolutionary change and speciation events (Woodward 1998; Beerling and Woodward 2001). In this scenario, falling atmospheric CO_2 levels would have led to falling water-use efficiencies, because of the high potential for water loss and the associated potential reductions in photosynthetic carbon gain. Under such conditions, plants essentially experience a 'physiological drought,' which can lead to a strong selective pressure favoring new physiological characteristics for enduring environmental constraints (Raven 1993). Artificial selection experiments with C_3 annuals support the scenario of low CO_2 by showing it is a more effective selection agent than is high CO_2 for traits determining plant fitness (Ward et al. 2000).

Phanerozoic patterns of vascular land-plant diversity inferred from the fossil record (Knoll and Niklas 1987) can be compared with the modeled CO_2 trajectory (Berner and Kothavala 2001) in the same way. It is noticeable that diversity was initially fairly stable for 200 million years but then increased exponentially over the past 100 million years (see Fig. 6.8). The exponential increase reveals the accumulation, over millions of years, of species originations exceeding species extinctions, and it becomes apparent that it occurred as atmospheric CO_2 levels declined. The fossil record indicates that under these circumstances, origination events are more strongly favored than are extinction events. For angiosperm species, therefore, declining (rather than rising) CO_2 levels played at least a circumstantial role in their diversification. Intriguingly, as with the evolution of megaphyll leaves, it seems to have been the plants themselves that were responsible the removal of carbon from the atmosphere (Beerling and Woodward 2001).

6.6 Summary

I sought to identify the influence of atmospheric CO_2 on evolutionary innovations in vascular land plants at a range of spatial scales. At the molecular scale, there appears to be little isotopic evidence that the low CO_2/high O_2 atmosphere characterizing the Permo-Carboniferous forced the evolution of C_4 photosynthesis earlier than the Tertiary. It is not yet clear why. In contrast, at the cellular

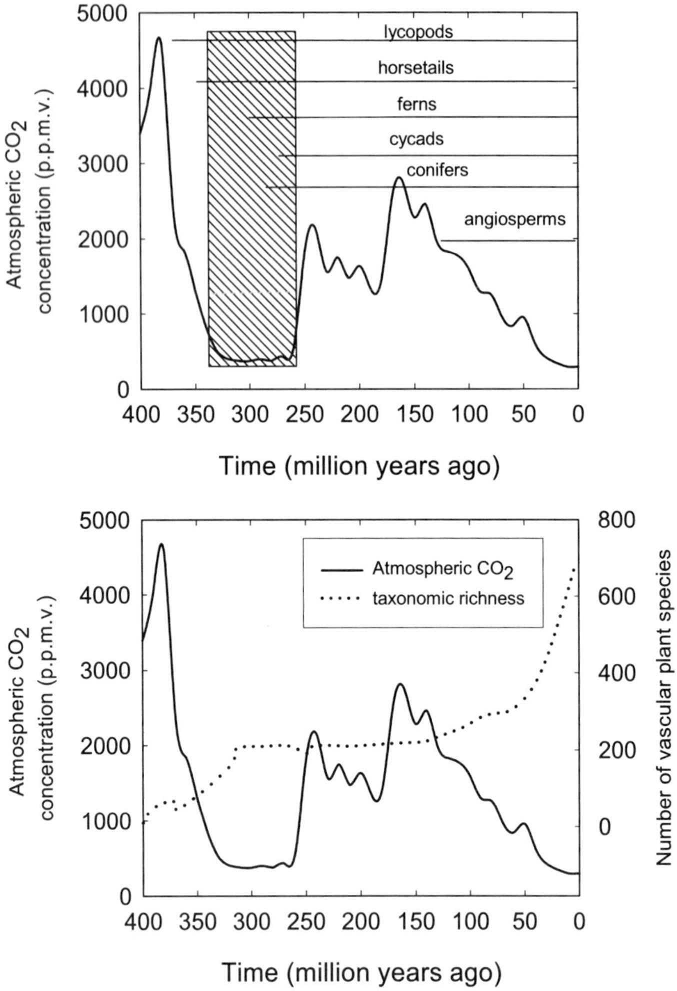

Figure 6.8. Changes in atmospheric CO_2 (Berner and Kothavala 2001) and the timing of the origination of major plant groups *(upper panel)* (Woodward 1998) and species richness of vascular land plants per epoch *(lower panel)* (Knoll and Niklas 1987). Broken lines indicate the margin of uncertainty.

level there is clear circumstantial evidence that atmospheric CO_2 exerted quite a strong degree of control on the stomatal characteristics of vascular land plants, with major downstream consequences for the course of evolution of organelles (leaves) and whole plants. Our understanding of how, if at all, ancient CO_2 variations influenced speciation events is highlighted as an area of major uncertainty. Of course, it is possible that CO_2 plays no role in determining patterns of species diversification seen in vascular plants and that, instead, the patterns are driven by their own internal dynamics (Hewzulla et al. 1999).

Central to the interpretation of the plant fossil record, the overarching framework of Phanerozoic atmospheric CO_2 history, itself a topic of research, is under continual revision. Moreover, a strong message to emerge from this review is that identifying evolutionary innovation driven by CO_2 changes depends on our understanding of how contemporary plants respond to CO_2. So it seems that as we better characterize the physiological mechanisms underpinning the observed effects of atmospheric CO_2 on modern plants and vegetation, we will be better placed to take a brighter look back into the dim geologic past.

Acknowledgments. I thank Colin Osborne for helpful comments on the manuscript and the editors for the invitation to speak at the Snowbird meeting. I gratefully acknowledge funding through a Royal Society University Research Fellowship and the Leverhulme Trust.

References

Algeo, T.J., and S.E. Scheckler. 1998. Terrestrial-marine teleconnections in the Devonian: Links between the evolution of land plants, weathering processes, and marine anoxic events. *Philosophical Transactions of the Royal Society* B353:113–30.

Barrett, P.M., and K.J. Willis. 2001. Did dinosaurs invent flowers? Dinosaur-angiosperm coevolution revisited. *Biological Reviews* 76:411–47.

Beerling, D.J. 2002. Palaeoclimatology: CO_2 and the end-Triassic mass extinction. *Nature* 415:386–87.

Beerling, D.J., and R.A. Berner. 2000. Impact of a Permo-Carboniferous high O_2 event on the terrestrial carbon cycle. *Proceedings of the National Academy of Sciences, USA* 97:12428–32.

———. 2002. Biogeochemical constraints on the Triassic-Jurassic carbon cycle event. *Global Biogeochemical Cycles* 16:doi/10.1029/2001GB001637.

Beerling, D.J., J.A. Lake, R.A. Berner, L.J. Hickey, D.W. Taylor, and D.L. Royer. 2002. Carbon isotope evidence implying high O_2/CO_2 ratios in the Permo-Carboniferous atmosphere. *Geochemica et Cosmochimica Acta* 66:3757–67.

Beerling, D.J., C.P. Osborne, and W.G. Chaloner. 2001. Evolution of leaf-form in land plants linked to atmospheric CO_2 decline in the Late Palaeozoic era. *Nature* 410:352–54.

Beerling, D.J., and F.I. Woodward. 1997. Changes in land plant function over the Phanerozoic: reconstructions based on the fossil record. *Botanical Journal of the Linnean Society* 124:137–53.

———. 2001. *Vegetation and the terrestrial carbon cycle. Modelling the first 400 million years.* Cambridge: Cambridge University Press.

Berner, R.A. 2001. Modeling atmospheric O_2 over Phanerozoic time. *Geochimica et Cosmochimica Acta* 65:685–94.

Berner, R.A., and Z. Kothavala. 2001. GEOCARB III: A revised model of atmospheric CO_2 over Phanerozoic time. *American Journal of Science* 301:182–204.

Berner, R.A., S.T. Petsch, J.A. Lake, D.J. Beerling, et al. 2000. Isotopic fractionation and atmospheric oxygen: Implications for Phanerozoic O_2 evolution. *Science* 287: 1630–33.

Cerling, T.E. 1999. Paleorecords of C_4 plants and ecosystems. In *Plant Biology,* ed. R.F. Sage, and R.K. Monson, 445–69. San Diego: Academic Press.

Cerling, T.E., J.M. Harris, B.J. MacFadden, M.G. Leakey, et al. 1997. Global vegetation change through the Miocene/Pliocene boundary. *Nature* 389:153–58.

Chaloner, W.G., and A. Sheerin. 1979. Devonian Macrofloras. In *The Devonian system,* ed. M.R. House, C.T. Scrutton, and M.G. Bassett (special paper in *Palaeontology* no. 23:145–61). London: Palaeontological Society.

Cornette, J.L., B.S. Lieberman, and R.H. Goldstein. 2002. Documenting a significant relationship between macroevolutionary origination rates and Phanerozoic pCO_2 levels. *Proceedings of the National Academy of Sciences, USA* 99:7832–35.
Crowley, T.J., and R.A. Berner. 2001. CO_2 and climate change. *Science* 292:870–72.
Edwards, D. 1998. Climate signals in Palaeozoic land plants. *Philosophical Transactions of the Royal Society* B353:141–57.
Ehleringer, J.R., T.E. Cerling, and B.R. Helliker. 1997. C_4 photosynthesis, atmospheric CO_2 and climate. *Oecologia* 112:285–99.
Ehleringer, J.R., R.F. Sage, L.B. Flanagan, and R.W. Pearcy. 1991. Climate change and the evolution of C_4 photosynthesis. *Trends in Ecology and Evolution* 6:95–99.
Falcon-Lang, H.J. 2000. Fire ecology of the Carboniferous tropical zone. *Palaeogeography, Palaeoclimatology, Palaeoecology* 164:339–55.
Farquhar, G.D., and S.C. Wong. 1984. An empirical model of stomatal conductance. *Australian Journal of Plant Physiology* 11:191–210.
Farquhar, G.D., J.R. Ehleringer, and K.T. Hubick. 1989. Carbon isotope discrimination and photosynthesis. *Annual Reviews of Plant Physiology and Plant Molecular Biology* 40:503–37.
Fowell, S.J., B. Cornet, P.E. Olsen. 1994. Geologically rapidlate Triassic extinctions: Palynological evidence from the Newark supergroup. In *Pangea: Paleoclimates, tectonics, and sedimentation during accretion, zenith, and breakup of a supercontinent,* ed. G.D. Klein, pp. 197–206. Colorado: Geological Society of America Species Paper 288.
Gale, J., S. Rachmilevitch, J. Reuveni, and M. Volokita. 2001. The high oxygen atmosphere toward the end-Cretaceous: A possible contributing factor to the K/T boundary extinctions and to the emergence of C_4 species. *Journal of Experimental Botany* 52: 801–809.
Haase, P., F.I. Pugnaire, S.C. Clark, and L.D. Incoll. 1999. Diurnal and seasonal changes in cladode photosynthetic rate in relation to canopy age structure in the leguminous shrub *Retama sphaerocarpa. Functional Ecology* 13:640.640–649.
Harris, T.M. 1937. The fossil flora of Scoresby Sound, East Greenland. Part 5. Stratigraphic relations of the plant beds. *Meddelelser Om Grønland* 112:1–111.
Hewzulla, D., M.C. Boulter, M.J. Benton, and J.M. Halley. 1999. Evolutionary patterns from originations and mass extinctions. *Proceedings of the Royal Society* B354:463–69.
Hibberd, J.M., and W.P. Quick. 2002. Characteristics of C_4 photosynthesis in stems and petioles of C_3 flowering plants. *Nature* 415:451–54.
Jones, T.P. 1994. ^{13}C enriched lower Carboniferous fossils from Donegal, Ireland: Carbon isotope constraints on taphonomy, diagensis and palaeoenvironment. *Review of Palaeobotany and Palynology* 97:39–50.
Keeley, J.M., C.B. Osmond, and J.A. Raven. 1984. *Stylites* a vascular land plant without stomata absorbs CO_2 via its roots. *Nature* 310:694–95.
Knoll, A.H., and K.J. Niklas. 1987. Adaptation, plant evolution and the fossil record. *Review of Palaeobotany and Palynology* 50:127–49.
Kuypers, M.M.M., R.D. Pancost, and J.S.S. Damste. 1999. A large and abrupt fall in atmospheric CO_2 concentration during Cretaceous times. *Nature* 399:342–45.
LaDeau, S.L., and J.S. Clark. 2001. Rising CO_2 levels and the fecundity of forest trees. *Science* 292:95–98.
Lenton, T.M. 2001. The role of land plants, phosphorus weathering and fire in the rise and regulation of atmospheric oxygen. *Global Change Biology* 7:613–29.
Lundblad, A.B. 1959. Studies in the Rhaeto-Liassic floras of Sweden. II. Ginkgophyta from the mining district of N.W. Scania. *Kungliga Svenska Vetenskapsakademiens Handlingar* 6:1–18.
McElwain, J.C., D.J. Beerling, and F.I. Woodward. 1999. Fossil plants and global warming at the Triassic-Jurassic boundary. *Science* 285:1386–90.

McElwain, J.C., and W.G. Chaloner. 1995. Stomatal density and index of fossil plants track atmospheric carbon dioxide in the Palaeozoic. *Annals of Botany* 76:389–95.

McLoughlin, S., A.N. Drinnan. 1997. Revised stratigraphy of the Permian Bainmedart coal measures, northern Prince Charles Mountains, East Antarctica. *Geological Magazine* 134:335–53.

Osborne, C.P., D.J. Beerling, B.H. Lomax, and W.G. Chaloner. 2004. Biophysical constraints on the origin of leaves inferred from the fossil record. *Proceedings of the National Academy of Sciences, USA* 101:10360–10362.

Osborne, C.P., W.G. Chaloner, and D.J. Beerling. 2002. Falling atmospheric CO_2 levels drove megaphyll leaf evolution. In *Evolution of plant physiology,* ed. A.R. Hemsley and I. Poole. London: Academic Press.

Parrish, J.T. 1993. Climate of the supercontinent Pangea. *Journal of Geology* 101:215–33.

Pant, D.D., and K.L. Gupta. 1968. Cuticular structure of some Indian lower Gondwana species of *Glossopteris* Brongniart. Part 1. *Palaeontographica* B132:45–81.

———. 1971. Cuticular structure of some Indian lower Gondwana species of *Glossopteris* Brongniart. Part 2. *Palaeontographica* B132:130–52.

Phillips, O.L., and A.H. Gentry. 1994. Increasing turnover through time in tropical forests. *Science* 263:954–58.

Phillips, T.L., and W.A. DiMichele. 1992. Comparative ecology and life-history biology of arborescent lycopsids in late Carboniferous swamps of Euramerica. *Annals of the Missouri Botanical Gardens* 79:560–88.

Prentice, I.C. et al. 2001. The carbon cycle and atmospheric carbon dioxide. In *Climate change 2001: The scientific basis,* ed. J.T. Houghton, Y. Ding, D.J. Griggs, M. Noguer, P.J. van der Linden, X. Dai, K. Maskell, and C.A. Johnson, 183–237. Cambridge: Cambridge University Press.

Raven, J.A. 1993. The evolution of vascular land plants in relation to quantitative function of dead-water conducting cells and of stomata. *Biological Reviews* 68:337–63.

Raven, J.A., and D. Edwards. 2001. Roots: evolutionary origins and biogeochemical significance. *Journal of Experimental Botany* 52:381–401.

Royer, D.L. 2001. Stomatal density and index as indicators of paleoatmospheric CO_2 concentration. *Reviews of Palaeobotany and Palynology* 114:1–28.

Royer, D.L., R.A. Berner, and D.J. Beerling. 2001. Phanerozoic atmospheric CO_2 change: Evaluating geochemical and paleobiological approaches. *Earth Science Reviews* 54: 349–92.

Sage, R.F. 2001. Environmental and evolutionary preconditions for the origin and diversification of the C_4 photosynthetic syndrome. *Plant Biology* 3:202–13.

Stewart, W.N., and G.W. Rothwell. 1993. *Palaeobotany and the evolution of plants.* New York: Cambridge University Press.

Tanner, L.H., J.F. Hubert, B.P. Coffey, and D.P. McInerney. 2001. Stability of atmospheric CO_2 levels across the Triassic/Jurassic boundary. *Nature* 411:675–77.

Thomas, B.A. 1970. Epidermal studies in the interpretation of *Lepidodendron* species. *Palaeontology* 13:29–34.

Valladares, F., and R.W. Pearcy. 1997. Interactions between water stress, sun-shade acclimation, heat tolerance and photoinhibition in the sclerophyll *Heteromeles arbutifolia. Plant, Cell, and Environment* 20:25–36.

Veizer, J. et al. 1999. $^{87}S/^{86}Sr$, $\delta^{13}C$ and δ^{18} O evolution of Phanerozoic seawater. *Chemical Geology* 161:59–88.

Visscher, H., and W.A. Brugman. 1981. Ranges of selected palynofloras in the Alpine Triassic of Europe. *Reviews of Palaeobotany and Palynology* 34:115–28.

von Caemmerer, S. 2002. *Biochemical models of leaf photosynthesis.* Australia: Commonwealth of Scientific Industrial Research Organization publishing.

Voznesenskaya, E., V.R. Franceschi, O. Kiirats, H. Freltag, and G.E. Edwards. 2001.

Kranz anatomy not essential for terrestrial C_4 plant photosynthesis. *Nature* 414:543–46.
Ward, J.K.W., J. Antonovics, R.B. Thomas, and B.R. Strain. 2000. Is atmospheric CO_2 a selective agent on model C_3 annuals? *Oecologica* 123:330–41.
Willis, K.J., and J.C. McElwain. 2002. *The evolution of plants.* Oxford: Oxford University Press.
Woodward, F.I. 1987. Stomatal numbers are sensitive to increases in CO_2 from pre-industrial levels. *Nature* 327:617–18.
———. 1998. Do plants need stomata? *Journal of Experimental Botany* 49:471–80.
Wright, V.P., and S.D. Vanstone. 1991. Assessing the carbon dioxide content of ancient atmospheres using palaeo-calcretes: Theoretical and empirical constraints. *Journal of the Geological Society of London* 148:945–47.

7. Cretaceous CO_2 Decline and the Radiation and Diversification of Angiosperms

Jennifer C. McElwain, K.J. Willis, and R. Lupia

7.1 Introduction

Determining how a projected doubling of atmospheric CO_2 concentration by the end of this century (IPCC 2001) will influence species composition and biodiversity remains a major scientific and political challenge (Loreau et al. 2001). Long-term CO_2 experiments using FACE (free air carbon dioxide enrichment) now yield invaluable results on how forests and crops/grassland vegetation respond in terms of their biomass allocation, productivity, light use, water use, and nutrient use efficiency in elevated CO_2 (500–700 ppm) (DeLucia et al. 1999; McLeod and Long 1999; Oren 2001). Furthermore, related experimental approaches are increasing our understanding of plant reproductive responses (Hussain, Kubiske, and Connor 2001) and competitive interactions (Bazzaz et al. 1995) in high CO_2 conditions. However, the current spatial and temporal scale of these experiments (<10 years) are limited and cannot yet take into consideration the floristic changes that may result from continuing increases in anthropogenic CO_2 at the ecosystem or biome level. Such floristic and macroecological data are readily available from the plant fossil record on timescales of millions of years and on spatial scales spanning whole continents. In this chapter we investigate the effects of long-term fluctuations in atmospheric CO_2 concentration during the Cretaceous period (145–65 Ma) (Tajika 1998; 1999) on patterns of ecological dominance and taxonomic diversity in Cretaceous fossil floras. In particular, we test the hypothesis of Teslenko (1967) and Robinson

(1994) that "CO_2 starvation" during the Cretaceous contributed to the taxonomic diversification and ecological radiation of angiosperms.

7.2 Background

7.2.1 Angiosperm Radiation and Diversification

Unequivocal evidence for early angiosperms in the geological record is based on fossil pollen, which appears for the first time in lower Cretaceous (Valanginian and Hauterivian; ~140–130 Ma) sediments of Israel and Europe (Hughes 1994; Brenner 1996) with additional supporting evidence from fossil flowers (Friis, Crane, and Pedersson 1999), fruits (Dilcher 1989), and leaves (Hughes 1994) by the Barremian and Aptian (127–112 Ma) (Willis and McElwain 2002). However, the time of origin of angiosperms remains controversial due to the recovery of Triassic and Jurassic fossils that are enigmatic (Cornet 1986) or contain some, but not all, of the characteristics necessary for assignment to angiosperms (Cornet 1993; Cornet and Habib 1992). More recently, *Archaefructus liaoningensis*, a flowering plant fruiting axis, was described from the Yixian Formation in the Liaoning Province of China (Sun et al. 1998). Unfortunately the exact chronostratigraphic age of the formation is still under debate, with age assignments ranging from the early Cretaceous (Barrett 2000) to the late Jurassic (Sun et al. 1998; 2002).

Angiosperms then underwent a rapid ecological radiation and taxonomic diversification starting in the Aptian-Albian (121–99 Ma) into the Campanian (83.5–71.3 Ma) (Crane and Lidgard 1989; Lupia, Lidgard, and Crane 1999). An explosive angiosperm radiation, originating in the tropics (Crane and Lidgard 1989), is supported by an increase in continent-level species richness from average values of $<5\%$ to $>40\%$ within approximately 40 million years (Lidgard and Crane 1988; Crane and Lidgard 1990; Lupia, Lidgard, and Crane 1999) (Fig. 7.1).

Similarly, marked increases in the relative abundance of angiosperm fossil pollen in North American (Lupia, Lidgard, and Crane 1999) and Australian (Nagalingum et al. 2002) fossil floras during the same interval are indicative of the rise of angiosperms to ecological dominance. Earlier analyses of large-scale floristic trends through the Cretaceous indicated that angiosperms competitively replaced all gymnosperm groups (Gothan and Remy 1957). However, this is not supported by more recent analyses demonstrating a more selective replacement of some but not all gymnosperm groups (Lidgard and Crane 1988). In particular, a coeval decline in the relative abundances of free-sporing plants (broadly, pteridophytes) in lower latitudes (below 42°N) and Cheirolepidaceae (an extinct group of arid-adapted conifers) in North America as the angiosperms radiated suggests that these groups were competitively replaced by angiosperms (Lupia, Lidgard, and Crane 1999).

In contrast with these patterns of decline and extinction among nonflowering

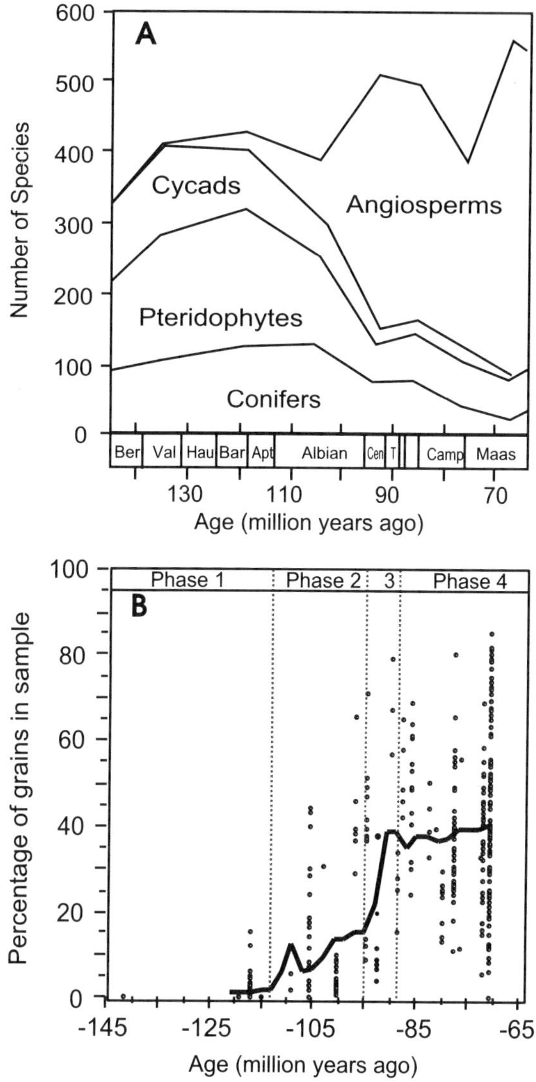

Figure 7.1. **(A)** Total species diversity changes during the Cretaceous illustrating the taxonomic radiation of angiosperms (redrawn from Lidgard and Crane 1988); **(B)** 7.5 million year moving average of angiosperm within flora relative abundance (providing an indication of ecological dominance) during the Cretaceous (redrawn from Lupia et al. 1999).

plant groups, fossil ephedroid (Gnetales) pollen diversity in lower latitudes increased synchronously with that of angiosperms (Crane and Lidgard 1989). This striking co-radiation of angiosperms and low latitude Gnetales during the early and middle Cretaceous led Crane and Lidgard (1990) to suggest that the evolution of both groups may have been strongly influenced or triggered by similar biological or abiotic factors. Unlike the angiosperms, however, the Gnetales did not continue to diversify and radiate into the late Cretaceous but underwent a precipitous decline in post-Cenomanian time (<90 Ma).

In summary, the Cretaceous period is characterized as a time of major floristic turnover and revolution, which in the case of North America (Lupia, Lidgard, and Crane 1999) and, in part, Australia (Nagalingum et al. 2002) can be divided into four distinct phases. In North America, the first phase, spanning from the Jurassic/Cretaceous boundary to the Aptian /Albian boundary (~145–112 Ma), gymnosperms (including conifers, cycads, bennettites, seed-ferns, and Ginkgos) and pteridophytes (all free-sporing plants) dominate the floristic diversity of Northern Hemisphere vegetation. The second phase is characterized by a gradual radiation and diversification of angiosperms between the early Aptian and mid to late Cenomanian (~95–93.5 Ma), followed by a third phase marked by a rapid rise to floristic and ecological dominance in the early Turonian (~93 Ma) to late Santonian (~84 Ma). A fourth macroecological phase is distinguished by an apparent stabilization of biome-level floristic composition to a new equilibrium with angiosperms making up over 40% of the floristic diversity and relative abundance; gymnosperms and pteridophytes shared the remaining 50 to 60%. This phase, however, is most probably an artifact of the data analysis, as the data series currently truncates at the boundary between the Cretaceous and the Tertiary. Extension of this type of analysis into the Tertiary would most likely reveal a continued radiation of angiosperms in the early Tertiary as demonstrated by previous largescale floristic studies (Lidgard and Crane 1990; Niklas, Tiffney, and Knoll 1983) and potentially by a second pulse of radiation in the latest Paleocene and earliest Eocene–a time when angiosperm-dominated forests extended beyond 60° paleolatitude in both poles (Upchurch, Otto-Bliesner, and Scotese 1999; Willis and McElwain 2002) and again in the Miocene, with the expansion of grass dominated ecosystems (Jacobs, Kingston, and Jacobs 1999, Willis and McElwain 2002). Owing to the lack of quantitative analysis of floral-level palynological and/or macrofossil records in the Tertiary, this chapter will focus on the Cretaceous record of angiosperm evolution only.

7.2.2 Environmental Trends and Events in the Cretaceous

The Cretaceous was the last entire period in Earth's history when a greenhouse climatic mode prevailed (Frakes 1999). Paleontological and geological indicators demonstrating significantly higher terrestrial and marine paleotemperatures than today are abundant (see Frakes 1999 for a review). Much higher paleotemperatures in the high polar latitudes are particularly striking in the Cretaceous. Both high greenhouse gas concentration (Arthur, Dean, and Schlanger 1985) and altered ocean circulation patterns (Herman and Spicer 1996) have been invoked to explain this phenomenon of Cretaceous polar warmth. However, despite a complete lack of evidence for polar ice throughout the entire period, the Cretaceous is no longer considered a period of uniformly warm climate (Barron and Washington 1984; Condie and Sloan 1998; Frakes 1999). Marked climatic variations, including short-term pulses of global cooling and warming, have been identified (Frakes 1999). In a recent climatic assessment of the Cretaceous based on a review of proxy evidence (e.g., Wolfe and Upchurch 1987; Parrish and Spicer 1988; Barrera et al. 1997), Frakes (1999) suggested that the early Cretaceous was generally cool, followed by a continuous warming through Albian to Cenomanian time, marked Cenomanian cooling, followed by even more marked Turonian warming, and then generally cooler global climates in the Santonian and Campanian (Fig. 7.2).

The Cretaceous period is also characterized as a period of substantial tectonic plate movements resulting in the final breakup of Pangea (Ziegler, Scotese, and Barrett 1982; Barron 1987; Scotese 1991). Most tectonic plate reconstructions show that breakup of major continental blocks—including Eurasia, Greenland, North America, South America, Africa, India, Australia, and Antarctica—had occurred by the end of the early Cretaceous (Ziegler, Scotese, and Barrett 1982; Barron 1987; Scotese 1991), but see Hay et al. (1999) for an alternative model, suggesting a late rather than early Cretaceous breakup. In particular, a period of rapid plate spreading was initiated in the Aptian (~124 Ma) until the Cenomanian (~83 Ma) (Sheridan 1997), resulting in changes in continental configuration and significantly increasing ocean crust production (Larson 1991; 1997). This period of enhanced ocean crust production and volcanic activity is thought to be primarily responsible for high global temperatures in the mid–Cretaceous, caused indirectly through increased volcanic release of greenhouse gases, such as CO_2 (Arthur, Dean, and Schlanger 1985) and possibly methane (Jahren et al. 2001). Large-scale changes in ocean circulation are also thought to have played an important role in the maintenance of high latitude polar warmth during the Aptian to Cenomanian interval (Herman and Spicer 1996; Zieglar 1998).

Three oceanic anoxic events (OAEs), characterized as intervals of enhanced organic carbon burial punctuated the Cretaceous period. These events are marked in the rock record by extensive deposits of black shale (Schlanger and Jenkyns 1976; Jenkyns 1980), which, based on stable carbon isotopic analysis, are interpreted to have resulted in major perturbations in the short-term (<1 million year) carbon cycle. Oceanic anoxic events are distinguished in the faunal record by major extinction events; although the causal mechanism or mechanisms for these events, as well as the extinctions that accompany them, are still hotly debated (Kaiho and Hasegawa 1994; Huber et al. 1999), there is a growing consensus that major perturbations in the composition of greenhouse gases (including methane and carbon dioxide) may be in some way involved (Kuypers, Pancost and Sinninghe-Damste 1999). Cretaceous OAE's occurred in the late Aptian to early Albian (OAE1), at the Cenomanian-Turonian boundary (OAE2)

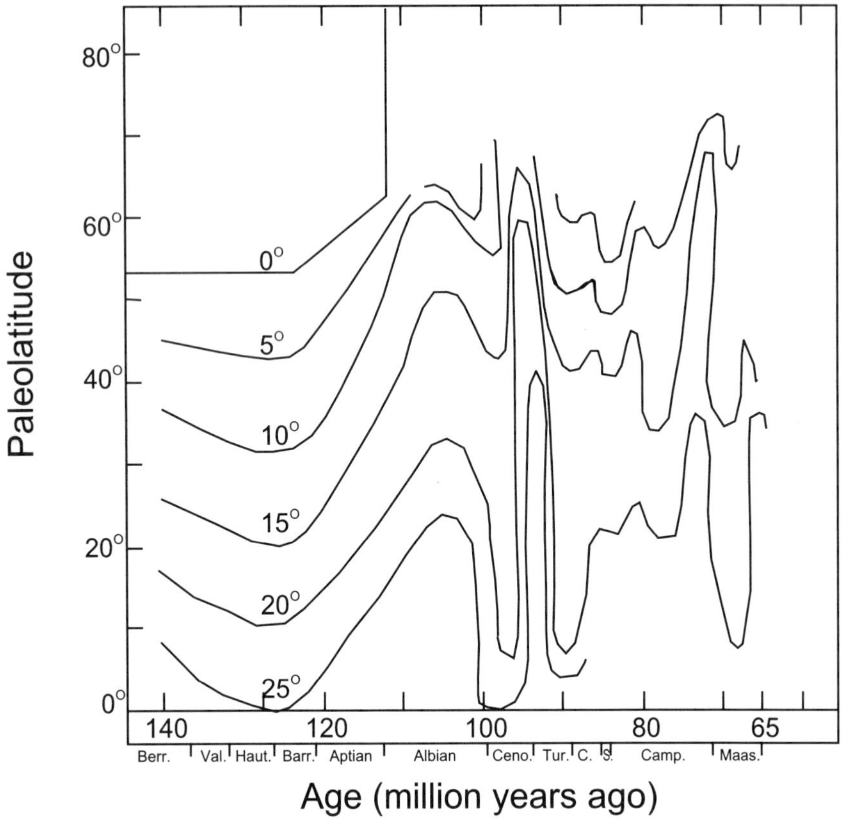

Figure 7.2. Cretaceous climate change (simplified from Frakes 1999) plotted as contours of marine paleotemperature on latitude versus age plot. Marked deviation of the paleotemperature contours into the higher latitudes indicate global warming intervals while deviations of the contours toward the lower latitudes provided an indication that cooler climates prevailed.

and during the Coniacian and early Campanian (OAE3) (Jenkyns 1980; Dean, Arthur, and Claypool 1986; Kauffman and Hart 1995; Arthur, Dean, and Pratt 1988). In summary, a combination of unusually high volcanic activity, large-scale changes in sea level and ocean circulation, with shorter term episodes of high organic carbon burial resulted in major atmospheric and climatic perturbations throughout the Cretaceous.

7.2.3 CO_2 Fluctuations and Trends in the Cretaceous

Biogeochemical models based on long-term processes influencing the carbon cycle (such as CO_2 production from volcanism and uptake from chemical weathering of silicate minerals, see Berner, this volume) predict that atmospheric CO_2

concentration underwent a long-term decline (Fig. 7.3) through the Cretaceous (Berner 1991; 1994; Berner and Kothavala 2001).

Estimates vary from 3.5 times to over 8 times preindustrial levels (300 ppm) for the early Cretaceous (Neocomian) and 1.8 to 4 times by the late Cretaceous (Senonian), according to the long-term carbon cycle models GEOCARB III (Berner and Kothavala 2001) and GEOCARB II (Berner 1994), respectively. In addition to solar and geological controls on the long-term carbon cycle, both models also take into account the role of such biological processes as the enhancement of chemical weathering by plants (see Berner this volume for further details).

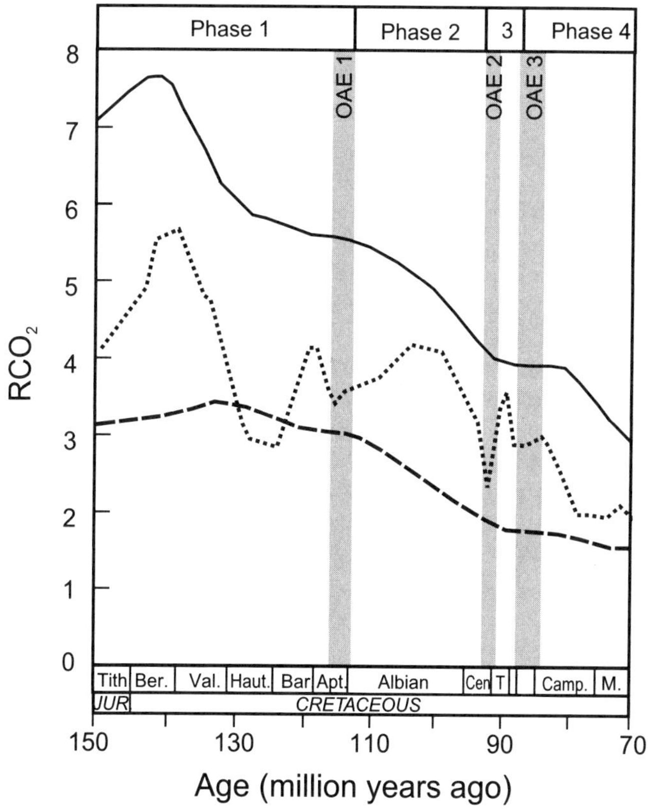

Figure 7.3. Compilation of Cretaceous CO_2 change and oceanic anoxic events. Solid (GEOCARB II, Berner 1994) and dashed (GEOCARB III, Berner and Kothavala 2001) lines represent estimates of CO_2 fluctuations from long-term models of the global carbon cycle. Dotted line represents CO_2 estimates from the carbon cycle model of Tajika (1998, 1999). Phase 1 to 4 represent distinct phases in angiosperms radiation and overall Cretaceous vegetation change as reflected in the data of Lupia et al. (1999) (see text for detailed description of phases).

It is noteworthy, however, that although the more recent and refined model, GEOCARB III, predicts lower concentrations in early and late Cretaceous CO_2 than does GEOCARB II, the general trends shown in both models are in good agreement and, in general, are well supported by independent proxy data (Chen et al. 2001; Robinson et al. 2002; Ekart et al. 1999). Long-term carbon cycle models do not take into account shorter scale processes influencing the carbon cycle, such as increased organic carbon burial at oceanic anoxic events, which may decrease concentrations of atmospheric CO_2 (Kuypers, Pancost, and Sinninghe-Damste 1999), or rapid methyle hydrate release, which hypothetically would increase CO_2 (Hesselbo et al. 2000; Jahren et al. 2001).

More recent advances in Cretaceous carbon cycle modeling now incorporate the influence of both long-term and short-term biogeochemical processes on atmospheric CO_2 (Tajika 1998; 1999). These higher resolution biogeochemical models predict a general pattern of declining CO_2 through the Cretaceous from an early Cretaceous high of nearly 6 times preindustrial levels (PIL) to concentrations in the region of 2 times PIL by the late Cretaceous (Tajika 1999), in good agreement with long-term CO_2 model estimates. Estimates of atmospheric CO_2 change from Tajika's model, as shown in Fig. 7.3, indicate relatively lower CO_2 levels in the Hauterivian (132–127 Ma), late Cenomanian (~95–93 Ma), and Campanian to Maastrichian (~85–65 Ma) in agreement with independent proxy CO_2 (Kuypers, Pancost, and Sinninghe-Damste 1999; Robinson et al. 2002; McElwain unpublished data) and climate (Frakes 1999) indicators for these times. Relatively higher CO_2 levels are estimated in the earliest Cretaceous (Tajika 1999). In the following sections we investigate how fluctuations in paleoatmospheric CO_2 concentration influenced the composition and macroecology of Cretaceous vegetation as reflected in North American fossil floras. In particular we present arguments and analyses in support of the original but as yet empirically untested hypotheses of Teslenko (1967) and Robinson (1994)—and noted by Beerling (1994) and Crane and Lidgard (1989)—that a reduction in atmospheric CO_2 concentration may have played an important role in the ecological radiation and taxonomic diversification of angiosperms (flowering plants) in the Cretaceous.

7.3 Material and Methods

Large-scale patterns of Cretaceous vegetation change were obtained from records of angiosperms, gymnosperms, and free-sporing plants and their relative floristic diversity and abundance in North American fossil floras (Fig. 7.4), compiled by Lupia, Lidgard, and Crane (1999).

This database contains pollen records from a total of 50 published papers incorporating 922 fossil pollen samples of relative floristic diversity (number of species in each group divided by total diversity, taken as an indicator of relative species richness) and 359 samples of relative abundance (number of specimens

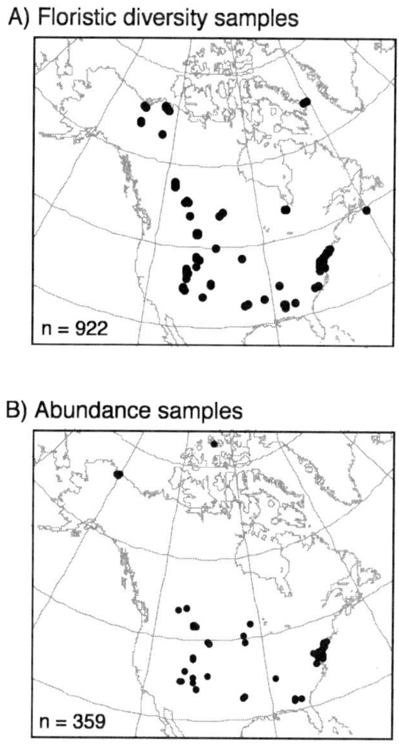

Figure 7.4. Maps of present-day locations of fossil pollen samples for **(A)** floristic diversity and **(B)** abundance in the North American Palynological Database (from Lupia, Lidgard, and Crane 1999, used with permission, The Paleontological Society).

of each group divided by number of specimens counted, taken as a measure of ecological dominance), spanning the entire Cretaceous period and covering 29°N to 84°N paleolatitude and 38° of paleolongitude. Further details on this database can be obtained from Lupia, Lidgard, and Crane (1999). In combination, the within-flora floristic diversity and abundance data serve as robust proxies for patterns of structure and ecological dominance within mainly local vegetation (see Lupia, Lidgard, and Crane 1999) thus enabling an assessment of the impact of CO_2 change on floristic composition through the Cretaceous period. Raw relative abundance and diversity data were then averaged per CO_2 concentration in order to investigate the role of CO_2 change in biome-level vegetation changes throughout the study interval. Long-term fluctuations in paleoatmospheric CO_2 concentrations were obtained from carbon cycle model predictions of Tajika (1998; 1999), GEOCARB II model (Berner 1994), and GEOCARB III (Berner and Kothavala 2001). Total relative floristic diversity and abundance data were regressed against atmospheric CO_2 concentration using simple linear and nonlinear correlation-regression analysis in order to evaluate the potential role of CO_2 change in vegetation dynamics and particularly in the angiosperm radiation and diversification.

7.4 Results and Discussion

7.4.1 CO_2 and Cretaceous Vegetation Composition

Our results indicate that CO_2 fluctuations may have played an important role in both the relative diversity (Table 7.1, Fig. 7.5D,E,F) and abundance (Fig. 7.5A,B,C) of angiosperms, gymnosperms, and pteridophytes in Cretaceous fossil floras.

At the broadest level of comparison, it is noteworthy that the termination or initiation, or both, of all four macroecological phases in North American Cretaceous vegetation change are temporally coincident with three major Cretaceous oceanic events (see Fig. 7.3), each of which was associated with a carbon cycle perturbation and likely CO_2 and climatic fluctuations.

Significant inverse correlations are observed between atmospheric CO_2 and angiosperm species richness and abundance at both the local level (within-flora floristic diversity) and the biome level (mean North American relative diversity), as shown in Fig. 7.6.

In comparison, significant positive correlations were observed between North American pteridophyte species richness and atmospheric CO_2 and also gymnosperm species abundance and atmospheric CO_2 (see Fig. 7.5, 7.6). The significance of both correlations is, however, largely dependent on the patterns of change in gymnosperm abundance and pteridophyte richness in lower latitudes

Table 7.1. Relationship between the relative abundance and diversity of angiosperms, gymnosperms, and pteridophytes with alternative carbon cycle model estimates of CO_2 change during the Cretaceous

	Carbon Cycle Models			
Relative within flora floristic diversity (Lupia et al. 1999)	Geocarb I (Berner 1991)	Geocarb II (Berner 1994)	Geocarb III (Berner, Kothavala 2001)	Tajika (1999)
Angiosperms (n = 938)	(−) ($r^2 = 0.5803$) ***	(−) ($r^2 = 0.6142$) ***	(−) ($r^2 = 0.6467$) ***	(−) ($r^2 = 0.6269$) ***
Gymnosperms (n = 938)	(0) ($r^2 = 0.095$) n.s.	(0) ($r^2 = 0.1562$) n.s.	(0) ($r^2 = 0.152$) n.s.	(0) ($r^2 = 0.0402$) n.s.
Pteridophytes (n = 938)	(+) ($r^2 = 0.3596$) *	(+) ($r^2 = 0.1562$) n.s.	(+) ($r^2 = 0.3617$) *	(+) ($r^2 = 0.4033$) *

r^2 = correlation coefficient

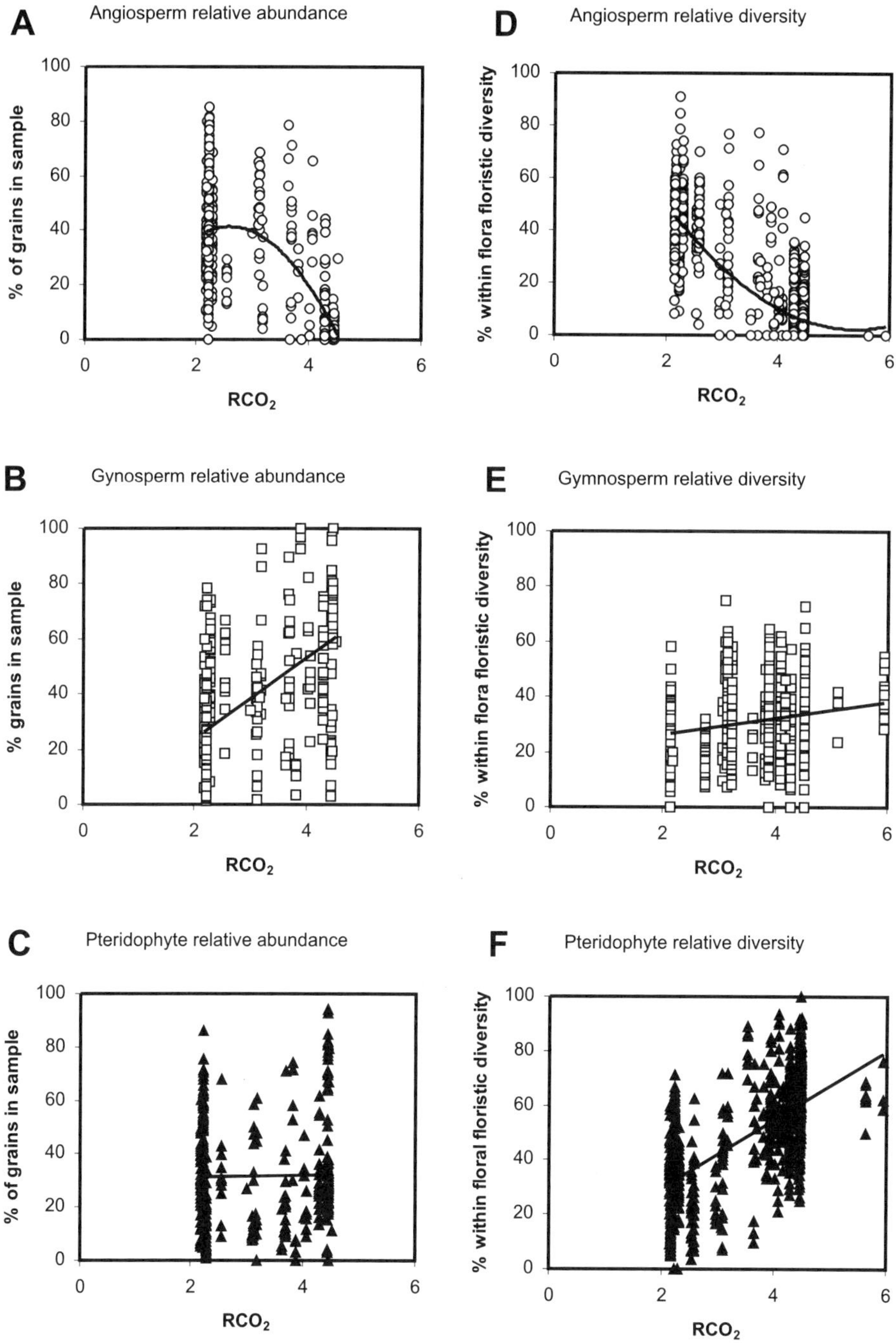

Figure 7.5. Relationship between relative abundance (dominance) and Cretaceous atmospheric CO_2 concentration for: **(A)** Angiosperms [$y = 4.1328x^2 - 44.503x + 122.1$ $r^2 = 0.6269$], **(B)** Gymnosperms [$y = 2.9474x + 20.369$ $r^2 = 0.0402$], and **(C)** Pteridophytes [$y = 12.547x + 4.5817$ $r^2 = 0.4033$]. Relationship between relative diversity (richness) and Cretaceous atmospheric CO_2 concentration for: **(D)** Angiosperms [$y=3.5689x^2 + 121.14$ $r^2 = 0.6513$], **(E)** Gymnosperms [$y= 2.0367x + 23.92$ $r^2 = 0.0466$], and **(F)** Pteridophytes [$y= -3.1461x^2 + 36.413x–40.195$ $r^2 = 0.5831$].

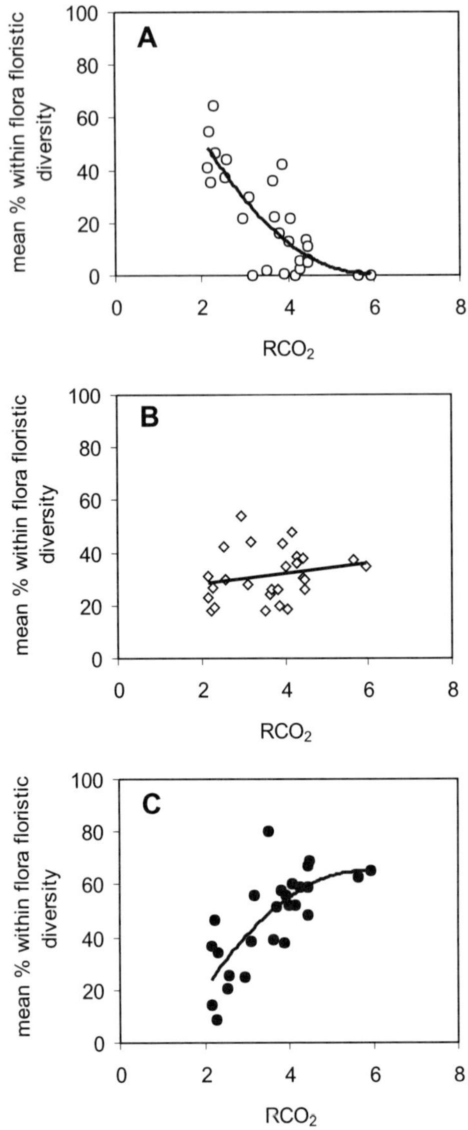

Figure 7.6. Relationship between mean relative diversity and Cretaceous atmospheric CO_2 concentration for: **(A)** Angiosperms [$y = -10.646x^2 + 55.092x–29.996$ R2 = 0.4506], **(B)** Gymnosperms [$y = 0.3769x + 30.553$ R2 = 0.0004], and **(C)** Pteridophytes [$y = 14.656x–5.5592$ $r_2 = 0.3188$].

of North America (below 42°N), suggesting that climatic and/or biogeographic factors, in addition to declining atmospheric CO_2, may have influenced the shifting pattern of gymnosperm diversity and ecology through the Cretaceous.

It has been suggested that key functional types rather than total biodiversity per se may exert the most important control on ecosystem functioning, including productivity, nutrient and water recycling, and so forth (Huston et al. 2000; Loreau et al. 2001). If this is the case, our observation that atmospheric CO_2 may have influenced the relative composition of angiosperms, gymnosperms, and pteridophytes in Cretaceous vegetation, particularly in the lower latitudes, may have extremely important implications for predicting the impact of future increases in atmospheric CO_2 on ecosystem functioning via its effect on large-scale compositional shifts in terrestrial vegetation.

Our results support the original but empirically untested hypotheses of Teslenko (1967) and Robinson (1994). These authors noted that a decrease in atmospheric CO_2 concentrations during the Cretaceous might have in some way triggered the remarkably rapid adaptive radiation of flowering plants. Although Teslenko (1967) had no way of testing his hypothesis, because no estimates of Cretaceous paleo-CO_2 concentration were available at that time, he speculated that angiosperms might have originated in higher elevations where they would have been accustomed and therefore pre-adapted to lower CO_2 partial pressures. The theory of high-elevation angiosperm origin (Axelrod 1952) has since been refuted due to both a lack of paleobotanical evidence (Doyle, Jardiné, and Doerenkamp 1982) and theoretical support from plant ecophysiological studies. Theoretical considerations, for instance, have shown that decreasing CO_2 partial pressure with elevation has little impact on overall carboxylation potential below 3000 meters, due to the counterbalancing effects of decreased O_2 partial pressure with elevation, resulting in reduced photorespiratory demand (Terashima et al. 1995) and the increased diffusivity of gases and hence conductance with elevation (Gale 1972).

By means of their rapid speciation rates—average speciation rates of > 0.35 Δs / Ma (new species per taxon per Ma) compared with <0.2 Δs / Ma for older mesophyte groups (Niklas, Tiffney, and Knoll 1983)—flowering plants, if not pre-adapted, likely would have been capable of rapidly evolving a suite of highly adapted vegetative characteristics, such as planate leaves with reticulate venation, fine-tuned stomatal regulation, highly efficient conductive tissue (vessels) that enabled them to optimize their physiology to the lowest CO_2 concentrations that had probably existed since the Carboniferous, approximately 100 million years earlier. Additionally, the almost unique ability of angiosperms to harness host-specific relationships with coevolving insects, may also have enabled angiosperm to survive as rarer species in isolated habitats, thereby reducing the high energetic cost associated with the manufacture of anti-herbivory toxins (Leigh and Vermeij 2002). Leaves that require a high investment of anti-herbivore defense compounds are more energetically expensive and therefore tend to have a longer leaf life span, which in turn is associated with lower maximum stomatal conductance and photosynthetic potential (Reich, Walters, and Ellsworth 1997)—traits that would not be favored under declining atmospheric CO_2.

7.4.2 Why Would a Decline in Atmospheric CO_2 Favor the Radiation and Diversification of Angiosperms?

7.4.2.1 Evidence from Comparative Plant Ecophysiology

From a photosynthetic perspective, although RUBISCO of most living angiosperm taxa today is saturated at CO_2 concentrations above 3 × PIL (preindustrial CO_2 level [PIL] taken as 300 ppm), i.e., ~900 ppm, it must be remembered that all gymnosperm and pteridophyte lineages in the early Cretaceous and the as yet unknown seed plant ancestor of basal angiosperms would have been photosynthetically adapted to much more elevated CO_2 concentrations (~>1500 ppm). Assuming therefore that Tajika's model estimates of paleo-CO_2 concentration are broadly correct, the sharp decline in atmospheric CO_2 concentrations from 5 × PIL in the Berriasian (144–137 Ma) to levels less than 3 times PIL by the Hauterivian (132–127 Ma) and perhaps as low as 1.67 times PIL by late Cenomanian time (Kuypers, Pancost and Sinninghe-Damste 1999) would have strongly selected plant anatomical, morphological, and physiological characteristics that enhanced water supply and improved water-use efficiency, while at the same time maintaining plant carbon balance. Plants maintain optimum photosynthetic efficiency by setting a constant CO_2 gradient of approximately 30% between the atmosphere and intercellular leaf space by increasing or decreasing their stomatal conductance (g_{max}) as CO_2 concentration decreases or increases, respectively (von Caemmerer and Evans 1991). Very high CO_2 gradients between atmosphere and leaf can be maintained with ease when atmospheric CO_2 concentration are very high, thereby enabling plants to improve their water-use efficiency (WUE) by either occluding the stomatal pore with papillae, decreasing stomatal density, or opening stomata only partially.

Plant stomata are kept open longer or wider, and/or stomatal number (density) increase (Beerling, McElwain, and Osborne 1998), when the concentration of atmospheric CO_2 decreases in order to maintain an adequate CO_2 gradient between the atmosphere and leaf. As a direct consequence of this, evaporative demand is increased and greater amounts of water are lost through transpiration. In a sense, therefore, reduced atmospheric CO_2 concentration imposes physiological drought (Street-Perrott et al. 1997). Significant long-term reductions in atmospheric CO_2 through the Cretaceous from typical levels of greater than 5 times PIL (Berner 1994; Ekart et al. 1999; McElwain 1998; McElwain, Beerling, and Woodward 1999), which had existed for the preceding approx 100 million years during the Triassic and Jurassic, would therefore select strongly for anatomical adaptations that enabled efficient transport of water to plant tissues and organs rapidly loosing water through increased evaporative demand.

The evolution of vessels in angiosperms from tracheids of a gymnospermous ancestor represents such an adaptation that would have readily enabled increasing conductive efficiency of xylem tissue as a whole (Table 7.2; Fig. 7.7).

Table 7.2. Generalized anatomical and morphological trends in the evolution of vessels in angiosperms from tracheids of a gymnospermous ancestor

Xylem Type	Type of bordered pit typically found in lateral xylem wall	Perforation plate	Taxa	First recorded occurrence of clade in fossil record	RCO_2
Simple tracheids (A) & (B) with circular bordered pits	Circular bordered pits of variable size and pit boarder thickness, complete pore membrane	Absent	Coniferales, Ginkgoales, Cycadales, Bennettittales	NA- predominant in Triassic and Jurassic (248 to 144 Ma)	>4 × PIL
Tracheids (C) with scalariform (ladder like) bordered pits on end walls only	Lateral wall bordered pits circular and alternately arranged	Absent	Winteraceae (eg Belliolum, Bubbia, Drimys, Tahkajania, Tasmania & Zygognum)	(121–99 Ma, Aptian-Albian)	3.62 to 4.43 × PIL
Tracheids (D) with scalariform bordered pits	Pits elliptical along horizontal axis pit membranes porous	Absent	[1]Trochodendron, [2]Tetracetron,	[1,2](118 Ma, Aptian),	4.47 × PIL
Vessels members with scalariform (E) or porous perforation plates of distinct morphology from some lateral pits	Some bordered pits of distinct morphology and porosity from perforation plates i.e. Nelumbo and some Nymphales lateral pits non porous	Present	[1]Chloranthaceae, [2]Illiciaceae [3]Nelumbo [1]Cornalean clade, [2]Ericalean clade, [3]Betulaceae, some [4]Saxifragaceae, [5]Caprifoliaceae,	[1](112–99 Ma, Albian), [2](? Aptian-Albian), [3](100 Ma, Albian), [4](69.5 Ma, Maastrichtian), [5](89.5 Ma, Turonian), [6](84 Ma, Santonian) [7](89.5 Ma), [8](53.2, Lwr. Eocene),	3.87 to 2.38 × PIL
Vessel members (F) with perforation plates intermediate between scalariform and porous	Multiple files of predominantly oppositely arranged bordered pits	Present	[1]Ericaceae, [2]Caprifoliaceae	[1](89.5 Ma Turonian), [2](53.2 Ma Lwr. Eocene)	3.07 to 2.38 × PIL

Table 7.2. *Continued*

Xylem Type	Type of bordered pit typically found in lateral xylem wall	Perforation plate	Taxa	First recorded occurrence of clade in fossil record	RCO_2
Vessel members (G) with simple perforation plates & strongly inclined end walls	Multiple files of opposite to alternately arranged circular bordered pits	Present	Boraginaceae, Rubiaceae, Oleaceae, Caprifoliaceae	(53.2 Ma Lwr Eocene)	2.38 × PIL
Wide (H) short vessel members with simple perforation plates & non inclined end walls	Multiple files of predominantly alternately arranged bordered pits	Present	[1]Proteaceae, [2]Capparales, [3]Rosaceae, [4]Myrtales, [5]Euphorbiaceae, [6]Polygalaceae, [7]Asteraceae, [8]Rubiaceae [9]Oleaceae, [10]Convolvulaceae, [11]Scrophulariaceae	[1](97 Ma, mid-Cretaceous), [2](89.5 Ma, Turonian), [3](44.3, Mid. Eocene), [4](84 Ma, late Santonian), [5,6](58.5 Ma, Upr. Paleocene), [7,8](29.3 Ma, Oligocene), [9,10](53.2 Ma, Lwr. Eocene), [11](37 Ma, Upr. Eocene)	4.27 to 2.07 × PIL

Xylem anatomy data sources (Bailey and Tupper, 1918; Carlquist and Schneider, 2002 and references therein;). RCO_2 = Range of estimated atmospheric CO_2 concentration for times of oldest occurrences of clades in fossil record (expressed as a ratio of past to pre industrial (PIL =300 ppm) CO_2 concentration (Cretaceous CO_2 data from Tajika, 1999, pre-Cretaceous CO_2 from Berner, 1994)

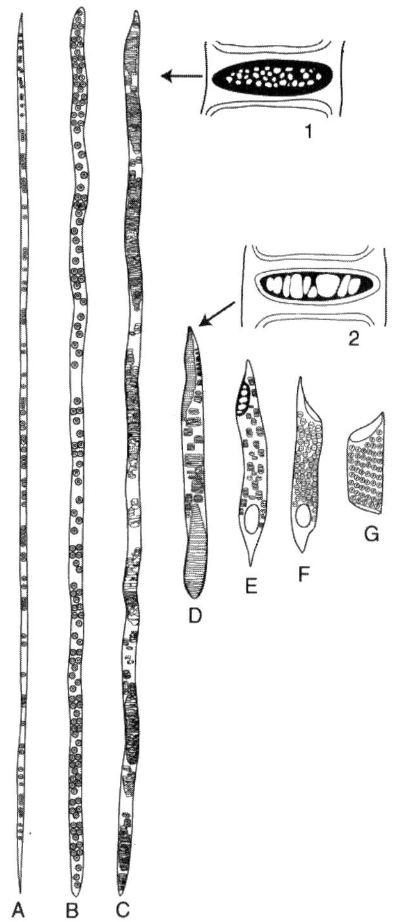

Figure 7.7. Diagrammatic illustration of the evolution of vessels from tracheids with simple bordered pits **(A, B)**, to elongated bordered pits in scalariform arrangement **(C)**, to vessel members **(D, E, F, G)** proposed by Bailey and Tupper (1918). **D** through **G** illustrate the proposed evolutionary trajectory from longer to shorter vessel members with reduced end-wall inclinations, from opposite to alternate pit arrangements, and a change from scalariform to simple perforation plates; **(A)** and **(B)** illustrate the evolutionary trajectory of increased porosity in the membranes of scalariform bordered pits in tracheids to those of scalariform perforation plates in primitive vessel members. See Table 7.2 for examples of living plant families possessing tracheid or vessel member anatomy illustrated as **A** through **G**. This xylem evolution model is broadly supported from observation of the plant fossil record of wood (Wheeler and Baas 1991). Redrawn from Bailey and Tupper (1918) with inset detail (1) and (2) from Carlquist and Schneider (2002).

Vessels are long (~10 mm to >15 feet) continuous water-conducting tubes made up of 'vessel elements' or 'vessel members' that are coalesced at their axial walls (Esau 1965; Zimmermann and Jeje 1981). Vessel members are elongated perforate cells with lignified secondary walls and no protoplast. They are characterized by the presence of perforated end walls, known as perforation plates, which enable water movement from member to member through the vessel. In contrast, tracheids, which is the only type of water-conducting cell found in most gymnospermous xylem (exceptions include Gnetales and Gigantopteridales, both of which have vessels), are imperforate cells. That is to say, tracheids do not have perforation plates in their end walls, and water flow from tracheid to tracheid occurs through the porous membrane of a pit depression in the secondary cell wall (Zimmermann 1983), which depending on the degree of porosity of the pit membrane, can moderately to greatly reduce water and solute flow, as shown in Fig 7.7A,B. In addition, tracheids are typically narrower in diameter and shorter in length than are vessels.

Such anatomical differences in the xylem of gymnosperms and angiosperms impact physiological function. For instance, comparative measurements of gymnosperm, angiosperm, and vessel-less angiosperm specific hydraulic conductive capacity (K_{sp}: conductivity per sapwood area) (Tyree and Ewers 1991; Becker, Tyree, and Tsuda 1999) and measures of whole plant conductivity (K_L stem hydraulic conductivity per unit leaf area) (Brodribb and Feild 2000) have demonstrated that conifers often have lower K_{sp} and K_L than do angiosperms or typically occupy the lowest ranges of angiosperm values but have similar or higher values than do vessel-less angiosperms (Brodribb and Hill 1999; Feild et al. 2001).

Although such comparative ecophysiological studies are rare and no such data are available for Pteridophytes (many of which, including most ferns, possess vessels), they suggest that the evolution of vessels in angiosperms and the evolutionary trajectory toward wider vessels (Bailey and Tupper 1918) with increased conductive area of vessel membrane end walls throughout the Cretaceous (see Table 7.2; Fig. 7.7D,G) (Carlquist and Schneider 2002) and on into the Tertiary (Wheeler and Baas 1991) would have enabled angiosperms to support greater leaf area and greater evaporative demand under progressively declining Cretaceous atmospheric CO_2, with relatively reduced investment in xylem tissue, compared to gymnosperms within the same communities and habitats. The work of Brodribb and Feild (2000) linking plant photosynthetic capacity with that of xylem conductivity also supports the suggestion that angiosperms may have been able to support greater photosynthetic capacity than gymnosperms within the same communities.

There is, however, a functional cost associated with increasing conductive capacity. The imperforate pit membrane of tracheids prevents movement of gas bubbles from tracheid to tracheid, thereby preventing the spread of air-filled, or embolized, tracheids. Air embolisms result in breakage of the water column within xylem tissue, and their spread can lead to catastrophic xylem dysfunction (Tyree and Sperry 1989). Air embolisms occur due to cavitations, which are sites, or nucleations, of vaporization within the water column. The main causes of cavitations in nature include: (1) air-seeding (Zimmermann 1983) due to water stress, where increasing negative pressure within the xylem can result in air being sucked into xylem through pores within the intertracheid or intervessel pit membranes—the larger the pores, the larger the bubbles (Tyree and Sperry 1989); (2) freeze-thaw cycles, where thawing of frozen water columns within xylem results in air bubbles being forced out of solution and nucleating a cavitation; and (3) sublimation of ice from frozen xylem leading to large bubbles on thawing (Tyree and Sperry 1989). Therefore, although the trajectory toward the evolution of vessel perforation plates in the axial walls of tracheids (see Fig. 7.7) would have enabled increased conductive capacity in basal angiosperm taxa (such as the scalariform bordered pits in the axial walls of *Trochodendron* and *Tetracetron*), by increasing the porosity of their pit membranes, total loss of pit membranes to form ever simpler perforation plates (see Table 7.2; Fig 7.7D,E,F,G transition) through the Cretaceous would have increased xylem vulnerability to both cavitation and spread of embolism in water-stressed and cold-stressed environments.

Taking these dual functions of xylem tissue into account, one would therefore expect the selection pressure for more efficient conductive tissue under declining Cretaceous atmospheric CO_2 to be greatest in seasonally dry but not permanently arid geographical areas of minimal temperature extremes with high evaporative demand (summer wet), and in light-limited habitats with ample water supply and limited temperature fluctuations, such as understory and subcanopy environments of tropical ever-wet forests. Plants in these habitats typically have lower maximum stomatal conductance (that is, they have higher resistance to CO_2 diffusion; Larcher 1995). In seasonally freezing climates (such as the cold temperate biome) and desert environments, retainment of tracheids or loss of vessels would provide protection against excessive cavitation and therefore embolism. An additional innovation that would have enhanced plant water supply under declining Cretaceous CO_2 was the evolution of high order reticulate venation with free-ending veinlets, which together with vessels evolved independently in Gnetales (*Gnetum*), early angiosperms, and Polypodiaceous ferns by the middle Cretaceous (Bailey 1944; Crane 1985; Trivett and Pigg 1996; Willis and McElwain 2002).

From a functional perspective, the models of Tyree and Sperry (1989) have demonstrated that xylem physiology and stomatal regulation must evolve as an integrated unit, as stomata play a vital role in the prevention of catastrophic xylem dysfunction from embolism. Robinson (1994) hypothesized that relatively lower CO_2 concentrations in the Cretaceous would also select strongly for plant groups with highly efficient stomatal control mechanisms and high potential maximum stomatal conductance (g_{max}). Although comparative statistics are few, the duration of stomatal opening and closing cycles is generally much faster in angiosperms than in gymnosperms and pteridophytes; angiosperms also tend to have higher maximum stomatal conductances than do gymnosperms and pteridophytes, respectively (see Robinson 1994 for review).

For instance, Gates (1968) has shown that the diffusive resistance in *Pteridium* is among the highest ever recorded. While Brodribb and Hill (1999) report typically stomatal opening times in excess of 2 hours for 16 Podocarpaceae and 7 Cupresaceae. Robinson (1994) argued therefore that angiosperms would be physiologically more optimized than gymnosperms and pteridophytes under conditions of "CO_2 starvation" as the more xerophyllous nature of gymnosperm leaves, which are often highly lignified and possess thick cuticle and lignified stomata, may have more inefficient stomatal control mechanisms since lignification may impede stomatal opening and closing responses (Robinson 1994). An alternative and perhaps more plausible explanation is that the degree of leaf lignification is positively correlated with leaf life span, which in turn is negatively correlated with leaf photosynthetic capacity (Reich, Walters, and Ellsworth 1997). The decline of overall gymnosperm diversity and abundance through the Cretaceous and perhaps the precipitous decline of Cheirolepidaceous conifers (many of which are characterized by xerophyllous leaves with deeply sunken stomata and overarching papillate epidermal cells) in post-Cenomanian time (Watson 1988) may be attributable to their inability to compete with higher photosynthetic efficiency among angiosperms under long-term declining atmospheric CO_2. By the middle Cretaceous, in marked contrast with gymnosperms, the leaves of angiosperms were larger and planate, relatively unlignified (based on comparative fossil anatomy), and presumably of shorter overall leaf life span, although this needs to be tested with fossil material using the leaf-life span method of Falcon-Lang (2000). Furthermore, angiosperm leaves were supported by a complex reticulate venation pattern and supplied with a highly conductive xylem tissue (vessels), which required less investment per unit leaf area than did their gymnospermous counterparts.

Estimates of maximum stomatal conductance of Cretaceous cheirolepidaceous conifer species range between 18.7 mmol m^{-2} s^{-1} (*Pseudofrenelopsis varians*) and 62.8 mmol m^{-2} s^{-1} (*Pseudofrenelopsis parceramosa*) (Cowie 1999). These are extremely low compared to values for modern angiosperm trees (^{150}–300 mmol m^{-2} s^{-1}; Larcher 1995) and herbs (150–700 mmol m^{-2} s^{-1}) and both deciduous and evergreen conifers, such as *Larix gmelinii* (365 mmol m^{-2} s^{-1}; Vygodskaya et al. 1997) and Pinaceae (200 mmol m^{-2} s^{-1}; Larcher 1995), respectively. The combination of potentially poor or slow stomatal control mechanisms and resultant lower stomatal conductance would be strongly selected against in a lower CO_2 world. Model and proxy estimates suggest that atmospheric CO_2 concentration may have reached levels of 2.5 × PIL (Tajika 1999) and potentially as low as 1.67 × PIL (Kuypers, Pancost, and Sinninghe-Damste 1999) due to high organic carbon burial flux at the C-T boundary. There is, therefore, a distinct possibility, although it remains to be tested, that CO_2 concentrations at the C-T boundary may have passed a critical lower threshold as a number of geographically widespread fossil taxa including numerous ephedroids, cheirolepidaceous conifers, and primitive clades producing the *Afropollis*-type pollen either suffered heavy extinctions or became extinct (Doyle, Jardiné, and Doerenkamp 1982; Crane and Lidgard 1989; Upchurch and Wolfe 1993).

The observation that both extinction (Cheirolepidaceae) and marked declines in relative abundances (*Afropollis* and ephedroid taxa) at the C-T boundary was most prevalent among xerophyllous taxa, as indicated by xeromorphic anatomy and morphology or by their association with arid indicators, such as evaporite sediments, is consistent with the prediction that low CO_2 would be more detrimental to plants of arid adapted habitats (Robinson 1994). A transient CO_2 drawdown for a total duration of approximately 100 K to 150 K years (Meyers, Sageman, and Hinnov 2001) at the Cenomanian-Turonian boundary, which is supported by geochemical proxy data (Kuypers, Pancost, and Sinninghe-Damste 1999) and by a biogeochemical model (Tajika 1999), may have conferred further competitive advantage to angiosperm clades since the North American pollen database (Lupia, Lidgard, and Crane 1999) shows the steepest increase in both relative richness and abundance during this interval. The same pattern of steeply increasing angiosperm abundance occurs synchronously in the Southern Hemisphere in Australian fossil floras (Nagalingum et al. 2002), again strongly supporting the role of a global abiotic trigger such as CO_2 change. High-resolution independent proxy CO_2 data, together with independent paleoclimate estimates spanning the Cenomanian-Turonian oceanic anoxic event, are now required in order to constrain the possible threshold CO_2 concentration or climatic perturbation that contributed to the rapid competitive replacement of cheirolepidaceae and ephedroids by flowering plants.

7.4.2.2 Supporting Evidence from the Plant Fossil Record

Reticulate venation has arisen independently and repeatedly in many different phylogenetic groups throughout the history of land plants (for a detailed review, see Trivett and Pigg 1996). At first view, therefore, there appears to be little adaptive significance of reticulate venation in relation to abiotic factors; it has instead been interpreted purely as an indicator of biological complexity or evolutionary status (Trivett and Pigg 1996). Highly ordered tertiary and quaternary reticulate venation does not appear in the fossil angiosperm record until the late Aptian and Cenomanian, as indicated by the classic work of Doyle and Hickey (1976), a time according to model estimates when paleo-CO_2 concentrations were declining sharply from >4 times to <3 times PIL. The only pre-Cretaceous occurrences of large compound or simple planate leaves with tertiary or quaternary reticulate venation and free-ending veinlets are restricted to two unique genera *Gigantoclea* and *Delnortea*, both of which belong to a group of extinct Permian seed ferns: Gigantopterids (Trivett and Pigg 1996). More importantly, Gigantopterids are the only pre-Cretaceous seed plants that possess true vessels within xylem tissue (Li, Taylor, and Taylor 1996; Li and Taylor 1999). It is noteworthy that the early Permian was the only other pre-Cretaceous time in Earth's history according to proxy CO_2 (McElwain and Chaloner 1996; Ekart et al. 1999) and model estimates (Berner and Kothavala 2001) when CO_2 concentrations reached levels lower that 3 times PIL. The number of shared vegetative characteristics between angiosperms, gigantopterids (Fig. 7.8), and

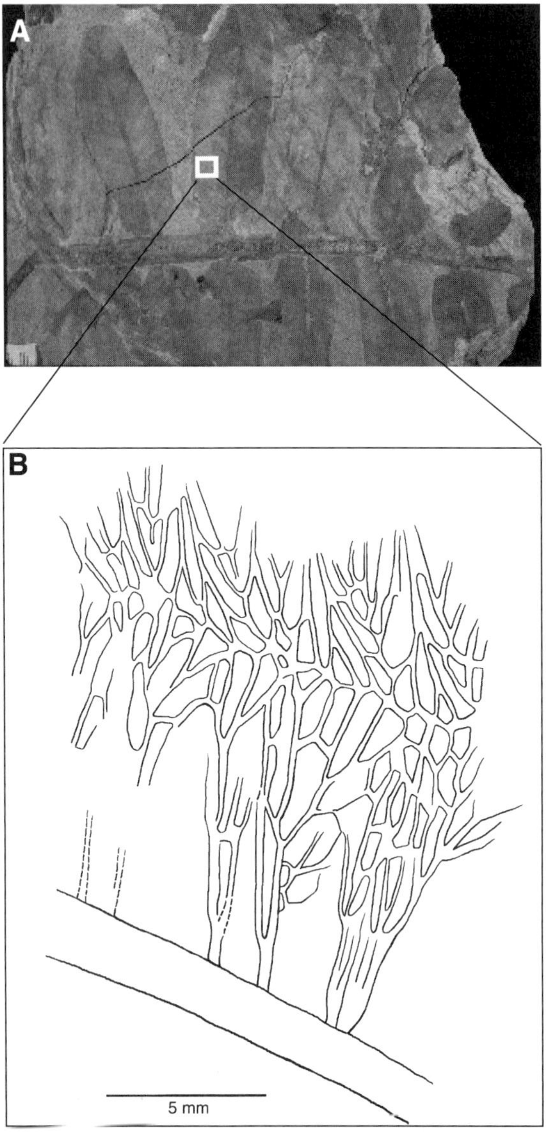

Figure 7.8. **(A)** Details of leaf morphology, and **(B)** venation of *Gingantoclea guizhouenensis* (PP34440), a Permian Gigantopterid seed fern from China.

Gnetum with respect to their venation and xylem are remarkable and may therefore represent functional convergence of morphology and anatomy to relatively lower atmospheric CO_2 concentrations.

However, similar patterns are not observed in spore-bearing plants as a number of pteridophytes, including lycopods (such as *Sellaginella,*), equisetum, and many ferns possessing both reticulate venation (Trivett and Pigg 1996) and true vessels in their rhizomes, roots, and/or stems (Carlquist and Schneider 2001) are not restricted to times of relatively low atmospheric CO_2 in Earth's history.

If the evolution of xylem vessels by angiosperms was a key innovation enabling them to competitively replace gymnosperms and pteridophytes in a lower CO_2 world, as we have argued, why have vessel-less angiosperms both persisted (e.g., Amborellaceae: Feild et al. 2001) and been secondarily derived (e.g., Winteraceae: Herendeen, Wheeler, and Baas 1999; Feild, Zwieniecki, and Holbrook 2000)? Recent comparative ecophysiological work has demonstrated that neither the presence of tracheids in primitive angiosperms nor the absence of vessels in conifers always imposes higher limitations on water transport than do plants with xylem vessels in the same environment (Feild and Holbrook 2000). In a recent review, Carlquist (1996) noted that "a reason for the persistence of primitive character states in wood may be the tendency for other aspects of the vegetative apparatus to evolve more readily, offering compensation." Carlquist (1996) argued that because leaf size and morphology are so plastic, xeromorphic leaf structure compensates readily for a primitive xylem configuration (e.g., microphylly) in Bruniaceae and Grubbiaceae, both of which have primitive wood and needle or scale leaves more typical of conifers. The evolution of true vessels, therefore, likely enabled angiosperms to attain much greater leaf sizes, which together with the structural support afforded by reticulate venation would have reduced the necessity for strongly xeromorphic leaf characteristics in angiosperms. As the high lignin content and structural rigidity of xeromorphic leaves may impede fine-tuned stomatal control (Robinson 1994) and tends to be associated with longer leaf life spans and lower photosynthetic capacity (Reich, Walters, and Ellsworth 1997), this in turn may have enabled the development of enhanced conductive capacity and photosynthetic potential in angiosperm leaves.

7.4.3 Role of Other Abiotic Triggers in the Angiosperm Radiation

Many competing hypotheses have been proposed for the remarkable radiation of angiosperms (see Taylor and Hickey 1996). It has long been held that global increases in aridity were the most probable trigger for the flowering plant radiation as angiosperms possess a number of innovations that may have made them more drought resistant and therefore competitively superior to other plants. These innovations include tough leathery leaves, a tough resistant seed coat that protected the young embryos from drying out, xylem vessel providing much more efficient water-conducting cells than in previous groups, and a deciduous habit. However, these traits are not unique to angiosperms. In fact, it seems paradoxical that cheirolepidaceous conifers, which were perhaps the most prominent fossil taxa found in association with aridity indicators, such as evaporite deposits, throughout the early and mid Cretaceous, became extinct if the trigger was increased aridity. Furthermore, the presence of tracheid elements rather than vessels is the preferred conducting tissue for plant taxa of the most extremely water-deficient environment because tracheids afford better protection from catastrophic embolism (Carlquist 1996) than do vessels. The possession of tracheids as well as vessels in many angiosperm evergreen shrubs of the Mediterranean region prevent embolism during the hottest part of the year when water availability reaches an annual minimum (Carlquist 1996). Also, the deciduous habit was not first invented by angiosperms. Indeed, supporting evidence suggests that many Paleozoic and Mesozoic gymnosperm taxa were deciduous (e.g., Glossopteridaceae, Ginkgoales, Taxodiaceae, and Cycadales) particularly in higher latitudes where seasonal climatic extremes and light limitation were more prevalent (Falcon-Lang 2000; Rees, Ziegler, and Valdes 2000). Finally, a gymnospermous seed would have afforded the same protection from desiccation that an angiospermous seed would have provided, and the foliage of gymnosperms, such as *Brachyphyllum* and *Pagiophyllum* (representing cheirolepidiaceous conifers at least in part), was eminently adapted to aridity.

It therefore appears highly unlikely that aridity alone, if at all, was an important driving influence on the radiation of angiosperms. We support the alternative (or at least complementary) explanation of Doyle and Donoghue (1986) that dramatically accelerated speciation rates, which are characteristic of angiosperms, simply led to an overwhelming diversity of adaptive morphologies (Lupia, Crane, and Lidgard 2000). However, we propose that a marked decline in atmospheric CO_2 through the Cretaceous imposed severe physiological limitation on gymnosperms and pteridophytes, particularly those of understory and seasonally dry habitats. Owing to the suite of adaptive vegetative characteristics (high potential stomatal conductance, high stomatal control, vessels, and reticulate venation), which in combination but not separately are unique to angiosperms, the reduction in CO_2 may have provided an essential abiotic trigger for the rapid evolutionary and ecological radiation of flowering plants.

7.4.4 Which Came First: Angiosperm Radiation or CO_2 Decline?

It has been proposed that the evolution of deciduous angiosperms in the Cretaceous played a significant role in the long-term decline in atmospheric CO_2 by accelerating chemical weathering of calcium and magnesium silicate rocks more than did the previous gymnosperm-dominated vegetation (Knoll and James 1987; Volk 1989). Because calcium and magnesium silicates are ultimately precipitated as carbonates in the oceans, silicate chemical weathering represents one of the most important sinks of atmospheric CO_2 in the long-term carbon cycle (see Berner, this volume). There are, however, very little comparative data on the relative weathering potential of angiosperms and gymnosperms, and based on the available studies to date, Volk's assumption that angiosperms have a higher chemical weathering potential remains controversial. For instance, studies in Iceland show that angiosperms have a significantly greater weathering potential on a per unit biomass basis than do gymnosperms (Berner 1997; Moulton, West, and Berner 2000; Berner and Kothavala 2001); the only other available study, however, shows the opposite (Quideau et al. 1996). Despite this uncertainty, it is important to consider whether declining atmospheric CO_2 may have led to the radiation and diversification of angiosperms as we have argued (see Table 7.1; Fig. 7.5), or whether the reverse is true: that the proliferation of angiosperms resulted in enhanced chemical weathering and Cretaceous CO_2 drawdown.

In order to address the problem of which came first, we have plotted the raw angiosperm relative abundance data from Lupia, Lidgard, and Crane (1999) on the same temporal scale as three paleo-CO_2 models (Fig. 7.9). It is clear from

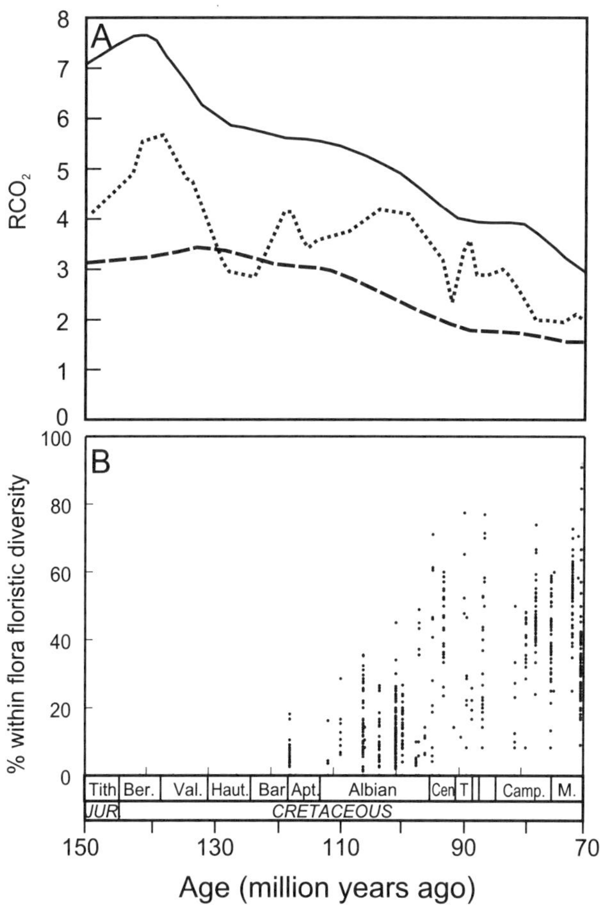

Figure 7.9. Illustrating a decline in atmospheric CO_2 in the Berriasian and Valaginian before the taxonomic radiation of angiosperms in the Aptian and Albian.

these comparisons that although angiosperms originated during an episode of high atmospheric CO_2, their taxonomic radiation and, more importantly, their rise to ecological dominance within Cretaceous fossil floras clearly followed an initial steep decline in CO_2 from $>5 \times$ PIL to $3 \times$ PIL in the Valanginian (137–Ma) (Tajika 1999). It is likely, therefore, that a long-term trend of CO_2 decline through the Cretaceous period may have played an extremely important role in the diversification of angiosperms, leading to their ecological dominance in almost every biome with the exception of the high latitude boreal realm by the Maastrichian.

7.5 Summary

The evolution of megaphyll leaves (Beerling, Osborne, and Chaloner 2001) and the C4 photosynthetic pathway (Ehleringer et al. 1991; Ehleringer, Cerling, and Helliker 1997) were both strongly influenced by declines in atmospheric CO_2 concentration in the Paleozoic and Cenozoic, respectively. In this chapter we present evidence that a long-term decline in atmospheric CO_2 concentration through the Cretaceous ($\sim$ 140–65 Ma) significantly influenced the floristic composition of Cretaceous vegetation. Highly significant inverse relationships are observed between both the relative abundance and the richness of angiosperms in Cretaceous fossil floras and estimated atmospheric Cretaceous CO_2 concentrations (Tajika 1998, 1999). In contrast, significant positive correlations are shown between paleo-CO_2 concentration and gymnosperm relative abundance and pteridophyte relative species richness. We propose that a combination of vegetative innovations, including high order reticulate venation, xylem vessels, and rapid stomatal control mechanisms enabled angiopserms to maintain high maximum stomatal conductance and CO_2 assimilation rates in a significantly lower CO_2 world. This combination of characteristics would have conferred a competitive advantage to angiosperms over gymnosperms and pteridophytes, particularly in seasonally but not permanently arid environments and light-limited habitats of the tropics, where adaptations to water and light limitations, respectively, likely severely impacted plant carbon balance. The ecophysiological differences among angiosperms, gymnosperms, and pteridophytes provide a mechanistic explanation for the observed relationships in the diversity and dominance patters of the three reproductive grades. Our results suggest that adaptation to CO_2 starvation, in addition to any advantages resulting from advanced reproductive biology in flowering plants, may have been an important trigger in the ecological radiation and taxonomic diversification of flowering plants through competitive replacement of less physiologically optimized gymnosperm and pteridophyte clades.

Our results indicate that an overall trend of declining CO_2 concentration through the Cretaceous differentially influenced each reproductive grade, favoring angiosperms over gymnosperms and pteridophytes in terms of their abundance and taxonomic diversity, respectively. It has been suggested from photosynthetic modeling studies on extant plants that angiosperm photosynthetic responses may be more optimized to low CO_2 concentrations than are gymnosperms and pteridophytes (Beerling 1994; Beerling and Woodward 1997). We propose that declining CO_2 concentrations during the Cretaceous may have provided a competitive physiological edge to angiosperms, which possessed reticulate veined leaves, vessel members, and sensitive stomatal control, thus contributing, along with their advanced reproductive strategy, to the competitive replacement of less physiologically optimized gymnosperm and free-sporing plant groups. However, the Gnetales remained an exception within the gymnosperms. Unlike all other gymnosperm clades, they show no significant reductions in relative diversity within Cretaceous fossil floras until post-Cenomanian time (Crane and Lidgard 1990), a point at which a critical minimum CO_2 threshold may have been reached resulting in widespread extinction of arid adapted clades. The fact that Gnetales, unlike all other Mesozoic gymnosperms, independently evolved two of the three key adaptations to optimize photosynthesis under relatively lower atmospheric CO_2 (planate leaves with high order reticulate venation and true vessels) helps to explain their initial co-radiation with angiosperms in response to declining CO_2 before the Cenomanian-Turonian OAE2. An explanation for the demise of many Gnetalean taxa at the C-T boundary, however, remains an interesting avenue for further research.

Our results, although preliminary and requiring further support from modern comparative plant ecophysiological studies, may have profound implications for projecting future effects of rapidly rising CO_2 on global plant biodiversity and the floristic composition of natural ecosystems. The predicted future doubling of CO_2 by the end of this century (IPCC 2001), based on our results, may significantly alter the percentage composition of angiosperms gymnosperms and pteridophytes in the world's biomes. Further work will aim to quantify CO_2 change throughout the Cretaceous period using the stomatal method (McElwain 1998) in order to define the critical threshold CO_2 level at which significant changes in vegetation composition occur, as current model estimates lack the precision to address this critical issue.

Acknowledgments. The authors thank two anonymous reviewers for their helpful comments and suggestions. Many thanks also to Marlene Donnelly, Scientific Illustrator at the Field Museum. McElwain thanks the Comer Science and Education Foundation (Grant no. 13) for funding this research.

References

Arthur, M.A., W.E. Dean, and L.M. Pratt. 1988. Geochemical and climatic effects of increased marine organic carbon burial at the Cenomanian/Turonian boundary. *Nature* 335:714–17.

Arthur, M.A., W.E. Dean, and S.O. Schlanger. 1985. Variations in the global carbon cycle during the Cretaceous related to climate, volcanism, and changes in atmospheric CO_2. *Geophysical Monograph* 32:504–29.

Axelrod, D.L. 1952. A theory of angiosperm evolution. *Evolution* 4:26–90.

Bailey, I.W. 1944. The development of vessels in angiosperms and its significance in morphological research. *American Journal of Botany* 31:421–28.
Bailey, I.W. and W.W. Tupper. 1918. Size variation in tracheary cells: I. A comparison between the secondary xylems of vascular cryptograms, gymnosperms and angiosperms. *Proceedings of the American Academy of Arts and Sciences* 54: 149–204.
Barrera, E., M. Savin, E. Thomas, and C.E. Jones. 1997. Evidence for themohaline-circulation reversals controlled by sea level change in the latest Cretaceous. *Geology* 25:715–18.
Barrett, P.M. 2000. Evolutionary consequences of dating the Yixian Formation. *Trends in Ecology and Evolution* 15:99–103
Barron, E.J. 1987. Cretaceous plate tectonics reconstructions. *Paleogeography, Paleoclimatology, Paleoecology* 59:3–29.
Barron, E.J., and W.M. Washington. 1984. The role of geographic variables in explaining paleoclimates: Results from Cretaceous climate model sensitivity studies. *Journal of Geophysical Research* 89:1267–79.
Bazzaz, F.A., M. Jasienski, S.C. Thomas, and P. Wayne. 1995. Microevolutionary responses in experimental populations of plants to CO_2-enriched environments: Parallel results from 2 model systems. *Proceedings of the National Academy of Sciences USA* 92:8161–65.
Becker, P., M.T. Tyree, and M. Tsuda. 1999. Hydraulic conductances of angiosperms versus conifers: similar transport sufficiency at whole-plant level. *Tree Physiology* 19: 445–52.
Beerling, D.J. 1994. Modeling palaeophotosynthesis: Late Cretaceous to present. *Philosophical Transactions of the Royal Society, London B* 346:421–32.
Beerling, D.J., J.C. McElwain, and C.P. Osborne. 1998. Stomatal responses of the 'living fossil' *Ginkgo biloba* L. to changes in atmospheric CO_2 concentrations. *Journal of Experimental Botany* 49: 1603–607.
Beerling, D.J., C.P. Osborne, and W.G. Chaloner. 2001. Evolution of leaf form in land plants linked to atmospheric CO_2 decline in the late Paleozoic era. *Nature* 410: 352–54.
Beerling, D.J., and F.I. Woodward. 1997. Palaeo-ecophysiological perspectives on plant responses to global change. *Trends in Ecology and Evolution* 11:20–23.
Berner, R.A. 1991. A model for atmospheric CO_2 over Phanerozoic time. *American Journal of Science* 291:339–75.
———. 1994. GEOCARB II: a revised model of atmospheric CO_2 over Phanerozoic time. *American Journal of Science* 294:56–91.
———. 1997. The Rise of Plants and Their Effect on Weathering and Atmospheric CO_2. *Science* 276:544–46.
Berner, R.A., and Z. Kothavala. 2001. GEOCARB III: A revised model of atmospheric CO_2 over Phanerozoic time. *American Journal of Science* 301:182–204.
Brenner, G.J. 1996. Evidence for the earliest stage of angiosperm pollen evolution: A paleoequatorial section from Israel. In *Flowering plant origin, evolution and phylogeny,* ed. D.W. Taylor and L.J. Hickey, 91–115. New York: Chapman and Hall.
Brodribb, T.J., and T.S. Field. 2000. Stem hydraulic supply is linked to leaf photosynthetic capacity: Evidence from New Caledonian and Tasmanian rainforests. *Plant, Cell, and Environment* 23:1381–87.
Brodribb, T.J., and R.S. Hill. 1999. The importance of xylem constraints in the distribution of conifer species. *New Phytologist* 143:365–72.
Carlquist, S. 1996. Wood anatomy of primitive angiosperms: New perspectives and syntheses. In *Flowering plant origin, evolution, and phylogeny,* ed. D.W. Taylor and L.J. Hickey, 68–91. New York: Chapman and Hall.
Carlquist, S., and E.L. Schneider. 2001. Vessels in ferns: Structural ecological and evolutionary significance. *American Journal of Botany* 88:1–13.
———. 2002. The tracheid-vessel element transition in angiosperms involves multiple

independent features: Cladistic consequences. *American Journal of Botany* 89:185–95.

Chen, L.-Q., C.-S. Li, W.G. Chaloner, D.J. Beerling, Q.-G. Sun. 2001. Assessing the potential for the stomatal characters of extant and fossil Ginkgo leaves to signal atmospheric CO_2 change. *American Journal of Botany* 88:1309–15.

Condie, K.C., and R.E. Sloan. 1998. *Origin and evolution of the Earth.* New Jersey: Prentice Hall.

Cornet, B. 1986. The reproductive structures and leaf venation of a late Triassic angiosperm *Sanmiguelia lewisii. Evolutionary Theory* 7:231–309.

Cornet, B. 1993. Dicot-like leaf and flowers from the late Triassic tropical Newark supergroup rift zone, U.S.A. *Modern Geology* 19:81–99.

Cornet, B., and D. Habib. 1992. Angiosperm-like pollen from the ammonite dated Oxfordian (Upper Jurassic) of France. *Review of Paleobotany and Palynology* 71:268–94.

Cowie, R.R. 1999. Gas exchange characteristics of an early cretaceous conifer, *Pseudofrenelopsis varians* (Cheirolepidiaceae), and its inferred paleoecology. Tex.: Southwest Texas State University. Master's Thesis.

Crane, P.R. 1985. Phylogentic alalysis of seed plants and the origin of angiosperms. *Annals of the Missouri Botanical Garden* 72:716–93.

Crane, P.R. and S. Lidgard. 1989. Angiosperm diversification and paleolatitudinal gradients in Cretaceous floristic diversity. *Science* 246:675–78.

———. 1990. Angiosperm radiation and patterns of Cretaceous palynological diversity. In *Major evolutionary radiations*, ed. P.D. Taylor and G.P. Larwood (Systematics Association Special) 42:377–407.

Dean, W.E., M.A. Arthur, and G.E. Claypool. 1986. Depletion of ^{13}C in Cretaceous marine organic matter: Source, diagenetic, or environmental signal? *Marine Geology* 70:119–57.

DeLucia, E.H., J.G. Hamilton, S.L. Naidu, R.B. Thomas, J.A. Andrews, A. Finzi, M. Lavine, R. Matamala, J.E. Mohan, G.R. Hendrey, and W.H. Schlesinger. 1999. Net primary productions of a forest ecosystem with experimental CO_2 enrichment. *Science* 285:1177–79.

Dilcher, D.L. 1989. The occurrence of fruits with affinities to Ceratophylaceae in lower and mid Cretaceous sediments. *American Journal of Botany* 76:162.

Doyle, J.A. and M.J. Donoghue. 1986. Seed plant phylogeny and the origin of angiosperms: an experimental cladistic approach. *Botanical Review* 52:321–431.

Doyle, J.A., and L.J. Hickey. 1976. Pollen and leaves from the mid Cretaceous Potomac Group and their bearing on early angiosperm evolution 139–206. In *Origin and early evolution of Angiosperms,* ed. C.B. Beck. New York: Columbia University Press.

Doyle, J., S. Jardiné, and A. Doerenkamp. 1982. *Afropollis*, a new genus of early angiosperm pollen, with notes on the Cretaceous palynostratigraphy and paleoenvironments of northern Gondwana. *Bulletin des Centres de Recherches Exploration-Proction Elf-Aquitaine* 6:39–117.

Ehleringer, J.R., T.E. Cerling, and B.R. Helliker. 1997. C-4 photosynthesis, atmospheric CO_2, and climate. *Oecologia* 112:285–99.

Ehleringer, J.R., R.F. Sage, L.B. Flanagan, and R.W. Pearcy. 1991. Climate change and the evolution of C4 photosyntheisis. *Trends in Ecology and Evolution* 3:95–99.

Ekart, D.D., T.E. Cerling, I.P. Montanez, and N.J. Tabor. 1999. A 400 million year carbon isotope record of pedogenic carbonate; implications for paleoatomospheric carbon dioxide. *American Journal of Science* 299:805–27.

Esau, K. 1964. *Plant anatomy.* New York: John Wiley. Falcon-Lang, H.J. 2000. A method to distinguish between woods produced by evergreen and deciduous coniferopsids on the basis of growth ring anatomy: A new palaeoecological tool. *Palaeontology* 43: 785–93.

Field, T.S., T. Brodribb, T. Jaffré, and N.M. Holbrook. 2001. Acclimation of leaf anatomy, photosynthetic light use, and xylem hydraulics to light in *Amborella trichopoda* (Amborellaceae). *International Journal of Plant Sciences* 162:999–1008.

Field, T.S., and N.M. Holbrook. 2000. Xylem sap flow and stem hydraulics of the vesselless angiosperm *Drimys granadensis* (Winteraceae) in a Costa Rican elfin forest. *Plant, Cell, and Environment* 23:1067–72.

Field, T.S., M.A. Zwieniecki, and N.M. Holbrook. 2000. Winteraceae evolution: An ecophysiological perspective. *Annals of the Missouri Botanical Gardens* 87:323–34.

Frakes, L.A. 1999. Estimating the global thermal state from Cretaceous sea surface and continental temperature data. In *Evolution of the Cretaceous ocean-climate system*, ed. E. Barrera and C.C. Johnson (special paper 332:49–57). Boulder, Colorado: Geological Society of America.

Friis, E.-M., P.R. Crane, and K.J. Pedersson. 1999. Early Angiosperm diversification: The diversity of pollen associated with angiosperm reproductive structures in early Cretaceous floras from Portugal. *Annals of the Missouri Botanical Garden* 86: 259–97.

Gale, J. 1972. Availability of carbon dioxide for photosynthesis as high altitudes: Theoretical considerations. *Ecology* 53:494–97.

Gates, D.M. 1968. Transpiration and leaf temperature. *Annual Review of Plant Physiology* 19:211–38.

Gothan, W., and W. Remy. 1957. *Steinkohlenpflanzen*. Essen: Glükauf.

Hay, W.W., R.M. DeConto, C.N. Wold, K.M. Wilson, S. Voigt, M. Schulz, A.R. Wold, W.-C. Dullo, A.B. Ronov, A.N. Balukhovsky, and E. Söding. 1999. Alternative global Cretaceous paleogeography. In *Evolution of the Cretaceous Ocean-Climate system*, ed. E. Barrera and C.C. Johnson (special paper 332:1–47). Boulder, Colorado: Geological Society of America.

Herendeen, P.S., E.A. Wheeler, and P. Baas. 1999. Angiosperm wood evolution and the potential contribution of paleontological data. *The Botanical Review* 65:278–300.

Herman, A.B., and R.A. Spicer. 1996. Paleobotanical evidence for a warm Cretaceous Arctic Ocean. *Nature* 380:330–33.

Hesselbo, S.P, D.R. Gröcke, H.C. Jenkyns, C.J. Bjerrum, P.L. Farrimond, H.S. Morgans-Bell, O. Green. 2000. Massive dissociation of gas hydrates during a Jurassic Oceanic anoxic event. *Nature* 406:392–95.

Huber, B.T., R.M. Leckie, R.D. Norris, T.J. Bralower, and E. CoBabe. 1999. Foraminiferal assemblage and stable isotopic change across the Cenomanian-Turonian boundary in the subtropical, North Atlantic. *Journal of Foraminiferal Research* 29:392–417.

Hughes, N.F. 1994. *The enigma of angiosperm origins*. Cambridge: Cambridge University Press.

Hussain, M., M.E. Kubiske, and K.F. Connor. 2001. Germination of CO_2-enriched *Pinus taeda L.* seeds and subsequent seedling growth responses to CO_2 enrichment. *Journal of Functional Ecology* 15:344–50.

Huston, M.A., et al. 2000. No consistent effect of plant diversity on productivity. *Science* 289:1255.

IPCC 2001. Climate change 2001: The scientific basis. Contributions of working group 1 to the third assessment report of the intergovernmental panel on climate change, ed. J.T. Houghton et al. Cambridge: Cambridge University Press.

Jacobs, B.F., J.D. Kingston, and L.L. Jacobs. 1999. The origin of grass dominated ecosystems. *Annals of the Missouri Botanical Garden* 86: 590–644.

Jahren, A.H., N.C. Arens, G. Sarmiento, J. Guerrero, and R. Amundson. 2001. Terrestrial record of methane hydrate dissociation in the Early Cretaceous. *Geology* 29: 159–62.

Jenkyns, H.C. 1980. Cretaceous anoxic events; from continents to oceans. *Journal of the Geological Society of London* 137: 171–88.

Kaiho, K., and T. Hasegawa. 1994. End-Cenomanian benthic foraminiferal extinctions and oceanic dysoxic events in the northwestern Pacific Ocean. *Palaeogeography, Palaeoclimatology, Palaeoecology* 111: 29–43.

Kauffman, E.G., and M.B. Hart. 1995. Cretaceous Bio-events. In *Global events and event stratigraphy in the Phanerozoic,* ed. O.H. Walliser. Berlin: Springer-Verlag.
Knoll, A.H., and W.C. James. 1987. Effect of the advent and diversification of vascular land plants on mineral weathering through geologic time. *Geology* 15: 1099–1102.
Kuypers, M.M.M., R.D. Pancost, and J.S. Sinninghe-Damste. 1999. A large and abrupt fall in atmospheric CO_2 concentration during Cretaceous times. *Nature* 399: 342–45.
Larcher, W. 1995. *Physiological plant ecology.* 3d ed. Berlin: Springer-Verlag.
Larson, R.L. 1991. Latest pulse of Earth: Evidence for a mid-Cretaceous superplume. *Geology* 19: 547–50.
———. Geological consequences of superplumes. *Geology* 19: 963–66.
Leigh, E.G., and G.J. Vermeij. 2002. Does natural selection organize ecosystems for the maintenance of high productivity and diversity? *Philosophical Transactions of the Royal Society London B* 357: 709–18.
Li, H., E.L. Taylor, and T.N. Taylor. 1996. Permian vessel elements. *Science* 271: 188–89.
Li, H., and D.W. Taylor. 1999. Vessel-bearing stems of Vsovinea tianii Gen. et Sp. Nov. (Gigantopteridales) from the upper Permian of Guizhou Province, China. *American Journal of Botany* 86: 1563–75.
Lidgard, S., and P.R. Crane. 1988. Quantitative analyses of the early angiosperm radiation *Nature* 331: 344–46
———. 1990. Angiosperm diversification and Cretaceous florisitc trends: a comparison of palynofloras and leaf macrofloras. *Paleobiology* 16: 77–93.
Loreau, M., S. Naeem, P. Inchausti, J. Bengtsson, J.P. Grime, A.D. Hector, U. Hooper, M.A. Huston, D. Raffaelli, B. Schmid, D. Tilman, and D.A. Wardle. 2001. Biodiversity and ecosystem functioning: Current knowledge and future challenges. *Science* 294: 804–808.
Lupia, R., S. Lidgard, and P.R. Crane. 1999. Comparing palynological abundance and diversity: Implications for biotic replacement during the Cretaceous angiosperm radiation. *Paleobiology* 25: 305–40.
———. 2000. Angiosperm diversification and Cretaceous environmental change. In *Biotic response to global change: The last 145 million years,* ed. S.J. Culver and P.F. Rawson, 223–43. Cambridge: Cambridge University Press.
McElwain, J.C. 1998. Do fossil plants signal palaeo-atmospheric CO_2 concentration in the geological past? *Philosophical Transactions of the Royal Society B* 353: 83–96.
McElwain, J.C., D.J. Beerling, and F.I. Woodward. 1999. Fossil plants and global warming at the Triassic-Jurassic boundary. *Science* 285: 1386–90.
McElwain, J.C., and W.G. Chaloner. 1996. The fossil cuticle as a skeletal record of environmental change. *Palaios* 11: 376–88.
McLeod, A.R., and S.P. Long. 1999. Free-air carbon dioxide enrichment (FACE) in global change research: A review. *Advances in Ecological Research* 28: 1–56.
Meyers, S., B. Sageman, and L. Hinnov. 2001. Integrated quantitative stratigraphy of the Cenomanian-Turonian Bridge Creek limestone member using evolutive harmonic analysis and stratigraphic modeling. *Journal of Sedimentary Research* 71: 627–43.
Moulton, K.L., J. West, and R.A. Berner. 2000. Solute flux and mineral mass balance approaches to the quantification of plant effects on silicate weathering. *American Journal of Science* 300: 539–70.
Nagalingum, N.S., A.N. Drinnan, R. Lupia, and S. McLoughlin. 2002. Fern spore diversity and abundances in Australia during the Cretaceous. *Review of Paleobotany and Palynology* 2431: 1–24.
Niklas, K.J., B.H. Tiffney, and A.H. Knoll. 1983. Patterns in vascular land plant diversification. *Nature* 303: 614–16.
Oren, R. 2001. Soil fertility limits carbon sequestration by forest ecosystems in a CO_2 enriched atmosphere. *Nature* 411: 469–72.

Parrish, J.T., and R.A. Spicer. 1988. Late Cretaceous terrestrial vegetation: A near polar temperature curve. *Geology* 16: 22–25.

Quideau, S.A., O.A. Chadwick, R.C. Graham, and H.B. Wood. 1996. Base cation biogeochemistry and weathering under oak and pine: A controlled long-term experiment. *Biogeochemistry* 35: 377–98.

Rees, P. McA., A.M. Ziegler, and P.J. Valdes. 2000. Jurassic phytogeography and climates: New data and model comparisons. In *Warm climates in Earth history,* ed. B.T. Huber, K.G. Macleod, and S.L. Wing. Cambridge: Cambridge University Press.

Reich, P.B., M.B. Walters, and D.S. Ellsworth. 1997. From tropics to tundra: Global convergence in plant functioning. *Proceedings of the National Academy of Sciences* 94: 13730–34.

Robinson, J.M. 1994. Speculations on carbon dioxide starvation, late Tertiary evolution of stomatal regulation and floristic modernization. *Plant Cell and Environment* 17: 1–10.

Robinson, S.A., J.E. Andrews, S.P. Hesselbo, J.D. Radley, P.F. Dennis, I.C. Harding, P. Allen. 2002. Atmospheric pCO_2 and depositional environment from stable-isotope geochemistry of calcrete nodules (Barremian, Lower Cretaceous, Wealden Beds, England). *Journal of the Geological Society London* 159: 215–24.

Schlanger, S.O., and H.C. Jenkyns. 1976. Creatceous oceanic anoxic events: Causes and consequences. *Geol. Mijnbouw* 55: 179–84.

Scotese, C.R. 1991. Jurassic and Cretaceous plate tectonic reconstructions. *Paleogeography, Paleoclimatology, Paleoecology* 87: 493–501.

Sheridan, R.E. 1997. Pulsation tectonics as a control on the dispersal and assembly of supercontinents. *Journal of Geodynamics* 23: 173–96.

Street-Perrott, F.A., Y. Huang, R.A. Perrott, G. Eglington, P. Barker, L.B. Khelifa, D.D. Harkness, and D.O. Olago. 1997. Impact of lower atmospheric carbon dioxide on tropical mountain ecosystems. *Science* 278: 1422–26.

Sun, Ge, D.L. Dilcher, S. Zheng, and Z. Zhou. 1998. In search of the first flower: A Jurassic angiosperm, *Archaefructus*, from Northeast China. *Science* 282: 1692–95.

Sun, G., Q. Ji, D.L. Dilcher, S. Zheng, K.C. Nixon, and X. Wang. 2002. *Archaefructaceae*, a New Basal Angiosperm Family. *Science* 296: 899–904.

Tajika, E. 1998. Climate change during the last 150 million years: Reconstruction from a carbon cycle model. *Earth and Planetary Science Letters* 160: 695–707.

———. 1999. Carbon cycle and climate change during the Cretaceous inferred from a biogeochemical carbon cycle model. *The Island Arc* 8: 293–303.

Terashima, I., T. Masuzawa, H. Ohba, and Y. Yokoi. 1995. Is photosynthesis suppressed at higher elevations due to low CO_2 pressure? *Ecology* 76: 2663–68.

Teslenko, Y.V. 1967. Some aspects of evolution of terrestrial plants. *Geologia i Geofizica (Novosibirsk)* 11: 58–64.

Trivett, M.L., and K.B. Pigg. 1996. A survey of reticulate venation among fossil and living plants. In *Flowering plant origin, evolution, and phylogeny,* ed. W.D. Taylor and L.J. Hickey, 8–12. New York: Chapman and Hall.

Tyree, M.T., and F.W. Ewers. 1991. The hydraulic architecture of trees and other woody plants. *New Phytologist* 119: 345–60.

Tyree, M.T., and J.S. Sperry. 1989. Vulnerability of xylem to cavitation and embolism. *Annual Review of Plant Physiological and Molecular Biology* 40: 19–38.

Upchurch, G.R., Jr., B.L. Otto-Bliesner, and C.R. Scotese. 1999. Terrestrial vegetation and its effect on climate during the latest Cretaceous. In *Evolution of the Cretaceous ocean-climate system,* ed. E. Barrera and C.C. Johnson, 406–26. Boulder, Colorado: Geological Society of America.

Upchurch, G.R., Jr., and J.A. Wolfe. 1993. Cretaceous vegetation of the Western Interior and adjacent regions of North America. *Geological Association of Canada Special Publications* 39: 243–81.

Vygodskaya, N.N., I. Milyukova, A. Varlagin, F. Tatarinov, A. Sogachev, K.I. Kobak, R.

Desyatkin, G. Bauer, D.Y. Hollinger, F.M. Kelliher, E.D. Schulze. 1997. Leaf conductance and CO_2 assimilation of *Larix gmelinii* growing in an eastern Siberian boreal forest. *Tree Physiology* 17: 607–15.

Volk, T. 1989. Rise of angiosperms as a factor in long-term climatic cooling. *Geology* 17: 107–10.

von Caemmerer, S., and J.R. Evans. 1991. Determination of the average partial pressure of CO_2 in chloroplasts from leaves of several C3 plants. *Australian Journal of Plant Physiology* 18: 287–306.

Watson, J. 1988. The Cheirolepidaceae. In *Origin and Evolution of Gymnosperms,* ed. C.B. Beck. New York: Columbia University Press.

Wheeler, E.A., and P. Baas. 1991. A survey of the fossil record for dicotyledonous wood and its significance for evolutionary and ecological wood anatomy. *International Association of Wood Anatomists Bulletin* 12: 275–332.

Willis, K.J., and J.C. McElwain. 2002. *The evolution of plants.* Oxford: Oxford University Press.

Wolfe, J.A., and G.R. Upchurch, Jr. 1987. North American non marine climates and vegetation during the late Cretaceous. *Paleogeography, Paleoclimatology, Paleoecology* 61: 33–77.

Ziegler, A.M., 1998. Warm Polar Currents. EOS Trans., AGU 78, Spring Meeting Suppl.

Ziegler, A.M., C.R. Scotese, S.F. Barrett. 1982. Mesozoic and Cenozoic paleogeographic maps. In *Tidal friction and the Earths rotation II,* Ed. P. Brosche, J. Sundermann. Berlin: Springer-Verlag.

Zimmermann, M.H. 1983. *Xylem structure and the ascent of sap.* Berlin: Springer-Verlag.

Zimmermann, M.H., and A.A. Jeje. 1981. Vessel-length distribution in stems of some American woody plants. *Canadian Journal of Botany* 59: 18.

8. Influence of Uplift, Weathering, and Base Cation Supply on Past and Future CO_2 Levels

Jacob R. Waldbauer and C. Page Chamberlain

8.1 Tectonics, Rock Weathering, and CO_2 Drawdown

The weathering of silicate minerals has long been recognized as one of the dominant controls on atmospheric CO_2. A pair of well-known reactions (Urey 1952) summarizes a series of biogeochemical processes that begin with the weathering of calcium and magnesium silicates and end with the burial of Ca-Mg carbonates in marine sediments:

$$CO_2 + CaSiO_3 \leftrightarrow CaCO_3 + SiO_2$$
$$CO_2 + MgSiO_3 \leftrightarrow MgCO_3 + SiO_2$$

The net result is that, for each mole of calcium or magnesium silicate weathered on the continents, one mole of carbon dioxide is sequestered in marine carbonate. Thermal decomposition of these carbonates releases the CO_2 back into the atmosphere. Along with fixation of CO_2 by plants and its release during oxidative weathering of organic matter, silicate weathering is the major process controlling atmospheric CO_2 levels on the multimillion-year timescale (Berner and Kothavala 2001).

In addition to causing this direct effect on atmospheric CO_2, rock weathering can influence carbon dioxide levels indirectly. In many ecosystems, the weathering of rock containing elements essential for plant growth (including Ca, Mg, K, and P) is a primary source of these nutrients. A deficit of cation nutrients can inhibit plant communities' capacity as CO_2 sinks by acting as the limiting factor in the synthesis of new biomass. Recently, increasing attention has been paid to declining levels of cation nutrients in forest ecosystems, particularly in the eastern United States (Federer et al. 1989; Bailey et al. 1996; Huntington et al. 2000). The prospect of large forests serving as long-term sinks for anthropogenic CO_2 is contingent upon the continued availability of base cations, a nutrient pool vulnerable to depletion by acid rain (Likens, Driscoll, and Buso 1996). In order to realistically gauge the potential for further forest growth, then, quantification of weathering rates and consequent base cation supplies is essential.

In discussions of rock weathering, there is an important distinction to be drawn between physical and chemical mechanisms. Broadly, physical weathering produces solids that are transported without significant chemical alteration as small water- or windborne particles. Chemical weathering, by contrast, releases dissolved solutes, the products of chemical alteration of primary minerals. The two types of processes reinforce each other. Chemical weathering disaggregates solid rock, increasing susceptibility to physical removal, while physical weathering fractures rock and increases the amount of mineral surface area accessible to water (White and Brantley 1995). Chemical weathering and physical weathering are quite distinct, however, in their effects on biological and atmospheric systems. Only chemical weathering draws down carbon dioxide from the atmosphere, and only chemical weathering directly produces soil nutrients, such as calcium, magnesium, and potassium, that are usable by plants.

A large number of variables exert influence on rates of rock weathering. Mineralogy of the bedrock is of primary importance, as the types of minerals present and their cation content will determine both how much CO_2 will be consumed by their weathering as well as how much nutrition the rock might supply to plants growing on its surface. The effect of climate on weathering rates has received significant attention, not least because weathering, in its role as a CO_2 sink, has a reciprocal effect on climate. White and Blum (1995) found a significant increase in weathering rates in granitic catchments with both temperature and precipitation, a trend that has been incorporated into global CO_2 models (Berner 1995). A persistent difficulty in quantifying the effects of climate on weathering has been isolating individual climatic variables from each other and from such factors as geomorphology, tectonics, vegetation, and lithology, a problem exacerbated by increasing the scale of investigation (White et al. 1999b).

The other major factor in determining long-term, large-scale weathering rates is tectonic uplift. Ultimately, it is uplift that exposes fresh continental material to weathering and so determines the total amount of weatherable material on Earth's surface. A problem in incorporating the role of uplift and mountain building into global CO_2 models is the lack of a clear paleoindicator of continental weathering rates. The strontium isotope composition of ocean water has been used as such a proxy (Raymo and Ruddiman 1992; Richter, Rowley, and DePaolo 1992), but the interpretation of the seawater Sr record is not so straightforward (Blum 1997; Berner and Kothavala 2001). The role of uplift in setting weathering rates, though incompletely quantified, is nevertheless essential to the global carbon cycle, as shown by the strong correlation of chemical and physical weathering rates in many world rivers (Gaillardet et al. 1999). Riebe et al. (2001) found that, for a group of granitic catchments in the Sierra Nevada Mountains, tectonics rather than climate is the decisive factor controlling long-term weathering rates.

In this paper, we present an analytical model for steady-state chemical weathering with an emphasis on the effect of uplift on the total rate of weathering in a depth profile. The model is parameterized in terms of the effective surface age, which allows comparisons of widely disparate tectonic regimes. From this model, we derive supply rates of base cation nutrients (Ca, Mg, and K) and discuss the implications for long-term CO_2 sequestration by forests. We also model the effect of the presence of disseminated hydrothermal calcite on the Sr isotope composition of weathered material and discuss how this affects interpretation of the seawater strontium record.

8.2 Modeling Steady-State Chemical Weathering

We sought to develop a model for steady-state chemical weathering for the case where uplift provides a constant supply of fresh rock to the bottom of the weathering zone and physical erosion removes incompletely weathered material from the surface at an equivalent rate. These conditions produce an unchanging total soil depth and stable flux of cations from the chemical weathering of parent rock. The situation is diagrammed in Fig. 8.1. Of particular interest is how variations in uplift velocity (and hence in supply rate of fresh material) affect the flux of cations from the weathering profile.

This balance of erosion and uplift occurs most clearly in active tectonic environments known as steady-state orogens (Koons 1990; Willet, Slingerland, and Hovius 2001). At convergent plate boundaries, thickening of the crust leads to high uplift rates due to isostasy. The rise of mountain belts results in an increase in erosion rates due to the additional relief (Ahnert 1970) and increased precipitation (Willet 1999). These two competing processes, uplift and erosion, reach a dynamic balance and create stable large-scale topography. Under these conditions, rock moves through into the weathering zone at relatively high velocities (on the order of millimeters per year), is partially chemically altered in the weathering zone, and is physically removed at the surface quickly enough to maintain a constant weathering profile. These steady-state orogens are found in essentially all active tectonic areas.

Beginning with unweathered rock of known composition, we modeled the chemical weathering rate of the various constituent minerals and the total release of cations from the weathering profile as material moves from the bottom of the weathering zone to the surface at constant velocity. Under semi-neutral conditions (pH ~5–8), chemical weathering rate is independent of pH and the kinetics of dissolution are zeroth order with respect to water (Sverdrup 1990).

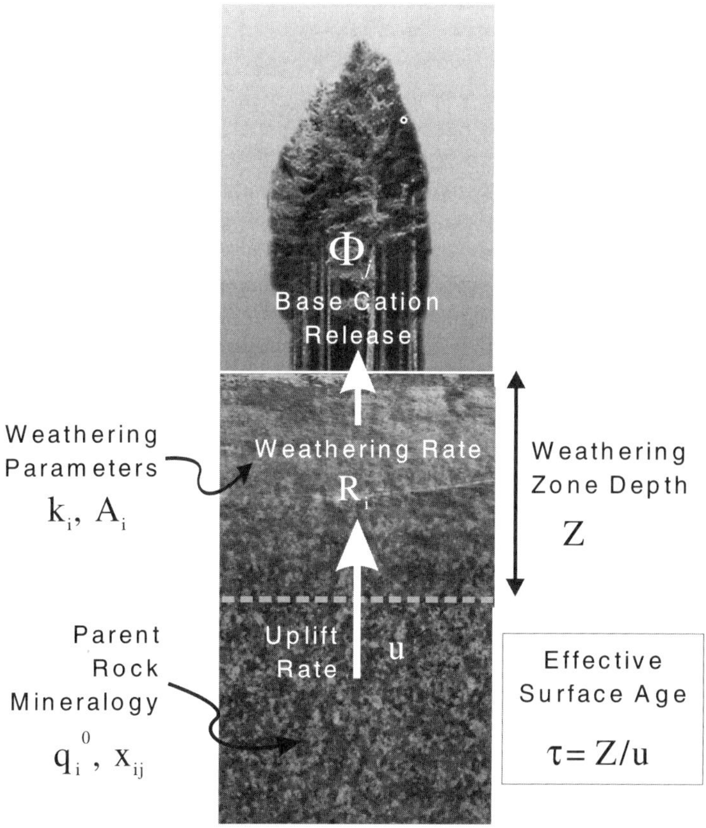

Figure 8.1. Diagram of steady-state chemical weathering model parameters and processes. Rock containing an initial (unweathered) concentration q_i^0 of mineral i is pushed upward with uplift velocity u into the weathering zone, which extends to a depth Z below the surface. In the weathering zone, the mineral weathers at a rate (R_i) dependent on the dissolution rate (k_i) and the mineral specific surface area (A_i). The model is parameterized in terms of the effective surface age, $\tau = Z/u$. Cation release rates (Φ_j) are calculated from mineral weathering rates and cation stoichiometry of the parent rock minerals (x_{ij}).

The change in concentration q_i (in mol/m^3) of a mineral i with time due to chemical weathering can be expressed as a function of the dissolution rate k_i (in units of mol/m^2/s) and the mineral specific surface area A_i (in m^2/mol):

$$\frac{dq_i}{dt} = -k_i A_i q_i \tag{8.1}$$

In the steady state described by this model, the weathering profile remains constant; that is, the composition of material as a function of depth within the weathering zone does not change with time. Hence the condition that $(\partial q_i\ \partial t)_z = 0$, and we may write:

$$\left(\frac{\partial q_i}{\partial t}\right)_z = 0 = -k_iA_iq_i - \frac{dz}{dt}\left(\frac{\partial q_i}{\partial z}\right)_t \tag{8.2}$$

In a state of steady uplift at a velocity u, $z = ut$ where $z = t = 0$ at the bottom of the weathering zone.

Thus $dz/dt = u$, and the solution of Equation 8.2 for the concentration as a function of distance from the bottom of the weathering zone is:

$$q_i(z) = q_i^0 \exp\left[-k_iA_i\frac{z}{u}\right] \tag{8.3}$$

where q_t^0 is the concentration of the mineral in the unweathered rock. Equation 8.3 describes how the concentration of a particular mineral changes as the host rock moves through the weathering zone. Of interest, however, is the total rate of chemical weathering in the entire weathering profile. This rate can be calculated by integrating this changing concentration over the depth of the weathering profile Z:

$$R_i' = -\frac{dq_i}{dt} = k_iA_i\int_0^z q_i(z)dz = q_i^0u\left[1 - \exp\left(-k_iA_i\frac{Z}{u}\right)\right] \tag{8.4}$$

Equation 8.3 gives the rate at which mineral i weathers as a function of weathering zone depth Z and uplift rate u. Note that R_i' is in units of mol/(m^2·s): i.e., it is a rate normalized to (ground) surface area. We can define an effective surface age τ, which is the amount of time it takes for rock to move from the lower boundary of the weathering profile to the surface, as

$$\tau = \frac{Z}{u} \tag{8.5}$$

The notion of effective surface age is a useful one, as it allows the weathering rate to be expressed as a function of a single variable. This can be done by normalizing the weathering rate instead to *volume* of weathering profile; we define $R_i = R_i'/Z$ so that the weathering rate R_i is now in units of mol/(m^3·s) and express the weathering rate of mineral i in terms of the effective surface age τ as:

$$R_i = \frac{q_i^0}{\tau}\left[1 - \exp(-k_iA_i\tau)\right] \tag{8.6}$$

Assuming complete and rapid physical weathering upon reaching the surface, the effective surface age is the timescale for chemical weathering of parent rock. The parameters that determine τ, the weathering zone depth Z (also termed the depth to bedrock), and the uplift velocity u are variables that are directly observable in the field. Furthermore, Z and u are regional quantities, whose values are generally uniform over relatively large areas of Earth's surface. This relationship makes the effective surface age a useful metric for comparing weathering rates in different tectonic regimes on a global scale.

It should be emphasized that effective surface ages vary greatly in differing tectonic environments. In an active collisional orogen like the Himalayas or Andes, uplift rates can be as high as 10 mm/yr and weathering zones are about 2 m deep, resulting in an effective surface age of 200 years. By contrast, a tectonically inactive craton in a tropical climate can be uplifting at only .01 mm/yr or less, while the weathering zone might be 30 m or deeper, so effective surface ages in these environments are in the range of 3–10 million years. This five-orders-of-magnitude range in effective surface ages is attributable more to the range of uplift rates across the globe (~four orders of magnitude) than to variations in the depth-to-bedrock (~two orders of magnitude); hence, effective surface age is more (though not exclusively) a measure of tectonic uplift than of weathering profile depth. It should also be recognized that uplift rate and weathering zone depth are not entirely independent, as very deep weathering profiles simply do not develop on rapidly uplifting bedrock, nor do stable cratons persist with only thin veneers of soils on top.

In order to recover quantitative estimates of chemical weathering rates from Equation 8.6, values of the specific surface area (A_i) for all the minerals of interest are required. Establishing accurate and consistent values for the specific surface area of minerals has been a difficult problem (see, e.g., Hochella and Banfield 1995; Brantley, White, and Hodson 1999; Gautier, Oelkers, and Schott 2001), and field and laboratory measurements are often at odds. For this model we chose to take the specific surface area to be the product of the geometric surface area for a given particle size and the surface roughness SR_i, defined as the ratio of the experimental BET surface area (measured by gas-adsorption methods) to the calculated geometric surface area (Sverdrup 1990; White 1995). This method, while perhaps an imprecise estimate of the true reactive surface area of a mineral grain, is the most direct method to obtain a self-consistent set of surface area values over a range of grain sizes and mineral types. Taking this approach, the weathering rate is:

$$R_i = \frac{q_i^0}{\tau}\left[1 - \exp\left(-\frac{6k_iSR_i}{\rho_i d_i}\tau\right)\right] \tag{8.7}$$

where ρ_i is the molar density of the mineral, d_i is the grain size (diameter of an assumed spherical grain), and SR_i is the surface roughness.

Estimates of the flux of base cation nutrients from a weathering profile can be made from this formula if the rock composition is known in detail. If a mole of mineral i contains x_{ij} moles of cation species j (e.g., for anorthite, $CaA1_2Si_2O_8$, $x_{An,Ca}=1$), then the flux, Φ_j, of that cation per cubic meter of weathering profile is:

$$\Phi_j = \sum_8 x_{ij}R_i \tag{8.8}$$

A value for the total flux from a catchment is obtained by multiplying the cation flux Φ_j by the volume of weathering profile in the catchment (its surface area times the depth of the weathering profile Z).

This model is admittedly simplified in several points. First, we assume that the weathering profile is in a steady state and does not change significantly in depth. This assumption certainly holds in the steady-state orogen described above and may be essentially valid for a broader range of circumstances (Heimsath et al. 1997). However, areas undergoing rapid physical erosion uncompensated by uplift, or where the soil profile is thickening significantly, will clearly fall outside its scope. Therefore, this model applies best to weathering-limited conditions (Stallard and Edmond 1983) where transport of weathered material is unhindered.

Second, we ignore the production of secondary minerals, such as clays, during the weathering of silicates. Since clay minerals contain very small amounts of the base cations of primary interest here, this step can be justified. However, the release of elements that are major constituents of clays, such as aluminum, cannot be calculated in the way described above, as they will be strongly retained in the soil profile. The availability of phosphorus, too, cannot be accurately predicted because of the complexation of phosphate with iron in secondary minerals. As shown by Walker and Syers (1976) in their study of New Zealand soils, weathering of apatite initially releases phosphate, but as the soil profile develops, more and more phosphate is complexed with Fe and becomes unavailable to plants. A third simplifying assumption is that all minerals begin weathering at the same depth and mineral surface reactivity remains essentially constant, which may not hold true in some environments (White et al. 2001). The effects of these assumptions do not, we feel, outweigh the overriding fact that uplift determines the rate at which fresh bedrock is exposed to weathering, and thereby tectonics plays a decisive role in weathering on a global scale.

8.3 Model Results: Comparison of CO_2 Drawdown and Base Cation Supply Between Tectonic Regimes

In Fig. 8.2 and Fig. 8.3, we show the results of this steady-state weathering model for the weathering of a rock of typical granitic composition, 20% quartz (SiO_2; $k_i=1 \times 10^{-15}$), 25% orthoclase feldspar ($KAlSi_3O_8$; $k_i=4 \times 10^{-13}$), 40% oligoclase feldspar ($Ca_2Na_8Al_{1.2}Si_{2.8}O_8$; $k_i=3 \times 10^{-12}$), and 15% biotite ($KMg_{1.2}Fe_{1.8}(AlSi_3O_{10})(OH)_2$; $k_i=6 \times 10^{-13}$). Indicated on the figures are ranges of effective surface ages characteristic of different tectonic environments.

The chemical weathering rate of the granitic substrate declines sharply with increasing surface age (Fig. 8.2A), dropping by more than an order of magnitude between typical active orogen effective surface ages and temperate craton conditions. For areas with $\tau > 100{,}000$ yr, the chemical weathering rate is very low, and for the oldest (i.e., slowest-uplifting, thickest-mantled) tropical cratons, the rate is approx 10^{-11} mol/($m^3 \cdot s$). These results are in general agreement with published measurements of weathering rates, as shown in Fig. 8.2B), which plots the observed annual silicon discharge from a range of world rivers against ef-

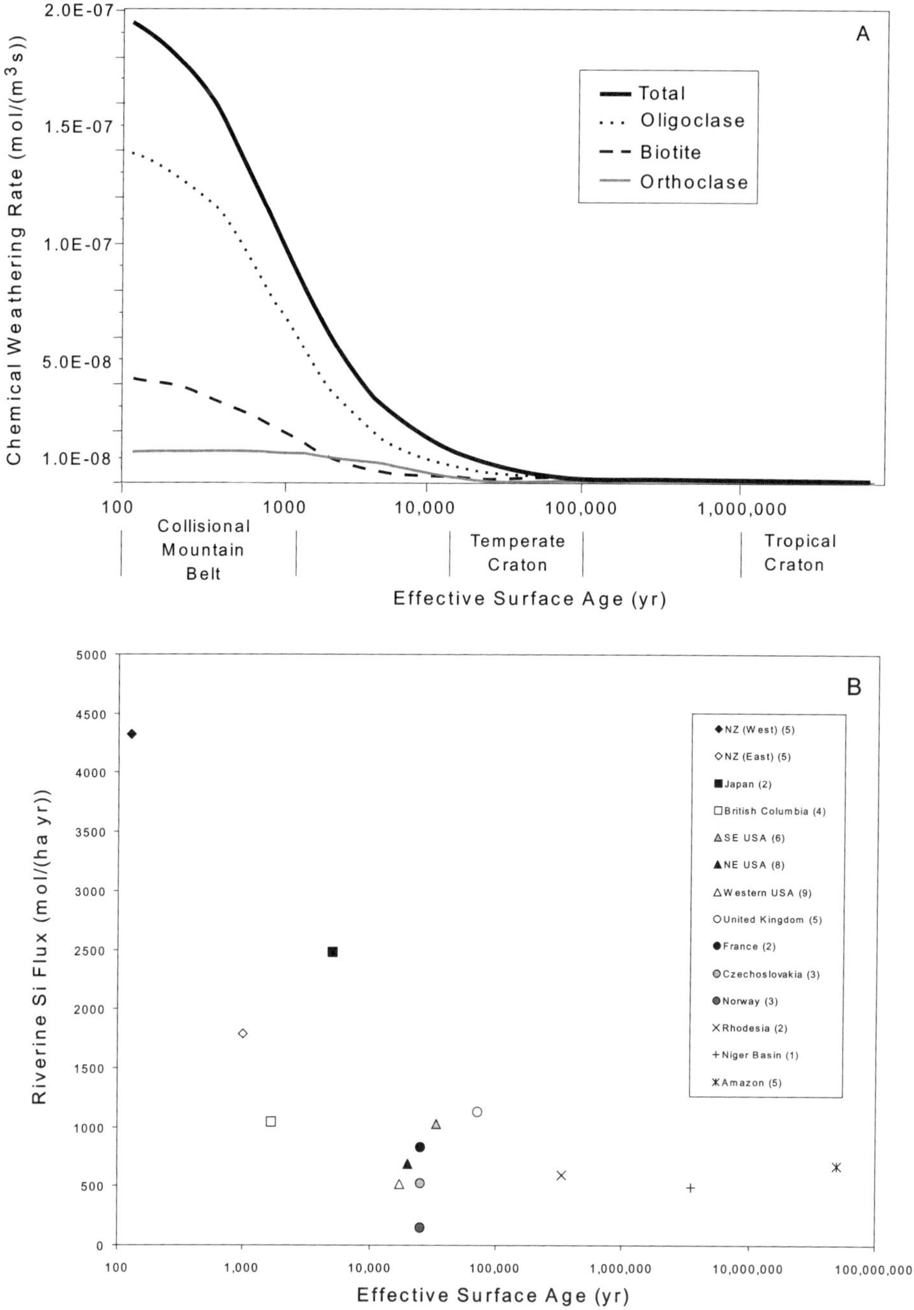

Figure 8.2. **(A)** Results of steady-state chemical weathering model for granitic material. Ranges of effective surface age values for different tectonic regimes are indicated. **(B)** Riverine silicon fluxes as a function of effective surface age for a number of world rivers. Numbers following labels in legend indicate number of measurements averaged to give the value for a particular area. Data from: Schaer et al. (1975); Kronberg et al. (1979); Miller and Lakatos (1983); Bierman et al. (1993); Tippet and Kamp (1993); Kelsey and Bockheim (1994); Simpson et al. (1994); White and Blum (1995); Gaillardet et al. (1997); Glasser and Hall (1997); Small et al. (1997); Boeglin and Probst (1998); Viers et al. (2000); Hiroshi et al. (2001); Jacobson et al. (2003).

fective surface age for the area in question. Riverine silicon is a useful field index for rock weathering because it derives almost entirely from the weathering of silicate minerals and its flux is not skewed by the presence of trace amounts of carbonates, oxides, or sulfides in the parent rock.

In Fig. 8.3 we show a similar comparison for release rates of base cations (Ca, Mg, and K). The cation release rate is very high in active orogen conditions and drops to negligible amounts on landscapes with $\tau > 100{,}000$ yr. As seen in Fig. 8.3B (note logarithmic scale), the observed cation fluxes do not decrease substantially for areas with greater surface ages. This is likely because of the lower bound on total flux due to atmospheric deposition, an input not taken into account in our model. The implication is that plant communities growing on high effective surface age landscapes are dependent primarily on atmospheric deposition to supply them with base cation nutrients, a suggestion supported by studies of forests growing on tropical cratons (Salati and Vose 1984; Markewitz et al. 2001).

The markedly higher weathering rates of low effective surface age environments mean that weathering in tectonically active areas is expected to draw down significantly more CO_2 than weathering in tectonically inactive areas. Hence, it is in the building of large mountain belts that the greatest amount of atmospheric carbon dioxide can be sequestered in marine carbonate. This result of the steady-state chemical weathering model is in good accord with the recent detailed field study of Mortatti and Probst (2003), who found that the Andes region accounts for 78% of CO_2 drawdown by silicate weathering in the Amazon basin. The largest orogens, for example the Himalayas, are likely to be recorded in the geochemical record; the interpretation of the strontium isotope composition of marine carbonates in this regard is discussed in the next section.

It should be remembered that this model calculates weathering rates normalized to volume of weathering profile. As a result, accurate comparison of large-scale weathering rates between environments depends on quantification of both exposed surface area and the depth at which weathering begins. In principle, one can use a model of this type—suitably extended to a variety of parent rock lithologies—to predict which areas of Earth's surface are the most active sinks for atmospheric CO_2. The importance of a large, tectonically active area such as the Himalayan chain is clear. Calculating the relative impact of weathering of a smaller orogen, such as the Southern Alps of New Zealand, is far more complicated. To obtain useful estimates of weathering in the disparate environments of these relatively small regions, a model would have to incorporate the nontrivial effects of climate (both temperature and precipitation), vegetation, and microbial activity, as well as factor in the geographical preponderance of weathering-limited versus transport-limited erosion regimes. For many of these variables, the requisite data simply do not exist on a global scale. As such, a comprehensive model of rock weathering and its effect on atmospheric CO_2 is still some ways off. It is clear from this model, though, that any attempt to calculate long-term, large-scale weathering rates must incorporate the effects of tectonic uplift.

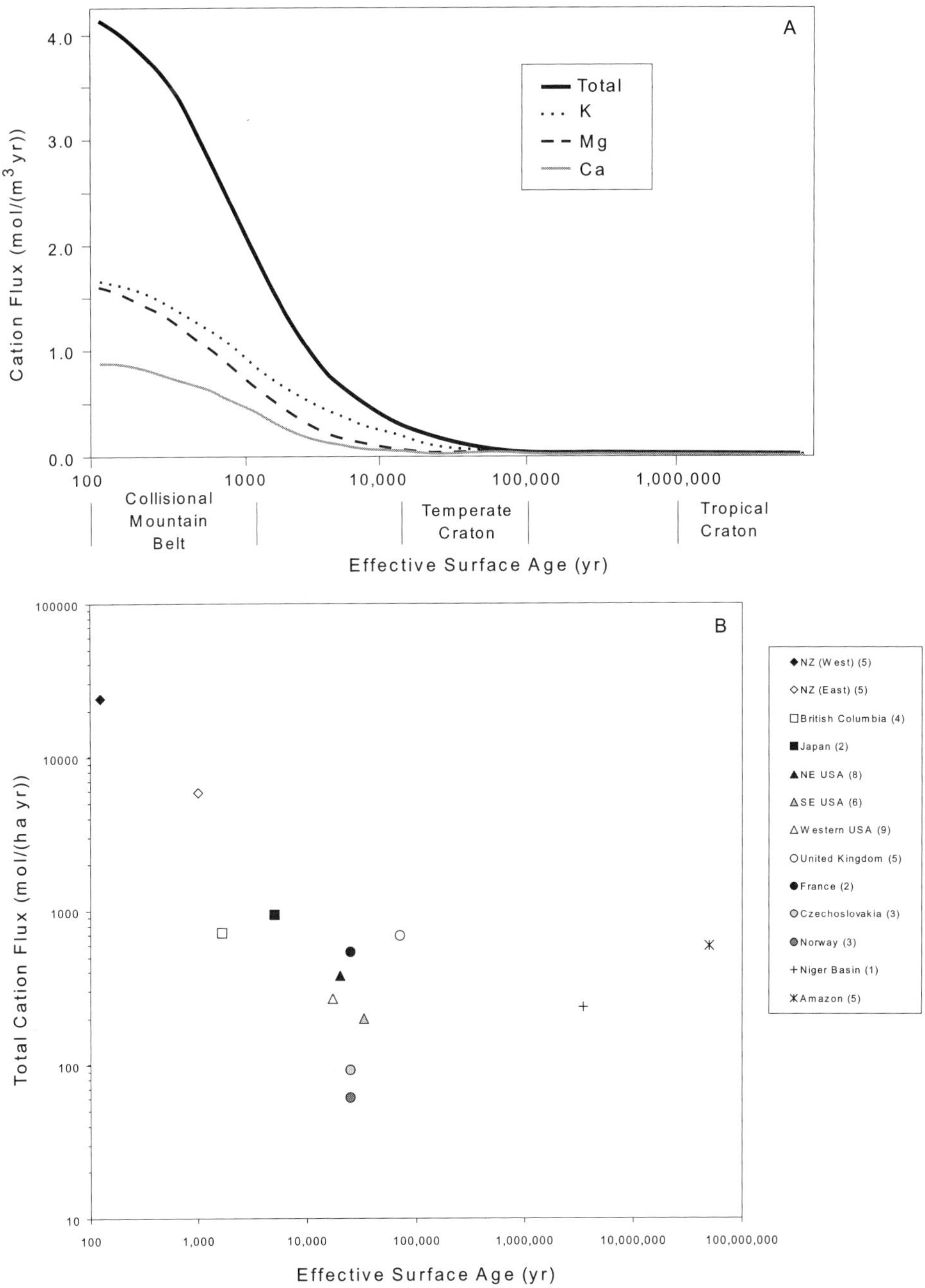

Figure 8.3. (A) Results of steady-state chemical weathering model for cation release from granitic material. Ranges of effective surface age values for different tectonic regimes are indicated. **(B)** Riverine cation fluxes as a function of effective surface age for a number of world rivers (note logarithmic scale). Numbers following labels in legend indicate number of measurements averaged to give the value for a particular area. Data from sources as in Fig. 8.2.

8.4 Interpretation of the Marine Strontium Isotope Record and Past CO_2 Drawdown

In order to reconstruct the history of atmospheric CO_2, it is necessary to have some information about the past continental rock weathering rates, as the weathering of silicate minerals serves as a major sink for atmospheric CO_2. One proxy that has been used to estimate past continental rock weathering is the strontium isotope composition of marine carbonates. The mechanism for the proxy is as follows: continental weathering of calcium silicates, rich in ^{87}Sr from the decay of ^{87}Rb, draws down CO_2 and produces dissolved carbonate (with a high $^{87}Sr/^{86}Sr$ ratio) in the oceans. This carbonate precipitates out of seawater to form carbonate minerals in marine sediments. An observed increase $^{87}Sr/^{86}Sr$ ratio of these marine carbonates, then, can be interpreted as an increase in continental weathering rates. The past 40 million years have seen a marked increase in the seawater $^{87}Sr/^{86}Sr$ ratio, a trend attributed to the India-Asia collision and consequent uplift of the Himalayas, which exposed large amounts of fresh granitic rock to weathering and possibly sequestered large amounts of atmospheric CO_2 in marine sediments (Edmond 1992; Richter, Rowley, and DePaolo 1992).

The strontium isotope composition of rock at Earth's surface is set by two factors: first, by the $^{87}Sr/^{86}Sr$ ratio of the constituent minerals at their closure temperature for strontium, which is typically about 300°C and achieved at about 15 km depth. At this point, strontium exchange between each mineral and any surrounding fluids stops. Second, the $^{87}Sr/^{86}Sr$ ratio of a mineral can increase due to the decay of ^{87}Rb to ^{87}Sr with a half-life of 48.8 Gyr. Uplift is critical for this second factor, because it determines how long it takes for a mineral to be brought from the closure depth to the surface where it can be weathered. Hence uplift will determine how much time ^{87}Rb has to decay and therefore the ultimate $^{87}Sr/^{86}Sr$ ratio of the weathered material. Taking a closure depth of 15 km and using the decay constant for ^{87}Rb, $\lambda = 1.42 \times 10^{-11}$ yr^{-1} (Steiger and Jäger 1978), the $^{87}Sr/^{86}Sr$ ratio of rock at the earth's surface can be written

$$(^{87}Sr/^{86}Sr)_{surface} = (^{87}Sr/^{86}Sr)_0 + (^{87}Rb/^{86}Sr)_0 (e^{\lambda(15 km/u)} - 1) \tag{8.9}$$

where the subscript 0 indicates the isotopic ratios at the closure temperature and u is the uplift velocity in km/yr. With knowledge of the initial ^{87}Sr and ^{87}Rb content of the rock, this equation provides a value for the strontium isotope composition of material weathering at the surface, which is the first step towards modeling how tectonics can influence the seawater Sr record.

A complication in the interpretation of the seawater strontium isotope record is the presence in many active orogen systems (including the Himalayas) of disseminated hydrothermal calcite. Large hydrothermal systems are a common feature of active collisional mountain belts (Koons and Craw 1991; Chamberlain et al. 1995). These hydrothermal systems are most vigorous in the areas of high uplift rates that have steep topography and an elevated brittle/ductile transition. A common feature of these hydrothermal systems is the abundant, secondary disseminated calcite within silicate rocks (Gazis et al. 1998; Templeton et al. 1998; White et al. 1999a). Recent work has shown that this calcite has elevated $^{87}Sr/^{86}Sr$ ratios as a result of equilibration with primary silicate minerals (Blum et al. 1998). Because calcite weathers far more rapidly than the host silicate minerals, the $^{87}Sr/^{86}Sr$ ratios of water in streams in these areas of rapid uplift are in large part dominated by weathering of calcite. It is, therefore, difficult if not impossible to distinguish calcite from silicate weathering with $^{87}Sr/^{86}Sr$ ratios alone.

Inclusion of even a small amount of interstitial calcite has a strong effect on the strontium flux from granitic rock due to weathering. This effect is shown in Fig. 8.4, Panel A, where 3% calcite ($CaCO_3$; $k_i = 1 \times 10^{-5}$) has been added to the granite composition described in the previous section, and $^{87}Rb/^{86}Sr$ ratios measured by Gazis et al. (1998) for a typical Himalayan granite included for the constituent minerals (oligoclase, .05; biotite, 441.4; orthoclase, 1.01). Representative strontium concentrations for minerals were obtained from Gazis et al. (1998) for oligoclase (128.2 ppm), biotite (3.5 ppm), and orthoclase (312.2 ppm) and from Jacobson and Blum (2000) for calcite (622.7 ppm). The calculated contribution of calcite to the total amount of strontium weathered from the rock is as high as 35% in tectonically active conditions, reflecting calcite's high Sr content and rapid weathering.

The calculated effect of the disseminated calcite on the isotopic composition of the weathered strontium is evident in Fig. 8.4B. The secondary, disseminated calcite dominates the strontium flux from low effective surface age areas, holding the $^{87}Sr/^{86}Sr$ ratio essentially constant over a range of tectonic conditions from active orogens to temperate cratons. As shown in the previous section, there is a tenfold decrease in the total chemical weathering rate over the same range. This result indicates that, where disseminated calcite is present, the isotopic composition of weathered Sr is insensitive to large changes in chemical weathering rate. Disseminated hydrothermal calcite has its largest effects on the Sr content of weathered material in tectonically active areas, which are precisely the places where weathering rates are highest and the most CO_2 is being drawn down. Furthermore, this model predicts that the most radiogenic strontium derives from regions of low total Sr flux, areas where silicate weathering consumes little atmospheric CO_2. Such a trend was reported by Lear et al. (2003) in their study of the Cenozoic benthic foraminiferal Sr/Ca record and reconstruction of global riverine Sr flux and isotopic composition. These results complicate interpretation of the seawater Sr isotope record, as they suggest that the isotopic composition of weathered strontium is not a clear index of the rate of continental silicate weathering.

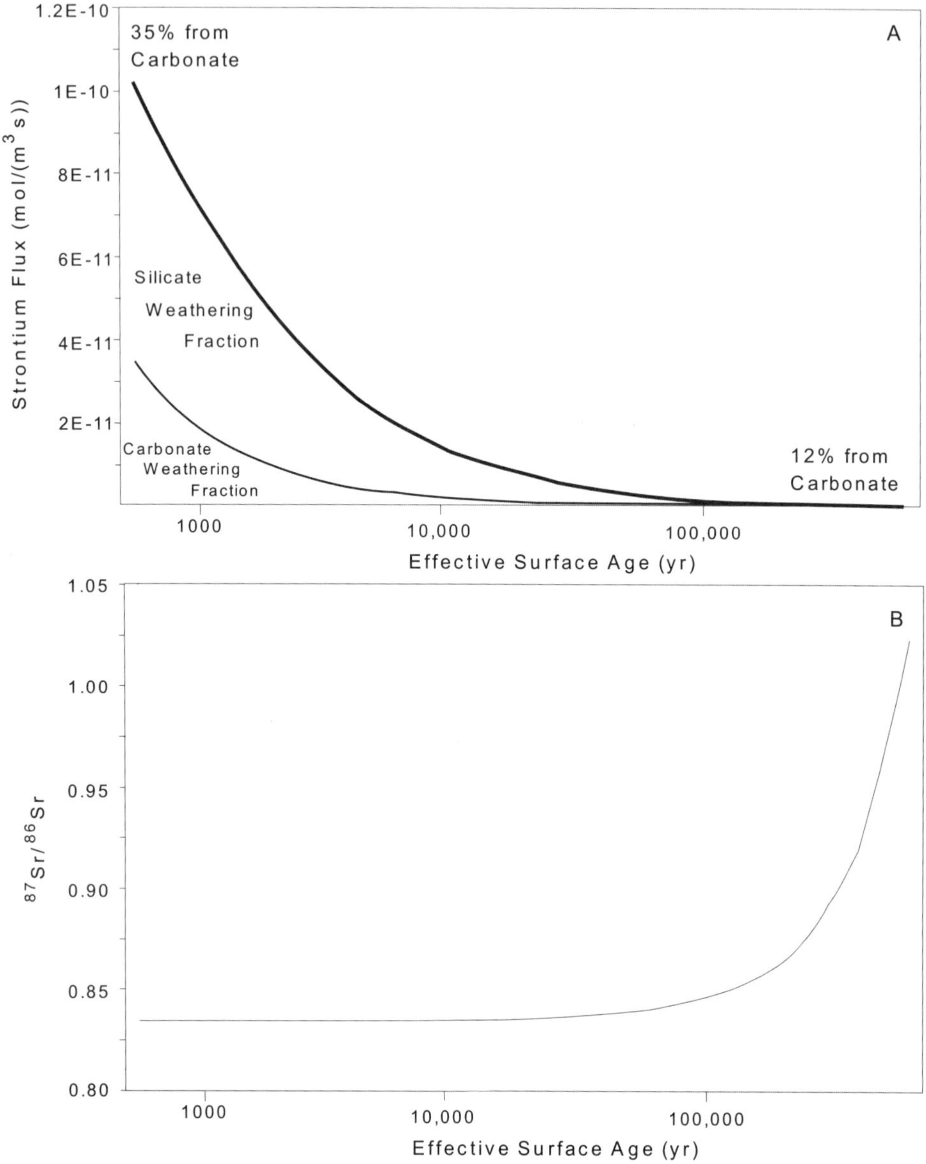

Figure 8.4. **(A)** Calculated Sr flux from weathering of granite with 3% disseminated calcite. The contribution of carbonate weathering to the total Sr flux ranges from 12% at τ = 650,000 yr to 35% at τ = 500 yr. **(B)** Calculated $^{87}Sr/^{86}Sr$ ratio of weathered material. The weathering of 3% calcite holds the isotopic composition essentially constant over a large range of effective surface ages; the total chemical weathering rate drops significantly over the same range. The most radiogenic Sr comes from regions with low total Sr flux.

8.5 Implications of Forest Base Cation Supply for Long-Term Sequestration of CO_2

The weathering of minerals containing base cations useful to plants (including Ca, K and Mg) is an important source of nutrients for many ecosystems (Waring and Schlesinger 1985). The stoichiometry for the growth of plant biomass by photosynthesis is approximately (Schnoor and Stumm 1985; Drever and Clow 1995):

$$\boxed{\begin{array}{ll} 800\,CO_2 & 1\,Al(OH)_2^{+} \\ 6\,NH_4^{+} & 1\,Fe^{2+} \\ 4\,Ca^{2+} & 2\,NO_3^{-} \\ 1\,Mg^{2+} & 1\,H_2PO_4^{-} \\ 2\,K^{+} & 1\,SO_4^{2-} \end{array}} + H_2O \xrightarrow[\textit{photosynthesis}]{\gamma} \text{Biomass} + 16\,H^{+} + 804\,O_2 \qquad (8.10)$$

At present atmospheric CO_2 levels, carbon dioxide is often a limiting nutrient for the synthesis of biomass in C_3 plants. Throughout the Phanerozoic, however, atmospheric CO_2 has repeatedly reached concentrations 10 or more times present levels (Berner and Kothavala 2001). At such high concentrations, CO_2 would cease to be a limiting nutrient and the supply of other species, principally nitrogen and the base cations, would determine plant growth rates. As anthropogenic inputs increase modern atmospheric CO_2 levels, there is the potential that an increase in total plant biomass—the greening of Earth's forests—might serve as a long-term sink for human-produced carbon dioxide. Such a sink will only be sustainable so long as the other requisite nutrients for plant growth are readily available and do not limit photosynthesis. Clearly, an understanding of how weathering supplies these nutrients to ecosystems is a prerequisite for both reconstructing how plants have responded to past changes in CO_2 levels and predicting how much carbon dioxide might be sequestered in present-day forests.

A growing body of evidence suggests that forest ecosystems may become cation-limited (and especially calcium-limited), particularly as a result of watershed acidification. Studies pointing to this conclusion have been carried out in both the northeastern (Federer et al. 1989; Bailey et al. 1996; Lawrence et al. 1997, 1999; Yanai et al. 1999) and southeastern United States (DeWalle et al. 1999; Huntington et al. 2000; Huntington 2000), as well as at several sites in Europe (Kram et al. 1997; Fichter et al. 1998). All of these areas would fall on the plots in Fig. 8.3 under the grouping of temperate cratons: i.e., areas with relatively low rates of base cation release due to weathering. Interestingly, human activity appears to have also significantly reduced atmospheric base cation deposition in these regions (Hedin et al. 1994).

As shown above, our model predicts much greater fluxes of base cations due to weathering under low effective surface age conditions. The implication is that base cations will not become limiting nutrients for forests growing in tectonically active areas. It is forests in these regions that are best suited to serve as long-term sinks for anthropogenic CO_2. These forests are also the least susceptible to soil nutrient leaching due to watershed acidification, since they receive a steady supply of nutrients from the weathering of fresh bedrock. Areas such as the eastern United States, on the other hand, are more vulnerable to losses due to acid rain, especially because a greater proportion of their forests' cation supplies comes from atmospheric deposition.

8.6 Summary

Tectonic uplift, through its control of chemical weathering rates—and, therefore, control of base cation supply rates to forests—exerts a powerful influence on atmospheric carbon dioxide. The analytical model presented in this paper quantifies the role of uplift in setting bulk weathering rates and demonstrates that areas with low effective surface ages (i.e., tectonically active areas) are likely to play the largest role as land-based CO_2 sinks. This is important both for reconstructing the history of atmospheric CO_2 and for predicting how and where human activity might have the most profound impacts on the global carbon dioxide cycle.

An interesting consideration in the discussion of the relationship between tectonics and weathering is the question of what it means for a landscape to be in a steady state. Over the longest periods, changes in plate motions will ultimately determine the condition of a particular tectonic environment, but shorter-timescale processes, especially glaciation, are also important. This question becomes particularly relevant to studies of atmospheric carbon dioxide because many of the northern latitude forest regions to which a great deal of CO_2 drawdown has been attributed have recently been glaciated. The forested temperate cratons, with effective surface ages in the 100,000-year range, sit on the border between weathering-dominated and atmosphere-dominated cation nutrient supplies.

Blum (1997) found that glacial soils in the Wind River Mountains younger than about 138,000 years had chemical weathering accelerated over steady-state rates. This span is close to the effective surface age, the amount of time the model presented here would predict it would take for material to move through the weathering profile in that environment. The implication is that forests growing on this terrain since the last glaciation have been supported by nutrient supplied from fresh rock exposed by the last glacial retreat, a supply that has essentially run out. If these northern latitude regions have in fact relatively recently reached a steady state with respect to weathering, they may actually become more cation limited in the future. It is worth noting that some northern areas, including parts of Canada and central Scandinavia, are still experiencing postglacial isostatic uplift. If the glacial-interglacial timescale is shorter than the time required to develop a constant weathering profile, then some areas may never actually reach what could be called a steady state.

The prospect of developing a global model for rock weathering that includes the combined effects of tectonics, climate, vegetation, and human activity along with glaciation, volcanism, and other geologic and geomorphic processes is a daunting one. The model presented here demonstrates, however, that a relatively simple framework can capture many of the main effects. Apparent areas for expansion of the scope of this model include quantifying the timescale for weathering profile development after glaciation, as well as the effects of the production and weathering of secondary minerals on release and retention of nutrients. Nevertheless, these additional factors will not mask the major role played by tectonic uplift and rock weathering in the dynamics of atmospheric CO_2 on a global scale.

Acknowledgments. The authors wish to thank Bob Berner and an anonymous reviewer for their helpful comments on this manuscript. This work was supported by the David and Lucile Packard Foundation.

References

Ahnert, F. 1970. Functional relationships between denudation, relief, and uplift in large mid-latitude drainage basins. *American Journal of Science* 268:243–63.

Bailey, S.W., J.W. Hornbeck, C.T. Driscoll, and H.E. Gaudette. 1996. Calcium inputs and transport in a base-poor forest ecosystem as interpreted by Sr isotopes. *Water Resources Research* 32:707–19.

Berner, R.A. 1995. Chemical Weathering and Its Effect on Atmospheric CO_2 and Climate. *Reviews in Mineralogy* 31:565–84.

Berner, R.A., and Z. Kothavala. 2001. GEOCARB III: A revised model of atmospheric CO_2 over Phanerozoic time. *American Journal of Science* 301:182–204.

Bierman, P.R., C.A. Massey, A.R. Gillespie, D. Elmore, and M. Caffe. 1993. Erosion rate and exposure age of granite landforms estimated with cosmogenic ^{36}Cl. *Geological Society of America Abstracts with Programs* 25:141.

Blum, J.D. 1997. The Effect of late Cenozoic glaciation and tectonic uplift on silicate weathering rates and the marine $^{87}Sr/^{86}Sr$ record. In *Tectonic uplift and climate change,* ed. W.F. Ruddiman, 260–289. New York: Plenum Press.

Blum, J.D., C.A. Gazis, A.D. Jacobson, and C.P. Chamberlain. 1998. Carbonate versus silicate weathering in the Raikhot watershed within the High Himalayan Crystalline Series. Geology 26:411–14.

Boeglin, J.-L., and J.-L. Probst. 1998. Physical and chemical weathering rates and CO_2 consumption in a tropical lateritic environment: the upper Niger basin. *Chemical Geology* 148:137–56.

Brantley, S.L., A.F. White, and M.E. Hodson. 1999. Surface area of primary silicate minerals. In *Growth, dissolution, and pattern formation in geosystems,* ed. B. Jamtveit and P. Meakin, 291–326. Dordrecht, Netherlands: Kluwer Academic Publishers.

Chamberlain, C.P., P.K. Zeitler, D.E. Barnet, D. Winslow, S.R. Poulson, T. Leahy, and J.E. Hammer. 1995. Active hydrothermal systems during the recent uplift of Nanga Parbat, Pakistan Himalaya. *Journal of Geophysical Research* 100:439–53.

DeWalle, D.R., J.S. Tepp, B.R. Swistock, W.E. Sharpe, and P.J. Edwards. 1999. Tree-ring cation response to experimental watershed acidification in West Virginia and Maine. *Journal of Environmental Quality* 28:299–309.

Drever, J.I., and D.W. Clow. 1995. Weathering rates in catchments. *Reviews in Mineralogy* 31:463–84.

Edmond, J.M. 1992. Himalayan tectonics, weathering processes, and the strontium isotope record in marine limestones. *Science* 258:1594–97.

Federer, C.A., J.W. Hornbeck, L.M. Tritton, C.M. Martin, R.S. Pierce, and C.T. Smith. 1989. Long-term depletion of calcium and other nutrients in Eastern US forests. *Environmental Management* 13:593–601.

Fichter, J., E. Dambrine, M.P. Turpault, and J. Ranger. 1998. Base cation supply in spruce and beech ecosystems of the Strengbach catchment (Vosges mountains, NE France). *Water Air and Soil Pollution* 104:125–48.

Gaillardet, J., B. Dupre, C.J. Allegre, and P. Negrel. 1997. Chemical and physical denudation in the Amazon River Basin. *Chemical Geology* 142: 141–73.

Gaillardet, J., B. Dupre, P. Louvat, and C.J. Allegre. 1999. Global silicate weathering and CO_2 consumption deduced from the chemistry of large rivers. *Chemical Geology* 159:3–30.

Gautier, J.-M., E.H. Oelkers, and J. Schott. 2001. Are quartz dissolution rates proportional to B.E.T. surface areas? *Geochimica et Cosmochimica Acta* 65:1059–70.

Gazis, C.A., J.D. Blum, C.P. Chamberlain, and M. Poage. 1998. Isotope systematics of granites and gneisses of the Nanga Parbat massif, Pakistan Himalaya. *American Journal of Science* 298:673–98.

Glasser, N.F., and A.M. Hall. 1997. Calculating Quaternary glacial erosion rates in northeast Scotland. *Geomorphology* 20: 29–48.

Hedin, L.O., L. Granat, G.E. Likens, T.A. Buishand, J.N. Galloway, T.J. Butler, and H. Rodhe. 1994. Steep declines in atmospheric base cations in regions of Europe and North America. *Nature* 367:351–54.

Heimsath, A.M., W.E. Dietrich, K. Nishiizumi, and R.C. Finkel. 1997. The soil production function and landscape equilibrium. *Nature* 388:358–61.

Hiroshi, S., O. Jun'ichi, N. Masao, and M. Yasuo. 2001. Holocene uplift derived from relative sea-level records along the coast of western Kobe, Japan. *Quaternary Science Reviews* 20:1459–74.

Hochella, M.F. Jr., and J.F. Banfield. 1995. Chemical weathering of silicates in nature: A microscopic perspective with theoretical considerations. *Reviews in Mineralogy* 31: 353–406.

Huntington, T.G. 2000. The potential for calcium depletion in forest ecosystems of southeastern United States: Review and analysis. *Global Biogeochemical Cycles* 14:623–38.

Huntington, T.G., R.P. Hooper, C.E. Johnson, B.T. Aulenbach, R. Capellato, and A.E. Blum. 2000. Calcium depletion in a Southeastern United States forest ecosystem. *Soil Science Society of America Journal* 64:1845–58.

Jacobson, A.D., and J.D. Blum. 2000. Ca/Sr and $^{87}Sr/^{86}Sr$ geochemistry of disseminated calcite in Himalayan silicate rocks from Nanga Parbat: Influence on river-water chemistry. *Geology* 28:463–66.

Jacobson, A.D., J.D. Blum, C.P. Chamberlain, D. Craw, and P.O. Koons. 2003. Climatic and tectonic controls on chemical weathering in the New Zealand Southern Alps. *Geochimica et Cosmochimica Acta* 67:29–46.

Kelsey, H.M., and J.G. Bockheim. 1994. Coastal landscape evolution as a function of eustasy and surface uplift rate, Cascadia margin, southern Oregon. *Geological Society of America Bulletin* 106:840–54.

Koons, P.O. 1990. The two-sided orogen: Collision and erosion from the sand box to the Southern Alps, New Zealand. *Geology* 18:679–82.

Koons, P.O., and D. Craw. 1991. Evolution of fluid driving forces and composition within collisional orogens. *Geophysical Research Letters* 280:935–38.

Kram, P., J. Hruska, B.S. Wenner, C.T. Driscoll, and C.E. Johnson. 1997. The biogeochemistry of basic cations in two forest catchments with contrasting lithology in the Czech Republic. *Biogeochemistry* 37:173–202.

Kronberg, B.I., W.S. Fyfe, O.H. Leonardos, and A.M. Santos. 1979. The chemistry of some Brazilian soils: Element mobility during intense weathering. *Chemical Geology* 24:211–29.

Lawrence, G.B., M.B. David, S.W. Bailey, and W.C. Shortle. 1997. Assessment of soil calcium status in red spruce forests in the northeastern United States. *Biogeochemistry* 38:19–39.

Lawrence, G.B., M.B. David, G.M. Lovett, P.S. Murdoch, D.A. Burns, J.L. Stoddard, B.P. Baldigo, J.H. Porter, and A.W. Thompson. 1999. Soil calcium status and the response of stream chemistry to changing acidic deposition rates. *Ecological Applications* 9:1059–72.

Lear, C.H., H. Elderfield, and P.A. Wilson. 2003. A Cenozoic seawater Sr/Ca record from benthic foraminiferal calcite and its application in determining global weathering fluxes. *Earth and Planetary Science Letters* 208:69–84.

Likens, G.E., C.T. Driscoll, and D.C. Buso. 1996. Long-term effects of acid rain: Response and recovery of a forest ecosystem. *Science* 272:244–46.

Markewitz, D., E.A. Davidson, R. de O. Figuelredo, R.L. Victoria, and A.V. Krusche. 2001. Control of cation concentrations in stream waters by surface soil processes in an Amazonian watershed. *Nature* 410:802–805.

Miller, D.S., and S. Lakatos. 1983. Uplift rate of Adirondack anorthosite measured by fission-track analysis of apatite. *Geology* 10:284–86.

Mortatti, J., and J.-L. Probst. 2003. Silicate rock weathering and atmospheric/soil CO_2 uptake in the Amazon basin estimated from river water geochemistry: seasonal and spatial variations. *Chemical Geology* 197:177–96.

Raymo, M.E., and W.F. Ruddiman. 1992. Tectonic forcing of late Cenozoic climate. *Nature* 359:117–22.

Richter, F.M., D.B. Rowley, and D.J. DePaolo. 1992. Sr isotope evolution of seawater: the role of tectonics. *Earth and Planetary Science Letters* 109:11–23.

Riebe, C.S., J.W. Kirchner, D.E. Granger, and R.C. Finkel. 2001. Strong tectonic and weak climatic control of long-term chemical weathering rates. *Geology* 26:511–14.

Salati, E., and P.B. Vose. 1984. Amazon basin: A system in equilibrium. *Science* 225: 129–38.

Schaer, J.P., G.M. Reimer, and G.A. Wagner. 1975. Actual and ancient uplift in the Gotthard region, Swiss Alps: A comparison between precise leveling and fission-track apatite age. *Tectonophysics* 29:293–300.

Schnoor, J.L., and W. Stumm. 1985. Acidification of aquatic and terrestrial ecosystems. In *Chemical processes in lakes,* ed. W. Stumm, 475–504. New York: Wiley.

Simpson, G.D.H., A.F. Cooper, and R.J. Norris. 1994. Late quaternary evolution of the Alpine fault zone at Paringa, South Westland, New Zealand. *New Zealand Journal of Geology and Geophysics* 37:49–58.

Small, E.E., R.S. Anderson, J.L. Repka, and R. Finkel. 1997. Erosion rates of alpine bedrock summit surfaces deduced from in situ ^{10}Be and ^{26}Al. *Earth and Planetary Science Letters* 150:413–25.

Stallard, R.F., and J.M. Edmond. 1983. Geochemistry of the Amazon 2: The influence of geology and weathering environment on the dissolved load. *Journal of Geophysical Research* 88:9688–9761.

Stallard, R.F. 1995. Relating chemical and physical erosion. *Reviews in Mineralogy* 31: 543–64.

Steiger, R.H., and E. Jäger. 1978. Subcommission on Geochronology: Convention on the use of decay constants in geochronology and cosmochronology. *Studies in Geology* 6:67–71.

Sverdrup, H.U. 1990. *The kinetics of base cation release due to chemical weathering.* Lund, Sweden: Lund University Press.

Templeton, A.S., C.P. Chamberlain, P.O. Koons, and D. Craw. 1998. Stable isotopic evidence for mixing between metamorphic and surface-derived waters during uplift of the Southern Alps, New Zealand. *Earth and Planetary Science Letters* 154: 73–92.

Tippet, J.M., and P.J.J. Kamp. 1993. Fission track analysis of the late Cenozoic vertical

kinematics of continental Pacific crust, South Island, New Zealand. *Journal of Geophysical Research* 98:16,119–48.
Urey, H.C. 1952. *The planets: Their origin and development.* New Haven: Yale University Press.
Viers, J., B. Dupre, J. Braun, S. Debert, B. Angeletti, J.N. Ngoupayou, and A. Michard. 2000. Major and trace element abundances, and strontium isotopes in the Nyong basin rivers (Cameroon): Constraints on chemical weathering processes and element transport mechanisms in humid tropical environments. *Chemical Geology* 169:211–41.
Walker, T.W., and J.K. Syers. 1976. The fate of phosphorus during pedogenesis. *Geoderma* 15:1–19.
Waring, R.H., and W.H. Schlesinger. 1985. *Forest ecosystems: Concepts and management.* San Diego: Academic Press.
White, A.F. 1995. Chemical weathering rates of silicate minerals in soils. *Reviews in Mineralogy* 31:407–61.
White, A.F., and A.E. Blum. 1995. Effects of climate on chemical weathering rates in watersheds. *Geochimica et Cosmochimica Acta* 59:1729–47.
White, A.F., A.E. Blum, T.D. Bullen, D.V. Vivit, M.S. Schulz, and J. Fitzpatrick. 1999b. The effect of temperature on experimental and natural weathering rates of granitoid rocks. *Geochimica et Cosmochimica Acta* 63:3277–91.
White, A.F., and S.L. Brantley. 1995. Chemical weathering rates of silicate minerals: An overview. *Reviews in Mineralogy* 31:1–22.
White A.F., T.D. Bullen, D.V. Vivit, and M.S. Schulz. 1999a. The role of disseminated calcite in the chemical weathering of granitoid rocks. *Geochimica et Cosmochimica Acta* 63:1939–53.
White, A.F., T.D. Bullen, M.S. Schulz, A.E. Blum, T.G. Huntington, and N.E. Peters. 2001. Differential rates of feldspar weathering in granitic regoliths. *Geochimica et Cosmochimica Acta* 65:847–69.
Willett, S.D. 1999. Orogeny and orography: The effects of erosion on the structure of mountain belts. *Journal of Geophysical Research* 104:28957–81.
Willett, S.D., R. Slingerland, and N. Hovius. 2001. Uplift, shortening, and steady state topography in active mountain belts. *American Journal of Science* 301:455–85.
Yanai, R.D., T.G. Siccama, M.A. Arthur, C.A. Federer, and A.J. Friedland. 1999. Accumulation and depletion of base cations in forest floors in the northeastern United States. *Ecology* 80:2774–87.

9. Atmospheric CO_2, Environmental Stress, and the Evolution of C_4 Photosynthesis

Rowan F. Sage

This paper is dedicated to the memory of Professor Vladimir Pyankov, an expert on C_4 dicots, who passed away unexpectedly in January of 2002.

9.1 Introduction

There are three modes of photosynthesis in terrestrial plants: C_3, C_4, and Crassulacean Acid Metabolism (CAM). C_3 photosynthesis is the oldest and most common of the three, having been present in the earliest land plants. It occurs in about 90% of all land plants and predominates in most life forms and taxonomic groups. CAM photosynthesis is the next most abundant in terms of species number and taxonomic distribution, with approximately 20,000 species (Winter and Smith 1996). C_4 photosynthesis is the least diverse of the three photosynthetic pathways, occurring in 17 families and some 7000 to 8000 species (Sage, Li, and Monson 1999). Despite the relatively small number of C_4 species, they have an ecological impact that far exceeds their numbers. C_4 grasses dominate warm-temperate to tropical grassland and savanna biomes, and tropical and subtropical marshes are often co-dominated by C_4 sedges (see Chapter 10) (Ueno and Takeda 1992; Knapp and Medina 1999;). C_4 plants are common in biomes that cover approximately 40% of Earth's land surface, and 20% to 25% of the global primary productivity occurs via the 3% of all plants species that use the C_4 pathway (Collatz, Berry, and Clark 1998; Sage, Wedin, and Li 1999).

While many consider the C_4 pathway as simply an interesting variety of photosynthesis, there are a number of reasons why C_4 plants deserve special attention. First, the biochemical technology of C_4 photosynthesis gives these species an ability to exploit habitats that are too hostile for C_3 species. As a result, large expanses of the world's deserts and salt flats support a vegetation cover that might be far more sparse in the absence of C_4 photosynthesis. Second, unique features of the C_4 pathway promoted the evolutionary radiation of numerous animal groups following the expansion of C_4 dominated biomes (MacFadden 1997) (see also Chapter 12). Third, because of its unique physiology, C_4 photosynthesis responds differently to climate and atmospheric CO_2 change than do C_3 plants. C_4 plants perform well relative to C_3 species at CO_2 levels below the current atmospheric CO_2 level, particularly in warm to hot climates (Johnson, Polley, and Mayeux 1993; Sage 1995). Thus, whenever there are pronounced changes in atmosphere and climate, the distribution of C_4 relative to C_3 vegetation is altered (Cerling et al. 1997; Huang et al. 2001; Schefuss et al. 2003). Understanding the interactions among climate, atmospheric change, and C_4 vegetation is, therefore, important for interpreting how the current biota of the planet came into being, as well as predicting how the future biota will respond to anthropogenic global change.

In recent years there have been important advances in our understanding of past climates, atmospheric composition, and the systematic diversification of C_4 photosynthesis, such that it is now possible to propose robust models of C_4 evolution. In this review, I present a comprehensive summary of how variation in atmospheric CO_2 content and environmental stress may have affected the evolutionary diversification of C_4 photosynthesis. Considerable emphasis is placed on the distribution of C_4 photosynthesis in families of dicotyledonous plants. C_4 photosynthesis arose very recently and on multiple occasions in the dicots, and many of the immediate C_3 ancestors and intermediate forms have been identified (Ehleringer, Cerling, and Helliker 1997; Sage 2001). Analysis of the C_4 dicot lineages thus facilitates understanding of the conditions leading to the rise of the C_4 pathway in a manner that may not be possible in the older, more diverse C_4 grass and sedge lineages.

9.2 A Brief Review of C_4 Photosynthesis

C_4 photosynthesis is not a single metabolic pathway but instead consists of distinct biochemical reactions that concentrate CO_2 into an internal compartment where the enzyme Ribulose (1,5) bisphosphate carboxylase/oxygenase (Rubisco) is localized (Fig. 9.1). This internal compartment is commonly termed the bundle sheath; however, it is often derived from cell layers other than the true bundle sheath tissue and for this reason is often termed the photosynthetic carbon reduction (PCR) tissue (Dengler and Nelson 1999). In addition to localizing Rubisco into an internal compartment, all C_4 plants share the initial enzymatic step of phosphoenol pyruvate (PEP) carboxylation to form oxaloacetic acid, a four-carbon compound.

PEP carboxylase is located in the mesophyll tissue of the leaf in what is termed the photosynthetic carbon assimilation (PCA) compartment (Dengler and

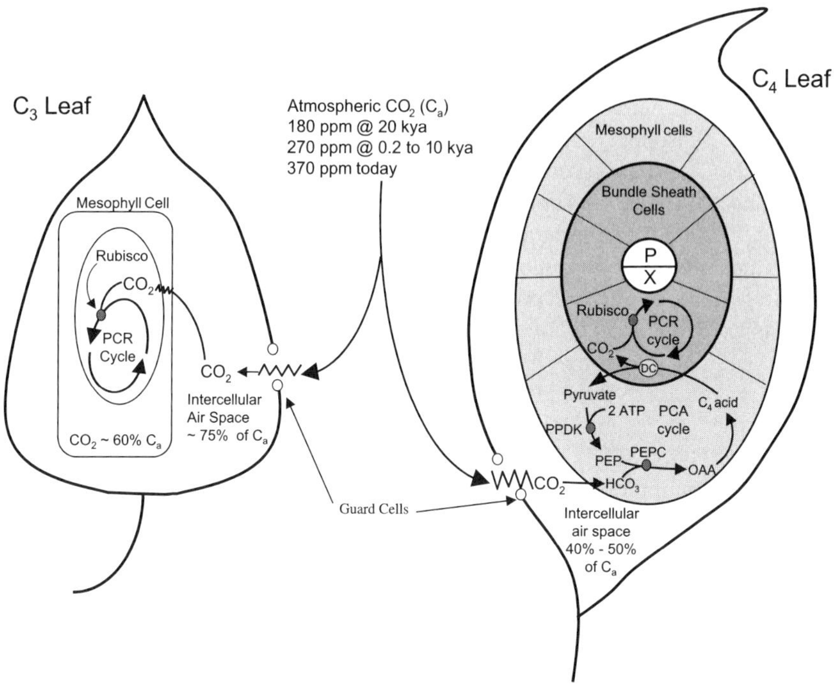

Figure 9.1. A schematic representation of photosynthesis in a C_3 and C_4 leaf. In the C_3 leaf, the outline of the biochemical path of the photosynthetic carbon reduction cycle (PCR) and photosynthetic oxidation cycle (PCO, or photorespiratory cycle) is shown for a single mesophyll cell, with the location of the Rubisco carboxylation and oxygenation steps highlighted. In the C_4 leaf, a cross section is shown for a single PCA-PCR unit comprised of adjacent mesophyll and bundle sheath cells, with a summary of key metabolic steps in the photosynthetic carbon assimilation (PCA) cycle. Atmospheric CO_2 levels (C_a) are shown for the late Pleistocene 20 thousand years ago (20 kya), in the Holocene before the industrial revolution (0.2–10 kya) and today. Other abbreviations: DC, the decarboxylase enzme; PEPC, phosphoenol pyruvate carboxylase; PPDK, pyruvate, phosphate-dikinase; p, phloem; x, xylem.

Nelson 1999). PCA cells usually surround PCR cells, creating a wreath-like arrangement termed Kranz anatomy (Metcalfe and Chalk 1979). The outer region of the PCA tissue is well exposed to the interecellular air species, while the PCR cells are positioned together along the inside walls of the PCA cells. This arrangement is important as it establishes a barrier that traps CO_2 in the PCR compartment.

Four-carbon acids produced by PEP carboxylation diffuse from the PCA to the PCR compartment. Once in the PCR compartment, they are dccarboxylated to release CO_2, which builds up to concentrations that saturate Rubisco and inhibit RuBP oxygenation (see Fig. 9.1) (Hatch 1987; Kanai and Edwards 1999). RuBP oxygenation is a side-reaction catalyzed by Rubisco that produces phosphoglycolate, a two-carbon compound that is of no use to the plant and is toxic in elevated concentrations (Jordan and Ogren 1984; Sage 1999). Plants must metabolize phosphoglycolate back to PCR cycle intermediates to avoid toxic accumulation of the byproducts of RuBP oxygenation. In doing this, reducing power and (ATP) are consumed, while previously fixed CO_2 is released (Douce and Heldt 2000). Together, the uptake of O_2 by Rubisco and the release of CO_2 during metabolism of phosphoglycolate are termed photorespiration. Because of the energy cost, the loss of previously fixed CO_2, and the competition from O_2 for Rubisco active sites, photorespiration is highly inhibitory for photosynthesis under conditions promoting high oxygenase activity (Sharkey 1988; Sage 1999). In the current atmosphere, Rubisco oxygenase activity becomes significant as atmospheric CO_2 levels decline below 500 ppm, but only at warmer temperature (Jordan and Ogren 1984). Rising temperature stimulates oxygenase activity and hence photorespiration in C_3 plants at CO_2 levels of the current atmosphere and below, such that C_3 photosynthesis is significantly (> 20%) inhibited by photorespiration above 25°C (see Chapter 10, Fig. 10.3). By accumulating CO_2 around Rubisco in the bundle sheath tissue, C_4 plants prevent significant levels of photorespiration at any temperature. The high CO_2 levels in the bundle sheath also allow Rubisco to operate near CO_2 saturation, so that its efficiency is maximized. In C_3 plants at warmer temperature, Rubisco operates at CO_2 levels well below saturation, and as a result its carboxylation potential is relatively low (Jordan and Ogren 1984; Sage 1999).

While all C_4 plants share certain features, such as PEP carboxylation, there are unique differences that demonstrate multiple evolutionary solutions to the challenge of concentrating CO_2 around Rubisco. For example, the decarboxylation of C_4 acids in the PCR compartment is catalyzed by one of three enzymes that vary between the different types of C_4 plants (Hatch 1987; Kanai and Edwards 1999). These are the NADP-malic enzyme (in NADP-ME type plants), the NAD-malic enzyme, and PEP carboxykinase (PCK). Differences also occur in the steps associated with metabolite transport between PCA and PCR cells, with the groups using the NADP-malic enzyme transporting C_4 acids as malate while the NAD-ME and PCK types transport C_4 acids as aspartate (Edwards and Walker 1983; Hatch 1987). Finally, while all C_4 plants segregate Rubisco and PEP carboxylation into physically distinct compartments, the source, location, and modifications of the PCA and PCR tissue vary substantially between different evolutionary lineages of C_4 photosynthesis. At least 8 different patterns of Kranz anatomy have been identified, based on the nature of the cellular arrangement, the ultrastructural patterns within photosynthetic cells, and the developmental lineage of the PCR tissue (Dengler and Nelson 1999). When all the differences in anatomy and decarboxylation types are accounted for, at least 16 different types of C_4 photosynthesis are apparent, and more appear likely as numerous groups of C_4 plants have yet to be studied.

Until recently, it was thought that Kranz anatomy was required for C_4 photosynthesis in order to trap CO_2 in the bundle sheath. This is no longer the case, because two species in the Suadeae tribe of the Chenopodiaceae have recently been shown to operate a complete C_4 pathway within single cells. One species, *Bienertia cycloptera*, localizes Rubisco into chloroplasts that reside in a cytoplasmic compartment in the center of barrel-shaped photosynthetic cells (Freitag and Stichler 2002; Voznesenskaya et al. 2002). PEP carboxylation occurs in cytoplasmic pockets located at the cell periphery and a large vacuole separates the central cytoplasmic region from the periphery. The second species is *Borszczowia aralocaspica*, a central-Asian shrub where photosynthetic cells form ranks of elongated columns surrounding the vascular tissue (Freitag and Stichler 2000; Voznesenskaya et al. 2001). Rubisco containing chloroplasts are located at the inner pole of the photosynthetic cells, while PEP carboxylation occurs at the outer end of the cell adjacent to the intercellular air spaces. As in *Bienertia*, *Borczscowia* has a large vacuole that separates the PCA and PCR regions of the cell and thus appears to serve as the diffusive barrier that limits CO_2 efflux.

9.3 The Taxonomic Distribution of C_4 Photosynthesis

C_4 photosynthesis in terrestrial plants is only found in the most advanced families of the angiosperms. It is absent in all of the more primitive plant groups (bryophytes, lycophytes, ferns, and gymnosperms), and in all the basal angiosperm orders, as shown in Fig. 9.2 (Kellogg 1999; Sage, Li, and Monson 1999).

C_4 photosynthesis occurs in 18 plant families in nine angiosperm orders, many of which are distantly related. Recent reports have identified C_4 photosynthetic metabolism in certain algae (for example, diatoms and the green algae *Egeria*; Reinfelder, Kraepiel, and Morel, 2000; Casati, Lara, and Andreo, 2000), but the issue of whether they constitute C_4 photosynthesis has not been fully resolved. Unlike terrestrial plants, C_4-like algae do not separate PCA and PCR functions into distinct compartments. Many C_3 plants also operate a C_4 cycle in vascular tissue of the stems (Hibberd and Quick 2002), but this is not considered to be C_4 photosynthesis because the C_4 metabolism is not linked to anatomical changes that facilitate CO_2 concentration around most of the Rubisco within the plant.

Most C_4 species are grasses (family Poaceae), with about 4600 of the estimated 10,000 grass species being C_4 (Sage, Li, and Monson 1999). Sedges (family Cyperaceae) are the next most abundant group with 1350 species, while the remaining 1200 or so C_4 species are from 15 families of dicots (Fig. 9.2 and Fig. 9.3).

In the dicots, 800 species are in the Chenopodiaceae and Amaranthaceae[1], 250 are in the Euphorbiaceae and 150 are scattered among 8 genera in the

---

1. Soltis et al. (2000) have merged the Chenopodiaceae into the Amaranthaceae based on molecular phylogeny data. Recent phylogenetic work with a larger number of species indicates separation of Chenopodiaceae and Amaranthaceae are justified (Kadereit et al. 2003). My treatment here recognizes the traditional split between the families.

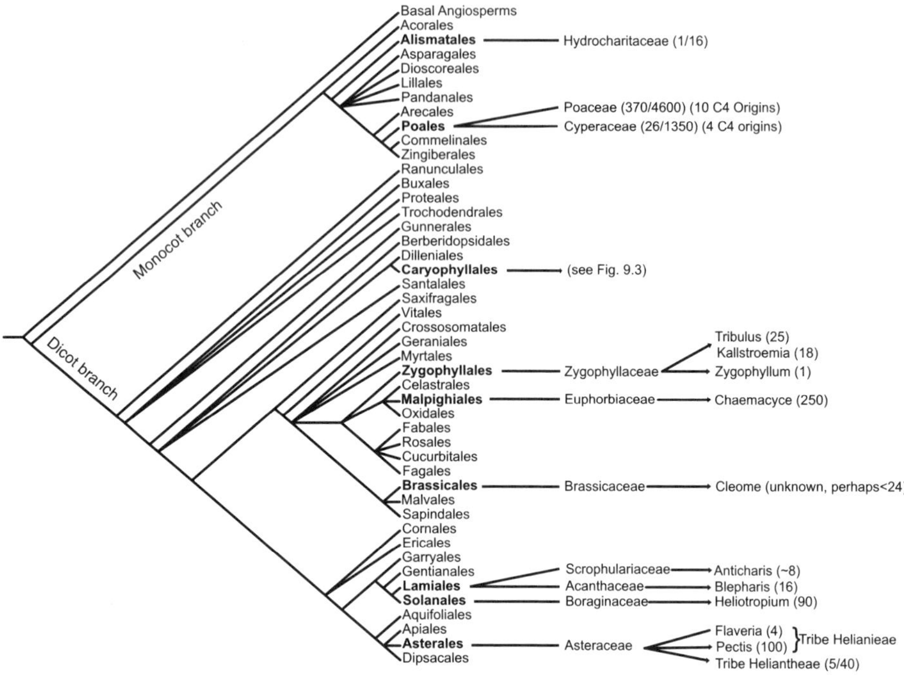

Figure 9.2. The distribution of C_4 photosynthesis in taxonomic orders of the angiosperms. Angiosperm orders containing C_4 species are shown in bold. Families containing C_4 species are shown for all orders except the Caryophyllales. Individual evolutionary lineages are shown for seven of the dicot families, with each lineage portrayed by the arrows connecting a family to the principle C_4 genera within the lineage. Genera/species numbers are shown in parentheses for grasses (Poaceae), sedges (Cyperaceae), the Hydrocharitaceae, and the tribe Heliantheae in the Astercacae. Estimated species numbers are shown for the genera in the other dicot lines. Based on the phylogeny of Stevens (2003). Estimates of genera and species numbers from Sage, Li, and Monson, 1999, as modified by recent carbon isotope surveys of photosynthetic pathway in Acanthaceae, Boraginaceae, Scrophulariaceae and Zygophyllaceae (Sage, unpublished).

Asteraceae. With the exception of the Chenopodiaceae and Amaranthaceae, the C_4 dicot species are relatively minor in their respective families. For example, the Asteraceae has 1530 genera and 23,000 C_3 species, but only 140 C_4 species scattered among 8 genera (Bremer 1994; Sage, Li, and Monson 1999). Ten of the 15 dicot families containing C_4 plants have three genera or less and each of these has fewer than 80 species. Many of the dicot genera with C_4 species also contain C_3 species, including a few (*Salsola, Heliotropium, Alternanthera, Moricandia, Flaveria*) that have species intermediate for C_3 and C_4 photosynthesis (Frohlich 1978; Monson 1989; Sage, Li, and Monson 1999; Pyankov et al.

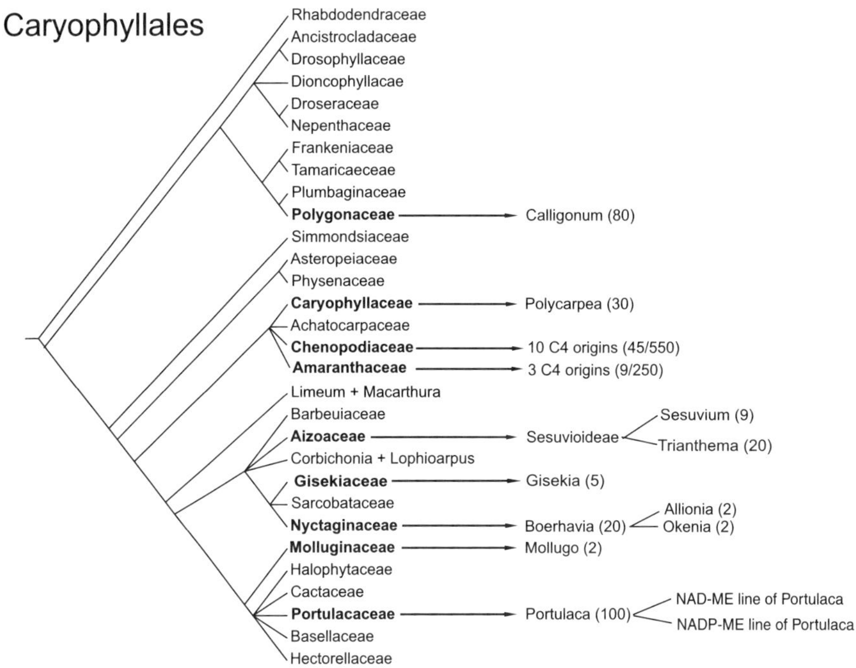

Figure 9.3. The distribution of C_4 photosynthesis in the Caryophyllales. Families with C_4 species are shown in bold, and species numbers corresponding to the individual genera are shown in parentheses. The genera beside the arrows are those where the C_4 pathway is postulated to have first arisen. Derived genera are shown for the smaller lineages. The systematic treatment follows Cuenoid et al. 2002. Species number estimates are from Sage, Li, and Monson, 1999, with modification based on carbon isotopic screens of herbarium materials (Sage, unpublished).

2001a,b). The presence of C_3/C_4 intermediates indicates that evolution from C_3 to C_4, or vice versa, is an ongoing process (Ehleringer and Monson 1993).

9.4 How Many Times Did C_4 Photosynthesis Evolve?

By comparing taxonomic affinity, biochemical traits, and anatomical features of the existing C_4 species, it is possible to identify distinct evolutionary origins of C_4 photosynthesis (Kellogg 1999; Pyankov et al. 2001a,b). The presence of C_4 photosynthesis in distantly related taxonomic families (see Fig. 9.2) clearly indicates distinct evolutionary lineages (Kellogg 1999). Within a given family, independent origin is indicated by differences in decarboxylation enzyme and Kranz anatomy type (Pyankov et al. 2001a,b). Recent phylogenetic analyses of the C_4-inclusive orders now allow for improved estimates of the number of origins of C_4 photosynthesis. For example, in the grass family, over 10 distinct C_4 origins are evident (see Fig. 9.2) if it is assumed that reversion from C_4 to C_3 photosynthesis is uncommon (GPWG 2001; Giussani et al. 2001). In the sedge family, four distinct evolutionary origins are indicated by phylogenetic treatments (Soros and Bruhl 2000; Sage 2001). In the dicots, 31 distinct evolutionary lines can be postulated (see Fig. 9.2, 9.3). Many independent origins in the dicots are obvious as they are single evolutionary lines within distinct families: for example, in the Acanthaceae (*Blepharis*), Boraginaceae (*Heliotropium*, section *Orthostachys*), Brassicaceae (*Cleome*, section *Gynandropsis*), Euphorbiaceae (*Chaemacyce*) and Scrophulariaceae (*Anticharis*) (Kellogg 1999; Sage, Li, and Monson 1999). Four distinct origins are evident in the Asteraceae: two in the tribe Helianieae (two in *Flaveria* and one in *Pectis*) and one in the Heliantheae, subtribe Coreopsideae (Karis and Ryding 1994; Korpiva et al. 1996; Kellogg 1999). Two lines are evident in the Zygophyllaceae, with one origin in *Tribulus* or *Kallstroemia*, and a second in *Zygophyllum* (Sheahan and Chase 1996; Kellogg 1999). In *Zygophyllum*, only one species, *Zygophyllum simplex*, is C_4 (Sage, unpublished based on a carbon isotope survey of the Zygophyllaceae).

After the grass order Poales, the most prolific order in terms of C_4 evolution is the dicot order Caryophyllales, with nine families containing C_4 species (see Fig. 9.3). C_4 species appear to derive from single evolutionary origins in six families of the Caryophyllales. Four independent origins are estimated in the Amaranthaceae (Kadereit et al. 2003; Kellogg 1999; and Sage unpublished). In the Portulacaceae, two independent lines are postulated, one using the NAD-malic enzyme type of C_4 photosynthesis, and a second using the NADP-malic enzyme type (Guralnick and Jackson 2001). The Chenopodiaceae is the most prolific dicot family in terms of producing independent C_4 lineages. Ten distinct C_4 origins are estimated based on a combination of taxonomic, anatomical, and physiological treatments (Kellogg 1999; Pyankov et al. 2001a,b; Freitag and Stichler 2002; Kadereit et al. 2003). Two genera in the Chenopodiaceae seem particularly adept at evolving C_4 species. C_3 *Salsola* species are postulated to have given rise to two C_4 lineages, one using NADP-ME as the primary decarboxylating enzyme, and a second line using NAD-ME (Pyankov et al. 2001a,b). The genus *Suada* in the subtribe Suadeae holds the generic record for distinct C_4 evolutionary origins, with four transitions from C_3 to C_4, including the only two known cases of single-celled C_4 photosynthesis in terrestrial plants (Freitag and Stichler 2002; Voznesenskaya et al. 2001, 2002; Schütze et al. 2003).

When the number of probable C_4 lineages is tabulated, more than 40 distinct evolutionary origins of C_4 photosynthesis appear likely. Given this high number of independent origins, it is apparent that the evolution of C_4 photosynthesis is relatively easy in certain groups of plants, such as grasses, sedges, and chenopods. However, the absence of C_4 photosynthesis in most groups of plants, including many families that share habitat or life form with C_4 plants, indicates there may be unique features in certain plant families that predisposed them to evolve C_4 photosynthesis on multiple occasions. The nature of these features is unknown, although it has been suggested that adaptations to aridity that affect bundle sheath size and vein spacing may be important (Ehleringer, Cerling, and Helliker 1997; Sage 2001). Enlarged bundle sheaths cells, reduced mesophyll cell number, and closer vein spacing may enhance leaf water status in environments with high evaporative demand (Sage 2002). These changes may, in turn, reduce the diffusive distance between mesophyll and bundle sheath cells, thus facilitating the exchange of metabolites between these tissues. Enhanced potential for metabolic exchange between mesophyll and bundle sheath cells is proposed to be an important early step in the evolution of C_4 photosynthesis (Monson 1999).

9.5 Where Did C_4 Photosynthesis Evolve?

The ability to resolve distinct C_4 evolutionary lineages in the taxonomic record allows for identification of probable centers of origin for C_4 photosynthesis, at least in the dicots. In the grasses, identifying centers of origin are more problematic because the origination events are older, the taxa more numerous, and the systematics more complicated than in the dicot lines where C_4 photosynthesis evolved. Usually, centers of diversity identify centers of origin. Using this approach, in combination with the location of the closest C_3 relatives and any C_3-C_4 intermediates, it is possible to postulate five general centers of origin for most of the C_4 dicot lineages (Fig. 9.4).

Centers of origin for the C_4 Asteraceae (*Pectis and Flaveria*), Boraginaceae, Euphorbiaceae, and Nyctaginaceae are proposed for the arid interior of Mexico, while arid South America is the center of origin of the C_4 species in the *Gomphrena Alteranthera* complex (Amaranthaceae), in *Alternanthera* (Amaranthaceae), and in the Helianteae tribe of the Asteraceae. The African center appears to be the home for many C_4 dicot lineages in Acanthaceae, Aizoaceae, Amaranthaceae (*Aerva*), Brassicaceae, Molluginaceae, Scrophulariaceae, and Zygophyllaceae (see Fig. 9.4). Central Asia is the likely home for all Chenopodicaceae, and *Calligonum* in the Polygonaceae. The two C_4 lines postulated for *Portulaca* (Portulcaceae) certainly originate in the Southern Hemisphere (Geesink 1969), but where exactly cannot be resolved until better phylogenies are produced.

In *Polycarpea* (Caryophyllaceae), C_3 and C_4 species co-occur in Africa and Arabia, indicating C_4 photosynthesis in this genus arose in here (Sage, unpublished). C_3-C_4 intermediates of *Heliotropium* are from Mexico, indicating an American center of origin seems likely (Frohlich 1978).

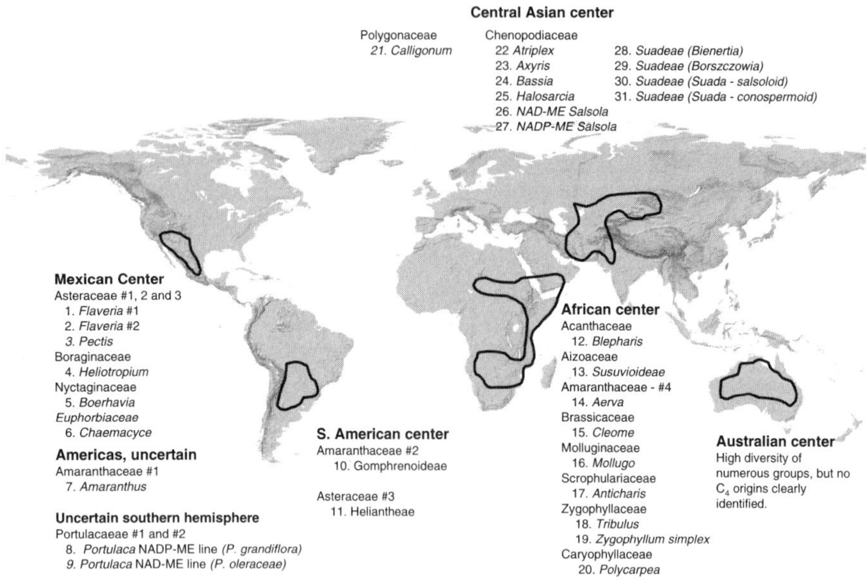

Figure 9.4. Postulated centers of origin for the C_4 dicot lineages. Centers of origin are based on (a) where the greatest diversity of C_4 taxa in a given lineage occur, (b) the distribution of any C_3-C_4 intermediates, and (c) the location of C_3 and C_4 species within a transitional genus. Key references for identifying distribution ranges include Volleson (2000, Acanthaceae); Hartmann (1993) for the Aizoaceae; Osmond, Björkman; and Anderson (1980), Kühn (1993), Pyankov et al. (2001a,b) and Townsend (1993) for the Amaranthaceae; Bremer (1994), Kadereit et al. (1996), and Powell (1978) for the Asteraceae; Johnston (1928), Frohlich (1978), and Hilger and Diane (2003) for the Boraginaceae; Bittrich (1993) for Caryophyllaceae; Webster, Brown, and Smith (1975) for the Euphorbiaceae; Endress and Bittrich (1993) for the Molluginaceae; Bittrich and Kühn (1993) for the Nyctaginaceae; Brandbyge (1993) for the Polygonaceae; Applequist and Wallace (2001) for the Portulaceae; plus Sage (unpublished, based on collection localities of over 1000 C_4 specimens housed in the herbaria of the Royal Botanical gardens at Kew, the Missouri Botanical Gardens, the New York Botanical Garden, and the Harvard University Herbarium).

9.6 How Old Is C_4 Photosynthesis?

9.6.1 The Evidence for Recent Evolution of C_4 Photosynthesis

Based on a number of criteria, the first C_4 ancestors of the existing C_4 flora most likely arose in recent geological time, with the earliest origin possibly occurring as far back as the mid-to-late Oligocene (23–30 Ma) (Kellogg 1999; Sage 2001). In rough terms, the greater the diversity in a C_4 lineage in terms of genera and species number, the older the lineage (Ehleringer, Cerling, and Helliker 1997). Using this criteria, the oldest C_4 lineage is in the grass family (370 genera, 4600 species), followed next by the sedge family (26 genera, 1350 species), and then the Chenopodiaceae (45 C_4 genera, 550 species), as shown in Fig. 9.5. Molluginaceae (one genus, less than 5 species), Scrophulariaceae (one genus, 6–10 species), and Brassicaceae (one genus with less than 20 species) would be very recent, possibly as late as the upper-Pleistocene (Ehleringer,

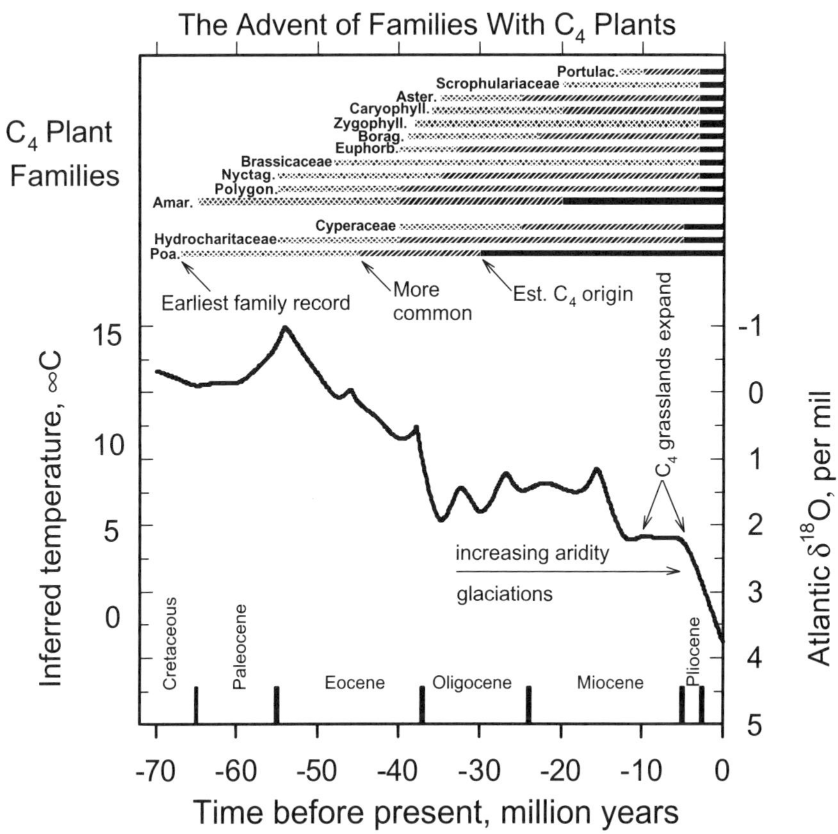

Figure 9.5. The occurrence of the families of plants containing C_4 photosynthesis in geologic time, with the corresponding inferred global temperature. Abbreviations correspond with the family names given in Fig. 9.2, 9.3. The solid portion of the histograms corresponds to estimated times for C_4 plant appearance in the respective families (after Kellogg, 1999, for grasses and Ehleringer, Cerling, and Helliker, 1997, and Kadereit et al. 2003 for dicots). Note: Amar refers to both the Amaranthaceae and Chenopodiaceae. Cyperaceae, Amaranthaceae, and Hydrocharitaceae are best guesses. The stippled middle portion of the histograms indicate times when family members become common in the fossil record, while the lightly stippled portions at the left end of the histogram indicate earliest reported origin (after Müller 1981 and Collinson, Boulter and Holmes, 1993) The inferred temperature is based on oxygen isotope ratios in Atlantic Ocean cores (Prothero 1994). From Sage (2001).

Cerling, and Helliker 1997). Most lines of C_4 dicots are suggested to originate in the Pleistocene, and it is possible that some of the less diverse grass and sedge lineages may date to the Pleistocene as well (see Fig. 9.5). For example, *Flaveria* is likely recent because there are only four C_4 species, and numerous C_3-C_4 intermediate species (Powell 1978; Kopriva, Chu, and Bauwe 1996). *Aerva* (Amaranthaceae) and *Halosarcia* (Chenopodiaceae) have multiple C_3 species, but these genera contain only one or two C_4 species, indicating a very recent transition to C_4 (Sage, unpublished data).

The earliest definitive evidence of C_4 photosynthesis in the geological record is represented by leaf samples that are 12 million years old in the Ricardo formation of California that have Kranz anatomy and C_4-like isotopic signatures (Nambudiri et al. 1978; Cerling 1999). Grass samples from Kenya that are 14.5 million years old have been suggested to be C_4, based on the presence of cuticles and other anatomical features that are similar to features that occur in modern C_4 grasses (Dugas and Retallack 1993). However, these samples have not been confirmed by isotopic screens or by Kranz anatomy (Cerling 1999). Isotopic signals from megafaunal remains and fossilized soils have been reported to indicate C_4 presence in East Africa as early as 15 million years ago (Kingston, Marino, and Hill 1994; Morgan, Kingston, and Marino 1994); however, C_4 species do not appear to dominate any ecosystems in East Africa prior to 10 million years ago (Cerling 1999). Beginning 10 million years ago, significant shifts in carbon isotope ratios from C_3 to C_4 values occur in many tropical sites, demonstrating ecosystems dominated by C_4 plants become widespread by the late Miocene (Cerling et al. 1997).

Evidence from molecular phylogenetic research with grasses indicates C_4 photosynthesis arose well before 12 to 15 million years ago. Using a molecular clock approach to analyze gene sequence differences between related grass genera that are both C_4 (hence the ancestor had to be C_4), Gaut and Doebley (1997) estimated distinct C_4 grass lineages were diverging by 25 million years ago. Kellogg and Russo (cited in GPWG 2001) estimated the C_4 grass *Danthoniopsis* diverged by 16 million years ago. Taking these observations into account, it is reasonable to propose that C_4 photosynthesis first arose in grasses by the mid Miocene, and possibly as early as the mid Oligocene. These species were relatively uncommon in most ecosystems until the late Miocene when they expanded across the globe, creating the first C_4 dominated biomes by 8 million years ago (Cerling 1999).

The Oligocene period (~34–23 Ma) is characterized by global environmental deterioration that left Earth colder, drier, and with an atmosphere depleted in CO_2 (see Fig. 9.5) (Prothero 1994; Zachos et al. 2001). Pagani, Freeman, and Arthur (1999); and Pearson and Palmer (2000) estimate that by 25 million years ago, high CO_2 levels (>500 ppm) of the Eocene and earlier had declined to below 300 ppm (see Chapter 3). This reduction correlates with the molecular clock data for C_4 divergence in major grass lineages.

In addition to the reduction in CO_2, the Oligocene was also a time when a number of other developments occurred that may have been essential for the rise of C_4 photosynthesis. First, the taxa and functional groups in which C_4 photosynthesis would appear showed marked diversification during this epoch (see Fig. 9.5). Although grasses pre-date the Oligocene by some 30 million years, they become common in the fossil record during the Oligocene (Kellogg 2000). Sedges also diversify about this time, as do most of the major dicot families that would one day contain C_4 plants (Collinson, Boulter, and Holmes 1993). In addition to the radiation of the taxonomic groups, herbaceous life forms become common in the fossil record only after 40 million years ago, while the annual plant life form, and drought-adapted features, becomes common after the early Oligocene (Wolfe 1997). In the large majority of the postulated C_4 lineages, the C_4 pathway primarily occurs in herbaceous species. Many of these are annual, and drought-adapted features prevail. In sum, the Oligocene is a time when a number of conditions favorable for C_4 photosynthesis first occurred. The biotic source material (advanced herbaceous angiosperm families) capable of evolving the pathway appears, atmospheric preconditions are met, and climate deterioration created extensive habitat favorable to C_4 species.

9.6.2 Did C_4 Photosynthesis Pre-Date the Oligocene Period?

A number of studies have proposed more ancient origins of C_4 photosynthesis than the Oligocene. Low CO_2 and high O_2 atmospheres favoring C_4 photosynthesis are postulated to have occurred in the Carboniferous period 300 Ma ago (Wright and Vanstone 1991), but clear evidence for C_4 plants in the Carboniferous has not yet been identified (Cerling 1999; Chapter 6). During the Carboniferous, angiosperms did not exist, and if C_4 plants were present, they would have occurred in ferns and related species, or gymnosperms. No evidence has ever been presented for C_4 photosynthesis in nonflowering plants. Kuypers, Pancost, and Damsté (1999) also propose C_4 plants were present during a possible low CO_2 excursion during an oceanic anoxic event 91 million years ago. This work is based on a sudden ($<$60 kya) isotopic shift in putative leaf waxes from marine deposits and is interpreted to indicate an expansion of C_4 community onshore in Northwest Africa. Without other supporting data, this interpretation is difficult to accept, as it would require either a ready source of C_4 species, or rapid evolution of C_4 photosynthesis, in taxa whose modern relatives contain no C_4 species.

9.7 Factors Promoting the Origin of C_4 Photosynthesis

Although C_4 plants have now radiated into a diverse range of habitats, including swamps, alpine tundra, and wet grasslands (Jones 1986; Long 1999), the general consensus is that they evolved in hot, arid, and often saline habitats (Ehleringer and Monson 1993). The greatest diversity of C_4 grasses corresponds to hot, dry environments (Hattersley 1992; Schulze et al. 1996); as shown in Fig. 9.4, the centers of diversity of the C_4 dicot lineages are arid interiors of continents. The main exception to these patterns may be the C_4 sedges, which tend to occur in wet habitats (Ueno and Takeda 1992). Ecological disturbance also may have played an important role, because many C_4 dicot species and C_3-C_4 intermediate species are common in recently disturbed habitats (Monson 1989).

High temperature is clearly important for the success of C_4 plants because C_4 photosynthesis has its greatest advantage over C_3 photosynthesis above 35°C (Pearcy, Tumosa, and Williams 1981; Osmond, Winter, and Ziegler 1982; Pearcy and Ehleringer 1984). For example, the temperature optimum for C_4 photosynthesis occurs at temperatures that are inhibitory for C_3 photosynthesis (Berry and Raison 1981; Sage and Pearcy 2000). At current CO_2 levels and below, the cause of the different temperature effects on C_3 versus C_4 photosynthesis is a high rate of photorespiration that occurs in C_3 plants above 30°C (see Chapter 10). By contrast, C_4 plants experience negligible inhibition from photorespiration at all temperatures (Kanai and Edwards 1999). Differences in photosynthetic performance also reflect ecological patterns. Studies comparing competitive abilities of C_3 and C_4 species in similar habitats demonstrate that C_4 plants dominate C_3 plants of similar growth form at elevated temperature (Pearcy, Tumosa, and Williams 1981; Sage and Pearcy 2000), and biogeographic surveys show C_4 grasses comprise over 90% of the biomass of grasslands at low latitude and altitude (Tiezen et al. 1997; Sage, Wedin, and Li 1999; Wan and Sage 2001).

The significance of aridity and salinity for C_4 evolution is threefold. First, the C_4 pathway provides clear advantages over the C_3 pathway in situations where the supply of freshwater is limited (Long 1999). C_4 plants have higher water use efficiency than do C_3 plants, a quality that gives them a greater capacity to produce biomass and to set seed on a limited amount of water (Osmond, Winter, and Ziegler 1982; Schulze and Hall 1982; Brown 1999). On saline soils, higher water use efficiency reduces the water flux through C_4 plants and, hence, the amount of salty water the plants encounter (Sage and Pearcy 2000). Second, drought and salinity reduce the density of plants, such that light levels remain high and surface heating of the ground maintains elevated temperatures near the plants. Third, drought and salinity stress reduce competitive interactions, so that transitional species with traits intermediate between C_3 and C_4 photosynthesis, and therefore potentially less efficient, may persist rather than being displaced by strong competitors.

9.7.1 The Role of Low Atmospheric CO_2

While important, the action of heat, drought, and salinity stress is probably not strong enough to select for the evolution of C_4 photosynthesis if atmospheric CO_2 levels are high. At elevated CO_2, photorespiration is low, Rubisco operates at high efficiency and photosynthesis in C_3 plants is typically greater than in ecologically similar C_4 plants (Berry and Raison 1981; Ehleringer et al. 1991). Without the added energy and nitrogen costs associated with the C_4 pump, C_3 plants at high CO_2 can also exhibit greater light, water, and nitrogen use efficiency than do C_4 plants, and the impact of drought, salinity, and heat stresses are reduced (Ehleringer, Cerling, and Helliker 1997; Hsiao and Jackson, 1999; Luo, Canadell, and Mooney 1999; Munns, Cramer, and Ball 1999; Sage and Cowling 1999). Furthermore, limitations associated with carbon supply become minor in high CO_2, and the control over productivity shifts from carbon supply to the availability of mineral nutrients, such as nitrogen and phosphorous (Agren, Shaver, and Rastetter 1999; Hungate 1999). Because of this, adaptations that improve the carbon economy of the plants, such as C_4 photosynthesis, have limited value in high CO_2 conditions.

At CO_2 levels less than those of today, C_4 plants show marked superiority over C_3 plants in terms of photosynthetic performance, growth, competitive ability, and reproductive output, particularly in warmer (>25°C) environments (Johnson, Polley, and Mayeux 1993; Dippery et al. 1995; Tissue et al 1995; Sage 1995). Furthermore, as CO_2 declines, the degree of inhibition associated with a given level of drought or heat stress substantially increases in C_3 species, potentially to the point where the survival of the plant is threatened (Sage and Cowling 1999). For example, in a variety of C_3 crops, biomass production at 200 ppm CO_2 and elevated temperature (35°C/29°C day/night) is reduced by 80% to 90% relative to plants at moderate temperature and current CO_2 levels (see Table 9.1). This level of yield reduction indicates plants exposed to warmer conditions in low CO_2 environments would have difficulty completing their lifecycle during a growing season.

A useful way to evaluate the ability of plants to survive at low CO_2 is to model the minimum CO_2 requirement for photosynthesis, vegetative growth, and reproduction (Fig. 9.6). If CO_2 supply is insufficient to support growth and reproduction, then the life cycle cannot be completed and plants will eventually disappear from a habitat. The minimum amount of CO_2 that is required to sup-

Table 9.1. Vegetative biomass (in grams) of three C_3 crop species grown in a 2 × 2 factorial experiment combining two growth CO_2 levels (380 and 200 ppm) and two growth temperature regimes (25°/19°C day/night and 35°/29°C day/night). The percent of the control value at 380 ppm and 25°/19° is shown in parentheses. The bean data is from Cowling and Sage (1998) and wheat and tobacco data from Sage and Cowling (1999). Means ± SE. DAP = days after planting.

		Growth Condition		
Species	380 ppm 25°/19°C	380 ppm 35°/29°	200 ppm 25°/19°	200 ppm 35°/29°
Phaseolus vulgaris	4.29 ± 0.15	2.74 ± 0.14	2.70 ± 0.12	0.82 ± 0.08
Black turtle bean at 23 DAP	(100%)	(64%)	(63%)	(19%)
Whole plant dry weight, N=15				
Nicotiana tabacum	1.20 ± 0.06	0.84 ± 0.06	0.51 ± 0.06	0.09 ± 0.01
Tobacco at 21 DAP	(100%)	(70%)	(43%)	(8%)
Shoot dry weight, N=6				
Triticum dicoccum	62 ± 2	14 ± 1	33 ± 3	5 ± 1
Emmer wheat at 41 DAP	(100%)	(23%)	(53%)	(8%)
Shoot fresh weight, N=3				

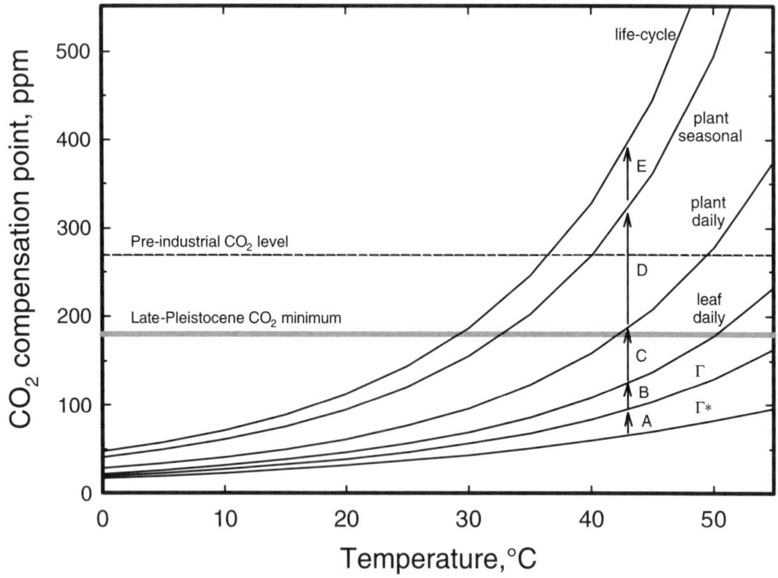

Figure 9.6. Hypothetical CO_2 compensation points for various processes in plants, as a function of temperature. Γ* is the CO_2 compensation point in the absence of mitochondrial respiration. Γ is the CO_2 compensation point for net photosynthesis, which incorporates the mitochondrial respiration rate in leaves in the light (indicated by arrow A). Leaf daily incorporates nighttime respiration of leaves (arrow B). Plant daily incorporates the respiration of nonphotosynthetic tissues over 24 hours (arrow C). Plant seasonal incorporates respiration and carbon construction costs associated with growth over the growing season (arrow D), and life cycle incorporates respiration and carbon costs for reproduction. Γ* and Γ responses modeled using Equations 2.37 and 2.32, and parameters in Table 2.3 from von Caemmerer (2000). These responses are experimentally supported but do not account for acclimation effects. To incorporate respiration and construction costs at higher levels than Γ, the R_d term in Equation 2.37 of van Caemmerer 2000 was increased to account for the additional respiratory carbon consumed during the added interval. Thus, for example, to account for nighttime respiration in leaves, R_d was doubled under the assumption that 50% of the leaf respiration occurs at night. For plant daily, R_d was doubled again on the assumption that 50% of the respiring plant biomass is nonphotosynthetic. Seasonal construction costs were assumed to be double maintenance costs, and R_d was increased another 25% to account for reproductive costs. Respiration costs at the whole plant level were loosely based on Lambers (1985) and are very approximate.

port this process is described in terms of the CO_2 compensation point. CO_2 compensation points vary with the frame of reference, rising as the scale increases from the biochemical to the whole plant level, and as the time frame increases from flux rates expressed per second to CO_2 exchange integrated over an entire life span. All versions of the CO_2 compensation point in C_3 plants are ultimately dependent on the ratio of oxygenation to carboxylation of RuBP, which is a property of the Rubisco enzyme (von Caemmerer 2000).

The CO_2 compensation point of gross photosynthesis in C_3 plants (Γ^*) occurs at the CO_2 level where RuBP carboxylation equals the rate of photorespiration. The CO_2 compensation point for net photosynthesis (Γ) occurs at the CO_2 level where the rate of carboxylation equals the sum of the photorespiration and mitochondrial respiration (van Caemmerer 2000). Because photorespiration and mitochondrial respiration increase with rising temperature at a faster rate than the rate of carboxylation, CO_2 compensation points also increase with temperature (Brooks and Farquhar 1985; von Caemmerer 2000).

For a plant to survive, carbon acquisition has to be greater than carbon loss through photorespiration and maintenance respiration, as well as having the carbon requirements to build new tissue. Respiratory losses are substantial, generally consuming 30% to 80% of the photosynthate gained by plants in the current atmosphere (Lambers 1985; van der Werf, Welschen, and Lambers 1992). For a leaf to have a positive photosynthesis rate over the course of a day, the daily gross photosynthesis rate has to exceed the sum of photorespiration and of mitochondrial respiration during both the light and dark periods of a diurnal cycle. The inclusion of the nighttime respiration costs yields the leaf-daily CO_2 compensation point, which is greater than the CO_2 compensation point for the instantaneous net photosynthesis (see Fig. 9.6). Similarly, CO_2 compensation points rise when respiration costs of nonphotosynthetic tissues are accounted for over a day (see the plant daily response) and an entire growing season (see the plant seasonal response).

The rise in the CO_2 compensation point for the plant over a growing season may be particularly pronounced because this response includes the construction costs to produce a mature vegetative plant. For a C_3 plant to complete its life cycle, the minimum carbon costs required to produce fruits and seeds must also be accounted for, and this further increases CO_2 compensation points. From an evolutionary perspective, the life cycle CO_2 compensation point is the most significant because it reflects the amount of CO_2 required for a population to avoid extinction.

In Fig. 9.6, the temperature response of the CO_2 compensation points for gross photosynthesis and net photosynthesis are experimentally well described (e.g., Brooks and Farquhar 1985). CO_2 compensation points at higher orders of complexity, such as the whole plant, have not been studied, and the curves presented are educated guesses. Despite this limitation, the hypothetical responses shown in Fig. 9.6 indicate that the rise in the CO_2 compensation point with increasing scale and temperature could be substantial enough to exceed prevailing atmospheric CO_2 levels during the Pleistocene, and possibly even the Holocene. As indicated in Fig. 9.6, the hypothesized rise in the CO_2 compensation point for completion of a life cycle may have exceeded the CO_2 level in the atmosphere of the late Pleistocene below 40°C. Above 40°C, the CO_2 compensation point for the life cycle is suggested to exceed the preindustrial CO_2 level of 270 ppm. If these responses are valid, they demonstrate that C_3 plants could reproduce neither in the late Pleistocene environments warmer than 30°C, nor in the Holocene environments above 40°C. The responses in Fig. 9.6 also indicate that vegetative C_3 plants could not grow above 40°C in late Pleistocene times, because the CO_2 compensation point for a positive daily carbon balance of whole plants is greater than atmospheric CO_2 levels of the time. Because few environments are consistently above 40°C, the importance of this may seem minor. However, seedlings experience temperatures that are well above air temperature in the boundary layer of the soil, and thus CO_2 requirements for establishment of C_3 seedlings could have been above what the prevailing climate conditions might indicate (Schulze et al. 1996; Sage and Sage 2002).

The situation would be even more perilous for C_3 plants if stomatal closure occurred to save water or to reduce salt loads, because this would effectively reduce the CO_2 availability to the plant, making it more likely that minimum CO_2 requirements were not met. As a result, it seems probable that many areas of the planet that support C_3 vegetation cover may have been unable to do so in low CO_2 environments of the past. These areas may thus have been wide open for colonization by genotypes expressing carbon-conservation traits that improved the carbon balance of the plant. Some of these carbon conservation mechanisms may have been the first steps in the evolution of C_4 photosynthesis.

9.7.2 Photorespiration: The Metabolic Bridge to C_4 Photosynthesis

In the origin of C_4 photosynthesis, there was no "final plan" for evolution to work toward. Instead, C_4 photosynthesis arose in a series of incremental steps from C_3 ancestors, and each step had to be adaptive in its own right (Monson and Rawsthorne 2000). Because each stage was likely necessary for subsequent stages to occur, the first stage may have been particularly important, as this would determine whether a population could even begin the C_4 evolutionary process (Sage 2002).

The first steps in C_4 evolution are thought to involve the localization of the enzyme glycine decarboxylase into the bundle sheath tissue (Monson 1999; Monson and Rawsthorne 2000). Glycine decarboxylase releases CO_2 in photorespiration, and its localization to the bundle sheath would force the photorespired CO_2 to diffuse through the mesophyll tissue, where it could be refixed by Rubisco. In addition, when photorespiration is high, any Rubisco in the bundle sheath could experience elevated CO_2 levels, allowing it to operate with greater efficiency. In this manner, the localization of glycine decarboxylase to the bundle sheath allows plants to operate a weak CO_2 concentrating mechanism that can partially offset the deleterious effects of high rates of photorespiration (Fig. 9.7). However, this will only be relevant when photorespiration is high, which is only in atmospheres of depleted CO_2.

The efficiency of a photorespiratory CO_2 pump can be increased in environments where photorespiration is elevated by enhancing Rubisco expression in the bundle sheath (Monson and Rawsthorne 2000). The next stage in the evolution toward C_4 photosynthesis may involve the enhancement of the CO_2 scavenging capacity by elevating PEP carboxylase activities in the mesophyll, and decarboxylase activity in the bundle sheath tissue (Monson 1999). Because C_3

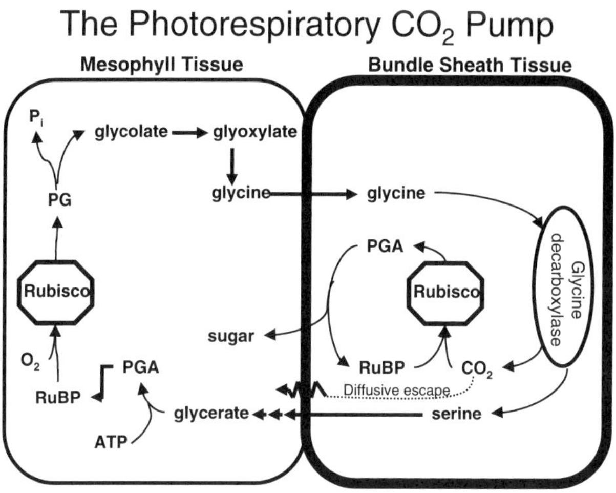

Figure 9.7. The photorespiratory CO_2 pump. In C_3-C_4 intermediate species, the first step on the evolutionary sequence to C_4 photosynthesis is proposed to be a shift in glycine decarboxylase expression from the mesophyll to the bundle sheath compartment (Monson 1999). Phosphoglycolate formed from RuBP oxygenation is converted to glycine in the mesophyll and then is transported to the bundle sheath where glycine decarboxylase metabolizes it to serine and CO_2. CO_2 levels in the bundle sheath become elevated during high rates of photorespiration and in doing so increase the efficiency of Rubisco use in the bundle sheath compartment. Derived from Sage 2001.

plants already have significant activities of decarboxylating enzymes in bundle sheath tissues (possibly to metabolize organic acids used in long-distance transport), this step may simply require the enhancement of existing patterns of gene expression (Hibberd and Quick 2002).

As the photorespiratory CO_2 pump develops, bundle sheath cells may enlarge to accommodate more Rubisco, while ratios of mesophyll to bundle sheath cells would further decline to reduce diffusion distances and to allow for close coordination of the various phases of the photorespiratory pump (Monson 1989). At some point, a recognizable PCA cycle would emerge as the primary role of PEP carboxylase shifts from refixation of photorespiratory CO_2 to assimilating CO_2 diffusing in from the atmosphere. In the final stages in C_4 evolution, Rubisco would be restricted to the bundle sheath cells, PEP carboxylase activity would be enhanced to meet the Rubisco capacity for CO_2 fixation, and enzyme regulation and cell ratios would be optimized to allow close coordination of the PCA and PCR cycles (Monson 1989).

In summary, the key initial step in the evolution of C_4 photosynthesis is thought to be the development of the photorespiratory CO_2 pump, which then makes the subsequent evolutionary stages possible. In this regard, photorespiration acts as the evolutionary bridge leading from C_3 photosynthesis to the intermediate stages where the PCA cycle can begin to develop (Sage 2001). To generate enough photorespiratory metabolites to allow a photorespiratory CO_2 pump to arise, low CO_2 is required because at high CO_2, Rubisco oxygenase activity is suppressed at all temperatures. Hence, for the evolutionary bridge to form, low CO_2 conditions must prevail.

9.8 Ecological Scenarios for C_4 Evolution

Once the preconditions for C_4 photosynthesis were met, how might the C_4 pathway have evolved in natural ecosystems? One scenario for C_4 evolution is indicated by the occurrence of the two single-celled C_4 species, *Bienertia cycloptera* and *Borszczowia aralocaspica* (Freitag and Stichler 2000, 2002). Both species occur on extremely saline soils in depressions, the shores of saline lakes, and along marine coastlines of central Asia; notably, they are found beyond the leading edge of the halophytic C_3 vegetation (Freitag and Stichler 2000; 2002). This distribution indicates C_4 photosynthesis may be the key trait enabling the single-celled C_4 species to exist on soils too saline for most C_3 competitors. If this is the case, the adaptive radiation of single-celled C_4 photosynthesis could have occurred on soils where the salinity is too high for C_3 competitors to occur. The process might have begun when individuals expressing the initial stages of C_3-C_4 intermediacy were able to survive on marginal soils just beyond the edge of the ancestral C_3 vegetation. Offspring of these colonists that had greater levels of intermediacy may then have been able to colonize soils of even greater salinity. In turn, some of their offspring may have exhibited a greater expression of C_3-C_4 intermediacy and, as a result, may have colonized soils further out on the salinity gradient. Eventually, the salinity gradient may have selected for the right combination of traits, in the right sequence, to allow for a fully developed C_4 pathway to evolve.

A second scenario describes how nonhalophytic C_4 species may have evolved along heat and aridity gradients, with selection at the seedling stage playing a major role (Fig. 9.8). In the hot deserts of the world, the establishment of seedlings is the stage of the life cycle when mortality is greatest. A particularly stressful feature of the establishment phase is that seedlings experience elevated temperature in the boundary layer of the soil. Insulation provided by the boundary layer retains heat from incoming solar radiation, causing temperatures near the soil surface to warm above air temperature (Oke 1987).

In the hot deserts of the world, daytime air temperatures commonly exceed 40°C (Walter, Harnickell, and Mueller-Dombois 1975), so boundary layer temperatures should be well into the 45°–55°C range. In low CO_2 environments, the high temperature of the boundary layer probably causes the growth CO_2 compensation point of the C_3 plants to approach the prevailing atmospheric CO_2 level, hindering the establishment of C_3 seedlings. Seedlings that localize glycine decarboxylase to the bundle sheath may experience enough improvement in carbon acquisition to be able to colonize marginally lethal microsites. In turn, their offspring with greater levels of C_3-C_4 intermediacy might be selected for

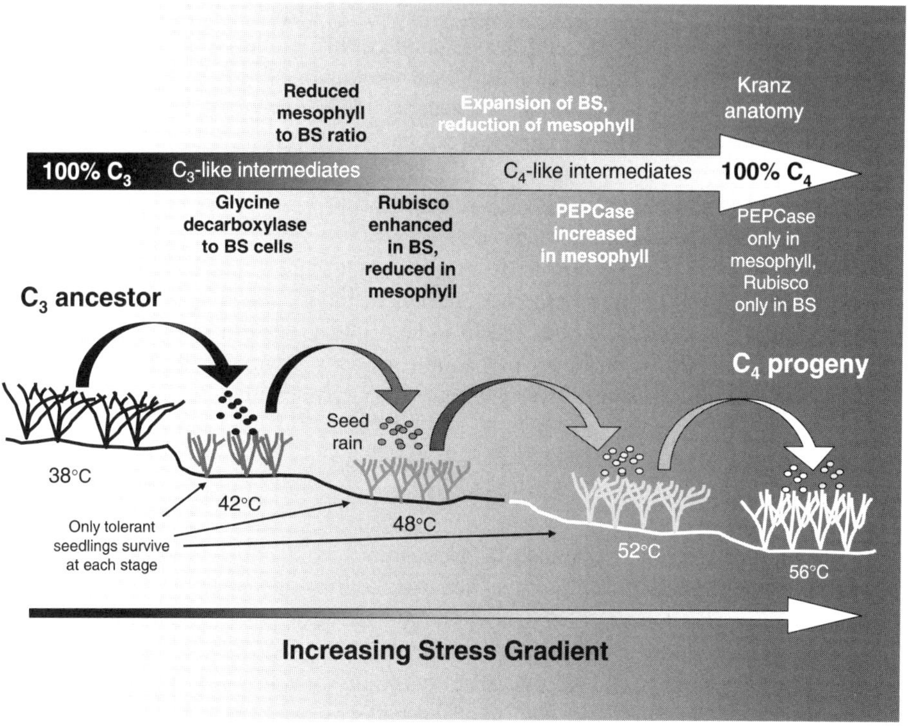

Figure 9.8. A conceptual model of how a stress gradient may select for varying stages of C_3-C_4 intermediacy and C_4 photosynthesis. Along the top of the figure are key steps in the evolutionary sequence of a complete C_4 pathway, which is based on recognized intermediate types in *Flaveria* species (Monson 1999). The temperatures at each stage represent possible temperatures in the boundary layer of desert soils along a microtopographic gradient. Where depressions form along the gradent, air movement slows, boundary layers thicken, and surface temperatures rise (Oke 1987). In the model, seed from C_3 progenitors will fall onto sites where the heat stress is slightly greater than can be tolerated by C_3 plants in a CO_2-depleted atmosphere. Most seedlings die, but seedlings with a weak carbon conservation mechanism, such as glycine decarboxylase localization to bundle sheath cells, will survive and establish. Seed from these individuals will rain onto sites of greater stress, and only offspring with a greater level of C_3-C_4 intermediacy will survive. In turn, only those offspring with an even greater degree of intermediacy will establish on the next most severe site, and so on until fully evolved C_4 progeny appear on sites with the most extreme stress.

progressively hotter sites, until eventually a full C_4 pathway evolved (see Fig. 9.8). Drought and salinity would act in concert with boundary layer heating, because other options for heat amelioration, such as opening stomates to transpirationally cool leaves, would not be possible where water is limited. As with salinity, in drought and high temperature scenarios, the inability of C_3 species to meet their carbon needs creates a gradient of open niches where experiments in carbon conservation and C_4 evolution could occur without interference from C_3 competitors. Once the C_4 pathway successfully evolved, the prototype C_4 plants would have a superior photosynthetic system that would allow them to spread back across the stress gradient and out of the C_4 so-called nursery. The novel C_4 species would then be able to radiate into C_3 dominated biomes and eventually adapt the wide range of habitats where C_4 photosynthesis is now successful, including swamps, wet grasslands, and river floodplains.

9.9 Summary

Significant advances in systematics, paleoecology, and climatology in recent years are now providing new perspectives that facilitate the generation of robust hypotheses of C_4 evolution. Fossil and isotopic evidence show when C_4 plants became common in the geological record, while phylogenetic evidence indicates when C_4 lineages began to diverge. The many independent origins of C_4 photosynthesis demonstrate the evolution of the pathway must be relatively easy in certain taxa; however, the restriction of C_4 photosynthesis to advanced angiosperm families indicates the evolution of these families may have been an important precondition that constrained the timing of C_4 evolution. The appearance of herbaceous and annual plant life forms in the Oligocene may also be significant, because C_4 photosynthesis is very successful in these groups, particularly in arid ecosystems (Sage 2001).

A plausible scenario indicated by the observations described here is that the superior reproductive system of the angiosperms allowed for the radiation of herbaceous species in hot, dry, and saline habitats. These plants in turn evolved drought adaptations that predisposed them to evolve C_4 photosynthesis once appropriate environmental conditions appeared. The full set of conditions promoting the success of C_4 photosynthesis may have first appeared during the Oligocene, when the global climate became drier, the tropics remained hot, and the atmospheric CO_2 levels declined below the high levels of the early Tertiary. In more recent geological time, climates deteriorated further, most notably during the late Miocene and early Pleistocene, allowing for widespread expansion of deserts and semi-arid landscapes. Although a further reduction in atmospheric CO_2 in the late Miocene has been debated (Cerling et al. 1997; Pagani, Freeman, and Arthur, 1999), it is clear from ice core records that record low CO_2 levels were present by late Pleistocene times (Petit et al. 1999; Zachos et al. 2001).

In summary, a picture emerges in which the Oligocene was a time when a number of critical factors coalesced to facilitate the origin of the first C_4 plants, and the phylogenetic evidence indicates major groups of C_4 grasses were diverging 20 to 30 million years ago. While there has been much emphasis on the first origin of C_4 plants, the current flora is the product of many separate origins, and many if not most of these occurred more recently, possibly following late Miocene and early Pleistocene climate and atmospheric change. A late Miocene pulse corresponds to when C_4 grasslands expanded across continents at low latitude, while the Pleistocene is associated with the radiation of C_4 photosynthesis in most dicot groups.

The relationship between CO_2 reduction and the late Miocene expansion of C_4 grasslands is uncertain; however, it is known that very low CO_2 levels were present by the late Pleistocene. These levels were so low that it is probable hot regions of the earth could not support C_3 plants because CO_2 requirements to complete a life cycle were greater than the prevailing atmospheric value. Under these conditions, widespread C_4 evolution could be expected in the dry tropics, because the heat, low humidity, limited water availability, and in many areas elevated salinity would have selected for traits that enhanced the carbon balance of the plant. The rate of photorespiration is high under these conditions. This would have promoted the rise of C_4 photosynthesis by limiting productive potentials in C_3 plants as well as by providing an opportunity to concentrate CO_2 by segregating glycine decarboxylase into the interior of a leaf. In so doing, high photorespiration is likely the stimulus that enabled C_4-evolutionary trends to commence in many taxa.

Acknowledgments. I thank Jaime Fraser, Danielle Way, James Havey, Seth Seegobin, and Dr. Tammy Sage for editorial assistance and substantive feedback. This work was supported by NSERC grant 0GP0154273 to RFS.

References

Agren, G.I., G.R. Shaver, and E.B. Rastetter. 1999. Nutrients: Dynamics and limitations. In *Carbon dioxide and environmental stress,* eds. Y. Luo and H.A. Mooney, 333–45. San Diego: Academic Press.

Applequist, W.L., and R.S. Wallace. 2001. Phylogeny of the Portulaceous cohort based on ndhF sequence data. *Systematic Botany* 26:406–19.

Berry, J.A., and J.K. Raison. 1981. Responses of macrophytes to temperature. In *Encyclopedia of plant physiology, new series, vol. 12A. Physiological plant ecology I. Water relations and carbon assimilation,* ed. O.L. Lange, P.S. Nobel, C.B. Osmond, and H. Ziegler, 277–338. Berlin: Springer-Verlag.

Bittrich, V. 1993. Caryophyllaceae. In *The families and genera of vascular plants. Vol II. Flowering Plants: Dicotyledons,* ed. K. Kubitzki, 206–36. Berlin: Springer-Verlag.

Bittrich, V., and U. Kühn. 1993. Nyctaginaceae. In *The families and genera of vascular plants. Vol II. Flowering plants: Dicotyledons*, ed. K. Kubitzki, 473–86. Berlin: Springer-Verlag.

Brandbyge, J. 1993. Polygonaceae. In *The families and genera of vascular plants. Vol II. Flowering plants: Dicotyledons,* ed. K. Kubitzki, 531–44. Berlin: Springer-Verlag.

Bremer, K. 1994. *Asteraceae: Cladistics and classification.* Portland: Timber Press.

Brooks, A., and G.D. Farquhar. 1985. Effect of temperature on the CO_2/O_2 specificity of ribulose-1,5-bisphosphate carboxylase/oxygenase and the rate of respiration in the light. *Planta* 165:397–406.

Brown, R.H. 1999. Agronomic implications of C_4 photosynthesis. In *C_4 plant biology*, ed. R.F. Sage and R.K. Monson, 473–507. San Diego: Academic Press.

Casati P., M.V. Lara, and C.S. Andreo. 2000. Induction of a C_4-like mechanism of CO_2 fixation in *Egeria densa*, a submersed aquatic species. *Plant Physiology* 123:1611–21.

Cerling, T.E. 1999. Paleorecords of C_4 plants and ecosystems. In *C_4 plant biology*, ed. R.F. Sage and R.K. Monson, 445–69. San Diego: Academic Press.

Cerling, T.E., J.M. Harris, B.J. MacFadden, M.G. Leacey, J. Quade, V. Eisenmann, and J.R. Ehleringer. 1997. Global vegetation change through the Miocene/Pliocene boundary. *Nature* 389:153–58.

Collatz, G.J., J.A. Berry, and J.S. Clark. 1998. Effects of climate and atmospheric CO_2 partial pressure on the global distribution of C_4 grasses: Present, past, and future. *Oecologia* 114:441–54.

Collinson, M.E., M.C. Boulter, and P.L. Holmes. 1993. Magnoliophyta ('Angiospermae'). In *The fossil record 2,* ed. M.J. Benton, 809–41. London: Chapman and Hall.

Cowling, S.A., and R.F. Sage. 1998. Interactive effects of low atmospheric CO_2 and elevated temperature on growth, photosynthesis and respiration in *Phaseolus vulgaris. Plant Cell and Environment* 21:427–35.

Cuénoud, P., V. Savolainen, L.W. Chatrou, M. Powell, R.J. Grayer, and M.W. Chase. 2002. Molecular phylogenetics of Caryophyllales based on nuclear 18S rDNA and plastid *rbc*L, *atp*B, and *mat*K DNA sequences. *American Journal of Botany* 89:132–44.

Dengler, N.G., and T. Nelson. 1999. Leaf structure and development in C_4 plants. In *C_4* plant biology, ed. R.F. Sage and R.K. Monson, 133–72. San Diego: Academic Press.

Diane, N., H. Förther, and H.H. Hilger. 2002. A systematic analysis of *Heliotropium, Tournefortia*, and allied taxa of the Heliotropiaceae (Boraginales) based on ITS1 sequences and morphological data. *American Journal of Botany* 89:287–95.

Dippery, J.K., D.T. Tissue, R.B. Thomas, and B.R. Strain. 1995. Effects of low and elevated CO_2 on C_3 and C_4 annuals. I. Growth and biomass allocation. *Oecologia* 101: 13–20.

Douce, R., and H.W. Heldt. 2000. In *Photosynthesis: Physiology and metabolism*, ed. R.C. Leegood, T.D. Sharkey, and S. von Caemmerer, 115–36. Dordrecht: Kluwer Academic Publishers.

Dugas, M.J., and G.J. Retallack. 1993. Middle Miocene fossil grasses from Fort Ternan, Kenya. *Journal of Paleontology* 67:113–28.

Edwards, G.E., and D.A. Walker. 1983. *C_3, C_4: Mechanism, and cellular and environmental regulation, of photosynthesis*. Oxford: Blackwell Scientific Publications.

Ehleringer, J.R., and R.K. Monson. 1993. Evolutionary and ecological aspects of photosynthetic pathway variation. *Annual Review of Ecology and Systematics* 24: 411–39.

Ehleringer, J.R., T.E. Cerling, and B.R. Helliker. 1997. C_4 photosynthesis, atmospheric CO_2 and climate. *Oecologia* 112:285–99.

Ehleringer, J.R., R.F. Sage, L.B. Flanagan, and R.W. Pearcy. 1991. Climate change and the evolution of C_4 photosynthesis. *Trends in Ecology and Evolution* 6:95–99.

Endress, M.E., and V. Bittrich. 1993. Molluginaceae. In *The families and genera of vascular plants. Vol II. Flowering plants: Dicotyledons,* ed. K. Kubitzki, 419–26. Berlin: Springer-Verlag.

Freitag, H., and W. Stichler. 2000. A remarkable new leaf type with unusual photosynthetic tissue in a central Asiatic genus of Chenopodiaceae. *Plant Biology* 2:154–60.

Freitag, H., and W. Stichler. 2002. *Bienertia cycloptera* Bunge ex Boiss., Chenopodiaceae: Another C_4 plant without Kranz tissues. *Plant Biology* 4:121–32.

Frohlich, M.W. 1978. Systematics of *Heliotropium* section *Orthistachys* in Mexico. Ph.D. Thesis. Cambridge: Harvard University.

Gaut, B.S., and J.F. Doebley. 1997. DNA sequence evidence for the segmental allotetraploid origin of maize. *Proceedings of the National Academy of Sciences U.S.A.* 94: 6809–14.
Geesink, R. 1969. An account of the genus Portulaca in Indo-Australia and the Pacific (Portulaceae). *Blumea* 17:275–301.
Giussani, L.M., J.H. Cota-Sánchez, F.O. Zuloaga, and E.A. Kellogg. 2001. A molecular phylogeny of the grass subfamily Panicoideae (Poaceae) shows multiple origins of C_4 photosynthesis. *American Journal of Botany* 88:1993–2012.
GPWG (Grass Phylogeny Working Group). 2001. Phylogeny and subfamilial classification of the grasses (Poaceae). *Annals of the Missouri Botanical Garden* 88:373–457.
Guralnick, L.J., and M.D. Jackson. 2001. The occurrence and phylogenetics of crassulacean acid metabolism in the Portulacaeae. *International Journal of Plant Science* 1162:257–62.
Hartmann, H.E.K. 1993. Aizoaceae. In *The families and genera of vascular plants. Vol II.* Flowering plants: Dicotyledons, ed. K. Kubitzki, 37–69. Berlin: Springer-Verlag.
Hatch, M.D. 1987. C_4 photosynthesis: a unique blend of modified biochemistry, anatomy and ultrastructure. *Biochemica Biophysica Acta* 895:81–106.
Hattersley, P.W. 1992. C_4 photosynthetic pathway variation in grasses (Poaceae): Its significance for arid and semi-arid lands. In *Desertified grasslands: Their biology and management, the Linnean Society of London*, 181–212. London: Academic Press.
Hibberd, J.M., and W.P. Quick. 2002. Characteristics of C_4 photosynthesis in stems and petioles of C_3 flowering plants. *Nature* 415:451–54.
Hsiao, T.C., and R.B. Jackson. 1999. Interactive effects of water stress and elevated CO_2 on growth, photosynthesis, and water use efficiency. In *Carbon dioxide and environmental stress,* ed. Y. Luo and H.A. Mooney, 3–31. San Diego: Academic Press.
Huang Y., F.A. Street-Perrott, S.E. Metcalfe, M. Brenner, M. Moreland, and K.H. Freemann. 2001. Climate change as the dominant control on glacial-interglacial variations in C_3 and C_4 plant abundance. *Science* 293:1647–51.
Hungate, B.A. 1999. Ecosystem response to rising atmospheric CO_2: feedbacks through the nitrogen cycle. In *Carbon dioxide and environmental stress,* ed. Y. Luo and H.A. Mooney, 265–85. San Diego: Academic Press.
Johnson, H.B., H.W. Polley, and H.S. Mayeux. 1993. Increasing CO_2 and plant-plant interactions: effects on natural vegetation. *Vegetatio* 104/105:157–70.
Johnston, I.M. 1928. Studies in the Boraginaceae,-VII: The South American species of *Heliotropium. The Gray Herbarium of Harvard University* 81:3–73.
Jones, M.B. 1986. Wetlands. In *Photosynthesis in contrasting environments,* ed. N.R. Baker and S.P. Long, 103–38. London: Elsevier.
Jordan, D.B., and W.L. Ogren. 1984. The CO_2/O_2 specificity of ribulose 1,5-bisphosphate carboxylase/oxygenase. *Planta* 161:308–13.
Kadereit, G., T. Borsch, K. Weising, and H. Freitag. 2002. Phylogeny of Amaranthaceae and Chenopodiaceae and the evolution of C_4 photosynthesis. *International Journal of Plant Science* 164:959–986.
Kanai, R., and G.E. Edwards. 1999. The biochemistry of C_4 photosynthesis. In *C_4 plant biology,* ed. R.F. Sage and R.K. Monson, 49–87. San Diego: Academic Press.
Karis, P.O., and O. Ryding. 1994. Tribe Heliantheae. In *Asteraceae: Cladistics and classification,* ed. K. Bremer, 559–624. Portland: Timber Press.
Kellogg, E.A. 1999. Phylogenetic aspects of the evolution of C_4 photosynthesis. In *C_4 plant biology,* ed. R.F. Sage and R.K. Monson, 411–44. San Diego: Academic Press.
———. 2000. The grasses: A case study in macroevolution. *Annual Review of Ecology and Systematics* 31:217–38.
Kingston, J.D., B.D. Marino, and A. Hill. 1994. Isotopic evidence for neogene hominid paleoenvironments in the Kenya Rift valley. *Science* 264:955–59.
Knapp, A.K., and E. Medina. 1999. Success of C_4 photosynthesis in the field: Lessons from communities dominated by C_4 plants. In *C_4 plant biology*, ed. R.F. Sage and R.K. Monson, 251–83. San Diego: Academic Press.

Kopriva, S., C.-C. Chu, and H. Bauwe. 1996. Molecular phylogeny of *Flaveria* as deduced from the analysis of nucleotide sequences encoding the H-protein of the glycine cleavage system. *Plant, Cell, and Environment* 19:1028–36.
Kühn, U. 1993. Chenopodiaceae. In *The families and genera of vascular plants. Vol II. Flowering plants: Dicotyledons,* ed. K. Kubitzki, 253–80. Berlin: Springer-Verlag.
Kuypers, M.M.M., R.D. Pancost, and J.S.S. Damsté. 1999. A large and abrupt fall in atmospheric CO_2 concentration during Cretaceous times. *Nature* 399:342–45.
Lambers, H. 1985. Respiration in intact plants and tissues: Its regulation and dependence on environmental factors, metabolism, and invaded organisms. In *Encyclopedia of plant physiology, New Series, Vol. 18: Higher plant respiration*, ed. R. Douce and D.A. Day, 418–73. Berlin: Springer-Verlag.
Long, S.P. 1999. Environmental responses. In *C_4 plant biology*, ed. R.F. Sage and R.K. Monson, 215–49. San Diego: Academic Press.
Luo, Y., J. Canadell, and H.A. Mooney. 1999. Interactive effects of carbon dioxide and environmental stress on plants and ecosystems: A synthesis. In *Carbon dioxide and environmental stress,* ed. Y. Luo and H.A. Mooney, 393–408. San Diego: Academic Press.
MacFadden, B.J. 1997. Origin and evolution of grazing guild in New World terrestrial mammals. *Trends in Ecology and Evolution* 12:182–87.
Metcalfe, C.R., and L. Chalk. 1979. *Anatomy of the dicotyledons. Vol. 1: Systematic anatomy of the leaf and stem,* 276. Oxford: Oxford Science Publishers.
Monson, R.K. 1989. On the evolutionary pathways resulting in C_4 photosynthesis and Crassulacean Acid Metabolism (CAM). *Advances in ecological research* 19:57–110.
———. 1999. The origins of C_4 genes and evolutionary pattern in the C_4 metabolic phenotype. In *C_4 plant biology,* ed. R.F. Sage and R.K. Monson, 377–410. San Diego: Academic Press.
Monson, R.K., and S. Rawsthorne. 2000. CO_2 assimilation in C_3-C_4 intermediate plants. In *Photosynthesis: Physiology and metabolism,* ed. R.C. Leegood, T.D. Sharkey, and S.C. von Caemmerer, 533–50. Dordrecht: Kluwer Academic Publihsers.
Morgan, M.E., J.D. Kingston, and B.D. Marino. 1994. Carbon isotopic evidence for the emergence of C_4 plants in the Neogene from Pakistan and Kenya. *Nature* 367:162–65.
Müller, J. 1981. Fossil pollen records of extant angiosperms. *Botanical Review* 47:1–146.
Munns, R., G.R. Cramer, and M.C. Ball. 1999. Interactions between rising CO_2, soil salinity and plant growth. In *Carbon dioxide and environmental stress,* ed. Y. Luo and H.A. Mooney, 139–67. San Diego: Academic Press.
Nambuduri, E.M.V., W.D. Tidwell, B.N. Smith, and N.P. Hebbert. 1978. A C_4 plant from the Pliocene. Nature 276:816–17.
Oke, T.R. 1987. *Boundary layer climates.* New York: Methuen.
Osmond, C.B., O. Björkman, and D.J. Anderson. 1980. Physiological processes in plant ecology: Toward a synthesis with *Atriplex.* In *Ecological studies: Analysis and synthesis, Vol. 36,* ed. W.D. Billings, F. Golley, O.L. Lange and J.S. Olson, 1–468. New York: Springer-Verlag.
Osmond, C.B., K. Winter, and H. Ziegler. 1982. Functional significance of different pathways of CO_2 fixation in photosynthesis. In *Encyclopedia of plant physiology, New Series Vol. 12B. Physiological plant ecology II. Water relations and carbon assimilation,* ed. O.L. Lange, P.S. Nobel, C. . Osmond, and H. Ziegler, 479–547. Berlin: Springer-Verlag.
Pagani, M., K.H. Freeman, and M.A. Arthur. 1999. Late Miocene atmospheric CO_2 concentrations and the expansion of C_4 grasses. *Science* 285:876–78.
Pearcy, R.W., and J. Ehleringer. 1984. Comparative ecophysiology of C_3 and C_4 plants. Plant, Cell, and Environment 7:1–13.
Pearcy, R.W., N. Tumosa, and K. Williams. 1981. Relationships between growth, photo-

synthesis and competitive interactions for a C_3 and a C_4 plant. *Oecologia* 48:371–76.

Pearson, P.N., and M.R. Palmer. 2000. Atmospheric carbon dioxide concentrations over the past 60 million years. *Science* 406:695–99.

Petit, J.R., J. Jouzel, D. Raynaud, N.I. Barkov, J.M. Barnola, I. Basile, M. Bender, J. Chappellaz, M. Davis, G. Delaygue, M. Delmotte, V.M. Kotlyakov, M. Legrand, V.Y. Lipenkov, C. Lorius, L. Pepin, C. Ritz, E. Saltzman, and M. Stievenard. 1999. Climate and atmospheric history of the past 420,000 years from the Vostok ice core, Antarctica. *Nature* 399:429–36.

Powell, A.M. 1978. Systematics of *Flaveria* (Flaveriinae-Asteraceae). *Annals of the Missouri Botanical Gardens* 65:590–636.

Prothero, D.R. 1994. *The Eocene-Oligocene transition: Paradise lost.* New York: Columbia University Press.

Pyankov, V.I., E.G. Artyusheva, G.E. Edwards, C.C. Black Jr., and P.S. Soltis. 2001a. Phylogenetic analysis of tribe Salsoleae (Chenopodiaceae) based on ribosomal ITS sequences: Implications for the evolution of photosynthesis types. *American Journal of Botany* 88:1189–98.

Pyankov, V.I., H. Ziegler, A. Kuz'min, and G. Edwards. 2001b. Origin and evolution of C_4 photosythesis in the tribe Salsoleae (Chenopodiaceae) based on anatomical and biochemical types in leaves and cotyledons. *Plant Systematics and Evolution* 230:43–74.

Reinfelder, J.R., A.M.L. Kraepiel, and F.M.M. Morel. 2000. Unicellular C_4 photosynthesis in a marine diatom. *Nature* 407:996–99.

Sage, R.F. 1995. Was low atmospheric CO_2 during the Pleistocene a limiting factor for the origin of agriculture? *Global Change Biology* 1:93–106.

———. 1999. Why C_4 Plants? In *C_4 plant biology,* ed. R.F. Sage and R.K. Monson, 3–16. San Diego: Academic Press.

———. 2001. Environmental and evolutionary preconditions for the origin and diversification of the C_4 photosynthetic syndrome. *Plant Biology* 3:202–13.

———. 2002. Are crassulacean acid metabolism and C_4 photosynthesis incompatible? *Functional Plant Biology* 29:775–85.

Sage, R.F., and S.A. Cowling. 1999. Implications of stress in low CO_2 atmospheres of the past: are today's plants too conservative for a high CO_2 world? In *Carbon dioxide and environmental stress,* ed. Y. Luo and H.A. Mooney, 289–308. San Diego: Academic Press.

Sage, R.F., M.-R. Li, and R.K. Monson. 1999. The taxonomic distribution of C_4 photosynthesis. In *C_4 plant biology,* ed. R.F. Sage and R.K. Monson, 551–84. San Diego: Academic Press.

Sage, R.F., and R.W. Pearcy. 2000. The physiological ecology of C_4 photosynthesis. In *Photosynthesis: Physiology and metabolism,* ed. R.C. Leegood, T.D. Sharkey, and S. von Caemmerer, 497–532. Dordrecht: Kluwer Academic.

Sage, R.F. and T.L. Sage. 2002. Microsite characteristics of *Muhlenbergia richardsonis* (Trin.) rydb., an alpine C_4 grass from the White Mountains, California. *Oecologia* 132:501–508.

Sage, R.F., D.A. Wedin, and M.-R. Li. 1999. The biogeography of C_4 photosynthesis: Patterns and controlling factors. In *C_4 plant biology,* ed. R.F. Sage and R.K. Monson, 313–73. San Diego: Academic Press.

Schefuss, E., S. Schouten, J.H.F. Jansen, and J.S.S. Damste. 2003. African vegetation controlled by tropical sea surface temperatures in the mid-Pleistocene period. *Nature* 422:418–21.

Schulze, E.-D., R. Ellis, W. Schulze, and P. Trimborn. 1996. Diversity, metabolic types and $\delta^{13}C$ carbon isotope ratios in the grass flora of Namibia in relation to growth form, precipitation and habitat conditions. *Oecologia* 106:352–69.

Schulze, E.-D., and A.E. Hall. 1982. Stomatal responses, water loss and CO_2 assimilation rates of plants in contrasting environments. In *Physiological plant ecology II: Water relations and carbon assimilation,* ed. O.L. Lange, P.S. Nobel, C.B. Osmond, and H. Ziegler, 181–230. Berlin: Springer-Verlag.

Schütze, P., H. Freitag, and K. Weising. 2003. An integrated molecular and morphological study of the subfamily Suaedoideae Ulbr. (Chenopodiaceae). *Plant Systematics and Evolution* 239:257–286.

Sharkey, T.D. 1988. Estimating the rate of photorespiration in leaves. *Physiologia Plantarum* 73:147–52.

Sheahan, M.C., and M.W. Chase. 1996. A phylogenetic analysis of Zygophyllaceae R. Br. based on morphological, anatomical and *rbcL* DNA sequence data. *Botanical Journal of the Linnean Society* 122:279–300.

Soltis, D.E., P.S. Soltis, M.W. Chase, M.E. Mort, D.C. Albach, M. Zanis, V. Savolainen, W.H. Hahn, S.B. Hoot, M.F. Fay, M. Aztell, S.M. Swensen, L.M. Prince, W.J. Kress, K.C. Nixon, and J.S. Farris. 2000. Angiosperm phylogeny inferred from 18S rDNA *rbc*L, and *atp*B sequences. *Botanical Journal of the Linnean Society* 133:381–461.

Soros, C.L., and J.J. Bruhl. 2000. Multiple evolutionary origins of C_4 photosynthesis in the Cyperaceae. In *Monocots: Systematics and evolution. Vol. 1.,* ed. K.L. Wilson and D.A. Morrison, 629–36. Collingwood, Australia: Commonwealth of Scientific Industrial Research Orgnization.

Stevens, P.F. 2003. Angiosperm phylogeny website. Version 4 May 2003. http://www.mobot.org/mobot//.

Tieszen, L.L., B.B. Reed, B.B. Bliss, B.K. Wylie, and D.D. DeJong. 1997. NDVI, C_3 and C_4 production, and distributions in Great Plains grassland land cover classes. *Ecological Applications* 7:59–78.

Tissue, D.T., K.L. Griffen, R.B. Thomas, and B.R. Strain. 1995. Effects of low and elevated CO_2 on C_3 and C_4 annuals. II. Photosynthesis and leaf biochemistry. *Oecologia* 101:21–28.

Townsend, C.C. 1993. Amaranthaceae. In *The families and genera of vascular plants. Vol. II. Flowering plants: Dicotyledons,* ed. K. Kubitzki, 70–91. Berlin: Springer-Verlag.

Ueno, O., and T. Takeda. 1992. Photosynthetic pathways, ecological characteristics, and the geographical distribution of Cyperaceae in Japan. *Oecologia* 89:195–203.

van der Werf, A., R. Welschen, and H. Lambers. 1992. Respiratory losses increase with decreasing inherent growth rate of a species and with decreasing nitrate supply: A search for explanations for these observations. In *Molecular, biochemical and physiological aspects of plant respiration,* ed. H. Lambers and L.H.W. van der Plas, 421–32. The Hague: SPB Publishing.

Vollesen, K. 2000. *Blepharis (Acanthaceae): A taxonomic revision.* Great Britain: The Board of Trustees of the Royal Botanic Gardens, Kew.

von Caemmerer, S. 2000. *Biochemical models of leaf photosynthesis.* Collingwood: Commonwealth Scientific Industrial Research Organization CSIRO Publishing.

Voznesenskaya, E.V., V.R. Franceschi, O. Kiirats, E.G. Artyusheva, H. Freitag, and G.E. Edwards. 2002. Evidence for C_4 photosynthesis without Kranz anatomy in *Bienertia cycloptera* (Chenopodiaceae). *Plant Journal* 31:649–62.

Voznesenskaya, E.V, V.R. Franceschi, O. Kiirats, H. Freitag, and G.E. Edwards. 2001. Kranz anatomy is not essential for terrestrial C_4 plant photosynthesis. *Nature* 414: 543–46.

Walter, H., E. Harnickell, and D. Mueller-Dombois. 1975. *Climate-diagram maps of the individual continents and the ecological climatic regions of the Earth, supplement to the vegetation monographs.* Berlin: Springer-Verlag.

Wan, C.S.M., and R.F. Sage. 2001. Climate and the distribution of C_4 grasses along the Atlantic and Pacific coasts of North America. *Canadian Journal of Botany* 79:474–86.

Webster, G.L., W.V. Brown, and B.N. Smith. 1975. Systematics of photosynthetic carbon fixation pathways in *Euphorbia. Taxon* 24:27–33.

Winter, K., and J.A.C. Smith. 1996. An introduction to crassulacean acid metabolism: biochemical principles and ecological diversity. In *Crassulacean acid metabolism: Biochemistry, ecophysiology, and evolution,* ed. K. Winter and J.A.C. Smith, 1–16. New York: Springer Verlag.

Wolfe, J.A. 1997. Relations of environmental change to angiosperm evolution during the late Cretaceous and Tertiary. In *Evolution and diversification of land plants,* ed. K. Iwatsuki and P.H. Raven, 269–90. New York: Springer-Verlag.

Wright, V.P., and S.D. Vanstone. 1991. Assessing the carbon dioxide content of ancient atmospheres using paleocalcretes: Theoretical and empirical constraints. *Journal of the Geological Society of London* 148: 945–47.

Zachos, J., M. Pagani, L. Sloan, E. Thomas, and K. Billups. 2001. Trends, rhythms and aberrations in global climate 65 Ma to present. *Science* 292:686–93.

10. The Influence of Atmospheric CO_2, Temperature, and Water on the Abundance of C_3/C_4 Taxa

James R. Ehleringer

10.1 Introduction

C_3 and C_4 photosynthesis are the two most common photosynthetic pathways contributing to global primary productivity, with Crassulacean Acid Metabolism (CAM), a third pathway found in succulents, being of limited importance to global productivity (Sage and Monson 1999; Still et al. 2003). Atmospheric CO_2 is the carbon source for all three pathways; one key distinguishing feature among them, however, is that both C_4 and CAM photosynthesis involve a CO_2-concentrating mechanism to create high CO_2 concentrations at the site of photosynthetic carboxylation. In contrast, C_3 photosynthesis relies solely on diffusion of CO_2 from the outside atmosphere to the sites of photosynthetic CO_2 fixation (Fig. 10.1). Hence, it is clear that changes in the atmospheric CO_2 levels will more likely influence the photosynthetic activities of C_3 than of C_4 and CAM plants. In this chapter, we focus on how changes in atmospheric CO_2 might favor plants with C_3 versus C_4 photosynthesis and, in particular, consider how environmental and ecological factors will affect these C_3/C_4 abundance relationships.

Quite surprisingly, C_4 photosynthesis is found only among the most advanced land plants. C_4 is known to occur among the Angiosperms but has not been reported among older taxonomic groups, such as Gymnosperms (e.g., pines and other coniferous trees) or the Pteridopyhta (e.g., ferns) (Ehleringer and Monson 1993; Sage and Monson 1999; Chapter 9). Within the Angiosperms, approx

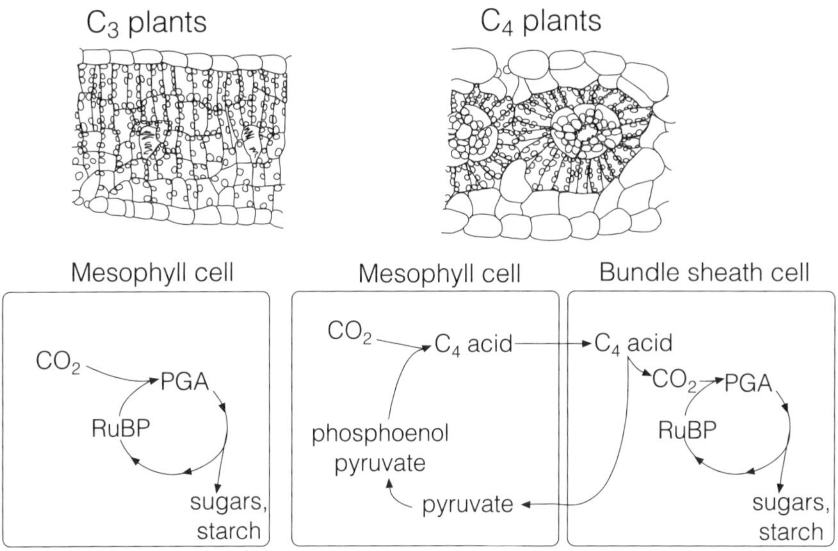

Figure 10.1. Cross-sections of leaves of C_3 and C_4 plants with the relevant features of the photosynthetic cycle for each pathway. Note that C_4 photosynthesis is essentially C_3 photosynthesis enclosed by a carbon-concentrating mechanism.

6000 of the approx 10,000 Monocotyledonae (hereafter monocots) have C_4 photosynthesis, whereas only approx 1500 of the approx 300,000 Dicotyledonae (hereafter dicots) have C_4 photosynthesis (Sage and Monson 1999). Within both monocots and dicots, C_4 photosynthesis has polyphyletic origins and appears to occur among only the most recently derived taxa, suggesting an evolutionary origin of only 10 to 15 million years ago.

Several general patterns emerge with respect to the distributions of C_3/C_4 taxa among different plant life forms (see Chapter 9). In today's world, C_4 taxa are more common among subtropical and tropical taxa; they are essentially absent from arctic and temperate taxa (Sage and Monson 1999). Furthermore, within both dicot annuals and annual/perennial monocots, C_4 taxa are typically active during the summer periods and far less common among the spring-active floras. Within the shrub life form, C_4 taxa are uncommon, but when they do occur they are largely halophytes (plants occurring on saline soils). Halophytes tend to be dicots instead of monocots, primarily in the family Chenopodiaceae. With perhaps one or two exceptions globally, C_4 photosynthesis is not found in trees. What emerges from these distribution patterns is that outside of saline soils, C_4 photosynthesis is found most commonly among herbaceous vegetation (both as annuals and as perennials). This may reflect an aspect of phylogenetic or evolutionary history, and it will certainly have ecological implications in the consideration of competition between low-stature C_4 herbaceous plants and tall C_3 trees.

The greater abundance of C_4 photosynthesis among the monocots may be a result of an anatomical pre-adaptation within this phylogenetic line. Ehleringer, Cerling, and Helliker (1997) and Sage (2001) discuss the importance of parallel venation in C_4 photosynthesis. Since the most common configuration of C_4 photosynthesis is a concentration of the C_3 cycle within the bundle sheath cells, having a parallel vein structure allows for a greater packing capacity of photosynthetic structures relative to the exposed surface area of a leaf blade. Dicots typically have a reticulate vein structure, which then results in a lower vein-packing density.

Beyond taxonomic and evolutionary inquiry, it is fair to question the value of understanding the dynamics of C_3 and C_4 taxa globally. For several key reasons, we should know the climatic and ecological factors controlling the abundances of C_4 taxa. First, in today's world, while C_4 taxa represent only approx 2% of the total species, they account for 25% to 30% of the global terrestrial primary productivity (Still et al. 2003). In a low-CO_2 world, such as during recent glacial periods, C_4 photosynthesis might have contributed an even greater proportion of global primary productivity (see Chapter 9). Second, there is ample evidence to suggest that herbivores (both insect and mammal) exhibit dietary preferences for one photosynthetic pathway over the other (summarized in Caswell et al. 1973; Ehleringer and Monson 1993; Sage and Monson 1999). Third, there is strong evidence to suggest that the relative abundances of C_3/C_4 ecosystems exhibit strong shifts over long temporal scales, such as between glacial-interglacial cycles, and that these changes may be related to atmospheric CO_2 (Talbot and Johannessen 1992; Aucour and Hillaire-Marcel 1994; Giresse, Maley, and Brenac 1994; Street-Perrott et al. 1997, 1998; Cowling and Sykes 1999; Huang et al. 1999, 2001; Boom et al. 2002; Harrison and Prentice 2003).

10.2 A Mechanistic Basis for the Advantage of C_4 Photosynthesis in a Low-CO_2 World

The enzyme Rubisco catalyzes the initial photosynthetic fixation of CO_2 in C_3 plants. Known more formally as ribulose bisphosphate carboxylase-oxygenase, Rubisco has both carboxylase and oxygenase activities:

$$\text{RuBP} + CO_2 \rightarrow 2\ \text{PGA} \tag{10.1}$$

$$\text{RuBP} + O_2 \rightarrow \text{PGA} + \text{PG} \tag{10.2}$$

where RuBP is ribulose bisphosphate, PGA is phosphoglycerate, and PG is phosphoglycolate. Subsequent oxidative metabolism of phosphoglycolate results in CO_2 loss and is known as photorespiration; the degree of oxygenase activity is a function of the atmospheric O_2/CO_2 ratio. Although Rubisco is thought to have originally evolved under anaerobic conditions, photosynthetic activity today in C_3 plants results in a relatively high photorespiration rate, simply because atmospheric CO_2 is low (~370 ppm) and atmospheric O_2 is high (~210,000 ppm). Roughly 25% to 30% of the Rubisco activity at 25°C is currently associated with oxygenase activity. Photorespiration rate is further enhanced by elevated temperature (Fig. 10.2). The photorespiratory increase with temperature occurs because Rubisco oxygenase activity is temperature sensitive, whereas carboxylase activity appears to be temperature insensitive.

The C_4 photosynthetic pathway represents a small but significant evolutionary change from C_3 photosynthesis. Biochemically, C_4 photosynthesis represents a CO_2 concentrating mechanism that achieves high CO_2 levels at the site of Rubisco activity (see Fig. 10.1). In C_3 plants, CO_2 diffuses into photosynthetic cells (resulting in lower-than-atmospheric values) and fixed by the enzyme Rubisco into phosphoglycerate (a 3C molecule). In C_4 plants, CO_2 (actually HCO_3-) is fixed more rapidly by PEP carboxylase into, initially, a 4C molecule. Since the activity of PEP carboxylase is faster than Rubisco, this leads to relatively high CO_2/O_2 levels inside bundle sheath cells, thereby eliminating photorespiration. Several C_4 photosynthetic subtypes exist (NADP-ME, NAD-ME, and PCK), with distribution following taxonomic lines (Sage and Monson 1999). The different photosynthetic subtypes represent different decarboxylation mechanisms.

The available evidence suggests that the atmospheric CO_2 levels 40 to 100 million years ago were higher than they are today (Ekart et al. 1999; see Chapter 2), and there is little likelihood that a C_4 photosynthetic carbon-concentrating

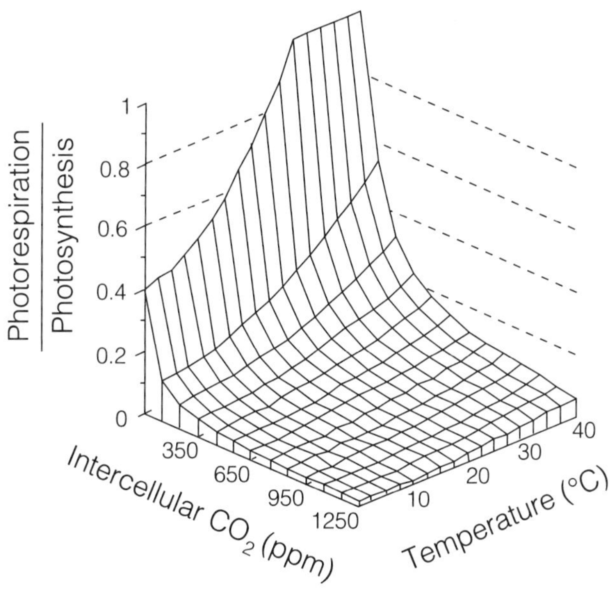

Figure 10.2. The relative proportion of photorespiration to photosynthesis in C_3 photosynthesis as a function of CO_2 and temperature. (Based on Ehleringer et al. 1991.)

mechanism would have been selectively advantageous under such conditions. Not too surprising, there is no fossil evidence to suggest that C_4 taxa were present under these high CO_2 conditions. In fact, since C_4 photosynthesis represents a CO_2 concentrating mechanism, it is likely that natural selection may not have favored the evolution of this pathway until CO_2 levels fell below some critical threshold (Ehleringer et al. 1991).

A common advantage of C_4 photosynthesis over C_3 photosynthesis is the lack of photorespiratory activity at the whole-leaf level. At low light levels, this difference is expressed as an enhanced photosynthetic quantum yield or light-use efficiency (Ehleringer and Björkman 1977; Ehleringer 1978; Ehleringer and Pearcy 1983; Sage 2001); at high light levels, photorespiration in today's atmosphere still results in a significant reduction in photosynthesis in C_3 plants. When individual leaves are exposed to higher light levels, the maximum quantum yield measured at low light levels may have little direct bearing on photosynthetic rate, but it is still an effective index of the relative inhibition of photosynthesis due to photorespiration (von Caemmerer 2000; Chapter 9). In this regard, the quantum yield serves a reliable indicator of the extent to which reduced CO_2 levels have resulted in lower photosynthetic rates in C_3 plants across a broad range of environmental conditions. The quantum yield of C_3 plants declined as temperatures increased (Fig. 10.3A), reflecting the negative impact of photorespiration on carbon gain.

Increasing atmospheric CO_2 levels increase quantum yield (reducing photorespiration), but the quantum yield still declines with increasing temperature. In contrast, the quantum yield of C_4 plants remains constant with temperature, since

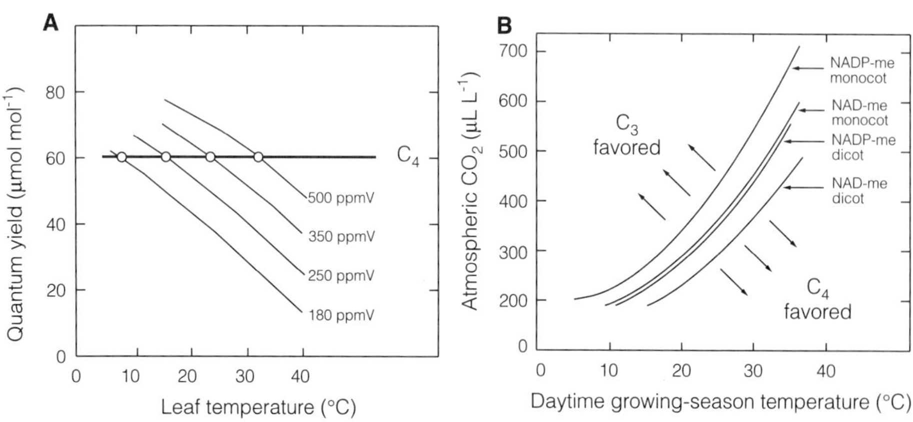

Figure 10.3. **(A)** The relationships between photosynthetic quantum yield (light-use efficiency) of C_3 and C_4 monocot taxa and temperature at several atmospheric CO_2 levels. **(B)** The prediction of where C_3 or C_4 taxa should predominate based on different combinations of atmospheric CO_2 and temperature. These curves are based on quantum yield relationships, Fig. 10.A. (Based on Ehleringer et al. 1997)

no photorespiratory activity is expressed at the leaf level. As a result of the different crossover points shown in Fig. 10.3A, we can construct a plot of those CO_2-temperature regimes where C_3 plants will have a photosynthetic advantage over C_4 plants and vice versa (Fig. 10.3B). The gap between the two central lines in the figure represents the intrinsic variations in quantum yield associated with the different C_4 subtypes.

Several distinct predictions can be made with respect to photosynthesis, temperature, and atmospheric carbon dioxide (Fig. 10.3B). First, at any point in Earth's history (and therefore at a single atmospheric CO_2 level), C_4 photosynthesis should be expected to occur largely in Earth's warmer regions with C_3 plants dominating the cooler regions. Second, as atmospheric CO_2 levels decreased in the past, the atmosphere of our planet crossed a threshold where C_4 photosynthesis became advantageous over C_3 photosynthesis, particularly at high growing-season temperatures. This necessarily implies warm temperatures during the growing season (therefore most likely a summer-wet or monsoon-type environment, but one in which grasses are not shaded out). While we cannot specify the CO_2 level at which C_4 first became advantageous, the results shown in Fig. 10.3B suggest that the crossover favoring expansion of C_4 plants occurred at atmospheric CO_2 levels that are higher than the present (i.e., >370 ppm). Atmospheric CO_2 will not be the sole factor determining a selective advantage to C_4 plants, but the differential photorespiratory responses of C_3 and C_4 photosynthetic physiologies certainly implies that C_3 plants will be sensitive to high temperature environments, particularly as atmospheric CO_2 levels decrease. Lastly, the quantum yield model predicts that C_4 photosynthesis should be expected to expand significantly under low CO_2 conditions, such as occurred during glacial periods. Under glacial CO_2 levels of 180 ppm (see Chapter 4), C_3 plants should be getting close to CO_2 starvation and photorespiration rates would be expected to be quite high (see Fig. 10.2).

10.3 C_4 Photosynthesis: An Adaptation to What?

While Fig. 10.3B implies that C_4 photosynthesis should be an adaptation to warm growing-season regions of a low CO_2 world, there are other potential ecological conditions where this pathway may exhibit a competitive advantage over C_3 plants (see Chapter 9). One obviously advantageous situation is a saline environment, where plants typically have reduced transpiration rates because stomatal conductances are reduced, thus limiting the inward diffusion of CO_2 (Pearcy and Ehleringer 1984; Sage and Monson 1999). Here it appears that C_4 dicots are most abundant, especially members of the Amaranthaceae and Chenopodiaceae families, such as *Atriplex* and *Salsola* (Sage and Monson 1999; Pyankov et al. 1999, 2000). Relatively few CO_2 monocot taxa occur in saline regions, although *Distichlis* and *Spartina* are clear examples of monocots that often predominate globally in many salt marsh ecosystems.

C_4 taxa have a wide distribution, largely centered in the tropical and subtropical regions (Fig. 10.4). Here the predominance of C_4 taxa as grasses and sedges becomes most evident in the savanna and grassland biomes. Although C_4 taxa are predicted to be common in tropical regions (Collatz, Berry, and Clark 1998), their distribution is restricted to open, disturbed spaces and some aquatic regions, where competition with taller life forms is minimal. In more temperate regions, C_4 taxa are far less common, occurring largely in halophytic regions or habitats of high disturbance (Sage and Monson, 1999; Chapter 9).

Today, C_4 monocots are most abundant in summer-wet, semi-arid ecosystems, where the ratio of precipitation to evaporation approaches unity (Fig. 10.5). We find that C_4 monocots (largely grasses and sedges) are quite common in both grassland and savanna ecosystems. C_4 taxa are not common to all arid land ecosystems, and it is difficult to attribute the current primary advantage of the C_4 pathway itself as an adaptation to drought. Many of the hot, summer-dry desert ecosystems, such as the Mohave Desert of North America and the Sahara Desert of Africa, are dominated by C_3 vegetation and not by C_4 vegetation, because rains fall during the winter-spring when temperatures are cool and not in the summer when temperatures are hot (Ehleringer and Monson 1993; Sage and Monson 1999).

While some paleontological studies assume that the presence of C_4 taxa is indicative of drought environments (e.g., Huang et al., 2001), there is no evidence of C_4 dominance in today's arid land ecosystems that lack summer rain (Ehleringer and Monson 1993; Collatz, Berry, and Clark 1998; Sage and

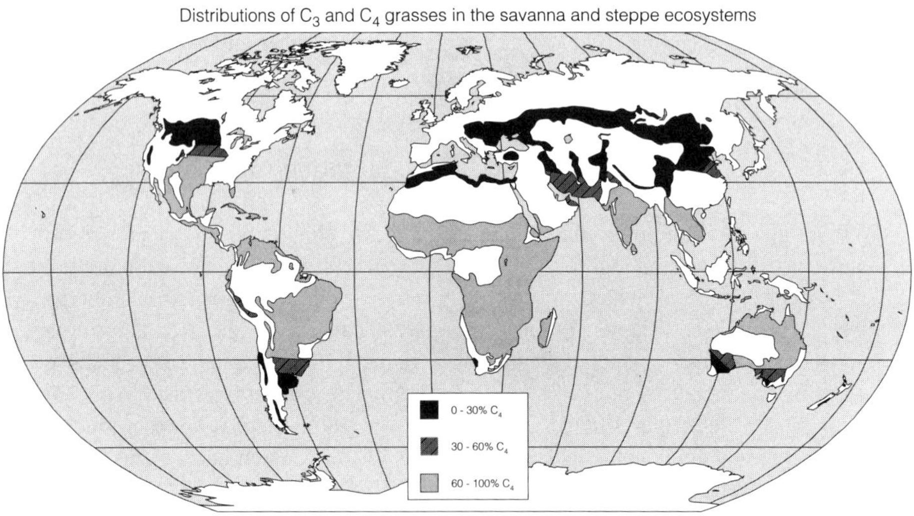

Figure 10.4. The distributions of C_3 and C_4 grasses in steppe and savanna ecosystems of the world. These are the two biomes wherein grasses are a significant fraction of the vegetation. (Adapted from Ehleringer and Cerling 2001.)

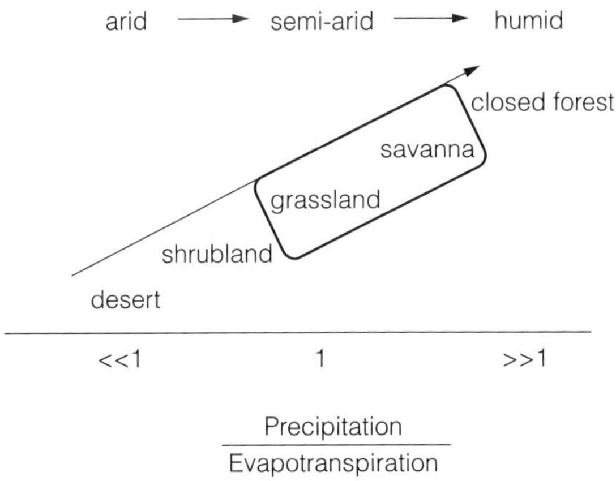

Figure 10.5. The distribution of different ecosystem types as a function of the ratio of precipitation to evapotranspiration. Circled are those ecosystems where C_4 monocot taxa are common.

Monson 1999). From a physiological perspective, there appears to be nothing inherent in the C_4 pathway that conveys a selective advantage to reduced plant water content, such as plants would typically experience during drought. Perhaps surprisingly, C_4 plants are not a large component of dry desert ecosystems where seasonal and interannual droughts are common (see Chapter 9). When they do occur in desert regions, C_4 monocots tend to be active only in the summer in regions with summer rains and then to be active only after the warm summer rains have arrived (Shreve and Wiggins 1964; Sage and Monson 1999).

As a consequence of their low stature, C_4 monocots can predominate only in open ecosystems where trees do not shade them out. This makes it particularly difficult in a historical sense to distinguish C_3/C_4 shifts that are associated with C_3/C_4 monocot changes (such as would occur in grasslands) from C_3/C_4 shifts associated with the gaps between tree canopies opening and closing, such as those that would occur in savannas in response to fire or grazing pressures (see Chapter 9). Bond, Midgley, and Woodward (2003) have recently modeled C_3 tree and C_4 grass responses to fire, a major disturbance of African grassland and savanna ecosystems. They found that the slow recovery of trees under a low-CO_2 environment favored the expansion of grasslands. This is similar to the conclusion of Cerling, Ehleringer, and Harris (1998) that C_3 taxa tend to approach CO_2 starvation under the low CO_2 levels of glacial periods. Thus, it is easier to interpret and understand the contributing factors impacting historical changes in local-to-regional C_3/C_4 abundances when comparisons are made within a single life form (grass-grass) rather than across life forms (tree-grass).

10.4 The Interactions Between Atmospheric CO_2, Temperature, and Precipitation: How They Affect C_4 Monocot Distribution

At any given atmospheric CO_2 level, the Ehleringer et al. (1997) model predicts that C_4 taxa should be most likely to occur in those habitats having a warmer growing season. Implicit in this model is that growing-season temperatures are distinguished from non–growing-season temperatures. Plants experiencing a cool winter-precipitation regime followed by a hot, dry summer (such as all Mediterranean climates) would not be considered warm-temperature ecosystems. Thus, Mediterranean climate and monsoonal climates would be distinct from each other, even though summertime temperatures in both climates might be similar. Teeri and Stowe (1976) were the first to show that C_4 grass distributions across the Great Plains of North America were linearly related to growing-season temperature. Ehleringer (1978) modeled current C_4 grass distributions under several ecological situations (geographic gradients in the Great Plains and biseasonal precipitation regimes) and showed that C_4 grass distributions followed the predicted temperature relationships shown in Fig. 10.3B. Recently, Ehleringer, Cerling, and Helliker (1997) reviewed more than 20 C_4 monocot studies and showed that in each case, % C_4 abundance was a linear function of growing-season temperature. Sage and Monson (1999) showed that % C_4 abundance on islands followed latitudinal gradients, with C_4 plants most common in warm regions. Collatz, Berry, and Clark (1998) predicted global C_4 abundances using the quantum yield model. Their results were constrained to consider only growing-season temperatures (i.e., monthly periods during which precipitation exceeded a threshold). In each of these studies, the regional and global C_3/C_4 distributions are consistent with predictions shown in Fig. 10.3B.

The quantum yield model predicts that at some time in Earth's recent history the atmospheric CO_2 declined to a point where a threshold was crossed and C_4 plants would be favored globally in the warmest growing-season habitats. Cerling et al. (1997) provided convincing evidence for a global expansion of C_4-dominated ecosystems approximately 6 to 8 million years ago during the late Miocene. At that time, C_4 ecosystems appeared nearly simultaneously in North and South America, Africa, and Asia. In each case, C_4 ecosystems appeared in warmer, lower latitudes and not in higher, cooler latitudes. Just exactly what that threshold CO_2 level was is unclear, but the threshold CO_2 would clearly have depended on the growing-season temperature (Fig. 10.3B). It is likely that the CO_2 level was somewhere in the range of 350 to 500 ppm. Firm estimates of the atmospheric CO_2 levels during the late Miocene are not well constrained. Some CO_2 proxies suggest that atmospheric CO_2 levels have been at near current levels since the Eocene (Pagani, Freeman, and Arthur 1999; Freeman and Colarusso 2001; see Chapter 3). Oxygen isotope ratio data suggest that increased monsoonal activity in southern Asia and the Indian subcontinent preceded the expansion of C_4 ecosystems by 1 to 2 million years (Quade, Cerling, and Bowman 1989). Such activity would have increased weathering of Himalayan uplift and could have been a driver for reducing atmospheric CO_2 levels (Ruddiman

1990). Until we have a clear high-resolution time course of CO_2 for the past 40 to 60 million years, the exact threshold that precipitated the expansion of C_4 ecosystems will remain unclear.

10.5 Was Atmospheric CO_2 a Driver for C_4 Expansion During Glacial Periods?

The changes in atmospheric CO_2 levels between 180 and 280 ppm over the past 420,000 years (Petit et al., 1999; see Chapter 4) should have impacted the abundances of C_3/C_4 taxa. The quantum yield model predicts that in ecosystems with warm growing-season temperatures the potential abundances of C_3/C_4 taxa fluctuated as CO_2 rose and declined between glacial and interglacial periods. There is now ample evidence from ecosystems in North America, South America, Central America, and Africa showing that C_4 taxa replaced C_3 taxa during glacial periods and that C_4 taxa often decreased in abundance following the last glacial maximum (e.g., Talbot and Johannessen, 1992; Aucour and Hillaire-Marcel 1994; Giresse, Maley, and Brenac 1994; Street-Perrott et al. 1997, 1998; Huang et al. 1999, 2001). Three depth-time profiles from different tropical regions clearly demonstrate a carbon isotopic shift in the historical record between glacial and interglacial periods (Fig. 10.6).

In each of these cases, the trends are consistent with increased C_4 abundance during the Last Glacial Maximum (LGM), followed by a decrease in the abundance of C_4 photosynthesis following deglaciation and the correlated increase in atmospheric CO_2. Recently, Boom et al. (2002) showed that the transition between C_3 and C_4 dominated ecosystems in the Bogota basin, Colombia, has repeatedly followed glacial-interglacial cycles for the past 400,000 years. During

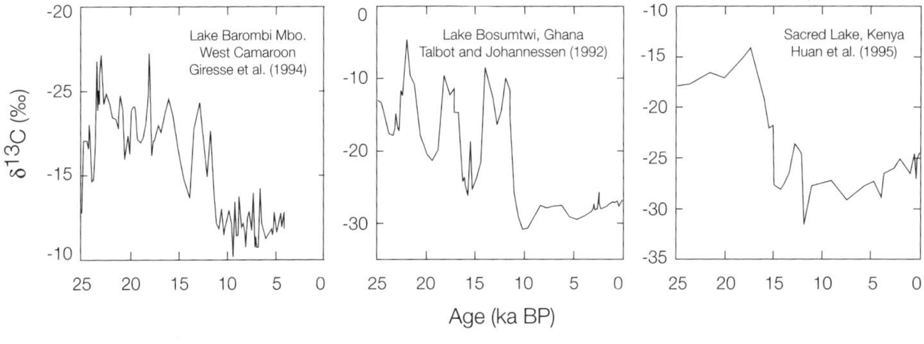

Figure 10.6. Chronological profiles of the carbon isotope ratio values of organic matter in *left* Lake Barombi Mbo, Cameroon (Giresse et al., 1994), *middle* Lake Bosumtwi, Ghana (Talbot and Johannessen (1992), and *right* Sacred Lake, Kenya (Huang et al., 1995).

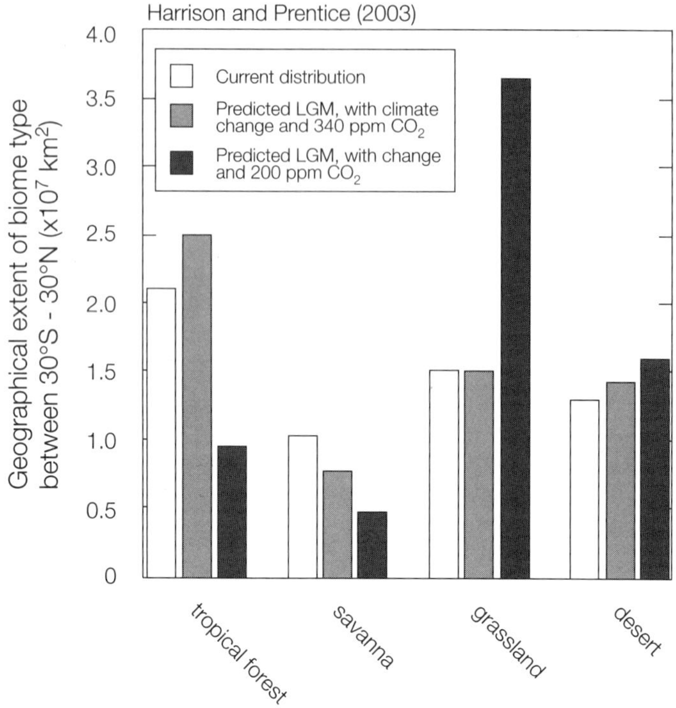

Figure 10.7. A comparison of the current geographic extent of tropical forest, savanna, grassland, and desert biomes between latitudes 30°S and 30°N to the geographical extent predicted from BIOME4 for these biomes types during the last glacial maximum under two environmental scenarios: (a) the last glacial maximum climate and today's atmospheric CO_2 level and (b) the last glacial maximum climate and a reduced CO_2 level of 200 ppm. Bars indicate the average geographic extent predicted based on 17 different climate simulations. (Based on data presented in Harrison and Prentice 2003.)

each glacial period (low CO_2), C_4 taxa expanded, whereas during interglacial periods, C_3 taxa predominated. Whether these C_3/C_4 shifts are associated with changes in the abundances of C_3/C_4 monocots or changes in the abundances of C_3 trees versus C_4 monocots is not always clear. Pollen evidence (when available) often indicates that the latter possibility is more likely. Other paleontological evidence indicates shifts in C_3/C_4 abundances that are correlated with the changes in atmospheric CO_2. In regions of western North America now dominated by C_3 taxa, tooth analyses of several extinct grazers (e.g., mastadonts, bison, and camels) indicate that C_4 species were a major dietary component (see Chapters 12 and 13 and Connin, Betancourt, and Quade 1998 for interpretation of tooth data). At the landscape scale, caliche carbonate analyses indicate that some portions of Arizona and New Mexico had a significant C_4 component even though these sites are C_3 dominated today (Cole and Monger 1994).

Recently, Harrison and Prentice (2003) have examined the extent to which the changes in global vegetation between the last glacial maximum and present have been driven by (a) changes in climate (temperature and precipitation) alone or (b) changes in both climate and atmospheric CO_2. They examined predictions of a global vegetation model driven by 17 different simulations of the paleoclimate at the LGM. Of considerable interest is that simulations with and without the direct consideration of atmospheric CO_2 on photosynthesis yielded quite different predictions (Fig. 10.7).

When compared to the present-day aerial extent of major biomes, the model predicted that with climate change alone (temperature and precipitation), there would be little change in the global extent of forest, savanna, and grassland vegetation. That predicted pattern is not consistent with the available pollen record. However, when the change in atmospheric CO_2 was also included with climate change in the model, then there were large predicted reductions in both the extent of tropical forest and savanna biomes (Fig. 10.7). This direct effect of CO_2 on biome-scale photosynthesis and hence on productivity and plant distribution is consistent with the predictions of the quantum yield model. C_4 grasslands apparently expanded as C_3 forests retreated during glacial periods. In effect, the C_3 forests in warm climates became much more CO_2 starved under the 180 to 190 ppm CO_2 of glacial periods.

10.6 Seasonality's Impact on When Plants Can Grow

Seasonality is an important ecological factor that is often difficult to extract from paleo-observations. When reconstructing C_3/C_4 abundance relationships, it is essential to recognize that changes in ocean circulation patterns can lead to changes in the seasonality of precipitation. For example, pack rat midden data indicate that the extent of summer rains in the western United States retreated southward over the past several thousand years (Betancourt et al. 1990). Bristlecone pine tree ring isotope data from the southern Sierra Nevada Range support the notion of the loss of summer rains in the western United States and suggest that some of these precipitation shifts could have been abrupt during the last millennium (Yapp and Epstein 1982). This possibility is also supported by lake-level increase indicating more winter precipitation in the area over the past 700 years (Stine 1994). In such cases, atmospheric CO_2 levels need not have decreased to favor expansion or contraction of C_4 monocots in the semi-arid western United States. Based on the quantum yield model, the critical factor would be the temperature during the growing season. In the absence of a summer growing season, the vegetation would likely be C_3 since temperatures are cool during the late winter and spring growing conditions.

Extending this analogy to present-day subtropical latitudes such as found in southwestern and western North America, winter precipitation regimes will favor C_3 monocots, whereas summer precipitation will favor C_4 monocots (Mulroy and Rundel 1977; Ehleringer, Cerling, and Helliker 1997; Sage and Monson

1999). This should also have been the case in the past, such as during glacial periods, although some regions could have been warm enough to favor C_4 monocots even during winter-spring periods. Huang et al. (2001) have attributed both seasonality and CO_2 as drivers in the C_3/C_4 shifts in Central American ecosystems, as predicted by the quantum yield model. In that study, Huang et al. hypothesized that shifts between summer monsoonal and winter frontal precipitation patterns likely contributed to an absence of C_4 taxa in northern Mexico (Chihuahua region) during the last glacial maximum, whereas C_4 taxa were quite common in a Guatemalan ecosystem. Ultimately, other factors could have contributed to the absence of C_4 taxa in this Chihuahuan ecosystem, since C_3 monocots are common there today (Shreve and Wiggins 1964; Huang et al. 2001) and since regions to the north were known to have been C_4 dominated during the last glacial maximum (cited above).

To illustrate the importance of seasonality favoring one photosynthetic pathway over another, consider the abundances of C_3/C_4 monocots in southern California and Florida today. Even though both locations are at similar latitudes, the C_4 abundance is high in Florida and low in southern California (Sage and Monson 1999). This is because the precipitation in California comes during the spring cool season, whereas the predominant precipitation in Florida is in the warm summer season. Accordingly, there is a high proportion of C_4 plants in Florida but not in southern California (Teeri and Stowe 1976; Sage and Monson 1999), as predicted by the quantum yield model shown in Fig. 10.3B. Similarly, in the southern United States and in northwestern Mexico where biseasonal precipitation patterns are common today, we see the same C_3/C_4 separations, with C_4 taxa active during the warm, wet summer periods (Shreve and Wiggins 1964; Mulroy and Rundel 1977). Today in these biseasonal precipitation ecosystems, C_3 grasses dominate the winter growing season, while C_4 grasses dominate the summer growing season. In order for C_4 grasses to have had a competitive advantage over C_3 grasses, the quantum yield model predicts that if current wintertime temperatures were maintained, the CO_2 level would have had to decrease substantially before C_4 grasses would be favored throughout the year. Thus, the seasonality of the precipitation event has a significant impact, influencing C_4 abundance because of temperature differences between winter and summer in temperate regions.

10.7 Ecological Factors Further Limit the Abundances of C_4 Monocot Taxa

Both climatic and nonclimatic factors will influence the abundances of C_4 taxa. Among the climatic factors already considered are atmospheric CO_2 and O_2 levels, temperature, and seasonality of precipitation (Fig. 10.8).

Yet consideration must also be given to nonclimatic drivers that will influence the abundances of C_3/C_4 taxa. Primary among these ecological considerations are factors that tend to open a C_3 forest and to allow expansion of the understory

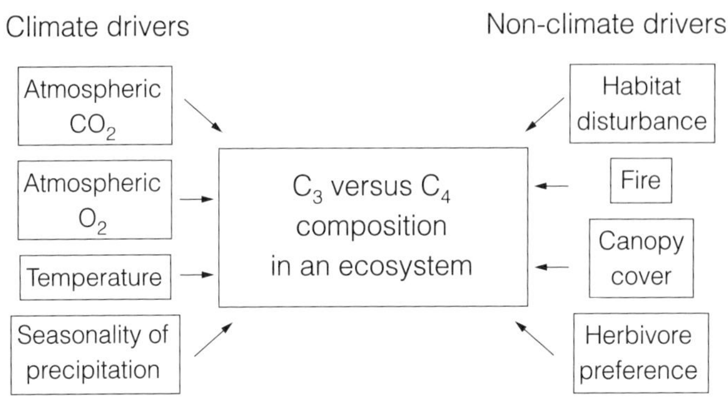

Figure 10.8. Climatic and nonclimatic drivers for changes in the abundances of C_3 and C_4 taxa in an ecosystem.

C_4 monocot components. These include habitat disturbance, fire, and herbivory, each of which affects the extent of tree canopy cover (Fig. 10.9).

In recent historical times, man is known to be a causal agent for such change as extensive deforestation, which leads to grass expansion. Consider four broad examples. First, forest destruction associated with the expansion of the Mayan civilization in Central America (Huang et al. 2001) and expansions in Ethiopia (Eshetu and Högberg 2000; Eshetu 2002) resulted in large changes in carbon isotope ratios of soil organic matter and lake sediments, allowing a time course reconstruction of the C_3/C_4 shifts. Second, recent conversions of tropical rainforests to pastures also result in profound shifts in the carbon isotope ratios of tropical soils (e.g., Neill et al. 1996). Third, large herbivores, such as elephants, are known to prefer C_3 trees and can have a profound effect on opening canopies (Cerling and Harris 1999). Fourth, fire will open forest canopies, allowing a C_3-to-C_4 shift (e.g., Bond 2000). In general, increased disturbance will make ecosystems appear more C_4-like, since C_3 trees are more likely to be impacted by disturbances such as fire and herbivory than are the C_4 grasses. Recovery from disturbance in these C_3/C_4 ecosystems will lead to a C_3 conversion as trees overtop and shade out the C_4 grasses (Bond 2000). If fine-scale paleontological isotope data were the only information available for recent time periods, it would be possible misinterpret the C_3/C_4 shifts as being climatically driven when, in fact, the C_3/C_4 shifts could have been driven by nonclimatic factors.

In Fig. 10.9 the climatic and nonclimatic drivers of change in C_3/C_4 ecosystems are integrated into a simple but mechanistically based model. Each of these ecosystems is semi-arid, with precipitation to evaporation ratios near unity. Shifts in C_3/C_4 abundances in tree-grass ecosystems are driven by both climatic (e.g., CO_2, temperature, precipitation) and nonclimatic factors (fire, disturbance). Within these C_3/C_4 shifts, C_4 NAD-ME grasses should occur on the drier sites, since these grasses have the lowest quantum yield values and may not out-

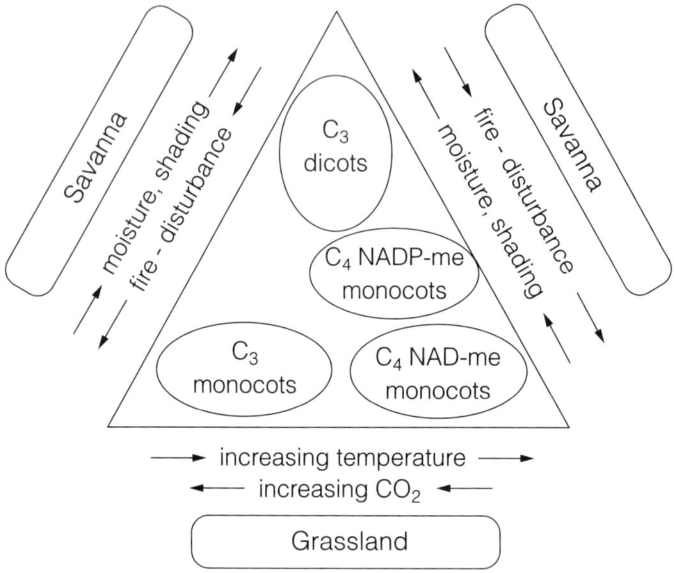

Figure 10.9. A plot of the environmental and ecological factors that govern shifts in the abundances of C_3 monocots and C_4 dicots within grassland and savanna ecosystems.

compete the more efficient NADP-ME grasses. In contrast, C_4 NADP-ME grasses are predicted to be more common on wetter sites than those occupied by NAD-ME plants. As moisture levels increase along this geographic cline, support of a higher leaf-area tree life form ultimately becomes possible. These predictions are consistent with observed temperate ecosystem distributions globally, such as the savannas, tall-grass prairie (NADP-ME dominated), and short-grass prairie (NAD-ME dominated) of North America. Similarly, in subtropical ecosystems globally, disturbance levels determine the shifts between open C_4 grass-dominated savanna and closed C_3 savanna forests in wet regions, while the C_3/C_4 savanna to C_4 grassland distributions are determined by precipitation levels. The second class of C_3/C_4 shifts occurs in grassland ecosystems, where life form remains constant but the growing-season temperatures dictate the extent to which photorespiratory costs favor the C_3 or C_4 photosynthetic pathway.

10.8 Summary

The abundance of plants having C_3 versus C_4 photosynthesis is strongly coupled to two environmental factors: atmospheric CO_2 and growing-season temperature. The mechanistic quantum yield predicts the combinations of CO_2 and temperature that result in photorespiratory changes that favor C_4 taxa over C_3 taxa. Warm temperatures and low atmospheric CO_2 levels favor C_4 taxa over C_3 taxa.

The abundance of C_4-dominated ecosystems is of only recent origin, with C_4-dominated ecosystems having expanded globally about 6 to 8 million years ago. Seasonality and disturbance regimes are important ecological considerations, which can contribute to the C_3/C_4 dominance in transition climate regimes under a constant atmospheric CO_2. Future increases in atmospheric CO_2 are likely to favor the expansion of C_3-dominated ecosystems over C_4-dominated ecosystems.

References

Aucour, A.-M., and C. Hillaire-Marcel. 1993. A 30,000 year record of ^{13}C and ^{18}O changes in organic matter from an equatorial bog. In: *Climate changes in continental isotopic records* (Geophysical Monograph 78), ed. P.K. Swart, K.C. Lohmann, J. McKenzie, S. Savin, 343–51. Washington, D.C.: American Geophysical Union.

———. 1994. Late quaternary biomass changes from ^{13}C measurements in a highland peatbog from equatorial Africa (Burundi). *Journal of Quaternary Research* 41:225–33.

Betancourt, J.L., T.R. Van Devender. et al., eds. 1990. Packrat middens: The last 40,000 years of biotic change. In *Chihuahuan desert: Vegetation and climate.* Tucson: University of Arizona Press.

Bond, W.J. 2000. A proposed CO_2-controlled mechanism of woody plant invasion in grasslands and savannas. *Global Change Biology* 6:865–69.

Bond, W.J., G.F. Midgley, and F.I. Woodward. 2003. The importance of low atmospheric CO_2 and fire in promoting the spread of grasslands and savannas. *Global Change Biology* 9:973–82.

Boom, A., R. Marchant, H. Hooghiemstra, and J.S.S. Damste. 2002. CO_2- and temperature-controlled altidudinal shifts of C_4- and C_3-dominated grasslands, allow reconstruction of palaeoatmospheric pCO_2. *Palaeogeography, Palaeoclimatology, and Palaeoecology* 177:151–68.

Caswell, H., F. Reed, S.N. Stephenson, and P.A. Werner. 1973. Photosynthetic pathways and selective herbivory: A hypothesis. *American Naturalist* 107:465–79.

Cerling, T.E., J.R. Ehleringer, and J.M. Harris. 1998. Carbon dioxide starvation, the development of C_4 ecosystems, and mammalian evolution. *Philosophical Transactions of the Royal Society of London Series B* 353:159–71.

Cerling, T.E., and J.M. Harris. 1999. Carbon isotope fractionation between diet and bioapatite in ungulate mammals and implications for ecological and paleoecological studies. *Oecologia* 120:347–63.

Cerling, T.E., J.M. Harris, B.J. MacFadden, M.G. Leakey, J. Quade, V. Eisenmann, and J.R. Ehleringer. 1997. Global vegetation change through the Miocene-Pliocene boundary. *Nature* 389:153–58.

Cole D.R., and H.C. Monger. 1994. Influence of atmospheric CO_2 on the decline of C_4 plants during the last deglaciation. *Nature* 368:533–36.

Collatz, G.J., J.A. Berry, and J.S. Clark. 1998. Effects of climate and atmospheric CO_2 partial pressure on the global distribution of C_4 grasses: present, past, and future. *Oecologia* 114:441–54.

Connin, S.L., J. Betancourt, and J. Quade. 1998. Late Pleistocene C_4 plant dominance and summer rainfall in the southwestern United States from isotopic study of herbivore teeth. *Quaternary Research* 50:179–93.

Cowling, S.A., and M.T. Sykes. 1999. Physiological significance of low atmospheric CO_2 for plant-climate interactions. *Quaternary Research* 55:140–49.

Ehleringer, J. 1978. Implications of quantum yield differences on the distributions of C_3 and C_4 grasses. *Oecologia* 31:255–67.

Ehleringer, J., and O. Björkman. 1977. Quantum yields for CO_2 uptake in C_3 and C_4 plants: dependence on temperature, CO_2 and O_2 concentration. *Plant Physiology* 59: 86–90.

Ehleringer, J.R., and T.E. Cerling. 2001. Photosynthetic pathways and climate. In *Global biogeochemical cycles in the climate system*, ed. E.-D. Schulze, M. Heimann, S.P. Harrison, E.A. Holland, J. Lloyd, I.C. Prentice, and D. Schimel, 267–77. San Diego: Academic Press.

Ehleringer, J.R., T.E. Cerling, and B.R. Helliker. 1997. C_4 photosynthesis, atmospheric CO_2, and climate. *Oecologia* 112:285–99.

Ehleringer, J.R., and R.K. Monson. 1993. Evolutionary and ecological aspects of photosynthetic pathway variation. *Annual Review of Ecology and Systematics* 24:411–39.

Ehleringer, J.R. and R.W. Pearcy. 1983. Variation in quantum yields for CO_2 uptake in C_3 and C_4 plants. *Plant Physiology* 73:555–59.

Ehleringer, J.R., R.F. Sage, L.B. Flanagan, and R.W. Pearcy. 1991. Climate change and the evolution of C_4 photosynthesis. *Trends Ecology and Evolution* 6:95–99.

Ekart, D.D., T.E. Cerling, I. Montanez, and N. Tabor. 1999. A 400 million year carbon isotope record of pedogenic carbonate: Implications for atmospheric carbon dioxide. *American Journal of Science* 299:805–17.

Eshetu, Z. 2002. Historical C_3/C_4 vegetation pattern of forested mountain slopes: its implication for ecological rehabilitation of degraded highlands of Ethiopia by afforestation. *Journal of Tropical Ecology* 18:743–58.

Eshetu, Z., and P. Högberg. 2000. Reconstruction of forest site history in Ethiopian highlands based on C-13 natural abundance of soils. *Ambio* 29:83–89.

Freeman, K.H., and L.A. Colarusso. 2001. Molecular and isotopic records of C_4 grassland expansion in the late Miocene. *Geochimica et Cosmochimica Acta* 65:1439–54.

Giresse, P., J. Maley, and P. Brenac. 1994. Late Quaternary palaeoenvironments in Lake Barombi Mbo (West Cameroon) deduced from pollen and carbon isotopes of organic matter. *Palaeogeography, Palaeoclimatology, and Palaeoecology* 107:65–78.

Harrison, S.P., and C.I. Prentice. 2003. Climate and CO_2 controls on global vegetation distribution at the last glacial maximum: analysis based on paleovegetation data, biome modeling and paleoclimate simulations. *Global Change Biology* 9:983–1004.

Huang, Y., K.H. Freeman, T.I. Eglington, and F.A. Street-Perrott. 1999. $\delta^{13}C$ analyses of individual lignin phenols in Quaternary lake sediments: A novel proxy for deciphering past terrestrial vegetation changes. *Geology* 27:471–74.

Huang, Y., F.A. Street-Perrott, S.E. Metcalfe, M. Brenner, M. Moreland, and K.H. Freeman. 2001. Climate change as the dominant control on glacial-interglacial variations in C_3 and C_4 plant abundance. *Science* 293:1647–51.

Mulroy, T.W., and P.W. Rundel. 1977. Annual plants: adaptations to desert environments. *BioScience* 27:109–14.

Neill, C., B. Fry, J.M. Melillo, P.A. Steudler, J.F.L. Moraes, and C.C. Cerri. 1996. Forest- and pasture-derived carbon contributions to carbon stocks and microbial respiration of tropical pasture soils. *Oecologia* 107:113–19.

Pagani, M., K.H. Freeman, and M.A. Arthur. 1999. Late Miocene atmospheric CO_2 concentrations and the expansion of C_4 grasses. *Science* 285:876–78.

Pearcy, R.W., and J. Ehleringer. 1984. Ecophysiology of C_3 and C_4 plants. *Plant Cell and Environment* 7:1–13.

Petit, J.R., J. Jouzel, D. Raynaud, et al. 1999. Climate and atmospheric history of the past 420,000 years from the Vostok ice core, Antarctica. *Nature* 399:429–36.

Pyankov, V.I., C.C. Black, Jr., E.G. Artyusheva, E.V. Voznesenskaya, M.S.B. Ku, and G.E. Edwards. 1999. Features of photosynthesis in *Haloxylon* species of Chenopodiaceae that are dominant plants in central Asian deserts. *Plant Cell Physiology* 40: 125–34.

Pyankov, V.I., et al. 2000. Occurrence of C_3 and C_4 photosynthesis in cotyledons and leaves of *Salsola* species (Chenopodiaceae). *Photosynthesis Research* 63:69–84.

Quade J., T.E. Cerling, and R.J. Bowman. 1989. Development of Asian monsoon revealed by marked ecological shift during the latest Miocene in Northern Pakistan. *Nature* 342:163–66.

Ruddiman, W.F. 1990. Changes in climate and biota on geologic time scales. *Trends in Ecology and Evolution* 5:285–88.

Sage, R.F. 2001. Environmental and evolutionary preconditions for the origin and diversification of the C_4 photosynthetic syndrome. *Plant Biology* 3:202–13.

Sage, R.F., and R.K. Monson. 1999. *C_4 plant biology.* San Diego: Academic Press.

Shreve, F., and I.R. Wiggins. 1964. *Vegetation and flora of the Sonoran Desert.* Palo Alto: Stanford University Press.

Still, C., J.A. Berry, G.J. Collatz, and R.S. Defries. 2003. Global distribution of C_3 and C_4 vegetation: Carbon cycle implications. *Global Biogeochemical Cycles* 17 (January): Article 1006.

Stine, S. 1994. Extreme and persistent drought in California and Patagonia during mediaeval time. *Nature* 369:546–49.

Street-Perrott, F.A., Y. Huang, R.A. Perrott, and G. Eglington. 1998. Carbon isotopes in lake sediments and peats of the last glacial age: Implications for the global carbon cycle. In *Stable isotopes,* ed. H. Griffiths, 381–96. Oxford: BIOS Scientific.

Street-Perrott, F.A., Y. Huang, R.A. Perrott, G. Eglington, P. Barker, L.B. Khelifa, D.D. Harkness, and D.O. Olago. 1997. Impact of lower atmospheric carbon dioxide on tropical mountain ecosystems. *Science* 278:1422–26.

Talbot, M.R., and R. Johannessen. 1992. A high resolution palaeoclimatic record for the last 27,500 years in tropical West Africa from the carbon and nitrogen isotopic composition of lacustrine organic matter. *Earth and Planetary Science Letters* 110:23–37.

Talbot, M.R., D.A. Livingstone, P.G. Palmer, J. Maley, J.M. Melack, J.M. Delibrias, and S. Gulliksen. 1984. Preliminary results from sediment cores from Lake Bosumtwi, Ghana. *Palaeoecology Africa* 16:173–92.

Teeri, J.A., and L.G. Stowe. 1976. Climatic patterns and the distribution of C_4 grasses in North America. *Oecologia* 23: 1–12.

von Caemmerer, S. 2000. *Biochemical models of leaf photosynthesis.* Collingwood: Commonwealth Scientific Industrial Research Organization Publishing.

Yapp, C.J., and S. Epstein. 1982. Climatic significance of the hydrogen isotope ratios in tree cellulose. *Nature* 297:636–39.

11. Evolution and Growth of Plants in a Low CO_2 World

Joy K. Ward

11.1 Introduction

11.1.1 Carbon Dioxide as a Plant Resource

CO_2 is unique among plant resources in that it is acquired from the atmosphere, whereas water and nutrients are taken up from the soil. During photosynthesis, plants produce organic carbon when RuBP (ribulose-1, 5-bisphosphate) is carboxylated via the enzyme Rubisco (ribulose bisphosphate carboxylase/oxygenase). The carbon dioxide concentration in the ambient air surrounding plants influences internal leaf CO_2 concentration (c_i), and this internal concentration influences the rate of carbon assimilation, or, CO_2 uptake per unit leaf area per unit time (see Sage 1994, Chapter 9) in the chloroplasts. Subsequently, the rate of carbon assimilation influences growth processes, plant functioning, and reproductive output (Curtis and Wang 1998; Makino and Mae 1999; Norby et al. 1999; Pritchard et al. 1999; Sage and Cowling 1999). In addition, the concentration of carbon dioxide relative to oxygen at the site of carboxylation of RuBP affects the rate of photorespiration, a process that reduces the efficiency of carbon uptake. Photorespiration is prevalent in plants with the C_3 photosynthetic pathway, whereas plants with the C_4 pathway have a CO_2-concentrating mechanism that essentially eliminates photorespiration (Ehleringer and Monson 1993, Chapter 10; Kanai and Edwards 1999). Therefore, changes in atmospheric CO_2 concentration would not be expected to affect all plant species equally and may have influenced plant competition and community structure within ecosystems over geologic timescales.

Although some photosynthetic microorganisms, such as *Chlamydomonas* species, can acclimate to CO_2 concentrations ranging from 10 ppm to as high as 10,000 ppm (Kaplan et al. 2001), land plants are adapted to operate within a more narrow range of atmospheric CO_2 concentrations. At CO_2 concentrations above the modern value of 370 ppm, C_3 plants often exhibit increased carbon assimilation and growth (Gunderson and Wullschleger 1994; Curtis and Wang 1998; Ward and Strain 1999), and generally become CO_2 saturated near 1200 ppm CO_2 (Bugbee et al. 1994). At extremely high CO_2 concentrations (10,000–25,000 ppm), plants may exhibit signs of CO_2 toxicity, including reduced seed yields (Bugbee et al. 1994) and leaf abnormalities (Schwarz and Strain 1990). These responses are likely due to starch accumulation in leaves and imbalances in carbon levels relative to other nutrients. At the other extreme, low CO_2 concentrations may be highly stressful on C_3 plants (95% of all plant species) and may reduce growth and reproductive output as a result of carbon limitations. During the late Pleistocene, CO_2 concentrations were 25% to 50% lower than at present, declining to values of 180 ppm during glacial periods (last occurring 18,000 to 20,000 years ago). These CO_2 concentrations may have been among the lowest that occurred during the evolution of land plants (Berner 1994; see Chapter 1) and likely reduced plant productivity (Polley et al. 1993a,b; Dippery et al. 1995; Sage 1995; Ward and Strain 1997; Sage and Coleman 2001), prevented reproduction in some species (Dippery et al. 1995), and altered evolutionary trajectories (Ward et al. 2000).

11.1.2 History of Low CO_2 Studies

Past studies addressing the effects of low CO_2 on plants can be used to determine mechanistic changes that may have occurred in ancient plants during periods of low CO_2, although this has not always been the direct goal of these studies. During the 1960s and 1970s, studies addressing plant responses to low CO_2 were often designed to improve crop productivity. These studies involved screening plants for high survival at highly limiting CO_2 concentrations that were often near the CO_2 compensation point (CO_2 concentration where carbon gain equals carbon loss; Cannel, Brun, and Moss 1969; Nelson, Asay, and Patton 1975; Sharma, Griffing, and Scholl 1979), with the hope that the surviving genotypes would exhibit unique properties of Rubisco, high rates of photosynthesis, or low rates of photorespiration that would be beneficial traits for crop selection. Unfortunately, most of these selection attempts met with little success in improving crop productivity. However, more recent selection experiments (reviewed in Medrano et al. 1995) with mutagenized tobacco resulted in increased growth at extremely low CO_2 concentrations (60–70 ppm) due to differences in biomass allocation, higher photosynthetic rates in older leaves, and lower respiration rates, whereas rates of photorespiration and Rubisco properties were unaffected. Although these studies were directed toward improving crop production, they also provide insights about the types of plant traits that may have been highly favored during past periods of low CO_2.

During the late 1980s, the role of low CO_2 studies transitioned from attempts at increasing agricultural yields (for which newer technologies were being applied) to understanding plant functioning and carbon sequestration within ancient ecosystems. This new interest was driven primarily by ice core data (see Chapter 4) indicating that CO_2 concentrations were as low as 180 ppm during glacial periods, ranged between 250 and 280 ppm CO_2 during interglacial periods, and were approximately 270 ppm just prior to the industrial revolution (Barnola et al. 1987; Petit et al. 1999). One of the first papers addressing the direct effects of CO_2 in this historical context was by Overdieck, Reid, and Strain (1988), who examined the effects of preindustrial CO_2 concentrations (270 ppm) on herbaceous annuals and found that growth of these annuals was reduced on average by 8% at preindustrial relative to modern CO_2 concentrations. This study suggested that human-induced increases in atmospheric CO_2 concentration may have already increased plant productivity at a global scale, particularly in ecosystems where other resources such as water and nutrients have a high availability. Then between 1990 and 1995 investigators began studying the effects of ancient CO_2 concentrations on a variety C_3 and C_4 annuals (Baker et al. 1990; Allen et al. 1991; Polley, Johnson, and Mayeux 1992; Sage and Reid 1992; Polley et al. 1993a,b; Dippery et al. 1995; Tissue et al. 1995). These studies clearly demonstrated that minimum CO_2 concentrations between 180 and 200 ppm, which last occurred 18,000 years ago (Last Glacial Maximum), were highly stressful on modern C_3 plants, whereas C_4 plants were barely affected (Ehleringer et al. 1991; Dippery et al. 1995). Because C_3 plants were so severely affected by low CO_2, more recent work has focused on the question of whether plants had the potential to evolve in response to low CO_2 and on the types of growth and physiological changes that may have been associated with these evolutionary responses (Ward and Strain 1997; Ward et al. 2000; Bunce 2001).

While plant ecologists were conducting growth studies involving low CO_2 treatments, paleoecologists and geologists were simultaneously applying carbon isotope analysis to investigate the effects of reduced CO_2 on plant community structure within ancient ecosystems (see Chapter 12). These studies often showed that C_4 plants became more prevalent within ancient ecosystems during periods when atmospheric CO_2 concentrations were low, whereas C_3 plants increased in frequency during periods of higher CO_2 (Cerling, Wang, and Quade 1993; Street-Perrott et al. 1997; Cerling et al. 1997; Cerling, Ehleringer, and Harris 1998; Connin, Betancourt, and Quade 1998; Boom et al. 2001; but also see Huang et al. 2001). Experimental studies (Dippery et al. 1995) and modeling efforts by Ehleringer, Cerling, and Helliker (1997) provided physiological and growth mechanisms to support the notion that C_4 species have higher competitive abilities at low CO_2 concentrations across a wide range of temperatures, whereas C_3 species are more competitive at higher CO_2 concentrations. In addition, other researchers were applying climate-vegetation models in order to predict the effects of low CO_2 on ecosystem functioning and carbon sequestration at larger spatial scales (Adams et al. 1990, Crowley 1995; Collatz, Berry, and Clark 1998; Cowling 1999; Levis and Foley 1999; Marchant, Boom, and Hooghiemstra 2002; Turcq et al. 2002). For example, Levis and Foley (1999) determined through modeling efforts that shifts from forests to C_4 grasslands during the Last Glacial Maximum (18,000–20,000 years ago) may have resulted in positive feedbacks associated with albedo effects, that resulted in warmer and drier ecosystems, particularly in the tropics.

11.1.3 Objectives

To better understand the effects of low CO_2 on ancient ecosystems, linkages between studies addressing questions at different spatial and temporal scales must be made. The primary goal of the present series is twofold: to form new connections between different disciplines in order to gain a broader understanding of the effects of low CO_2 in the past, and to determine how increasing CO_2 may influence future ecosystems. The specific objectives of this chapter are (a) to review studies on plant growth responses to low CO_2, (b) to describe the interactive effects of low CO_2 and other stresses, (c) to compare the effects of low CO_2 on the growth of C_3 versus C_4 species (little is known about responses of CAM species), and (d) to discuss how C_3 plants may have evolved in response to low CO_2. Because CO_2 concentration has fluctuated over geologic time, this chapter will emphasize plant responses to minimum atmospheric CO_2 concentrations that occurred during glacial periods of the late Pleistocene, although responses to other historical CO_2 concentrations will also be discussed occasionally.

11.2 Growth of C_3 Plants at Low CO_2

11.2.1 Biomass Production

From studies incorporating a range of past-through-predicted CO_2 concentrations, it is clear that an absolute change in CO_2 concentration *below* the modern CO_2 value has a much greater effect on plant growth than the same absolute change in CO_2 concentration *above* the modern value, as shown in Fig. 11.1 (Polley et al. 1993b; Dippery et al. 1995). Therefore, even small changes in global CO_2 concentrations during glacial periods may have greatly affected primary productivity.

Furthermore, unlike the majority of responses at elevated CO_2, reductions in growth during long-term exposure to low CO_2 are closely associated with reductions in the rate of carbon assimilation at the leaf level (Tissue et al. 1995; Sage 1995; but also see Cowling and Sage 1998 where responses are decoupled

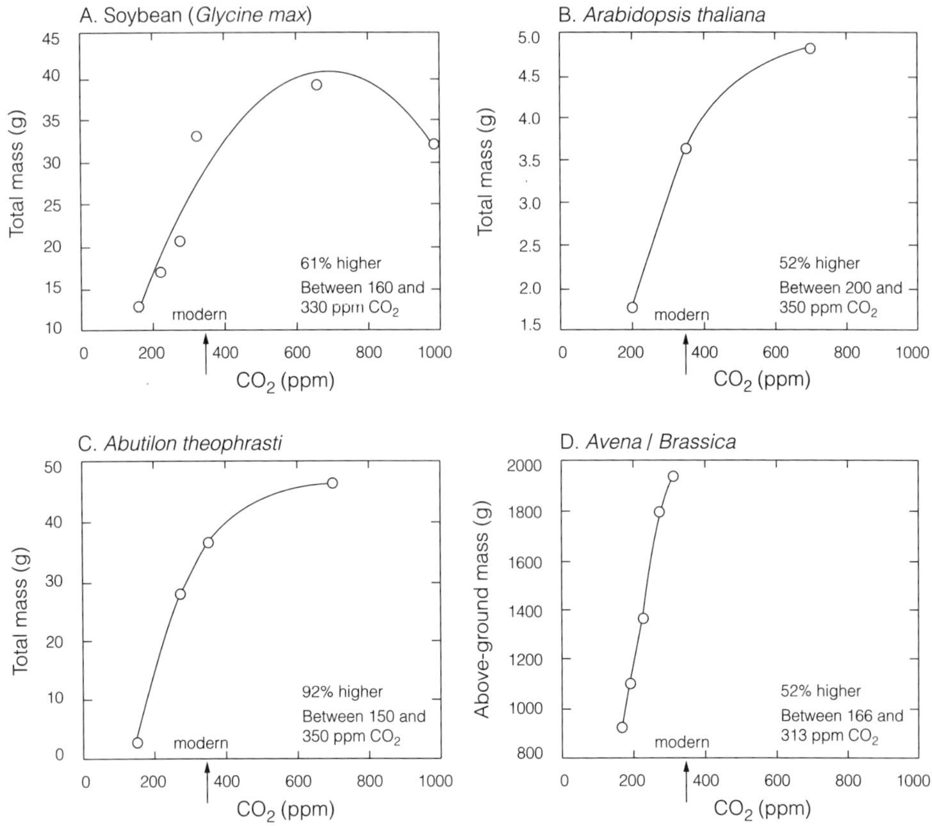

Figure 11.1. Effects of growth CO_2 concentrations ranging from past through predicted future values on the biomass production of several C_3 annuals. The modern CO_2 concentration is indicated with an arrow. Data are from **(A)** Allen et al. 1991, **(B)** Ward and Strain 1997, **(C)** Dippery et al. 1995, and **(D)** Polley, Johnson, and Mayeux 1992.

at high temperature). On average, the growth of modern C_3 plants is reduced by approximately 50% at glacial (180–220 ppm CO_2) versus modern CO_2 concentrations (350–380 ppm), with other conditions being optimal (Sage and Coleman 2001), and this is similar to reductions in carbon assimilation rate at the leaf level (see Chapter 9). There is, however, a range in this growth reduction that varies among species (see Fig. 11.1). As CO_2 concentrations are further decreased to 150 ppm, growth may be reduced by as much as 92% as has been observed for *Abutilon theophrasti,* shown in Fig. 11.1C (Dippery et al. 1995). In this case, carbon assimilation was reduced to such an extent that reproduction failed, indicating that 150 ppm CO_2 may be near the threshold where some C_3 plants may not have adequate carbon uptake and storage for completion of the life cycle.

11.2.2 Development Rate

Plant development rate is an important characteristic of life history that affects responses to stress, competitive ability, and long-term fitness. At elevated CO_2, C_3 species most commonly exhibit similar or faster development rates, with decreases in the time to transition from vegetative to floral stages (St. Omer and Horvath 1983; Garbutt and Bazzaz 1984; Cook et al. 1998; Ward and Strain 1997), although slower development rates have also been reported in rare cases (Carter and Peterson 1983). Much less is known about the effects of low CO_2 on the development rate of C_3 plants. To the author's knowledge, this measurement has only been reported for field-collected genotypes of *Arabidopsis thaliana* (hereafter *Arabidopsis*), which on average required 9 more days to initiate reproduction when grown at 200 versus 350 ppm CO_2 (Ward and Strain 1997). This represented a proportionally large shift in developmental timing considering that the life cycle of these *Arabidopsis* genotypes was only 60 days on average. In this regard, Sage and Coleman (2001) hypothesized that increasing the amount of carbon storage prior to reproduction by delaying developmental transitions would be beneficial for survival at low CO_2. Such a delay in the initiation of reproduction would allow for greater accumulation of stored reserves that could be allocated to reproduction, resulting in increased fitness under low CO_2 conditions. It should be noted, however, that climates were cooler during glacial periods; earlier frost events, therefore, may have limited seed production in many regions, rendering a delayed development rate as maladaptive in some annual species. In contrast, perennial species would have had an advantage in that reproduction could be delayed until sufficient resources were accumulated (Cowling 2001), and these species may have grown more slowly during years when other stresses (e.g., drought) were prominent in addition to low CO_2.

11.2.3 Allometry: Above- Versus Below-Ground Biomass

It has been postulated that an increase in the allocation of biomass to leaves versus roots and higher investment in photosynthetic machinery may improve carbon uptake under low CO_2 conditions (Sage and Coleman 2001). Similar to this prediction, Dippery et al. (1995) reported that *Abutilon theophrasti* partitioned a greater proportion of biomass to leaves relative to roots at 150 versus 350 ppm CO_2 (root/shoot mass = 0.17 ± 0.05 vs. 0.34 ± 0.02) when measured after the same growth period. Allometric analyses also confirmed that partitioning to leaves was increased at low CO_2 when differences in overall plant size and development rate were controlled for (see Samson and Werk 1986 for details on this analysis). In this same study, however, lower root production reduced nitrogen uptake and resulted in lower Rubisco production that further compounded the effects of limiting CO_2 availability (Tissue et al. 1995). Furthermore, the *Abutilon* plants grown at 150 ppm CO_2 had higher leaf area relative to whole-plant biomass (leaf area ratio) compared to plants developed at higher CO_2 concentrations. This response likely contributed to the low, but constant, relative growth rates (change in biomass per unit time per total dry mass) observed for plants grown at 150 ppm CO_2, whereas plants produced at higher CO_2 concentrations exhibited declining relative growth rates over time, due to a reduction in the proportion of carbon assimilating structures (Dippery et al. 1995). Interestingly, Cowling and Sage (1998) observed that *Phaseolus vulgaris* also had reduced allocation of biomass to roots under low, glacial CO_2 concentrations and optimal temperatures, but this difference was associated with increased stem biomass as opposed to leaf biomass. This difference may be associated with direct effects of CO_2 on biomass allocation, or to indirect effects on plant ontogeny. Furthermore, neither Dippery et al. (1995) nor Bunce (2001) found differences in the amount of leaf area relative to whole-plant biomass in *Abutilon theophrasti* and other C_3 annuals grown at preindustrial (270 ppm) versus modern (350–370 ppm) CO_2 concentrations, indicating that increases in CO_2 over the past hundred years may not have been large enough to induce allometric shifts in C_3 annuals.

11.2.4 Allometry: Allocation to Reproduction

The effects of CO_2 on the allocation of biomass to reproduction is an important topic because it describes the relative amount of maternal resources invested into future progeny. This allocation pattern would have been critical during periods of low CO_2 when the potential for carbon gain was greatly reduced. However, relatively little is known about the effects of low CO_2 on reproduction, and what is known has focused solely on annuals. Dippery et al. (1995) showed that low CO_2 (150 ppm) could prevent reproduction in annuals, namely because all flower buds aborted prior to anthesis, presumably due to carbon limitations. Others have reported large reductions in absolute reproductive biomass under glacial CO_2 concentrations relative to modern values, with reductions averaging 64% among *Arabidopsis* genotypes and 60% for soybean, indicating similar reductions among species. When allometric patterns were considered, Ward and Strain (1997) found relatively similar allocation of biomass to reproduction in the annual *Arabidopsis* grown at glacial (200 ppm) versus modern (350 ppm) and elevated (700 ppm) CO_2 concentrations when the effects of plant size and development stage were taken into account. Thus, reduced final reproductive mass at 200 ppm CO2 was due to overall smaller plant size, and not to decreased allocation to reproduction. Furthermore, for *Arabidopsis* (Ward and Strain 1997) and soybean (Allen et al. 1991), absolute reductions in seed production between modern and glacial CO_2 concentrations were due primarily to a lower number of fruits, and not to changes in seed number per fruit, which tends to be more highly conserved.

The reproductive responses of perennial species to low CO_2 are largely unknown. It would be interesting to observe the timing of reproduction in species that could potentially delay the initiation of reproduction for long periods of time when starved for carbon. Furthermore, it would be important to determine if reproduction occurs at the same plant size (but after a longer growing period) at low CO_2 relative to modern values, or alternatively if low CO_2 alters the minimum size required for initiation of reproduction. Such studies may shed light on how perennials functioned during glacial periods, and may further our understanding of the effects of resource limitations on reproduction in plants with different life histories.

11.3 C_3 Plant Response to Low CO_2 and Other Stresses

Exposure of C_3 plants to elevated CO_2 above the modern value often ameliorates the negative effects of other stressful factors, such as drought, low nutrient availability, reduced light, and herbivore damage (Tolley and Strain 1984; Wray and Strain 1986; Clifford et al. 1993; Tschaplinski et al. 1995; but also see Fajer, Bowers, and Bazzaz 1991). Conversely, it is predicted that under low CO_2 conditions, stressful factors may have more pronounced effects on plant growth and functioning (Cowling and Sage 1998; Cowling and Sykes 1999). This is because plant growth would be reduced from low carbon availability, and excess carbon would not be available in storage for replacing lost structures or for increasing carbon-based secondary compounds to compensate for stress effects.

11.3.1 Temperature

Sage and Cowling (1999) investigated the effects of low CO_2 and high temperature on the physiology and growth of several C_3 species (*Phaseolus vulgaris*, wheat, and tobacco). They predicted a priori that high temperatures in combination with low CO_2 would greatly reduce plant carbon gain due to both reduced CO_2 substrate for carbon assimilation and higher rates of photorespiration induced by high temperature and reduced CO_2 relative to O_2. They found that under warm conditions (35 day/ 29 night °C), plant size was reduced by 75% to 95% at 200 relative to 380 ppm CO_2, whereas at cool conditions (25 day/ 20 night °C) plant size was only reduced by 40% to 60%. However, reduced biomass at warm temperatures could not be attributed to lower rates of carbon assimilation on a leaf area basis, mainly due to increased leaf nitrogen. Instead, lower growth was a result of highly reduced leaf area (proportional to reductions in total mass) occurring from lower leaf expansion and less leaf initiation (Sage and Cowling 1999). Thus, changes in leaf development and production have been implicated as the primary response to thermal stress at low CO_2. Moreover, these results suggest that C_3 plants occurring in regions with relatively warm growing seasons in the past may have been more negatively affected by low CO_2 compared to plants occurring in cooler regions (Sage and Cowling 1999).

11.3.2 Water and Nitrogen

With regard to water relations, it is predicted that low CO_2 concentrations during glacial periods would have increased the water consumption of C_3 plants relative to interglacial periods when CO_2 concentrations were much higher (Baker et al. 1990; Polley, Johnson, and Mayeux 1992; Beerling and Woodward 1993; Polley et al. 1993a,b; Sage 1995; Ward et al. 1999). This may be due to several factors, such as relatively greater production of leaf biomass and area (see "Allometry: Above- Versus Below-Ground Biomass," above), higher stomatal conductance, and greater stomatal density that would serve to enhance CO_2 uptake on a whole-plant basis, but would produce greater water loss (Sage 1995; Beerling and Chaloner 1993). The efficiency of water use (carbon assimilation/stomatal conductance, A/g), determined from both instantaneous gas exchange measures and more integrated carbon isotopes values, decreased in a variety of C_3 annuals developed under low CO_2, and these reductions were proportional to reductions in CO_2 concentration, as shown in Fig. 11.2 (Polley, Johnson, and Mayeux 1995). Decreases in the efficiency of nitrogen use (biomass produced/plant N) were not as pronounced as for water, particularly when plants were droughted (*Triticum aestivum, Bromus tectorum*), indicating that low CO_2 increases the requirement for water to a greater extent than nitrogen (Polley, Johnson, Mayeux 1995).

Stomatal conductance has been shown to increase between modern and glacial CO_2 concentrations for all C_3 plants measured to date, and this increase ranges between 35% and 50% depending upon the species (Polley et al. 1993b; Ward et al. 1999). In field studies, it has been determined that such physiological responses have translated into greater depletion of soil water by a C_3/C_4 grass assemblage grown at low CO_2 concentrations compared to those grown above the modern value (Polley, Johnson, and Derner 2002). In addition, stomatal density (number of stomates/leaf area) was found to be higher in fossil leaves of ancient plants that occurred during past periods of low CO_2 relative to those that occurred during more recent periods when atmospheric CO_2 concentrations were higher (Beerling et al. 1993; Beerling and Royer 2002). However, it should also be noted that recent studies with modern C_3 plants (*Solanum dimidiatum, Bromus japonicus*) indicate that stomatal density may be reduced at low CO_2 (Maherali et al. 2002), and that larger pore size may be an alternative response for increasing carbon uptake. This discrepancy between modern and ancient studies may be due to the ability to control the growth environment for modern studies that is not possible with ancient material, or it may be due to differences between short-term acclimation to low CO_2 in modern genotypes compared to longer-term evolutionary responses in ancient material. Despite evidence indicating greater water demand at low CO_2, precipitation was much lower during glacial maxima on a global basis (by as much as one-half of the present amount), as indicated by higher amounts of dust within portions of ice cores corresponding to glacial periods (Yung et al. 1996). Such a combination of low water availability and low CO_2 may have been extremely stressful on C_3 plants and may have greatly reduced C_3 plant productivity at a global scale.

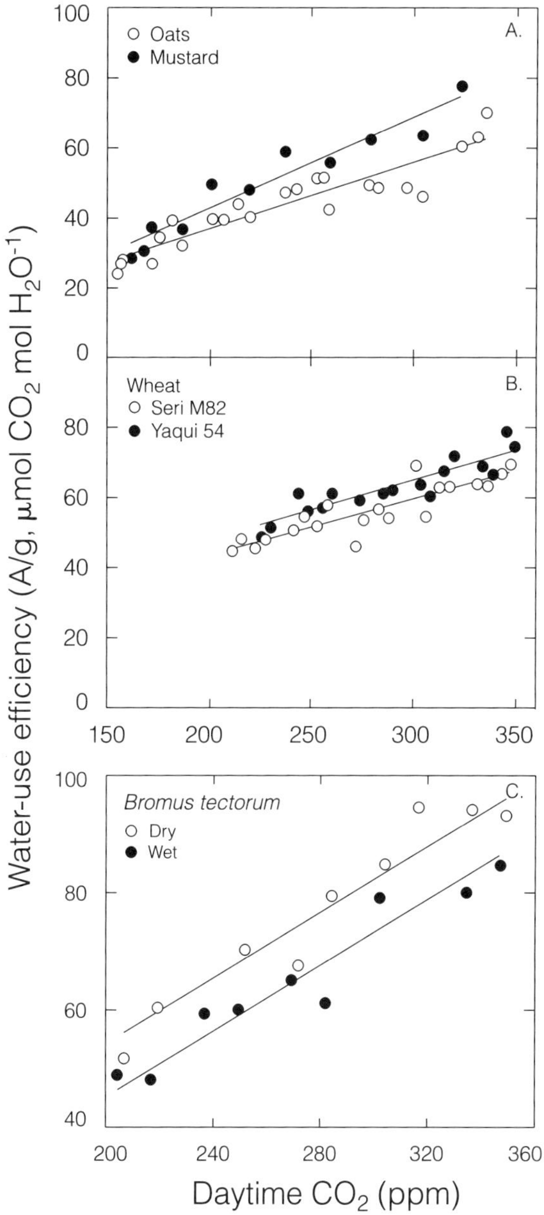

Figure 11.2. Relationship between CO_2 concentration and water-use efficiency (carbon assimilation/stomatal conductance, A/g) calculated from $\delta^{13}C$ values of leaves for several C_3 annuals. Graphs are **(A, B)** reprinted with permission from Nature (Polley et al. 1993a), copyright 1993 MacMillan Magazines Limited, and **(C)** Polley, Johnson, and Mayeux 1995 (copyright Blackwell Science Ltd.).

11.3.3 Symbionts

Although very little is known about the effects of low CO_2 on plant symbionts, it is predicted that low CO_2 would have influenced these relationships. These organisms are present in almost all ecosystems, and they enhance the movement of elements from the soil or atmosphere to plants (Schlesinger 1997). Both mychorrizal fungi and nitrogen fixing bacteria derive carbohydrate resources from the host plant and, in turn, provide nutrients (phosphorus and nitrogen) to the host. Therefore, maintaining symbionts can be expensive since they consume a substantial proportion of photosynthate; plants generally reduce these associations when soil nutrients are nonlimiting (Hetrick, Wilson, and Todd 1996). Similarly, housing symbionts may not be as beneficial under carbon limitations induced by low CO_2, due to the high cost of maintaining such organisms, and nutrients may not be as limiting relative to carbon. As a result of these issues, symbiotic relationships with mychorrizal fungi and nitrogen fixing bacteria may have been less prevalent during glacial periods when CO_2 concentrations were low relative to modern time periods (Sage 1995). In support of this possibility, Polley and co-workers (1994) found that nitrogen fixation was reduced for plants grown at glacial versus modern CO_2 concentrations with the same availability of soil nitrogen. Presumably, however, these symbiotic relationships were still maintained during past periods of low CO_2, and coevolutionary responses may have adjusted these interactions to be mutually beneficial in a low CO_2 world.

11.4 Comparisons of C_3 and C_4 Responses to Low CO_2

11.4.1 Low CO_2

Several studies have compared the responses of C_3 and C_4 species grown at elevated CO_2 (Coleman and Bazzaz 1992; Arp et al. 1993; Dippery et al. 1995), and many of these have indicated that C_3 species will become more competitive as CO_2 continues to rise in the future. However, few studies have focused on comparisons of these functional types grown at low CO_2 below the modern value (Polley, Johnson, and Mayeux 1992; Polley, Johnson, and Mayeux 1994; Dippery et al. 1995; Sage 1995; Tissue et al. 1995). As described above, it is known that C_3 species exhibit decreased carbon assimilation rates and growth with decreasing atmospheric CO_2 concentration, whereas C_4 species are generally less affected by CO_2 (Sage 1994; Tissue et al. 1995; Ward et al. 1999). Furthermore, it is clear from modeling efforts involving physiological parameters (Ehleringer, Cerling, and Helliker 1997, Chapter 10), measures of carbon isotopes in ancient material (see Chapter 12), and observed growth responses (Dippery et al. 1995; Tissue et al. 1995), that C_4 species would have had a strong competitive advantage over C_3 species in many regions of the world during glacial periods when CO_2 concentrations were at a minimum. This is due in most part to the CO_2-concentrating mechanism of C_4 plants that eliminates photorespiration and increases carboxylation efficiency, resulting in high growth potential, even under very reduced CO_2 concentrations. Furthermore, in light of these responses, low CO_2 has been implicated as a major driver for the evolution of the C_4 photosynthetic pathway during reductions in atmospheric CO_2 that began during the Cretaceous and continued until the Miocene (Ehleringer et al. 1991).

Although C_4 plants are much less affected by low CO_2 compared to C_3 plants, this is not to say that they are completely unaffected. For example, Ward et al. (1999) found that the C_4 annual *Amaranthus retroflexus* (hereafter *Amaranthus*) exhibited physiological responses to CO_2 that were unexpected based on theoretical predictions. Between 180 and 700 ppm CO_2, the C_4 species exhibited gradual increases in carbon assimilation and decreases in stomatal conductance and transpiration that were similar to the response patterns observed for the C_3 species (*Abutilon theophrasti*), as shown in Fig. 11.3, indicating that the C_4 species was not CO_2-saturated at 180 ppm CO_2.

However, these physiological responses did not alter the leaf area and biomass production of the C_4 species, whereas the C_3 species exhibited reduced biomass production at 180 relative to 700 ppm CO_2 (Fig. 11.4). Similar to these findings, Polley, Johnson, and Mayeux (1994) observed that *Schizachyrium scoparium* (C_4) exhibited increased stomatal conductance and had lower rates of carbon assimilation when CO_2 concentrations were decreased from 340 to 200 ppm, but plant growth was unaffected.

11.4.2 Low CO_2 and Drought

The interactive effects of low CO_2 and low water availability during glacial periods (Petit et al. 1999) may have further altered the relative performance of C_3 versus C_4 species. Based on physiological responses, it has been predicted that C_3 species may have been more negatively affected by the combination of low CO_2 and drought than were C_4 species (Ward et al. 1999). This is expected because C_3 species would need to maintain higher stomatal conductance for increasing carbon uptake, but this response would simultaneously facilitate water consumption and loss (Baker et al. 1990; Polley et al. 1993a; Polley, Johnson, and Mayeux 1995). To investigate this issue experimentally, *Amaranthus* (C_4) and *Abutilon* (C_3) were grown under low CO_2 and drought conditions (Ward et al. 1999). In response to severe drought, C_3 plants dropped a large amount of leaf area and maintained higher leaf water potential in remaining leaves. In contrast, C_4 plants retained greater leaf area, but at a lower leaf water potential. Furthermore, C_3 plants grown at 180 ppm CO_2 delayed the reduction of stomatal conductance after the initiation of drought and retained greater leaf area (relative to total biomass) compared to C_3 plants grown at the modern CO_2 value. As a result, when plants were re-watered, C_3 and C_4 plants showed similar relative recovery from rapid drought at 180 ppm CO_2 when the biomass of droughted

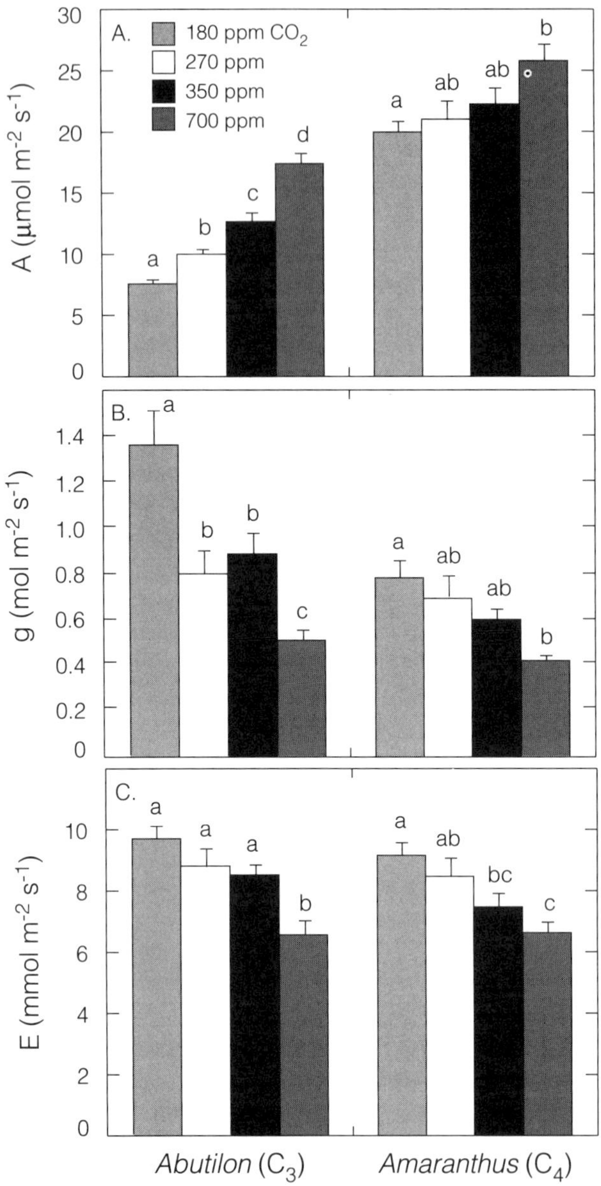

Figure 11.3. **(A)** Carbon assimilation rate **(B)** stomatal conductance, and **(C)** transpiration of *Abutilon theophrasti* (C_3) and *Amaranthus retroflexus* (C_4) after 28–31 days of growth at CO_2 concentrations ranging from past through predicted future values. Values are means (n = 16–20) ± SE. Different letters within a species indicate statistically different responses to CO_2 at $P < 0.05$. Graphs are taken from Ward et al. 1999 (copyright Blackwell Science Ltd.).

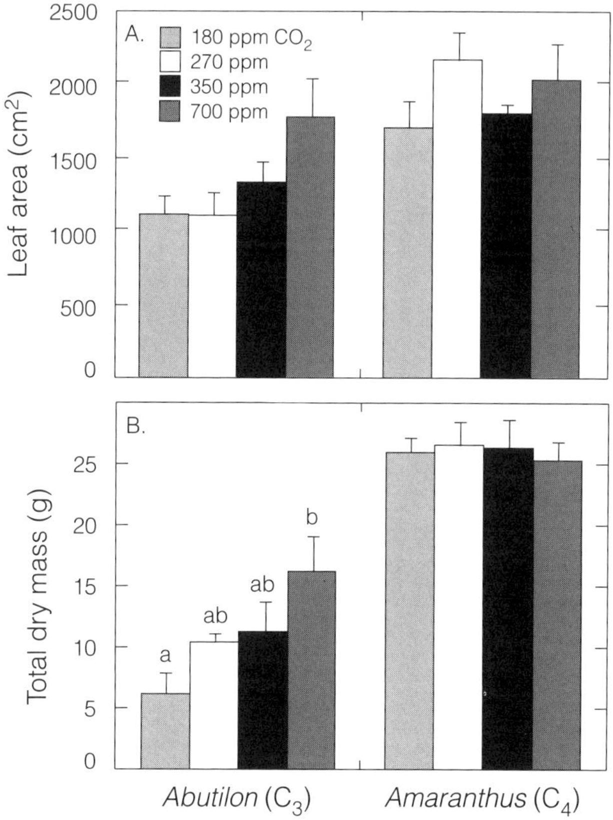

Figure 11.4. **(A)** Total leaf area and **(B)** total biomass production of *Abutilon theophrasti* (C_3) and *Amaranthus retroflexus* (C_4) after 38 days of growth at CO_2 concentrations ranging from past through predicted future values. Values are means (n = 4) ± SE. Different letters within a species indicate statistically different responses to CO_2 at $P < 0.05$. Graphs are taken from Ward et al. 1999 (copyright Blackwell Science Ltd.).

and well-watered plants was compared. In this case, growth responses superseded physiological expectations, producing the result that C_3 plants grown under low CO_2 conditions were no more relatively affected by rapid and severe drought than were C_4 plants. However, it is important to note that low CO_2 greatly reduced the growth of the C_3 plant on an absolute basis, and therefore the C_4 plant would likely have been more competitive against the C_3 plant if this had been a competitive scenario. Moreover, this study points out the need for more growth studies dealing with the interactive effects of low CO_2 and other factors (such as low temperature) on C_3 and C_4 species.

11.5 Evolution of C_3 Plants at Low CO_2

11.5.1 Relevant Questions

Studies with modern plants described in the previous sections indicate that C_3 species may have been severely affected by low CO_2 concentrations that last occurred 18,000 years ago. In response to this work, the compelling question has become: How did C_3 plants continue to persist during past periods of low CO_2, when modern plants are so negatively affected by this factor?

The first consideration to make when attempting to answer this question is that low CO_2 and low temperatures were tightly coupled throughout interglacial/glacial cycles of the late Pleistocene (Petit et al. 1999). Theoretically, reduced temperatures would have improved the carbon balance of C_3 plants, because photorespiration is reduced at cold temperatures (Ehleringer and Björkman 1977; Ehleringer, Cerling, and Helliker 1997; but also see Sage and Cowling 1999). Despite this effect, it is unlikely that the amount of temperature reduction in most regions would have negated the very dominating and negative effects of low CO_2 on C_3 physiology and growth (Ehleringer, Cerling, and Helliker 1997). In addition, C_4 species would have been relatively unaffected by reduced CO_2 concentrations and may have contributed additional competitive stress, particularly in warm regions (Dippery et al. 1995).

Clearly, the potential for evolution of plants in response to low CO_2 must be considered when determining how C_3 plants persisted during glacial periods (Strain 1991; Ward and Strain 1997; Ward et al. 2000; Bunce 2001; Beerling and Chaloner 1993). Like most stressful factors, it is likely that low CO_2 acted as a strong selective agent on C_3 plants due to stress imposed on physiology, growth, and reproduction. Furthermore, declining CO_2 concentrations at the onset of glacial periods occurred over tens of thousands of years (Petit et al. 1999), providing sufficient time for evolution of both annual and perennial species. Below are listed several critical questions that emerge as being particularly relevant for understanding the effects of low CO_2 on the evolution of C_3 plants:

1. *Do plants evolve in response to low CO_2?* This question addresses whether reduced atmospheric CO_2 concentrations act as a selective agent on C_3 plants.
2. *Can we extrapolate modern plant responses back in time to predict responses of ancient vegetation?* This question addresses whether plants have developed adaptations to modern atmospheric CO_2 concentrations or are better adapted to CO_2 concentrations occurring in the past. The rate of evolutionary response is dependent on generation time, and therefore the answer to this question may vary depending on whether annuals or long-lived perennials are being considered (Sage and Cowling 1999).
3. *What plant traits may be affected as a result of evolutionary responses to low CO_2?* This question is critical as it may shed light on how C_3 plants were able to persist during glacial periods when CO_2 concentrations were at a minimum (Ward et al. 2000).
4. *How might plant adaptation to low CO_2 have influenced ecosystem function-*

ing? Information pertaining to this question would improve our predictions of the turnover rates of carbon and rates of water flux in ancient ecosystems. Furthermore, possible changes in tissue quality and biomass partitioning also may indicate that herbivores were indirectly affected by low CO_2.

Although these questions are only beginning to be addressed, they provide a conceptual framework for the important advances to be made in the future for understanding how C_3 plants might have adapted to low CO_2 and the possible effects of this at the ecosystem scale. A summary of the advances that have been made to date in understanding how C_3 plants may have evolved in response to low CO_2 concentrations follows below:

When designing evolutionary studies to determine if low CO_2 might have acted as a selective agent, the issue of whether modern genotypes are representative of those occurring in the past must be addressed. It has been suggested that modern plants are adapted to reduced CO_2 of the past, due to slower rates of evolution relative to the rate of CO_2 increase (Strain 1991; Sage and Cowling 1999; Tissue et al. 1995). However, recent studies indicate that annuals, at least, might have adapted to more recent CO_2 concentrations (Bunce 2001); Ward et al. (2000) demonstrated that it is possible for evolution in response to low CO_2 to proceed very rapidly. Therefore, if current plants have adapted to modern CO_2 concentrations, single generation experiments may provide only a baseline for understanding how plants respond to low CO_2 as a novel environmental factor rather than indicating how plants functioned in the past under these conditions. Therefore, studies incorporating ancient plant material and modern plant genotypes are both necessary for fully understanding how plants persisted in response to low atmospheric CO_2 during glacial periods.

11.5.2 Genetic Variation of Modern Plants at Low CO_2

In order to determine if low CO_2 acts as a selective agent, an assessment must be made of the amount of genetic variation (intraspecific variation) for responses to low CO_2, which must be present in order for evolution to occur (Bradshaw and McNeilly 1991). In response to CO_2 concentrations spanning modern through predicted future values, C_3 plants have been shown to exhibit genetic variation for carbon assimilation rate (Curtis et al. 1996), stomatal responses (Case, Curtis and Snow 1998), growth (Norton, Firbank, and Watkinson 1995; Zhang and Lechowicz 1995), and reproduction (Curtis, Snow, and Miller 1994; Bazzaz et al. 1995). These results suggest that C_3 plants will undergo changes in genetic composition in response to increasing CO_2 concentrations, so long as antagonistic genetic correlations do not decrease evolutionary potential, as has been observed for other global change factors (Etterson and Shaw 2001). Low CO_2 of the past may have been an even stronger selective agent on plants, because genetic variation is often enhanced in response to stressful conditions (Bazzaz et al. 1992). Therefore, more evolutionary work involving low CO_2 concentrations is needed to better predict the responses of plants to changes in atmospheric CO_2 over geologic timescales.

In order to investigate levels of genetic variation in response to CO_2 availability, Ward and Strain (1997) conducted a norm-of-reaction experiment using field-collected genotypes of *Arabidopsis* to determine the potential for evolution of modern plants at past and predicted future CO_2 concentrations. The authors focused on variation for important components of fitness, including total biomass production and reproductive output. As with growth responses, seed production exhibited greater response to changes in CO_2 below the modern value relative to above the value (Fig. 11.5).

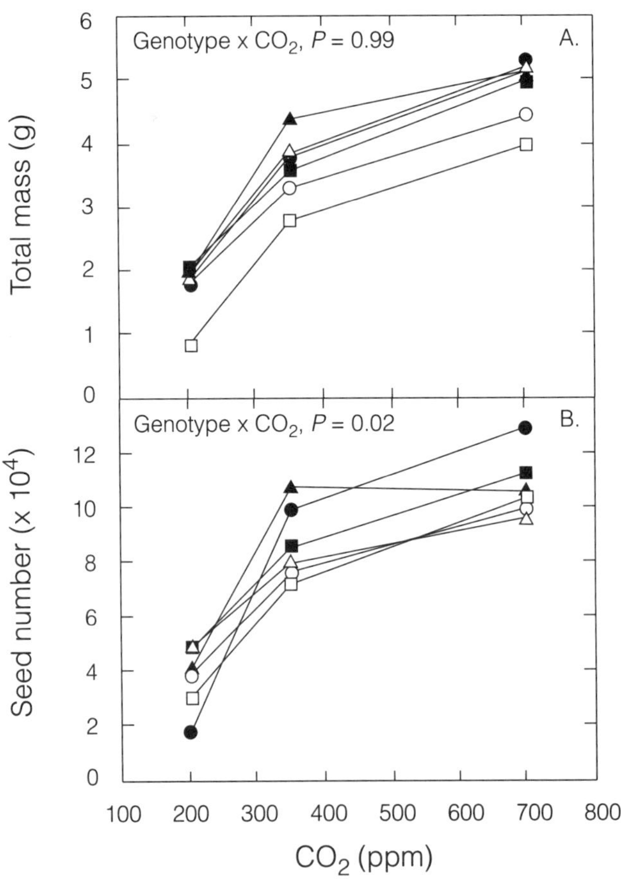

Figure 11.5. Effects of growth CO_2 concentration on **(A)** biomass production and **(B)** total seed number for six field-collected genotypes of *Arabidopsis thaliana*. *P*-values indicate the level of significance for the interaction between the effects of CO_2 and genotype. Data are from Ward and Strain 1997. Prior to this experiment, seeds were collected from plants grown under common conditions to avoid maternal effects. In addition, initial seed mass did not vary between genotypes.

With regard to intraspecific responses, genotypes did not exhibit variation in response to CO_2 for total biomass production (Fig. 11.5A), but they did show significant variation for seed production, a response that is closely associated with fitness (Fig. 11.5B). More specific analyses revealed that reduced seed production between 350 and 200 ppm CO_2 was variable among genotypes, ranging from a 38% to 81% reduction. However, there were no significant differences in seed number (although a trend toward increase was observed) between 350 and 700 ppm CO_2. Therefore, with regard to genetic variation, modern C_3 annuals may have greater potential for evolution in response to the low CO_2 that occurred during glacial periods relative to the elevated CO_2 predicted for the future (Ward and Strain 1997).

11.5.3 Selection Responses of Modern Plants to Low CO_2

Studies investigating genetic variation in response to novel CO_2 concentrations are necessary to determine whether it is possible for evolution to occur in various conditions. Indicating the specific mechanisms on which selection will act is a separate issue. To begin identifying these mechanisms, Ward et al. (2000) reported on a five-generation selection experiment using *Arabidopsis* selected at both low (200 ppm) and elevated (700 ppm) CO_2. Seeds used for selection were derived from random crosses between field-collected genotypes, allowing for selection on variation resulting from recombination and segregation. Selection was conducted for high seed number (25% truncation at each generation), which is a major component of fitness (Primack and Antonovics 1981), and changes in physiology, biomass production, and development rate were measured throughout selection. At 200 ppm CO_2, selection plants had significantly higher seed number than did the control plants after only three generations of selection (with no change in seed mass or quality), indicating that selection responses to low CO_2 were very strong in this species (Fig. 11.6A). Furthermore, this rate of selection under limiting CO_2 availability is one of the fastest rates reported for novel environments induced by a global change factor (Bone and Farres 2001). In addition, plants selected for high seed number had, on average, 26% higher total biomass compared to control plants at the fifth generation of selection when responses of selection and control populations are aggregated into overall means (Fig. 11.6C). The increased biomass was attributed to an extended period of productivity during which plants initiated reproduction later (see Fig. 11.6B) and began senescing later than control plants (data not shown). Furthermore, selection and control plants had similar rates of carbon accumulation prior to the initiation of reproduction, suggesting that photosynthetic capacity might not have been affected by selection, at least during early stages of growth (see Ward et al. 2000). Thus, this study revealed that evolution in response to low CO_2 could result in increased lifetime carbon gain, due to shifts in development rate as opposed to changes in physiological processes. The study also indicated that selection responses can improve plant growth at low CO_2 as a result of genetic changes, and therefore plant responses produced during single-

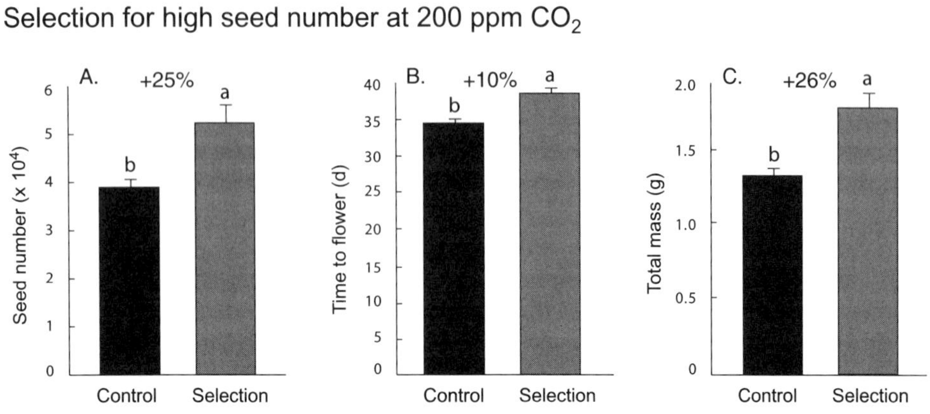

Figure 11.6. Responses of total seed number, time to first flower, and total biomass production for control (random selection) and selection plants that were selected for high seed number at 200 ppm CO_2 (glacial value). Values are from overall means of all selection and control population replicates (spanning different growth chambers) taken at the fifth and final generation of selection. Upper values indicate percent difference and direction of change between control and selection plants. Different letters within a measurement indicate significant differences at the $P < 0.05$ level.

generation experiments might not be fully representative of those that occurred in the ancient past.

In follow-up to this selection experiment, a reciprocal transplant study was conducted to determine if low CO_2 was indeed the selective agent that was acting on these plants (Ward et al. 2000). It was found that plants selected for high seed number at 200 ppm CO_2 had 30% higher seed number than plants selected at 700 ppm CO2 when both were grown at 200 ppm CO_2 ($P = 0.0015$). Because genetic adaptation can be defined as higher performance under the conditions of selection (Tousignant and Potvin 1996), this result indicated that plants selected at 200 ppm CO_2 exhibited true adaptive responses to low CO_2. In addition, overall changes in biomass production and development rate occurred in opposite directions during selection at 700 relative to 200 ppm CO_2 (Ward et al. 2000, data not shown), supporting the notion that low CO_2 acted as a selective agent. Thus, further work with these selection populations will continue to shed light on possible mechanisms by which C_3 annuals may evolve in response to changes in CO_2 concentration over geologic and contemporary timescales.

In addition to developmental responses, it is likely that modifications in seed size and quality might also have been influenced during evolution at low CO_2 concentrations. The carbon balance of seedlings is tightly coupled to atmospheric CO_2 availability, especially following depletion of seed reserves. Therefore, seedling mortality might have been very high at low CO_2 concentrations because of the inability of seedlings to maintain a positive carbon balance during the physiological transition between using stored carbon from seed reserves versus carbon from the atmosphere. During the vulnerable seedling stage, maintaining a positive carbon balance at low CO_2 would require sufficient leaf area that would most effectively be produced from stored seed reserves originating from the mother plant. Furthermore, a greater amount of seedling resources could also increase root production that would enhance nitrogen uptake. This would ultimately contribute to greater Rubisco production in leaves and would increase carbon uptake during critical growth stages of young seedlings. Thus, there might also have been strong selection pressure for increased seed size (presumably at the expense of seed number) during glacial periods when carbon availability was limiting.

11.6 Summary

Low CO_2 concentrations during glacial periods of the late Pleistocene were among the lowest that have occurred during the evolution of land plants. Modern C_3 plants grown at low CO_2 are highly stressed, as indicated by large reductions in carbon assimilation rates and biomass production. Furthermore, reproduction of C_3 plants is also reduced at low CO_2, and these responses are closely associated with plant fitness and evolution. In contrast, C_4 plants are much less affected by low CO_2 and generally do not exhibit changes in growth between modern and glacial CO_2 concentrations. Understanding how C_3 plants adapted to low CO_2 has become a critical question for recent low CO_2 work. Within a single generation, C_3 plants may acclimate to low CO_2 by allocating a higher proportion of biomass to leaves and maintaining higher stomatal conductance to increase carbon uptake when CO_2 is limiting. In response to longer-term selection over multiple generations, C_3 plants have been shown to exhibit slower development rates that increase resource availability for reproduction. A list of observed and predicted responses to low CO_2 is summarized in Fig. 11.7; these responses incorporate both short-term (during a single generation of exposure to low CO_2) and long-term (evolutionary timescales) responses. More work involving the effects of low CO_2 needs to be conducted to more fully understand how C_3 plants persisted during low CO_2 periods. For example, new technologies are now available for studying responses at the genome level (e.g., microarray technology), and these should be employed to understand how gene expression varies at modern versus low CO_2 concentrations. Future work such as this will increase our understanding of how land plants adapted during critical periods in geologic history when carbon resources were limiting; it also may indicate how responses to ancient conditions influenced the evolution and functioning of modern plants.

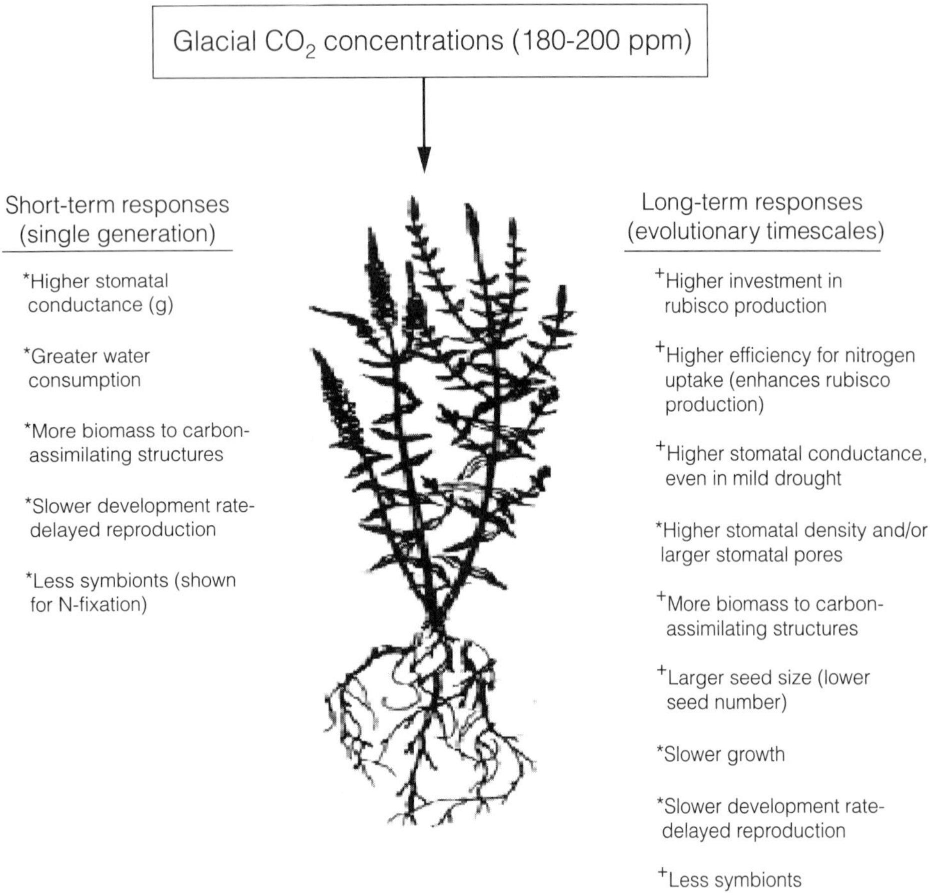

Figure 11.7. A summary of the responses of C_3 plants to low atmospheric CO_2 concentrations of glacial periods that have been observed (*) or proposed (+) to occur over short- and long-term timescales.

Acknowledgments. The author would like to thank Jim Ehleringer, Thure Cerling, and Denise Dearing for their excellent efforts in organizing the symposium "History of CO_2 and Responses of Plants, Animals, and Ecosystems," from which the present book originated.

References

Adams, J.M., H. Faure, L. Faure-Denard, J.M. McGlade, and F.I. Woodward. 1990. Increases in terrestrial carbon storage from the Last Glacial Maximum to the present. *Nature* 348:711–14.

Allen, L.H., Jr., E.C. Bisbal, K.J. Boote, and P.H. Jones. 1991. Soybean dry matter allocation under subambient and superambient levels of carbon dioxide. *Agronomy Journal* 83:875–83.

Arp, W.J., B.G. Drake, W.T. Pockman, P.S. Curtis, and D.F. Whigham. 1993. Interactions between C_3 and C_4 salt marsh plant species during four years of exposure to elevated atmospheric CO_2. *Vegetatio* 104/105:133–43.

Baker, J.T., L.H. Allen, Jr., K.J. Boote, P. Jones, and J.W. Jones. 1990. Rice photosynthesis and evapotranspiration in subambient, ambient, and superambient carbon dioxide concentrations. *Agronomy Journal* 82:834–40.

Barnola, J.-M., D. Raynaud, Y.S. Korotkevich, and C. Lorius. 1987. Vostok ice core provides 160,000-year record of atmospheric CO_2. *Nature* 329:408–18.

Bazzaz, F.A., D.D. Ackerly, F.I. Woodward, and L. Rochefort. 1992. CO_2 enrichment and dependence of reproduction on density in an annual plant and a simulation of its population dynamics. *Journal of Ecology* 80:643–51.

Bazzaz, F.A., M. Jasienski, S.C. Thomas, and P. Wayne. 1995. Microevolutionary responses in experimental populations of plants to CO_2-enriched environments: Parallel results from two model systems. *Proceedings of the National Academy of Sciences, USA* 92:8161–65.

Beerling, D.J., and W.G. Chaloner. 1993. Evolutionary responses of stomatal density to global CO_2 change. *Biological Journal of the Linnean Society* 48:343–53.

Beerling, D.J., W.G. Chaloner, B. Huntley, J.A. Pearson, and M.J. Tooley. 1993. Stomatal density responds to the glacial cycle of environmental change. *Proceedings of the Royal Society of London* 251:133–38.

Beerling, D.J., W.G. Chaloner, B. Huntley, J.A. Pearson, M.J. Tooley, and F.I. Woodward. 1992. Variations in the stomatal density of *Salix herbacea* L. under the changing atmospheric CO_2 concentrations of late- and post-glacial time. *Philosophical Transactions of the Royal Society of London* 336:215–24.

Beerling, D.J., and D.L. Royer. 2002. Reading a CO_2 signal from fossil stomata. *New Phytologist* 153:387–97.

Beerling, D.J., and F.I. Woodward. 1993. Ecophysiological responses of plants to global environmental change since the Last Glacial Maximum. *New Phytologist* 125:641–48.

Berner, R.A. 1994. Geocarb II: A revised model of atmospheric CO_2 over Phanerozoic time. *American Journal of Science* 294:56–91.

Bone, E., and A. Farres. 2001. Trends and rates of microevolution in plants. *Genetica* 112/113:165–82.

Boom, A., G. Mora, A.M. Cleef, and H. Hooghiemstra. 2001. High altitude C_4 grasslands in the Northern Andes: Relicts from glacial conditions? *Review of Palaeobotany and Palynology* 115:147–60.

Bradshaw, A.D., and T. McNeilly. 1991. Evolutionary response to global climatic change. *Annals of Botany* 67:5–14.

Bugbee, B., B. Spanarkel, S. Johnson, O. Monje, and G. Koerner. 1994. CO_2 crop growth enhancement and toxicity in wheat and rice. In *Life sciences and space research XXV, Vol. 14,* 257–67. Oxford: Pergamon Press.

Bunce, J.A. 2001. Are annual plants adapted to the current atmospheric concentration of carbon dioxide? *International Journal of Plant Science* 162:1261–66.

Cannel, R.A., W.A. Brun, and D.N. Moss. 1969. A search for high net photosynthetic rate among soybean genotypes. *Crop Science* 9:840–41.

Carter, D.R., and K.M. Peterson. 1983. Effects of a CO_2-enriched atmosphere on the growth and competitive interaction of a C_3 and C_4 grass. *Oecologia* 58:188–93.

Case, A.L., P.S. Curtis, and A.A. Snow. 1998. Heritable variation in stomatal responses to elevated CO_2 in wild radish, *Raphanus raphanistrum* (Brassicaceae). *American Journal of Botany* 85:253–58.

Cerling, T.E., J.R. Ehleringer, and J.M. Harris. 1998. Carbon dioxide starvation, the development of C_4 ecosystems, and mammalian evolution. *Philosophical Transactions of the Royal Society of London* 353:159–71.

Cerling, T.E., J.M. Harris, B.J. MacFadden, M.G. Leakey, J. Quade, V. Eisenmann, and J.R. Ehleringer. 1997. Global vegetation change through the Miocene/Pliocene boundary. *Nature* 389:153–58.

Cerling, T.E., Y. Wang, and J. Quade. 1993 Expansion of C_4 ecosystems as an indicator of global ecological change in the late Miocene. *Nature* 361:344–45.

Clifford, S.C., I.M. Stronach, A.D. Mohamed, S.N. Azam-Ali, and N.M.J. Crout. 1993. The effects of elevated atmospheric carbon dioxide and water stress on light interception, dry matter production and yield in stands of groundnut (*Arachis hypogaea* L.). *Journal of Experimental Botany* 44:1763–70.

Coleman, J.S., and F.A. Bazzaz. 1992. Effects of CO_2 and temperature on growth and resource use of co-occurring C_3 and C_4 annuals. *Ecology* 73:1244–59.

Collatz, G.J., J.A. Berry, and J.S. Clark. 1998. Effects of climate and atmospheric CO_2 partial pressure on the global distribution of C_4 grasses: present, past, and future. *Oecologia* 114:441–54.

Connin, S.L., J. Betancourt, and J. Quade. 1998. Late Pleistocene C_4 plant dominance and summer rainfall in the southwestern United States from isotopic study of herbivore teeth. *Quaternary Research* 50:179–93.

Cook, A.C., D.T. Tissue, S.W. Roberts, and W.C. Oechel. 1998. Effects of long-term elevated [CO_2] from natural CO_2 springs on *Nardus stricta*: Photosynthesis, biochemistry, growth, and phenology. *Plant, Cell, and Environment* 21:417–25.

Cowling, S.A. 1999. Simulated effects of low atmospheric CO_2 on structure and composition of North American vegetation at the Last Glacial Maximum. *Global Ecology and Biogeography* 8:81–93.

———. 2001. Plant carbon balance, evolutionary innovation and extinction in land plants. *Global Change Biology* 7:231–39.

Cowling, S.A., and R.F. Sage. 1998. Interactive effects of low atmospheric CO_2 and elevated temperature on growth, photosynthesis and respiration in *Phaseolus vulgaris*. *Plant, Cell, and Environment* 21:427–35.

Cowling, S.A., and M.T. Sykes. 1999. Physiological significance of low atmospheric CO_2 for plant-climate interactions. *Quaternary Research* 52:237–42.

Crowley, T.J. 1995. Ice age terrestrial carbon changes revisited. *Global Biogeochemical Cycle* 9:377–89.

Curtis, P.S., D.J. Klus, S. Kalisz, and S.J. Tonsor. 1996. Intraspecific variation in CO_2 responses in *Raphanus raphanistrum* and *Plantago lanceolata*: Assessing the potential for evolutionary change with rising atmospheric CO_2. In *Carbon dioxide, populations, and communities,* ed. C. Körner and F.A. Bazzaz, 13–22. San Diego: Academic Press.

Curtis, P.S., A.A. Snow, and A.S. Miller. 1994. Genotype-specific effects of elevated CO_2 on fecundity in wild radish (*Raphanus raphanistrum*). *Oecologia* 97:100–105.

Curtis, P.S., and X. Wang. 1998. A meta-analysis of elevated CO_2 effects on woody plant mass, form, and physiology. *Oecologia* 113:299–313.

Dippery, J.K., D.T. Tissue, R.B. Thomas, and B.R. Strain. 1995. Effects of low and elevated CO_2 on C_3 and C_4 annuals. I. Growth and biomass allocation. *Oecologia* 101: 13–20.

Ehleringer, J., and O. Björkman. 1977. Quantum yields for CO_2 uptake in C_3 and C_4 plants. Dependence on temperature, CO_2, and O_2 concentration. *Plant Physiology* 59: 86–90.

Ehleringer, J.R., T.E. Cerling, and B.R. Helliker. 1997. C_4 photosynthesis, atmospheric CO_2, and climate. *Oecologia* 112:285–99.

Ehleringer, J.R., and R.K. Monson. 1993. Evolutionary and ecological aspects of photosynthetic pathway variation. Annual Review of Ecology and Systematics 24:411–39.

Ehleringer, J.R., R.F. Sage, L.B. Flanagan, and R.W. Pearcy. 1991. Climate change and the evolution of C_4 photosynthesis. *Trends in Ecology and Evolution* 6:95–99.

Etterson, J.R., and R.G. Shaw. 2001. Constraint to adaptive evolution in response to global warming. *Science* 294:151–54.
Fajer, E.D., M.D. Bowers, and F.A. Bazzaz. 1991. Performance and allocation patterns of the perennial herb, *Plantago lanceolata*, in response to simulated herbivory and elevated CO_2 environments. *Oecologia* 87:37–42.
Garbutt, K., and F.A. Bazzaz. 1984. The effects of elevated CO_2 on plants. III. Flower, fruit and seed production and abortion. *New Phytologist* 98:433–46.
Gunderson, C.A., and S.D. Wullschleger. 1994. Photosynthetic acclimation in trees to rising atmospheric CO_2: A broader perspective. *Photosynthesis Research* 39: 369–88.
Hetrick, B.A.D., G.W.T. Wilson, and T.C. Todd. 1996. Mycorrhizal response in wheat cultivars: Relationship to phosphorus. *Canadian Journal of Botany* 74:19–25.
Huang, Y., F A. Street-Perrott, S.E. Metcalfe, M. Brenner, M. Moreland, and K.H. Freeman. 2001. Climate change as the dominant control on glacial-interglacial variations in C_3 and C_4 plant abundance. *Science* 293:1647–51.
Kanai, R., and G.E. Edwards. 1999. The biochemistry of C_4 photosynthesis. In *C_4 plant biology,* ed. R.F. Sage and R.K. Monson, 49–87. San Diego: Academic Press.
Kaplan, A., Y. Helman, D. Tchernov, and L. Reinhold. 2001. Acclimation of photosynthetic microorganisms to changing ambient CO_2 concentration. *Proceedings of the National Academy of Science, USA* 98:4817–18.
Levis, S., and J.A. Foley. 1999. CO_2, climate, and vegetation feedbacks at the Last Glacial Maximum. *Journal of Geophysical Research* 104:31191–98.
Maherali, H., C.D. Reid, H.W. Polley, H.B. Johnson, and R.B. Jackson. 2002. Stomatal acclimation over a subambient to elevated CO_2 gradient in a C_3/C_4 grassland. *Plant, Cell, and Environment* 25:557–66.
Makino, A., and T. Mae. 1999. Photosynthesis and plant growth at elevated levels of CO_2. *Plant and Cell Physiology* 40:999–1006.
Marchant, R., A. Boom, and H. Hooghiemstra. 2002. Pollen-based biome reconstructions for the past 450 000 yr from the Funza-2 core, Colombia: comparisons with model-based vegetation reconstructions. *Palaeogeography, Palaeoclimatology, Palaeoecology* 177:29–45.
Medrano, H., A.J. Keys, D.W. Lawlor, M.A.J. Parry, J. Azcon-Bieto, and E. Delgado. 1995. Improving plant production by selection for survival at low CO_2 concentrations. *Journal of Experimental Botany* 46:1389–96.
Nelson, C.J., K.H. Asay, and L.D. Patton. 1975. Photosynthetic responses of tall fescue to selection for longevity below the CO_2 compensation point. *Crop Science* 15:629–33.
Norby, R.J., S.D. Wullschleger, C.A. Gunderson, D.W. Johnson, and R. Ceulemans. 1999. Tree responses to rising CO_2 in field experiments: Implications for the future forest. *Plant, Cell, and Environment* 22:683–714.
Norton, L.R., L.G. Firbank, and A.R. Watkinson. 1995. Ecotypic differentiation of response to enhanced CO_2 and temperature levels in *Arabidopsis thaliana. Oecologia* 104:394–96.
Overdieck, D., C. Reid, and B.R. Strain. 1988. The effects of preindustrial and future CO_2 concentrations on growth, dry matter production and the C/N relationship in plants at low nutrient supply: *Vigna unguiculata* (Cowpea), *Abelmoschus esculentus* (Okra) and *Raphanus sativus* (Radish). *Angew, Botanik* 62:119–34.
Petit, J.R., J. Jouzel, D. Raynaud, N.I. Barkov, J.-M. Barnola, I. Basile, M. Benders, J. Chappellaz, M. Davis, G. Delaygue, M. Delmotte, V.M. Kotlyakov, M. Legrand, V.Y. Lipenkov, C. Lorius, L. Pépin, C. Ritz, E. Saltzman, M. Stievenard. 1999. Climate and atmospheric history of the past 420,000 years from the Vostok ice core, Antarctica. *Nature* 399:429–36.
Polley, H.W., H.B. Johnson, and J.D. Derner. 2002. Soil- and plant-water dynamics in a

C_3/C_4 grassland exposed to a subambient to superambient CO_2 gradient. *Global Change Biology* 8:1118–29.
Polley, H.W., H.B. Johnson, B.D. Marino, and H.S. Mayeux. 1993a. Increase in C_3 plant water-use efficiency and biomass over glacial to present CO_2 concentrations. *Nature* 361:61–64.
Polley, H.W., H.B. Johnson, and H.S. Mayeux. 1992. Carbon dioxide and water fluxes of C_3 annuals and C_4 perennials at subambient CO_2 concentrations. *Functional Ecology* 6:693–703.
———. 1994. Increasing CO_2: Comparative responses of the C_4 grass *Schizachyrium* and grassland invader *Prosopis*. *Ecology* 75:976–88.
———. 1995. Nitrogen and water requirements of C_3 plants grown at glacial to present carbon dioxide concentrations. *Functional Ecology* 9:86–96.
Polley, H.W., H.B. Johnson, H.S. Mayeux, and S.R. Malone. 1993b. Physiology and growth of wheat across a subambient carbon dioxide gradient. *Annals of Botany* 71: 347–56.
Primack, R.B., and J. Antonovics. 1981. Experimental ecological genetics in *Plantago*. V. Components of seed yield in the ribwort plantain Plantago lanceolata L. *Evolution* 35:1069–79.
Pritchard, S.G., H.H. Rogers, S.A. Prior, and C.M. Peterson. 1999. Elevated CO_2 and plant structure: A review. *Global Change Biology* 5:807–37.
Sage, R.F. 1994. Acclimation of photosynthesis to increasing atmospheric CO_2: The gas exchange perspective. *Photosynthesis Research* 39:351–68.
———. 1995. Was low atmospheric CO_2 during the Pleistocene a limiting factor for the origin of agriculture? *Global Change Biology* 1:93–106.
Sage, R.F., and S.A. Cowling. 1999. Implications of stress in low CO_2 atmospheres of the past: Are today's plants too conservative for a high CO_2 world? In *Carbon dioxide and environmental stress,* ed. Y. Luo and H.A. Mooney, 289–308. San Diego: Academic Press.
Sage, R.F., and C.D. Reid. 1992. Photosynthetic acclimation to sub-ambient CO_2 (20 Pa) in the C_3 annual *Phaseolus vulgaris* L. *Photosynthetica* 27:605–17.
Sage, R.F., and J.R. Coleman. 2001. Effects of low atmospheric CO_2 on plants: More than a thing of the past. *Trends in Plant Science* 6:18–24.
Samson, D.A., and K.S. Werk. 1986. Size-dependent effects in the analysis of reproductive effort in plants. *The American Naturalist* 127:667–80.
Schlesinger, W.H. 1997. *Biogeochemistry.* San Diego: Academic Press.
Schwarz, N., and B.R. Strain. 1990. Carbon- a plant nutrient, deficiency and sufficiency. *Journal of Plant Nutrition* 13:1073–78.
Sharma, R.K., B. Griffing, and R.L. Scholl. 1979. Variations among races of *Arabidopsis thaliana* (L.) Heynh for survival in limited carbon dioxide. *Theoretical and Applied Genetics* 54:11–15.
St. Omer, L., and S.M. Horvath. 1983. Potential effects of elevated carbon dioxide levels on seed germination of three native plant species. *Botanical Gazette* 144:477–80.
Strain, B.R. 1991. Possible genetic effects of continually increasing atmospheric CO_2. In *Ecological genetics and air pollution,* ed. G.E. Taylor, Jr., L.F. Pitelka, and M.T. Clegg, 237–44. Berlin: Springer-Verlag.
Street-Perrott, F.A., Y. Huang, R.A. Perrott, G. Eglinton, P. Barker, L.B. Khelifa, D.D. Harkness, and D.O. Olago. 1997. Impact of lower atmospheric carbon dioxide on tropical mountain ecosystems. *Science* 278:1422–26.
Tissue, D.T., K.L. Griffin, R.B. Thomas, and B.R. Strain. 1995. Effects of low and elevated CO_2 on C_3 and C_4 annuals. II. Photosynthesis and leaf biochemistry. *Oecologia* 101:21–28.
Tolley, L.C., and B.R. Strain. 1984. Effects of CO_2 enrichment and water stress on growth of *Liquidambar styraciflua* and *Pinus taeda* seedlings. *Canadian Journal of Botany* 62:2135–39.

Tousignant, D., and C. Potvin. 1996. Selective responses to global change: Experimental results on *Brassica juncea* (L.) Czern. In *Carbon dioxide, populations, and communities,* ed. C. Körner and F.A. Bazzaz, 23–30. San Diego: Academic Press.

Tschaplinski, T.J., D.B. Stewart, P.J. Hanson, and R.J. Norby. 1995. Interactions between drought and elevated CO_2 on growth and gas exchange of seedlings of three deciduous tree species. *New Phytologist* 129:63–71.

Turcq, B., R.C. Cordeiro, A. Sifeddine, F.F.L. Simões Filho, A.L.S. Albuquerque, and J.J. Abrão. 2002. Carbon storage in Amazonia during the Last Glacial Maximum: Secondary data and uncertainties. *Chemosphere* 49:821–35.

Ward, J.K., J. Antonovics, R.B. Thomas, and B.R. Strain. 2000. Is atmospheric CO_2 a selective agent on model C_3 annuals? *Oecologia* 123:330–41.

Ward, J.K., and B.R. Strain. 1997. Effects of low and elevated CO_2 partial pressure on growth and reproduction of *Arabidopsis thaliana* from different elevations. *Plant, Cell, and Environment* 20:254–60.

———. 1999. Elevated CO_2 studies: Past, present, and future. *Tree Physiology* 19:211–20.

Ward, J.K., D.T. Tissue, R.B. Thomas, and B.R. Strain. 1999. Comparative responses of model C_3 and C_4 plants to drought in low and elevated CO_2. *Global Change Biology* 5:857–67.

Wray, S.M., and B.R. Strain. 1986. Response of two old field perennials to interactions of CO_2 enrichment and drought stress. *American Journal of Botany* 73:1486–91.

Yung, Y.L., T. Lee, C. Wang, and Y. Shieh. 1996. Dust: A diagnostic of the hydrologic cycle during the last glacial maximum. *Science* 271:962–63.

Zhang, J., and M.J. Lechowicz. 1995. Responses to CO_2 enrichment by two genotypes of *Arabidopsis thaliana* differing in their sensitivity to nutrient availability. *Annals of Botany* 75:491–99.

12. Environmentally Driven Dietary Adaptations in African Mammals

Thure E. Cerling, John M. Harris, and Meave G. Leakey

12.1 Introduction

From the late Miocene to early Pliocene, global terrestrial ecology underwent a major change, from a world where plants used C_3 photosynthesis almost exclusively to one where C_4 photosynthesis was a major component (Cerling et al. 1997b). This is especially true in the tropics, where C_4 photosynthesis might have comprised 50% of the net primary productivity. Evidence for this change is found in changes in the $^{13}C/^{12}C$ ratios in soil organic matter paleosol carbonates, and in reconstructions of the diets of large mammals (Quade and Cerling 1995; Cerling et al. 1997b; Freeman and Colarusso 2001). This change has so far been well documented in Africa, Asia, North America, and South America. Europe and Antarctica are at latitudes where C_4 photosynthesis does not occur in significant amounts. Comparable events in Australia might have occurred slightly later in time and are currently being studied by Ayliffe and co-workers. The principal method for documenting changes of this nature in the geological past is the determination of carbon isotope ratios, because the $^{13}C/^{12}C$ ratios are significantly different for plants using the C_3 versus the C_4 photosynthetic pathways.

The abrupt change from the C_3-world to the C_4-world makes it possible to study the response of mammals to a major change in food resources. In tropical Africa, dicotyledonous plants (hereafter, dicots) use C_3 photosynthesis almost exclusively, whereas monocotyledonous plants (hereafter, monocots) use the C_4

photosynthetic pathway. This means that most browsers, because they consume dicots, will have a dietary record of C_3 plants, whereas grazers will have a record of a C_4 plant diet preserved in their tissues. This isotopic distinction is very useful for quantifying dietary differences in mammals: grasses and dicots differ in chemical composition (hence, in nutrition content, toxin content, digestibility, etc.) as well as preferred habitat (e.g., openness of canopy, height above ground level, etc.). The paleontological record has many examples of morphological changes in mammals that are attributed to dietary shifts from browsing to grazing. Therefore, stable isotope studies coupled with observations on morphological change in mammals can test some of these hypotheses of dietary change.

In this study we use the well-dated, and very fossiliferous, collections from the Turkana basin in Northern Kenya to document evolutionary change in mammalian lineages, considering in turn equids, proboscideans, suids, and giraffids.

12.2 Methods and Materials

Fossil teeth from various localities in the Lake Turkana basin of Northern Kenya were sampled in the field and from the paleontological collections of the National Museums of Kenya, Nairobi. We use the stratigraphic and geochronological scheme of Brown, Fiebel, and McDougall (Feibel, Brown, and McDougall 1989; McDougall et al. 1992; Brown 1995; McDougall and Feibel 1999) to place fossils in their chronological order. Teeth of extant species from various African localities were sampled in the field and from the osteological collections of the National Museums of Kenya, Nairobi.

The stable isotopic composition of tooth apatite was determined using standard methods and is reported using the standard per mil (‰) isotopic notation:

$$\delta^{13}C = (R_{sample}/R_{standard} - 1)\ 1000 \qquad (12.1)$$

where R_{sample} and $R_{standard}$ are the $^{13}C/^{12}C$ ratios in the sample and standard, respectively. For carbon the standard is PDB. The apparent isotope enrichment factor ε^* between diet and enamel is 14.1‰ where:

$$\Box\varepsilon^* = [(1000 + \delta_{enamel})/(1000 + \delta_{diet}) - 1] * 1000 \qquad (12.2)$$

(Cerling and Harris 1999). In this paper we are assuming that the isotope enrichment is the same for all large ungulate mammals.

The $\delta^{13}C$ composition of the tooth enamel of mammalian herbivores reflects the proportion of C_3 versus C_4 vegetation that the animal was ingesting at the time the tooth germ was being formed. For mammals living in warm climates with rain during the growing season (a prerequisite for C_4 vegetation), it is thus possible to infer whether an individual was a browser, grazer, or mixed feeder.

12.3 Mammalian Tooth Enamel as a Paleoenvironmental Recorder

The isotope enrichment from diet to bioapatite is about 14‰ for ruminant bovids and is similar for other large mammals (Cerling and Harris, 1999). Therefore, the average $\delta^{13}C$ of diet can be estimated from the $\delta^{13}C$ of bioapatite. Enamel is the preferred substance to work with for paleodiet studies because it is the least susceptible to recrystallization during diagenesis (Ayliffe, Chivas, and Leakey 1994). In addition, because of the high selectivity in dietary preferences, studies of the isotopic composition of tooth enamel amplify the ecosystem-level isotope signal. The presence of a significant C_4 component in one species indicates that it would have been available for all other extant species.

Although C_3 and C_4 plants have average $\delta^{13}C$ values of about −27% and −13‰, respectively (Bender 1971), there is considerable variation about the mean. C_4 plants tend to have slightly more positive values in mesic environments and more negative values in xeric environments, whereas C_3 plants have the opposite trend. For tropical ecosystems, Cerling, Harris, and Passey (2003) found $\delta^{13}C$ values averaging between −31% to −34‰ for closed canopy forest understory plants; in African tropical savanna and bushland ecosystems, the $\delta^{13}C$ for C_3 plants averaged between −29% and −25‰ while the $\delta^{13}C$ value for C_4 plants averaged between −14% to −12‰. The isotope separation between average C_3 and C_4 plants at the ecosystem scale was as high as 17‰ in some mesic environments and as low as 11‰ in certain xeric environments (see examples in Cerling et al. 2003). The isotope composition of plants is also related to the $\delta^{13}C$ of the atmosphere from which plants derive CO_2. The isotopic composition of the atmosphere has changed by 1.5‰ since 1850 because of fossil fuel burning, and it is likely that the $\delta^{13}C$ of the atmosphere has undergone natural variations of a few ‰ over the past several million years. Therefore, in the following discussion we distinguish only the broad classes of a predominantly C_4-diet, a mixed C_3-C_4 diet, and a predominantly C_3-diet, with $\delta^{13}C$ values of >-2‰, −2 to −8‰, and <-8‰, respectively. Because of the uncertainty in the $\delta^{13}C$ of the local ecosystem mixing endmembers and the uncertainty in the $\delta^{13}C$ of the atmosphere, an enamel $\delta^{13}C$ value of −8‰ is treated as a cutoff for indicating a measurable C_4 component in the diet as recorded by tooth enamel. Thus a value of −8‰ could indicate any dietary component between 0 and about 25% C_4 biomass.

For the past 7 to 8 million years, the C_3 and C_4 distinction in the tropics is between C_3-dicots and C_4-grasses, as it is today. Therefore the isotope distinction that is made is between browsers and grazers, where browsing animals consume dicots and grazers consume grasses. We note that this distinction is lost in regions where the growing season is at cooler temperatures (< 15–20°C) because C_3-grasses dominate in such regions. This includes most of Europe, some Mediterranean climates, and alpine environments including those above about 3000 m in the tropics. Prior to 8 million years ago C_4 plants had very low abundances even in tropical and subtropical regions where they are common today (Cerling et al. 1997b). Therefore, before 8 million years ago it is not possible to use the isotope composition of enamel to indicate the fraction of browse versus grass in the diet, because the isotope distinction is lost. Cerling et al. (1997b) found that equids in southern North America, Pakistan, and Africa all had a predominantly C_3 diet before 8 million years ago; this could have been either C_3 browse or C_3 grass.

12.4 Case Histories of Diet Change in the Turkana Basin

In this section we look at four different mammalian lineages found in the Turkana basin: equids, proboscideans, suids, and giraffids. The Turkana basin is ideally suited for studying mammalian evolution associated with diet change. Neogene sediments of the region are exceptionally fossiliferous, and the frequency of volcanic ash layers throughout the basin has allowed a detailed chronology to be established. The principal fossiliferous localities in the Lake Turkana basin sample the following times: early Miocene (Buluk, Kalodirr), late Miocene (Lothagam), and Pliocene through early Pleistocene (Kanapoi, Koobi Fora, West Turkana, lower Omo Valley). We also include in our discussion data from the nearby Samburu Hills (Namurungule Formation) and Nakali, as well as isolated samples from the Baringo basin, Laetoli, and South Africa.

12.4.1 Equids

The evolution of the horse—from the small, multi-toed, browsing forest-dwellers of the Eocene to the large, single-toed, open country grazers whose undomesticated representatives can still be found in Africa and Asia—has long been construed as a classic example of evolution in action. Equids underwent much of their early evolution in North America and first appear in East Africa at about 10.5 million years ago, when three-toed hypsodont hipparionines became widespread throughout the Old World. There were evidently two episodes of hipparionine migration into Africa from Eurasia, the earlier characterized by species of *Hippotherium* and the later by *Eurygnathohippus* species (Bernor and Harris 2003). We analyzed samples of equids from Nakali and from the Namurungule Formation in the Samburu Hills, in addition to those from the Nawata and Nachukui formations at Lothagam and the Koobi Fora Formation. The Samburu Hills sequence overlaps, in part, with the lowest part of the Lower Nawata Member. The sediments at Nakali predate both the Samburu Hills and the Lothagam sequences.

The crown height of the teeth of Nakali equids, which are the earliest equids from East Africa, was sufficient to invoke a grazing habit, but analyzed teeth have $\delta^{13}C$ values ranging from -9.2 to -11.5‰ and indicate a pure C_3 diet (Fig. 12.1).

Other equids from Ngeringeriwe (ca. 9.5 Ma) in the Tugen Hills have $\delta^{13}C$ values between -9.5 and -11‰ indicating little, if any, C_4 contribution to their diet (Kingston, Marino, and Hill 1994). Transition to a C_4 diet took place during

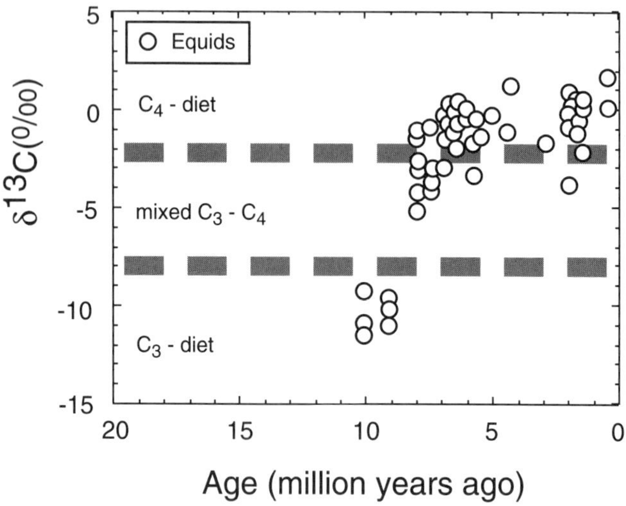

Figure 12.1. $\delta^{13}C$ of tooth enamel versus age for fossil and modern equids from East Africa. The isotope dietary classification is based on an enrichment factor for enamel-diet of 14.1‰.

an interval represented within the Samburu Hills succession and by the lowest part of the Nawata Formation. A mixed C_3/C_4 diet, with $\delta^{13}C$ values between −3 and −5‰ is found in about half (3 of 7) of the samples from the Samburu Hills and in the lower 82 meters of the Nawata Formation (3 of 6 samples). All samples, except one, from later in the Lower Nawata or from younger horizons in the Koobi Fora Formation (n = 14), have $\delta^{13}C$ values more positive than −2‰ (see Fig. 12.1).

Equids appear to have made the transition from a C_3 to a C_4 diet quickly both in Africa and elsewhere, although we have yet to identify a lengthy stratigraphic section where we can document the transition continuously over time. The transition appears to have occurred at about 8 million years ago. Grazing rhinoceroses (*Ceratotherium* species) appear to have begun exploiting C_4 grasses at about the same point in time.

12.4.2 Proboscideans

Three different clades of proboscidean occurred in the African Neogene and are represented by the Deinotheriidae, Gomphotheriidae, and Elephantidae. Deinotheres have low-crowned, lophodont teeth superficially similar to those of tapirs. The lophs have beveled cutting edges that are maintained, though at different angles, throughout the life of the individual for processing the food prior to digestion. Deinothere teeth remain essentially unchanged in morphology throughout the early Miocene to early Pleistocene; however, there is an overall increase in size from the earliest representatives of *Prodeinotherium* to the latest representatives of *Deinotherium.* In contrast, the early gomphotheres have low-crowned bunodont cheek teeth that are adapted for crushing and grinding rather than cutting. There is a general tendency for increase in complexity of the crown pattern and increase in the length of the teeth in different lineages of gomphotheres. This led to a pattern of delayed eruption of the posterior cheek teeth (horizontal tooth replacement), which both allowed the accommodation of larger teeth in the jaw and helped prolong the use of the teeth and, hence, the life of the individual.

The elephantids emerged from gomphothere stock near the end of the Miocene. Morphological differences between gomphotheres and early elephantids suggest a major change in the method of chewing, together with increases in length and hypsodonty of elephantid cheek teeth (Maglio 1973). The teeth of elephantids contain a large number of transverse plates instead of the paired trifoliate bunes characteristic of gomphotheres. Horizontal tooth replacement was further refined so that normally only one cheek tooth was erupted and functional in each side of the jaw. The earliest elephantids (*Primelephas* species) have low-crowned teeth, but the ensuing *Loxodonta*, *Elephas*, and *Mammuthus* lineages have hypsodont teeth; each lineage demonstrates an increase in crown height and plate number and a decrease in enamel thickness through time. The difference in crown height between gomphotheres and elephantids appears to indicate exploitation of different kinds of vegetation; the sequential changes seen in the cheek teeth of elephantid lineages have long been interpreted as progressive modifications for a primarily graminiferous (grazing) diet.

Deinotheres maintain a C_3 diet from their earliest record until their extinction in the early Pleistocene. Early gomphotheres from Maboko (ca. 17 Ma), Buluk (ca. 15 Ma), and Fort Ternan (14 Ma) all have $\delta^{13}C$ values that indicate a pure C_3 diet (Fig. 12.2). Gomphotheres from the Namurungule Formation in the Samburu Hills have $\delta^{13}C$ values between -8 and $-9‰$ indicating a diet dominated by (or solely based on) C_3 plants. Gomphotheres and elephantids, however, evidently changed from a C_3 dominated diet before 8 million years ago, to one that is C_4-dominated by 7 million years ago.

Unfortunately, the gomphotheres sampled from the Samburu Hills assemblages are documented from different sites than the sampled equids, so that in the absence of information about the size and nature of the different localities, it cannot be determined if the transition to C_4 diet in equids preceded that in gomphotheres. The transition from a C_3 to a C_4 diet was rapid in gomphotheres and elephantids (see Fig. 12.2). Elephantid teeth subsequently underwent a series of adaptations to more effectively process an abrasive graminiferous diet, but those of gomphotheres did not; that family became extinct in Africa at about 4 million years ago.

Elephantids have a predominantly grazing diet from about 7 million years ago until less than one million years ago (Cerling et al. 1999). Both fossil *Loxodonta* and *Elephas* were grazers, yet the modern *Loxodonta* in Africa is predominantly a browser as is *Elephas* in Asia (Cerling et al. 1999). Elephants are widely

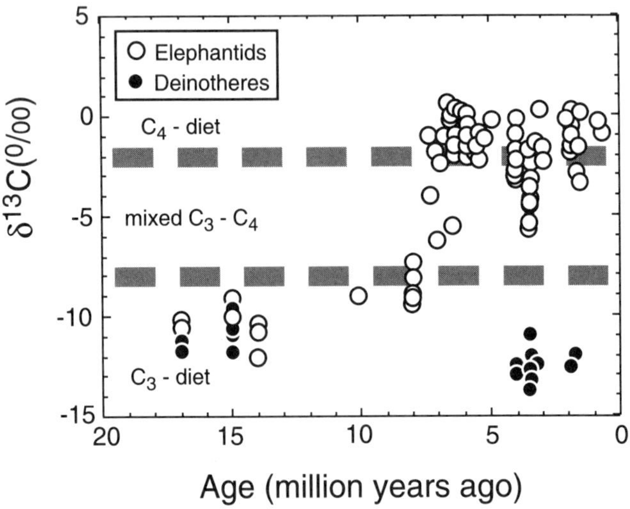

Figure 12.2. $\delta^{13}C$ of tooth enamel versus age for fossil and modern proboscideans from East Africa. The isotope dietary classification is based on an enrichment factor for enamel-diet of 14.1‰.

thought to have an enormous influence on the promotion of grasses in Africa because of their destructive feeding on trees. Their influence on the landscape must have been different in the past than it is today.

12.4.3 Suids

Suid assemblages from sub-Saharan Africa underwent a major transformation at the end of the Miocene. Genera representing the subfamilies Kubanochoerinae, Listriodontinae, and Paleochoerinae were replaced by *Nyanzachoerus* species (*Ny. syrticus*, *Ny. devauxi*) that comprised immigrant representatives of the Eurasian Tetraconodontinae. The derived tetraconodontine *Notochoerus jaegeri* appeared in the early Pliocene, and its daughter species *Not. euilus* characterizes the later Pliocene together with *Nyanzachoerus pattersoni*, *Ny. kanamensis*, and *Ny. australis*. The middle Pliocene saw a second wave of immigration from Eurasia that founded the *Potamochoerus*, *Kolpochoerus,* and *Metridiochoerus* lineages. Today these three lineages are represented, respectively, by the bush pig (*Potamochoerus porcus)*, the forest hog (*Hylochoerus meinertzhageni*), and the warthog (*Phacochoerus aethiopicus*).

Four of these suid genera—(*Nyanzachoerus*, *Notochoerus*, *Kolpochoerus*, and *Metridiochoerus*)—were characterized by increases in size, hypsodonty, and in the complexity of the third molar. As in elephantids, these trends were previously interpreted to represent adaptations from a browsing to a grazing diet. This interpretation was confirmed for *Nyanzachoerus* and *Notochoerus* species. The earliest nyanzachoeres had shorter third molars and lower $\delta^{13}C$ values than did the later nyanzachoeres and notochoeres (Fig. 12.3), thereby supporting previous hypotheses that the lengthening of the third molar represented an adaptation to a more abrasive diet. However, of the four species of *Metridiochoerus* recognized in the Plio-Pleistocene portion of the Koobi Fora succession—*M. an-*

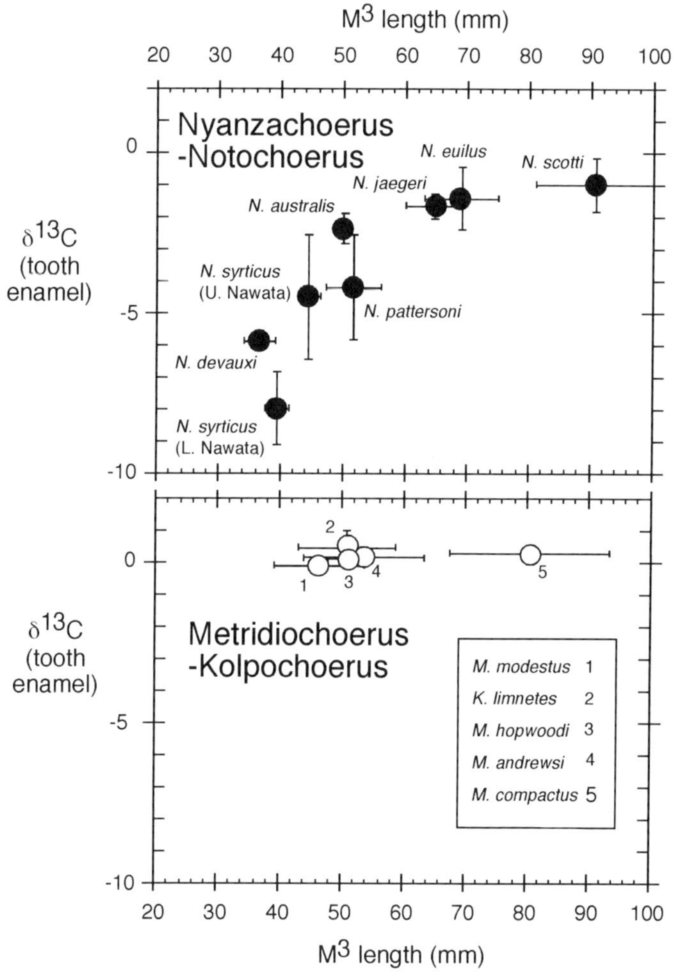

Figure 12.3. $\delta^{13}C$ of tooth enamel versus length of M^3 for fossil suids in the Turkana Basin. The *Nyanzachoerus-Notochoerus* lineage shows increasing length of the M^3 as more C_4 grass is incorporated into their diet. *Metridochoerus* and *Kolpochoerus* do not show any change in the fraction of C_4 biomass in their diet.

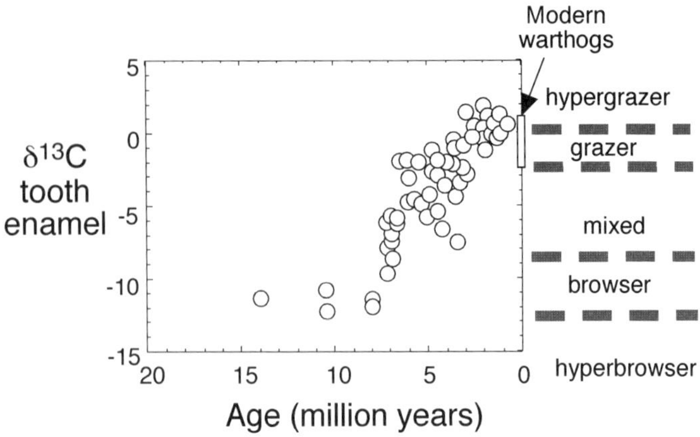

Figure 12.4. $\delta^{13}C$ of tooth enamel versus age for fossil and modern suids from East Africa. The isotope dietary classification is based on an enrichment factor for enamel-diet of 14.1‰.

drewsi, M. modestus, M. hopwoodi, and *M. compactus* (Harris and White 1979; Harris 1983)—all were found to be dedicated C_4 grazers throughout their known history, as were representatives of the *Kolpochoerus* lineage. Third molars of the *M. andrewsi* and *K. limnetes* lineages display a progressive increase through time in the length, height, and number of pillars of the third molars, so it was somewhat surprising to find that even the earliest and least progressive representatives from the Nachukui Formation were dedicated C_4 grazers (and that there is no relationship between M^3 length and $\delta^{13}C$). In this case, it appears that even the earliest kolpochoeres and metridiochoeres had adapted to grazing and that the progressive lengthening of the molars does not reflect an overall change in diet but, as with the elephantids, an adaptation for increased efficiency in processing a graminiferous diet. Overall, suids show a gradual increase in the fraction of C_4 grass in their diets over several million years, becoming 100% grazers by about 3 million years ago (Fig. 12.4).

12.4.4 Giraffids

The Giraffidae is today represented by only two living species, although approximately 30 fossil species have been recognized from the fossil record (Solounias et al. 2000). The family is characterized by the presence of a bifurcate lower canine and the presence of skin-covered ossicones; most representatives are of large size and have elongate legs and necks. Although seldom comprising significant elements of the African biota, giraffe relatives are represented by *Prolibytherium* and *Zaraffa* in the early Miocene of North Africa, by *Climacocerus*, *Paleotragus,* and *Samotherium* species in the middle to late Miocene of East Africa, and by *Sivatherium* and *Giraffa* species during the past five million years.

Despite the diversity of extinct forms, most have small and low-crowned teeth, and until recently the consensus had been that giraffids were committed browsers. However, Solounias and Dawson-Saunders (1988) used tooth microwear analysis to demonstrate that representatives of the Eurasian genus *Samotherium* were mixed feeders and hypothesized that *Samotherium* species seasonally alternated between browsing and grazing. A grazing component of the diet was further substantiated by the proportionately wide nature of the premaxilla (Solounias and Dawson-Saunders 1988; Solounias and Moelleken 1993). More recently, Solounias et al. (2000) have expanded their database of extant ruminant tooth wear and in so doing were able to recognize in the Giraffidae five dietary categories: grazers, facultative grazers, seasonal mixed feeders, nonseasonal mixed feeders, and browsers.

In sub-Saharan Africa, the middle Miocene giraffoids, *Climacoceras* and *Paleotragus primaevus,* both seem to have had a pure or nearly pure C_3 diet (Fig. 12.5). However, the $\delta^{18}O$ value for *Paleotragus* teeth indicated a more evaporated water source suggesting that, like extant giraffes, *Paleotragus* may have obtained much of the water it required from its food (Cerling et al. 1997a). Specimens of *Paleotragus* from the Samburu Hills are slightly more positive than those from Fort Ternan, although specimens from both localities plot as browsers. The difference may reflect a slight increase in the C_4 dietary component but could suggest that the later and more northerly locality was drier and more water-stressed.

Several species of *Giraffa* have been recognized in the East African Plio-

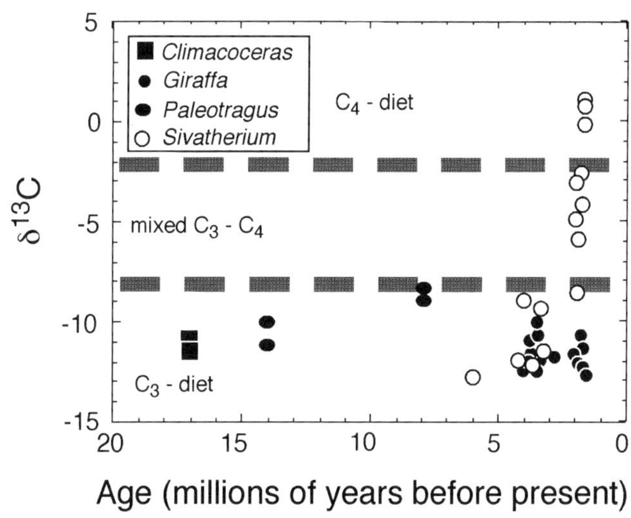

Figure 12.5. $\delta^{13}C$ of tooth enamel versus age for fossil and modern giraffids from East Africa. The isotope dietary classification is based on an enrichment factor for enamel-diet of 14.1‰.

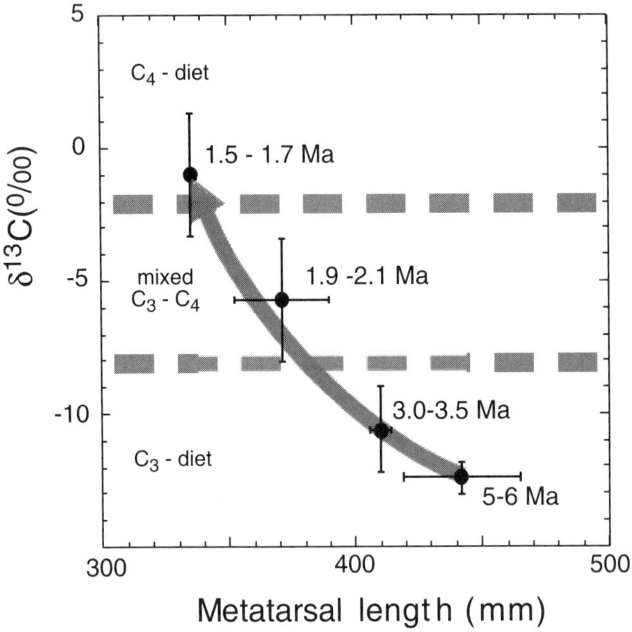

Figure 12.6. $\delta^{13}C$ in tooth enamel compared to metatarsal length in Sivatheres. The metatarsal becomes shorter as more C_4 grass is incorporated into the diet.

Pleistocene. All seem to plot out isotopically as dedicated browsers. Although the samples are small, specimens of *Giraffa jumae* and *G. stillei* from Laetoli appear to be less negative in $\delta^{13}C$ and less positive in $\delta^{18}O$ than specimens from the Lake Turkana basin; this may suggest that the Serengeti specimens drank more water but ate more grass. The findings could also indicate, however, that browse vegetation in the Laetoli region was stressed from lack of water at the time the giraffes were feeding on it.

Two species of *Sivatherium* have been recognized from sub-Saharan Africa: *Sivatherium hendeyi* from Langebaanweg and the more widely distributed *Sivatherium maurusium*. Isotopically, *Sivatherium hendeyi* plots out as a browser, as do the samples of *S. maurusium* from Kanapoi and from the lower portions of the Koobi Fora Formation (Fig. 12.6). However, a change in diet is indicated by specimens from the Upper Burgi Member; one of the specimens plots as a browser, but the other one plots as a mixed feeder. Specimens from the upper part of the Koobi Fora Formation plot as grazers.

It is intriguing that African sivatheres adopted a grazing diet several millions of years after proboscideans, perissodactyls, suids, and bovids exploited C_4 grasses in sub-Saharan Africa. Corroboration of the dietary change is seen in the lengths of the metatarsals: The metatarsal length in the Miocene is considerably longer than those from the younger Pleistocene.

12.5 Discussion

Carbon isotopic analysis is a very powerful tool for determining dietary adaptations of mammals from the tropics of Africa, Asia, and the Americas and can be used for fossil as well as extant specimens. The technique supplements and, in some instances, surpasses the accuracy of dietary information derived from direct observation of extant individuals. The technique may be applied to most, if not all, tissues from living specimens; dentine and bone provide believable results from late Pleistocene fossils, but tooth enamel is the most reliable source of isotopic information for fossil specimens because of possible complications from diagenetic alteration. In particular, stable isotopes can be used to test hypotheses about anatomical characteristics and their relationship to diet if the question can be cast in terms of a C_3-diet versus and C_4-diet.

One drawback of the methodology is that enamel germs are formed only at specific times during the development of young individuals, and thus the dietary information from a specific tooth reflects the diet of the individual (or that of its mother) only when that tooth was being formed. In contrast, hair and horn grow continuously throughout life, but these tissues are seldom fossilized. Microscopic pits and scratches on the wear surfaces of teeth record abrasion by dietary items ingested by the sampled specimen during the course of the previous few months. Thus microscopic wear analysis provides a useful method for refining dietary information provided by isotopic analysis.

This study shows the dietary response of four different mammalian lineages to ecological change. Equids, elephantids, suids, and giraffids/sivatheres responded in different ways to the expansion of C_4 biomass in the East Africa ecosystem. Between 8 million and 7 million years ago, C_4 grasses became abundant in East Africa. Equids responded very quickly to this new diet and became exclusive, or almost so, C_4-grazers by about 7 million years ago. From 7 million years ago to the present, C_4 grasses have been the dominant food source for equids in Africa. The proboscidean response was similar; very shortly after C_4 grasses became abundant, the diet of elephantids was dominated by C_4-grass. However, they have experienced a recent change so that modern elephants are browsers, unlike their fossil relatives.

Suids provide an interesting test case for dietary adaptation as they underwent significant morphological change during this time period. Stable isotopes show this change is accompanied by diet change for only one lineage, while two other lineages underwent similar change (elongation of molars) yet maintained a predominantly C_4 diet throughout. Late Miocene and early Pliocene sivatheres have isotopically similar tooth enamel to giraffines but became grazers in the late Pliocene; this change to grazing was accompanied by a shortening of their limb bones. Among the large herbivores from the African fossil record, giraffines and deinotheres have consistently negative $\delta^{13}C$ tooth enamel values, indicative of a pure C_3 browsing diet throughout their history.

12.6 Summary

The Turkana basin provides an excellent study area to examine dietary adaptations to ecological change. C_4 biomass increased significantly in East Africa between 7 million and 8 million years ago, and the Turkana basin is well dated. It contains abundant fossils, which have been placed in their stratigraphic and chronological context. Because most grasses in sub-Saharan Africa use the C_4 method of photosynthesis, grazing mammals may be readily distinguished from browsing mammals by the isotopic composition of their body tissues. The isotopic composition of tooth enamel distinguishes fossil mammals that fed on C_4 plants from those that exploited C_3 vegetation, thereby documenting dietary changes associated with the late Miocene radiation of C_4 grasses.

Hipparionine horses with hypsodont teeth entered Africa at about 10 Ma but did not start exploiting C_4 grasses until about 8 million years ago. By 7 million years ago they had a predominantly C_4 diet, which continues today in the equids of Africa. Both gomphotheriid and elephantid proboscideans started to exploit C_4 grasses between 8 million and 7 million years ago. Elephantids subsequently developed longer and taller teeth that had more plates with thinner enamel; these represent progressive adaptations to an increasingly graminiferous diet. However, modern African elephants (*Loxodonta*) are predominantly browsers, unlike the fossils from the Turkana basin and elsewhere in Africa. Gomphotheriid teeth did not display comparable changes, and this group became extinct at about 4 million years ago. Deinotheriid proboscideans remained C_3 browsers throughout their known history. Tetraconodontine suids arrived in Africa at about 8 million years ago. The third molars of *Nyanzachoerus* and *Notochoerus* species display changes in length, height, and complexity through time, and such changes were associated with a change in diet from browse to graze as recorded by the carbon isotopes.

The suines *Metridiochoerus* and *Kolpochoerus* migrated into Africa after 4 million years ago; changes in the morphology of their third molars (M^3), like comparable changes seen in elephantid teeth, represent adaptation to a graminiferous diet. The third molar increased in length over several million years although their diet was predominantly C_4 throughout. Giraffids were present in sub-Saharan Africa from the middle Miocene. Whereas giraffines had small and low-crowned teeth and remained browsers throughout their known history, sivatheres started exploiting C_4 grasses during the late Pliocene (i.e., several million years after other African mammals had exploited C_4 vegetation) and became grazers by about 2 million years ago. The adaptation in sivatheres was a shortening of both the fore and hind legs as C_4 grass was increasingly used as a diet resource.

References

Ayliffe, L.K., A.R. Chivas, and M.G. Leakey. 1994. The retention of primary oxygen isotope composition of fossil elephant skeletal phosphate. *Geochimica et* Cosmochimica Acta 58:5291–98.

Bender, M.M. 1971. Variations in the $^{13}C/^{12}C$ ratios of plants in relation to the pathway of photosynthetic carbon dioxide fixation. *Phytochemistry* 10:1239–45.

Bernor, R.L., and J.M. Harris. 2003. Systematics and evolutionary biology of the Late Miocene and Early Pliocene hipparionine equids from Lothagam, Kenya. In *Lothagam: The dawn of humanity in Africa,* ed. M.G. Leakey and J.M. Harris. New York: Columbia University Press.

Brown, F.H. 1995. The potential of the Turkana Basin for paleoclimatic reconstruction in East Africa. In *Paleoclimate and evolution,* ed. E.S. Vrba, G.H. Denton, T.C. Partridge, and L.H. Burkle, 319–30. New Haven: Yale University Press.

Cerling, T.E., and J.M. Harris. 1999. Carbon isotope fractionation between diet and bioapatite in ungulate mammals and implications for ecological and paleoecological studies. *Oecologia* 120:347–63.

Cerling, T.E., J.M. Harris, S.H. Ambrose, M.G. Leakey, and N. Solounias. 1997a. Dietary and environmental reconstruction with stable isotope analyses of herbivore tooth enamel from the Miocene locality of Fort Ternan, Kenya. *Journal of Human Evolution* 33:635–50.

Cerling, T.E., J.M. Harris, and M.G. Leakey. 1999. Browsing and grazing in modern and fossil proboscideans. *Oecologia* 120:364–74.

Cerling, T.E., J.M. Harris, B.J. MacFadden, M.G. Leakey, J. Quade, V. Eisenmann, and J.R. Ehleringer. 1997b. Global change through the Miocene/Pliocene boundary. *Nature* 389:153–58.

Cerling, T.E., J.M. Harris, and B.H. Passey. 2003. Dietary preferences of East African Bovidae based on stable isotope analysis. *Journal of Mammalogy* 84:456–70.

Feibel, C.S., F.H. Brown, and I. McDougall. 1989. Stratigraphic context of fossil hominids from the Omo Group deposits; northern Turkana Basin, Kenya and Ethiopia. *American Journal of Physical Anthropology* 78:595–622.

Freeman, K.H., and L.A. Colarusso. 2001. Molecular and isotopic records of C_4 grassland expansion in the late Miocene. *Geochimica et Cosmochimica Acta* 65:1439–54.

Harris, J.M. 1983. Family Suidae. In *The fossil ungulates:Proboscidea, Perissodactyla, and Suidae,* ed. J.M. Harris. Koobi Fora Research Project Monograph Series 2:215–302. Clarendon Press, Oxford, UK.

Harris, J.M., and T.D. White. 1979. Evolution of the Plio-Pleistocene African Suidae. *Transactions of the American Philosophical Society* 69:1–128.

Kingston, J.D., B.D. Marino, and A. Hill. 1994. Isotopic evidence for Neogene hominid paleoenvironments in the Kenya Rift valley. *Science* 264:955–59.

Maglio, V.J. 1973. Origin and evolution of the Elephantidae. *Transactions of the American Philosophical Society* 63(part 3):148.

McDougall, I., F.H. Brown, T.E. Cerling, and J.W. Hillhouse. 1992. A reappraisal of the geomagnetic polarity time scale to 4 Ma using data from the Turkana Basin, East Africa. *Geophysical Research Letters* 19:2349–52.

McDougall, I., and C.S. Feibel. 1999. Numerical age control for the Miocene-Pliocene succession at Lothagam, a hominoid-bearing sequence in the northern Kenya Rift. *Journal of the Geological Society of London* 156:731–45.

Quade, J., and T.E. Cerling. 1995. Expansion of C_4 grasses in the late Miocene of northern Pakistan: Evidence from stable isotopes in paleosols. *Palaeogeography, Palaeoclimatology, Palaeoecology* 115:91–116.

Solounias, N., and B. Dawson-Saunders. 1988. Dietary adaptations and paleoecology of the late Miocene ruminants from Pikermi and Samos in Greece. *Palaeogeography, Palaeoclimatology, Palaeoecology* 65:149–172.

Solounias, N., W.S. McGraw, L. Hayek, and L. Werdelin. 2000. The paleodiet of the Giraffidae. In *Antelopes, deer, and relatives: Fossil record, behavioral ecology, systematics, and conservation*, ed. E.S. Vrba and G.B. Schaller, 83–95. Yale University Press, New Haven.
Solounias, N., and S.M.C. Moelleken. 1993. Determination of dietary adaptation in an archaic antelope through tooth microwear and premaxillary analysis. *Lethaia* 26:261–268.

13. Terrestrial Mammalian Herbivore Response to Declining Levels of Atmospheric CO_2 During the Cenozoic: Evidence from North American Fossil Horses (Family Equidae)

Bruce J. MacFadden

13.1 Introduction

The fossil record preserves a wonderfully rich sequence spanning 65 million years of Cenozoic mammals that lived in ancient terrestrial ecosystems. During this time in Earth history, major global climate changes undoubtedly affected the course of mammalian evolution. Similarly, countless biotic interactions, such as competition, contributed to the struggle for existence of particular species. One of the interesting challenges for the paleontologist is to understand the relative role of physical/climatic factors versus those of biological interactions in driving evolution. This challenge has been with us since Darwin, and is unlikely to be solved at any time in the near future. As more data and interpretations accumulate from the fossil record, we will have an enhanced understanding of these determinants of evolution.

The paleoecology of Cenozoic mammals has long been studied, mostly because these animals are abundant and many have living close relatives, which yield modern-day data that can be compared with paleontological data in order to reconstruct ancient ecosystems. With regard to the physical and climatic factors that structure ancient ecosystems, rainfall and seasonality are, classically, the best known. Other climatic parameters also have undoubtedly affected Cenozoic mammalian evolution. We now know, for example, that early Cenozoic Earth was a far more humid hothouse than it is today and that during a relatively short interval of several million years in the Eocene, this dominant climatic regime transformed into an "icehouse," that is, with significant global cooling that apparently affected the rate of extinctions (Prothero 1994). It has been asserted that, in a similar way, the late Cenozoic has undergone significant global change manifested by aridification, higher seasonality, and general cooling, particularly over the past 5 million years with the onset of Pleistocene glacial stages (Raymo 1994).

Over the past decade, studies of stable isotopes of ancient soils and fossil herbivore teeth (Cerling, Wang, and Quade 1993, Cerling et al. 1997) have demonstrated a major global change in the relative abundance of C_3 and C_4 photosynthesis in terrestrial ecosystems during the late Miocene. Prior to approx 7 million years ago, terrestrial ecosystems, even including tropical and temperate grasslands, were fundamentally C_3-based. After that time, approx the analogs of the older grasslands became C_4-based, which set the stage for modern C_4-based grassland biomes.

The purpose of this paper is to ask the question: How did mammalian herbivores living in late Cenozoic terrestrial ecosystems respond to declining levels of atmospheric CO_2? The short answer to the question might be that these herbivores were oblivious to this global climate change. However, as will be developed below, the long answer to this question is both more complex and certainly more interesting. As mammalian herbivores ate plants, and plants responded in fundamental ways to declining atmospheric CO_2, the primary consumers were indirectly affected by this global change. It should be noted that similar studies have been presented for browsing (Janis, Damuth, and Theodor 2000) and grazing ungulates (Janis, Damuth, and Theodor 2002).

Fossil horses (Family Equidae) from North America are chosen here as a model, or case example, because (1) this group was very abundantly represented in North America from the Eocene, approx 57 million years ago, until the last surviving genus *Equus* became extinct on this continent during the late Pleistocene, approx 10,000 years ago, (2) this group is a classic textbook example of morphological evolution related to adaptive change (e.g., Simpson 1951, 1953; MacFadden 1992), and (3) Recent work that analyzes stable isotopes of North American fossil horse teeth allow testing of previous morphological hypotheses concerning the dietary evolution related to the adaptive shift from browsing to grazing (Wang, Cerling, and MacFadden 1994; MacFadden and Cerling 1994; Cerling, Harris, and MacFadden 1998a; MacFadden, Solounias, and Cerling 1999).

This paper assumes that levels of atmospheric CO_2 declined during the Cenozoic and that this change, either fundamentally or partially (i.e., other factors such as increased aridity may also have been involved) affected the photosynthetic pathways used by grasses. There are some different interpretations about the exact timing of decreased levels of atmospheric CO_2 (e.g., Boucot and Gray 2001). Most authors (e.g., Cerling 1991; Berner 1991; 1994; Ekart et al. 1999; Pearson and Palmer 2000; Wallmann 2001) indicate low, or declining, levels to a point sometime during the Miocene that would have potentially favored C_4

photosynthesis, i.e., below approx 500 ppm (for an exception, see Pagani, Arthur, and Freeman 1999; Pagani, Freeman, and Arthur 1999).

13.2 Ancient Grasslands and Paleobotanical Evidence for C_3 and C_4 Plants During the Cenozoic

In the broad scheme of life on this planet, with land plants known to have existed at least since the Devonian, approx 300 million years ago (Behrensmeyer et al. 1992), grasses are a relatively recent arrival on the global ecological landscape. Today, grasslands constitute a major ecological biome type and cover about 25% of the Earth's land. They exist in a wide variety of climatic regimes ranging from mean annual temperatures between 0 and 20°C and between 250 and 1000 mm of annual precipitation. Grassland types vary from tropical savannas, to temperate tall and mixed-grass prairies, shrub steppe, high-latitude cold steppes and arid short-grass steppes, and montane grass communities (Shantz 1954; Lauenroth 1979). Grasses, therefore, contribute greatly to primary productivity and represent a considerable food resource for terrestrial herbivores.

With regard to the origin of Poaceae, Kellogg (1999, p. 431) states: "We can comfortably date the origin of the grasses at approximately 65 mya based on pollen grains from the Upper Maastrichtian" (Linder 1987). The earliest evidence of grasses during the Cenozoic comes from Paleocene pollen, approx 65 to 55 million years ago (Muller 1981), and the earliest plant macrofossil comes from the Paleocene/Eocene of Tennessee, approx 55 million years ago (Crepet and Feldman 1991). Thereafter, the fossil record lacks good evidence of grasses until the middle Cenozoic (Jacobs, Kingston, and Jacobs 1999). Based on the abundance of precocious (relative to, e.g., horses in North America; Shockey 1997) high-crowned, presumed grazing, mammalian herbivores during the late Eocene and Oligocene in South America, Stebbins (1981) suggested that grasses may have arisen, or spread, earlier there than in the Northern Hemisphere; fossil evidence for this assertion, however, is scanty. In the Northern Hemisphere, the next definitive evidence for fossil grasses includes paleofloras of the Great Plains that are approx 15 million years old and come from middle Miocene (MacGinitie 1962; Gabel, Backlund, and Hafner 1998) and dumb-bell shaped (panicoid) phytoliths (Fig.13.1), preserved from the 14-million-year-old Fort Ternan hominoid site in East Africa (Dugas and Retallack 1993; Jacobs, Kingston, and Jacobs 1999). It is not otherwise known whether middle Miocene grasses were principally C_3 or C_4, but proxy evidence from fossil high-crowned, presumed grazing, herbivores suggests that they were mostly C_3 (Wang, Cerling, and MacFadden 1994; MacFadden, Cerling, and Prado 1996; Cerling et al. 1997, Cerling, Ehleringer, and Harris 1998b).

The first definitive evidence for C_4 grasses, that is, with preserved Krantz anatomy, comes from the late Miocene (~12.5 Ma) Ricardo Formation from the

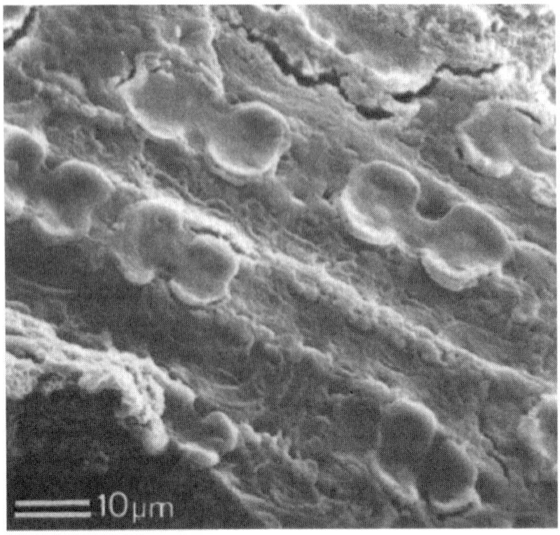

Figure 13.1. Middle Miocene dumb-bell shaped (panocoid) phytoliths from Fort Ternan, Kenya. From Dugas and Retallack (1993) and reproduced with permission of the Paleontological Society.

Mojave desert in southern California (Tidwell and Nambudiri 1989). There also is a late Miocene occurrence (~7 Ma) from Kansas (Thomasson, Nelson, and Zakrzewski 1986). Considerable discussion in the literature addresses the extent of and biogeographic distribution of C_4 grasses prior to the late Miocene (see Cerling 1999; Kellogg 1999). Based on carbon isotopic evidence from fossil herbivores and ancient soil carbonates, it seems that if C_4 grasses were present prior to this time, they were relatively rare in terrestrial ecosystems. A large body of carbon isotopic data from tooth enamel and ancient soil carbonates during the late Miocene (Fig. 13.2) demonstrates that C_4-based ecosystems became widespread globally in a relatively short period of time after about 7 million years ago (Cerling et al. 1997; 1999).

Studies of the relative proportions of C_3 and C_4 plant foods eaten by mammalian herbivores in the New World during the latest Miocene through late Pleistocene indicate that latitudinal gradients of C_3 and C_4 distributions, as we know them from the present-day, were established during this time, as shown in Fig. 13.3 (MacFadden, Cerling, and Prado 1996; MacFadden et al. 1999). In the context of the present discussion, the proximate cause for the advent of C_4 grasses during the late Miocene approx 7 million years ago seems to have been the decrease in atmospheric CO_2 to below levels (<400 ppm) that would favor C_4 photosynthesis as an adaptive strategy for grasses (Ehleringer, Cerling, and Harris 1997; Cerling 1999).

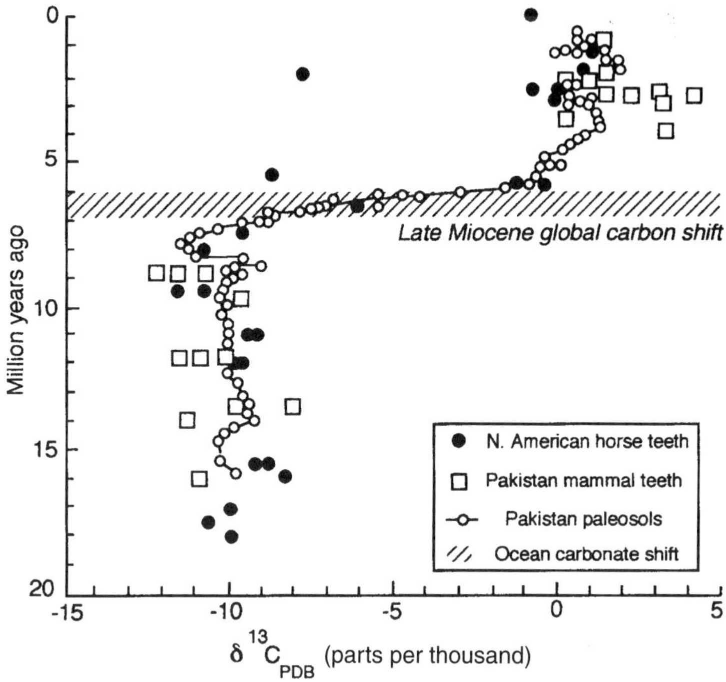

Figure 13.2. Late Miocene global carbon shift based on data from fossil teeth, paleosols, and the deep-sea carbonate record. The carbon isotopic content is expressed in $\delta^{13}C$ (parts per thousand) = $(R_{sample}/R_{standard} - 1) \times 100$, where R = $^{13}C/^{12}C$, sample is unknown and standard is PBD Cretaceous belemnite. Modified from Cerling, Wang and Quade (1993) and reproduced with permission of Macmillan.

13.3 Reconstruction of Ancient Diets: General Background for Mammalian Herbivores

Traditionally, the diets of ancient herbivores have been interpreted through study of dental morphology. This stems from a classic study of fossil horses from Europe (Kowalevsky 1873) but is also based on an extensive body of modern literature (e.g., Janis and Fortelius 1988). Recent studies also have added paleodietary information from the study of cranial morphology (Janis and Ehrhardt 1988; Solounias and Moelleken 1993) and enamel-wear patterns (Solounias and Hayek 1993; Fortelius and Solounias 2000; Solounias and Semprebon 2001) of fossil mammalian herbivores.

Based on study of extant mammalian herbivores with known diets and habitat preferences, the feeding adaptations of corresponding extinct species can be modeled from the fossil record. For example, extant mammals such as the white-tailed deer (*Odocoileus virginianus*) and tapir (*Tapirus*) have a feeding mor-

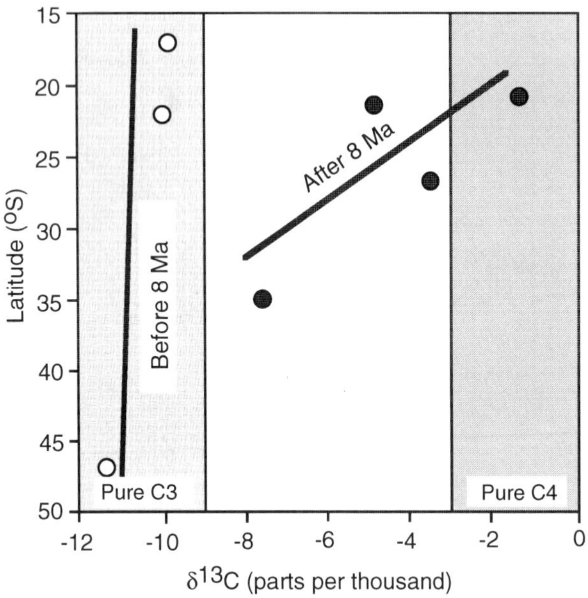

Figure 13.3. Variation in tooth enamel carbonate $\delta^{13}C$ of terrestrial mammalian grazers from South America, showing the development of latitudinal gradients in the distribution of C_3 and C_4 grasses after 8 million years ago (Ma). Modified from MacFadden, Cerling and Prado (1996) and reproduced with permission of the Society for Sedimentary Geology.

phology that includes short-crowned (brachydont) teeth (Fig. 13.4) and a rounded incisor cropping mechanism (Janis and Fortelius 1988; Janis and Ehrhardt 1988; Solounias and Moelleken 1993). These species are known to be principally forest-dwelling browsers, that is, feeding on relatively soft leafy plant foods (Walker 1975).

In contrast, the extant North American Bison, *Bison bison* or African zebras (e.g., *Equus burchelli*) have high-crowned (hypsodont) teeth and relatively straight incisor cropping mechanisms. These species are known to be principally open-country grazers, that is, feeding principally (>90%, Janis and Ehrhardt 1988) on grasses. The primary adaptive explanation for the increased crown height in grazers is the accelerated tooth wear rate resulting from chewing abrasive grasses, which contain phytoliths—silica plant parts that function to deter herbivory (McNaughton et al. 1985; Piperno 1988). To a lesser extent, another factor potentially contributing to tooth wear could be the introduction of contaminant grit in herbivore diets, either from feeding close to the ground, or from abrasive dust particles on plant foodstuffs.

The radiation of Cenozoic mammals during the past 65 million years is characterized by increasing taxonomic and morphological diversification and by re-

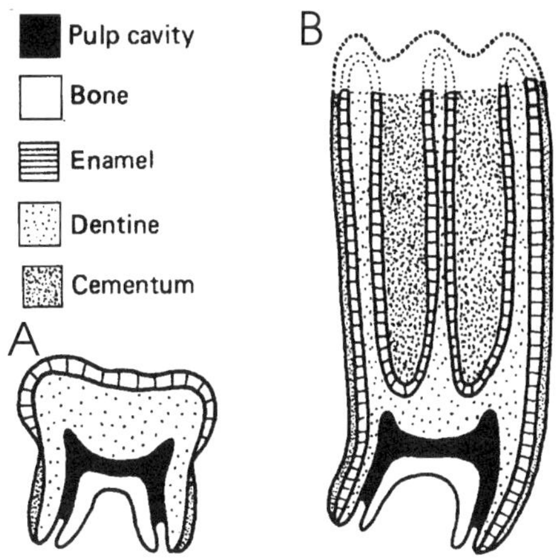

Figure 13.4. Comparisons of tooth crown height represented by cross-sectioned teeth. A. a short-crowned (brachydont) molar from a human. B. High-crowned (hypsodont) molar from a horse. From Janis and Fortelius (1988) and reproduced with permission of Cambridge University Press.

placement, since some clades radiated into new "adaptive zones" (*sensu* Simpson 1953), or guilds (*sensu* Root 1967). With a few very rare exceptions (e.g., archaic mammalian orders Taeniodontia and Tillodontia), early Tertiary (Paleocene through Oligocene) terrestrial mammalian herbivores are universally characterized by short-crowned, or brachydont, teeth, as shown in Fig. 13.4 (Janis 2000). This morphology is interpreted as a generalized browsing adaptation, either for folivory or for frugivory. This is also consistent with paleoecological reconstructions in which Earth's terrestrial ecosystems during the early Tertiary had a relatively high proportion of forested and woodland biomes available as food resources to these ancient browsers (Webb 1977).

During the Miocene there was a widespread change in the dental morphology of terrestrial mammalian herbivores. Notwithstanding what occurred in South America (to be discussed below), there was a widespread and geologically simultaneous (i.e., all within a period of a several million years) increase in crown height in these mammals. This dramatic evolution of hypsodont, or high-crowned, teeth is interpreted as a response to the advent of widespread grassland communities during the Miocene and corresponding adaptive radiation of terrestrial mammalian grazers. Correspondingly, many mammal groups that during the earlier Tertiary were represented by brachyodont species (e.g., Artiodactyla and Perissodactyla) evolved into clades consisting of grazers.

Although this rapid and widespread (both taxonomically and geographically) increase in hypsodonty occurred throughout Holarctica and Africa during the Miocene, a different pattern is observed for extinct South American mammalian herbivores. Mammalian faunas from the late Eocene, and particularly the De-

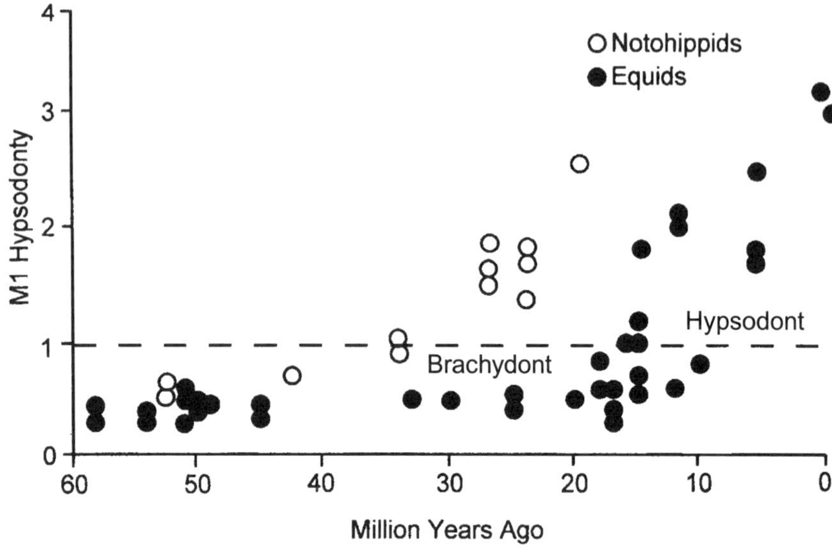

Figure 13.5. The timing of increased crown height (hypsodonty) in South American notoungulates (Family Notohippidae) as compared to North American horses (Family Equidae). Hypsodonty, or the hypsodonty index (HI), is the ratio of molar occlusal length to unworn molar crown height. Mammals with HIs of <1 are short-crowned, or brachydont, whereas those with HIs of >1 are high-crowned, or hypsodont. From Shockey (1997) and reproduced with permission of the Society of Vertebrate Paleontology.

seadan late Oligocene, approx 30 million years ago, show a large diversity of the high-crowned, endemic notoungulates. The stage of evolution seen in the dentition of these terrestrial mammalian herbivores is comparable to that of North American clades, such as horses approx 20 million years ago, thus giving rise to the concept of "precocious hypsodonty" in South America (Fig. 13.5). Several workers have asserted that the advent of precocious hypsodonty indicates an earlier origin of grasses in South America than in the Northern Hemisphere (Patterson and Pascual 1972; Stebbins 1981; Jacobs, Kingston, and Jacobs 1999) by approx 10 million years ago.

13.4 Fossil Horses, Diets, and Climate: Case Example from North America

Fossil horses (Family Equidae) have an abundant fossil record in North America spanning most of the Cenozoic, that is, for the past 57 million years. For this reason, and probably because of their close association with humans, fossil horses have been a classic example of long-term macroevolution (e.g., Simpson

1951, 1953; MacFadden 1992). In this section, fossil horses will be presented as a case example of patterns of morphological evolution, extinction, and possible climatic influences, particularly during the late Cenozoic, that is, the past 25 million years, within the Family Equidae (Fig. 13.6).

Like other early Cenozoic terrestrial mammalian herbivores, Eocene through early Miocene horses are characterized by brachydont (short-crowned) dentitions. During the middle Miocene, approx 20 million years ago, horses underwent a dramatic morphological evolution of their dentitions (MacFadden and Hulbert 1988), resulting in a rapid diversification of hypsodont (high-crowned) taxa. This Great Transformation has been interpreted to represent the advent of grass-dominated ecosystems and the corresponding onset of grazing by horses and other herbivores (Simpson 1951; Webb 1977; MacFadden 1992). Given that the carbon isotopic composition of plant foods are incorporated into the enamel mineral lattice of teeth, Wang, Cerling, and MacFadden (1994) investigated the carbon isotopic composition of fossil horse teeth to determine if there was an isotopic shift corresponding to a dietary shift from C_3 to C_4 plant foods during the Great Transformation.

Interestingly, the results were somewhat unexpected in terms of the original hypothesis to be tested. Wang, Cerling, and MacFadden (1994) found that although the increase in hypsodonty occurs during the middle Miocene, starting at approx 20 million years ago, the isotopic shift from C_3 to C_4 plant foods did not occur until the late Miocene, approx 7 million years ago (Fig. 13.7).

Several subsequent studies have confirmed that the carbon isotopic shift in diets from C_3 to C_4 plants recorded in extinct terrestrial grazing-mammal teeth also did not occur until the late Miocene in South America, Africa, and Eurasia (Cerling 1999; Cerling et al. 1997, 1998b), thus indicating a global phenomenon (see Fig. 13.2). This observation has direct relevance here because this global carbon shift preserved in fossil grazing-mammal teeth is a response to changes in grass photosynthesis, that is, predominantly C3 prior to approx 7 million years ago to predominantly C_4 thereafter (in temperate and tropical grassland biomes). Although other factors (such as increased aridity) may have been involved, this change in photosynthesis seems related to declining atmospheric CO_2 to levels less than approx 500 ppm—the threshold below which C_4 photosynthesis becomes advantageous as a physiological adaptation (Ehleringer et al. 1991; Ehleringer, Cerling, and Helliker 1997).

The macroevolution of fossil horses during the browsing/grazing transition was part of the widespread change in overall standing diversity of terrestrial mammalian herbivores (Janis, Damuth, and Theodor 2000). During the middle Miocene, this terrestrial herbivore diversity reached its peak in the Clarendonian Chronofauna (Webb 1977; 1984) between approx 15 million to 12 million years ago. This time is characterized by high-productivity terrestrial biomes that supported a rich ancient biodiversity, including both browsing and grazing herbivore taxa. For example, a dozen genera, many of which were sympatric, of fossil horses lived in North America during this time. Beginning 12 million years ago,

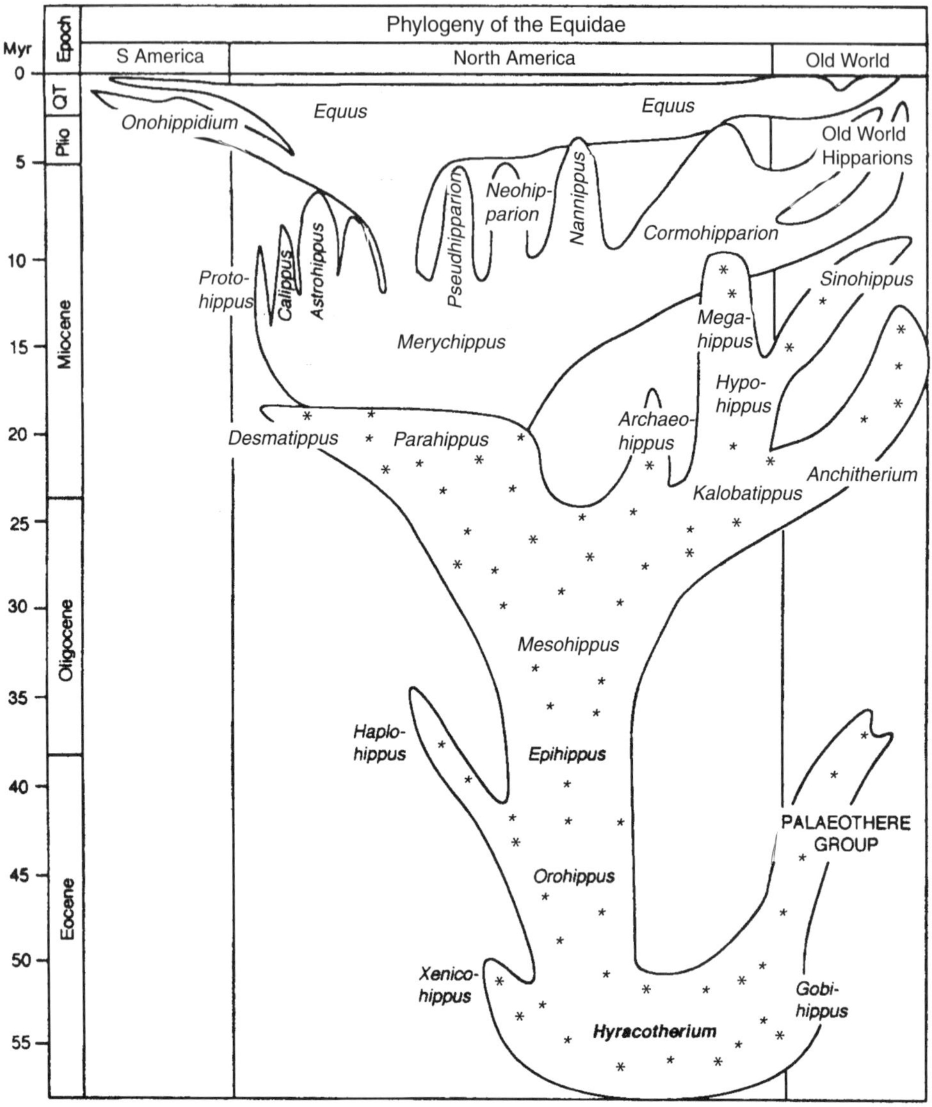

Figure 13.6. Phylogeny of the Equidae, with emphasis on the North American radiation. Principally browsing taxa are represented by the region of the phylogeny between 57 and 20 million years ago with "*" symbols. Modified from MacFadden (1992) and reproduced with permission of Cambridge University Press.

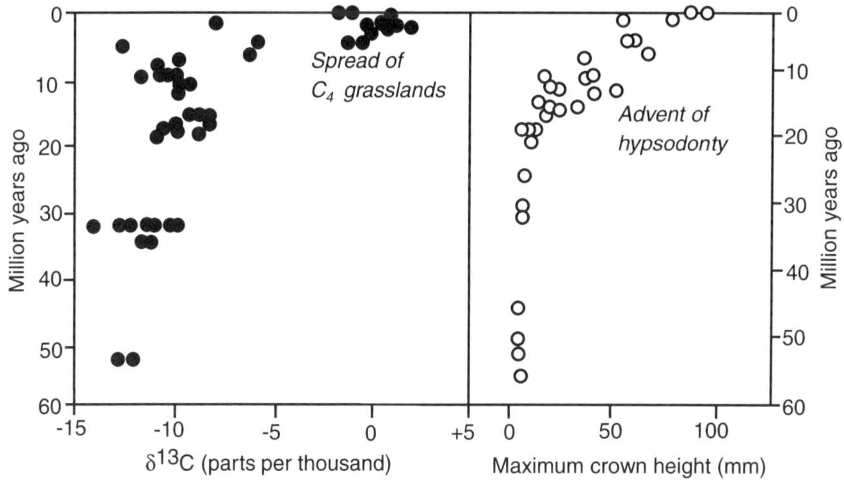

Figure 13.7. The shift in tooth enamel carbonate $\delta^{13}C$ (left) versus increased crown height (right) in Cenozoic horses from North America. From Wang, Cerling and MacFadden (1994) and reproduced with permission of Elsevier.

terrestrial mammalian herbivore diversity has declined, so that today it is the lowest it has been during the late Cenozoic. During this interval, the decrease in herbivore diversity occurred within different feeding guilds and apparently for different reasons.

The relatively high diversity of browsing taxa during the early and middle Cenozoic correlates with widespread and abundant forested and woodland communities, providing extensive browsing plant food resources (Webb 1977; Janis, Damuth, and Theodor 2000). By the middle Miocene, forested and woodland biomes decreased as grasslands expanded. Correspondingly, the number of browsing mammals declined and grazers increased in their diversity (Webb 1977), although Janis, Damuth, and Theodor (2000, 2002) assert that these changes were unrelated. Thereafter, overall standing herbivore diversity declined steadily, and this is even seen within the grazing mammals such as horses (Fig. 13.8) and therefore cannot be simply explained by the decline of browsers.

The interpretation for this decline in the diversity of grazing herbivore has been a global deterioration of climate, including increased aridity, higher seasonality, and general global cooling resulting in lower productivity ecosystems supporting lower biodiversity (Webb 1977; MacFadden 1997; Janis, Damuth, and Theodor 2000). While these previously recognized climatic parameters undoubtedly affected the diversity of grazing mammals during the late Cenozoic, the decrease in atmospheric CO_2 also undoubtedly played a part in this equation, as will be discussed in the next section.

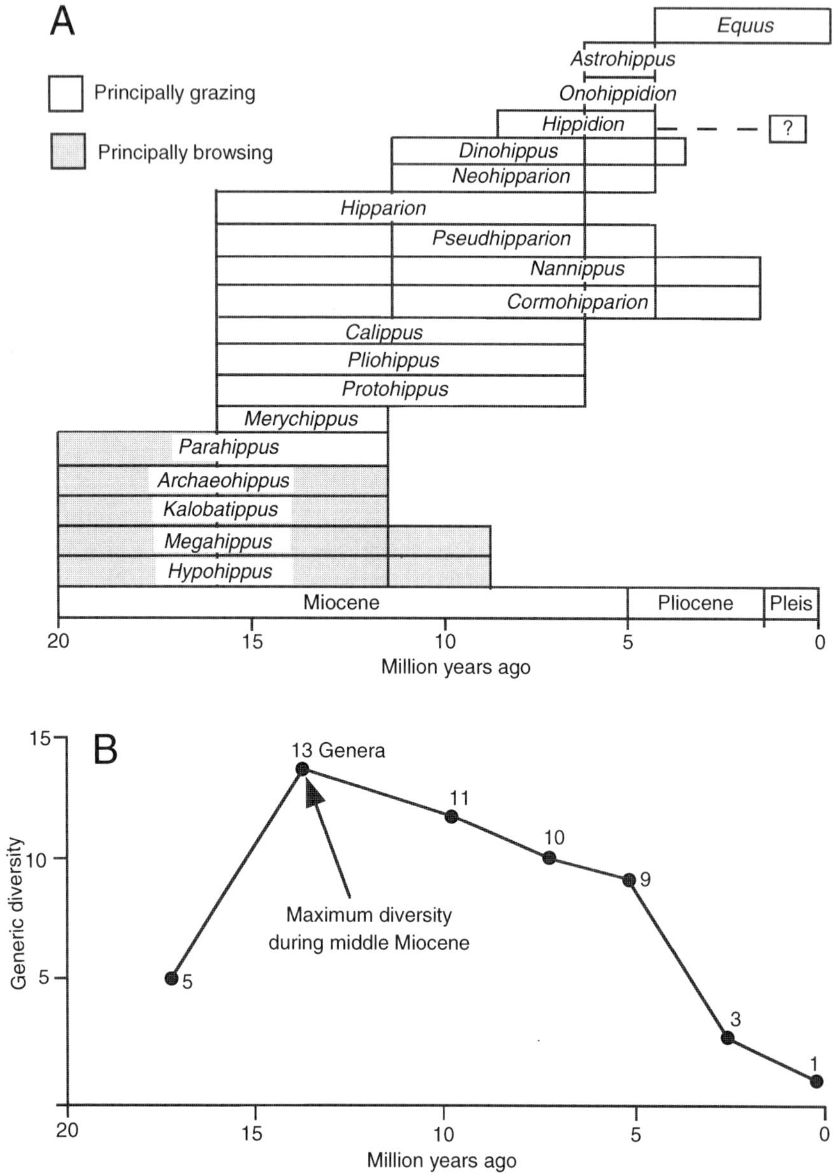

Figure 13.8. Decline in horse diversity since the middle Miocene, approx 15 Ma ago. **(A)** Time ranges of principally browsing and grazing genera. **(B)** Plot of generic diversity. Modified from MacFadden (1992) and reproduced with permission of Cambridge University Press.

13.5 The Present Is the Key to the Past

The proximate result of the decrease in atmospheric CO_2 during the late Miocene to levels below approx 500 ppm (the C_3/C_4 adaptive crossover) was the proliferation of C_4 grasses at the expense of C_3 grasses in temperate and tropical regions. Using modern analogs, we might ask what are the differences between C_3 versus C_4 grasses as food resources for grazers? The following potential factors are relevant in this context:

1. Lower ecosystem productivity. Studies have demonstrated that many C_4 grassland ecosystems generally have lower productivity than do corresponding C_3 grasslands. This does not result from an inherent lowered productivity of the grass species, per se, but rather because they exist in climates that support lower productivity. Along with this, lower productivity ecosystems support a lower biomass and lower biodiversity (Dyer et al. 1982; Lauenroth 1979; McNaughton et al. 1989), including the consuming herbivores. While this may be a general rule for grasslands, an exception is the higher mammalian herbivore diversity seen in some C_4 grasslands, for example, African savannas.
2. Lower nutritive value. C_4 grasses have a lower percentage of digestible plant parts than do C_3 grasses: for example, a greater percentage of bundle sheaths (Heckathorn, McNaughton, and Coleman 1999). Herbivore species feeding on this low-nutritive, low-digestibility food resource will "need" to evolve adaptive strategies to extract more nutrition from their diet per unit volume. This could be done by one or more ways, including higher digestibility and/or increased intake of plant foods.
3. Greater plant defenses. Plants potentially have the following dominant antiherbivory adaptations: physical, that is, the presence of abrasive materials (phytoliths) and or other structures, such as thorns; and chemical, that is, toxic or unpalatable compounds within their leaves. Grasses, characteristically have neither secondary chemical compounds (Heckathorn, McNaughton, and Coleman 1999) nor thorns. The primary antiherbivory mechanism in grasses, therefore, consists of internal silica phytoliths. Phytoliths vary significantly between C_3 and C_4 grasses (Piperno 1988; Twiss 1992; Piperno and Pearsall 1998). These differences may have affected the consuming herbivores in two ways: C_4 grasses have a greater phytolith content than do C_3 grasses; and generally C_4 grasses have more elongated (panicoid) phytoliths, whereas in C_3 grasses these structures are more rounded or equidimensional (e.g., pooid). These differences potentially affected evolution because C_4 grass foodstuffs would therefore have accelerated tooth wear relative to C_3 grasses. In the context of late Cenozoic mammalian herbivores, this antiherbivory adaptation of C_4 grasses that spread after approx 7 million years ago in terrestrial ecosystems could have affected grazing mammals by increasing the rate of tooth wear. This potentially would have "upped the ante" in the

coevolutionary arms race between the plant foods and the mammalian grazing species that consumed the foods.

13.6 How Were the Cenozoic Mammalian Herbivores Affected?

Decreased atmospheric CO_2 and the corresponding increase in C_4 plants resulted in lower productivity ecosystems, lower digestibility food resources available to terrestrial mammalian herbivores, and significantly increased rates of tooth wear. These parameters can now be examined in a context of patterns of mammalian herbivore morphology and macroevolution observed from the fossil record over the past 20 million years, that is, the part of the Cenozoic that seems to have been most affected by changes in threshold levels of atmospheric CO_2.

Lower productivity grassland biomes produce less biomass (Lauenroth 1979; Dyer et al. 1982; McNaughton et al. 1989) and consequently can support less overall species diversity up the food chain. It has been clearly demonstrated that after the peak diversity of the Clarendonian Chronofauna in North America (Webb 1977, 1984; Janis, Damuth, and Theodor 2000), overall herbivore diversity dropped steadily from the Miocene through the late Pleistocene. This is evidenced by the fact by approx 2 million years ago, terrestrial herbivore diversity in North America was about one-third of what it previously had been at 15 million years ago (Janis, Damuth, and Theodor 2000). The potential explanation for this decline in diversity is complex and involves several different phases. The first phase involved the reduction of browsing herbivore diversity during the late middle Miocene (~15–12 Ma) when the forested and woodland biomes decreased as grasslands spread. Thereafter, however, herbivore diversity continued to decline, as is exemplified by Equidae (see Fig. 13.8). This drop in diversity has mostly been explained by a global climate change resulting in increased aridity, seasonality, and cooling in lower productivity grassland ecosystems (Webb 1977, 1984; Webb and Opdyke 1995; Janis 1993). In the context of this discussion, decline in atmospheric CO_2 also potentially contributed to this drop in diversity.

Undoubtedly, the quality of food resources eaten by mammalian herbivores affected evolution. Species that transition from higher- to lower-quality plant foods require a different kind of digestive physiology. Feeding on lower-quality forage tends to be correlated to increased retention time in the gut, which yields more time for extracting nutrition from the plant foodstuffs. This is seen in larger perissodactyls, such as horses and grazing rhinos, and large ruminant artiodactyls, for example, bison (Janis 1976). Retention time of food in the gut of mammalian herbivores is proportional to body size, so it has been suggested that one of the contributing factors for the increased body size (Fig. 13.9) seen in some clades (e.g., horses)was the change to lower digestibility plant foods, such as would be seen by a change from feeding on C_3 to C_4 grasses (McNaughton 1991; Owen-Smith 1988; Janis, Damuth, and Theodor 2000).

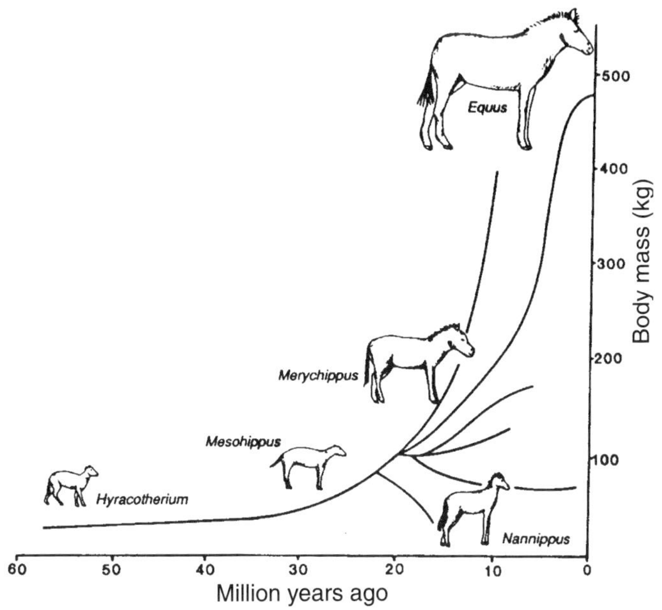

Figure 13.9. Body size evolution in horses. The first half of equid evolution is characterized by relative stasis, whereas the second half is characterized by diversification of body size since the late Miocene. The clade with *Equus* has included some of the largest extinct species, some with estimated body masses between 450 and 500 kg. From MacFadden (1992) and reproduced with permission of Cambridge University Press.

How does the evolution of body size in horses and other terrestrial grazers potentially relate to changes in atmospheric CO_2? At first glance this may seem like a stretch, but actually, an indirect mechanistic relationship can be argued, particularly for those examples of increased body size. Previous explanations for increased body size in mammalian clades have included such factors as more adaptive physiology (Eisenberg 1981), increased energy input into species (Brown 1995), more efficient defense against predators in the coevolutionary arms race (Janis 1993), and/or larger home ranges (McNab 1963) in which to forage for seasonally available plant food resources (Owen-Smith 1988). To these can be added the factor that C_4 grasses, with their lower digestibility and lower per unit volume nutritive value, could potentially favor increased body size for longer retention times in the gut to increase digestive efficiency (Owen-Smith 1988; Janis, Gordon, and Illius 1994).

Grazers that feed on more abrasive plant foods with increased phytolith content will concomitantly increase tooth wear rates. It therefore follows that those late Miocene grazing species that fed on C_4 grasses after 7 million years ago that had relatively more hypsodont teeth would be adaptively favored. Mac-

Fadden, Solounias, and Cerling (1999) studied six species of 5-million-year-old sympatric horses from the latest Miocene (late Hemphillian) of Florida. This diversity was just before a major extinction event at the Hemphillian/Blancan boundary at 4.5 million years ago. Of the six species clades that coexisted and divided up the available plant food resources prior to the end of the Hemphillian, only three survived after 4.5 million years ago. The three surviving Pilocene (Blacan) clades demonstrate a range in adaptive characteristics, most notably body size and relative tooth crown height (hypsodonty index [HI], higher index indicates increased relative crown height), as follows:

1. The "dwarf" *Nannippus* (species *aztecus* to *peninsulatus*), with body masses estimated to be approx 65 to 75 kg and HIs of approx 2.5 to 3.
2. *Cormohipparion* (single late-surviving species *emsliei*), with body mass estimated to be approx 105 kg and a relatively low HI of approx 2.
3. *Dinohippus/Equus* (*D. mexicanus* to *E. simplicidens*, to several later *Equus* species) with body masses ranging from approx 275 to 450 kg and HIs from 2.3 to >>3 (MacFadden 1987; MacFadden, Solounias, and Cerling 1999).

Of the three equid clades that persisted into the Pliocene and earliest Pleistocene, only one, *Equus*, continues through the late Pleistocene after approx 1.5 million years ago when it further diversifies into several apparently distinct species and is exceedingly abundant throughout North America, South America, Eurasia, and Africa. During the late Pleistocene *Equus* becomes geographically restricted to the Eurasian steppe after about 10,000 years ago, when this genus became extinct elsewhere throughout its previous range.

The foregoing discussion predicts that the equid clade with increased body size and relatively more hypsodont teeth would be adaptively favored in a regime of more abrasive and less nutritious C_4 grasses. This is the pattern that is observed: *Equus* was the latest surviving and most widespread of the late Cenozoic equid clades observed in the fossil record. Its adaptability stems not only from some critical adaptations that indirectly result from changes in atmospheric CO_2 and the spread of C_4 grasses, but also from the fact that *Equus* has been a generalist and can shift to a more mixed diet when other hypergrazers invade, or are dominant in, a local grassland community: for example, bison (*Bison bison*) in North America (MacFadden and Cerling 1996; Feranec and MacFadden 2000) or wildebeest (*Connochaetes taurinus*) in Africa (Owen-Smith 1988).

13.7 Summary

Changes in atmospheric CO_2 potentially had a profound, albeit indirect, effect on the evolution of terrestrial mammalian grazers during the late Cenozoic. As evidenced from fossil horses (Family Equidae), the larger-bodied and higher-crowned clade that included *Dinohippus* and *Equus* persisted through the series of extinctions that occurred since the middle Miocene. This clade survival is consistent with an adaptive strategy for feeding on more abrasive and less nutritive C_4 grasses that spread in tropical and temperate regions after the late Miocene global carbon shift approx 7 million years ago.

This observation can be further tested through future studies of molar wear rates in clades such as *Dinohippus/Equus* that fed on C_4 grasses. The prediction is that relative to other clades that fed mostly on relatively less abrasive C_3 grasses, those feeding on C_4 grasses would demonstrate increased molar wear rates. Although such a study has not yet been done with this goal in mind, it is theoretically possible given the available specimens from the fossil record.

Earth underwent a profound global change during the late Cenozoic, including the advent of the glacial stages over the past 5 million years. This "climatic deterioration" has traditionally been explained by increased aridity, seasonality, and decreased mean annual temperature, all of which undoubtedly affected the global floras and faunas. To this array of physical and climatic determinants now can be added decreased atmospheric CO_2 to a level below which C_4 photosynthesis in grasses was favored as an adaptive strategy; the change in atmospheric CO_2 ultimately affected the terrestrial mammalian grazers of the world, as exemplified by late Cenozoic horses from North America.

Acknowledgments. This research was supported by NSF grant EAR 9901896. I thank J. Ehleringer for his encouragement in submitting this paper and the two reviewers for their comments. This is University of Florida Contribution to Paleobiology 544.

References

Behrensmeyer, A.K., J.D. Damuth, W.D. DiMichele, R. Potts, H.-D. Sues, and S.L. Wing, eds. 1992. *Terrestrial ecosystems through time: Evolutionary paleoecology of terrestrial plants and animals.* Chicago: University of Chicago Press.

Berner, R.A. 1991. A model for atmospheric CO_2 over Phanerozoic time. *American Journal of Science* 291:339–76.

———. 1994. GEOCARB II: A revised model of atmospheric CO_2 over Phanerozoic time. *American Journal of Science* 294:56–91.

Boucot, A.J., and J. Gray. 2001. A critique of Phanerozoic climatic models involving changes in the CO_2 content of the atmosphere. *Earth-Science Reviews* 56:1–159.

Brown, J.H. 1995. *Macroecology.* Chicago: University of Chicago Press.

Cerling, T.E. 1991. Carbon dioxide in the atmosphere: Evidence from Cenozoic and Mesozoic paleosols. *American Journal of Science* 291:377–400.

———. 1999. Paleorecords of C_4 plants and ecosystems. In *C_4 plant biology,* ed. R.F. Sage and R.K. Monson, 445–69. San Diego: Academic Press.

Cerling, T.E., J.R. Ehleringer, and J.M. Harris. 1998. Carbon dioxide starvation, the development of C_4 ecosystems, and mammalian evolution. *Philosophical transactions Royal Society London. B.* 353:159–71.

Cerling, T.E., J.M. Harris, B.J. MacFadden, M.G. Leakey, J. Quade, V. Eisenmann, and J.R. Ehleringer. 1997. Global vegetation cahnge through the Miocene/Pliocene boundary. *Nature* 389:153–58.

———. 1998. Carbon isotopes, diets of North American equids, and the evolution of North American grasslands. In *Stable isotopes and the integration of biological, ecological, and geochemical processes,* ed. H. Griffiths, D. Robinson, and P. Van Gardingen, 363–79. Oxford: BIOS Scientific Publishers.

Cerling, T.E., Y. Wang, and J. Quade. 1993. Global ecological change in the late Miocene: Expansion of C_4 ecosystems. *Nature* 361:344–45.
Crepet, W.L., and G.D. Feldman. 1991. The earliest remains of grasses in the fossil record. *American Journal of Botany* 78:1010–14.
Dugas, D.P. and G.J. Retallack. 1993. Middle Miocene fossil grasses from Fort Ternan, Kenya. *Journal of Paleontology* 76:113–28.
Dyer, M.I., J.K. Detling, D.C. Coleman, and D.W. Hilbert. 1982. The role of herbivores in grasslands. In *Grasses and grasslands: Systematics and ecology,* ed. J.R. Estes, R.J. Tyrl, and J.N. Brunken, 253–95. Norman: University of Oklahoma Press.
Ehleringer, J.R., T.E. Cerling, and B.R. Helliker. 1997. C_4 photosynthesis, atmospheric CO_2, and climate. *Oecologia* 112:285–89.
Ehleringer, J.R., R.F. Sage, L.B. Flanagan, and R.W. Percy. 1991. Climate change and the evolution of C_4 photosynthesis. *Trends in Ecology and Evolution* 6:95–99.
Eisenberg, J.F. 1981. *The mammalian radiations: An analysis of trends in evolution, adaptation, and behavior.* Chicago: University of Chicago Press.
Ekart, D.D., T.E. Cerling, I.P. Montanez, and N.J. Tabor. 1999. A 400 million year carbon isotope record of pedogenic carbonate: Implications for paleoatmospheric carbon dioxide. *American Journal of Science* 10:805–27.
Feranec, R.S., and B.J. MacFadden. 2000. Evolution of the grazing niche in Pleistocene mammals: Evidence from stable isotopes. *Palaeogeography, Palaeoclimatology, Palaeoecology* 162:155–69.
Fortelius, M., and N. Solounias. 2000. Functional characterization of ungulate molars using the abrasion-attrition wear gradient: A new method for reconstructing paleodiets. *American Museum Novitates* 3301:1–36.
Gabel, M.L., D.C. Backlund, and J. Hafner. 1998. The Miocene macroflora of the northern Ogallala Group, northern Nebraska and southern South Dakota. Journal of Paleontology 72:388–97.
Heckathorn, S.A., S.J. McNaughton, and J.S. Coleman. 1999. C_4 plants and herbivory. In *C_4 plant biology,* ed. R.F. Sage and R.K. Monson, 285–312. San Diego: Academic Press.
Jacobs, B.F., J.D. Kingston, and L.L. Jacobs. 1999. The origin of grass-dominated ecosystems. *Annals Missouri Botanical Garden* 86:590–643.
Janis, C.M. 1976. The evolutionary strategy of the Equidae and the origins of rumen and caecal digestion. *Evolution* 30:757–74.
———. 1993. Tertiary mammal evolution in the context of changing climates, vegetation, and tectonic events. *Annual Review of Ecology and Systematics* 24:467–500.
———. 2000. Patterns in the evolution of herbivory in large terrestrial mammals: The Paleogene of North America. In *Evolution of herbivory in terrestrial vertebrates,* ed. H.-D. Sues, 168–222. New York: Cambridge University Press.
Janis, C.M., J. Damuth, and J.M. Theodor. 2000. Miocene ungulates and terrestrial primary productivity: Where have all the browsers gone? *Proceedings National Academy of Science* 97:7899–7904.
———. 2002. The origins and evolution of the North American grassland biome: The story from hoofed mammals. *Palaeogeography, Palaeoclimatology, Palaeoecology* 177:183–98.
Janis, C.M., and D. Ehrhardt. 1988. Correlation of relative muzzle width and relative incisor width with dietary preference in ungulates. *Zoological Journal Linnean Society* 92:267–84.
Janis, C.M., and M. Fortelius. 1988. On the means whereby mammals achieve increased functional durability of their dentitions, with special reference to limiting factors. *Biological Reviews* 63:197–230.
Janis, C.M., I.J. Gordon, and A.W. Illius. 1994. Modelling equid/ruminant competition in the fossil record. *Historical Biology* 8:15–29.
Kellogg, E.A. 1999. Phylogenetic aspects of the evolution of C_4 photosynthesis. In

C_4 plant biology, ed. R.F. Sage and R.K. Monson, 411–44. San Diego: Academic Press.
———. 2000. The grasses: A case study in macroevolution. *Annual Review of Ecology and Systematics* 31:217–38.
Kowalevsky, V. 1873. Sur l'Anchitherium aurelianense Cuv. Et sur l'histoire paléontologique des chevaux. (Prèmiere Partie.) Mémoire Academie Imperiale Scientifique St. Pétersbourg, Série 7, 20(5):1–73.
Lauenroth, W.K. 1979. Grassland primary production: North American grasslands. In *Perspectives in grassland ecology,* ed. N. French, 3–24. New York: Springer-Verlag.
Linder, H.P. 1987. The evolutionary history of the Poales/Restionales—a hypothesis. *Kew Bulletin* 42:297–318.
MacFadden, B.J. 1987. Fossil horses from "Eohippus" (*Hyracotherium*) to *Equus*: Scaling, Cope's law, and the evolution of body size. *Paleobiology* 12:355–69.
———. 1992. *Fossil horses: Systematics, paleobiology, and evolution of the family equidae.* New York: Cambridge University Press.
———. 1997. Origin and evolution of the grazing guild in New World terrestrial mammals. *Trends in Ecology and Evolution* 12:182–87.
MacFadden, B.J., and T.E. Cerling. 1994. Fossil horses, carbon isotopes and global change. *Trends in Ecology and Evolution* 12:481–86.
MacFadden, B.J., T.E. Cerling, J.M. Harris, and J. Prado. 1999. Ancient latitudinal gradients of C_3/C_4 grasses interpreted from stable isotopes of New World Pleistocene horse (*Equus*) teeth. *Global Ecology and Biogeography* 8:137–49.
MacFadden, B.J. and R.C. Hulbert, Jr. 1988. Explosive speciation at the base of the adaptive radiation of Miocene grazing horses. *Nature* 336:466–68.
MacFadden, B.J., T.E. Cerling, and J. Prado. 1996. Cenozoic terrestrial ecosystem evolution in Argentina: Evidence from carbon isotopes of fossil mammals. *Palaios* 11: 319–27.
MacFadden, B.J., N. Solounias, and T.E. Cerling. 1999. Ancient diets, ecology, and extinction of 5-million-year-old horses from Florida. *Science* 283:824–27.
MacGinitie, H.D. 1962. The Kilgore flora: A late Miocene flora from northern Nebraska. *University of California Publications Geological Sciences* 35:67–158.
McNab, B.K. 1963. Bioenergetics and the determination of home range size. *American Naturalist* 97:133–40.
McNaughton, S.J. 1991. Evolutionary ecology of large tropical herbivores. In *Plant-animal interactions: Evolutionary ecology in tropical and temperate regions,* ed. P.W. Price, T.M. Lewisohn, G. Wilson Fernandez, and W.W. Benson, 509–22. New York: Wiley.
McNaughton, S.J., M. Oesterheld, D.A. Frank, and K.J. Williams. 1989. Ecosystem-level patterns of primary productivity and herbivory in terrestrial habitats. *Nature* 341:142–44.
McNaughton, S.J., J.L. Tarrants, M.M. McNaughton, and R.H. Davis. 1985. Silica as a defense against herbivory and a growth promoter in African grasses. *Ecology* 66:528–35.
Muller, J. 1981. Fossil pollen records of extant angiosperms. *Botanical Review* 45:1–145.
Owen-Smith, N. 1988. *Megaherbivores: The influence of very large body size on ecology.* Cambridge: Cambridge University Press.
Pagani, M., M.A. Arthur, and K.H. Freeman. 1999. Miocene evolution of atmospheric carbon dioxide. *Paleooceanography* 14:273–92.
Pagani, M., K.H. Freeman, and M.A. Arthur. 1999. Late Miocene atmospheric CO_2 concentrations and the expansion of C_4 grasses. *Science* 285:876–69.
Patterson, B., and R. Pascual. 1972. The fossil mammal fauna of South America. In *Evolution, mammals, and southern continents,* ed. A. Keast, F.C. Erk, and B. Glass, 247–309. Albany: State University of New York Press.

Pearson, P.N., and M.R. Palmer. 2000. Atmospheric carbon dioxide concentrations over the past 60 million years. *Nature* 406:695–99.
Piperno, D. 1988. *Phytolith analysis: An archaeological and geological perspective.* San Diego: Academic Press.
Piperno, D., and D.H. Pearsall. 1998. The silica bodies of tropical American grasses: Morphology, taxonomy, and implications for the grass systematics and fossil phytolith identification. *Smithsonian Contributions to Botany* 85:40.
Prothero, D.R. 1994. *The Eocene-Oligocene transition: Paradise lost.* New York: Columbia University Press.
Raymo, M.E. 1994. The initiation of northern hemisphere glaciation. *Annual Review Earth Planetary Science Letters* 22:353–83.
Root, R.B. 1967. The niche exploitation pattern of the blue-gray gnatcatcher. *Ecological Monographs* 37:317–50.
Shantz, H.L. 1954. The place of grasslands in the earth's cover of vegetation. *Ecology* 35:142–45.
Shockey, B.J. 1997. Two new notoungulates (family Notohippidae) from the Salla Beds of Bolivia (Deseadan: late Oligocene): Systematics and functional morphology. *Journal of Vertebrate Paleontology* 17:584–99.
Simpson, G.G. 1951. *Horses: The story of the horse family in the modern world and through sixty million years of history.* Oxford: Oxford University Press.
———. 1953. *The major features of evolution.* New York: Columbia University Press.
Solounias, N., and L.-A. Hayek. 1993. New methods of tooth microwear analysis and application to dietary determination of two extinct antelopes. *Journal of Zoology* 229: 421–45.
Solounias, N., and S.M.C. Moelleken. 1993. Dietary adaptation of some extinct ruminants determined by premaxillary shape. *Journal of Mammalogy* 74:1059–71.
Solounias, N., and G. Semprebon. 2001. Advances in the reconstruction of ungulate ecomorphology with application to early fossil equids. *American Museum of Natural History Novitates* 3366:1–31.
Stebbins, H.L. 1981. Coevolution of grasses and herbivores. *Annals Missouri Botanical Garden* 68:75–86.
Thomasson, J.R., M.E. Nelson, and R.J. Zakrzewski. 1986. A fossil grass (Gramineae: Chloridoideae) from the Miocene with Krantz anatomy. *Science* 233:876–78.
Tidwell, W.D., and E.V.M. Nambudiri. 1989. *Tomlinsonia thomassonii*, gen. et sp. nov., a permineralized grass from the Upper Miocene Ricardo Formation, California. *Reviews Paleobotany Palynology* 60:165–77.
Twiss, P. 1992. Predicted world distribution of C_3 and C_4 grass phytoliths. In *Phytolith systematics,* ed. G. Rapp, Jr., and S.C. Mulholland, 113–28. New York: Plenum Press.
Walker, E.P. 1975. *Mammals of the World,* 3d ed. Baltimore: Johns Hopkins University Press.
Wallmann, K. 2001. Controls on the Cretaceous and Cenozoic evolution of seawater composition, atmospheric CO_2 and climate. *Geochimica et Cosmochimica Acta* 65: 3005–25.
Wang, Y., T.E. Cerling, and B.J. MacFadden. 1994. Fossil horses and carbon isotopes: new evidence for Cenozoic dietary, habitat, and ecosystem changes in North America. *Palaeogeography, Palaeoclimatology, Palaeoecology* 107:269–79.
Webb, S.D. 1977. A history of savanna vertebrates in the New World. Part I. North America. *Annual Review of Ecology and Systematics* 8:355–80.
———. 1984. Ten million years of mammal extinctions in North America. In *Quaternary extinction: A prehistoric revolution,* ed. P.S. Martin and R.G. Klein, 189–210. Tucson: University of Arizona Press.
Webb, S.D. and N.D. Opdyke. 1995. Global climatic influence on Cenozoic land mammal faunas. In *Effects of past global change on life,* 184–208. Washington: National Academy Press.

14. CO_2, Grasses, and Human Evolution

Nicholaas J. van der Merwe

14.1 Introduction

Hominins (i.e., humans and our ancestors) have evolved over the past 7 million years in a world with relatively low concentrations of CO_2 in the atmosphere. During recent geological time, the concentration of CO_2 in the atmosphere dropped steadily: for example, in the Jurassic, concentrations were several thousands of ppmV (parts per million by volume), then, in the late Cretaceous, they were about 1250 ppmV. By 8 million years ago, the concentrations of CO_2 may have dropped below 500 ppmV for the first time. These conditions favored the productivity of C_4 plants, which evolved during the past 20 million years in response to declining CO_2 concentrations. In contrast, C_3 plants grow best at higher CO_2 levels. The interplay between CO_2 levels and the productivity of C_3 versus C_4 plants had profound effects during different stages of human evolution. In the current and future stages of human evolution, CO_2 concentrations may rise to levels not seen since the early Miocene; the possible effects of CO_2 concentrations are of interest to scientists and other observers worldwide.

Most (but not all) C_4 plants are subtropical grasses. The emphasis in this chapter will be on the importance of C_4 and C_3 grasses in human evolution, but attention will also be paid to the role of other plants. Humans are members of the genus *Homo* and their evolution includes biological, cultural, and social elements. For the purposes of this chapter, cultural evolution refers primarily to technology: how humans went about making a living and feeding themselves.

Social evolution refers to the way that groups of humans have organied their societies.

Between 8 million and 7 million years ago, C_4 plants expanded rapidly around the world. The earliest hominins evolved in Africa between 7 million and 6 million years ago, when C_4 grassland and woodland savannas were already widespread in the continent and around the world. By 2 million years ago, when the first representatives of the genus *Homo* emerged, this "C_4 world" was well established, and all hominin species were extensively engaged with it. Hominins ate plants that were for the most part probably of the C_3 photosynthetic type; their dietary involvement with C_4 plants probably involved not only the plants, but also the insects and animals that ate C_4 plants.

At a much later stage of human evolution, well after the emergence of modern *Homo sapiens*, humans developed the technology to cultivate plants and proceeded to domesticate an array of C_3 and C_4 plants. It is of particular interest, and almost counterintuitive, that the first plants to be domesticated were of the C_3 type and that this event occurred shortly after the Last Glacial Maximum, when atmospheric CO_2 concentration had been at an all-time low of 180 ppmV. The domestication of C_3 grasses, such as rye, wheat, and barley, was completed in the Near East by 11,000 years ago and resulted from the fortuitous confluence of a small but critical rise in the CO_2 level above 200 ppmV (due to the melting of continental ice sheets), warmer climate, and an inventive population of humans at the cultural crossroads of the world at that time. The domestication of C_4 grasses followed soon thereafter, with two species of millet leading the way in northern China by about 9000 years before the present. Maize, another C_4 plant, was domesticated by about 7000 years before the present in Mexico, followed by a suite of other C_4 grasses in Africa (sorghum, millets, fonio, and tef), perhaps after 6000 years before the present. Other important domesticates that became staple food items included rice, a C_3 plant, which was domesticated in the tropical Yangtze River area of China at about the same time as millets but became an important crop of northern Chinese civilization only after the development of complex irrigation systems. Elsewhere in the world, C_3 tubers were staple crops of early agricultural societies; these included cassava (tropical South America), potatoes (the Andes), and yams (Central America and West Africa), among others.

The development of agriculture, wherever it occurred, had profound effects on the size of human populations, settlement patterns, and social organization. In several parts of the world, it led to the development of urban communities with social and political stratification. The civilizations of the Middle East and Egypt were built on C_3 cultigens, while those of China, Africa outside Egypt, and the Americas were built primarily on C_4 grains. In the past 1000 years, the dispersal around the world of two C_4 crops, maize and sugar cane, produced major economic and societal changes; they included the expansion of empires and the dislocation of millions of people through the widespread practice of slavery. These events are examined below (for a more expansive version, particularly with reference to the role of C_4 plants in social evolution, see van der Merwe and Tschauner (1999).

14.2 C_4 Plants and Hominin Evolution

Modern apes and humans evolved from Miocene apes that lived in forests and ate primarily forest plants, including leaves, fruits, seeds, and tubers. In the course of their evolution, hominins became bipedal, evolved larger brains, and developed the dexterity and cultural knowledge to produce an array of artifacts. The latter are assumed to have been used primarily for the purpose of acquiring and processing food, some of which was meat. These different elements did not evolve at the same rate (bipedality preceded larger brains by more than 2 million years), but together they characterize the genus *Homo*. Larger brain size is of particular importance in this context, since it provided the capacity for the evolution of human culture. In turn, brain size is related to diet: for example, the "expensive tissue hypothesis" of Aiello and Wheeler (1995) holds that the development of a larger brain would have required a larger proportion of high-nutrient foods, including meat, since, in their view, the gut became smaller as the brain grew larger. In the process, hominins developed larger teeth and thicker enamel to deal with lipid-rich seeds as well as gastrointestinal systems that could metabolize lipid-rich and protein-dense foods (see Schoeninger et al. 2002).

Dietary hypotheses play an important role in understanding evolutionary scenarios pertaining to hominins. The central question is, How and when did tree-dwelling, apelike creatures with diets of forest plants evolve into bipedal savanna foragers who came to depend on meat for a substantial part of their diet? One approach that provides some answers to this question involves measurements of the ^{13}C contents of early hominin fossils, specifically their tooth enamel. Since C_4 plants have higher ^{13}C contents than do C_3 plants, the C_4 component in hominin diets can be determined. The question then becomes, To what extent is this C_4 dietary component supplied by C_4 plants, or by insects and mammals that ate the plants? ^{13}C measurements alone cannot answer this question; additional information is required from environmental reconstructions, plant fossils, tooth-wear studies, and artifacts.

The earliest fossil that has been described as a hominin was found in Chad and may be as old as 7 million years (Brunet et al. 2002), while fossil specimens of the genus *Homo* are known from East and South Africa, starting at around 2 million years ago. Whether one takes 7 million or 2 million years ago as the starting date for humanity, hominins have spent 99% of their time on Earth as foragers of plants and meat in their natural environment. Modern hunter-gatherers are adapted to a wide range of environments, from the semifrozen Arctic to tropical forests. The early hominins of Africa, however, had a choice of tropical forests, C_4 savannas, lakes, and river systems. The earliest hominins evolved fairly soon after the expansion of C_4 savannas, and it is tempting to see this evolution as a response to the changes in the environment. This notion has been called the "savanna hypothesis," and it has been placed under a critical spotlight in recent years. The basis of this criticism is some evidence, at this point not very well developed, that before about 3 million years ago, our ancestors may have lived in forested environments. It is nevertheless the case that hominins of 3 million years ago and later were fully engaged with C_4-based foods in the savannas of both East and South Africa, as shown by their ^{13}C contents. The validity of the savanna hypothesis really depends on *when*, rather than *if*.

The global expansion of C_4 plants has been demonstrated by measurements of ^{13}C contents in the tooth enamel of grazing animals (Cerling, Wang, and Quade 1993; Cerling et al. 1997b). This event took place at the Miocene/Pliocene boundary, between about 8 million and 5 million years ago, with some differences in the chronology from one area to another. In Kenya, C_4 plants were present in the diets of some grazing animals by about 15 million years ago (Morgan, Kingston, and Marino 1994), but their major expansion occurred between about 8 million and 7 million years ago (Cerling et al. 1997a; Cerling 1999). The timing of this event has not been established for South Africa, but we know that grazing animals of the plateau interior of this region had fully C_4 diets by 3 million years ago (Sponheimer and Lee-Thorp 1999; van der Merwe et al. 2003a).

The reduction of atmospheric CO_2 content has been suggested as the trigger for the global expansion of C_4 plants (Cerling et al. 1997b), with the critical threshold at about 500 ppmV (Ehleringer et al. 1991; Ehleringer, Cerling, and Helliker 1997; Ehleringer and Björkman 1997). The details of this hypothesis are the subject of debate in this volume. If we accept it as a working hypothesis, then atmospheric CO_2 concentration played an important role in the adaptive radiation(s) of the hominin line, whether at the very beginning or somewhat later in hominin evolution. This role may have been at the remove of several trophic levels, with hominins including the consumers of C_4 plants in their diets. They may have also eaten some seeds and rhizomes of grasses themselves (as baboons are known to do on a seasonal basis), as well as the starchy tubers of C_4 sedges.

All modern humans belong to a single species, *Homo sapiens*. This is an unusual situation, because the fossil record of our early ancestry includes a variety of species. A detailed account of these species is not appropriate in this chapter; a broad outline will suffice. During the past decade, hominin fossils older than 3 million years have been discovered in Chad (Brunet et al. 2002), Ethiopia (White, Suwa, and Asfaw 1994; White et al. 2000), Kenya (Leakey et al. 2001; Pickford and Senut 2001), and at Sterkfontein in South Africa (Clarke 1998). These fossils are in some cases fragmentary, and the taxonomic labels that have been attached to them may very well change. The East African specimens have been described as living in forested environments, although the evidence for this assertion is as yet incomplete. The South African specimen, known as "Little Foot," is represented by what may turn out to be a complete skeleton, but it is still being excavated from the rock-hard breccia. Its footbones have been described as appropriate to a treeclimber. These finds present us with an incomplete and tantalizing picture of a stage in human evolution when hominins were making the transition from forest to savanna habitats. At this point, nothing is known about their behavior or diet.

By about 3 million years ago, the picture becomes more detailed (see Asfaw et al. 1999). Small-brained, but fully bipedal hominins were present in the savannas of East and South Africa. These were the early australopithecines, represented in East Africa by, among others, *Australopithecus garhi* and *A. afarensis,* and in the south by *A. africanus.* By about 2.5 million years ago, *A. garhi* of Ethiopia produced stone tools to deflesh the bones of large mammals and cracked their long bones to extract the marrow (de Heinzelin et al. 1999). In South Africa, *A. africanus* also produced stone tools by this time, while the 13C content of its tooth enamel shows that it was extensively engaged with C_4-based foods (details below).

By 2 million years ago, representatives of the genus *Homo*, with increasingly larger brains, were established in East and South Africa, with various fossil specimens known as *Homo habilis*, *H. rudolfensis,* and (somewhat later) *H. ergaster*. At the same time, a more robust form of australopithecine had evolved, represented by *A. aethiopicus* and *A. boisei* in East Africa and *A. robustus* (also known as *Paranthropus robustus*) in South Africa. These robust australopithecines co-existed with members of the *Homo* lineage from about 2 million to 1 million years ago.

A. robustus has been described as a specialized herbivore, who used its oversized back molars to chew large quantities of plant foods (Robinson 1954). In contrast, members of the genus *Homo* are generally considered as omnivores. The idea that *A. robustus* was an herbivore has been reinforced by microscopic studies of its teeth, which show extensive pitting (Grine 1981; Grine and Kay 1988); these were ascribed to the chewing of hard nuts. This contrasting view of the diets of *A. robustus* and early species of *Homo* is contradicted, at least in part, by isotopic evidence.

Our knowledge of hominin diets at about 2 million years ago is limited. Most of the animal bones found in the same deposits as hominin remains came to be there through natural accumulation or the action of carnivores, for whom the hominins were also prey. The demonstrable exceptions are bones with cutmarks from stone tools, which have been found at several sites from this time period, in both East and South Africa. In addition, use-wear marks on bone tools from the South African sites of Swartkrans and Drimolen (ca. 2–1 Ma) have been shown to be the result of digging into termite mounds (Backwell and d'Errico 2000). Most (but not all) of the hominin fossils from these two sites are of the species *A. robustus*. This suggests the tantalizing possibility (but does not prove) that the robust australopithecines ate termites and that, unlike chimpanzees, they did not wait for the short annual season when termites make openings in their mounds for the departure of nuptial flights. Nevertheless, the logistics of gathering enough termites to serve as more than a snack was probably beyond the capabilities of a hominin. The aardwolf (*Proteles cristatus*), a specialized termite forager of about 15 kg body weight, eats an estimated 300,000 termites every night and requires a territory with about 3000 termite mounds to accomplish this task (Smithers 1983; Bothma 2001).

The carbon isotope ratios of herbivore tooth enamel can provide us with a clear indication of whether particular animals consumed C_3 plants (e.g., browsers like giraffe and bushbuck), C_4 plants (e.g., grazers like wildebeest), or were mixed feeders who consumed both (e.g., impala). Tooth enamel preserves the original biological ^{13}C signal of these animals, because, unlike bone, it is highly crystalline and resists chemical alteration in the burial environment (Lee-Thorp and van der Merwe 1987). The carbon isotope ratios of carnivores cannot tell us which species provided their favorite prey but can significantly narrow the range of possibilities. Lions, for example, usually have isotope signatures that resemble those of grazing animals, because they prefer medium to large grazers as prey. Leopards, by contrast, usually have isotope signatures that resemble those of browsers, because they hunt in woodland environments. This approach has been used to identify the prey possibilities of several carnivores at the hominin site of Swartkrans (Lee-Thorp, Thackeray, and van der Merwe 2000).

The carbon isotope ratios of the tooth enamel of potential omnivores, such as hominins, cannot provide us with a detailed menu of their dietary preferences, but it can provide important constraints to arguments regarding what omnivores might have eaten. To investigate the diets of fossil hominins, it is necessary to establish the C_3 and C_4 "end members" among herbivores that lived at the same time and place. These end members may vary over time and space, depending on the ^{13}C content of the atmosphere and on local environmental conditions (e.g., aridity and temperature) that cause slight alterations in the ^{13}C contents of plants. Burning of fossil fuels (a C_3 carbon source) during the industrial era has significantly diluted the ^{13}C content of the atmosphere. The waxing and waning of glacial ice sheets in the past trapped and released large amounts of CO_2 in concert with global temperature changes, producing changes in atmospheric CO_2 levels. These changes did not greatly affect atmospheric ^{13}C content, however, since the same CO_2 was being recycled. On average, the ^{13}C content of the preindustrial atmosphere was quite stable for the past 3 million years or more. In the delta notation of carbon isotope measurements, the preindustrial $\delta^{13}C$ values of C_3 and C_4 plants were about $-25‰$ (minus 25 per mil) and $-11‰$, respectively, relative to the PDB marine carbonate standard (which has a value of zero). The tooth enamel $\delta^{13}C$ values of South African herbivores were, on average, about $-13‰$ for browsers and $+1‰$ for grazers. By comparing these C_3 and C_4 end members for herbivores with the isotopic signatures of hominins, the relative amounts of C_3 and C_4 based foods in hominin diets can be established. Whether these foods were plants or the consumers of plants cannot be determined, but a carbon isotope measurement provides a fixed point from which to consider the possibilities in a given environment. These possibilities can be based on the fossil remains and carbon isotope ratios of plants and animals in the landscape in which the hominin lived, cutmarks on bone, tooth wear, artifacts, and other lines of evidence that provide environmental and dietary information. At the moment, published isotope measurements for multiple specimens of tooth enamel are available for three species of early hominins from South Africa.

Fourteen specimens of *A. africanus* have been analyzed isotopically: 4 from the site of Makapansgat (Sponheimer and Lee-Thorp 1999) at about 3 million years ago and 10 specimens from Sterkfontein, Member 4, dating between 3 million and 2 million years ago (van der Merwe et al.2003a). The results show a surprisingly wide variation in C_4-based foods between individuals. Among the 4 specimens from Makapansgat, the C_4 dietary component varies from near zero to about 55%; among the 10 individuals from Sterkfontein, the C_4 component varies from 20 to 50%. This is an extraordinarily wide variation in diet for any animal species. *A. africanus* was evidently a generalized feeder who was exceptionally opportunistic in its dietary behavior; it was probably an omnivore, but we cannot prove that it was.

The other published isotope results from South Africa are for 2 species of hominins from the site of Swartkrans, Members 1 and 2 (ca. 1.8–1 Ma), including, 3 individuals of *Homo* sp. (*H. habilis* or *H. ergaster*) and 8 individuals of *A. robustus* (Lee-Thorp and van der Merwe 1993; Lee-Thorp, Thackeray, and van der Merwe 2000). The 2 species had identical C_4 components in their diets of about 25%, on average. The results for both species show fairly low variation; they were much more constrained in their diets than *A. africanus*. The fact that both species had about 25% C_4-based foods in their diets does not mean that their diets were identical, but it does indicate that they were generalized feeders and probably omnivores. The latter hypothesis is reinforced by the occurrence at Swartkrans (in all strata) of stone tools, cutmarks on bone, and bone tools for digging into termite mounds (various authors in Brain 1993; Backwell and d'Errico 2000). It is not possible, however, to associate this evidence with only one or both of the hominin species.

Unpublished isotope data are now available for 2 hominin species from Tanzania (van der Merwe et al. in prep.): 3 specimens of *H. habilis* and 2 specimens of *A. boisei,* dated to about 1.8 million years ago. Both species had substantial C_4 dietary components.

The C_4-based foods in the diets of early hominins from South and East Africa could have included parts of C_4 grasses and sedges. Hominins did not have the type of teeth that are required for eating large amounts of grass, however, nor did their teeth acquire the signs of wear that graminivory produces on tooth enamel. Grine's microscopic study of the teeth of *A. africanus* led him to the assertion that this species was a frugivore and folivore (Grine 1987); the isotopic data do not support the conclusion that *A. africanus* was an exclusive C_3 plant feeder. If C_4 plants were directly involved in the diets of these early hominins, C_4 sedges are more likely possibilities, since some sedge species produce starchy underground storage organs of substantial size. There is no evidence for the harvesting of sedge tubers by early hominins, but aboriginal Australians have been reported to do so (Cane 1989), while charred plant remains from the late

Pleistocene site of Wadi Kubbaniya near Aswan show that *Cyperus rotundus* tubers were the staple food of Palaeolithic hunter-gatherers who lived along the Nile (Hillman, Madeyska, and Hathor 1989). Sedges grow only in wetlands, thus constraining their distribution to lake regions and river systems.

Other C_4-based foods, of course, are the consumers of C_4 plants. In Africa, the most obvious candidates were grazing animals that provided a plentiful and ready supply of meat for hominin scavenging or hunting. Also abundant were grass-eating insects, such as locusts (available in vast swarms on a seasonal basis) and termites. Of the latter, some species eat wood (C_3 plants) to excrete in their nests as a substrate for raising fungus gardens, while others eat grass. The termites that are currently available in the Sterkfontein Valley (*Trinervitermes* sp.) are grass eaters. These are distributed only in relatively arid savanna regions. Termites that are gathered by chimpanzees in the wetter parts of the continent (*Macrotermes* sp.) are primarily wood eaters and have C_3 isotopic signatures.

We can conclude from the carbon isotope signatures of hominins who lived after 3 million years ago that they were fully engaged with C_4 foods, i.e., they were creatures familiar with the savanna. The question of when early hominins became engaged with this C_4 world is still unanswered. It will remain so until hominins who lived before 3 million years ago have been documented more extensively and carbon isotope measurements on their tooth enamel have been carried out.

14.3 Evolution of Food Production

14.3.1 Modern Humans and Plants

All humans in the world today belong to a single species (or even subspecies) known as anatomically modern *Homo sapiens sapiens* (amHss). They are genetically very similar and are apparently descended from a small group of female breeders at some genetic bottleneck in the past (Harpending et al. 1993, 1998; Jorde et al. 1995, 1997; Rogers and Jorde 1995). This species evolved in Africa during the past 200,000 years: the earliest securely dated skeletal evidence comes from Ethiopia and is 160,000 years old (White et al. 2003). During the past 100,000 years or so, modern *H. sapiens* spread from Africa to the other continents and replaced all the human groups they encountered there. This account, so simply provided here, constitutes a hypothesis that is variously referred to as Mitochondrial Eve (after the DNA of a single female ancestor), African Garden of Eden, or Out of Africa. This does not mean that it is a fantasy, but it is certainly the subject of vigorous debate. Depending on how the genetic evidence is interpreted, the case can be made for an African origin (Vigilant et al. 1991; Wilson and Cann 1992) or for a multiregional origin (Thorne and Wolpoff 1992) of modern humans. Relethford (2001) has extensively and critically reviewed this debate, and no attempt will be made to settle it here.

By 15,000 years ago, humans were present on all the continents except Antarctica. The Americas were the last to be settled, sometime before 15,000 years ago, with humans thinly populating both continents within a relatively short time (Gamble 1994). The retreat of the glaciers and global increases in temperature at this time provided an environmental setting for a new departure in human subsistence, resulting in the domestication of animals and plants. In the account that follows, the emphasis is on plants, particularly on C_3 and C_4 grasses, including a number of cultigens that are not well known outside their areas of domestication. Geographically, the emphasis is on those parts of the world where significant domestication events occurred: the Near East, China, Africa, and Central and South America. The chronology is based on radiocarbon dates, commonly used by archaeologists. These have been calibrated for variations in atmospheric ^{14}C content to read in calendar years (Stuiver, Reimer, and Reimer 2000) and are given in years Before Present (BP); the present is AD 1950, according to radiocarbon convention. A radiocarbon measurement of about 10,000 years calibrates to a calendar age of about 12,000 BP so the chronology provided here may differ substantially from those in the original archaeological reports.

The melting of polar ice caps and continental glaciers put a large amount of CO_2 back into circulation, raising the atmospheric concentration from below 200 to about 270 ppmV between 15,000 and 12,500 BP. Coupled with a temperature rise, this caused a 25% to 50% increase in the productivity of C_3 plants (Sage 1995). Of particular interest here are C_3 grasses, which probably grew in dense stands at middle altitudes in the so-called Fertile Crescent: the mountainous region from the Taurus Mountains in Anatolia to the Zagros Mountains of Mesopotamia. Dense stands of wild wheat, barley, rye, and oats can still be seen on some of the foothills of these mountains. During this post-glacial period, the Near East seems to have been a busy crossroads of human activity and a centre of innovation. Hunter-gatherers in this region harvested wild grasses and ground the seeds into flour with stone mortars and pestles (Hillman, Colledge, and Harris 1989). They also hunted larger numbers of gazelle, but the animal resources were abundant enough to make seasonal movements unnecessary. They settled in villages and developed the early stages of more complex social organisation. These conditions were repeated at various times in different parts of the world; in several such cases, they gave rise to the domestication of animals and plants (Gebauer 1995; Watson 1995). The conditions for such domestication are, of course, related: abundant resources make sedentism and social complexity possible. Accordingly, domestication of wild plants and animals usually occurred in coastal or riverine settings with an abundant diversity of resources. One such area was the Levant (roughly the lowlands along the east coast of the Mediterranean), where the earliest plant domestication took place (Fig. 14.1).

14.3.2 Domestication of C_3 Plants in the Levant

A number of C_3 plants were domesticated in the Levant, including rye, wheat, barley, and oats. Among the pulses, there were peas, lentils, and chickpeas. Nuts

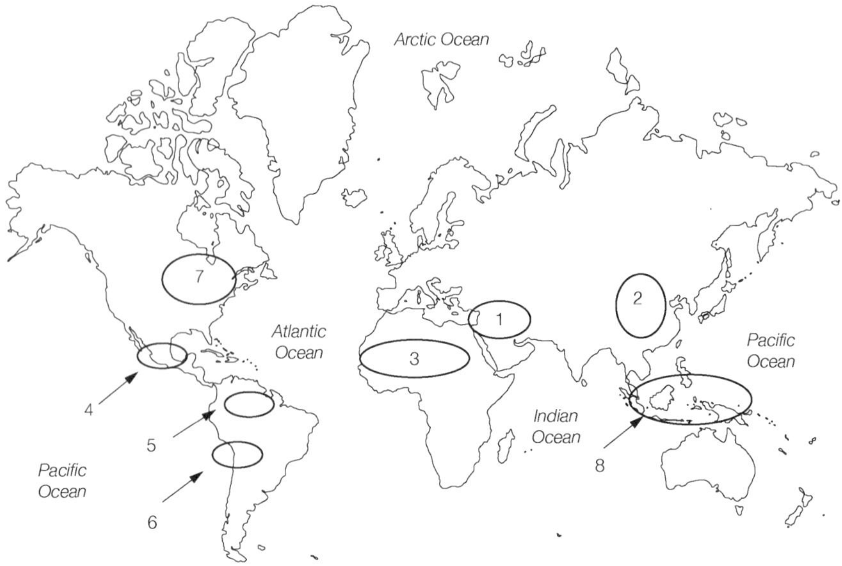

Figure 14.1. Major centers of plant domestication. The circles indicate the general area, not the precise geographical location: **(1)** Near East: wheat, barley, rye, oats, lentils, pulses, various fruits; **(2)** China: foxtail and broomcorn millet, Asian rice, soybeans; **(3)** Africa: sorghum, pearl and finger millet, fonio, tef, African rice, yam, oil palm, melons; **(4)** Mesoamerica: maize, beans, squashes, chili, yams, tomato; **(5)** Tropical America: cassava, yams, pineapple, avocado, cashew and Brazil nuts; **(6)** Andes: potatoes, quinoa, kiwicha; **(7)** Northeast America: sunflower, sumpweed, goosefoot, squashes; **(8)** Southeast Asia: sugar cane, yams, taro.

and fruits included almonds, olives, pomegranates, apples, grapes, and various soft fruits. These were presumably domesticated at different times over several thousand years. For most of these plants, it is not possible to tell the difference between wild and early domestic forms. In the case of wheat and barley, the distinction can be made when the brittle-rachised wild forms were replaced by tough-rachised domesticates. This is the result of sickle-harvesting and selection for mutants that do not scatter their seeds during the reaping process (Harris 1998).

According to the archaeological evidence, the first occurrence of plant domestication in the world happened in the Levant. It involved wheat and barley, and the initial process was completed by about 11,500 BP (Bar-Yosef and Belfer-Cohen 1991; Bar-Yosef and Meadow 1995; Harris 1998). The impetus for this domestication is ascribed to a period of climatic deterioration known as the Younger Dryas, when global temperatures decreased and glaciers re-advanced between 12,000 and 11,600 BP. This aberrant period of cooling was the result of ice melting: for example, the St. Lawrence Seaway in North America opened and discharged large amounts of glacial meltwater from the continent into the North Atlantic, which cooled the oceans worldwide (Broecker 1987). The effect on the Near East was to make the mountains of the Fertile Crescent inhospitable to grasses and the humans who had been harvesting them. To maintain food supplies, people at lower altitudes in the Levant increased the manipulation of cereals they had come to rely on, saving some of the harvest to plant during the next growing season. By 11,500 BP, when the climate was warming once more, this manipulation can be shown to have resulted in domesticated forms. The archaeological evidence comes from three sites in the south central Levant, between Lake Tiberias and the Dead Sea in the Jordan Valley (Harris 1998). These are dated between 11,800 and 10,500 BP, a chronological and cultural period referred to by archaeologists as the Pre-Pottery Neolithic. The evidence is very meager, but includes confirmed domesticated emmer wheat and barley at the site of Iraq ed-Dubb and emmer wheat at Jericho and Aswad I. Other plant remains from the three sites are wild or of uncertain domestic status. At Iraq ed-Dubb, these remains included wild barley, wild einkorn, wild or domestic lentil, and legume of the genus *Vicia* (possibly fava bean). At Jericho, some charred grains of possible domestic two-row barley were found, as were some very fragmentary remains of einkorn, emmer, lentil, and chickpea. At Tell Aswad, remains of two-row barley, lentil, and field pea were found, all of uncertain status. The confirmed domesticates, therefore, are emmer wheat at three sites and barley at one. At the same time, it should be noted that wild einkorn is not native to this area (Bar-Yosef and Meadow 1998), which means that it had been brought into the area for cultivation. Wild or domesticated, its reproduction was already under the control of humans.

Although the cultivation of domesticated forms may have been practiced in only a few places at the end of the Younger Dryas, it became the subsistence strategy of most people in the Near East by about 10,500 BP (Harris 1998). Within the next 5,000 years, these C_3 domesticates came to provide the staple diet of the earliest urban civilizations of the world in Mesopotamia and Egypt.

14.3.3 Domestication of C_4 Plants in China

Two species of millet, foxtail (*Setaria italica*) and broomcorn (*Panicum miliaceum*), were the first C_4 plants to be domesticated. This domestication took place in the loess plains of North China and along the Yellow River; the process had apparently been completed by about 10,000 BP (Chang 1980, 1986). These cereals provided the staple foods on which the classic Bronze Age civilizations of the Hsia, Shang, and Chou dynasties were built (ca. 5500–2000 BP). Rice (*Oryza sativa*), a C_3 plant, so intimately associated with things Chinese, did not contribute to the subsistence base of the earliest Chinese civilizations. It was domesticated further to the south, along the Yangtze River (Cohen 1998) at about the same time that millets were domesticated in the north. Rice cultivation did not move north into drier conditions from its tropical origins until complex terrace irrigation systems were developed.

The archaeological evidence for millet domestication in China is still fairly sparse (Chang 1986; Cohen 1998; Crawford 1992a; Underhill 1997). About 12,000 years ago, China was populated by small groups of people who subsisted by hunting, fishing, and plant collecting. By 9500 to 9000 years ago, sedentary villages of agriculturalists were already established along the Yellow River (or Huanghe) in the North China plain. Archaeologists group these villagers together as the Peiligang culture, after the site of Peiligang in Henan province. The villages were as large as two hectares, with round houses of 2 to 3 m in diameter (Smith 1995). Skeletal remains of domesticated animals include dogs, pigs, and chickens, but hunting and fishing also brought in substantial numbers of deer, pheasant, carp, turtle, and river shellfish. The identifiable remains of wild plant foods included walnuts, hazelnuts, and hackberry (Rodwell 1984).

Botanical evidence that people of the Peiligang culture grew domesticated forms of millets is lacking but can be deduced from their artifacts and storage practices. Their grave goods included stone axes for felling trees, stone hoes for tilling the soil, stone mortars with feet for grinding cereal grains, and serrated sickles for harvesting. Storage pits between the houses were common: at the site of Cishan, 80 such pits were excavated, some with 1 m thick deposits of rotting grain in them (Rodwell 1984). Botanical identification of these grains can be problematic (Crawford 1992a), but it appears that foxtail millet was grown at Cishan and broomcorn millet at Peiligang (Ren 1995). After 8000 BP, people of the Yangshao culture grew both species of millet over much of North China (Chang 1986). During the time of the Bronze Age Shang dynasty (after 4500 BP), separate Chinese ideographs for foxtail and broomcorn millet were written on oracle bones (Chang 1980) and their use as staples continued into the historical period. Chinese millet became widely distributed in prehistoric times. Both species were grown in Japan by 3500 BP (Crawford 1992b). During the Neolithic and Early Bronze Age (ca. 7000–3500 BP), millets were grown great distances from their origin in East Asia, across Eurasia and as far north as the Alps (Kroll 1981, 1983). At the Macedonian site of Kastanas, broomcorn millet was an important crop during the Late Bronze Age and Early Iron Age, while foxtail millet was present in small quantities. At the Hallstatt Iron Age site of Magdalenska gora in Slovenia (ca. 3000 BP), the average carbon isotope ratio of human skeletal collagen indicates that some 35% of the human diet was dependent on C_4 plants, presumably millets (Murray and Schoeninger 1988).

The wild ancestor of foxtail millet (*Setaria italica*) is probably *S. viridis* (green bristle grass), with which it can produce fertile hybrids. The wild species occurs in both Asia and Europe and may have been independently domesticated in the latter area, as the species name for the domesticated form indicates (Rodwell 1984). No wild ancestor for broomcorn millet has been identified, but China is its center of botanical diversity.

14.3.4 C_4 Crops from Africa

Most of Africa is unsuitable for dry-land cultivation of C_3 grains such as wheat and barley, which require rain during a cool growing season. These conditions are available only in the winter rainfall areas along the Mediterranean coast of North Africa and at the southwestern tip of the continent in South Africa. Wheat and barley could also be grown by early agriculturalists along the Nile Valley; they sowed the seeds on the wet floodplain after the summer floodwaters receded. Africa did, however, produce some C_3 domesticates that are indigenous to the forest-savanna ecotone, along the northern margins of the equatorial forests: these were yams, cow peas, oil palm, and African rice, *Oryza glaberrima* (Harlan 1992). From the highlands of Ethiopia also came domesticated noog, an oil plant, and ensete, a shrub with bananalike leaves and an edible, starchy stem (Harlan 1992). These African domesticates have not achieved much wider distribution than the regions in which they were domesticated. An exception is the watermelon, from West Africa.

For agriculture to become established in the savanna regions of Africa, it was necessary to domesticate indigenous C_4 grasses. The best known of these are sorghum (*Sorghum bicolor*) and pearl millet (*Pennisetum glaucum*, which includes *P. typhoides and P. americanum*). Somewhat lesser known are finger millet (*Eleusine coracana*), tef (*Eragrostis tef*), and fonio (including white fonio, *Digitaria exilis*, and black fonio. *D. iburna*). Sorghum and pearl millet were domesticated in the Sahel and/or Sahara; tef and finger millet originated in the highlands of Ethiopia and Uganda; and fonio is of West African origin.

The archaeological evidence for the domestication of African plants is essentially nonexistent, because focused research has not been done in the areas where domestication may have taken place. During the domestication process, cereal grains increase in size, and a tougher rachis develops, thus increasing productivity and making it easier to harvest the seed heads. Such botanical evidence can be convincing only if a sizeable number of plant parts are recovered from an archaeological deposit by means of fine screening. In addition, impressions of grains in fired clay (pottery, bricks) can be studied under the microscope. Such impressions can be misleading, however, since seeds may expand during the firing and leave impressions that are larger than the original. A critical synthesis of the archaeobotanical evidence is provided by Wetterstrom (1998), who cites misidentification of wild forms as a reason for eliminating several claims for the early presence of domesticated African cereals.

During the 1990s, Fred Wendorf and a team of co-investigators carried out an exemplary archaeological investigation of early plant use in Egypt. This project focused on the site of Nabta Playa, some 100 km west of the Nile in the southwest desert of Egypt, and other related sites in the area (Wendorf et al. 1992; Wasylikowa et al. 1993; Wasylikowa et al. 1995). Wetter conditions in the past made human occupation possible in this area from 10500 BP to 6300 BP, interrupted by several hiatuses during the dry periods and ending in final abandonment. The area has been a barren desert ever since. The archaeobotanical evidence from this project is impressive, including some 20,000 specimens, and the period of most importance is at Nabta Playa around 9000 BP. A small village of huts with bell-shaped storage pits was located near the bottom of a seasonally flooded basin, and the inhabitants herded cattle and hunted small game. Plant remains included parts of the underground storage organs of water-edge plants, such as *Cyperus rotunda*, a C_4 sedge, and the C_3 plants *Typha* sp. (cattail) and *Nymphaea* sp. (water lily) (Hathor 1995). Grass seeds were well represented, including small numbers of *Digitaria*, *Setaria*, *Brachiara*, and *Urochloa*, as well as larger numbers of *Echinocloa* cf. *colono, Panicum* cf. *turgidum* (a wild millet), and sorghum (Wasylikowa et al. 1995). These C_4 grasses would have been available for harvest in the autumn around the edges of the seasonal lake in the Nabta Playa basin and around other pools in the area. Occupation was apparently in the winter dry season, because the village would have been under water during the summer rains.

The grasses that were gathered by the people of Nabta Playa were all wild forms, including the sorghum. The sorghum grains are small and the rachis is of the brittle wild form (Dahlberg and Wasylikowa 1996). The plant remains from later sites in the area are also those of wild grasses (Close 1992), so sorghum domestication did not take place in the Egyptian Sahara. Environmental conditions similar to those at Nabta Playa would also have pertained in the Sahara, which had lakes and streams during wet phases that occurred between about 10500 and 9000 BP and between 7500 and 5000 BP (Grove 1995; Muzzolini 1995). The wild forms of sorghum and pearl millet occur today in the Sahel, a broad band of savanna across Africa between about 5° and 15° North (Harlan 1992, 1995). During the wetter periods in the past, they probably grew in the Sahara. Archaeological evidence shows that sedentary communities were present at water sources in the Sahara and Sahel from 9000 BP, where they depended for food on fishing, hunting, and gathering wild grasses (Muzzolini 1995). These are the kinds of conditions that led to domestication in the Levant and in China.

A plausible scenario for early domestication of African grains and their dispersal to other regions can be constructed from a few suggestive pieces of evidence. The data consist of a sorghum grain impression, claimed to be domesticated and dated to about 5000 BP, from the archaeological site of Adrar Bous in the central Sahara (Clark, Williams, and Smith 1973); charred remains of African cereal grains in India at Rodji, in peninsular Gujarat, including finger millet at about 5000 BP and sorghum at about 3000 BP (Weber 1991); and impressions of sorghum grains in clay bricks, dated to about 5000 BP, from the site of Hili in the Oman peninsula (Cleuziou and Constantini 1980). This suggests an Arabian bridge for the transmission of crop plants from Africa to India. The data points in this scenario, however, do not stand up well under close scrutiny (Wetterstrom 1998; Meadow 1998). The sorghum grain imprint from Adrar Bous has been re-identified as a wild form (Stemler 1990); the identification of *Sorghum bicolor* at Hili is not unequivocal; and the identification of finger millet and sorghum at Rodji is not secure at species level. Nevertheless, the African domesticates finger millet, sorghum, and pearl millet *did* travel to India at a very early stage and became important crops in that region. In the case of sorghum, for example, this happened early enough for a defined race to be developed in India. *Durra* sorghum, the most derived race of the plant, has been re-exported from India to Africa, where it is grown along the edges of the Sahara by Islamic peoples (Harlan 1995).

The earliest archaeological evidence for a domesticated African cereal comes from pottery impressions of pearl millet at various sites along the Tichitt escarpment (Dhar Tichitt) of central Mauretania. This evidence has been the subject of vigorous debate, regarding both the botanical identification and the radiocarbon dating (Munson 1971, 1976, 1989; Jacques-Félix 1971; Brunken et al. 1977; Holl 1985, 1986). Two potsherds from different Dhar Tichitt sites with confirmed impressions of domesticated pearl millet have been directly dated by radiocarbon to about 3800 BP (Amblard 1996).

It is clear that the early phases of African cereal domestication still await archaeological discovery. Linguistic evidence and the botanical geography of wild and domesticated forms do provide some approaches to a reconstruction of African domestication (Harlan 1992, 1995). In doing so, however, one must recognize that the reconstruction may be based on events of relatively recent vintage that mask those of earlier times. A single, well-founded piece of archaeological evidence may serve to rewrite the whole account.

Sorghum (*S. bicolor*) was domesticated from the wild species *S. verticilliform*, which still grows in dense stands in parts of the Sudan. This cereal has been in cultivation long enough for five distinct races to develop; these are associated with particular African language groups (Harlan 1995). *Durra* sorghum has already been mentioned; it was apparently developed in India from African *S. bicolor* and then re-introduced to Africa with the spread of Islam during the past 1000 years. The *guinea* race is mostly West African, requiring high rainfall, and is grown by people who speak Niger-Congo languages. People of the Chari-Nile language family grow the *caudatum* race from Lake Chad to the Ethiopian border. In southern Africa, speakers of Bantu languages (a subgroup of Niger-Congo) who moved into this part of the continent nearly 2000 years ago grow a drought-resistant *southern* race. The *bicolor* race closely resembles the wild form and is found nearly everywhere that sorghum is grown. Today sorghum (Table 14.1) is grown in 30 countries, which produce about 60 million metric tons annually to provide the staple food of 500 million people (National Research Council 1996; Brown 1999).

Pearl millet (*Pennisetum glaucum*) is highly drought resistant and is probably the most modified of all domesticated cereals (Harlan 1995). The wild forms of pearl millet grow in the Sahel and deep into the Sahara. The wild seed heads are up to 100 mm long, while the domesticated seed heads may be as long as 2 m. About 10 million tons are currently produced per year, half of it in India (National Research Council 1996). The Indian name is *bajra* and it is used to make the unfermented bread *roti* and fermented *kisra*. In Africa, it is produced across the Sahel from Senegal to Somalia and in drought-ridden parts of southern Africa, including the edges of the Kalahari desert (Harlan 1995). It is made into porridge, while sprouted millet is used to make beer.

Finger millet (*Eleusine coracana*) was domesticated from the wild *E. africana*, which occurs in the highlands of Ethiopia, Uganda, and Kenya (Harlan

Table 14.1. Estimated global production of food crops in 2000. Sugar cane is represented by wet cane weight; cereals, beans, and pulses by dry kernel weight; and tubers by wet tuber weight. Source: Food and Agriculture Organization website:http://apps.fao.org.

Crop	Area harvested (Ha × 10^6)	Production (Mt × 10^6)
C_4 plants		
Sugar cane	19	1260
Maize	139	594
Sorghum	42	58
Millets	36	27
Fonio	0.35	0.25
C_3 plants		
Wheat	213	584
Barley	54	132
Oats	13	26
Quinoa	0.08	0.06
Soybeans	74	161
Lentils	3	3
Potatoes	20	321
Sugar beets	6	245
Sweet potatoes	9	142
Yams	4	38

1995). It is the major crop plant of Uganda and is grown on a smaller scale in East and southern Africa, as well as in India (National Research Council 1996). It is variously used to make porridge, malt, beer, and distilled *arake* (in Ethiopia). Current production is about 4.5 million metric tons per year.

Fonio (*Digitaria exilis*, white fonio, and *D. iburna*, black fonio) was formerly the major crop plant of West Africa, where it was presumably domesticated (National Research Council 1996, Harlan 1995). It is still the staple food in parts of Mali, Burkina Faso, Guinea, and plateau Nigeria. Its most common use is to make couscous, which is reputed to taste better than the same dish made with wheat. The latter has to be processed into small grains for the purpose, a procedure that may have been developed to imitate the small seeds of fonio. Two other species of fonio were domesticated in India and Europe at an unknown time in the past, of which the European species (*D. sanguinalis*) was planted until quite recently (Harlan 1995). Current production of fonio is about 250 thousand metric tons per year.

Tef (*Eragrostis tef*) is the staple grain of Ethiopia but is almost unknown elsewhere (National Research Council 1996). It is used to make *injera*, a fermented flat bread like a large pancake (more than 0.5 m in diameter), on which various spicy stews are served to diners. Tef requires the hot days and cool nights of the highlands to produce well; about half the hectarage under cereal cultivation in Ethiopia is planted in tef. Its wild form is probably *E. pilosa*, a harvested wild grass of the Sudan; this distribution suggests that wild tef was brought to Ethiopia and domesticated there (Harlan 1995). Tef is also grown as animal fodder on the high plateau interior of South Africa, where Kew Botanical Gardens introduced it experimentally in 1886. The spread of Ethiopian restaurants in North America has produced a demand for tef, which is supplied by a single producer, Wayne Carlson of Caldwell, Idaho (located in a high altitude desert). He learned about tef as a community health worker in Ethiopia, returned to the United States after the Marxist revolution in that country, and imported the seed from South Africa (Carlson, personal communication). This serendipitous transplantation of a lesser-known grain is highlighted here because it may well exemplify at least one mechanism by which some crop plants spread beyond their areas of origin in the past.

In many parts of Africa, maize has replaced the local C_4 cereals as the staple crop; sorghum and millets are frequently grown for the making of beer alone. Maize, however, is not well adapted to drought-stricken areas or high altitudes. If global warming should cause the desiccation of African savannas, a resurgence of interest in indigenous cereals is to be expected.

C_4 plants played a crucial role in the development of African urban societies, but the archaeobotanical evidence for the species involved in any particular case is largely lacking. The earliest and best known urban society of the African continent was that of Dynastic Egypt (ca. 5000–2500 BP), but this civilization was built on the staple crops of the eastern Mediterranean: wheat and barley. The urban societies of the Nubian Nile are less well known: the kingdoms of Kerma (3500–2900 BP), Napata (2900–2400 BP), and Meroe (2400–1800 BP) (Shaw et al. 1995). These kingdoms had close connections with Egypt and probably followed the Egyptian lead in agriculture, but direct evidence is lacking. The Kingdom of Aksum in Ethiopia (1800–1400 BP), a civilization with a written Semitic language and many cultural practices derived from Arabia, may have relied exclusively on local domesticates for food, given Ethiopia's unusual environmental setting. Tef, finger millet, and other Ethiopian plants probably dominated the Aksumite menu.

In West Africa, the earliest documented urban society was that of Jesse-Jenno (2000–600 BP), located in Mali in the resource-rich inner delta of the Niger River (McIntosh and McIntosh 1995). African rice, sorghum, and millet (species not given) were the important cereals here. Other West African states included the kingdoms of Ghana (1600–800 BP), Mali (800–500 BP), and Songhay (500–400 BP).

Finally, sorghum and millets were brought to southern Africa by Bantu-speaking farmers of the Iron Age, who settled most of the southern subcontinent after 2000 BP. The earliest-known archaeological occurrences of both pearl and finger millet are from the site of Ndondwane in Kwazulu-Natal, South Africa, dated to about 1250 BP (Maggs and Ward 1984), but farmers were already present in this area by 1700 BP. Urban societies developed in southern Africa after

1000 BP, the best known being Mapungubwe on the Limpopo (ca. 1000–800 BP) and the kingdom of Great Zimbabwe (Garlake 1973; Hall 1987; Huffman 1996).

14.3.5 Maize in the Americas

Maize (*Zea mays*) was domesticated some 5000 years later than wheat but now beats its C_3 rival by a short lead in global production (see Table 14.1). The wild ancestor of maize is teosinte (*Zea mexicana* spp. *parviglumis* or *mexicana*), which occurs naturally in the west-central highlands of Mexico. Teosinte looks so much unlike domesticated maize that the connection between them has been the subject of much debate; it was only settled by genetic studies (Iltis 1983; Galinat 1985a,b; Benz and Iltis 1990; Dobley 1990). As yet, no archaeobotanical evidence has been produced for the origin of maize domestication, which should presumably be in the same area where teosinte grows.

Richard MacNeish set out in the 1960s to document the origins of maize domestication. He chose the arid Tehuacán Valley in southwestern Mexico and excavated a series of dry caves with well-preserved plant remains. The earliest deposits with small maize cobs in them produced radiocarbon dates of about 6500 BP on charcoal samples (MacNeish 1967, 1981). Since then, however, some of these maize cobs have been radiocarbon dated directly by accelerator mass spectrometry (AMS) and found to be 5500 years old (Long et al. 1989). Earlier evidence for the presence of maize is available from elsewhere in South America. These early occurrences are not from the area where teosinte grows, so the origins of maize domestication still await archaeological documentation. Conclusive proof would require the discovery of cobs or kernels, or both, that are recognizably those of some early form of maize.

Outside the region where teosinte grows, the occurrence of maize pollen and phytoliths (silica skeletons) can attest to its presence. These have usually been found in sediment cores from lake beds. Early occurrences of maize pollen or phytoliths are from Colombia at approx 7500 BP (Bray et al. 1987), western Panama at approx 5800 BP (Piperno et al. 1985; Piperno, Bush, and Colinvaux 1990), and coastal Ecuador by approx 6000 BP (Bush, Piperno, and Colinvaux 1989). Stable carbon isotope data for human skeletal collagen from Panama show that maize was present in the local diet during the period of approx 7800 to 5300 BP (Norr 1995).

The presence of maize in a given area does not mean that it was an important crop. In most parts of the American tropics, maize was introduced to people who probably already practiced swidden (slash-and-burn) cultivation of tropical forest plants like yams and cassava, plus assorted fruits and nuts (Lathrap 1987). The same pollen cores that produced early occurrences of maize pollen and phytoliths also contained high concentrations of charcoal (from forest burning to make fields); these are already present in earlier parts of the cores (Piperno 1990). It is likely that maize was planted as an exotic garden vegetable and that it took a long time to develop successful methods for growing it in quantity.

The Valdivia culture of coastal Ecuador provides an example of such a slow adaptation (Raymond 1983). This is one of the earliest "Formative" cultures of the Americas, with settled villages, long distance trade, and ceramics. Maize phytoliths have been identified in this area by at least 6000 BP and perhaps even earlier (Pearsall and Piperno 1990; Piperno 1991). Carbon and nitrogen isotope measurements on human skeletons of the Valdivia people indicate, however, that their food sources during the early phases (ca. 6000 BP) were the ocean, rivers, and tropical forest (van der Merwe, Lee-Thorp, and Raymond 1993). During the Middle Formative (ca. 4000 BP), maize remains occur in the archaeological deposits (Lippi et al. 1984) and the isotopic data attest that maize had become an important dietary item. Maize was also cultivated farther south at this time on the coast of Peru, by people who did not have ceramics yet (the Peruvian Preceramic) and who had no prior agriculture.

In North America, maize agriculture was adopted in the southwest between about 3000 and 2000 BP (Wills 1988), but the presence of foods other than maize with C_4 signatures (bison, amaranth, and various cactuses) makes it difficult to document this fact by isotopic analysis (Schoeninger 1995). The pace of its spread to other parts of the continent varied, as new varieties were developed to cope with shorter growing seasons. Some maize remains have been found in the Mississippi River Valley and dated to 2000 BP (Riley et al.1994); a few maize remains also have been found in Tennessee and Ohio that are as old as 1800 BP (Fritz 1990; Wagner 1994). In the lower Great Lakes area, maize formed an important part of the diet by 1400 BP (Crawford, Smith, and Bowyer 1997), but the isotopic data for much of the northeastern woodlands show that it did not become an important dietary item until after 1200 BP (van der Merwe and Vogel 1978; see Schoeninger and Moore 1992 for review). In southern Ontario, maize reached a peak of production (about 60% of the diet) around 700 BP (Williamson and Pfeiffer 2003; van der Merwe et al. 2003b), but in coastal New England it did not achieve importance until after European colonization (Little and Schoeninger 1995).

14.3.6 Maize and the Development of Complex Societies

Maize has been an extraordinary agent for social change in many parts of the world. In several well-known cases it replaced indigenous plants as the staple crop and provided the wellspring for the development of complex societies with economic and political stratification. Under the right conditions, it has the highest grain yield of all the major cereals (Doswell, Paliwal, and Cantrell 1996). It has a high content of protein (10% or more), which can be a liberating agent for a population constrained by low-protein plant foods and hunted animal protein. Its downside is a deficiency in the essential amino acids tryptophan and lysine, which leads to health problems in maize-dependent populations. Maize is also an exacting crop, requiring fertile soils, and it is sensitive to cold, excessive heat, insufficient sunshine, drought, and excessive humidity. It grows best in temperate environments with warm days and cool nights (Doswell, Paliwal, and Cantrell 1996). There is a high risk of frost in such situations, requiring careful planning of the time to plant and harvest, even before the cycle of frost, rain, and sun has been established for the season (Kirkby 1973; Sanders et al. 1979; Wills 1988).

A high-risk crop with high yield potential provides a fertile field for entrepreneurial activity and the rise of an elite social group. Based on numerous ethnographic accounts from the Americas, Hayden (1990, 1995) has described how such social stratification takes place. Given a stable population in which each family can feed itself by hunting, fishing, and the cultivation of local plants, sanctions against competitive behavior are relaxed. This allows ambitious "aggrandizers" to enhance their status in the community by sponsoring competitive feasts, which are usually based on the beer made from maize. This creates social debts and alliances, leading to control of labor to grow more of the foreign crop. As maize becomes more important in a society, those aggrandizers who are able to organize production most effectively gain high economic and political rank. For maize to become a staple crop requires control of the agricultural cycle by means of calendars, as among the Maya (Forde 1931; Earls 1986, 1989, 1991); irrigation and preplanting irrigation (Sanders et al. 1979; Hastorf 1993); terraces in the highlands (Earls 1989); and ridged fields, as found in eastern North America and the Andes (Erickson 1986; Riley 1987; Woods 1987; Knapp 1991). Populations increased in response to a large supply of maize. In the floodplains of the lower Orinoco, for example: the area covered by habitation sites increased fifteenfold within a few centuries (Roosevelt 1980). Such population increases lead to the formation of kingdoms that are socially and politically stratified and eventually to the emergence of complex urban societies.

This scenario played itself out in several parts of the Americas. After its dispersal from western Mexico into the forested, humid lowlands of Mesoamerica, maize was gradually adopted during the early Formative period (4700–3200 BP; Blake et al. 1992a,b). A subsequent dramatic increase in maize consumption was accompanied by the emergence of La Venta on the Gulf Coast of Mexico; this was a major ceremonial center of the Olmec civilization. In the Maya lowlands, detailed isotopic data for human burials show a clear trend of maize consumption from a low of about 30% at Preclassic Cuello in Belize, a town with a small pyramid at the center (starting ca. 3500 BP), to a nearly total dependence on maize and beans at the major Classic ceremonial center of Copan in Honduras (1700–1100 BP) (White and Schwarcz 1989; White et al. 1993; Gerry 1994; Reed 1994; L.E. Wright 1994; Tykot et al. 1996; van der Merwe et al. 2000). Maize, beans, and squash, plus chilli peppers for flavor, became a defining feature of the cultures of Mesoamerica (Kirchhof 1943). It continues to be so in the cuisines we think of as Mexican and Tex-Mex.

The Mesoamerican food complex was adopted as a package by Archaic hunter-gatherers of the southwestern United States, where maize became the staple food crop and led to the development of pueblo societies (Ford 1985; Wills 1988; Minnis 1992). In Northeast America, however, a local crop complex based on C_3 plants was already in place. It included domesticated sunflower

(*Helianthus annuus*), sumpweed (*Iva annua*), goosefoot (*Chenopodium bushianum*), maygrass (*Phalaris caroliniana*), and several kinds of squash (Ford 1985; Asch and Asch 1985). After its introduction in this area around 1800 BP, maize was grown on a small scale as a special crop. It was apparently consumed in community ceremonies and social gatherings (Johannessen 1993ab; Wymer 1994). Between 1200 and 1000 BP, there was an explosive increase in maize consumption, as attested by isotopic data from human burials (van der Merwe and Vogel 1978; Bender et al. 1981; Buikstra et al. 1988; Lynott et al. 1986; Rose et al. 1991; Tuross et al. 1994). This increase was accompanied by a change in settlement pattern as people moved out of the river valleys to plant upland fields (Riley 1987; Woods 1987; Scarry 1993). Here they lived in scattered hamlets under the control of newly emerged elites at ceremonial mound centers. Of the latter, the sites of Moundville, Alabama, and Cahokia, Illinois, are well-known examples.

In South America, maize also replaced indigenous crops after a long period of being regarded as an exotic plant. People lived in the tropical lowlands of the Amazon and Orinoco drainages from about 10000 BP (Roosevelt et al. 1991). They developed a subsistence system based on the cultivation of cassava as a staple, with yields from fishing and hunting supplying the protein. Evidence for maize at about 3000 to 2500 BP has been excavated at Parmana on the Orinoco (Roosevelt 1980). By 1600 BP, maize had become the staple crop of this region, according to carbon isotope measurements on human skeletons (van der Merwe et al. 1981); earlier skeletons are not available, so it is not clear when maize achieved importance. Ethnohistoric sources from the period of early European contact describe complex chiefdoms based on maize agriculture in the river floodplains (Roosevelt 1980, 1987). These presumably developed before 1600 BP, during the period when maize became an important crop and the population increased dramatically. The paramount chiefs of the region collected a share of the maize crop from their subjects and held elaborate religious ceremonies, with music, dancing, and lots of maize beer.

The early inhabitants of the Andes developed a system of subsistence agriculture based on many varieties of potato and tubers, plus the cereals quinoa (*Chenopodium quinoa*, a C_3 plant) and kiwicha (*Amaranthus caudatus*, a C_4 plant) as minor crops (National Research Council 1989). Maize appeared on the Peruvian coast for the first time during the late Preceramic period (5000–4000 BP) and was depicted in the artworks of the Chavin civilization in the highlands, about 4000 to 3000 BP (Lathrap 1970). Carbon isotope measurements on human and llama burials associated with the temple at Chavin provide intriguing evidence: the humans had about 20% C_4 carbon in them, while the llamas had as much as 40% (Burger and van der Merwe 1990). Maize was obviously not the dietary staple at this time; the evidence suggests that the guardians of the temple acquired their C_4 carbon content from maize beer consumed during religious ceremonies, while the sacrificial llamas were fed on maize fodder or grain, or both. By the time of the Inca Empire (after 800 BP), maize had become an important crop, with maize beer (chichi) a substance of great prestige and ceremonial significance (Hastorf 1990a,b; Hastorf and Johannessen 1993). Early Spanish colonial documents (Netherly 1978) relate that the Inca authorities could successfully recruit labor for public works projects only if they sponsored festivals and supplied large amount of maize beer.

14.4 Sugar and Maize in the Colonial Era

If we consider, in less serious moments, that maize beer was a major driving force in the development of American civilizations, then rum can be considered to have served a similar purpose in the rise of European colonial powers and their merchant navies. Next to maize, sugar cane (*Saccharum officinarum*) has been the C_4 plant that unleashed the strongest demographic forces on a global scale.

14.4.1 Sugar Cane

There are two wild forms of sugar cane (Artschwager and Brandes 1958). *S. robustum* has its center of diversity in New Guinea and grows from Borneo to the New Hebrides. *S. spontaneum* has a range from Nigeria, across the Sudan, South Asia, and the Indian subcontinent, to Taiwan, Japan, and the Solomon Islands; its center of diversity is India. Most domesticated cultivars have 80 chromosomes, such as *S. robustum*, but several varieties have admixtures of *S. spontaneum* germ plasm (82–142 chromosomes). The two species have probably been hybridized several times in different places. Most domesticated varieties of sugar cane are sterile and are reproduced by cloning or ratooning (i.e., letting the cane resprout after cutting).

The development of sugar cane cultivation and sugar production has a history that circles the globe (see Sauer 1993: 236–250). It is thought that the plant was domesticated in New Guinea (Warner 1962), where archaeological evidence as old as 5000 years has been excavated (Daniels and Daniels 1993); originally, it was probably used just as something sweet to chew. From this starting point, sugar cane spread eastward through the Pacific Islands to Hawaii. From New Guinea, sugar cane also spread westward through Southeast Asia to India.

The planting of sugar cane is referred to in Sanskrit literature as early as 3000 BP (Sauer 1993). It was probably rendered into brown loaves called *gur* at this time, produced by boiling the juice until it crystallized. By 2300 BP, two forms of more refined sugar are mentioned: *sarkara* (English "sugar") and *khanda* ("candy"). From India, the sugar-making process spread eastward through Southeast Asia to China. Here it was considered important enough for the emperor to send a study mission to India in 1303 BP (i.e., AD 647) to observe that country's superior technology.

Sugar was known in Europe at the time of Alexander the Great (2300 BP) and during Roman times, but it was an exotic import and very expensive. The spread of sugar production from India to the West was almost entirely due to the expansion of Islam (Watson 1974). After the death of Mohammed in 632 AD, Abu Bakr launched from Medina and Mecca the jihad that engulfed Pakistan, Iraq, Iran, Syria, North Africa, and Spain within a century. After 1000 AD, Muslim Turks added Anatolia and parts of the Balkans to the area under Islamic influence. Irrigation systems were developed to grow sugar cane during the dry summer season, and the foundations of the sugar plantation system were laid. Conquest destroyed the traditional patterns of land ownership, and entrepreneurs could acquire large estates to produce sugar for export. Sugar production required large labor forces, which were usually imported or enslaved locally. The word "slave" derives from Slav (a person whose native tongue is a Slavic language), and the political ramifications from ethnic hostilities engendered by sugar production can still be observed in such places as Kosovo and Macedonia.

The sugar plantation system with its workforce of slaves was "... to grow monstrous during the European colonial expansion" (Sauer 1993: 239). In the course of their encounters in the Mediterranean, Christians had several centuries to learn about sugar production from Muslims. In the wake of the Crusades, they adopted the system wholesale (Galloway 1977). Venetian merchant bankers invested in large sugar plantations on Cyprus and Crete in the fourteenth century. They imported slaves from Russia or bought African slaves from the Arabs in North Africa. By 1500 AD, Portugal and Spain had introduced sugar production to the uninhabited islands of the Atlantic (the Canaries, Azores, Cape Verdes, and São Tomé), importing the slave workforce from Africa. Sugar and slavery became inextricably linked on these islands (Greenfield 1977, 1979).

Columbus was married to the daughter of a sugar planter on Madeira (in the Canary Islands), and he introduced the plant to Hispaniola (now Haiti and the Dominican Republic) during his second voyage of exploration (Sauer 1993). The Spanish were not very successful with sugar in the Caribbean, but the Portuguese achieved great success with it in the tropics of northeast Brazil (Greenfield 1979). They brought slaves and settlers from the Atlantic islands and by 1600 AD had turned a region that was otherwise economically worthless to them into one of the most profitable colonies in the world. In competition with the Portuguese, the British, Dutch, and French introduced sugar production and slaves to their Caribbean possessions and all over the Americas. As the demand from Europe increased dramatically, sugar became the most important agricultural import from the colonies. In the course of these developments, sugar cane came to be a major demographic force, causing the migration of millions of people around the globe: some willing, but the majority not (Mintz 1959). The sugar cane agroindustry, which is capital-intensive and requires factory-like mills, also proved to be a major precursor of European industrialization (Mintz 1985).

14.4.2 Maize

European explorers and maritime traders were the agents who spread maize from its American heartland. Within a century of colonial conquests in the Americas, the plant had been introduced to most parts of the world. By the end of the sixteenth century, maize had become a field crop that was widely grown in Spain, France, and Italy. Portuguese traders introduced it to Africa in the 1500s; during the 1700s, it became the staple crop in several parts of the continent and provided the food base for the development of the kingdoms of the Congo, Maravi (in Malawi), and the Nguni (Zulu), among others. Portuguese and Arab traders introduced maize to the west coast of India, and from there it spread into the interior via the Silk Route. In the seventeenth century, maize became well established in Indonesia, the Philippines, and Thailand. In seventeenth-century China, maize cultivation led to the opening up of new agricultural lands outside the rice-growing area to cope with rapid population increase (Norman et al. 1995; Doswell et al. 1996).

The adoption of maize as a staple in preference to indigenous crop complexes took many centuries in several parts of the Americas, as already described. In many parts of the world, however, this adoption took only one or two centuries. There has been at least one exception to this rule: in South Africa, in the realm of Modjadji the Rain Queen. This story reads like a fantasy, but it is very much based on fact and will serve to bring the 7-million-year saga of C_4 plants to an end. The village of Modjadji, Rain Queen of the Lovedu people, is situated on the edge of the Drakensberg escarpment near the town of Tzaneen, in what used to be called the Transvaal. Here the introduction of maize was restricted for centuries. As late as the 1950s, the stamping and grinding of maize was forbidden in the village (Davison 1984), because it was believed that this foreign crop could bring disruptive influences that would interfere with the rainmaking process. The Modjadji lineage derives from the royal dynasties of Great Zimbabwe (Krige and Krige 1943) and has been very powerful in the past. The reclusive queen never marries but may have a large number of wives herself. Her oldest daughter, the offspring of an incestuous union, is trained in the secrets of rainmaking and succeeds her mother after the queen commits ritual suicide. Modjadji II, a powerful Rain Queen of the Victorian era, is said to have inspired the character of She-who-must-be-obeyed, a mysterious Sheba-like African queen, who was immortalized by novelist H. Rider Haggard (in *She*, first published 1886; for those not familiar with Victorian best-sellers, Ursula Andress played "She" in the 1965 Hollywood production by that name). The prohibition against maize faded along with much of the power of Modjadji V, who succeeded to the throne in 1984. Irrigation systems and agribusiness lapped up to the edges of her small queendom, and maize flour became readily available at the trading store. On 28 June 2001, Modjadji V died of renal failure and cardiac congestion. Her followers believe that she died of a broken heart, just three days after the death of her eldest daughter, Crown Princess Makheala (after a "short illness," presumably AIDS-related). Cabinet ministers of the South African government attended the joint funeral of the queen and princess, a testimony to the lingering power of the Modjadjis. With no trained successor available, the tribal elders were unable to appoint a new Rain Queen until May 2003. The year 2001 proved to be the wettest on record in many parts of South Africa, only to be followed by a drought; many people of the area attribute these phenomena to the absence of a Rain Queen who could control the rain.

In the meantime, what has happened to maize? In 1994, global production of maize exceeded that of wheat for the first time. The lead was reversed every year for the next three, but maize has been in first place since 1998. Maize is indeed a highly productive crop, with a global yield in 2000 (see Table 14.1) of 4.2 metric tons per hectare, while wheat yielded 2.2 Mt/Ha and sorghum 1.4 Mt/Ha. Given these figures, the message of the Modjadji story is that a group of people may choose for many years not to grow a crop for reasons of tradition, even when everyone else is doing so for reasons of productivity. This is the rational, scientific interpretation of the data; additionally, the extremes of high rainfall and drought since the death of Modjajdji V are due to the onset of a new climatic regime, associated with global warming. At yet another level of interpretation, the story of her passing and the accession of Modjajdi VI, a hurriedly and incompletely trained Rain Queen, should perhaps be seen as a metaphor for our time: the old order is changing in a disorderly fashion, and so is the climate.

14.5 Summary

Humans evolved in a C_4 world, after the global expansion of C_4 plants some 8 million to 7 million years ago. It has been argued that this expansion was triggered by the reduction of atmospheric CO_2 content below about 500 ppmV. Stable carbon isotope measurements on tooth enamel are available for hominid fossils of about 3 million to 1 million years ago, including specimens of the genus *Homo*. All of these early hominids had substantial components of C_4 carbon in their diets, acquired by eating C_4 plants and/or the consumers of C_4 plants.

For more than 99% of their time on Earth, humans have lived a hunter-gatherer lifestyle, hunting for meat and foraging for wild plant foods. Among the latter, grasses did not play an important role. Modern humans became more directly involved with grasses after the end of the last glacial event, from about 15,000 years ago. This took place in the Near East, a center of cultural innovation at that time. The grasses were wild forms of wheat and barley, domesticated along with other C_3 plants of this winter rainfall region. Their productivity was substantially increased by warmer temperatures and by the postglacial increase in atmospheric CO_2 content, which is of special importance to C_3 plants. After a period of intensive harvesting of wild grasses in this region, wheat and barley were domesticated by about 11,500 years ago. By about 10,500 years ago, these storable grains had come to provide the food base for the emergence of the first dynastic civilizations in Mesopotamia and Egypt, which took place about 5,000 years later.

The first C_4 grasses to be domesticated were two species of millet: foxtail and broomcorn. This took place in northern China by about 10,000 years ago and gave rise to the development of Chinese civilization some 5,000 years later. The other major C_4 domesticates are sorghum and various millets from Africa and maize from Mexico. All of these may have been domesticated by 7,000 years ago, but the archaeological evidence is completely lacking in Africa and is insecure in Mexico before 5,500 years ago. In both Africa and the Americas, these C_4 grains provided the staples for the development of complex societies.

Maize has been a particularly strong force for the development both of an elite social group and of social complexity. In the Americas, it became the staple food by replacing several indigenous crop complexes based on C_3 plants. These include the tropical forest system of South America (based on cassava and other tubers), the Andean crop complex (potatoes and other tubers plus the C_3 grain quinoa), and the crops of the eastern United States (sunflower and various chenopodia). During European colonial expansion, maize was dispersed to most parts of the world; it became the most productive grain on a global scale during the 1990s, slightly ahead of wheat. Another C_4 domesticate, sugar cane, has an unsavory reputation as a major demographic force. During the expansion of Islam and the European colonial powers, sugar production became synonymous with slavery on a large scale, causing the dislocation of millions of people.

What of the future? Will rising CO_2 levels and global warming prove to be the nemesis of human beings, who evolved in a world with low CO_2 and plenty of C_4 plants? The answer is nothing as apocalyptic: a change in the status quo, certainly, but extinction certainly not.

Let us consider a world with 750 ppmV CO_2 in the atmosphere, global temperatures about 3°C to 5°C higher than today, and sea level about 1 m to −2 m higher. Would we be able to breathe? If one suddenly entered a warm room full of people and a high CO_2 content, it would seem very stuffy and your breathing rate would rise. After a while, however, you would adapt your breathing to the oxygen content of the air, instead of the CO_2. Regarding the level of the ocean: beachfront property is obviously a bad investment and low-lying coastal cities will have problems, and yet a sea-level rise of 1 m to 2m is not unusual in human history.

The big question, of course, is what about plants? Will C_3 plants crowd out C_4 plants and will wheat replace maize as the world's staple crop? This is unlikely, although higher CO_2 levels do favor C_3 plants. Forests will grow faster, but not necessarily bigger. Trees may invade some grasslands and turn them into woodland savannas; bush encroachment is already a recognized problem in African savannas. C_4 plants, however, are not that sensitive to higher CO_2 levels, and C_3 plants are sensitive to *low* CO_2 levels. The areas of the world where crops of the two photosynthetic types grow best are delineated by the seasonality of rainfall or, failing that, the availability of water for out-of-season irrigation. These variables will change, with some areas gaining at the expense of others. The disruptions are likely to be economic and political, rather than in global productivity. Wars may be fought over access to water, rather than oil.

One of the most unpredictable variables in futuristic scenarios is human behavior. The choices people make about the plants they grow are not necessarily based on plant biology, but on tradition and local economics (Brown 1999). To this one might add serendipity, as in tef moving from Ethiopia (where it is the staple food for humans) to South Africa (where it is valued as fodder for race horses) to Idaho (to supply Ethiopian restaurants in the United States). Human fickleness and fashion cause Africans to eat white maize and to feed yellow maize to livestock; in some other parts of the world, the opposite is true. Considerations of power caused Napoleon, for example, to promote the cultivation of sugar beets (to counter the English trade in cane sugar) or Modjadji to outlaw maize, so as to keep her rainmaking rituals pristine. Finally, religious beliefs promote certain foods and declare others to be taboo. As Damon Runyan (author of *Guys and Dolls*) would sum up the situation: in the affairs of man the odds are always six to five against.

Acknowledgments. I thank several colleagues who provided information included in this chapter and commented on portions of the text. They are Wilma Wetterstrom, Richard Meadow, and Ofer Bar-Yosef of Harvard University; Rebecca Ackerman, Julia Lee-Thorp, and Judith Sealy of the University of Cape Town; and one anonymous reviewer.

References

Aiello, L.C., and P. Wheeler. 1995. The expensive tissue hypothesis. *Current Anthropology* 36:199–221.

Amblard, S. 1996. Agricultural evidence and its implications on the Dhars Tichitt and Oualata, southeastern Mauretania. In *Aspects of African archaeology,* ed. G. Pwiti and R. Soper, 421–27. Harare: University of Zimbabwe Publications.

Artschwager, E., and E.W. Brandes. 1958. *Sugar cane: Origin, classification characteristics, and description of representative clones*. U.S. Department of Agriculture Handbook no. 122. Washington, D.C.: Government Printing Office.

Asch, D.L., and N.E. Asch. 1985. Prehistoric plant cultivation in west-central Illinois. In *Prehistoric food production in North America,* ed. R.I. Ford, 149–203. Anthropological Papers no. 75, Museum of Anthropology. Ann Arbor: University of Michigan.

Asfaw, B., T. White, O. Lovejoy, B. Latimer, S. Simpson, and G. Suwa. 1999. *Australopithecus garhi*: A new species of early hominid from Ethiopia. *Science* 284:629–35.

Backwell, L.R., and F. d'Errico. 2000. First evidence for termite foraging by Swartkrans early hominids. Proceedings of the National Academy of Sciences, USA 98:1358–63.

Bar-Yosef, O., and A. Belfer-Cohen. 1991. From sedentary hunter-gatherers to territorial farmers in the Levant. In *Between bands and states*, ed. S.A. Gregg, 181–202. Occasional Paper, Center of Archaeological Investigations, no. 9. Carbondale: Southern Illinois University.

Bar-Yosef, O., and R.H. Meadow. 1995. The origins of agriculture in the Near East. In *Last hunters, first farmers: New perspectives on the prehistoric transition to agriculture*, ed. T.D. Priceand A.B. Gebauer, 39–49. Santa Fe, N.M.: School of American Research.

Bender, M.M., D.A. Bareis, and A.L. Stevenson. 1981. Further light on carbon isotopes and Hopewell agriculture. *American Antiquity* 46:346–53.

Benz, B.F., and H.H. Iltis. 1990. Studies in archaeological maize I: The "wild" maize from San Marcos cave re-examined. *American Antiquity* 55:500–511.

Blake, M., B. Chisholm, J.E. Clark, and K. Mudar. 1992a. Non-agricultural staples and agricultural supplements: Early formative subsistence in the Soconusco region, Mexico. In *Transitions to agriculture in prehistory*, ed. A.B. Gebauer and T.D. Price, 133–51. Madison, Wis.: Prehistory Press.

Blake, M., B. Chisholm, J.E. Clark, B. Voorhies, and M.W. Love. 1992b. Prehistoric subsistence in the Soconusco region. *Current Anthropology* 33:83–94.

Bothma, J. du P. 2001. Die aardwolf-Proteles cristatus. *S.A. Game and Hunt* 7:6–7.

Brain, C.K., ed. 1993. Swartkrans: A cave's chronicle of early man. Pretoria: Transvaal Museum Monographs.

Bray, W., L. Herrera, C. Schrimpff, and J.G. Monslave. 1987. The ancient agricultural landscape of Calima, Colombia. *In Pre-Hispanic agricultural fields in the Andean region*, vol. 2, ed. W.M. Denevan, K. Matthewson, and G. Knapp, 443–81. Oxford: BAR International Series 359.

Broecker, W.S. 1987. *The biggest chill.* Natural History 96:74–82.

Brown, R.H. 1999. Agronomic implications of C_4 photosynthesis. In *C_4 plant biology*, ed. R.F. Sage and R.K. Monson, 473–508. London and New York: Academic Press.

Brunet, M., F. Guy, D. Pilbeam, H. Mackaye, A. Likius, D. Ahountas, A. Beauvilain, C. Blondel., H. Bocherens, J. Boisserie, L. De Bonis, Y. Coppens, J. Dejax, C. Denys, P. Duringer, V. Eisenmann, G. Fanone, P. Fronty, D. Geraads, T. Lehmann, F. Lihoreau, A. Louchart, A. Mahamat, G. Merceron, G. Mouchelin, O. Otero, P. Campomanes, M. Ponce De Leon, J. Rage, M. Sapanet, M. Schuster, J. Sudre, P. Tassy, X. Valentin, P. Vignaud, L. Viriot, A. Zazzo, and C. Zollikofer. 2002. A new hominid from the upper Miocene of Chad, Central Africa. *Nature* 418:145–51.

Brunken, J., J.M.J. de Wet, and J.R. Harlan. 1977. The morphology and domestication of pearl millet. *Economic Botany* 31:163–74.

Buikstra, J.E., W. Autry, E. Breitburg, L. Eisenberg, and N.J. van der Merwe. 1988. Diet and health in the Nashville Basin: H.human adaptation and maize agriculture in middle Tennessee. In *Diet and subsistence: Current archaeological perspectives*, ed. B.V. Kennedy and G.M. LeMoine, 243–59. Proceedings of the 19th Annual Chacmool Conference. Calgary, Alberta: University of Calgary.

Burger, R.L., and N.J. van der Merwe. 1990. Maize and the origin of highland Chavin civilization: An isotopic perspective. *American Anthropologist* 92:85–95.

Bush, M.B., D.R. Piperno, and P.A. Colinvaux. 1989. A 6000 year history of Amazonian maize cultivation. *Nature* 340:303–305.

Cane, S. 1989. Australian aboriginal seed grinding and its archaeological record: A case study from the western desert. In *Foraging and farming: The evolution of plant exploitation*, ed. D.R. Harris and G.C. Hillman, 99–119. London: Unwin Hyman.

Cerling, T.E. 1999. Paleorecords of C_4 plants and ecosystems. In *C_4 plant biology*, ed. R.F. Sage and R.K. Monson, 445–72. London and New York: Academic Press.

Cerling, T.E., J.M. Harris, S.H. Ambrose, M.G. Leakey, and N. Solounias. 1997a. Dietary and environmental reconstruction with stable isotope analyses of herbivore tooth enamel from the Miocene locality of Fort Ternan, Kenya. *Journal of Human Evolution* 33:635–50.

Cerling, T.E., J.M. Harris, B.J. MacFadden, M.G. Leakey, J. Quade, V. Eisenmann, and J.R. Ehleringer. 1997b. Global vegetation change through the Miocene and Pliocene. *Nature* 389:153–58.

Cerling, T.E., Y. Wang, and J. Quade. 1993. Expansion of C_4 ecosystems as an indicator of global ecological change in the Miocene. *Nature* 361:344–45.

Chang, K.C. 1980. *Shang civilization.* New Haven: Yale University Press.

———. 1986. *The archaeology of Ancient China*, 4th ed. New Haven: Yale University Press.

Clark, J.D., M.A.J. Williams and A.B. Smith. 1973. The geomorphology and archaeology of Adrar Bous, central Sahara: preliminary report. *Quaternaria* 18:245–97.

Clarke, R.J. 1998. First ever discovery of a well-preserved skull and associated skeleton of *Australopithecus, South African Journal of Science* 94:463.
Cleuziou, S., and L. Constantini. 1980. Premiers elements sur l'agriculture protohistorique de l'Arabie orientale. *Paléorient* 6:245–51.
Close, A.E. 1992. Holocene occupation of the eastern Sahara. In *New light on the Northeast African past*, ed. F. Klees and R. Kuper, 155–83. Köln: Heinrich Barth Institut.
Cohen, D.J. 1998. The origins of domesticated cereals and the Pleistocene-Holocene transition in East Asia. *Review of Archaeology* 19:22–29.
Crawford, G.W. 1992a. Prehistoric plant domestication in East Asia. In *The origins of agriculture*, ed. C.W. Cowan and P.J. Watson, 7–38. Washington, D.C.: Smithsonian Institution Press.
———. 1992b. The transition to agriculture in Japan. In *Transitions to agriculture in prehistory*, ed. A.B. Gebauer and T.D. Price, 117–32. Madison, Wis.: Prehistory Press.
Crawford, G.W., D.G. Smith, and V. Bowyer. 1997. Dating the entry of corn (*Zea mays*) to the Lower Great Lakes region. *American Antiquity* 62:112–19.
Dahlberg, J.A., and K. Wasylinkowa. 1996. Image and statistical analyses of early sorghum remains (8000 BP) from the Nabta Playa archaeological site in the western desert, southern Egypt. *Vegetation History and Archaeobotany* 5:293–99.
Daniels, J., and C. Daniels. 1993. Sugarcane in prehistory. *Archaeologia Oceania* 28:1–7.
Davison, P. 1984. *Lobedu material culture*, vol. 94, no. 3. Cape Town: Annals of the South African Museum.
De Heinzelin, J., J.D. Clark, T. White, W. Hart, P. Renne, G. WoldeGabriel, Y. Beyene, and E. Vrba. 1999. Environment and behavior of 2.5-million-year-old Bouri hominids. *Science* 284:625–29.
Dobley, J.F. 1990. Molecular evidence and the evolution of maize. *Economic Botany* 44: 6–27.
Doswell, C.R., R.L. Paliwal, and R.P. Cantrell. 1996. *Maize in the Third World*. Boulder: Westview Press.
Earls, J. 1986. Evolución de administración ecológia Inca. In *Andenes y Camellons en el Perú Andino: Historia, Presente y Futuro*, ed. C. de la Torre and M. Burga, 23–57. Lima: Consejo Nacional de Ciencia y Technologia.
———. 1989. *Planificación agrícola Andina: Bases para un Manejo Cibernético de Sistemas de Andenes*. Lima: Centro de Investigación, and Ediciones (Corporación Financiera del Desarrollo).
———. 1991. *Ecología y Agronomia en los Andes*. La Paz: Hisbol.
Ehleringer, J.R., and O. Björkman. 1977. Quantum yields for CO_2 uptake in C_3 and C_4 plants: Dependence on temperature, CO_2 and O_2 concentrations. *Plant Physiology* 59: 86–90.
Ehleringer, J.R., T.E. Cerling, and B.R. Helliker. 1997. C_4 photosynthesis, atmospheric CO_2 and climate. *Oecologia* 112:285–99.
Ehleringer, J.R., R.F. Sage, L.B. Flanagan, and R.W. Percy. 1991. Climate change and the evolution of C_4 photosynthesis. *Trends in Ecological Evolution* 6:95–99.
Erickson, C.L. 1986. Waru-waru: una technología agrícola de altiplano prehispánico. In *Andenes y Camellons en el Perú Andino: Historia, Presente y Futuro*, ed. C. de la Torre and M. Burga, 59–84. Lima: CONCYTEC.
Ford, R.I. 1985. Patterns of prehistoric food production in North America. In *Prehistoric food production in North America*, ed. R.I. Ford, 341–64. Anthropological Papers no. 75, Museum of Anthropology. Ann Arbor: University of Michigan.
Forde, C.D. 1931. Hopi agriculture and land ownership. *Journal of the Royal Anthropological Institute* 61:357–405.
Fritz, G.J. 1990. Multiple pathways to farming in precontact eastern North America. *Journal of Prehistory* 4:387–435.
Galinat, W.C. 1985a. The missing links between teosinte and maizc: A review. *Maydica* 30:137–60.

———. 1985b. Domestication and diffusion of maize. In *Prehistoric food production in North America*, ed. R.I. Ford, 245–78. Anthropological Papers no. 75, Museum of Anthropology. Ann Arbor: University of Michigan.

Galloway, J.H. 1977. The Mediterranean sugar industry. *Geographic Review* 67:177–92.

Gamble, C. 1994. *Timewalkers: The prehistory of global colonization.* Cambridge: Harvard University Press.

Garlake, P.S. 1973. *Great Zimbabwe.* London: Thames and Hudson.

Gerry, J.P. 1994. *Diet and status among the classic Maya: An isotopic perspective.* Doctoral dissertation, Harvard University. Ann Arbor: University Microfilms.

Greenfield, S.M. 1977. Madeira and the beginnings of New World sugar cultivation and plantation slavery: A study in institution building. In *Comparative perspectives on slavery in New World plantation societies*, ed. V. Rubin and A. Tuden. Annals of the New York Academy of Sciences no. 292. New York: New York Academy of Sciences.

———. 1979. Plantations, sugar cane and slavery. In *Roots and branches: Current directions in slave studies*, ed. M. Craton, 85–119. Toronto: Bergamon Press.

Grine, F.E. 1981. Trophic level differences between "gracile" and "robust" australopithecines: A scanning electron microscope analysis of occlusal events. *South African Journal of Science* 77:203–30.

———. 1987. The diet of South African Australopithecines based on a study of dental microwear. *L'Anthropologie* 91:467–82.

Grine, F.E., and R.F. Kay. 1988. Early hominid diets from quantitative image analysis of dental microwear. *Nature* 333:765–68.

Grove, A.T. 1995. Africa's climate in the Holocene. In *Archaeology of Africa: Food, metals and towns,* ed. T. Shaw, P. Sinclair, B. Andah, and A. Okpoko, 32–42. London: Routledge.

Hall, M. 1987. *The changing past: Farmers, kings, and traders in Southern Africa 200–1860.* Cape Town: David Phillip.

Harlan, J.R. 1992. Indigenous African agriculture. In *The origins of agriculture: An international perspective,* ed. W. Cowan and P.J. Watson, 59–70. Washington, D.C.: Smithsonian Institution Press.

———. 1995. The tropical African cereals. In *The archaeology of Africa: Food, metals, and towns,* ed. T. Shaw, P. Sinclair, B. Andah, and A. Okpoko, 53–60. London and New York: Routledge.

Harpending, H.C., M.A. Batzer, M. Gurven, L.B. Jorde, A.R. Rogers, and S.T. Sherry. 1998. Genetic traces of ancient demography. *Proceedings of the National Academy of Sciences, USA* 95:1961–67.

Harpending, H.C., S.T. Sherry, A.R. Rogers, and M. Stoneking. 1993. The genetic structure of ancient human populations. *Current Anthropology* 34:483–96.

Harris, D.R. 1998. The origins of agriculture in southwest Asia. *Review of Archaeology* 19:5–11.

Hastorf, C.A. 1990a. The effect of the Inca state on Sausa agricultural production and crop consumption. *American Antiquity* 55:262–90.

———. 1990b. One path to the heights: Negotiating political inequality in the Sausa of Peru. In *The evolution of political systems*, ed. S. Upman, 146–76. Cambridge: Cambridge University Press.

———. 1993. *Agriculture and the onset of political inequality before the Inca.* Cambridge: Cambridge University Press.

Hastorf, C.A., and S. Johannessen. 1993. Pre-Hispanic political change and the role of maize in the Central Andes of Peru. *American Anthropologist* 95:115–38.

Hathor, J. 1995. Parenchymatous tissues from the early Neolithic site E-75-6 at Nabta Playa, western desert, south Egypt. *Acta Palaeobotanica* 35:157–62.

Hayden, B. 1990. Nimrods, piscators, pluckers, and planters: The emergence of food production. *Journal of Anthropological Archeology* 9:31–69.

———. 1995. Pathways to power: principles for creating socio-economic inequalities. In *Foundations of social inequality*, ed. T.D. Price and G.M. Feinman, 15–86. New York: Plenum Press.

Hillman, G.C., S. Colledge, and D.R. Harris. 1989. Plant food economy during the epi-Paleolithic period at Tell Abu Hureyra, Syria: Dietary diversity, seasonality, and modes of exploitation. In *Foraging and farming: The evolution of plant exploitation,* ed. G.C. Hillman and D.R. Harris, 240–66. London: Unwin Hyman.

Hillman, G., E. Madeyska, and J. Hathor. 1989. Wild plant foods and diet at Late Paleolithic Wadi Kubbaniya: The evidence from charred remains. In *The prehistory of Wadi Kubbaniya, vol.2: Stratigraphy, paleoeconomy, and environment,* ed. F. Wendorf, R. Schild, and A.E. Close, 162–242. Dallas: Southern Methodist University Press.

Holl, A. 1985. Subsistence patterns of the Dhar Tichitt Neolithic, Mauretania. *African Archaeological Review* 3:131–62.

———. 1986. *Économie et Société Néolithique du Dhar Tichitt (Mauritánie).* Paris: Editions Recherce sur les Civilisations.

Huffman, T.N. 1996. *Snakes and crocodiles: Power and symbolism in Ancient Zimbabwe.* Johannesburg: University of the Witwatersrand Press.

Iltis, H.H. 1983. From teosinte to maize: the catastrophic sexual transmutation. *Science* 222:886–94.

Jacques-Félix, H. 1971. Appendix K, Grain impressions. In *The Tichitt Tradition: A late prehistoric occupation of the Southwestern Sahara* (P.J. Munson, doctoral dissertation). University of Illinois at Champaign-Urbana. Ann Arbor: University Microfilms.

Johannessen, S. 1993a. Farmers of the late woodland. In *Foraging and farming in the eastern woodlands*, ed. C.M. Scarry, 57–77. Gainesville: University of Florida Press.

———. 1993b. Food, dishes, and society in the Mississippi Valley. In *Foraging and farming in the eastern woodlands*, ed. C.M. Scarry, 182–205. Gainesville: University of Florida Press.

Jorde, L.B., M.J. Bamshad, W.S. Watkins, R. Zenger, A.E. Fraley, P.A. Krakowiak, K.D. Carpenter, H. Soodyall, R. Jenkins, and A.R. Rogers. 1995. Origins and affinities of modern humans: A comparison of mitochondrial and nuclear genetic data. *American Journal of Human Genetics* 57:523–38.

Jorde, L.B., A.R. Rogers, M. Bamshad, W.S. Watkins, P. Krakowiak, S. Sung, J. Kere, and H.C. Harpending. 1997. Microsatellite diversity and the demographic history of modern humans. *Proceedings of the National Academy of Sciences, USA* 94:3100–3103.

Kirkby, A.V.T. 1973. *The use of land and water resources in the past and present valley of Oaxaca.* Memoirs of the Museum of Anthropology no.5. Ann Arbor: University of Michigan.

Kirchoff, P. 1943. Mesoamerica. *Acta Americana* 1:92–107.

Knapp, G. 1991. *Andean ecology: Adaptive dynamics in Ecuador.* Dellplain Latin American Studies no. 27. Boulder: Westview Press.

Krige, E.J., and J.D. Krige. 1943. *Realm of a rain queen: A study of the pattern of Lovedu Society.* London: Oxford University Press.

Kroll, H. 1981. Thessalische Kulturpflanzen. *Zeitschrift für Archäologie* 15:97–103.

———. 1983. Kastanas: Ausgrabungen in einem Siedlungshügel der Bronze- und Eisenzeit Makedoniens 1975–1979. *Die Pflanzenfunde Prähistorische Archäologie in Südost-europa 2.* Berlin: Spiess.

Lathrap, D.W. 1970. *The upper Amazon.* London: Thames and Hudson.

———. 1987. The introduction of maize in prehistoric eastern North America: The view from Amazonia and the Santa Elena peninsula. In *Emergent horticultural economies of the eastern oodlands*, ed. W.F. Keegan, 345–71. Occasional Paper no. 7, Center for Archaeological Investigations. Carbondale: Southern Illinois University.

Leakey, M.G., F. Spoor, F.H. Brown, P.N. Gathogo, C. Kiarie, L.N. Leakey, and I. McDougall. 2001. New hominin genus from eastern Africa shows diverse middle Pliocene lineages. *Nature* 410:433–40.

Lee-Thorp, J.A., J.F. Thackeray, and N.J. van der Merwe. 2000. The hunters or the hunted revisited. *Journal of Human Evolution* 39:565–76.

Lee-Thorp, J.A, and N.J. van der Merwe. 1987. Carbon isotope analysis of fossil bone apatite. *South African Journal of Science* 83:712–15.

———. 1993. Stable carbon isotope studies of Swartkrans fossils. In *Swartkrans: A cave's chronicle of early man* (monograph no. 8), ed. C.K. Brain, 251–56. Pretoria: Transvaal Museum.

Lippi, R.D., R.M. Bird, and D.M. Stemper. 1984. Maize recovered at La Ponga: An early Ecuadorian site. *American Antiquity* 49:118–24.

Little, E.A., and M.J. Schoeninger. 1995. The late woodland diet on Nantucket Island and the problem of maize in coastal New England. *American Antiquity* 60:351–68.

Long, A.B., B.F. Benz, D.J. Donahue, A.J.T. Tull, and L.J. Toolin. 1989. First direct AMS dates on early maize from Tehuacan, Mexico. *Radiocarbon* 31:1030–35.

Lynott, M.J., T.W. Boutton, J.E. Price, and D.E. Nelson. 1986. Stable carbon isotopic evidence for maize agriculture in southeast Missouri and northeast Arkansas. *American Antiquity* 51:51–65.

Maggs, T. and V. Ward. 1984. Early Iron Age sites in the Muden area of Natal. *Annals of the Natal Museum* 26:105–40.

MacNeish, R.S. 1967. A summary of the subsistence. In *The prehistory of the Tehuacan Valley, vol.1: Environment and subsistence*, ed. D.S. Byers, 290–309. Austin: University of Texas Press.

———. 1981. Tehuacan's accomplishments. In *Supplement to the handbook of South American Indians, vol. 1: Archaeology,* ed. J.A. Sabloff and V.R. Bricker, 31–47. Austin: University of Texas Press.

McIntosh, S.K., and R.J. McIntosh. 1996. Cities without citadels: Understanding urban origins along the middle Niger. In *The archaeology of Africa: Food, metals, and towns*, ed. T. Shaw, P. Sinclair, B. Andah, and A. Okpoko, 622–41. London and New York: Routledge.

Meadow, R.H. 1998. Pre- and proto-historic agricultural and pastoral transformations in northwestern south Asia. *Review of Archaeology* 19:12–21.

Minnis, P. 1992. Earliest plant cultivation in the desert borderlands of North America. In *The origins of agriculture: An international perspective*, ed. C.W. Cowan and P.J. Watson, 121–41. Washington, D.C.: Smithsonian Institution Press.

Morgan, M.E., J.D. Kingston, and B.D. Marino. 1994. Carbon isotopic evidence for the emergence of C_4 plants in the Neogene from Pakistan and Kenya. *Nature* 367:162–65.

Mintz, S.W. 1959. The plantation as a sociocultural type. In *Plantation systems of the New World*, 42–50. Social Science Monographs 7. Washington, D.C.: Pan American Union.

———. 1985. *Sweetness and power: The place of sugar in modern history*. New York: Elizabeth Sifton Books.

Munson, P.J. 1971. *The Tichitt tradition: A late Prehistoric occupation of the southwestern Sahara* (doctoral dissertation, University of Illinois at Champaign-Urbana). Ann Arbor: University Microfilms.

———. 1976. Archaeological data on the origins of cultivation in the southwestern Sahara and their implications for West Africa. In *Origins of African plant domestication*, ed. J.R. Harlan, J.M. J. de Wet, and A. Stemler, 187–209. The Hague and Paris: Mouton.

———. 1989. About 'Économie et Société Néolithic du Dhar Tichitt (Mauritanie)'. *Sahara* 2:106–108.

Murray, M., and M.J. Schoeninger. 1988. Diet status, and complex social structure in

Iron Age central Europe: Some contributions of bone chemistry. In *Tribe and polity in late prehistoric Europe,* ed. D.B. Gibson and M.N. Geselowitz, 155–76. New York: Plenum Press.

Muzzolini, A. 1995. The emergence of food-producing economy in the Sahara. In *The archaeology of Africa: Food, metals, and towns*, ed. T. Shaw, P. Sinclair, B. Andah, and A. Okpoko, 227–39. London and New York: Routledge.

National Research Council. 1989. *Lost Crops of the Incas: Little-known plants of the Andes with promise for worldwide cultivation.* Washington, D.C.: National Academy Press.

———. 1996. *Lost crops of Africa, vol. 1: Grains.* Washington, D.C.: National Academy Press.

Netherly, P.J. 1978. *Local level lords on the North Coast of Peru* (doctoral dissertation, Cornell University). Ann Arbor: University Microfilms.

Norman, M.J.T., C.J. Pearson, and P.G.E. Searle. 1995. *Ecology of tropical food crops.* Cambridge: Cambridge University Press.

Norr, L. 1995. Interpreting dietary maize from bone stable isotopes in the American tropics: The state of the art. In *Archaeology in the lowland American tropics,* ed. P.W. Stahl, 198–223. Cambridge: Cambridge University Press.

Pearsall, D.M. and D.R. Piperno. 1990. Antiquity of maize cultivation in Ecuador: Summary and re-evaluation of the evidence. *American Antiquity* 55:324–37.

Pickford, M., and B. Senut. "Millennium ancestor," a 6-million year old bipedal hominid from Kenya. *South African Journal of Science* 97:22.

Piperno, D.R. 1990. Aboriginal culture and land usage in the Amazon basin, Ecuador. *Journal of Archaeological Science* 17:665–77.

Piperno, D.R., M.B. Bush, and P.A. Colinvaux. 1990. Paleoenvironments and human occupation in the late glacial Panama. *Quaternary Research* 33:108–16.

Piperno, D.R., K.H. Clary, R.G. Cooke, A.J. Ranere, and D. Weiland. 1985. Preceramic maize in central Panama: Phytolith and pollen evidence. *American Anthropologist* 87: 871–78.

Price, T.D., and A.B. Gebauer, eds. 1995. *Last hunters, first farmers: New perspectives on the prehistoric transition to agriculture.* Santa Fe, N.M.: School of American Research.

Raymond, J.S. 1988. Subsistence patterns during the early Formative in the Valdivia Valley, Ecuador. In *Diet and subsistence: Current archaeological perspectives*, ed. B.V. Kennedy and G.M. LeMoine, 159–64. Proceedings of the 19th Chacmool Conference. Calgary Alberta: University of Calgary.

Reed, D.M. 1994. Ancient Maya diet at Copán, Honduras, as determined through the analysis of stable carbon and nitrogen isotopes. In *Paleonutrition: The diet and health of prehistoric Americans,* ed. K.D. Sobolik, 210–21. Occasional Paper no. 22, Center for Archaeological Investigation. Carbondale: Southern Illinois University.

Relethford, J.H. 2001. *Genetics and the search for modern human origins.* New York: Wiley-Liss.

Ren, S. 1995. Several important achievements in Neolithic culture before 5000 B.C. *Kaogu* 1995(1):37–49.

Riley, T.J. 1987. Ridged-field agriculture and the Mississippian agricultural pattern. In *Emergent horticultural economies of the eastern woodlands*, ed. W.W. Keegan, 295–304. Occasional Paper no. 7, Center for Archaeological Investigations. Carbondale: Southern Illinois University.

Riley, T., G.R. Walz, C.J. Baereis, A.C. Fortier, and K.E. Parker. 1994. Accelerator mass spectrometry (AMS) dates confirm early *Zea mays* in the Mississippi River valley. *American Antiquity* 59:490–98.

Robinson, J.T. 1954. Prehominid dentition and hominid evolution. *Evolution* 8:324–34.

Rodwell, S. 1984. China's earliest farmers: The evidence from Cishan. *Bulletin of the Indo-Pacific Prehistory Association* 5:55–63.

Rogers, A.R., and L.B. Jorde. 1995. Genetic evidence on modern human origins. *Human Biology* 67:1–36.
Roosevelt, A.C. 1980. *Parmana: Prehistoric maize and manoic subsistence along the Amazon and Orinoco*. New York: Academic Press.
Roosevelt, A.C., R.A. Housley, M. Imazio de Silveira, S. Maranca, and R. Johnson. 1991. Eighth millennium pottery from a prehistoric shell midden in the Brazilian Amazon. *Science* 254:1621–24.
Rose, J.C., M.K. Marks, and L.L. Tieszen. 1991. Bioarchaeology and subsistence in the central and lower portions of the Mississippi Valley. In *What mean these bones?*, ed. M.L. Powell, P.S. Bridges, and A.M. Mires, 7–21. Tuscaloosa: University of Alabama Press.
Sage, R.F. 1995. Was low atmospheric CO_2 during the Pleistocene a limiting factor for the origin of agriculture? *Global Change Biology* 1:93–106.
Sanders, W.T., J.R. Parsons, and R.S. Santley. 1979. *The Basin of Mexico: Ecological processes in the evolution of civilization*. New York: Academic Press.
Sauer, J.D. 1993. *Historical geography of crop plants*. Boca Raton, Ann Arbor, London, and Tokyo: CRC Press.
Scarry, C.M. 1993. Agricultural risk and the beginning of the Moundville chiefdom. In *Foraging and farming in the eastern woodlands*, ed. C.M. Scarry, 157–81. Gainesville: University of Florida Press.
Schoeninger, M.J. 1995. Stable isotope studies in human evolution. *Evolutionary Anthropology* 4:83–98.
Schoeninger, M.J., H.T. Bunn, S.S. Murray, T. Pickering, and J. Moore. 2001. Meat-eating by the fourth African ape. In *Meat-eating and human evolution*, ed. C.B. Stanford and H.T. Bunn, 179–95. New York: Oxford University Press.
Schoeninger, M.J. and J. Moore. 1992. Bone stable isotope studies in archaeology. *Journal of World Prehistory* 6:247–96.
Shaw, T., P. Sinclair, B. Andah, and A. Okpoko, eds. 1995. *The archaeology of Africa: Foods, metals, and towns*. London and New York: Routledge.
Smith, B.D. 1995. *The emergence of agriculture*. Scientific American Library. New York: W.H. Freeman.
Smithers, R.H.N. 1983. *The mammals of the Southern African ubregion*, 347–51. Pretoria: University of Pretoria Press.
Sponheimer, M., and J.A. Lee-Thorp. 1999. Isotopic evidence for the diet of an early hominid, *Australopithecus africanus*. *Science* 283:368–70.
Stemler, A. 1990. A scanning electron microscopic analysis of plant impressions from the sites of Kadero, El Zakiab, Um Direiwa, and El Kadada. *Archéologie du Nil Moyen* 4:87–105.
Stuiver, M., P.J. Reimer, and R. Reimer. 2000. CALIB Radiocarbon Calibration, HTML version 4.2. Website http://depts.washington.edu/qil/calib/index.html.
Thorne, A.G., and M.H. Wolpoff. 1992. The multiregional evolution of humans. *Scientific American* 266:76–83.
Tuross, N., M.L. Fogel, L. Newsom, and G.H. Doran. 1994. Subsistence in the Florida Archaic: The stable-isotope and archaeobotanical evidence from the Windover site. *American Antiquity* 59:288–303.
Tykot, R.H., N.J. van der Merwe, and N. Hammond. 1996. Stable isotope analysis of bone collagen, bone apatite, and tooth enamel in the reconstruction of human diet: A case study from Cuello, Belize. In *Archaeological chemistry: Organic, inorganic, and biochemical analysis*, ed. M.V. Orna, 355–65. Washington, D.C.: American Chemical Society.
Underhill, A.P. 1997. Current issues in Chinese Neolithic archaeology. *Journal of World Prehistory* 11:103–61.
van der Merwe, N.J., J.A. Lee-Thorp, and J.S. Raymond. 1993. Light stable isotopes and

the subsistence of Formative cultures at Valdivia, Ecuador. In *Prehistoric human bone at the molecular level,* ed. J.B. Lambert and G. Grupe, 63–98. Berlin: Springer-Verlag.
van der Merwe, N.J., J.F. Thackeray, J.A. Lee-Thorp, and J. Luyt. 2003a. The isotopic ecology and diet of *Australopithecus africanus* at Sterkfontein, South Africa. *Journal of Human Evolution* 44:581–97.
van der Merwe, N.J., and H. Tschauner. 1999. C_4 plants and the development of human societies. In *C_4 plant biology,* ed. R.F. Sage and R.K. Monson, 509–50. London and New York: Academic Press.
van der Merwe, N.J., R.H. Tykot, N. Hammond, and K. Oakberg. 2000. Diet and animal husbandry of the Preclassic Maya at Cuello, Belize: Isotopic and zooarchaeological evidence. In *Biogeochemical approaches to Paleodietary analysis*, ed. S.H. Ambrose and M.A. Katzenberg, 23–38. New York: Kluwer Academic/ Plenum Publishers.
van der Merwe, N.J., and J.C. Vogel. 1978. ^{13}C content of human collagen as a measure of prehistoric diet in woodland North America. *Nature* 276:815–16.
van der Merwe, N.J., R.F. Williamson, S. Pfeiffer, S.C. Thomas, and K.O. Allegretto. 2003b. The Moatfield ossuary: Isotopic dietary analysis of an Iroquoian community, using dental tissue. *Journal of Anthropological Archaeology*, in press.
Vigilant, L., M. Stoneking, H. Harpending, K. Hawkes, and A.C. Wilson. 1991. African populations and the evolution of human DNA. *Science* 253:1503–1507.
Wagner, G.E. 1994. Corn in eastern woodlands late prehistory. In *Corn and culture*, ed. S. Johannessen and C.A. Hastorf, 335–46. University of Minnesota Publications in Anthropology 5. Boulder: Westview Press.
Warner, J.N. 1962. Sugar cane: An indigenous Papuan cultigen. *Ethnology* 1:405–11.
Wasylikowa, K., J.R. Harlan, J. Evans, F. Wendorf, R. Schild, A.E. Close, H. Królik, and R.A. Housley. 1993. Examination of botanical remains from early Neolithic houses at Nabta Playa, western desert, Egypt, with special reference to sorghum grains. In *The archaeology of Africa: Food, metals, and towns*, ed. T. Shaw, P. Sinclair, B. Andah, and A. Okpoko, 163–226. London and New York: Routledge.
Wasylikowa, K., R. Schild, F. Wendorf, H. Królik, L. Fubiak-Martens, and J.R. Harlan. 1995. Archaeobotany of the early Neolithic site E-75-6 at Nabta Playa, western desert, south Egypt. *Acta Palaeobotanica* 35:135–55.
Watson, A.M. 1974. The Arab agricultural revolution and its diffusion, 700–1100. *Journal of Economic History* 34:8–35.
Watson, P.J. 1995. Explaining the transition to agriculture. In *Last hunters, first farmers*, ed. T.D. Price and A.B. Gebauer, 21–37. Santa Fe, N.M.: School of American Research.
Weber, S.A. 1991. *Plants and Harappan subsistence*. New Delhi: Oxford and IBH.
Wendorf, F., A.E. Close, R. Schild, K. Wasylinkowa, R.A. Housley, J.R. Harlan, and H. Królik. 1992. Saharan exploitation of plants 8,000 years b.p. *Nature* 359:721–24.
Wetterstrom, W. 1998. The origins of agriculture in Africa, with particular reference to sorghum and pearl millet. *Review of Archaeology* 19:30–46.
White, T.D, B. Asfaw, D. DeGusta, H. Gilbert, G.D. Richards, G. Suwas, and F.C. Howell. 2003. Pleistocene *Homo sapiens* from Middle Awash, Ethiopia. *Nature* 423:747–52.
White, C.D., P.F. Healy, and H.P. Schwarcz. 1993. Intensive agriculture, social status, and Maya diet at Pacbitun, Belize. *Journal of Anthropological Research* 49:347–75.
White, C.D., and H.P. Schwarcz. 1989. Ancient Maya diet as inferred from isotopic and elemental analysis of human bone. *Journal of Archaeological Science* 16:451–74.
White, T.D., G. Suwa, and B. Asfaw. 1994. *Australopithecus ramidus*, a new species of early hominid from Aramis, Ethiopia. *Nature* 371:306–12.
White, T.D., G. Suwa, S. Simpson, and B. Asfaw. 2000. Jaws and teeth of *A. afarensis* from Maka, Middle Awash, Ethiopia. *American Journal of Physical Anthropology* 111:45–68.

Williamson, R.F., and S. Pfeiffer, eds. 2003. Bones of the ancestors: Archaeology and osteobiography of the moatfield ossuary, in press. Canadian Museum of Civilization, Mercury Series.

Wills, W.H. 1988. *Early prehistoric agriculture in the American Southwest*. Santa Fe, N.M.: School of American Research.

Wilson, A.L., and R. Cann. 1992. The recent African genesis of humans. *Scientific American* 266:68–73.

Woods, W.I. 1987. Maize agriculture and the late prehistoric: a characterization of settlement location strategies. In *Emergent horticultural economies of the eastern woodlands*, ed. W.W. Keegan, 275–294. Occasional Paper no. 7, Center for Archaeological Investigations. Carbondale: Southern Illinois University.

Wright, L.E. 1994. *The sarifice of the Earth? Diet, health, and inequalities in the Pasión Maya Lowlands* (doctoral dissertation University of Chicago). Ann Arbor: University Microfilms.

Wymer, D.A. 1994. The social context of early maize in the mid-Ohio Valley. In *Corn and culture*, ed. S. Johannessen and C.A. Hastorf, 411–26. University of Minnesota Publications in Anthropology 5. Boulder: Westview Press.

15. The Carbon Cycle over the Past 1000 Years Inferred from the Inversion of Ice Core Data

Cathy Trudinger, Ian Enting, David Etheridge,
Roger Francey, and Peter Rayner

15.1 Introduction

Atmospheric CO_2 has increased by almost 30% since preindustrial times as a result of anthropogenic activities. The anthropogenic CO_2 perturbation sits on top of the natural variability of the carbon cycle that occurs on a range of timescales. To understand CO_2 variations over the past 1000 years, we need to consider exchanges between the atmosphere, the oceans and the terrestrial biosphere, and, in recent centuries, the anthropogenic source of CO_2 from the burning of fossil fuels. Climate is the most important driver of natural variations in exchange between the reservoirs. In the oceans, temperature and wind-speed influence CO_2 solubility, air-sea gas exchange rates, biological activity, and ocean circulation, all of which influence CO_2 fluxes. Temperature and precipitation influence photosynthesis and respiration for the terrestrial biosphere.

The anthropogenic CO_2 perturbation is due to fossil fuel burning, cement production, and changes in land use: for example, biomass burning, harvest of forests, clearing of natural forests for agriculture, and afforestation (Houghton 1995; see also Kammen and Marino 1993). The oceans and terrestrial biosphere have responded to the higher levels of CO_2 in the atmosphere by taking up some of the extra CO_2. In the case of oceans, this response is due to the partial pressure differences across the air-sea interface; for the terrestrial biosphere, it is due to the CO_2 fertilization effect (Wullschleger, Post, and King 1995). To what extent this response will continue into the future is the subject of ongoing research.

Measurements from Antarctic ice cores are an important source of CO_2 data prior to continuous high-quality direct atmospheric measurements, which began in 1958. Air trapped in bubbles in polar ice is extracted and measured to reconstruct past levels. The processes involved in storing air in the ice, which are mainly bubble trapping and diffusion through the porous firn layer, influence the trapped concentrations (see Chapter 4). They cause fractionation and smoothing of the concentration record in time relative to the actual atmospheric variations, so that high-frequency variations are not recorded in the ice. The degree of smoothing and fractionation depends on the snow accumulation rate and the depth of the firn layer and can be quantified with a numerical model of the firn processes (Schwander, Stauffer, and Sigg 1988; Trudinger et al. 1997). As Antarctica is far from most CO_2 source regions, and CO_2 is well mixed in the atmosphere on timescales of more than a year, the ice core measurements reflect global changes.

The minor stable isotope ^{13}C in CO_2 can be used to distinguish between the uptake of atmospheric CO_2 by the oceans and terrestrial biosphere due to the different degrees of isotopic fractionation associated with exchange with each reservoir. Uptake of CO_2 by plants and the oceans both discriminate against ^{13}C, but the discrimination associated with the uptake by plants is about a factor of 10 greater than that for the oceans (Keeling et al. 1989). Isotopic composition is usually expressed in the $\delta^{13}C$ notation, defined as

$$\delta^{13}C = \left(\frac{^{13}C/^{12}C_{\text{sample}}}{^{13}C/^{12}C_{\text{standard}}} - 1 \right) \times 1000 \tag{15.1}$$

in units of ‰ (permil). The more negative the $\delta^{13}C$ value, the more depleted the sample is in ^{13}C relative to ^{12}C. Preindustrially, atmospheric $\delta^{13}C$ was about −6.5‰. Present levels are below −8.0‰ and continue to decrease due to the addition of fossil fuel CO_2, which is depleted in ^{13}C. While $\delta^{13}C$ can tell us more about carbon fluxes than is possible with just CO_2 alone, there are complications.

The budget of CO_2 (in GtC yr^{-1}, where GtC is gigatonnes carbon, 1 Gt = 10^{15} g) can be expressed as

$$\begin{aligned} \frac{d}{dt}C_a &= F_f + F_{ba} - F_{ab} + F_{oa} - F_{ao} \\ &= F_f + F_b + F_o \end{aligned} \tag{15.2}$$

where C_a is the atmospheric CO_2 content, F_f is the source due to fossil fuel burning, and F_{ab}, F_{ba}, F_{ao}, and F_{oa} denote the (one-way) *gross* fluxes from the atmosphere to biosphere, biosphere to atmosphere, atmosphere to oceans, and oceans to atmosphere, respectively. The gross fluxes, of order 50–100 GtC yr^{-1}, are due to photosynthesis and respiration for the biosphere and molecular diffusion across the air-sea interface for the oceans, and they are almost balanced annually. The *net* fluxes leaving the atmosphere, today of the order of a few

GtC yr^{-1}, are given by the differences between the gross fluxes, $F_o = F_{oa} - F_{ao}$ for the oceans and $F_b = F_{ba} - F_{ab}$ for the terrestrial biosphere.

The ^{13}C budget can be expressed as

$$\begin{aligned}\frac{d}{dt}(C_a\delta_a) &= C_a\frac{d}{dt}\delta_a + \delta_a\frac{d}{dt}C_a \\ &\approx F_f(\delta_f - \delta_a) + (F_{ba} - F_{ab})\varepsilon_{ab} + F_{ba}(\delta_a^b - \delta_a) \\ &+ (F_{oa} - F_{ao})\varepsilon_{ao} + F_{oa}(\delta_a^o - \delta_a)\end{aligned} \tag{15.3}$$

(Tans, Berry, and Keeling 1993), where ε_{ij} (in permil) is the isotopic shift that occurs for a flux from reservoir i to reservoir j, and δ_a^b and δ_a^o are the isotopic ratios that the atmosphere would have if it were in isotopic equilibrium with the terrestrial biosphere and ocean, respectively. The addition of ^{13}C-depleted anthropogenic CO_2 to the atmosphere disturbs the isotopic equilibrium with the oceans and terrestrial reservoirs. The gross CO_2 fluxes reduce this disequilibrium. Terms in the atmospheric ^{13}C budget describing this effect, F_{ba} $(\delta_a^b - \delta_d)$ and F_{oa} $(\delta_a^o - \delta_a)$, are often referred to as *isofluxes* (isotopic disequilibrium fluxes). The degree of isotopic disequilibrium between the reservoirs is difficult to measure but can be modeled. Other processes can also disturb the isotopic equilibrium between the reservoirs. For example, isotopic fractionation can vary with sea-surface temperature (Heimann and Keeling 1989) or in response to water stress on plant stomata (Randerson et al. 2002). These second-order effects are quite uncertain on the global average, particularly on decadal timescales, and have generally been neglected. They will also be neglected here, but they are an aspect of ^{13}C interpretation that certainly needs further investigation.

We will use a simple box model to investigate an ice core record of CO_2 and δ^{13}C from Law Dome in Antarctica. Three modeling approaches have traditionally been used to study variations in CO_2 over recent centuries. The first is a *forward* calculation, where the model determines variations in CO_2 and δ^{13}C due to specified fluxes. The other two methods, the *single deconvolution* and *double deconvolution*, are inverse methods, and as such they deduce fluxes that are required to match the changes measured in the ice core record. In a single deconvolution, anthropogenic inputs are specified, CO_2 uptake by the ocean calculated, and terrestrial uptake deduced to match the observed variations in CO_2. In a double deconvolution, the anthropogenic fluxes are specified, and the net ocean and terrestrial fluxes determined simultaneously to match both CO_2 and δ^{13}C. All three approaches will be used here to study different aspects of the Law Dome ice core record. By using these different approaches in combination, we can compare our current, somewhat limited, knowledge of the fluxes and the processes responsible for them with what the concentration and isotope data tell us. This will help us to gain a better understanding of the processes and to develop better models of the carbon cycle, which are required for predictions of future levels of CO_2. This chapter is a summary of work presented in Trudinger et al. (1999, 2002b) and Trudinger (2000).

15.2 CO_2 and $\delta^{13}C$ Over the Past 1000 Years

The Law Dome CO_2 ice core record (Fig. 15.1) from Etheridge et al. (1996) consists of ice core measurements from three cores on Law Dome, East Antarctica: DE08, DE08-2, and DSS, as well as firn measurements from DE08-2. The three cores were drilled using different techniques and dated independently. The snow accumulation rate at these sites is very high, giving the ice core record high time resolution and low smoothing due to the bubble trapping process. The smoothing is about 10 years for DE08 and DE08-2 and about 18 years for DSS (Trudinger 2000). The ice core record overlaps the modern CO_2 record, providing confirmation that the record represents atmospheric concentrations.

The most obvious feature in the CO_2 record is the strong increase in CO_2 since about 1800 due to the anthropogenic CO_2 source. In addition there is significant variability on a range of timescales, mainly due to the effects of climate on exchange between the atmosphere, oceans, and terrestrial biosphere. Between 1550 and 1800, CO_2 was about 6 ppm lower than it was during the

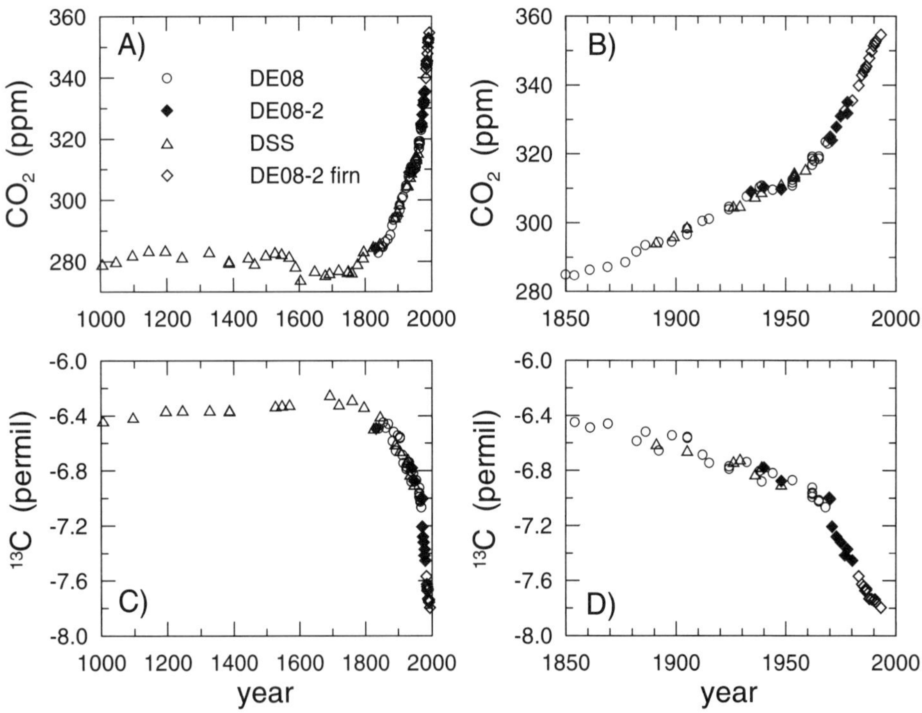

Figure 15.1. **(A)** Law Dome CO_2 ice core record over the past 1000 years (Etheridge et al. 1996); **(B)** CO_2 record, shown again since 1850; **(C)** Law Dome $\delta^{13}C$ ice core record (Francey et al. 1999); and **(D)** $\delta^{13}C$ record shown since 1850.

rest of the preindustrial period. This low CO_2 occurs during the Little Ice Age (LIA), a period of cool, dry conditions, particularly in the Northern Hemisphere (Lamb 1982; Grove 1988). The ice core shows a period of CO_2 stabilization or possibly even a decrease in CO_2 in the 1940s, at a time when the fossil fuel source was significant. The Law Dome ice core record of $\delta^{13}C$ (Francey et al. 1999) is also shown in Fig. 15.1. The record shows how $\delta^{13}C$ has decreased over the industrial period. $\delta^{13}C$ was at its highest during the LIA period, and seems not to have mirrored the flattening seen in CO_2 during the 1940s.

There is generally consistency between the Law Dome record and other ice core records of CO_2 and $\delta^{13}C$ over the past 1000 years (Neftel et al. 1985; Raynaud and Barnola 1985; Friedli et al. 1986; Pearman et al. 1986; Etheridge, Pearman, and de Silva 1988; Siegenthaler et al. 1988; Nakazawa et al. 1993; Barnola et al. 1995; Kawamura et al. 2000). However, two main differences between CO_2 records are worth noting. First, the Law Dome record does not show the increase in CO_2 in the thirteenth century that Siegenthaler et al. 1988 and Barnola et al. 1995 saw in their records. Second, the sharp decrease around 1600 seen in the Law Dome record is not seen in other records. Analytical problems associated with the earlier records may help to explain some of the differences. In addition, the Law Dome record has less smoothing in time than do the other records due to the very high accumulation rate; some differences might be expected due to the different smoothing at each site. The differences between $\delta^{13}C$ records are often larger than their measurement uncertainties. However, the Law Dome ice core record involved detailed study of systematic influences on the measurements, which seems not to have been done for the other records. Sponge records of $\delta^{13}C$ provide a method to reconstruct ocean mixed-layer $\delta^{13}C$ over the past 1000 years. The sponge records of Böhm et al. (2002) compare well with the ice core records. Due to actual differences between atmospheric and mixed-layer $\delta^{13}C$, the sponge record cannot easily be used to confirm or reject particular ice core records.

The ice core records mentioned above are all from Antarctic sites. CO_2 has also been measured in Greenland ice cores, but they show CO_2 levels that are up to 20 ppm higher than Antarctic levels (Anklin et al. 1995), with lower $\delta^{13}C$ (Francey et al. 1997). These differences are too large to be real interhemispheric gradients; the current interhemispheric gradient caused by the input of fossil fuel CO_2 (mainly into the Northern Hemisphere) is only 3 to 4 ppm. Based on measurements of CO_2, $\delta^{13}CO_2$, and CO, it is likely that the Greenland ice cores are contaminated with CO_2 derived from organic impurities (Francey et al. 1997; Haan and Raynaud 1998).

While the possibility of minor CO_2 contamination in Antarctic ice cannot be completely ruled out, Law Dome CO_2 measurements show no evidence of contamination within the analytical reproducibility of about 1 ppm. Comparison between measurements of ice core air, firn air, and atmospheric CO_2 levels, between adjacent samples of the same core with the same air age, and between CO_2 records from three Law Dome ice cores at different locations (with different accumulation and thus age-depth relations, and different chemical impurity levels) show no unexplained differences (Etheridge et al. 1996; Etheridge 1999). The Law Dome ice samples were all relatively shallow (100s of meters) and thus had low bubble pressure and no evidence of clathrated air; they were also measured typically within only six years of drilling. The level of organic material in Antarctic ice is 10 to 100 times lower than in Greenland. Impurity levels are generally very low in the high accumulation snowfall of Law Dome, and CO_2 levels do not correlate with impurity levels (Etheridge 1999). Tschumi and Stauffer (2000) discussed in situ production of CO_2 due to contamination in ice. However, the Antarctic ice they studied (Byrd) was drilled many years ago and was very deep (and possibly clathrated) ice, so their results are unlikely to be applicable to Law Dome CO_2 over the past 1000 years. We therefore take the Law Dome record as an accurate representation of (smoothed) atmospheric concentrations, and we determine below what the variations in concentration and isotopic ratio imply about the net fluxes.

15.3 Modeling: Forward

The next three sections focus on interpreting features in the Law Dome record using a simple carbon cycle model (Enting and Lassey 1993; Lassey, Enting, and Trudinger 1996; Trudinger et al. 1999; Trudinger 2000). The model shown in Fig. 15.2 has 2 atmospheric boxes (troposphere and stratosphere), an ocean mixed-layer box, an Oeschger et al. (1975) diffusive deep ocean (18 layers), and a 2-box terrestrial biosphere. In addition to CO_2, the model includes the isotopes ^{13}C and ^{14}C. The model parameters are calibrated to give a good fit to measurements of CO_2 and $^{14}CO_2$.

We ran the model in forward mode, using best estimates of the CO_2 net fluxes over recent centuries, and compared the calculated CO_2 and $\delta^{13}C$ with the ice core measurements. The fluxes used were: (a) the source due to fossil fuel burning (Keeling 1997; Marland and Boden 1997); (b) the net source due to land-use change from Houghton (1995); (c) enhanced terrestrial uptake due to CO_2 fertilization (using the formulation by Allen et al. (1987) with parameters calibrated as part of the model's calibration routine); and (d) ocean uptake determined with the model.

For the calculated CO_2, $\delta^{13}C$ and the net fluxes for the forward calculation, see Fig. 15.3. The model has been tuned to match the overall industrial change in CO_2. It also matches the overall change in $\delta^{13}C$ very well ($\delta^{13}C$ was not used for calibration of the model parameters, but tuning to ^{14}C data establishes parameters that are also important for ^{13}C). The average net oceanic and terrestrial uptake in the 1980s is similar to generally accepted estimates of these quantities (Schimel et al. 1995). However, some decadal, multidecadal, and century-scale variations in the ice core record are not captured by this calculation. These variations will be the focus of the remainder of this chapter. We will look first at the LIA period.

Although the Law Dome record shows sustained low levels of CO_2 for a

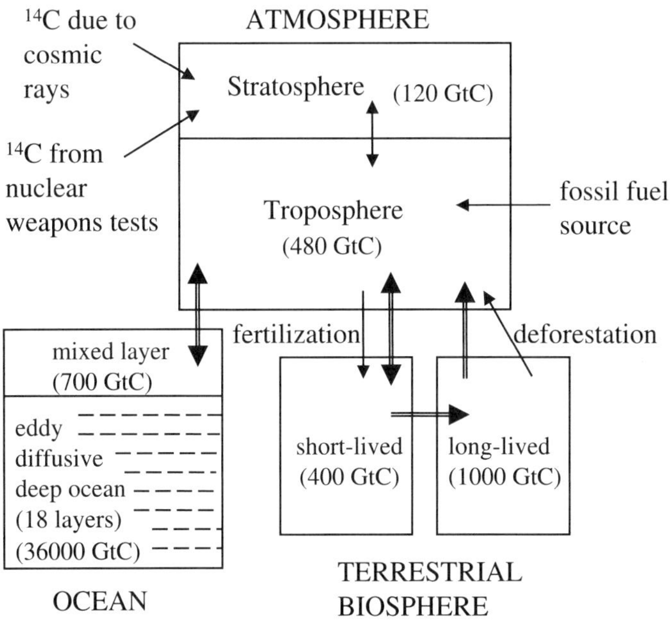

Figure 15.2. Structure of the box diffusion model, showing the fluxes between reservoirs and typical reservoir sizes.

number of centuries during the LIA period, the LIA was not a prolonged, globally synchronous climate event (Jones and Bradley 1992). Most of the well-known Northern Hemisphere climate reconstructions do show colder conditions on average over the years 1600 to 1800 compared to conditions at other times during the past 1000 years (Briffa and Osborn 2002). Although the climate reconstructions differ in their timing of the onset of cooling, the timing of the decrease in CO_2 is very clear. We will explore simple mechanisms to try to explain the high $\delta^{13}C$ and low CO_2 with temperature as a driver. The decrease in atmospheric CO_2 is unlikely to have been the primary cause of the LIA cooling (Etheridge et al. 1996). It is more likely that reduced temperatures, probably driven by volcanic and solar forcing, affected the carbon exchange between the different reservoirs, changing atmospheric CO_2 levels. The feedback of reduced CO_2 on temperature would have been small (Etheridge et al. 1996).

The forward model calculation was repeated with a hypothetical temperature record of lower temperatures during the LIA (Fig. 15.4A) and either an ocean response or a terrestrial response to this temperature variation (Trudinger et al. 1999). The hypothetical temperature record was very idealized and created so that modeled CO_2 would roughly match changes in the CO_2 ice core record. The ocean response included temperature dependence of the ocean buffer factor (Revelle and Suess 1957) and isotopic fractionation (Heimann and Keeling 1989),

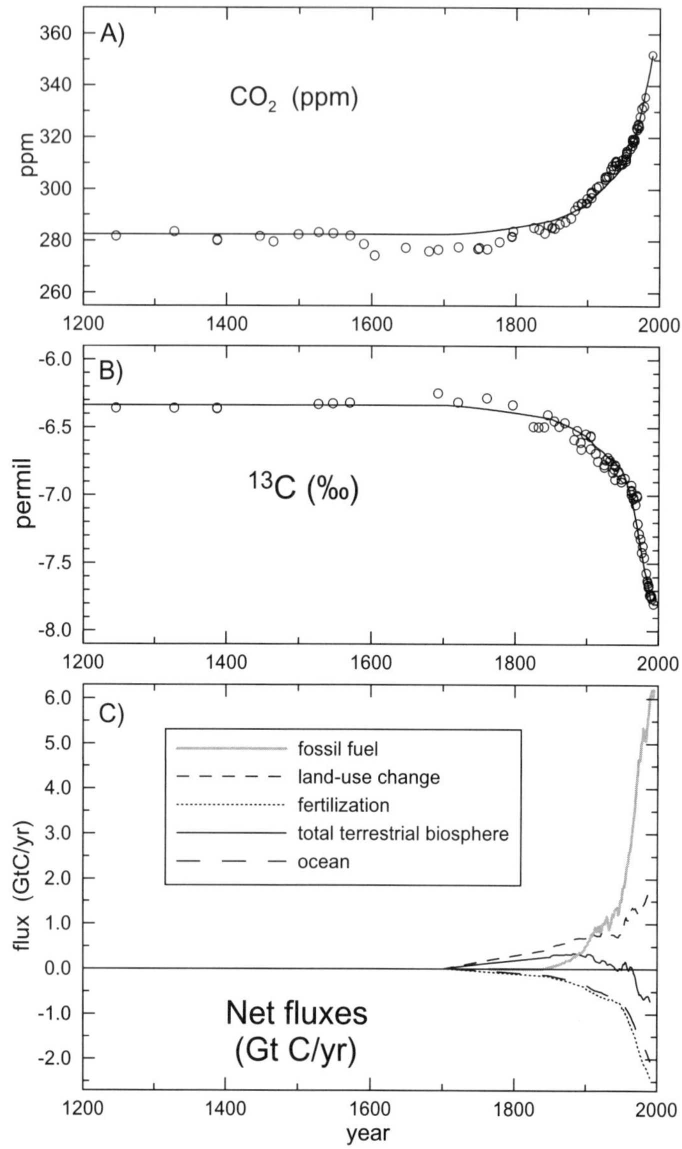

Figure 15.3. **(A)** CO_2 and **(B)** $\delta^{13}C$ calculated with the fluxes in **(C)** run forward in the carbon cycle model.

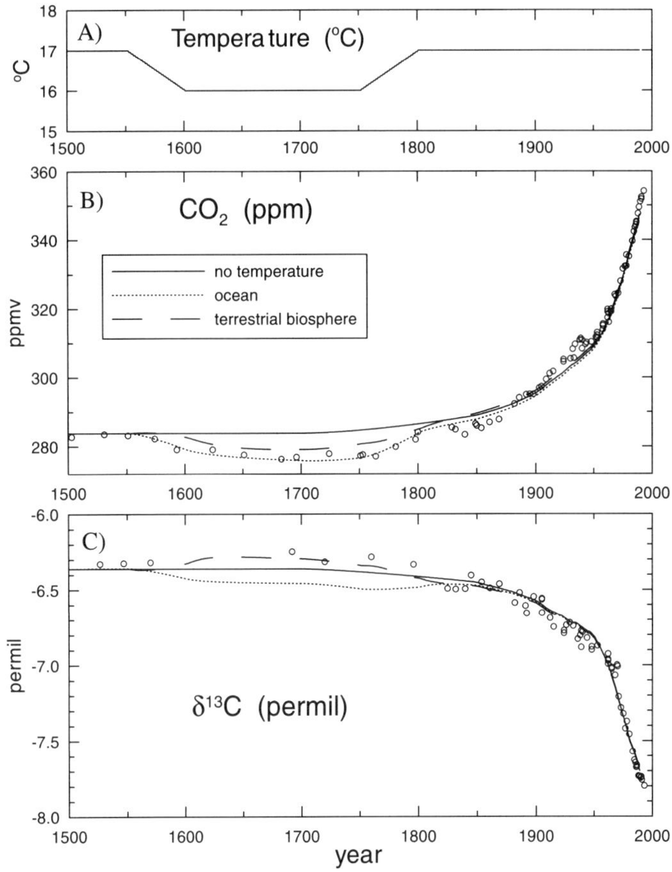

Figure 15.4. **(A)** Hypothetical temperature record used in the carbon cycle model; **(B)** Modeled CO_2; and **(C)** $\delta^{13}C$ from a forward calculation with temperature dependence of either the oceans (dotted lines) or the terrestrial biosphere (dashed lines).

and it gave a decrease in both CO_2 and $\delta^{13}C$ (dotted lines in Fig. 15.4). The terrestrial response was modeled using typical Q_{10} factors to reflect the greater sensitivity of respiration than of NPP to temperature variations (Q_{10} is the factor by which a CO_2 flux increases per 10°C of warming) (Harvey 1989). The terrestrial response to the lower temperature was a decrease in CO_2 and an increase in $\delta^{13}C$ (dashed lines in Fig. 15.4), similar to the changes observed. With such a simple model, the magnitude of the temperature response may be uncertain, but the sense of the changes should be reliable.

In the model, the effect of a temperature decrease on both terrestrial and oceanic exchange would be addition of the CO_2 decrease and partial cancellation of the $\delta^{13}C$ changes. A scenario consistent with the changes measured in the ice core record is that global sea surface temperatures were not significantly different during the LIA, but that major terrestrial biomes experienced significant cooling. Many of the temperature records that show the cooling during this period are from Northern Hemisphere land sites, and information for the Southern Hemisphere is more limited. Etheridge et al. (1998) measured a decrease in CH_4 through this period, and a change in the interpolar difference of CH_4, consistent with lower land temperatures reducing the extent and magnitude of CH_4 emissions from wetlands in the Northern Hemisphere.

15.4 Modeling: Single Deconvolution

Global CO_2 concentrations are known better than global CO_2 fluxes. The single deconvolution calculation uses CO_2 concentration data to infer global terrestrial fluxes, with ocean fluxes determined with an ocean model. Fig. 15.5 shows CO_2, $\delta^{13}C$ and the net ocean and terrestrial fluxes from the single deconvolution. The calculation is run from AD 1200, as that is the earliest relatively steady period in the ice core record, and the model starts at steady state.

The terrestrial flux deduced in the single deconvolution minus the Houghton (1995) land-use change flux is shown by the dotted line in Fig. 15.5 Panel C. This flux varies around zero until about 1930, after which time there is a significant terrestrial uptake. Friedlingstein et al. (1995) calculated the fertilization sink with a gridded biosphere model. The time evolution of their flux was different from that suggested by a deconvolution of ice core CO_2 data: the fertilization sink, which depends on the increase of atmospheric CO_2, was larger than the deconvolution sink up to 1950, then smaller between 1960 and 1980. They suggested that other processes, such as nitrogen deposition, climate variability, or mid-latitude forest regrowth, were required to match the atmospheric CO_2.

The $\delta^{13}C$ that was calculated in the single deconvolution matches more of the variations in the ice core $\delta^{13}C$ record than $\delta^{13}C$ calculated in the forward calculation, but it is still not an exact match. Therefore more information is available from $\delta^{13}C$ about exchange of CO_2, and this will be used in the double deconvolution in the next section.

15.5 Modeling: Double Deconvolution

The forward and single deconvolution calculations use the $\delta^{13}C$ record for validation, whereas the double deconvolution uses it directly to deduce the net fluxes. Previous double deconvolutions (Francey et al. 1995; Joos and Bruno 1998; Joos et al 1999) have used the change in CO_2 and $\delta^{13}C$ through time with the budget equations of Tans, Berry, and Keeling (1993) to solve for the net fluxes by mass balance. This requires CO_2 and $\delta^{13}C$ to be known at each time step, so smoothing splines have been used to interpolate the ice core data. How-

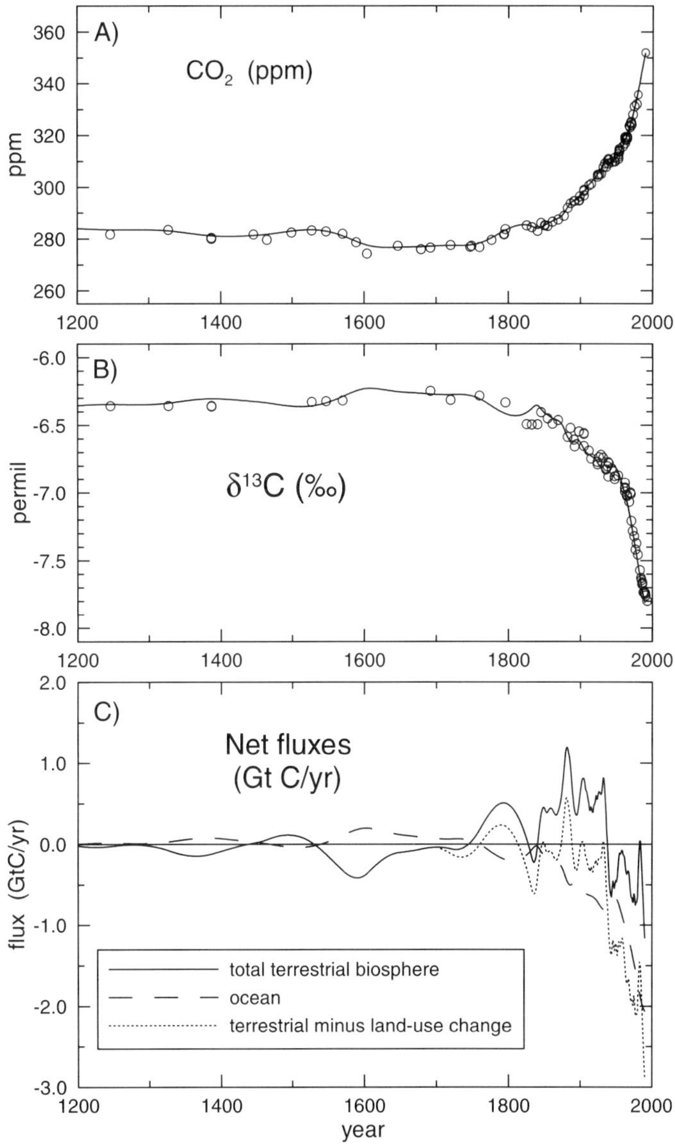

Figure 15.5. **(A)** CO_2 and **(B)** $\delta^{13}C$ and **(C)** fluxes from the single deconvolution calculation (Trudinger et al. 1999).

ever, the degree of smoothing used for the splines can have important implications for the inferred sources and their variability.

The double deconvolution described here uses the Kalman filter (Kalman 1960) to incorporate statistical analysis into the carbon cycle modeling (Trudinger 2000; Trudinger et al. 2002b). The Kalman filter is a state-space modeling technique. Data are processed sequentially to give optimal estimates of the state of the system. The carbon cycle model used with the Kalman filter is described by the budget equations of Tans, Berry, and Keeling (1993), which were given in the Introduction, with the terrestrial and oceanic components of the box diffusion model (see Fig. 15.2) used for calculation of the isofluxes. The Kalman filter double deconvolution uses the CO_2 and $\delta^{13}C$ ice core measurements, with their measurement uncertainties, to estimate the net terrestrial and oceanic fluxes.

Because the ice core record is a smoothed representation of the actual atmospheric variations, we also use a numerical model of firn diffusion and bubble trapping (Trudinger et al. 1997) to determine how concentrations are likely to vary in the firn-smoothed ice core record. Part of the variability in the ice core record is treated as signal, and part of it is treated as noise. The measurement uncertainties tell us how much can be noise, and from (off-line) calculations with the firn model we estimate how much is likely to be signal. This helps us set parameters in the Kalman filter to specify what we treat as signal versus noise (Trudinger, Enting, and Rayner 2002a). With statistical tests such as the chi-squared test (Tarantola 1987) we make sure that all of the variability in the measurements is accounted for with our statistical model.

An important feature of the calculation is that it determines the uncertainty in the estimated fluxes due to uncertainty in the measurements and evolution of the fluxes. We use a random walk model for flux evolution (Mulquiney et al. 1995). This is a weak constraint on the fluxes, and the uncertainties represent upper bounds. Uncertainty in the total net CO_2 flux and its partition into ocean and terrestrial components varies throughout the calculation, depending on the ice core data density and uncertainty, with the carbon cycle model relating CO_2, $\delta^{13}C$, the fluxes, and their uncertainties in a consistent way. For example, the uncertainty in the terrestrial flux is often lower than the uncertainty in the ocean flux, because $\delta^{13}C$ constrains the terrestrial flux better than CO_2 constrains the ocean flux with the data uncertainties we have used. In general, the flux uncertainty increases when there are long gaps in the data, which seems intuitively reasonable. Full details of the Kalman filter deconvolution calculations are given in Trudinger et al. (2002b).

Fig. 15.6 shows the results of the Kalman filter double deconvolution calculation over the past 800 years. This particular calculation has parameters chosen to give variability on century timescales (higher frequency variations such as those on the decadal timescale are treated as noise because the data density for much of the preindustrial period is insufficient to resolve them). Also shown are two temperature reconstructions: the Bradley and Jones (1993) Northern Hemisphere summer temperature anomaly record and the Mann, Bradley, and

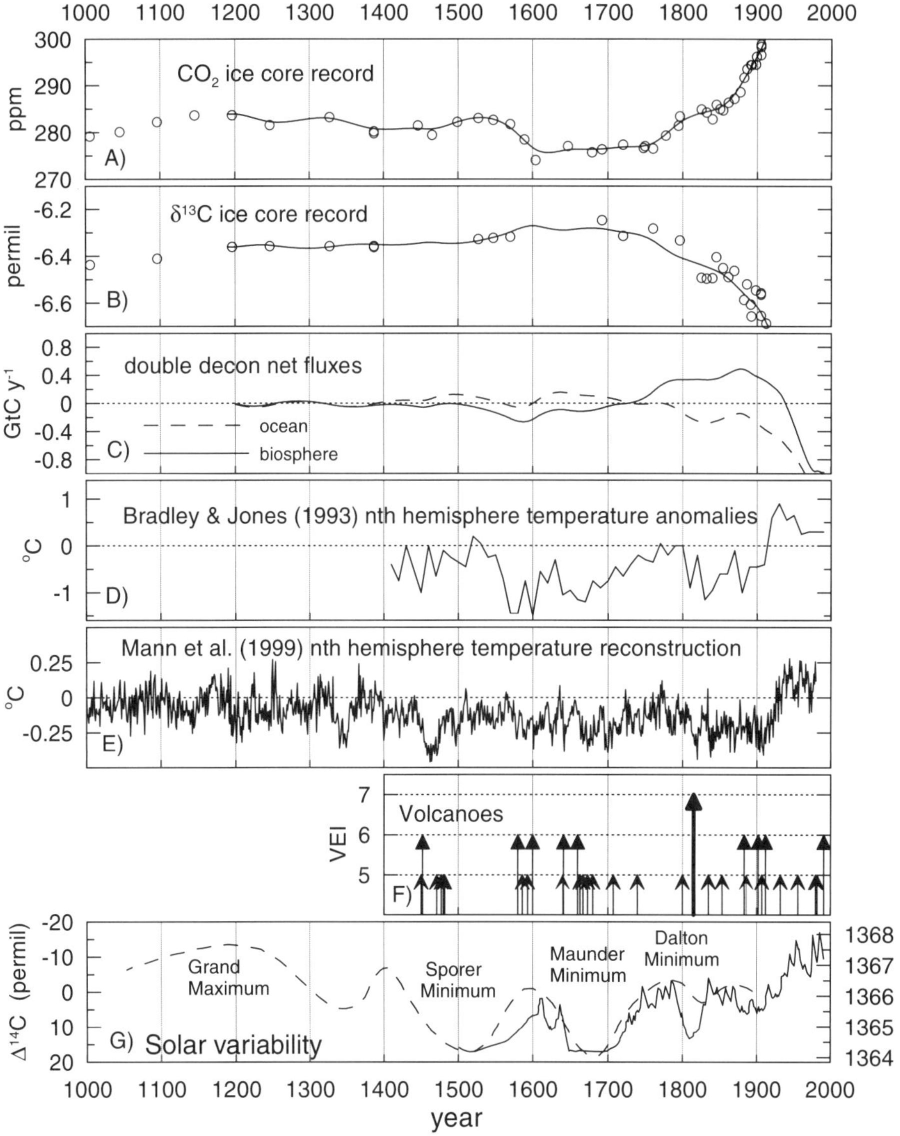

Figure 15.6. **(A)** Law Dome CO_2 measurements with the Kalman filter double deconvolution results for the century timescale (Trudinger et al. 2002b); **(B)** Law Dome $\delta^{13}C$ and the double deconvolution results; **(C)** Net fluxes from the double deconvolution; **(D)** Bradley and Jones Northern Hemisphere summer temperature anomalies; **(E)** Mann, Bradley, and Hughes (1999) Northern Hemisphere temperature reconstruction; **(F)** Volcanic Explosivity Index for the largest explosive volcanic eruptions (Briffa et al. 1998); **(G)** *solid line*, *right axis* solar irradiance reconstruction from Rind, Lean, and Healey (1999) and *dashed line*, *left axis* smoothed ^{14}C record from Eddy (1976).

Hughes (1999) temperature reconstruction. Explosive volcanic eruptions and solar forcing are believed to have an important effect on climate (Crowley 2000), and measures of their variability over the recent centuries are shown in Fig. 15.6F,G.

The Kalman filter double deconvolution and a mass balance double deconvolution by Joos et al. (1999) using the same ice core record both suggest biospheric uptake as well as an ocean source during the LIA period. Due to gaps in the data, it is not known whether $\delta^{13}C$ increased when CO_2 decreased around 1600, although the sponge record of ocean mixed-layer $\delta^{13}C$ of Böhm et al. (2002), with an almost linear increase in $\delta^{13}C$ between 1550 and 1670, suggests that it may not have increased at that time. The Kalman filter takes into account the fact that there are CO_2 but no $\delta^{13}C$ measurements around 1600, and the uncertainty in the total CO_2 flux is less than the uncertainty in the partition, as would be expected intuitively.

Biospheric uptake between about 1500 and 1750 from the Kalman filter deconvolution looks roughly like the Bradley and Jones (1993) Northern Hemisphere temperature record. The lowest temperature in the Bradley and Jones record occurs between about 1560 and 1600, when CO_2 decreased to its lowest level for the past 1000 years. The Mann, Bradley, and Hughes (1999) record has the lowest temperature for the past 1000 years at about 1460. The CO_2 measurement near 1460 is one of the lowest measurements in the record, not counting the period between 1550 and 1800 already discussed. After 1750 the biospheric flux becomes a source, and at least part (if not most) of this would have been due to land-use change.

Fig. 15.7 shows the results of the Kalman filter double deconvolution calculation since AD 1850. We have not used DSS data after 1830 in this calculation because the DSS data are inherently more smoothed in time than the DE08/DE08-2 data, and it may be inconsistent to use data with different degrees of firn smoothing without taking this into account. The estimated total flux (ocean + biosphere), minus the fluxes from the forward calculation described in Section 15.3 (i.e., land-use change, uptake due to fertilization, and modeled ocean uptake/release) is shown in Fig. 15.7C. We interpret this as the flux not already explained by the forward fluxes. Fig. 15.7D,E shows the terrestrial and oceanic fluxes, minus the terrestrial and oceanic forward fluxes. Uncertainties (1σ) from the Kalman filter are shown. This particular calculation uses a random walk parameter based on the firn model calculation and the data uncertainties given by Francey et al. (1999) for $\delta^{13}C$, and CO_2 uncertainties of 0.8 ppm.

Although Etheridge et al. (1996) suggested CO_2 uncertainties of 1.2 ppm, we choose to use the smaller CO_2 uncertainties based on the statistics of the residuals in the Kalman filter calculation. With these parameters, the Kalman filter tracks much of the decadal variation in the ice core record. This calculation pushes the ice core data to their limits, and further confirmation of features in the ice core record is needed (and planned) from new measurements. In an alternative calculation with larger data uncertainties of 1.2 ppm for CO_2 and 0.04‰ for $\delta^{13}C$, the Kalman filter essentially does not follow much of the decadal variation, as the data are thought to be too uncertain and the decadal

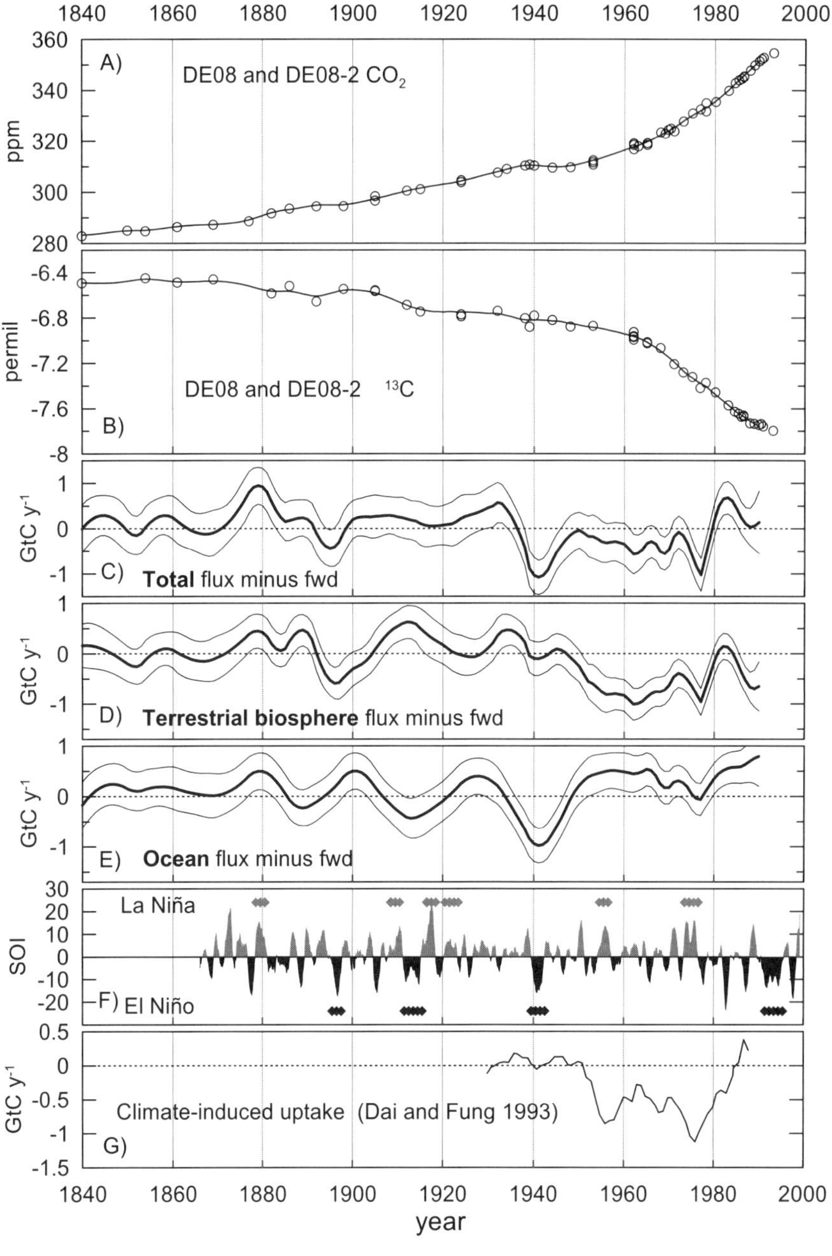

Figure 15.7. **(A)** Law Dome CO_2 measurements from 1840 with the Kalman filter double deconvolution results for the decadal timescale; **(B)** Law Dome $\delta^{13}C$ measurements from 1840 and the double deconvolution; **(C)** total CO_2 fluxes (ocean + biosphere) from the double deconvolution minus the fluxes from the forward calculation (see text) with 1σ uncertainties; **(D)** terrestrial biospheric fluxes and **(E)** oceanic fluxes, minus the forward fluxes, from the double deconvolution; **(F)** Southern Oscillation Index (11 month moving average). The diamond symbols show the extended El Niño and La Niña sequences identified by Allan and D'Arrigo (1999); and **(G)** Climate-induced biospheric CO_2 uptake from Dai and Fung (1993).

variation treated as noise rather than signal. Trudinger et al. (2002b) has more details of these calculations.

In general, the temporal structure of anomalies in the CO_2 net fluxes from our deconvolution over the industrial period is similar to that deduced by Joos et al (1999) in their mass balance double deconvolution on the same ice core record. This is not surprising given that both methods solve mass balance for the same ice core measurements. However, our deconvolution calculation suggests that natural variations in the net fluxes may be up to about 1 GtC yr^{-1} (peak-to-peak) on timescales of slightly less than a decade up to centuries, whereas the Joos et al. (1999) calculation found that natural variations were usually below ± 0.2 GtC yr^{-1} on timescales of decades to centuries. These results depend very much on the degree of smoothing used, and how variability is partitioned between signal and noise. Our calculation has assumed that many of the decadal features represent real atmospheric changes, and the firn model has shown that features such as these can survive firn smoothing. New ice core measurements will confirm which of the features are real.

Panel F of Fig. 15.7 shows the Southern Oscillation Index (11-month moving average) from http://www.dar.csiro.au/information/soi.html. The extended El Niño and La Niña events identified by Allan and D'Arrigo (1999) are indicated by the diamond symbols. Etheridge et al. (1996) noted that the persistent El Niño sequences identified by Allan and D'Arrigo (1999) around 1895 to 1898, 1911 to 1916, and 1939 to 1943 coincided with decreases in the growth rate of CO_2. We believe that the time resolution of air stored in the ice at the DE08 site (i.e., considering the smoothing by firn processes) should be sufficient to give some insight into the effect of SOI on CO_2 fluxes for the extended El Niño sequences (but not El Niño events of shorter duration). The deconvolution suggests biospheric uptake around 1895 to 1898, ocean uptake and a biospheric source around 1911 to 1916, and strong ocean uptake around 1939 to 1943. Except for the ocean uptake from 1911 to 1916, these results are all statistically significant with the data uncertainties we have used. However, the $\delta^{13}C$ variation on this timescale is sometimes defined by individual ice core measurements, which is not the ideal situation. We therefore treat these results as suggestive rather than conclusive, until the features in the ice core record can be confirmed by further measurements. The effect of ENSO on CO_2 is potentially complicated as it can involve both oceanic and terrestrial processes on a range of timescales.

We investigated the CO_2 and $\delta^{13}C$ variations corresponding with one of these extended El Niño sequences in more detail. As mentioned earlier, the CO_2 stabilization in the 1940s is a prominent feature in the Law Dome CO_2 record. The CO_2 variation is not mirrored in the $\delta^{13}C$ measurements, suggesting mainly ocean uptake. Due to the effects of firn smoothing, the real atmospheric variation would need to have been even more extreme than the ice core measurements show. We used the firn model to determine the atmospheric CO_2 variations that would have been required to leave the measured CO_2 variation trapped in the ice. This CO_2 variation was then used in the double deconvolution, with the measured $\delta^{13}C$, to see what fluxes are suggested. Large uncertainties were used for the $\delta^{13}C$ measurements in this calculation (4 times the standard uncertainties), because firn smoothing has been accounted for in relation to CO_2 but not for $\delta^{13}C$. The calculation suggests that CO_2 uptake of up to 3 GtC yr^{-1} is required in the mid-1940s. This uptake may be up to about one-third biospheric but is mainly oceanic. If the uptake were one-third biospheric, there would have been a peak in atmospheric $\delta^{13}C$ that would not be detectable in the ice after smoothing by the firn processes. If the uptake were more than one-third biospheric, the peak in atmospheric $\delta^{13}C$ would have been too large to be consistent with the ice core measurements when smoothed. Similarly, the $\delta^{13}C$ data are not consistent with the CO_2 decrease being due to a strong decrease in the fossil fuel source at this time. This calculation suggests a cumulative uptake of up to 16 GtC between 1938 and 1951. Deconvolution of the data without taking into account the firn smoothing gives a cumulative ocean uptake of about 8 GtC between 1934 and 1949. Joos et al. (1999) found a cumulative ocean uptake of 11 GtC for this period, but their smoothing spline widens the feature compared to the amount found in our calculation.

The double deconvolution allows us to estimate terrestrial and oceanic CO_2 fluxes, but it does not immediately tell us about the processes responsible for them. That would require a process model of CO_2 exchange, and the logical next step of this work is to link the ice core inversions to a more process-based model. For now we can compare our estimated fluxes with those predicted by a process model. Fig. 15.7G shows the net terrestrial CO_2 flux due to climate variations (temperature and precipitation) calculated by Dai and Fung (1993) in a gridded model of the terrestrial biosphere. The temporal evolution and magnitude of this flux is remarkably similar to the net terrestrial flux from the double deconvolution.

15.6 Summary

In this chapter, we have investigated various features in the Law Dome CO_2 and $\delta^{13}C$ ice core record over the past 1000 years. The ice core record shows, with high precision and time resolution, the natural variability in the carbon cycle and the large anthropogenic perturbation. A simple one-dimensional carbon cycle model was used in both forward and inverse approaches to interpret the observed variations in terms of net fluxes of CO_2. As well as the standard forward and single deconvolution calculations, we used a Kalman filter double deconvolution method (Trudinger et al. 2002b). This is a new method that incorporates a rigorous statistical analysis into the carbon cycle modeling and allows quantification of the uncertainties in the inferred oceanic and terrestrial fluxes.

The main results of this work are:

- The low levels of CO_2 and high levels of $\delta^{13}C$ during the Little Ice Age period (approximately 1550 to 1800) suggest a predominantly biospheric response to reduced temperatures rather than an oceanic response.

- The double deconvolution suggests natural variability in the net CO_2 fluxes of about 1 GtC yr^{-1} on timescales of slightly less than decades up to centuries. This is greater than that suggested by a previous study and very dependent on what variation in the ice core record is considered real.
- The CO_2 flattening in the 1940s requires almost 3 GtC yr^{-1} uptake in the mid-1940s. The corresponding $\delta^{13}C$ variations suggest that most of this uptake was oceanic.
- The double deconvolution suggests variations in biospheric uptake of CO_2 between 1950 and 1980 that are very similar in both magnitude and temporal evolution to the terrestrial sink estimated by Dai and Fung (1993) using regional climate records to force globally gridded ecosystems.

The forward and inverse modeling approaches are complementary, and both are useful. The inverse approach has the advantage that global concentration is known better than global fluxes, but it has the disadvantage that the flux estimates are not linked with the processes that cause their variability. Future work will involve the use of more process-based models for flux evolution in the double deconvolution, perhaps using the Kalman filter. Confirmation of the ice core records, ideally with finer sampling density, is also needed.

References

Allan, R.J., and R.D. D'Arrigo. 1999. 'Persistent' ENSO sequences: How unusual was the 1990–1995 El Niño? *The Holocene* 9:101–18.

Allen, L.H., K.J. Boote, J.W. Jones, P.H. Jones, R.R. Valle, B. Acock, H.H. Rogers, and R.C. Dahlman. 1987. Response of vegetation to rising carbon dioxide: Photosynthesis, biomass, and seed yield of soybean. *Global Biogeochemical Cycles* 1:1–14.

Anklin, M., J.-M. Barnola, J. Schwander, B. Stauffer, and D. Raynaud. 1995. Processes affecting the CO_2 concentrations measured in Greenland ice. *Tellus* 47B:461–70.

Barnola, J.-M., M. Anklin, J. Porcheron, D. Raynaud, J. Schwander, and B. Stauffer. 1995. CO_2 evolution during the last millennium as recorded by Antarctic and Greenland ice. *Tellus* 47B:264–72.

Böhm, F., A. Haase-Schramm, A. Eisenhauer, W. Dullo, M.M. Joachimski, H. Lehnert, and J. Reitner. 2002. Evidence for preindustrial variations in the marine surface water carbonate system from coralline sponges. *Geochemistry, Geophysics, Geosystems* 3: 10.1029/2001GC000264.

Bradley, R.S., and P.D. Jones. 1993. 'Little Ice Age' summer temperature variations: Their nature and relevance to recent global warming trends. *The Holocene* 3:367–76.

Briffa, K.R., P.D. Jones, F.H. Schweingruber, and T.J. Osborn. 1998. Influence of volcanic eruptions on Northern Hemisphere summer temperature over the past 600 years. *Nature* 393:450–55.

Briffa, K.R., and T.J. Osborn. 2002. Blowing hot and cold. *Science* 295:2227–28.

Crowley, T.J. 2000. Causes of climate change over the past 1000 years. *Science* 289: 270–77.

Dai, A., and I.Y. Fung. 1993. Can climate variability contribute to the "missing" CO_2 sink. *Global Biogeochemical Cycles* 7:599–609.

Eddy, J.A. 1976. The maunder minimum. *Science* 192:1189–1202.

Enting, I.G., and K.R. Lassey. 1993. *Projections of future CO_2*. Technical paper no. 27. Aspendale, Australia: CSIRO Division Atmospheric Research.

Etheridge, D.M. 1999. Natural and anthropogenic changes in atmospheric carbon dioxide and methane over the last 1000 years (Ph.D. thesis). Melbourne, Australia: University of Melbourne.

Etheridge, D.M., G.I. Pearman, and F. de Silva. 1988. Atmospheric trace-gas variations as revealed by air trapped in an ice core from Law Dome, Antarctica. *Annals of Glaciology* 10:28–33.

Etheridge, D.M., L.P. Steele, R.J. Francey, and R.L. Langenfelds. 1998. Atmospheric methane between 1000 AD and present: evidence for anthropogenic emissions and climate variability. *Journal of Geophysical Research* 103D:15979–93.

Etheridge, D.M., L.P. Steele, R.L. Langenfelds, R.J. Francey, J.-M. Barnola, and V.I. Morgan. 1996. Natural and anthropogenic changes in atmospheric CO_2 over the last 1000 years from air in Antarctic ice and firn. *Journal of Geophysical Research* 101D: 4115–28.

Francey, R.J., C.E. Allison, D.M. Etheridge, C.M. Trudinger, I.G. Enting, M. Leuenberger, R.L. Langenfelds, E. Michel, and L.P. Steele. 1999. A 1000-year high precision record of $\delta^{13}C$ in atmospheric CO_2. *Tellus* 51B:170–93.

Francey, R.J., E. Michel, D.M. Etheridge, C.E. Allison, M.L. Leuenberger, and D. Raynaud. 1997. The preindustrial difference in CO_2 from Antarctic and Greenland ice. In *Extended abstracts of the 5th International Carbon Dioxide Conference,* 211–12. Cairns, Australia.

Francey, R.J., P.P. Tans, C.E. Allison, I.G. Enting, J.W.C. White, and M. Trolier. 1995. Changes in oceanic and terrestrial carbon uptake since 1982. *Nature* 373:326–30.

Friedli, H., H. Loetscher, H. Oeschger, U. Siegenthaler, and B. Stauffer. 1986. Ice core record of the $^{13}C/^{12}C$ ratio of atmospheric carbon dioxide in the past two centuries. *Nature* 324:237–38.

Friedlingstein, P., I. Fung, E. Holland, J. John, G. Brasseur, D. Erickson, and D. Schimel. 1995. On the contribution of CO_2 fertilisation to the missing biospheric sink. *Global Biogeochemical Cycles* 9:541–56.

Grove, J.M. 1988. *The Little Ice Age.* New York: Methuen.

Haan, D., and D. Raynaud. 1998. Ice core record of CO variations during the last two millennia: Atmospheric implications and chemical interactions within the Greenland ice. *Tellus* 50:253–62.

Harvey, L.D.D. 1989. Effect of model structure of the response of terrestrial biosphere models to CO_2 and temperature increases. *Global Biogeochemical Cycles* 3:137–53.

Heimann, M., and C.D. Keeling. 1989. A three-dimensional model of atmospheric CO_2 transport based on observed winds: 2. Model description and simulated tracer experiments. In *Aspects of climate variability in the Pacific and the Western Americas,* Geophysical Monograph 55, ed. D. Peterson, 237–75, American Geophysical Union, Washington, D.C. (USA).

Houghton, R.A. 1995. Effects of land-use change, surface temperature and CO_2 concentration on terrestrial stores of carbon. In *Biotic feedbacks in the global climatics system: Will the warming feed the warming?* Ed. G.M. Woodwell and F.T. Mackenzie, 333–50. Oxford: Oxford University Press.

Jones, P.D., and R.S. Bradley. 1992. Climate variations over the last 500 years. In *Climate since A.D. 1500,* ed. R.S. Bradley and P.D. Jones, 649–65, New York: Routledge.

Joos, F., and M. Bruno. 1998. Long-term variability of the terrestrial and oceanic carbon sinks and the budgets of the carbon isotopes ^{13}C and ^{14}C. *Global Biogeochemical Cycles* 12:277–95.

Joos, F., R. Meyer, M. Bruno, and M. Leuenberger. 1999. The variability in the carbon sinks as reconstructed for the last 1,000 years. *Geophysical Research Letters* 26:1437–40.

Kalman, R.E. 1960. A new approach to linear filtering and prediction problems. *Journal of Basic Engineering* 82D:35–45.

Kammen, D.M., and B.D. Marino. 1993. On the origin and magnitude of pre-industrial anthropogenic CO_2 and CH_4 emissions. *Chemosphere* 26:69–86.

Kawamura, K., T. Nakazawa, T. Machida, S. Morimoto, S. Aoki, A. Ishizawa, Y. Fujii, and O. Watanabe. 2000. Variations of the carbon isotope ratio in atmospheric CO_2 over the last 250 years recorded in an ice core from H15, Antarctica. *Polar Meteorology and Glaciology* 14:47–57.

Keeling, C.D. 1997. Global historical CO_2 emissions. In *Trends: A compendium of data on global change.* Carbon Dioxide Information Analysis Center, Oak Ridge, Tennessee U.S.A.

Keeling, C.D., R.B. Bacastow, A.F. Carter, S.C. Piper, T.P. Whorf, M. Heimann, W.G. Mook, and H. Roeloffzen. 1989. A three-dimensional model of atmospheric CO_2 transport based on observed winds: 1. Analysis of observational data. In *Aspects of climate variability in the Pacific and the Western Americas*, Geophysical Monograph 55, ed. D.H. Peterson, 165–236, American Geophysical Union, Washington, D.C. (U.S.A).

Lamb, H.H. 1982. *Climate, history and the modern world.* New York: Methuen.

Lassey, K.R., I.G. Enting, and C.M. Trudinger. 1996. The Earth's radiocarbon budget: A consistent model of the global carbon and radiocarbon cycles. *Tellus* 48B:487–501.

Mann, M.E., R.S. Bradley, and M.K. Hughes. 1999. Northern hemisphere temperatures during the past millennium: Inferences, uncertainties and limitations. *Geophysical Research Letters* 26:759–62.

Marland, G., and T.A. Boden. 1997. Global, regional and national CO_2 emissions. In *Trends: A compendium of data on global change*, Carbon Dioxide Information Analysis Center, Oak Ridge, Tennessee, U.S.A.

Mulquiney, J.E., J.P. Norton, A.J. Jakeman, and J.A. Taylor. 1995. Random walks in the Kalman filter: Implications for greenhouse gas flux deductions. *Environmetrics* 6:473–78.

Nakazawa, T., T. Machida, M. Tanaka, Y. Fujii, S. Aoki, and O. Watanabe. 1993. Atmospheric CO_2 concentrations and carbon isotope ratios for the last 250 years deduced from an Antarctic ice core, H15. In *Extended abstracts of the 4th International Carbon Dioxide Conference*, 193–96. Carqueiranne, France.

Neftel, A., E. Moor, H. Oeschger, and B. Stauffer. 1985. Evidence from polar ice cores for the increase in atmospheric CO_2 in the past two centuries. *Nature* 315:45–47.

Oeschger, H., U. Siegenthaler, U. Schotterer, and A. Gugelmann. 1975. A box diffusion model to study the carbon dioxide exchange in nature. *Tellus* 27B:168–92.

Pearman, G.I., D.M. Etheridge, F. de Silva, and P.J. Fraser. 1986. Evidence of changing concentrations of atmospheric CO_2, N_2O and CH_4 from air bubbles in Antarctic ice. *Nature* 320:248–50.

Randerson, J.T., G.J. Collatz, J.E. Fessenden, A.D. Munoz, C.J. Still, J.A. Berry, I.Y. Fung, N. Suits, and A.S. Denning. 2002. A possible global covariance between terrestrial gross primary production and ^{13}C discrimination: Consequences for the atmospheric ^{13}C budget and its response to ENSO. *Global Biogeochemical Cycles* 16 doi:10.1029/2001GB001845.

Raynaud, D., and J.-M. Barnola. 1985. An Antarctic ice core reveals atmospheric CO_2 variations over the past few centuries. *Nature* 315:309–11.

Revelle, R., and H.E. Suess. 1957. Carbon dioxide exchange between the atmosphere and ocean and the question of an increase of atmospheric CO_2 during the past decades. *Tellus* 9:18–27.

Rind, D., J. Lean, and R. Healy. 1999. Simulated time-dependent climate response to solar radiative forcing since 1600. *Journal of Geophysical Research* 104:1973–90.

Schimel, D., I.G. Enting, M. Heimann, T.M.L. Wigley, D. Raynaud, D. Alves, and U. Siegenthaler. 1995. CO_2 and the carbon cycle. In *Climate change 1994: Radiative forcing of climate change and an evaluation of the IPCC emission scenarios*, 35–71. Cambridge: Cambridge University Press.

Schwander, J., B. Stauffer, and A. Sigg. 1988. Air mixing in firn and the age of the air at pore close-off. *Annals of Glaciology* 10:141–45.

Siegenthaler, U., H. Friedli, H. Loetscher, E. Moor, A. Neftel, H. Oeschger, and B. Stauffer. 1988. Stable-isotope ratios and concentrations of CO_2 in air from polar cores. *Annals of Glaciology* 10:151–56.

Tans, P.P., J.A. Berry, and R.F. Keeling. 1993. Oceanic $^{13}C/^{12}C$ observations: A new window on ocean CO_2 uptake. *Global Biogeochemical Cycles* 7:353–68.

Tarantola, A. 1987. *Inverse problem theory: Methods for data fitting and parameter estimation*. Amsterdam: Elsevier.

Trudinger, C.M. 2000. *The carbon cycle over the last 1000 years inferred from inversion of ice core data*. PhD Thesis, Monash University. [Available online at http://www.dar.csiro.au/publications/trudinger_2001a0.htm].

Trudinger, C.M., I.G. Enting, D.M. Etheridge, R.J. Francey, V.A. Levchenko, L.P. Steele, D. Raynaud, and L. Arnaud. 1997. Modeling air movement and bubble trapping in firn. *Journal of Geophysical Research* 102D:6747–63.

Trudinger, C.M., I.G. Enting, R.J. Francey, D.M. Etheridge, and P.J. Rayner. 1999. Long-term variability in the global carbon cycle inferred from a high precision CO_2 and $\delta^{13}C$ ice core record. *Tellus* 51B:233–48.

Trudinger, C.M., I.G. Enting, and P.J. Rayner. 2002a. Kalman filter analysis of ice core data: 1. Method development and testing the statistics. *Journal of Geophysical Research* 107D doi:10.1029/2001JD001111.

Trudinger, C.M., I.G. Enting, P.J. Rayner, and R.J. Francey. 2002b. Kalman filter analysis of ice core data: 2. Double deconvolution of CO_2 and $\delta^{13}C$ measurements. *Journal of Geophysical Research* 107D doi:10.1029/2001JD001112.

Tschumi, J., and B. Stauffer. 2000. Reconstructing past atmospheric CO_2 concentration based on ice-core analysis: open questions due to in situ production of CO_2 in the ice. *Journal of Glaciology* 46:45–53.

Wullschleger, S.D., W.M. Post, and A.W. King. 1995. On the potential for a CO_2 fertilisation effect in forests: Estimates of the biotic growth factors based on 58 controlled-exposure studies. In *Biotic feedbacks in the global climate system: Will the warming feed the warming?*, ed. G.M. Woodwell and F.T. Mackenzie, 85–107. Oxford University Press, U.K.

16. Remembrance of Weather Past: Ecosystem Responses to Climate Variability

David Schimel, Galina Churkina, Bobby H. Braswell, and James Trenbath

> . . . the biosphere is . . . a highly complex system with ten million to thirty million different kinds of working parts and myriad feedback systems positive and negative, each with its own lag time, of which we have little ken.
>
> —Thomas E. Lovejoy 1992

16.1 Introduction

The future of the carbon cycle is one of the great uncertainties in projecting the future of climate. The response of terrestrial ecosystems remains largely unknown, and qualitative disagreement persists regarding the influence of climate change on ecosystem carbon storage. There has been a sustained effort to understand how ecosystems respond to climate and carbon dioxide (Prentice et al. 2001) using observations, experiments, models, and the paleorecord. Despite this massive research effort, and considerable effort to synthesize its results, considerable uncertainty remains. We do not have a clear paradigm for interpreting observations of ecosystem climate responses in time and space and for translating that paradigm into predictive models. Many of the disagreements occur because it is difficult to separate the effects of processes operating on different timescales. For example, low cloud cover, high solar radiation, and consequent high temperatures may increase photosynthesis (A) in the short term. On slightly longer timescales, these conditions may favor allocation of new photosynthate to roots, reducing leaf area relative to a cooler year. The *flux* in units of mass per area depends on both the specific rate of photosynthesis (mass per unit Leaf Area Index (LAI)) and the LAI, multiplicatively. The ecosystem-level change can be predicted only by knowing the impact of physical variables on a range of parameters, some of which respond in seconds (for example, A) and others that respond over days to months (e.g., allocation to leaf area).

This type of complexity pervades the terrestrial carbon cycle, in which fluxes depend on both rate constants (e.g., water, temperature, or light dependence) and substrate (e.g., leaf area or detrital carbon) in a first-order (multiplicative) fashion. Biogeochemists have tended to extrapolate the consequences of environmental change from the direction and relative magnitude of the immediate rate constants (e.g., photosynthesis vs. respiration), since we have much less direct information about the "slower" physiological rate constants, such as those controlling allocation of photosynthate to leaves and litter quality and microbial processes controlling soil carbon. These slower processes tend to affect the state variable component of first-order regulation of fluxes and so transmit the integrated history of influences on the ecosystem to the short-term responses.

The growing attention paid to slower processes partly reflects developments in methodology. Early efforts to model whole-ecosystem carbon exchange (Detling, Parton, and Hunt 1978) were based on scaling-up leaf-level of chamber measurements, with fluxes measured in area-specific rates, multiplied by estimates of the relevant substrate (leaf area). Currently, eddy covariance provides a measure of the flux on a true area basis, but the specific rates and relevant substrate (leaf area, litter, etc.) amounts are more difficult to ascertain because of the high heterogeneity in vegetation and soil properties within the footprint of flux towers. This poses new challenges for modeling (Thornton et al. 2002). Existing observational time series intrinsically capture the higher frequencies best, but eddy covariance time series are now long enough that the role of interannual changes to state variables, such as LAI, litter mass, or soil water amount, should be evident.

Developing an observational basis for understanding long-term carbon dynamics from flux observations is challenging. The problem can be described as detecting the signal of slow processes through the "noise" of daily and seasonal changes. Fluxes vary strongly and predictably with physical forcing. Interannual changes to carbon uptake may be driven by climate but do not scale simply with the integral or average of short-term forcing (Vukicevic, Braswell, and Schimel 2001; Wofsy personal communication). In this paper, we use a model to simulate observations of ecosystem flux and state variables and analyze the interaction between short-term climate effects and slower state variable responses, as reflected in both "fast fluxes" and slow pools. We do not address the longer timescales of sand and soil development.

The existence of processes operating on different timescales in the carbon system is well known. Both formal and empirical approaches to unraveling the dynamics of multiscales processes are just beginning to be applied in biogeochemistry (Katul et al. 2001a), although mathematical scaling has long been the subject of formal analysis in atmospheric science and hydrology (Famiglietti and Wood 1994). While current models simulate dynamics on various timescales (Collatz et al. 1991; Rosenbloom, Doney, and Schimel 2001), observations and experiments intrinsically provide more information about "fast" than "slow" processes, although as time series of observations in flux networks and long-term ecological studies lengthen, long-term dynamics become more apparent

(Goulden et al. 1996; Thornton et al. 2002). Creative use of long operational time series (Caspersen et al. 2000) and isotopic measurements (Gaudinski et al. 2000) also provide insight into long-term processes.

In the past decade, the amount and sophistication of terrestrial biogeochemical data has increased dramatically (Canadell et al. 2000). At the same time, ecosystem models have generally increased in complexity and in their ability to accurately predict observations (Jenkins, Birdsey, and Pan 2001; Schimel et al. 1997). Furthermore, the number of measured variables available for comparison to simulated quantities has expanded greatly, even from the 1990s (Parton et al. 1993) to the present (Kelly et al. 2000). However, validation of the time-dependence of many modeled processes remains elusive, and it is still not clear why terrestrial models do not show closer agreement with each other (McGuire et al. 2001; Schimel et al. 2001). The abundance of data and the increasing sophistication of models therefore pose operational problems for the biogeochemical community. How can models be evaluated against observations of multiple timescales simultaneously and objectively? We address this through a simple and instructive model experiment.

16.2 Methods

16.2.1 Model

We used the model Biome-BGC to simulate flux and process observations. Biome-BGC has been extensively described in the literature (Running et al. 1994) and is a comprehensive model of ecosystem physiology. A detailed and recent comparison of Biome-BGC to eddy covariance measurements of Net Ecosystem Exchange (NEE) is found in Thornton et al. (2002), along with a description of the model. In short, the Biome-BGC model simulates daily photosynthesis and respiration as a function of physical variables and nutrient availability, and decomposition as a function of soil microclimate and nutrient dynamics. Plant growth and allocation to leaves, roots, and wood are simulated at the whole-plant level and respond dynamically to the environment. In Biome-BGC as in other state-of-the-art ecosystem models water and nitrogen couple autotrophic and heterotrophic processes together (Schimel et al. 1997). Biome-BGC was chosen for this study because of its relatively mechanistic canopy model, and because it has been extensively compared to flux measurements.

16.2.2 Experimental Design

An ensemble of experiments was performed with Biome-BGC and is referred to as the *ecosystem memory* experiment. In the ecosystem memory experiment, parallel simulations were run to create an ensemble. The simulations were integrated using long-term weather until the systems were a statistical steady state (year-to-year carbon fluxes varied with the weather, but decadal fluxes were stable). This results in systems with relatively low Net Ecosystem Exchange and

Net Primary Productivity. We chose this state because it makes the sensitivity of disparate systems comparable, and to maximize sensitivity of the analysis. The experiments were identical in every respect until 1975, which we chose arbitrarily as the base year. Note that the base year includes daily, seasonal, and interannual variability although most figures show annual results (Fig. 16.1). Simulations were integrated in which the weather from each of the 10 years prior to and including 1975 were substituted for 1975 (Table 16.1). After 1975, the inputs to all runs were once again identical. Table 16.1 shows the differences in annual meteorological parameters between the long-term mean, the base year of 1975, and each year substituted for 1975 to create the ensemble. The simu-

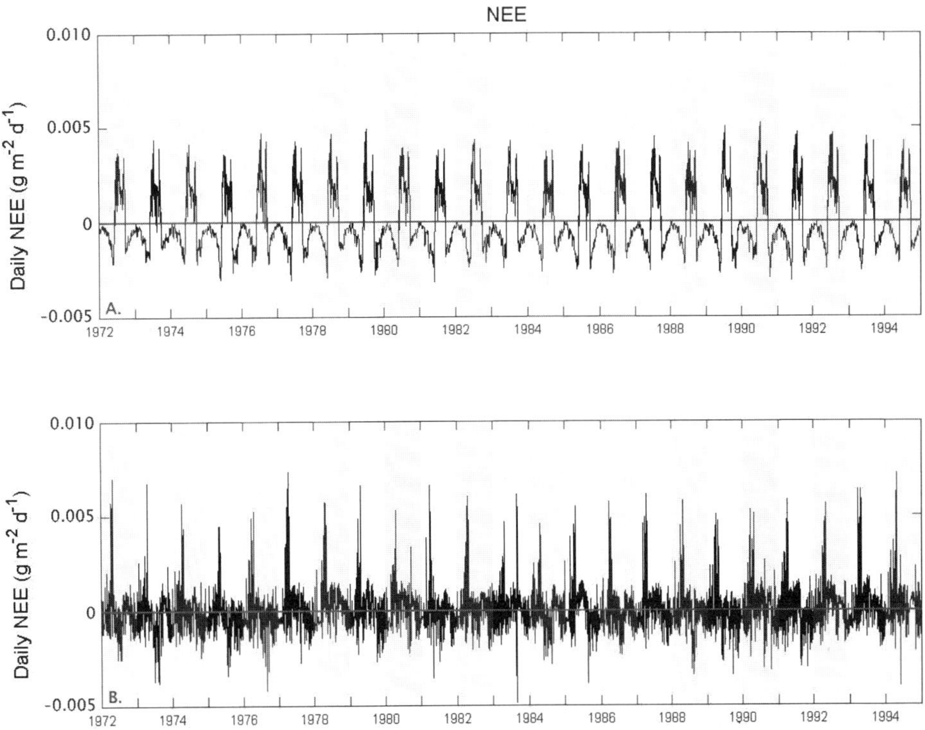

Figure 16.1. Simulated daily NEE for the Harvard Forest stand base cases. This figure illustrates the amount of variability on all timescales inherent in the simulations and the significant differences in the character of fluxes between growth forms. The Harvard Forest is shown to visualize the amount and scales of variability. Hyytiala's simulations have comparable daily, seasonal and interannual variability in fluxes. To produce Figs. 16.2 through 16.7, the annual values of the perturbed simulations were subtracted from the annual values derived from the base simulation to visualize the relative change introduced by direct and delayed effects. **(A)** Harvard Deciduous; **(B)** Harvard Coniferous. Subsequent plots are based on annual rather than daily values.

Table 16.1. Location and meteorology for the Harvard Forest and Hyytiala sites. Annual meterological parameters are shown for the mean, the base year of 1975, and the years substituted for 1975 to create the ensemble. Note that the model is run using daily weather files and these are computed average values.

Year	Mean T°	**SD**	**Mean annual precip.**	**SD**	**Mean vapor pressure deficit**	**SD**	**Mean Incident shortwave**	**SD**
Hyvitiala, Finland, 61° 51' N								
1965–1994 (annual)	4.67	1.0	70.52	10.7	414.55	32.70	162.27	5.05
1965	4.17	9.8	65.53	—	413.5	396.4	163.0	135.6
1966	3.27	11.8	68.43	—	415.7	463.3	161.17	140.2
1967	4.83	11.0	83.4	—	400.7	391.9	157.2	135.9
1968	3.9	10.8	61.78	—	411.9	421.3	161.5	137.0
1969	3.89	11.3	61.18	—	461.99	471.3	171.8	143.9
1970	4.09	10.7	67.24	—	411.14	436.6	159.6	139.4
1971	4.63	9.93	60.32	—	479.2	513.2	168.69	141.0
1972	5.77	10.0	75.5	—	417.4	432.2	160.7	136.6
1973	4.92	10.3	59.24	—	442.91	462.4	162.17	135.2
1974	6.03	7.7	97.85	—	389.26	375.5	158.5	136.3
1975 (Base)	6.4	8.7	60.09	—	468.4	445.1	167.32	141.2
Havard Forest USA, 42° 30' N								
1965–1994 (annual mean)	9.8	0.7	111.7	19.4	639.4	36.5	316.33	8.2
1965	9.1	10.9	62.5	—	647	474.2	321.5	124.2
1966	9.78	10.2	80.88	—	684.59	489.1	320.52	121.6
1967	9.0	10.5	112.1	—	606.5	426.9	324.28	121
1968	9.41	11.2	87.06	—	648.44	454.6	324	122.2
1969	9.56	10.5	103.01	—	610.47	424.9	314.67	123.6
1970	9.26	11.4	86.33	—	629.83	454	327.04	119.8
1971	9.51	10.9	99.84	—	635.21	451.3	325.24	127.4
1972	8.77	10.3	122.84	—	550.73	395	308.94	123.9
1973	10.7	10.1	123.59	—	635.93	412.3	313.65	112.5
1974	9.51	10	107.93	—	604.46	419.1	320.74	118.5
1975 (Base)	10.0	10.4	135.7	—	622.27	437.9	309.63	128.7

lations of 1965 to 1994 differed only during simulated 1975. By comparing the perturbed runs to the base case (1975 weather was the actual weather for that year), we could assess how long ecosystem pools and fluxes would be influenced by past weather conditions.

We substituted natural variability in climate forcing as an alternative to a systematic factorial sensitivity analysis in order to (1) probe realistic levels of ecosystem response to the degree of natural variability experienced at each site, and (2) because, as a practical matter, it is difficult to distribute a systematic perturbation over a daily weather year, while maintaining realistic autocorrelation, correlations between climate variables, event structures, and other characteristics of the weather that affect ecosystems. As Knapp and Smith (2001) show, ecosystem characteristics are closely adapted to both the local mean and natural variability of the site. Our ensemble approach produces a model estimate of responses scaled to representative behavior at each site. A systematic sensitivity experiment may produce exaggerated results in some cases, while underestimating the response in others, depending on the relationship of the perturbations applied to a site's natural variability.

16.2.3 Sites

Two sites were used for the ecosystem memory experiment: Hyytiala, Finland, and Harvard Forest, United States. At each site, defined in the model by latitude, climate (see Table 16.1), and soils, deciduous and coniferous stands were simulated. Meteorological data were derived from local weather station records, as both sites had available daily meteorological data. Both deciduous and coniferous growth forms were simulated: first, because the two regions contain both growth forms (the Hyytiala region contains birch and deciduous understory vegetation, although the flux site footprint is pine); and second, to examine the effect of differences between growth forms in leaf turnover time and foliar chemistry. Note that for the Harvard Forest site, by chance, 1975 was the wettest and one of the warmest years in the period from 1965 to 1975. As a consequence, at that site, the ensemble members differ monotonically and systematically from the base case.

16.3 Results and Discussion

16.3.1 Responses of Net Ecosystem Exchange

We analyzed the initial and long-term responses of multiple ecosystem variables. Plotting the difference between the base case and perturbed ensemble members allows visualization of how long weather effects in a single year cause indirect ecosystem responses. Fig. 16.2 shows the difference in NEE resulting from a single year's perturbation. NEE fluxes are significantly affected by weather in prior years. The effect is large for several years and can continue for periods of decades.

The differences in NEE were often larger in the year after the perturbed year (1975), even though the weather for that subsequent year was identical for all simulations. NEE slowly returned to the base case and generally approached the baseline after 3 to 5 years. The rates and pattern of this "relaxation" were ecosystem specific. Note that because these experiments were analyzed while systems were at a mature NEE/NPP state, the sensitivity expressed as a fractional change is amplified relative to systems rapidly accumulating carbon. Experience

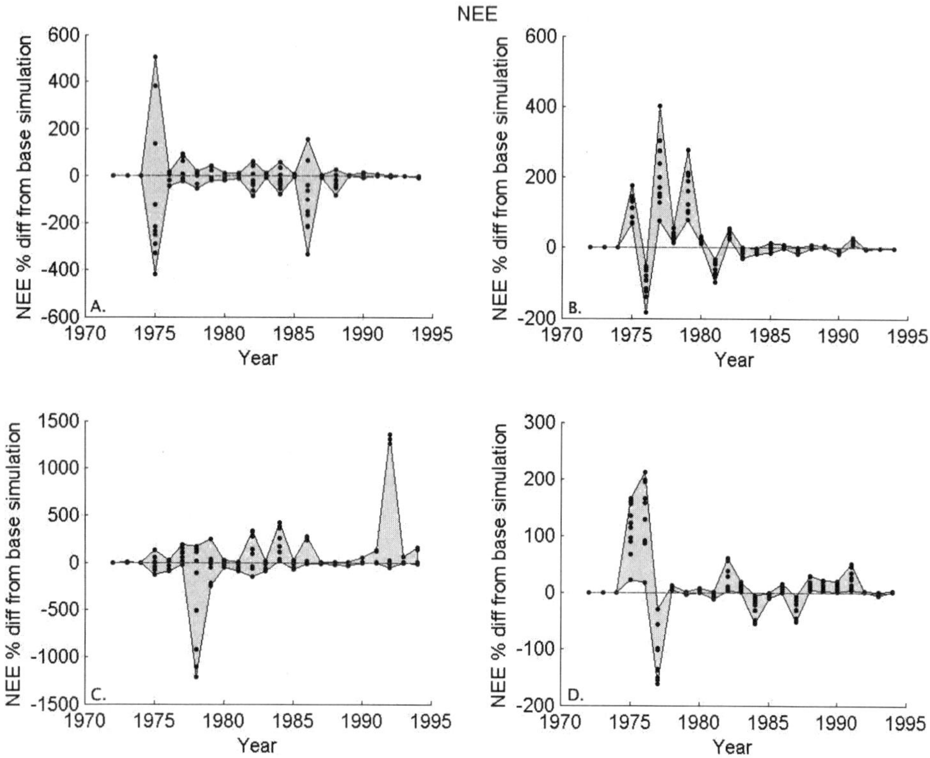

Figure 16.2. Net ecosystem exchange for the model ensemble. All results are shown as annual % difference from the base case. The shaded envelope brackets the ensemble of results; individual ensemble results are shows as points within the envelope. **(A)** the Hyytiala site coniferous stand; **(B)** the Harvard Forest coniferous stand; **(C)** Hyytiala decidous; and **(D)** Harvard Forest deciduous stand.

with this model suggests that the various effects described below remain significant and are on the order of 10% to 50% of the mean NEE flux in accumulating stands (Thornton et al. 2002).

The effects on NEE do not decay away smoothly but decline and recur, depending on the subsequent year's weather and variations in litter carbon and leaf area. Ecosystem sensitivity to climate variability in one year is contingent on responses to the weather in previous years. Since fluxes per se have no memory in the model, these lags arise because of changes to model state variables, including carbon, nitrogen, or water pools. Responses to weather in one year can predispose ecosystem component fluxes to respond more or less strongly to weather in subsequent years. This type of contingency, though expected from ecosystem theory, is rarely accounted for in analyses of biogeochemical flux time series (Goulden et al. 1996; Kelly et al. 2000).

16.3.2 State Variable Dynamics

Plotting state variable changes over time provides insights into the mechanisms causing lagged responses in NEE (Fig. 16.3). The responses of maximum leaf area, soil water, soil nitrogen, and litter mass are shown for the Harvard Forest Coniferous simulation; other sites and growth forms show qualitatively similar results. Foliage, soil water, and nitrogen cycles were all affected by the weather perturbation and then recovered on different timescales. Soil water generally responds quickly (1–2 years) and quickly returns to the base case, which is expected since the modeled turnover time for soil water is a little less than a year. Leaf area also has a short turnover time (depending on coniferous or deciduous growth habit) but is affected through water and nitrogen effects on productivity for many years. Since the nitrogen cycling response is controlled by decomposition, it responds slowly, as does litter carbon. Litter responds to

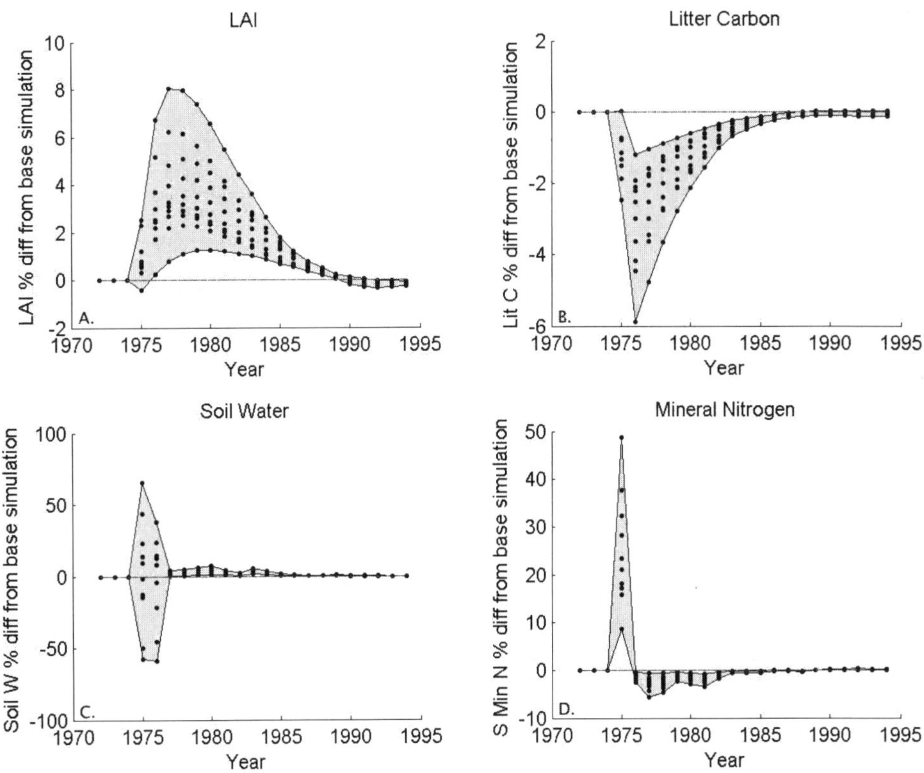

Figure 16.3. Effects on key state variables. All results are shown as annual % difference from the base case. Example results are shown for the Harvard Forest Coniferous stand. **(A)** Maximum leaf area index; **(B)** Litter carbon; **(C)** Soil water; and **(D)** soil mineral nitrogen (plant available N).

altered NPP and then returns to quasi-equilibrium slowly with a rate determined by litter decomposition rates and changes to inputs from NPP. Nitrogen plays a key role in coupling plant and soil time scales together. Rates of N cycling are modified by the initial water and litter C perturbations. Altered N cycling then affects productivity, litter production, and so feeds back again to N cycling and so on.

Plant growth forms affect response times of state variables. The magnitude of effects arising from growth forms is at least as large as the differences between latitudes, based on this limited comparison. Fig. 16.4 shows a comparison of leaf area responses between growth forms at the two sites. Sites and growth forms differ significantly in their responses. The difference in peak proportional response between the deciduous and coniferous responses is large (6 vs. 18% for the northern site) compared to the mean site-to-site differences (9 vs. 12%).

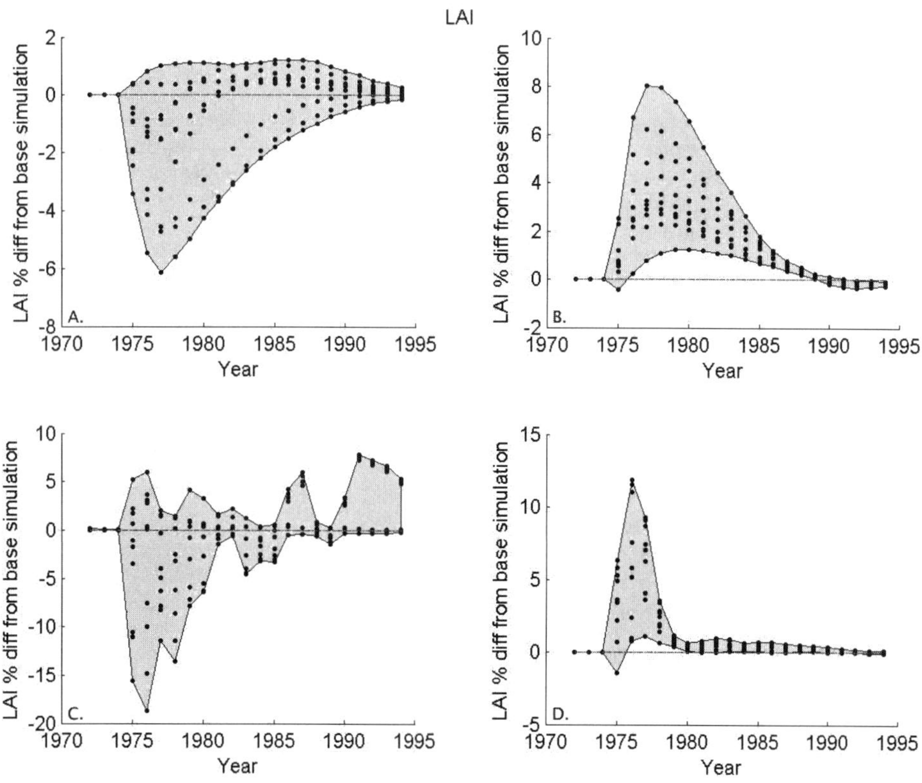

Figure 16.4. Effect of growth form and site on state variables: maximum leaf area index. All results are shown as annual % difference from the base case. This figure compares leaf area index responses of coniferous and deciduous stands at the two sites. **(A)** Hyytiala Coniferous; **(B)** Harvard Coniferous; **(C)** Harvard Deciduous; and **(D)** Hyytiala Deciduous.

For three of the simulations, the timescale of the LAI perturbation is about 15 years, while at the Harvard Forest deciduous site, the timescale is only about 4 years. In the case of the northern deciduous simulation, the extremely long timescale appears to result from the long-lived perturbation to the nitrogen cycle, which in this cold forest is controlled by slow decomposition rates. The differences are indicative of the additional parameterization and validation needed to accurately model "slow" responses globally.

16.3.3 GPP and Ecosystem Respiration Responses

Ecosystem state variables have direct effects on the component fluxes Gross Primary Production and Ecosystem Respiration (ER = microbial respiration + plant growth and maintenance respiration). Fig. 16.5 shows proportional changes in GPP and ER for the coniferous sites. The proportional changes in GPP and ER are larger than those in the state variables, although much smaller than the

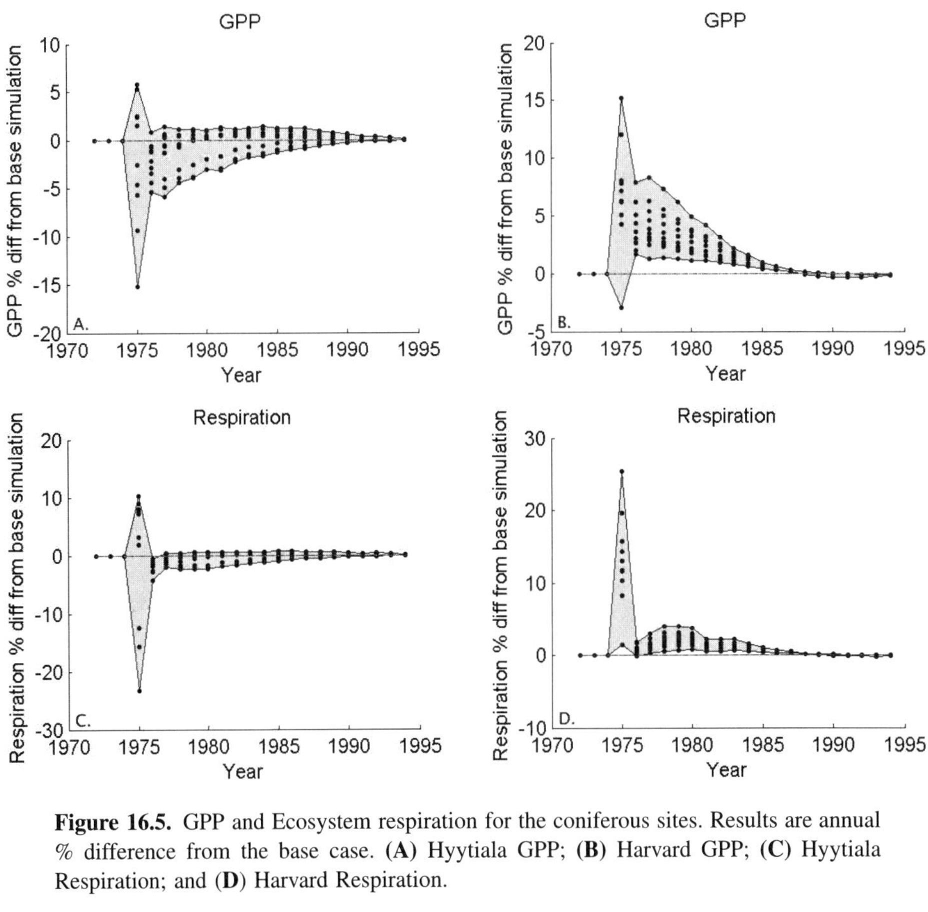

Figure 16.5. GPP and Ecosystem respiration for the coniferous sites. Results are annual % difference from the base case. **(A)** Hyytiala GPP; **(B)** Harvard GPP; **(C)** Hyytiala Respiration; and **(D)** Harvard Respiration.

changes in NEE (see Fig. 16.2). The initial responses and recovery of GPP and ER are smoother than those of NEE. The highly discontinuous responses of NEE occur because GPP and ER change out of phase with one another, on slightly different timescales, causing the difference to vary nonlinearly. Growth form also affects the responses of GPP and ER. Fig. 16.6 compares the GPP perturbations for the Harvard Forest coniferous and deciduous simulations. The GPP effects are longer lasting in the coniferous stand, due mainly to the multi-year needle retention time, and secondarily to coupled effects resulting from slower decomposition rates.

NEE is more sensitive to lagged effects than are GPP and ER separately. Large effects on NEE occur when GPP and ER responses are uncorrelated in time. A variety of ecosystem processes can cause uncorrelated changes in ER and GPP. For example, the production of a large litter cohort in one year may

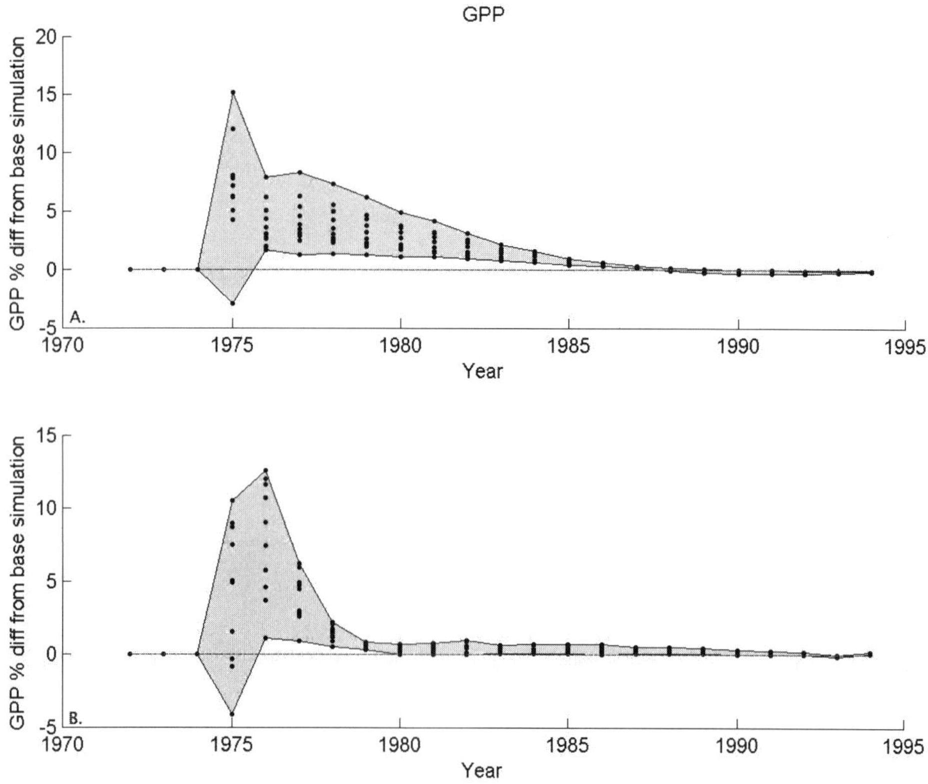

Figure 16.6. GPP perturbations (annual % difference from the base case) for the Harvard Forest coniferous and deciduous simulations, showing the effects of growth form on a component flux response. Note that ER in the model is computed from the sum of plant growth respiration maintenance respiration and heterotrophic respiration, all of which are computed separately. **(A)** Coniferous; **(B)** Deciduous.

increase ER the next year without a corresponding impact on state variables affecting photosynthesis. Changes in nitrogen availability driven by changes in litter mass and heterotrophic activity may also affect GPP. The effects of climate perturbations tend to be largest the year after the perturbation. During the anomalous year, changes to temperature and moisture affect both autotrophic and heterotrophic processes. The effects on GPP and ER are largest initially but may cancel each other out. In subsequent years, the anomalies in GPP and ER are smaller but uncorrelated, leading to larger effects on NEE than in the year when GPP and ER respond the most.

16.3.4 Multiple Steady States in Total Carbon Storage

State variables are important controls in ecosystem models because of the strong first-order dependence of most fluxes. Because the state variables change slowly relative to fluxes and change as a consequence of integrated fluxes, they contribute low-frequency behavior to the model. Differences in the values of state variables that are small relative to measurement error, and too small to simulate accurately, will affect model solutions. This is a precondition for complex responses or even chaotic behavior (May 2001). Fig. 16.7 shows that the model, examined in terms of total carbon storage, has multiple steady states, at least on decadal timescales. Each realization of the model equilibrates at a slightly different level of total carbon. Initially, the total carbon responses diverge, a precondition for chaotic behavior. For the model to be chaotic, these responses would have to diverge exponentially as a function of an initial infinitesimal perturbation. The plots show modest quantitative differences between model states, but given that the variable plotted is total carbon, these small differences are significant. Given the timescale of the perturbation, the differences would be entirely between much smaller active pools (e.g., litter, microbial C, leaf area) and so significant for fluxes. These small differences in state can "precondition" the system for later, altered sensitivity as can be seen 5 to 15 years after the perturbation for some site/growth form combinations.

16.3.5 Significance of Lagged Effects

Ecosystem responses to the environment are a complex mixture of immediate responses to environmental variability and indirect effects mediated through biological processes. Ecosystems therefore respond to any given forcing (e.g., weather) on multiple timescales. This is a characteristic of systems that contain components that respond with different time-constants. The carbon cycle literature has long identified multiple timescales of response as a key aspect of ecosystem behavior (Moore and Braswell 1994; Schimel et al. 1997). Recent papers have begun to address this question as well, as time series grow longer (Katul et al. 2001a,b; Baldocchi, Falge, and Wilson 2001). Ecosystem memory—the effects of prior disturbance and weather forcing of ecosystem responses—exerts a significant effect on net carbon exchange (Vukicevic, Braswell, and Schimel 2001).

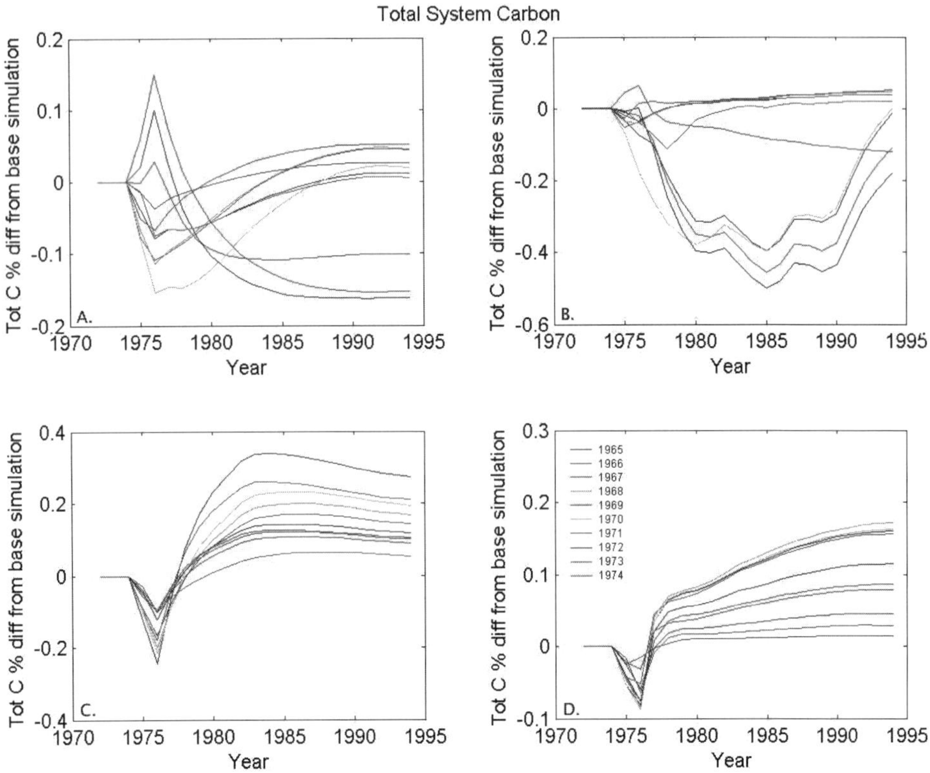

Figure 16.7. Total ecosystem carbon (plant plus soil stocks) for both sites and growth forms, shown as annual % difference from the base case. Each individual ensemble member is shown as a separate trajectory and is color-coded. To create the ensemble members, a year previous to 1975, chosen from 1965 to 1974, is substituted in place of 1975; the substituted year is indicated in the legend in Panel C. Note the variety of dynamical responses of total C illustrated by the NEE trajectories. All simulations show parallel behavior of ensemble members as they approach new steady states (possibly converging on very long timescales). **(A)** the Hyytiala site coniferous stand; **(B)** the Harvard Forest coniferous stand; **(C)** Hyytiala deciduous; and **(D)** Harvard Forest deciduous stand.

Braswell et al. (1997) and Vukicevic, Braswell, and Schimel (2001) argue that the atmospheric carbon dioxide response to temperature variability suggests that ecosystems respond initially to increased temperature by releasing carbon, but that a delayed increase in uptake occurs on average 1 to 3 years after the initial release. They hypothesized that this occurs because increased respiration in the initial year causes increased nutrient availability, and hence increased productivity in subsequent years, without a balancing increase in respiration. There is some evidence for this pattern in the ecosystem memory experiment, as this is the pattern shown in the Harvard Forest deciduous simulation. Other sites and forest types show other responses. As in Schimel et al. (1996) and Braswell et al. (1997), Biome-BGC suggests that ecosystem characteristics significantly influence the response to climate variability. Large responses in the atmosphere must result when spatially coherent climate anomalies trigger responses across ecosystems that respond in similar ways. Vukicevic, Braswell, and Schimel (2001) show that a dominant mode of temperature variability on land occurs in the Northern Hemisphere mid-latitudes, a region that contains most temperate deciduous forests. Because effects of weather in prior years can significantly alter NEE, GPP, and ER from average values, we caution that space-for-time analyses where short flux records are plotted against mean or current year's climate (Valentini et al. 2000) may be significantly affected by lagged effects. As a result, space for time substitution can give extremely misleading results for environmental controls over fluxes.

16.3.6 The Response Hierarchy and State Space Estimation

We observed a hierarchy in the impact of indirect effects from state variables to component fluxes to the integrated (NEE) flux. The ecosystem memory experiment shows that lagged effects are discernible in ecosystem state variables such as leaf area, soil water, soil mineral nitrogen, and litter mass. The magnitude of these effects is of the order 1% to 18% difference from the base case, depending on variable and time since disturbance. The effects persist for 3 to 15 years. The ecosystem component fluxes (GPP and ER) likewise show lagged effects and have delayed responses of up to 25% to 50%. The component fluxes show their largest responses soon after the initiation of the ensemble and decay away smoothly. Delayed differences relative to the base case arise from lags in the ecosystem state variables, because, unlike fluxes, they are able to carry information from one time step to the next. NEE is the difference between GPP and ER and is, therefore, typically a small fraction of either flux. Small changes in either GPP or ER can cause large changes in NEE. The magnitude of the differences between ensemble members in NEE is of the order 50% to 1500%, far larger than the proportional differences in state variables or the component fluxes. At each level of control, the effects of the perturbation are amplified. Accurate simulation of NEE on the interannual timescale depends on highly accurate simulation of state variables and component fluxes of GPP and ER.

Note that the sensitivity of modeled fluxes to key control variables, such as leaf area and litter fall, has profound implications for measurement strategies. It has long been known that errors in GPP scales nearly linearly with errors in leaf area, and as a result major efforts have gone into developing in situ and satellite estimates of leaf area (Running et al. 1994). The current state of the art allows for precision in LAI measurements at the stand scale of 10% to 25% of the mean. Model predictions of leaf area can typically be verified to within about 10% to 25% of the true value. This may result in acceptable errors in GPP and NPP, which are also typically known to within about 25% (Parton et al. 1993) but can result in very large errors in NEE (including a possible sign change). Similar scaling applies to errors in ecosystem respiration, which, however, is rarely measured with comparable accuracy and precision to NPP or GPP.

Until the relatively recent advent of long, quality-controlled NEE time series from eddy covariance, model verification was usually based on comparison to state variables, and NPP, estimated from state variables. Accurate simulation of NEE will require models whose simulation of key state variables is optimized as carefully as are parameter values. While correct model structure and accurate parameters will help with simulation of state variables, other techniques may also be applied. State space estimation, or adjustment of initial and simulated state variable values to minimize the difference between observed and simulated fluxes, while accounting for uncertainties in all variables, is a clear alternative to estimation and tuning or model parameters. This approach is widely used in geophysical models and in weather forecasting and is well justified in systems where model evolution depends sensitively on poorly known state variables (Kalnay, in press).

The sensitivity of NEE to small changes in state variables may explain some of the discrepancies among ecosystem models (Melillo et al. 1995). Ecosystem models with similar structures and parameterizations generally perform similarly under today's climate and compare similarly against NPP data (Schimel et al. 2000; Jenkins, Birdsey, and Pan 2001). However, these same models tend to diverge when simulating climate change (Melillo et al. 1995). Given the sensitivity of NEE to small differences in LAI and other state variables, model biases too small to detect with today's data can have large impacts on modeled NEE. This provides an alternate perspective on model-model disagreements and suggests that far more detailed model intercomparisons will be required to diagnose the cause of differences, such as detailed diagnoses of interannual variability. Models validated against state variable measurements can have major errors in NEE, and equally "valid" models can differ substantially in predictions. Small errors, undetectable with current measurements, in allocation, litter chemistry, soil water, and other key states can accumulate and produce large errors in the long-term carbon balance. This is especially true when models represent aggregated properties (stands or pixels) and average parameter values must be estimated. Long-term modeling of NEE can easily require more accuracy than sampling error permits! Correct models may not even be identifiable with current process study techniques. The solution to this dilemma is not purely reductionist, but rather through the use of novel estimation strategies, as discussed above.

16.3.7 Complex Dynamics in the Carbon System

It is well known that model systems can exhibit stable or chaotic behavior, depending on parameter values (May 2001). In the state and parameter domain examined in this study, Biome-BGC exhibits multiple steady states but not chaotic dynamics. As ecosystem models and data are analyzed under more and more conditions, we should pay close attention to appropriate diagnostics to identify whether systems are entering chaotic domains. The existence of chaotic behavior in biogeochemical systems would have a major impact on both observing system design and modeling strategy. Currently, monitoring and analysis systems for chaotic systems (e.g., the atmosphere, some animal populations) exist, but most biogeochemical observations have tacitly assumed a high degree of linearity between forcing and response (Schimel et al. 1997). If this were to change with alterations in either the biota or the climate system, management of carbon, water quality, and other ecological goods and services would become even more challenging.

Katul et al. (2001b) suggested that ecosystems were dissipative systems, creating some degree of order from chaotic atmospheric forcing (this amounts to an operational definition of life, couched in terms of fluxes). That hypothesis remains tenable. Our results suggest that one mechanism of dissipation is to "rectify" chaotic forcing from the atmosphere into longer timescales. These model results show that interannual variability in climate forcing of ecosystems has impacts on timescales up to decadal. Ecosystem responses may be proportional in the integral to the chaotic portion of forcing but spread over a long period of time, thus limiting the change to average or instantaneous rates. Because NEE is a small difference between large opposing fluxes, its dynamical characteristics are different from the variables ecosystem modelers have focused on for decades (NPP, biomass, and other state variables). Our study addresses intermediate timescales just reached in the short time series that Katul et al. (2001b) analyzed, but it has implications for behavior over a longer term. Ecosystems may either damp or amplify high-frequency chaotic forcing into chaotic fluxes on longer timescales, depending on whether ecosystems are chaotic systems themselves. Models and data about the dynamical characteristics of biogeochemical systems have just reached the point where questions about complexity, nonlinearity, and predictability can be concretely addressed.

16.4 Summary

Understanding the contribution to ecosystem change from processes with different time constants is extremely difficult using short observational time series. We used a model to simulate observations of ecosystem flux and state variables and then linked behavior in the simulated observations to underlying mechanisms and timescales as a model for developing new strategies for observing and analyzing ecosystems. We performed a simple model experiment, using a widely used and well-tested ecological mode, Biome-BGC. We ran many parallel multidecadal simulations, which differed only in the weather during one year (simulated 1975). One simulation used 1975 weather during 1975 while the others had different weather years substituted in that year. Following the 1975 perturbation, the inputs were again identical. This allowed us to analyze both the rate of recovery of ecosystem variables after a one-year perturbation and the mechanisms that caused delayed responses. We observed a hierarchy in the impact of indirect effects arising from state variables and component fluxes upon the integrated (NEE) flux. State variables such as leaf area, soil water, and litter mass were affected by 1% to 15% difference from the base case. The effects persisted for 3 to 15 years. The ecosystem component fluxes (GPP and ecosystem respiration) had responses of up to 25% to 50%, which persisted for 3 to 5 years. The component fluxes show their largest responses soon after the initiation of the ensemble and decay away smoothly. Net Ecosystem Exchange was altered by 50% to 1500% with lagged effects persisting for many years. The fluxes themselves have no memory per se and so lagged differences from the base case arise from anomalies in the ecosystem state variables. The model suggests that a large fraction of interannual variation in NEE occurs when subtle lagged changes in GPP and Ecosystem Respiration occur out of phase with each other. The sensitivity of NEE to small changes in state variables may explain some of the discrepancies amongst ecosystem models. Model biases too small to detect with today's data can have large impacts on modeled NEE. Models validated against state variable measurements can have major errors in NEE, and equally "valid" models can differ substantially in predictions. Aggregated state variables (ecosystem carbon storage) showed multiple steady states but not chaotic dynamics. Biogeochemical systems may be, however, on the "edge" of chaos and future observations, and models should consider the possibility of highly nonlinear and chaotic dynamics arising as novel climate and ecological changes occur.

Acknowledgments. This research was supported by the European Commission's CARBODATA project, by a NASA grant, by the Max Planck Institute for Biogeochemistry, the National Science Foundation, and the U.S. Department of Energy. The authors would like to thank Dennis Baldocchi, Steve Wofsy, Tomi Vukicevic, Peter Thornton, and Steve Running for various forms of help and dispute. The National Science Foundation sponsors the National Center for Atmospheric Research.

References

Baldocchi, D., E. Falge, and K. Wilson. 2001. A spectral analysis of biosphere-atmosphere trace gas flux densities and meteorological variables across hour to multiyear time scales. *Agricultural and Forest Meteorology* 107:1–27.

Braswell B.H., Schimel D.S., Linder E, Moore B. 1997. Science 278 (5339): 870–872.

Canadell, J.G., H.A. Mooney, D.D. Baldocchi, J.A. Berry, J.R. Ehleringer, C.B. Field, S.T. Gower, D.Y. Hollinger, J.E. Hunt, R.B. Jackson, S.W. Running, G.R. Shaver, W. Steffen, S.E. Trumbore, R. Valentini, and B.Y. Bond. 2000. Carbon metabolism of the terrestrial biosphere: A multitechnique approach for improved understanding. *Ecosystems* 3:115–30.

Caspersen, J.P., S.W. Pacala, J.C. Jenkins, G.C. Hurtt, P.R. Moorcroft, and R.A. Birdsey. 2000. Contributions of land-use history to carbon accumulation in US forests. *Science* 290:1148–51.

Collatz, G.J., J.T. Ball, C. Grivet, and J.A. Berry. 1991. Physiological and environmental regulation of stomatal conductance, photosynthesis, and transpiration: A model that includes a laminar boundary-layer. *Agricultural and Forest Meteorology* 54:107–36.

Detling, J.K., W.J. Parton, and H.W. Hunt. 1978. An empirical model for estimating CO_2 exchange of *Bouteloua gracilis* (H.B.K.) Lag. in the shortgrass prairie. *Oecologia* 33: 137–47.

Famiglietti, J.S., and E.F. Wood. 1994. Multiscale modeling of spatially variable water and energy-balance processes. *Water Resources Research* 30:3061–78.

Gaudinski, J.B., S.E. Trumbore, E.A. Davidson, and S.H. Zheng. 2001. Soil carbon cycling in a temperate forest: radiocarbon-based estimates of residence times, sequestration rates, and partitioning of fluxes. *Biogeochemistry* 52:113–14.

Goulden, M.L, J.W. Munger, S.M. Fan, B.C. Daube, and S.C. Wofsy. 1996. Exchange of carbon dioxide by a deciduous forest: Response to interannual climate variability. *Science* 271:1576–78.

Jenkins, J.C., R.A. Birdsey, and Y. Pan. 2001. Biomass and NPP estimation for the mid-Atlantic region (USA) using plot-level forest inventory data. *Ecological Applications* 11:1174–93.

Kalnay, E. [*in press*] Atmospheric modeling, data assimilation, and predictability. Cambridge: Cambridge University Press.

Katul, G.G., C.T. Lai, J.D. Albertson, B. Vidakovic, K.V.R. Schafer, C.I. Hsieh, and R. Oren. 2001b. Quantifying the complexity in mapping energy inputs and hydrologic state variables into land-surface fluxes. *Geophysical Research Letters* 28:3305–3307.

Katul, G., C.T. Lai, K. Schafer, B. Vidakovic, J. Albertson, D. Ellsworth, and R. Oren. 2001a. Multiscale analysis of vegetation surface fluxes: From seconds to years. *Advances in Water Resources* 24:1119–32.

Kelly, R.H., W.J. Parton, M.D. Hartman, L.K. Stretch, D.S. Ojima, and D.S. Schimel. 2000. Intra-annual and interannual variability of ecosystem processes in shortgrass steppe. *Journal of Geophysical Research-Atmospheres* 105:20093–20100.

Knapp, A.K., and M.D. Smith. 2001. Variation among biomes in temporal dynamics of aboveground primary production. *Science* 291:481–84.

Lovejoy, T. 1992. Preface. In *Global warming and biological diversity,* ed. R.L. Peters and T.E. Lovejoy. New Haven: Yale University Press.

May, R.M. 2001. *Stability and complexity in model ecosystems.* Princeton, New Jersey: Princeton University Press.

McGuire, A.D., S. Sitch, J.S. Clein, R. Dargaville, G. Esser, J. Foley, M. Heimann, F. Joos, J. Kaplan, D.W. Kicklighter, R.A. Meier, J.M. Melillo, B. Moore, I.C. Prentice, N. Ramankutty, T. Reichenau, A. Schloss, H. Tian, L.J. Williams, and U. Wittenberg. 2001. Carbon balance of the terrestrial biosphere in the twentieth century: Analyses of CO_2, climate and land use effects with four process-based ecosystem models. *Global Biogeochemical Cycles* 15:183–206.

Melillo, J.M. J. Borchers, J. Chaney, H. Fisher, S. Fox,A. Haxeltine, A. Janetos, D.W. Kicklighter, T.G.F. Kittel, A.D. McGuire, R. McKeown, R. Neilson, R. Nemani, D.S. Ojima, T. Painter, Y. Pan, W.J. Parton, L. Pierce, L. Pitelka, C. Prentice, B. Rizzo, N.A. Rosenbloom, S. Running, D.S. Schimel, S. Sitch, T. Smith, I. Woodward. 1995. Vegetation/Ecosystem Modeling and Analysis Project (VEMAP): continental-scale study of terrestrial ecosystem responses to climate change and CO_2 doubling. Global Biogeochemical Cycles 9(4): 407–437.

Moore, B., and B.H. Braswell. 1994. The lifetime of excess atmospheric carbon-dioxide *Global Biogeochemical Cycles* 8:23–38.

Parton, W.J., J.M. O. Scurlock, D.S. Ojima, T.G. Gilmanov, R.J. Scholes, D.S. Schimel, T. Kirchner, J.C. Menaut, T. Seastedt, E.G. Moya, A. Kamnalrut, J.I. Kinyamario. 1993. Observations and modeling of biomass and soil organic-matter dynamics for the grassland biome worldwide. *Global Biogeochemical Cycles* 7:785–809.

Prentice, I.C., G.D. Farquhar, M.J.R. Fasham, M.L. Goulden, M. Heimann, V.J. Jaramillo, H.S. Kheshgi, C. Le Quere, R.J. Scholes and D.W.R. Wallace. In: J.T. Houghton, Y. Ding, D.J. Griggs, M. Noguer, P.J. van der Linden, X. Dai, K. Maskell, C.A. Johnson (eds) Climate Change 2001: The Scientific Basis. Contribution of Working Group I to the Third Assessment Report of the Intergovernmental Panel on Climate Change. Cambridge University Press, Cambridge, U.S.A. pp. 183–237.

Rosenbloom, N.A., S.C. Doney, and D.S. Schimel. 2001. Geomorphic evolution of soil texture and organic matter in eroding landscapes. *Global Biogeochemical Cycles* 15: 365–81.

Running, S.W., C.O. Justice, V. Salomonson, D. Hall, J. Barker, Y.J. Kaufmann, A.H. Strahler, A.R. Huete, J.P. Muller, V. Vanderbilt, Z.M. Wan, P. Teillet, and D. Carneggie. 1994. Terrestrial remote-sensing science and algorithms planned for EOS MODIS. *International Journal of Remote Sensing* 15:3587–3620.

Schimel, D.S., B.H. Braswell, R. McKeown, D.S. Ojima, W.J. Parton, and W. Pulliam. 1996. Climate and nitrogen controls on the geography and timescales of terrestrial biogeochemical cycling. *Global Biogeochemical Cycles* 10:677–92.

Schimel, D.S., W. Emanuel, B. Rizzo, T. Smith, F.I. Woodward, H. Fisher, T.G.F. Kittel, R. McKeown, T. Painter,N. Rosenbloom, D.S. Ojima, W.J. Parton, D.W. Kicklighter, A.D. McGuire, J.M. Melillo, Y. Pan, A. Haxeltine, C. Prentice, S. Sitch, K. Hibbard, R. Nemani, L. Pierce, S. Running, J. Borchers, J. Chaney, R. Neilson, and B.H. Braswell. 1997. Continental scale variability in ecosystem processes: Models, data, and the role of disturbance. *Ecological Monographs* 67:251–71.

Schimel D.S., J.I. House, K.A. Hibbard, P. Bousquet, P. Ciais, P. Peylin, B.H. Braswell, M.J. Apps, D. Baker, A. Bondeau, J. Canadell, G. Churkina, W. Cramer, A.S. Denning, C.B. Field, P. Friedlingstein, C. Goodale, M. Heimann, R.A. Houghton, J.M. Melillo, B. Moore, D. Murdiyarso, I. Noble, S.W. Pacala, I.C. Prentice, M.R. Raupach, P.J. Rayner, R.J. Scholes, W.L. Steffen, C. Wirth. 2001. Recent patterns and mechanisms of carbon exchange by terrestrial ecosystems. *Nature* 414 (6860): 169–172

Thornton, P.E., B.E. Law, H.L. Gholz, K.L. Clark, E. Falge, D.S. Ellsworth, A.H. Goldstein, R.K. Monson, D. Hollinger, M. Falk, J. Chen, and J.P. Sparks. 2002. Modeling and measuring the effects of disturbance history and climate on carbon and water budgets in evergreen needleleaf forests. *Agricultural and Forest Meteorology* 113:185–222.

Valentini, R., G. Matteucci, A.J. Dolman, E.D. Schulze, C. Rebmann, E.J. Moors, A. Granier, P. Gross, N.O. Jensen, K. Pilegaard, A. Lindroth, A. Grelle, C. Bernhofer, T. Grunwald, M. Aubinet, R. Ceulemans, A.S. Kowalski, T. Vesala, U. Rannik, P. Berbigier, D. Loustau, J. Guomundsson, H. Thorgeirsson, A. Ibrom, K. Morgenstern, R. Clement, J. Moncrieff, L. Montagnani, S. Minerbi, P.G. Jarvis. 2000. Respiration as the main determinant of carbon balance in European forests. *Nature* 404:861–65.

Vukicevic, T., B.H. Braswell, and D. Schimel. 2001. A diagnostic study of temperature controls on global terrestrial carbon exchange. *Tellus* B53:150–70.

17. Effects of Elevated CO_2 on Keystone Herbivores in Modern Arctic Ecosystems

Scott R. McWilliams and James O. Leafloor

Current atmospheric CO_2 enrichment represents a dietary change of the biosphere not precedented in the past 0.5 million years.

—Körner 2000

17.1 Introduction

A central theme of this chapter is that elevated atmospheric CO_2 will directly alter growth and nutrient quality of arctic plants and will indirectly change climate, especially at high latitudes. Determining how this environmental change will affect arctic ecosystems requires an understanding of not only the physiological ecology of arctic biota but also the strength and nature of biotic interactions. For example, changes in plant abundance and quality associated with elevated atmospheric CO_2 could negatively affect herbivores and so alter the competitive interplay between species and, thus, the ecological community.

Since much has recently been written about the effect of elevated CO_2 on the physiology and ecology of arctic plant communities (Chapin et al. 1992; Oechel and Vourlitis 1996; Oechel et al. 1997; Dormann and Woodin 2002), in this chapter we primarily focus on the potential effects of these environmental changes on key consumers in arctic ecosystems, specifically herbivores. Thus, a second theme of this chapter is that elevated CO_2 will affect the interaction between herbivores in arctic ecosystems and their plant food supply. We primarily focus on keystone herbivores in extant arctic ecosystems because they provide important insights into the effect of changes in plant quality and quantity on herbivore populations, specifically, and on arctic ecosystems, in general.

17.2 Elevated Atmospheric CO_2 Will Change Climate Especially at High Latitudes

Carbon availability to plants has increased at least 30% during the past 150 years due to burning of fossil fuels and forest destruction; although the same level of increase in carbon availability has occurred historically, it took place over about 500,000 years (Petit et al. 1999). This acceleration of CO_2 enrichment with industrialization is projected to continue so that by the end of the twenty-first century the concentration of CO_2 in Earth's atmosphere will have doubled compared to preindustrial concentrations (IPCC 1996).

Increased availability of carbon affects plant growth (via fertilization) and plant quality, as discussed below. Besides this fertilization effect of CO_2 enrichment, increased atmospheric CO_2 will also contribute to global warming because CO_2 is one of the greenhouse gases (Crowley 2000). The effect of global warming on arctic ecosystems, the region north of the latitudinal treeline, is difficult to ascertain because many other components of the climate system besides the greenhouse gases will also change: for example, hydrological processes (Körner 2000). However, current predictions suggest that global warming will vary across the globe with relatively small changes in temperature near the equator and the most substantial changes (>4°C) occurring at high latitudes such as the Arctic. These types of large-scale increases in temperature in the Arctic would change the seasonal phenology of arctic biota (Penuelas and Filella 2001; Lucht et al. 2002), cause substantial changes in distribution and abundance of plants (Davis and Shaw 2001) and animals (Walker, Gould, and Chapin 2001), and perhaps alter the significant amount of carbon now stored in arctic soils, thus further elevating atmospheric CO_2 (Oechel and Billings 1992).

Changes in seasonal phenology would have profound effects on the fauna in arctic ecosystems. For example, the breeding season of many arctic herbivores coincides with periods of maximal availability of young plant tissue that is most nutritious (e.g., Raveling 1978; Sedinger and Raveling 1986; Prop and de Vries 1993; Gunn and Skogland 1997). Increased ambient temperature would increase plant growth rate and so not only reduce the time period during which high-quality immature forage is available (Ayres 1993) but also directly affect the nutrient quality of forage plants (Tolvanen and Henry 2001). This can create a phenological mismatch between consumer and consumed that negatively affects reproduction of consumers (Dunn and Winkler 1999; Both and Visser 2001) and, hence, their populations (Boyd and Madsen 1997; Prop et al. 1998; Inouye et al. 2000; Price and Root 2000; Lenart et al. 2002; Root and Schneider 2002).

The primary prediction regarding changes in plant distribution and abundance with climate change is that shrub-tundra and taiga communities will spread northward and replace tussock-tundra (Bliss and Matveyeva 1992). Such a change in plant communities would decrease the abundance of tundra graminoids that, in turn, would adversely affect populations of arctic grazers. For example, large-scale climatic variability associated with the North Atlantic Oscillation between 1928 and 1977 was significantly related to plant phenology and interannual changes in growth, fecundity, and population trends for seven species of northern ungulates (Post and Stenseth 1999). If habitat preferences of contemporary herbivores remain fixed as plant distribution and abundance change, then browsing mammals currently inhabiting taiga (e.g., moose, snowshoe hare) will likely replace browsing mammals currently inhabiting shrub-tundra (e.g., caribou) as well as vertebrate grazers (e.g., voles, lemmings, geese) in the Arctic (Bryant and Reichardt 1992; Kerr and Packer 1998; Walker, Gould, and Chapin 2001). Such a change in animal communities would seriously affect the economies of several human populations in the Arctic, especially those that depend on caribou (e.g., Inuit and Cree) and geese (e.g., Yupik Eskimo).

In the remainder of this chapter, we primarily focus on the direct effect of CO_2 enrichment on arctic ecosystems and not on the indirect effects of CO_2 enrichment, such as global warming (Chapin et al. 1992; Arft et al. 1999).

17.3 Effects of Elevated CO_2 on Growth and Nutrient Quality of Arctic Plants

Our understanding of how elevated CO_2 affects plant growth and quality is based largely on studies of plants that are endemic to mid-latitudes and not high-latitudes (Bazzaz 1990; Mooney et al. 1991). However, some studies (Table 17.1) have focused on the direct effects of elevated CO_2 on arctic plants in intact arctic ecosystems (reviewed by Oechel and Billings 1992; Dormann and Woodin 2002). Elevated CO_2 has a short-term (days to weeks) positive effect on plant productivity, but over longer timescales (months to years) there is often a complete homeostatic adjustment that results in similar rates of photosynthesis in plants exposed to ecologically relevant but different levels of ambient CO_2 (Tissue and Oechel 1987; Oechel and Vourlitis 1996). In general, these studies reveal that elevated CO_2 has relatively minor direct effects on photosynthesis and net primary productivity of arctic plants, with the level of plant response primarily depending on temperature, and water, and nutrient availability (Oechel and Billings 1992; Oechel and Vourlitis 1997).

Although elevated CO_2 has relatively minor direct effects on photosynthesis and net primary productivity of arctic plants, elevated CO_2 significantly changes plant quality (Bryant and Reichardt 1992). In general, the carbon/nitrogen ratio in plants increases with CO_2 enrichment (Ayres 1993; Cotrufo, Ineson, and Scott 1998; Körner 2000). Increased carbon availability is often used by plants to boost the amount of carbon-based secondary compounds, such as tannins, especially in C_4 woody plants (Jonasson et al. 1986; Bryant and Reichardt 1992), or fiber, as in C_3 plants (Ayres 1993). Nitrogen limitation is often associated with CO_2 enrichment so that plants have less tissue protein (Berendse and Jonasson 1992; Cotrufo, Ineson, and Scott 1998). Thus, from an herbivore's perspective, increased plant fiber or secondary compounds and decreased protein in plants grown in conditions of elevated CO_2 directly reduce the nutritional

Table 17.1. Effects of elevated atmospheric CO_2 on characteristics of arctic plants and the arctic ecosystem depend on timescale (short-term is days to months, intermediate is months to years, and long-term is decades to centuries). Symbols denote whether the plant or ecosystem response to elevated CO_2 was higher (+), lower (−), or the same (0) compared to a control (i.e., the plant or ecosystem exposed to ambient CO_2). Data are from Oberbauer et al. (1986), Tissue and Oechel (1987), Grulke et al. (1990), Mooney et al. (1991), and Oechel and Vourlitis (1996).

	Timescale of response		
Plant or Ecosystem Effect	Short-term	Intermediate	Long-term
Plant Effect			
Growth and Production			
Photosynthesis	+	0	0
Transpiration	+	0	0
Biomass production	+	+/0	?
Shoot growth	+	0	?
Root:shoot ratio	+	?	+
Tillering	+	+	?
Tissue composition			
Nitrogen concentration	−	−	−
Carbon:nitrogen ratio	+	+	+
Starch content	+	+	+
Ecosystem Effect			
Evapotranspiration	+	0	0
Net ecosystem flux	+	0	?
Net carbon storage	+	+/0	+/0
Soil enzyme activity	+	+/−	+/−
Soil solution nitrogen	−	−/0	−/0

quality of the plants (Ayres 1993). As discussed in the following sections, these changes in plant resource allocation associated with elevated CO_2 are predicted to affect growth and condition of herbivores, herbivore population dynamics, and reciprocal interactions between arctic herbivores and the plant communities upon which they depend.

Although we focus in this review on the direct effects of elevated CO_2 on plant quality and how this will affect certain herbivores, it is worth emphasizing that elevated CO_2 will directly and indirectly influence interactions between plants and herbivores. For example, the increased carbon availability to plants with elevated CO_2 will likely interact with the effect of elevated CO_2 on ambient temperature, precipitation, and nitrogen deposition and their effects on plant quality (Oechel and Billings 1992; Shaw et al. 2002). Increased ambient temperature would likely increase nutrient availability in the arctic ecosystem so that plants may then route available carbon to growth rather than the production of carbon-based fiber or secondary compounds. Increased ambient temperature would also directly affect herbivore energy and nutrient requirements, development, and survival. Thus, elevated CO_2 will certainly affect plant-herbivore interactions in several ways (Ayres 1993). We focus here on one dramatic effect of elevated CO_2 on plant-herbivore interactions: how changes in plant quality can directly affect keystone herbivores in arctic ecosystems.

17.4 Top-Down and Bottom-Up Processes: The Importance of Keystone Herbivores in Arctic Ecosystems

Understanding how elevated CO_2 will affect arctic ecosystems requires knowledge of how elevated CO_2 will directly influence plant growth and quality, and how these changes in plant growth and quality will, in turn, influence consumers in higher trophic levels. Biotic communities in arctic ecosystems are primarily organized in linear food chains with distinct trophic levels (i.e., "trophic ladders") rather than in food webs with less distinct trophic levels (i.e., "trophic webs") (Strong 1992). Since arctic ecosystems have relatively low species diversity (Hansell et al. 1998) and low annual primary production (Bliss 1986) and are organized in multiple trophic ladders, predicting how changes in primary producers will influence consumers should be relatively straightforward. However, even for the relatively simple arctic ecosystem, there are multiple controls on ecosystem processes that complicate our ability to predict the effect of such perturbations as elevated CO_2.

According to Drent and Prins (1987), the distribution and abundance of arctic herbivores is largely determined by "bottom up" processes, such as nutrient availability to plants. Thus, increased plant growth associated with elevated CO_2 would increase primary productivity, which in turn would support more herbivores. However, the simultaneous decrease in plant quality associated with elevated CO_2 would nutritionally challenge some arctic herbivores and likely compromise their ability to numerically respond to the predicted increase in plant growth (see Section 17.5).

There are also "top down" processes, such as herbivory and predation, in arctic ecosystems that are important in influencing how the biota responds to elevated CO_2. Arctic ecosystems are unusual in that some herbivores, such as geese, are "keystone" species (Kerbes, Kotanen, and Jefferies 1990; Jefferies 1999), which means that they have a significant and disproportionately large impact on the ecosystem (Paine 1966; Paine 1969; Power et al. 1996). Keystone species in other ecosystems are usually predators (e.g., sea otters, starfish) near the top of the food chain rather than mid-level consumers such as herbivores (Strong 1992; Bond 1994).

The role of geese as keystone species in arctic ecosystems is described in the next section. The important point here is that the population density of keystone herbivores in arctic ecosystems affects the relative strength of top-down and bottom-up forces. At high population densities, keystone herbivores exert strong top-down influences on forage plants that can lead to runaway consumption and destruction of the arctic vegetation (Jefferies 1997; Mulder and Ruess 1998; Jefferies 1999; Mulder and Ruess 2001; Handa, Harmsen, and Jefferies 2002). In contrast, at lower herbivore population densities, resources available to plants may limit their productivity, which then will exert strong bottom-up control of herbivore populations (Drent and Prins 1987). In short, keystone herbivores in arctic ecosystems are of central ecological importance and so must be considered when determining how elevated CO_2 affects arctic ecosystems.

17.5 Geese as Keystone Herbivores in Arctic Ecosystems and Trophic Cascades

Herbivores in extant arctic ecosystems include relatively small populations of large mammals (e.g., muskox, caribou), large and often cyclic populations of small mammals (e.g., voles, lemmings), resident galliformes (e.g., grouse, ptarmigan), and migratory waterfowl (e.g., geese and a few duck species) (Bliss 1975; Batzli et al. 1980; Bliss 1986; Jefferies et al. 1992). Muskox and caribou are the only two ungulate species that are resident in the northernmost regions of the world (Klein 1986). Muskox are relatively unselective, bulk feeders that can eat plants high in fiber, whereas caribou are selective grazers whose performance is strongly affected by forage quality (White 1983; White and Fancy 1986; Lenart et al. 2002). Voles and lemmings are also selective grazers, their population dynamics are directly influenced by plant quality and quantity, and their grazing can have dramatic affects on the habitat (Batzli et al. 1980; Batzli 1983; Oksanen 1983; Bliss 1986). Grouse and ptarmigan, as well as medium-sized mammals such as arctic hares, do not occur at sufficiently high enough densities in arctic ecosystems to have major impacts on the vegetation (Batzli et al. 1980).

For the purposes of this review, we have chosen to focus on geese that inhabit primarily coastal salt marsh and adjacent uplands in arctic ecosystems for three main reasons: (1) avian herbivores such as geese require relatively high-quality forage and so changes in plant quality, like those predicted to occur with elevated CO_2, should rapidly affect their distribution and abundance; (2) coastal regions, in general, and those in the arctic, in particular, are especially vulnerable to changes in sea level like those predicted to occur with elevated CO_2 and global warming; and (3) geese are keystone species in this system (Kerbes, Kotanen, and Jefferies 1990; Jefferies 1999) that can consume as much as 90% of the net aboveground primary production (Cargill and Jefferies 1984). Thus, we focus on arctic-nesting geese because these keystone herbivores and their primary habitat are both likely to respond sensitively to the environmental change predicted with elevated atmospheric CO_2.

Body size is thought to put limits on the quality of foods that can be eaten by herbivores. In general, as body size of herbivores declines so does the fiber level of the selected diet and the proportion of energy requirements satisfied by energy yield from fermentation (Illius and Gordan 1993; Cork 1994). This pattern of decline in reliance on fiber with decreasing body size in vertebrate herbivores is thought to occur because of unequal allometries of energy requirements and gut capacity (Demment and Van Soest 1985). The relative rarity of avian herbivores (i.e., only 3% of extant bird species) suggests these allometric constraints are especially difficult to circumvent for birds (McWilliams 1999). True geese are exceptional birds in that they eat leaves throughout their life cycle and they are strong flyers with associated high metabolic rates. Thus, adult geese and especially their goslings are expected to respond sensitively to changes in plant quality and quantity because they require a relatively high quality and quantity of forage to satisfy their relatively high energy and nutrient requirements (McWilliams 1999).

Evidence for geese as "keystone herbivores" in arctic ecosystems (Kerbes, Kotanen, and Jefferies 1990; Jefferies 1999) includes the effect of their grazing on plant primary production (Cargill and Jefferies 1984; Bazely and Jefferies 1985; Jefferies 1988a; Hik and Jefferies 1990; Gauthier et al. 1995; Person, Babcok, and Ruess 1998) and on plant succession (Bazely and Jefferies 1986; Drent and van der Wal 1999); further evidence includes the dramatic effect of changing density of geese on the plant community (Mulder, Ruess, and Sedinger 1996; Srivastava and Jefferies 1996; Zacheis, Hupp, and Ruess 2001; Person et al. 2003) and the ecosystem properties (Ruess, Hik, and Jefferies 1989; Bazely and Jefferies 1997; Jefferies 1999).

Overgrazing by lesser snow geese has caused large-scale, long-term changes in quality and quantity of preferred arctic plants in coastal salt marshes throughout southern Hudson Bay and James Bay, Canada (Kerbes, Kotanen, and Jefferies 1990; Jefferies 1997; Handa, Harmsen, Jefferies 2002). This large-scale impact of geese on the vegetation is primarily the result of improved survival and habitat on the wintering areas (in the mid-southern U.S.), which has effectively removed what historically was the primary, density-dependent limitation on arctic-nesting goose populations (Batt 1997). These large-scale effects of overabundant geese on arctic plant communities has, in turn, negatively affected goslings attempting to grow up in arctic breeding areas with reduced plant quality and quantity (Cooch et al. 1991; Cooch, Rockwell, and Brault 2001).

The positive feedback responses between grazing goslings and their forage plants that occur at low to moderate grazing intensities break down at higher grazing intensities (Jefferies 1999). Destructive removal of below-ground parts of plants and overgrazing of above-ground plant parts by geese creates bare patches of sediment (Fig. 17.1), which then become hypersaline during summer due to increased rates of soil evaporation (Srivastava and Jefferies 1996). The hypersaline soil conditions depress growth and inhibit recolonization of the salt-marsh graminoids so that an alternative vegetational state develops that is dominated by *Salicornia borealis*, a salt-tolerant plant species that is not eaten by geese and persists for many decades. This process is called a "trophic cascade" by ecologists and refers to runaway consumption and dominance by species at higher trophic levels (in this case by geese) that has dramatic effects through the food chain (Jefferies 1999). Trophic cascades are rare in terrestrial systems, but when they do occur they are found where species diversity is low and food chains are linear: for example, in arctic ecosystems (Strong 1992).

Clearly, understanding how elevated CO_2 will affect arctic ecosystems requires

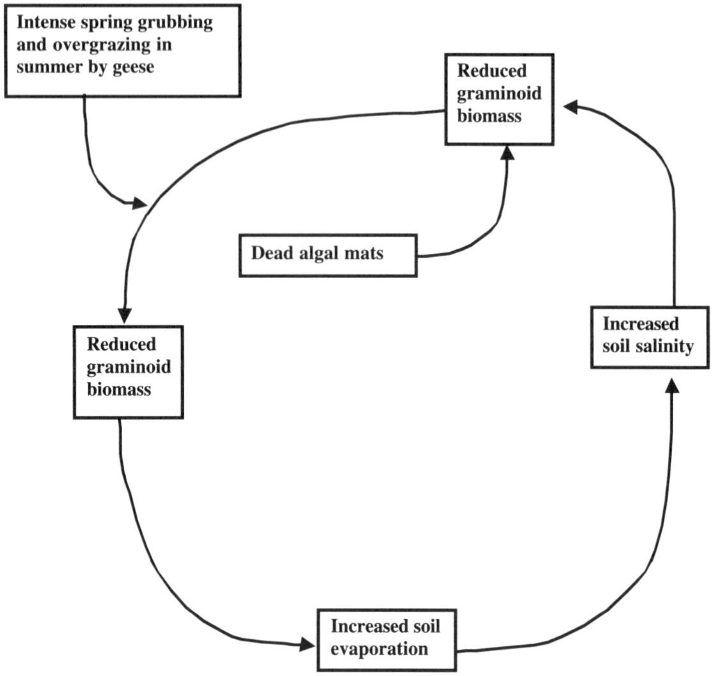

Figure 17.1. Overgrazing and grubbing by arctic-nesting geese in the low arctic salt-marsh ecosystem of Hudson Bay, Canada, has led to reductions in plant cover, increased soil evaporation, and salinity and thus to further reduction and degradation of the salt-marsh ecosystem. This trophic cascade has led to dramatic, large-scale changes in coastal salt-marsh vegetation from a vegetation community that supported large populations of vertebrate herbivores to a system with few if any plants palatable to herbivores. Geese are clearly keystone herbivores in this system and so must be considered when determining the effects of elevated CO_2 on arctic ecosystems. Modified from Srivastava and Jefferies (1996) and Jefferies (1999).

knowledge of how changes in plant growth and quality affect the keystone herbivores that dominate this ecosystem. We expect keystone herbivores such as geese to respond sensitively to changes in plant growth and quality because of the allometric constraints associated with being a bird that flies and eats leaves.

17.6 Reciprocal Interactions Between Plants and Herbivores in Arctic Ecosystems

In this section and the next, we discuss the interplay between arctic herbivores and their food plants and emphasize how changes in food quality affect plant-herbivore interactions. The reciprocal interactions between arctic herbivores and their food plants demonstrate how changes in plant quality, like those predicted to occur with elevated CO_2, can affect key consumers in arctic ecosystems.

In arctic ecosystems, plant chemistry (e.g., secondary compounds) controls herbivory on woody vegetation (Bryant and Kuropat 1980; Batzli 1983), and deciduous woody plants that are browsed respond to herbivory by changing their plant chemistry (Bryant and Reichardt 1992). Vertebrate herbivores such as the arctic hare browse deciduous woody plants (i.e., they eat shoots and leaves) whereas insect herbivores eat primarily just the leaves. When deciduous woody plants are browsed, more nutrients (e.g., nitrogen) are available for the remaining shoots so that leaf growth increases and available carbon is used to support increased plant growth rather than the production of carbon-based secondary compounds. In contrast, when leaves are repeatedly eaten from deciduous woody plants by insect herbivores, fewer nutrients are available for new leaf growth. This limitation of nutrients causes a surplus of available carbon that can be used to support increased production of secondary compounds or fiber. Thus, woody plants that are browsed are higher in quality (i.e., more nitrogen, less secondary compounds) than are woody plants that are defoliated by insects, and these differences in quality subsequently influence future herbivory (e.g., Danell and Huss-Danell 1985).

The interplay between arctic plants and herbivores is also evident for vertebrate grazers that eat herbaceous (nonwoody) plants. For example, flocks of arctic-nesting brent geese revisit certain areas of coastal salt marsh at intervals of time that maximize plant quality and quantity—a phenomena known as "cyclic grazing" (Prop 1991; Rowcliffe et al. 1995; Drent and van der Wal 1999)). Given the time required for plants to recover from being grazed and the intake rates achieved by brent geese on swards of certain biomass, the frequency of revisitation by these geese coincided with that which maximized their intake on a given sward (Fig. 17.2). Thus, cyclic grazing by brent geese effectively increased the amount of high-quality forage plants available to the geese within a season (Drent and van der Wal 1999).

Moderate grazing by geese increases the quality of forage because newer, younger plant tissue is higher in nitrogen and lower in fiber than is older tissue (Fig. 17.3). For example, grazing by lesser snow goose goslings increased net aboveground primary productivity by up to 100% and increased nitrogen concentration in forage plants, in large part because the droppings produced by feeding goslings increased available nitrogen for plant growth (Cargill and Jefferies 1984; Bazely and Jefferies 1985). Thus, goslings modulate nitrogen flow in this arctic ecosystem by their grazing and so increase the quality and quantity of available forage (Jefferies 1988a; Jefferies 1988b; Jefferies 1999).

17.7 Reduced Plant Quality like that Predicted to Occur with Elevated CO_2: How Keystone Herbivores Respond

As discussed above, elevated atmospheric CO_2 consistently causes changes in plant quality that generally reduce the plant's nutritional value for herbivores.

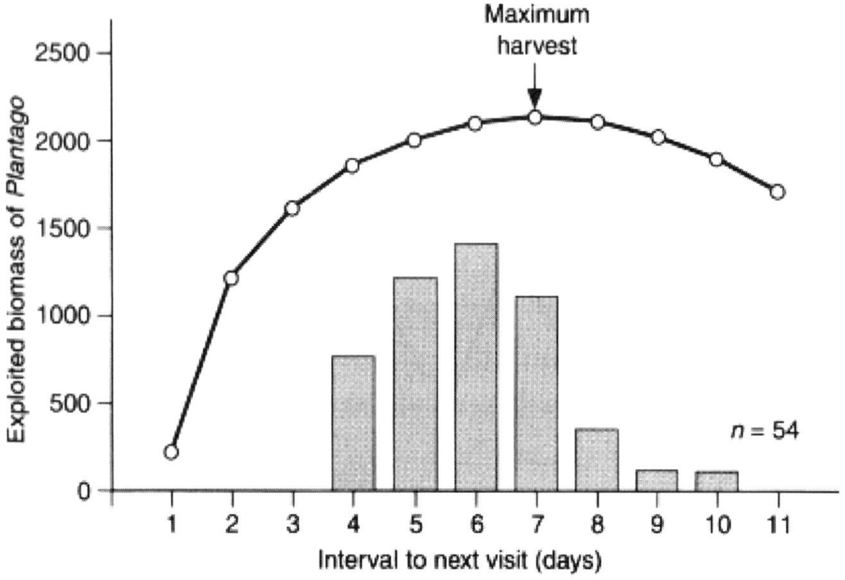

Figure 17.2. Keystone herbivores such as geese sensitively respond to changes in plant quantity. Available biomass of a preferred forage plant (*Plantago maritima*) initially increases with less grazing by brent geese (*Branta bernicla*), as indexed by how often they graze the sward: for example, every day, every other day, and so forth. Maximum harvest by the geese is achieved when they return to a given sward every 7 days, since this allows the plant enough recovery time to allow the geese to maximize their intake rate. Too little grazing of the sward by geese allows the plants to grow beyond the optimal size for maximum harvest, based on a model by Prop (1991). The measured frequency of visits by brent goose flocks to specific swards (shown by the histogram) coincides with maximum harvest but is skewed toward slightly earlier harvest (Prop 1999). From Drent and van der Wal (1999). Reprinted with permission from Blackwell Science.

Since the effects of CO_2 enrichment on the forage quality of plants is predicted to be especially evident in C_3 plants, which predominate in arctic ecosystems, the effect of reduced forage quality on arctic herbivores is particularly pertinent yet poorly understood (Dormann and Woodin 2002). Most research on the response of herbivores to plants treated with CO_2 has focused on insect herbivores inhabiting temperate systems (reviewed by Lincoln, Fajer, and Johnson 1993; Lindroth 1996). We are unaware of any research that focuses on the response of arctic herbivores to changes in plant quality directly caused by elevated atmospheric CO_2. Thus, we will focus here on how arctic herbivores respond to forage with increased plant fiber and decreased protein since these are the expected changes in quality of plants grown under conditions of elevated CO_2.

During the breeding season in the Arctic, adult geese and goslings select plant

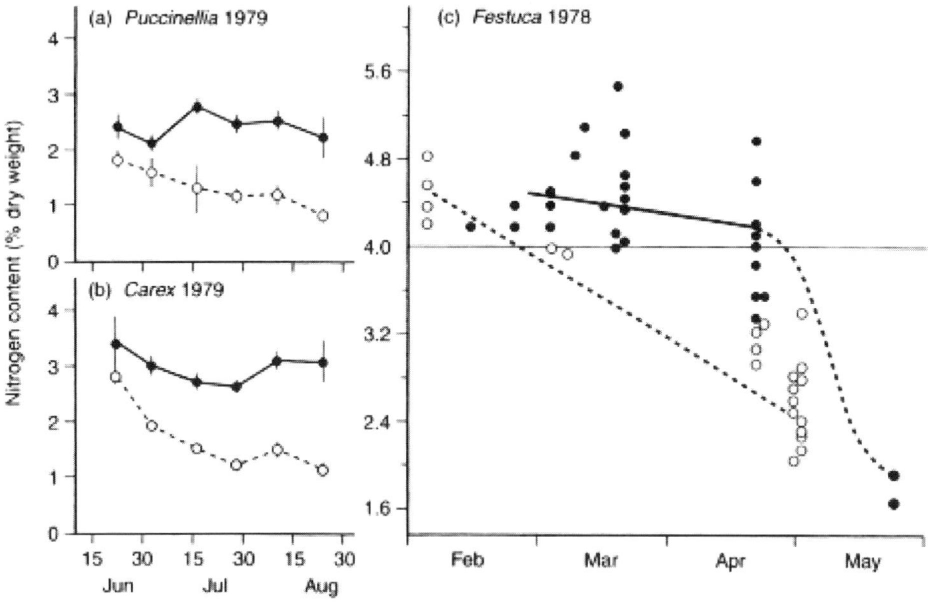

Figure 17.3. Grazing by keystone herbivores such as geese improves plant quality. Nitrogen content of three arctic plant species **(A)** *Puccinellia phryganodes*, **(B)** *Carex subspathacea*, and **(C)** *Festuca rubra* normally decreases during the growing season when plants are ungrazed (*open circles*). Moderate grazing by geese (*solid circles and lines*) delays the decline in nitrogen content of the three plant species and hence extends the period of time with good forage quality. During the summer breeding season, more days with high-protein, low-fiber forage increases growth, survival, and productivity of geese. Data for 1978 are from Cargill and Jefferies (1984), and the data for 1979 are from Ydenberg and Prins (1981). From Drent and van der Wal (1999). Reprinted with permission from Blackwell Science.

species and plant parts that are high in protein and low in fiber (Sedinger and Raveling 1984; Gauthier and Bedard 1990; Krapu and Reinecke 1992; Fox 1993; Laing and Raveling 1993; Manseau and Gauthier 1994). High-quality forage (low fiber, high protein) is needed by geese and goslings during this time to satisfy the high nutrient requirements for breeding and growth, respectively (Sedinger and Raveling 1986; Sedinger 1992). Protein content is an important nutrient because nitrogen is typically a limiting nutrient for herbivores (Mattson 1980), especially during growth and reproduction (Robbins 1993), and geese are relatively inefficient at retaining dietary protein (Sedinger 1997). Fiber content is important because it largely determines digestibility of forage and may limit food intake (Sedinger and Raveling 1988; Sedinger 1997).

The requirement for high-quality forage should make it difficult for avian herbivores such as geese to survive when plant quality and quantity diminishes as predicted with elevated CO_2. Few studies have directly tested how wild herbivores respond to reduced forage quality. Small mammalian herbivores (e.g., voles) increase intake and passage rates and reduce fiber digestibility when eating higher fiber diets (Batzli, Broussard, and Oliver 1994; Cork 1994; Owl and Batzli 1998). This potentially important adjustment in intake and digestibility in response to changes in dietary fiber has rarely been studied in avian herbivores (McWilliams 1999). In one of the few such studies, Japanese quail increased daily food intake with increased dietary fiber, as predicted, although digestibility did not change (Savory and Gentle 1976a,b). More recent studies showed that quail increased their food intake when fed diets with 15% neutral detergent fiber (NDF) compared to diets with less fiber, but they decreased their food intake when fed higher-fiber diets (Starck 1999).

To address how extant avian herbivores respond to reduced-quality forage, we fed Canada goose and lesser Snow goose goslings from hatch until at least 100 days old on one of six diets that were a factorial combination of two levels of fiber and three levels of protein. All birds received the same dietary fiber levels (either 45% or 30% NDF) throughout growth. Levels of dietary protein were high (18%), medium (14%), or low (10%) for goslings after 21 days old. The levels of dietary protein and fiber were chosen based primarily on ecological relevance and nutritional adequacy. In short, the lower level of dietary fiber was representative of selected foods of wild goslings, whereas only a few selected wild foods contained as much fiber as those in our higher level of dietary fiber (Sedinger 1992). We expected this higher level of dietary fiber to increase processing time and thus potentially limit food intake of goslings. The levels of dietary protein were at or below expected protein requirements (ca. 18%) of young growing goslings based on studies of domestic waterfowl (NRC 1984).

All diets were formulated from the same five ingredients: beet sugar (*Beta vulgaris altissima*), soybean (*Glycine max*) meal, orchard grass (*Dactylis glomerata*) hay, oat (*Avena sativa*) hulls, and wheat (*Triticum aestivum*) middlings (a flour byproduct). Use of these ingredients insured that the dietary fiber was primarily from grasses as it is for wild goslings (Sedinger 1992). The following measurements were made on each gosling every 2 to 3 days after hatch until they were at least 100 days old or until growth was complete: body mass, culmen, tarsus, head length, and length of ninth primary (Dzubin and Cooch 1992).

Diet quality affected survival of Snow goslings especially early in growth, whereas diet quality had less effect on survival of Canada goslings (Fig. 17.4). For Snow goslings but not Canada goslings, survival was consistently lower for birds fed the two 10% protein diets and for birds fed the higher fiber diet at each level of dietary protein.

There were also important differences in survival between Snow and Canada goslings fed the same diets. Survival of Snow goslings fed the two 10% protein diets was lower than Canada goslings fed the same diets (LP/HF diet:$\chi 2 = 16.0$, d.f.$=1$, $p=0.0001$; LP/LF diet:$\chi 2 = 5.6$, d.f.$=1$, $p=0.02$). Similarly, survival

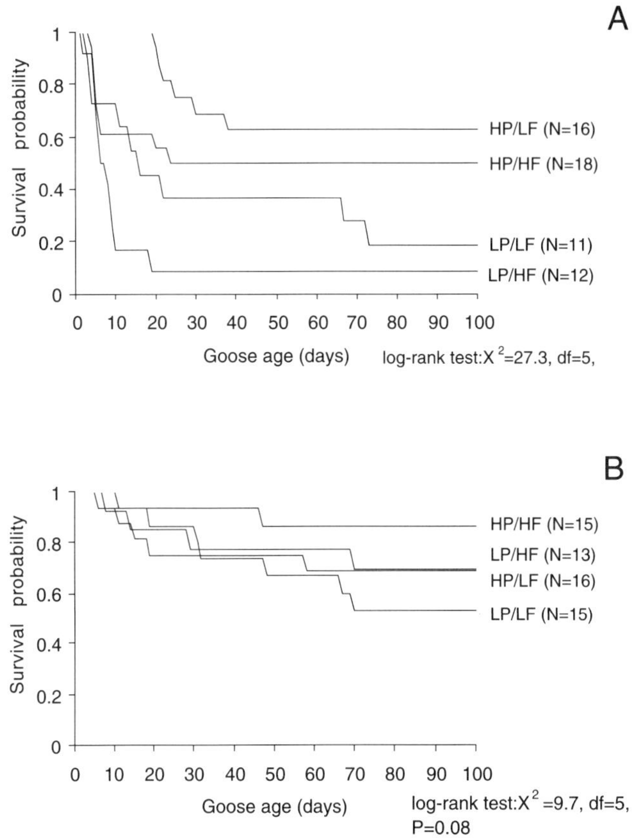

Figure 17.4. Survival probability of Snow geese **(A)** and Canada geese **(B)** fed grass-based diets with either high protein and high fiber (HP,HF), high protein and low fiber (HP,LF), low protein and high fiber (LP,HF), or low protein and low fiber (LP,LF). All geese were fed these diets from hatch until they were at least 100 days old. Survival of Snow goslings was significantly affected by both dietary protein and fiber, whereas survival of Canada goslings was less affected by diet (Kaplan Meir estimates of survival were compared using a log-rank test).

of Snow goslings fed the high-protein, high-fiber diet was lower than Canada goslings fed the same diet ($\chi 2 = 5.3$, d.f.=1, p=0.02). These results suggest that Snow geese have higher protein requirements during growth (>10%) than Canada geese, and that Canada goslings can survive better than Snow goslings when eating diets with more fiber. These results suggest that Snow goslings require a higher quality foraging environment than Canada goslings for normal growth.

Diet quality also affected growth rate and size of Snow and Canada geese (Fig. 17.5). Growth rate and structural size, as indexed by tarsus length, were

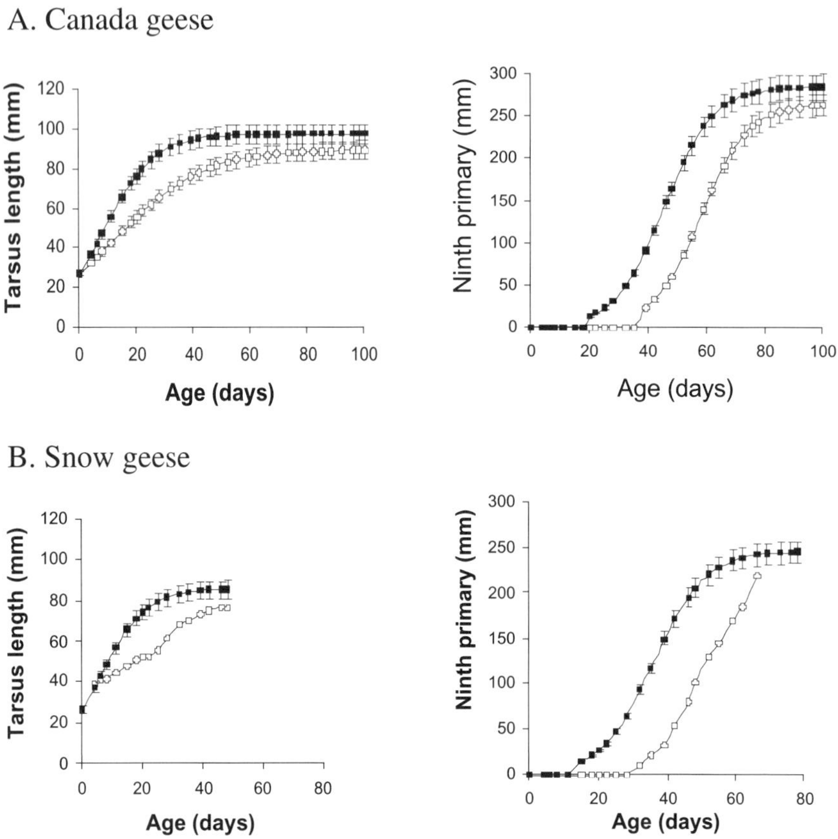

Figure 17.5. Body size (as indexed by tarsus length) and feather growth (as indexed by length of ninth primary feather) of growing Canada geese **(A)** and lesser Snow geese **(B)** fed grass-based diets with either high protein (*solid circles*) or low protein (*open circles*). All geese were fed these diets from hatch until they were at least 100 days old. There was no significant effect of dietary fiber on growth rates (not shown), while geese of both species fed low dietary protein grew slower than geese fed high dietary protein. No standard errors are shown for Snow geese fed low dietary protein because only one Snow gooese survived to at least 60 days when fed that diet.

significantly reduced for both Canada and Snow geese fed the low protein diets. Feather growth, as indexed by length of ninth primary, was also delayed for both goose species fed the low protein diets. We detected no significant effects of dietary fiber on growth of goslings, suggesting geese were able to compensate for this aspect of reduced diet quality.

For arctic-nesting geese, slower growth, smaller body size, and delayed feather growth of goslings is associated with reduced survival and fitness

(Cooke, Rockwell, and Lank 1995; Sedinger, Flint, and Lindberg 1995; Cooch, Rockwell, and Brault 2001) and so these effects of reduced diet quality can have major impacts on goose population dynamics. If elevated CO_2 generally reduces forage quality of arctic plants as expected, then the interspecific differences in nutritional requirements and growth rates of Snow and Canada goslings would suggest that Snow geese would be more affected by reduced forage quality than Canada geese. Thus, the response(s) of such avian herbivores to changes in plant quality and quantity associated with elevated CO_2 is likely to be strongly negative and species-specific.

17.8 Genetic Variation and Phenotypic Plasticity Influence an Organism's Response to Environmental Change

The ability of plant and animal species to respond to environmental change depends on the level of genetic variability within species as well as the ability of specific genotypes to respond flexibly as the environment changes. Arctic plants exhibit substantial ecotypic variation, which increases the chances that some populations will survive when the environment changes (reviewed by McGraw and Fetcher 1992). Arctic herbivores such as Canada geese, *Branta canadensis*, also have substantial ecotypic variation with at least 11 or >100 recognized subspecies depending on the authority (Bellrose 1980; Hanson 1997; Shields and Cotter 1998) and considerable phenotypic variation within subspecies (Leafloor and Rusch 1997).

Phenotypic plasticity (phenotypic variability within single genotypes) may also enable individuals to respond favorably to environmental change (Piersma and Drent 2003). For example, environmental conditions during gosling growth determine adult body size, which, in turn, affects survival and future fecundity (Sedinger, Flint, and Lindberg 1995; Leafloor, Ankney, and Rusch 1998; Cooch, Rockwell, and Brault 2001). In addition, recent work with arctic-nesting geese has demonstrated considerable phenotypic flexibility (reversible phenotypic change within single individuals) in their digestive system that has important ecological implications (McWilliams and Karasov 2001; Piersma and Drent 2003). For example, Snow and especially Canada goslings increase their food intake with increased dietary fiber to compensate for the decrease in digestibility of the higher-fiber diet (Fig. 17.6). A reduction in diet quality is also associated with an increase in digestive capacity (Fig. 17.7), which allows goslings to eat more of the lower quality diet while maintaining digestibility and metabolizability of the diet relatively constant (McWilliams et al. 2000). These same patterns of digestive response in relation to diet quality have been documented for other relatively small herbivores (Cork 1994; Iason and Wieren 1999).

We know little about the extent of phenotypic plasticity and phenotypic flexibility in different species living in arctic ecosystems and its ecological implications. Presumably, species that can flexibly respond are more likely to adapt

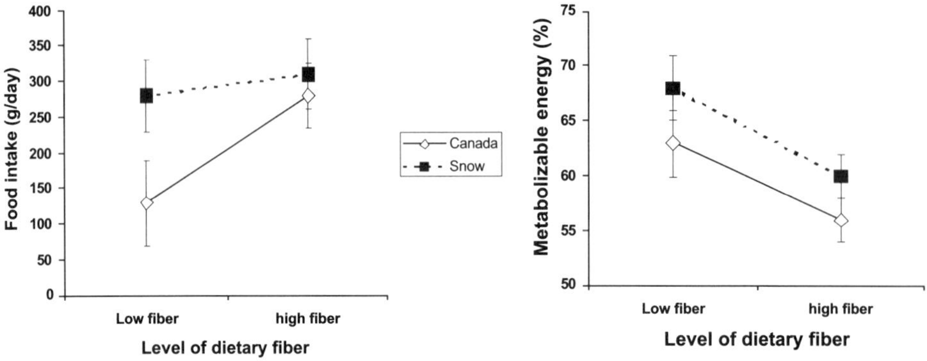

Figure 17.6. Daily food intake (adjusted for goose body mass using ANCOVA) and apparent metabolizable energy of 8-week-old Canada geese and lesser Snow geese fed grass-based diets with either high fiber or low fiber. All geese were fed these diets from hatch until the total collection digestibility trials were conducted when they were 8 weeks old. Canada and Snow geese digested less of the available dietary energy and ate relatively more when fed higher fiber diets. However, the Snow geese with smaller body size increased their food intake only 11% while the larger Canada geese increased their food intake 115% when fed the low-fiber compared to the high-fiber diets. This suggests that food intake of Snow geese may be constrained when they consume high-fiber forage.

successfully to environmental changes compared to other species that are inflexible (Pigliucci 1996; Piersma and Lindstrom 1997). Our results indicate that arctic-nesting geese have significant phenotypic flexibility in their digestive system, yet there is clearly a limit to this flexibility as indicated by their slow growth and reduced survival when fed diets high in fiber and low in dietary protein. For most birds, maximum size of the digestive tract is likely limited by constraints associated with flying (McWilliams and Karasov 2001, 2003). Thus, arctic-nesting geese can flexibly respond to short-term changes in diet quality, but the extent of change in forage quality expected with elevated CO_2 may well exceed their capacity.

17.9 Migration and Mobility Influence an Organism's Response to Environmental Change

Migration of both plants and animals will play a role in the redistribution of arctic biota in response to elevated CO_2 (Davis and Shaw 2001). In general, evidence from past climatic change demonstrates that species migrate individualistically rather than as assemblages, so that autoecological studies should be useful for predicting response(s) of extant biota to future environmental change (Cronin and Schneider 1990; Graham and Grimm 1990; Davis and Shaw 2001).

Arctic herbivores vary tremendously in their mobility with the most sedentary

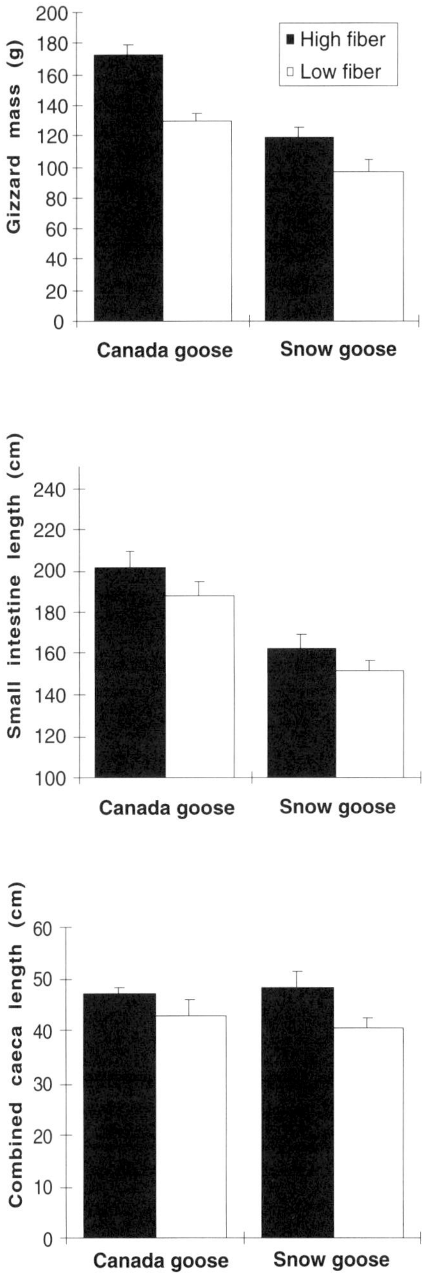

Figure 17.7. Phenotypic flexibility in digestive organs of Canada and lesser Snow geese in response to dietary fiber. Gizzard mass, small intestine length, and the combined length of both caeca increased when geese were fed high-fiber diets compared to low-fiber diets (Factorial ANOVA; Goose species effect: $P<0.0001$ for gizzard and small intestine, $P=0.96$ for caeca; Dietary fiber effect: $P<0.0001$ for gizzard, $P=0.019$ for small intestine, $P=0.004$ for caeca). There was no significant effect of dietary protein on digestive organs (not shown).

(e.g., voles and lemmings) exhibiting dramatic, and often cyclic, population fluctuations, whereas the most mobile (e.g., migratory geese) apparently are able to find preferred habitat when it is available (Boyd and Madsen 1997). If responding to changes in plant quantity and quality simply involves spatially and temporally tracking preferred habitat, then animal species that can outpace the rate of environmental change will survive and those that cannot will perish. Arctic ecosystems are characterized by short growing seasons with high daily rates of plant production (but low annual productivity) even when ambient temperatures are only slightly above freezing (Bliss 1986). In general, birds use their mobility afforded by flight to capitalize on the brief pulse of productivity during the arctic summer—an impressive 15% of the world's bird species breed in the Arctic but then spend the nonbreeding period in more hospitable temperate or tropical climes (Hansell et al. 1998). Thus, arctic-nesting birds provide good examples of highly mobile animals that can effectively track their preferred habitat over relatively large spatial scales (Boyd and Madsen 1997).

However, major changes in plant quantity and quality, and interactions between predators, herbivores, and primary producers are expected with elevated CO_2 that may affect arctic wildlife populations even if they are highly mobile (Root and Schneider 2002). For example, breeding during the short arctic summer requires relatively fast growth rates for especially larger animal species such as geese. Faster growth rates of goslings increase short-term nutritional requirements. Thus, although arctic-nesting geese are quite mobile, their life history is such that large-scale reductions in plant quality and quantity or changes in plant phenology would negatively affect gosling growth and survival unless adaptation occurred. Undoubtedly, the responses of arctic plants and animals to the effects of elevated CO_2 will involve the interplay of both adaptation and migration (Davis and Shaw 2001).

17.10 Summary

Elevated atmospheric CO_2 generally decreases plant quality (i.e., lower protein, increased fiber), especially of C_3 plants such as those that dominate arctic ecosystems. Predicting how changes in the quality and distribution of plants will affect higher trophic levels in arctic ecosystems requires knowledge of the reciprocal interactions between keystone herbivores and their food plants. We review contemporary studies of one group of keystone herbivores in arctic ecosystems, arctic-nesting geese, and their response to changes in plant quality and quantity.

Arctic-nesting geese are good indicators of the health of arctic ecosystems because the ecological limitations associated with being an avian herbivore require them to respond sensitively to changes in plant quality and quantity. Field studies have demonstrated that goslings that grow up in areas with poor habitat quality are smaller as adults and have reduced survival and fitness compared to geese in good quality habitat. Recent captive-rearing experiments with goslings have elucidated some of the important physiological responses of geese to poor quality forage. This recent work emphasizes the importance of phenotypic flexibility in the digestive system of keystone herbivores, and of digestive constraints in determining the lowest quality of food eaten by such herbivores. Thus, predicting the effects of elevated atmospheric CO_2 on arctic ecosystems requires understanding the interplay between phenotypic flexibility, adaptation, and migration of the arctic biota.

References

Arft, A.M., M.D. Walker, D. Gurevitch, J.M. Alatalo, M.S. Bret-Herte, M. Dale, M. Diemer, F. Gugerli, G. Henry, M. Jones, and others. 1999. Response patterns of tundra plant species to experimental warming: A meta-analysis of the International Tundra Experiment. *Ecological Monographs* 69:491–511.

Ayres, M.P. 1993. Plant defense, herbivory, and climate change. In *Biotic interactions and global change*, ed. P.M. Kareiva, J.G. Kingsolver, and R.B. Huey, 75–94. Sunderland, Mass.: Sinauer Associates.

Batt, B., ed. 1997. *Arctic ecosystems in peril: Report of the arctic goose habitat working group*. Washington, D.C.: U.S. Fish and Wildlife Service, Canadian Wildlife Service.

Batzli, G.O. 1983. Responses of arctic rodent populations to nutritional factors. *Oikos* 40:396–406.

Batzli, G.O., A.D. Broussard, and R.J. Oliver. 1994. The integrated processing response in herbivorous small mammals. In *The digestive system in mammals: Food, form, and function*, ed. D.J. Chivers and P. Langer, 324–36. Cambridge: Cambridge University Press.

Batzli, G.O., R.G. White, S.F. MacLean, Jr., F.A. Pitelka, and B.D. Collier. 1980. The herbivore-based trophic system. In *An arctic ecosystem: The coastal tundra at Barrow, Alaska,* ed. J. Brown, P.C. Miller, L. Tieszen, and F.L. Bunnell, 335–410. Stroudsburg, Penn.: Dowder, Hutchinson, and Ross.

Bazely, D.R., and R.L. Jefferies. 1985. Goose faeces: A source of nitrogen for plant growth in a grazed salt marsh. *Journal of Applied Ecology* 22:693–703.

———. 1986. Changes in the composition and standing crop of salt marsh communities in response to removal of a grazer. *Journal of Ecology* 22:693–703.

———. 1997. Trophic interactions in arctic ecosystems and the occurrence of a terrestrial trophic cascade. In *Ecology of arctic environments*, ed. S.J. Woodin and M. Marquiss, 183–208. Oxford: Blackwell Science Ltd.

Bazzaz, F.A. 1990. The response of natural ecosystems to the rising global CO_2 levels. *Annual Review of Ecology and Systematics* 21:167–96.

Bellrose, F.C. 1980. *Ducks, geese, and swans of North America.* Harrisburg, Penn.: Stackpole Books.

Berendse, F., and S. Jonasson. 1992. Nutrient use and nutrient cycling in northern ecosystems. In *Arctic ecosystems in a changing climate: An ecophysiological perspective*, ed. F.S. Chapin III, R.L. Jefferies, J.F. Reynolds, G.R. Shaver, and J. Svoboda, 337–56. New York: Academic Press.

Bliss, L.C. 1975. Tundra, grasslands, herblands, and shrublands and the role of herbivores. *Geoscience and Man* 10:51–79.

———. 1986. Arctic ecosystems: their structure, function and herbivore carrying capacity. In *Grazing research at northern latitudes*, ed. O. Gudmundsson, 5–26. New York: Plenum Press.

Bliss, L.C., and N.V. Matveyeva. 1992. Circumpolar arctic vegetation. In *Arctic ecosys-*

tems in a changing climate: An ecophysiological perspective, ed. F.S. Chapin III, R.L. Jefferies, J.F. Reynolds, G.R. Shaver, and J. Svoboda, 59–89. New York: Academic Press.

Bond, W.J. 1994. Keystone species. In *Biodiversity and ecosystem function*, ed. E.D. Schulze and H.A. Mooney, 237–53. Berlin: Springer-Verlag.

Both, C., and M.E. Visser. 2001. Adjustment to climate change is constrained by arrival date in a long-distance migrant bird. *Nature* 411:296–98.

Boyd, H., and J. Madsen. 1997. Impacts of global change on arctic-breeding bird populations and migration. In *Global change and arctic terrestrial ecosystems*, ed. W.C. Oechel, T. Callaghan, T. Gilmanov, J.I. Holten, B. Maxwell, U. Molau, and B. Sveinbjornsson, 201–17. New York: Springer-Verlag.

Bryant, J.P., and P.J. Kuropat. 1980. Selection of winter forage by subarctic browsing vertebrates: The role of plant chemistry. *Annual Review of Ecology and Systematics* 11:261–85.

Bryant, J.P., and P.B. Reichardt. 1992. Controls over secondary metabolite production by Arctic woody plants. In *Arctic ecosystems in a changing climate: An ecophysiological perspective,* ed. F.S. Chapin III, R.L. Jefferies, J.F. Reynolds, G.R. Shaver, and J. Svoboda, 377–90. New York: Academic Press.

Cargill, S.M., and R.L. Jefferies. 1984. The effects of grazing by lesser snow geese on the vegetation of a sub-arctic salt marsh. *Journal of Applied Ecology* 21:669–86.

Chapin, F.S. III, R.L. Jefferies, J.F. Reynolds, G.R. Shaver, and J. Svoboda, eds. 1992. *Arctic ecosystems in a changing climate: An ecophysiological perspective*. New York: Academic Press.

Cooch, E.G., D.B. Lank, R.F. Rockwell, and F. Cooke. 1991. Long-term decline in body size in a snow goose population: evidence of environmental degradation? *Journal of Animal Ecology* 60:483–96.

Cooch, E.G., R.F. Rockwell, and S. Brault. 2001. Retrospective analysis of demographic responses to environmental change: An example in the lesser snow goose. *Ecological Monographs* 71:377–400.

Cooke, F., R.F. Rockwell, and D.B. Lank. 1995. *The snow geese of La Perouse Bay: Natural selection in the wild.* New York: Oxford University Press.

Cork, S.J. 1994. Digestive constraints on dietary scope in small and moderately small mammals: How much do we really understand? In *The digestive system in mammals: Food, form, and function,* ed. P. Langer, 337–69. Cambridge: Cambridge University Press.

Cotrufo, M.F., P. Ineson, and A. Scott. 1998. Elevated CO_2 reduces the nitrogen concentration of plant tissues. *Global Change Biology* 4:43–54.

Cronin, T.M., and C.E. Schneider. 1990. Climatic influences on species: Evidence from the fossil record. *Trends in Ecology and Evolution* 5:275–79.

Crowley, T.J. 2000. Causes of climate change over the past 1000 years. *Science* 289: 270–77.

Danell, K., and K. Huss-Danell. 1985. Feeding by insects and hares on birches earlier affected by moose browsing. *Oikos* 44:75–81.

Davis, M.B., and R.G. Shaw. 2001. Range shifts and adaptive responses to Quaternary climate change. *Science* 292:673–79.

Demment, M.W., and P.J. Van Soest. 1985. A nutritional explanation for body-size patterns of ruminant and nonruminant herbivores. *American Naturalist* 125:641–72.

Dormann, C.F., and S.J. Woodin. 2002. Climate change in the arctic: Using plant functional types in a meta-analysis of field experiments. *Functional Ecology* 16:4–17.

Drent, R.H., and H.H.T. Prins. 1987. The herbivore as prisoner of its food supply. In *Disturbance in grasslands: Causes, effects, and processes,* ed. J. van Andel, J.P. Bakker, and R.W. Snaydon, 131–48. Dordrecht, Netherlands: Kluwer Academic Publishers.

Drent, R.H., and R. van der Wal. 1999. Cyclic grazing in vertebrates and the manipulation

of the food resource. In *Herbivores: Between plants and predators*, ed. H. Olff, V.K. Brown, and R.H. Drent, 271–99. Oxford: Blackwell Science Ltd.
Dunn, P.O., and D.W. Winkler. 1999. Climate change has affected the breeding date of tree swallow throughout North America. Proceedings of the Royal Society of London 266:2487–90.
Dzubin, A., and E.G. Cooch. 1992. *Measurements of geese: General field methods*. Sacramento, Calif.: California Waterfowl Association.
Fox, A.D. 1993. Pre-nesting feeding selectivity of Pink-footed geese *Anser brachyrhynchus* in artificial grasslands. *Ibis* 135:417–23.
Gauthier, G., and J. Bedard. 1990. The role of phenolic compounds and nutrients in determining food preference in Greater Snow Geese. *Oecologia* 84:553–58.
Gauthier, G., R.J. Hughes, A. Reed, J. Beaulieu, and L. Rochefort. 1995. Effect of grazing by greater snow geese on the production of graminoids at an arctic site (Bylot Island, NWT, Canada). *Journal of Ecology* 83:653–64.
Graham, R.W., and E.C. Grimm. 1990. Effects of global climate change on the patterns of terrestrial biological communities. *Trends in Ecology and Evolution* 5:289–92.
Grulke, N.E., G.H. Riechers, W.C. Oechel, U. Hjelm, and C. Jaeger. 1990. Carbon balance in tussock tundra under ambient and elevated CO_2. *Oecologia* 83:485–94.
Gunn, A., and T. Skogland. 1997. Responses of caribou and reindeer to global warming. In *Global change and arctic terrestrial ecosystems*, ed. W.C. Oechel, T. Callaghan, T. Gilmanov, J.I. Holten, B. Maxwell, U. Molau, and B. Sveinbjornsson, 189–200. New York: Springer-Verlag.
Handa, I.T., R. Harmsen, and R.L. Jefferies. 2002. Patterns of vegetation and the recovery potential of degraded areas in a coastal marsh system of the Hudson Bay lowlands. *Journal of Ecology* 90:86–99.
Hansell, R.I.C., J.R. Malcolm, H. Welch, R.L. Jefferies, and P.A. Scott. 1998. Atmospheric change and biodiversity in the Arctic. *Environmental Monitoring and Assessment* 49:303–25.
Hanson, H.C. 1997. *The giant Canada goose*, 2d ed. Carbondale, Ill.: Southern Illinois University Press.
Hik, D.S., and R.L. Jefferies. 1990. Increases in net above-ground primary production of a salt-marsh forage grass: A test of the predictions of the herbivore-optimization model. *Journal of Ecology* 78:180–95.
Iason, G.R., and S.E.V. Wieren. 1999. Digestive and ingestive adaptations of mammalian herbivores to low-quality forage. In *Herbivores: Between plants and predators,* ed. H. Olff, V.K. Brown, and R.H. Drent, 337–69. Oxford: Blackwell Science Ltd.
Illius, A.W., and I.J. Gordan. 1993. Diet selection in mammalian herbivores: Constraints and tactics. In *Diet selection: An interdisciplinary approach to foraging behavior,* ed. R.N. Hughes, 157–81. Oxford: Blackwell Scientific Publishers.
Inouye, D.W., B. Barr, K.B. Armitage, and B.D. Inouye. 2000. Climate change is affecting altitudinal migrants and hibernating species. *Proceedings of the National Academy of Sciences* 97:1630–33.
IPCC. 1996. Climate change. The second assessment report of the Intergovernmental Panel for Climate Change (IPCC) working group I. Cambridge: Cambridge University Press.
Jefferies, R.L. 1988a. Vegetational mosaics, plant-animal interactions, and resources for plant growth. In *Plant evolutionary biology,* ed. L. Gottlieb and S.K. Jain, 341–69. London: Chapman and Hall.
———. 1988b. Pattern and process in arctic coastal vegetation in response to foraging by lesser snow geese. In *Plant form and vegetation structure, adaptation, plasticity, and relation to herbivory,* ed. M.J.A. Werger, P.J. M.van der Aart, H.J. During, and J.T.A. Verhoeven, 281–300. The Hague: SPB Academic.
———. 1997. Long-term damage to sub-arctic coastal ecosystems by geese: ecological indicators and measures of ecosystem dysfunction. In *Disturbance and recovery in*

arctic lands, ed. R.M.M. Crawford, 151–65. Boston, Mass.: Kluwer Academic Publishers.

———. 1999. Herbivores, nutrients and trophic cascades in terrestrial environments. In *Herbivores: Between plants and predators,* ed. H. Olff, V.K. Brown, and R.H. Drent, 301–330. Oxford: Blackwell Science.

Jefferies, R.L., J. Svoboda, G. Henry, M. Raillard, and R. Ruess. 1992. Tundra grazing systems and climatic change. In *Arctic ecosystems in a changing climate: An ecophysiological perspective,* ed. F.S. Chapin III, R.L. Jefferies, J.F. Reynolds, G.R. Shaver, and J. Svoboda, 391–412. New York: Academic Press.

Jonasson, S., J.P. Bryant, F.S. Chapin III, and M. Andersson. 1986. Plant phenols and nutrients in relation to variations in climate and rodent grazing. *American Naturalist* 128:394–408.

Kerbes, R.H., P.M. Kotanen, and R.L. Jefferies. 1990. Destruction of wetland habitats by lesser snow geese: A keystone species on the west coast of Hudson Bay. *Journal of Applied Ecology* 27:242–58.

Kerr, J., and L. Packer. 1998. *The impact of climate change on mammal diversity in Canada.* Boston: Kluwer Academic Publishers.

Klein, D.R. 1986. Latitudinal variation in foraging strategies. In *Grazing research in northern latitudes,* ed. O. Gudmundsson, 237–57. New York: Plenum Press.

Körner, C. 2000. Biosphere responses to CO_2 enrichment. *Ecological Applications* 10: 1590–1619.

Krapu, G.L., and K.J. Reinecke. 1992. Foraging ecology and nutrition. In *Ecology and management of breeding waterfowl,* ed. B.D.J. Batt, A.D. Afton, M.G. Anderson, C.D. Ankney, D.H. Johnson, J.A. Kadlec, and G.L. Krapu, 1–29. Minneapolis: University of Minnesota Press.

Laing, K.K., and D.G. Raveling. 1993. Habitat and food selection by Emperor Goose goslings. *Condor* 95:879–88.

Leafloor, J.O., C.D. Ankney, and D.H. Rusch. 1998. Environmental effects on body size of Canada Geese. *Auk* 115:26–33.

Leafloor, J.O., and D.H. Rusch. 1997. Clinal size variation in Canada Geese: Implications for the use of morphometric discrimination techniques. *Journal of Wildlife Management* 61:183–90.

Lenart, E.A., R.T. Bowyer, J.V. Hoef, and R.W. Ruess. 2002. Climate change and caribou: Effects of summer weather on forage. *Canadian Journal of Zoology* 80:664–78.

Lincoln, D.E., E.D. Fajer, and R.H. Johnson. 1993. Plant-insect herbivore interactions in elevated CO_2 environments. *Trends in Ecology and Evolution* 8:64–68.

Lindroth, R.L. 1996. Consequences of elevated atmospheric CO_2 for forest insects. In *Carbon dioxide, populations, and communities,* ed. C. Körner and F.A. Bazzaz, 347–61. San Diego: Academic Press.

Lucht, W., I.C. Prentice, R.B. Myneni, S. Sitch, P. Friedlingstein, W. Cramer, P. Bousquet, W. Buermann, and B. Smith. 2002. Climatic control of the high-latitude vegetation greening trend and Pinatubo effect. *Science* 296:1687–89.

Manseau, M., and G. Gauthier. 1994. Interactions between Greater Snow Geese and their rearing habitat. *Ecology* 74:2045–55.

Mattson, W.J. 1980. Herbivory in relation to plant nitrogen content. *Oecologia* 81:186–91.

McGraw, J.B., and N. Fetcher. 1992. Response of tundra plant populations to climate change. In *Arctic ecosystems in a changing climate: An ecophysiological perspective,* ed. F.S. Chapin III, R.L. Jefferies, J.F. Reynolds, G.R. Shaver, and J. Svoboda, 359–376. New York: Academic Press.

McWilliams, S.R. 1999. Digestive strategies of avian herbivores. In *Proceedings of the 22d International Ornithological Congress,* ed. N. Adams and R. Slotow, 2198–2207. Durban, South Africa: University of Natal.

McWilliams, S.R., and W.H. Karasov. 2001. Phenotypic flexibility in digestive system structure and function in migratory birds and its ecological significance. *Comparative Biochemistry and Physiology,* Part A, 128:579–93.

———. 2003. Migration takes guts: Digestive physiology of migratory birds and its ecological significance. In *Birds of two worlds,* ed. P. Marra and R. Greenberg. Washington, D.C.: Smithsonian Institution Press.

McWilliams, S.R., W.H. Karasov, and E. Caviedes-Vidal. 2000. Digestive adjustments in geese to reduced forage quality and its ecological implications. *American Zoologist* 40:1127.

Mooney, H.A., B.G. Drake, R.J. Luxmoore, W.C. Oechel, and L.F. Pitelka. 1991. Predicting ecosystem responses to elevated CO_2 concentrations. *BioScience* 41:96–104.

Mulder, C.P.H., and R.W. Ruess. 1998. Effects of herbivory on arrowgrass: Interactions between geese, neighboring plants, and abiotic factors. *Ecological Monographs* 68: 275–93.

———. 2001. Long-term effects of changes in goose grazing intensity on arrowgrass populations: A spatially explicit model. *Journal of Ecology* 89:406–17.

Mulder, C.P.H., R.W. Ruess, and J.S. Sedinger. 1996. Effects of environmental manipulations on *Triglochin palustris*: Implications for the role of goose herbivory in controlling its distribution. *Journal of Ecology* 84:267–78.

National Research Council. 1984. *Nutrient requirements of poultry.* Washington, D.C.: National Academy of Science.

Oberbauer, S.O., N. Sionit, S.J. Hastings, and W.C. Oechel. 1986. Effects of carbon dioxide enrichment on growth, photosynthesis, and nutrient concentration of Alaskan tundra plant species. *Canadian Journal of Botany* 64:2993–98.

Oechel, W.C., and W.D. Billings. 1992. Effects of global change on the carbon balance of Arctic plants and ecosystems. In *Arctic ecosystems in a changing climate: An ecophysiological perspective,* ed. F.S. Chapin III, R.L. Jefferies, J.F. Reynolds, G.R. Shaver, and J. Svoboda, 139–68. New York: Academic Press.

Oechel, W.C., A.C. Cook, S.J. Hastings, and G.L. Vourlitis. 1997. Effects of CO_2 and climate change on arctic ecosystems. In *Ecology of arctic environments,* ed. S.J. Woodin and M. Marquiss, 255–73. Oxford: Blackwell Science.

Oechel, W.C., and G.L. Vourlitis. 1996. Direct effects of elevated CO_2 on arctic plant and ecosystem function. In *Carbon dioxide and terrestrial ecosystems,* ed. G.W. Koch, and H.A. Mooney, 163–76. New York: Academic Press.

———. 1997. Climate change in northern latitudes: Alterations in ecosystem structure and function and effects on carbon sequestration. In *Global change and arctic terrestrial ecosystems,* ed. W.C. Oechel, T. Callaghan, T. Gilmanov, J.I. Holten, B. Maxwell, U. Molau, and B. Sveinbjornsson, 381–401. New York: Springer-Verlag.

Oksanen, L. 1983. Trophic exploitation and arctic phytomass patterns. *American Naturalist* 122:45–52.

Owl, M.Y., and G.O. Batzli. 1998. The integrated processing response of voles to fibre content of natural diets. *Functional Ecology* 12:4–13.

Paine, R.T. 1966. Food web complexity and species diversity. *American Naturalist* 100: 65–75.

———. 1969. A note on trophic complexity and community stability. *American Naturalist* 103:91–93.

Penuelas, J., and I. Filella. 2001. Responses to a warming world. *Science* 294:793–95.

Person, B.T., C.A. Babcock, and R.W. Ruess. 1998. Forage variation in brood-rearing areas used by pacific black brant geese on the Yukon-Kuskokwim delta, Alaska. *Journal of Ecology* 86:243–59.

Person, B.T., M.P. Herzog, R.W. Ruess, J.S. Sedinger, R.M. Anthony, and C.A. Babcock. 2003. Feedback dynamics of grazing lawns: Coupling vegetation change with animal growth. *Oecologia* 135:583–92.

Petit, J.R., J. Jouzel, D. Raynaud, N. Barkov, J.-M. Barnola, I. Basile, M. Bender, J. Chappellaz, M. Davis, G. Delaygue, M. Delmotte, V. Kotlyakov, M. Legrand, V. Lip-

enkov, C. Lorius, L. Pepin, C. Ritz, E. Saltzman, and M. Stievenard. 1999. Climate and atmospheric history of the past 420,000 years from the Vostok ice core, Antarctica. *Nature* 399:429–36.
Piersma, T., and J. Drent. 2003. Phenotypic flexibility and the evolution of organismal design. *Trends in Ecology and Evolution* 18:228–33.
Piersma, T., and A. Lindstrom. 1997. Rapid reversible changes in organ size as a component of adaptive behavior. *Trends in Ecology and Evolution* 12:134–38.
Pigliucci, M. 1996. How organisms respond to environmental changes: From phenotypes to molecules (and vice versa). *Trends in Ecology and Evolution* 11:168–73.
Post, E., and N.C. Stenseth. 1999. Climatic variability, plant phenology, and northern ungulates. *Ecology* 80:1322–39.
Power, M.E., D. Tilman, J.A. Estes, B.A. Menge, W.J. Bond, L.S. Mills, G. Daily, J.C. Castilla, J. Lubchenco, and R.T. Paine. 1996. Challenges in the quest for keystones. *BioScience* 46:609–20.
Price, J.T., and T.L. Root. 2000. Focus: Effects of climate change on bird distributions and migration patterns. In *Preparing for a changing climate: The potential consequences of climate variability and change,* ed. P.J. Sousounis and J.M. Bisanz, 65–68. Ann Arbor: University of Michigan.
Prop, J. 1991. Food exploitation patterns by brent geese *Branta bernicla* during spring staging. *Ardea* 79:331–42.
Prop, J., J.M. Black, P. Shimmings, and M. Owen. 1998. The spring range of barnacle geese *Branta leucopsis* in relation to changes in land management and climate. *Biological Conservation* 86:339–46.
Prop, J., and J. de Vries. 1993. Impact of snow and food conditions on the reproductive performance of Barnacle Geese *Branta leucopsis. Ornis Scandinavica* 24:110–21.
Raveling, D.G. 1978. The timing of egg laying by northern geese. *Auk* 95:294–303.
Robbins, C.T. 1993. *Wildlife feeding and nutrition,* 2d ed. New York: Academic Press.
Root, T.L., and S.H. Schneider. 2002. Climate change: Overview and implications for wildlife. In *Wildlife responses to climate change,* ed. S.H. Schneider and T.L. Root, 1–56. Washington, D.C.: Island Press.
Rowcliffe, J.M., A.R. Watkinson, W.J. Sutherland, and J.A. Vickery. 1995. Cyclic winter grazing patterns in brent geese and the regrowth of salt-marsh grass. *Functional Ecology* 9:931–41.
Ruess, R.W., D.S. Hik, and R.I. Jefferies. 1989. The role of lesser snow geese as nitrogen processors in a sub-arctic salt marsh. *Oecologia* 79:23–29.
Savory, C.J., and M.J. Gentle. 1976a. Effects of dietary dilution with fibre on the food intake and gut dimensions of Japanese quail. *British Journal of Poultry Science* 17: 561–70.
———. 1976b. Changes in food intake and gut size in Japanese quail in response to manipulation of dietary fiber content. *British Journal of Poultry Science* 17:571–80.
Sedinger, J.S. 1992. Ecology of prefledging waterfowl. In *Ecology and management of breeding waterfowl,* ed. B.D.J. Batt, A.D. Afton, M.G. Anderson, C.D. Ankney, D.H. Johnson, J.A. Kadlec, and G L. Krapu, 109–27. Minneapolis: University of Minnesota Press.
———. 1997. Adaptations to and consequences of an herbivorous diet in grouse and waterfowl. *Condor* 99:314–26.
Sedinger, J.S., P.L. Flint, and M.S. Lindberg. 1995. Environmental influence on life-history traits: Growth, survival, and fecundity in Black Brant (*Branta bernicla*). *Ecology* 76:2404–14.
Sedinger, J.S., and D.G. Raveling. 1984. Dietary selectivity in relation to availability and quality of food for goslings of Cackling Geese. *Auk* 101:295–306.
———. 1986. Timing of nesting by Canada geese in relation to the quality and availability of their food plants. *Journal of Animal Ecology* 55:1083–1102.

———. 1988. Foraging behavior of cackling Canada goose goslings: implications for the roles of food availability and processing rate. *Oecologia* 75:119–24.

Shaw, M.R., E.S. Zavaleta, N.R. Chiariello, E.E. Cleland, H.A. Mooney, and C.B. Field. 2002. Grassland responses to global environmental changes suppressed by elevated CO_2. *Science* 298:1987–90.

Shields, G.F., and J.P. Cotter. 1998. Phylogenies of North American geese: The mitochondrial DNA record. In *Biology and management of Canada geese,* ed. D.H. Rusch, M.D. Samuel, D.D. Humburg, and B.D. Sullivan, 405–11. Milwaukee: Proceedings of International Canada Goose Symposium.

Srivastava, D.S., and R.L. Jefferies. 1996. A positive feedback: Herbivory, plant growth, and salinity and the desertification of an arctic salt marsh. *Journal of Ecology* 84:31–42.

Starck, J.M. 1999. Phenotypic flexibility of the avian gizzard: rapid, reversible and repeated changes of organ size in response to changes in dietary fibre content. *Journal of Experimental Biology* 202:3171–79.

Strong, D.R. 1992. Are trophic cascades all wet? Differentiation and donor-control in speciose ecosystems. *Ecology* 73:747–54.

Tissue, D.T., and W.C. Oechel. 1987. Response of *Eriophorum vaginatum* to elevated CO_2 and temperature in the Alaskan tussock tundra. *Ecology* 68:401–10.

Tolvanen, A., and G.H.R. Henry. 2001. Reponses of carbon and nitrogen concentrations in high arctic plants to experimental increased temperature. *Canadian Journal of Botany* 79:711–18.

Walker, M.D., W.A. Gould, and F.S. Chapin III. 2001. Scenarios of biodiversity changes in arctic and alpine tundra. In *Global biodiversity in a changing environment: Scenarios for the 21st century,* ed. F.S. Chapin III, O.E. Sala, and E. Huber-Sannwald, 83–100. New York: Springer-Verlag.

White, R.G. 1983. Foraging patterns and their multiplier effects on productivity of northern ungulates. *Oikos* 40:377–84.

White, R.G., and S.G. Fancy. 1986. Nutrition and energetics of indigenous northern ungulates. In *Grazing research in northern latitudes,* ed. O. Gudmundsson, 259–70. New York: Plenum Press.

Ydenberg, R.C., and H.H.T. Prins. 1981. Spring grazing and the manipulation of food quality by barnacle geese. *Journal of Applied Ecology* 18:443–53.

Zacheis, A., J.W. Hupp, and R.W. Ruess. 2001. Effects of migratory geese on plant communities of an Alaskan salt marsh. *Journal of Ecology* 89:57–71.

18. Modern and Future Forests in a Changing Atmosphere

Richard J. Norby, Linda A. Joyce, and Stan D. Wullschleger

18.1 Introduction

Through direct observation and clever use of proxy signals, we can reconstruct the atmospheric conditions of past years, and that allows us to analyze the interactions between plants and the atmosphere in the near and distant past. Although predicting future conditions will always be fraught with uncertainty, we can state with a very high degree of confidence that trees grew in a lower CO_2 concentration in past decades and will experience a higher concentration of atmospheric CO_2 in the future. The inexorable increase in atmospheric CO_2 since the onset of industrialization in the eighteenth century is a global phenomenon that has been altering the environment of forests worldwide.

Since all C_3 plants, including trees, respond to atmospheric CO_2 concentration, forests also should respond to this altered environment. The primary response is straightforward: CO_2 is the substrate of photosynthesis, and as the substrate concentration increases so does the reaction rate. This analysis, however, is highly simplistic. Plants can adjust to changing [CO_2] such that the relationship between photosynthesis and [CO_2] changes (Stitt 1991). These adjustments are themselves responses to CO_2 that can alter the water or nutrient cycles of the forest or its vulnerability to biotic or climatic stresses (Fig.18.1).

In addition, myriad secondary and tertiary adjustments, feedbacks, and interactions affect how trees respond to CO_2, and thereby the response of a forest (Bazzaz 1990; Field et al. 1992). The net result can be nearly impossible to

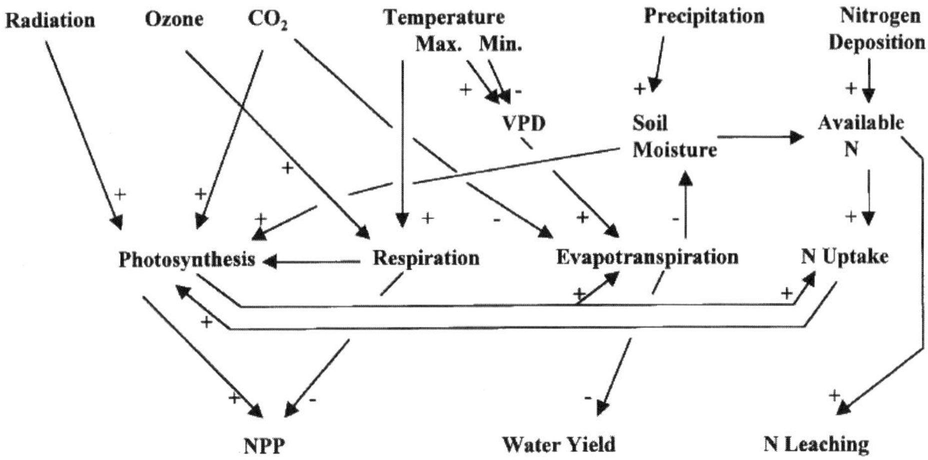

Figure 18.1. The primary response of plants to increased CO_2 concentration starts with an increase in photosynthesis, and this can lead to increased plant production. The outcome of this primary effect, however, is moderated by numerous secondary effects and the influences of other environmental variables. Figure is from Aber et al. 2001; copyright, American Institute of Biological Sciences.

predict or detect. As the primary responses to CO_2 become obscured, so too does our understanding and confidence in predicting how forest structure or function will change as the atmospheric [CO_2] continues to rise. These adjustments and feedbacks are perhaps the most important aspect of forest response to CO_2, and it is critical that analyses of forest response take a whole-system approach that considers CO_2 in relation to multiple interacting environmental variables and biotic factors. The greatest challenge is to account for issues of scale imposed by the inescapable constraints of the large size and long life span of trees. It is no easy matter to relate the responses of a leaf in a cuvette over a matter of minutes to those of a forest over decades, but it also is impossible to maintain a forest for decades in a cuvette to observe it. In this chapter we will consider how these seemingly incompatible needs and overwhelming complications are considered as we analyze how rising atmospheric [CO_2] has been influencing modern forests, and how forests will respond in the near future as humans continue to cause [CO_2] to increase.

18.2 CO_2 Influences on Modern Forests

When a 100-year-old tree in the forest today started out its life, the global average atmospheric [CO_2] was 295 ppm. The 25% increase in [CO_2] since then should have brought about an increased rate of photosynthesis and growth. Our modern forests may look substantially different today and be holding significantly more C because of the human-caused increase in [CO_2] during the twentieth century. Finding clear evidence of a past effect of rising [CO_2] is difficult and indirect. Analyses of the global C cycle (Schimel 1995), including anthropogenic sources of CO_2 (e.g., fossil fuel burning, cement manufacturing, and land clearing), the modeled sink of CO_2 into the ocean, and the record of atmospheric [CO_2] leave an unaccounted fraction (missing sink) that is presumed to have been sequestered by the terrestrial biosphere, of which forests are a major component. Some of this terrestrial sink is ascribed to the regrowth of forests after earlier land clearing; some may be a response to perturbation of the nitrogen cycle; and the rest, as much as 1.0 ± 0.5 Gt yr^{-1}, is attributed to increased growth in response to rising [CO_2], or, CO_2 fertilization (Schimel 1995). Controlled experiments in which plants are grown in subambient concentrations of CO_2 support the premise that plant growth is stimulated over the range of 200 to 360 ppm CO_2 (Mielnick et al. 2001), but there have been few such experiments with woody plants and none with forests. Hence, we must rely on indirect evidence, such as tree-ring chronologies and forest inventory analysis, in which any effect of CO_2 is confounded by many other uncontrolled variables.

18.2.1 Tree-Ring Evidence

In 1984 LaMarche and colleagues reported that they had detected possible evidence of a CO_2 fertilization effect in the annual rings of subalpine conifers growing at high altitude sites in the southwestern United States (LaMarche et al. 1984). By that time, growth rates of the trees had increased significantly since the mid-nineteenth century. The positive growth trend was attributed to rising warm-season temperatures until about 1960, but there were no apparent climatic trends that could explain the more recent growth increases. The recent growth increases were thought to be consistent in magnitude with the increase in atmospheric [CO_2], leading to speculation that increased CO_2 uptake and storage could be occurring in the high-altitude habitats and that growth enhancement from elevated CO_2 should be included in global C balance models (LaMarche et al. 1984).

The LaMarche evidence was difficult to interpret, in part because their trees were highly stressed and slow growing, but primarily because of the absence of quantitative analysis of climatic records from nearby weather stations. In contrast, quantitative correlation of tree-ring chronologies with climatic records was conducted for conifers in the Cascade Mountains of western Washington (Graumlich, Brubaker, and Grier 1989) and in the Sierra Nevada Mountains of California (Graumlich 1991). In Washington, annual production was significantly correlated with long-term variation in summer temperature and short-term variation in annual precipitation, and it was uncorrelated with atmospheric [CO_2]. In three of the five Sierra Nevada sites, twentieth-century growth variation could be modeled as a function of climatic variation. Elevated [CO_2] was a possible explanation for unexplained growth increases at two sites, but other possible explanations also were offered. Regardless of whether the growth increases at the sites studied by LaMarche were indeed a response to CO_2 fertilization, there is no evidence for a general effect of rising [CO_2] on the forests of the western United States.

While tree-ring chronologies collected elsewhere seemed to show an unexplained trend of increasing growth that was consistent with (but certainly not diagnostic of) a CO_2 fertilization effect (West et al. 1993; Graybill and Idso 1993), other tree-ring investigations have concluded that tree growth patterns were not related to increasing [CO_2] (Kienast and Luxmoore 1988; D'Arrigo and Jacoby 1993). Jacoby and D'Arrigo (1997) concluded that the evidence for CO_2 fertilization was inconclusive for trees growing in natural settings, where there can be many other limiting and interacting factors. The absence of clear evidence in tree-ring chronologies of a CO_2 fertilization effect should not be taken to mean that rising [CO_2] has had no effect on modern forests, or by extension that an effect in the future is unlikely; rather, the ambiguous record is probably a reflection of the aforementioned problem of detection in the face of multiple feedbacks and interacting environmental and biological factors.

18.2.2 Forest Inventory Analysis

On a global scale, forests account for a large fraction of C absorbed annually by the terrestrial biosphere. About half of the C absorbed in photosynthesis (gross primary production, GPP) is lost in plant respiration in the process of synthesizing organic matter (net primary production, NPP), and much of the rest is equaled by soil microbial (heterotrophic) respiration. The small excess in assimilation over total ecosystem respiration (net ecosystem production, NEP), however, represents C sequestration, which is an important factor in the relationship between fossil fuel emissions and the increase in atmospheric [CO_2]. Sequestration in forests can be due to CO_2 fertilization or a result of land use history (Houghton, Hackler, and Lawrence 1999). The current forests of the northeastern United States, for example, mostly began after the land was abandoned from agriculture in the late 1800s. Since the stands are young, they are sequestering C, but as the stands get older, the amount of C sequestered annually will decline. Estimation of the size of the terrestrial C sink in the future based on observations of twentieth-century forests requires that CO_2 fertilization and land use change be accounted for separately, because their relative contributions to the C sink will change in the future.

Caspersen et al. (2000) used forest inventory analysis to estimate the relative contributions of growth enhancement and historical changes in land use to the C accumulation in forests in the eastern United States. Two successive inventories between the late 1970s and mid-1990s provided data on aboveground biomass, changes in biomass due to growth and mortality, and the age structure of the plot. Since the biomass of a stand is the cumulative result of growth and mortality rates over the age of the stand, Caspersen et al. (2000) could construct stand growth curves based on current rates of growth. The predicted standing biomass was then compared to the actual standing biomass; and discrepancies were ascribed to changes in growth rates. The researchers concluded that C accumulation since 1930 was overwhelmingly due to forest regrowth rather than growth enhancement. The fraction of regional-scale aboveground net ecosystem production (dry matter accumulation) that could be attributed to growth enhancement was 2.0 ± 4.4%, with an upper limit of 7.0%. (The upper limit is not inconsistent with experimental data on tree responses to CO_2 enrichment, given the change in atmospheric [CO_2] that occurred between 1930 and 1995.) Growth enhancement might include CO_2 fertilization, nitrogen deposition, and offsetting factors such as ozone pollution or calcium depletion. The Caspersen et al. (2000) analysis has been challenged because their analytical approach was not sufficiently robust to estimate small changes in growth (Joos, Prentice, and House 2002). The problem again is one of detecting a response to CO_2 that is independent of other environmental influences, stand developmental history, and regional-scale land-use patterns.

18.3 Forests in a Future Atmosphere

Understanding the influence of CO_2 on modern forests is of interest in part because of our general desire to understand how our planet works. More importantly, however, responses of forests today may provide insight into the responses of future forests. Our perspective is a time frame of about 30 to 100 years, during which [CO_2] is expected to rise from the current (2001) level of 372 ppm to about 450 ppm in 2030 and 700 in 2100, based on a middle-range scenario of fossil fuel emissions (IPCC 1992). Individual trees and forests living today, and the people who enjoy them, will experience this important environmental change. Hence, this time frame is relevant to public policy decisions about energy use and the environment.

Surprisingly, perhaps, there is much better evidence describing the responses of trees to future CO_2 levels than to past levels. It is much easier to add CO_2 to the air in an experimental system than it is to remove CO_2, and therefore, there is a wealth of data from controlled, replicated experiments investigating various responses to elevated [CO_2] (Eamus and Jarvis 1989; Ceulemans and Mousseau 1994; Curtis and Wang 1998). We can also make use of observational data from stands of trees near natural springs that have been creating a high [CO_2] environment for many decades. Dendrochronological studies of the drought-impacted trees at the CO_2 springs in Tuscany, Italy, however, have yielded conflicting results (Hättenschwiler et al. 1997; Tognetti, Cherubini and Innes 2000) depending on which trees were measured and how the data were analyzed (Norby et al. 1999). These conflicts illustrate both the difficulty of finding unambiguous expressions of growth response and the overriding importance of confounding environmental factors. Experimental approaches to investigating tree and forest responses to future CO_2 concentrations circumvent many of the problems associated with observations of natural systems. The CO_2 concentration can be precisely controlled and monitored, and responses can be compared to those of comparable plants under ambient conditions. Other environmental variables can be controlled or at least equalized across treatments so that problems of confounding factors are minimized. Interactions between [CO_2] and temperature, N availability, soil moisture, or ozone can be explored. Over the past two decades of research there has been a gradual increase in scale and complexity of the experiments: from potted seedlings in growth chambers, to saplings grown in field chambers for several growing seasons, to forest stands exposed in the open air.

18.3.1 Experimental Studies with Young Trees

Our questions of interest concern the responses of forests in future decades, yet it is impossible to devise a reasonable experiment that exposes an entire forest for a significant fraction of its life cycle. Instead, researchers interested in forest responses must investigate individual components and specific mechanisms of forests with the hope that their answers will be valid and relevant at the larger scale of interest. Early experiments were of short duration (weeks) with young tree seedlings in pots in growth cabinets, but as the need to address the role of forests in the global C cycle became clearer (Kramer 1981), experiments were directed toward defining mechanisms of responses that would be relevant to unmanaged forests, such as nutrient interactions (Norby, O'Neill, and Luxmoore 1986) and competition (Tolley and Strain 1984). Experiments in field chambers allowed investigations over several growing seasons with the influence of multiple, interacting, and fluctuating resources and provided a context for investigating potential important ecological feedbacks such as litter decomposition (Cotrufo, Ineson, and Rowland 1994), herbivory (Lindroth et al. 1997) and nitrogen mineralization (Zak et al. 2000).

While we continue to struggle to understand how the primary responses to [CO_2] are altered by various adjustments, feedbacks, and interactions in the forest environment, we can be confident that we have a good understanding of the primary responses themselves. Norby et al. (1999) reviewed the responses of freely rooted trees exposed to elevated [CO_2] in field experiments and concluded that for the most part the earlier experiments with seedlings were correct. The responses are as follows:

- Photosynthesis is increased approx 60% with an increase of 300 ppm CO_2, and there is little evidence of the long-term loss of sensitivity to CO_2 that was suggested by earlier experiments with potted seedlings.
- The relative effect of [CO_2] on aboveground biomass is highly variable in experiments, but this static measure of response is inappropriate for characterizing a dynamic process that is confounded by the developmental pattern of the plant. When aboveground growth is normalized to constant leaf area, as is appropriate for addressing responses in a closed-canopy forest, annual wood increment per unit leaf area increased 26%, with a fairly consistent response across different tree species and experimental systems (Fig. 18.2).

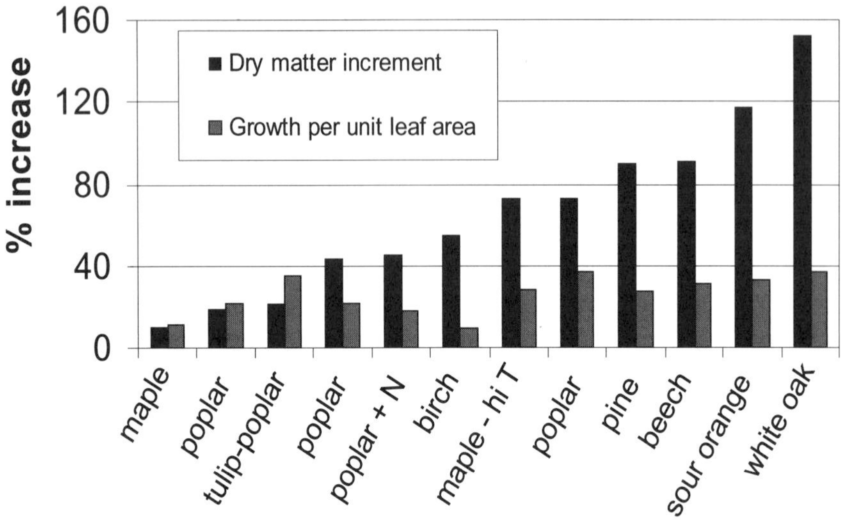

Figure 18.2. The relative effect of elevated [CO_2] (600–700 ppm) on the aboveground dry matter increment of tree seedlings and saplings grown in field chambers. The wide range in response does not represent inherent differences between species but rather the confounding effect of increasing leaf area and exponential growth. Expressing annual stem growth per unit leaf area adjusts for exponential growth pattern and may be more relevant to growth responses that may occur in a closed-canopy forest. The experiments represented in this graph and the conceptual basis for the analysis are in Norby (1996) and Norby et al. (1999).

- Root-to-shoot mass ratio does not typically change (after correcting for developmental effects), but the production of physiologically active fine roots may be enhanced.
- Foliar nitrogen concentrations generally are lower in CO_2-enriched trees.

Additional analyses have indicated the following conclusions:

- Stomatal conductance is lower in field-grown trees exposed to elevated [CO_2] for more than 1 year (Medlyn et al. 2001).
- There probably are no direct effects of CO_2 on leaf respiration (Amthor 2000).
- Litter nitrogen is only slightly lower and this does not result in demonstrable effects on decomposition (Norby et al. 2001a).

These conclusions all derive from experiments with small trees grown individually or in small groups for one to several years. It is encouraging that there is a general concurrence between these field studies and older studies with seedlings, but there are still many differences between young trees in open-top chambers and mature forest trees (Lee and Jarvis 1995). It is important to recognize the limitations of the data before using the results to predict the future state of forests. In particular, we must account for the overriding influence of tree and stand development, ecological feedbacks that could not be replicated in a small group of saplings, and nonlinearity in the physiological responses to [CO_2].

18.3.2 Stand-Level Responses

The application of free-air CO_2 enrichment (FACE) methodology to forest systems (Hendrey et al. 1999) has permitted a new series of experiments in which new hypotheses and critical uncertainties can be explored at the forest stand scale (Karnosky et al. 2001). Important scale-related questions identified in previous open-top chamber experiments that can be addressed in these FACE experiments include the following:

- Does maximum stand leaf area index (LAI) increase in elevated [CO_2]?
- Does growth per unit leaf area remain enhanced after canopy closure?
- Is fine root turnover increased by CO_2 enrichment after a soil is fully exploited by root systems?
- Do feedbacks through the nitrogen cycle lead to down regulation of the tree growth response?
- Do effects of CO_2 on stomatal conductance translate to lower stand water use?
- Do differential effects of CO_2 on competing species alter stand structure?

Many of these questions are components of a more general one posed by Strain and Bazzaz (1983): "The initial effect of elevated CO_2 will be to increase NPP [net primary productivity] in most plant communities. . . . A critical question is the extent to which the increase in NPP will lead to a substantial increase in plant biomass. Alternatively, increased NPP could simply increase the rate of turnover of leaves or roots without changing plant biomass" (Strain and Bazzaz 1983). The fate of increased C absorbed by a future forest in a CO_2-enriched atmosphere is one of the defining questions that must be answered if we are to understand how future forest metabolism will be different from that of today and what the implications are for forest management strategies to reduce the rate of increase in [CO_2] in the atmosphere through enhanced C sequestration.

18.3.3 Oak Ridge Experiment on CO_2 Enrichment of Sweetgum

Four FACE experiments have been initiated in forest ecosystems: two in young, expanding stands and two in established plantations of evergreen or deciduous trees (Karnosky et al. 2001). The objective of the FACE experiment in Oak Ridge, Tennessee, has been to understand how the eastern deciduous forest will be affected by CO_2 enrichment of the atmosphere, and what the feedbacks are from the forest to the atmosphere. The forest stand under study is a sweetgum (*Liquidambar styraciflua*) plantation that was established in 1988. Beginning in 1998, the forest canopy was exposed to an elevated concentration of atmospheric CO_2 ($\sim$ 550 ppm) by emitting CO_2-enriched air from vent pipes that surround the 25 m diameter experimental plots (Norby et al. 2001b).

The responses of this simple forest stand to three years of CO_2 enrichment should inform us about the metabolism of forest stands in 50 to 100 years or should at least reveal the range of responses that might occur. Measurements of a wide range of processes clearly show that the trees responded to the higher CO_2 in predictable ways. Photosynthesis was stimulated, the stimulation was consistent throughout the canopy and with varying weather conditions, and it has not declined through time (Gunderson et al. 2002). Leaf area index was not altered by CO_2 enrichment (Norby et al. 2001b), so it can be presumed that gross primary productivity, or the total amount of C absorbed by the stand, increased. Net primary productivity (NPP) increased 21% during the first four years of CO_2 enrichment (Fig. 18.3), a response that represents a fundamental change in the metabolism of this forest stand that could have ramifications throughout the C cycle (Norby et al. 2002). In many respects the responses to CO_2 enrichment in a stand of loblolly pine (*Pinus taeda*) in North Carolina, of similar age and stature, were similar: increased photosynthesis, no change in leaf area index, higher GPP, and a 27% increase in NPP (Luo et al. 2001; DeLucia, George, and Hamilton. 2002; Hamilton et al. 2002). Although N limitations might eventually impose a constraint on response (Oren et al. 2001) and the stimulation of photosynthesis declines in older needles (Rogers and Ellsworth 2002), the increase in NPP has been sustained for at least 4 years (Hamilton et al. 2002).

What was (or will be) the fate of the additional C that was removed from the atmosphere by these stands each year? As Strain and Bazzaz (1983) asked, Does the increased production result in a greater accumulation of tree biomass, or does the additional fixed C turn over faster with no net accumulation in the plant? In the first year of the sweetgum study, increased NPP resulted in a

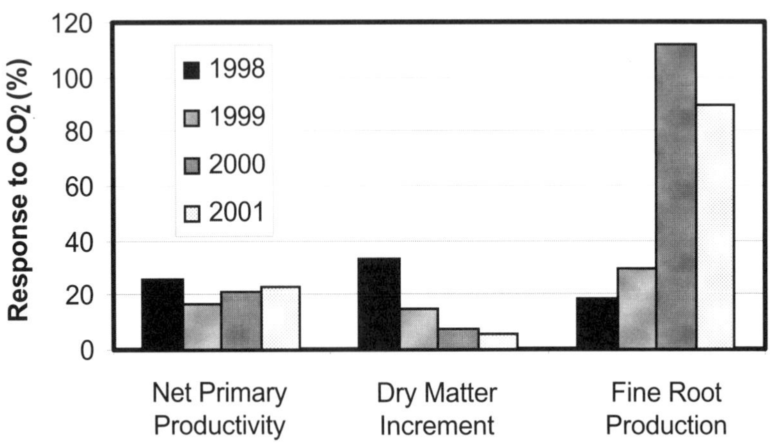

Figure 18.3. The relative response to elevated CO_2 (550 ppm) of NPP, dry matter increment (wood production), and fine root production in a sweetgum forest stand in a FACE experiment in Oak Ridge, Tennessee. (Norby et al. 2002.)

substantial (33%) increase in aboveground dry matter production (Norby et al. 2001b), similar to the 25% increase in relative basal area increment in the first full year of the loblolly pine experiment (DeLucia et al. 1999). Caution is needed, however, in the interpretation of these data: the first-year response of a tree to a sudden increase in [CO_2] from 360 to 550 ppm is not analogous to a tree growing 50 years in the future with a relatively constant (or gradually increasing) CO_2 concentration. As the sweetgum experiment progressed, the response of aboveground dry matter increment rapidly diminished to only a 6% increase, whereas the aboveground growth response was sustained in the pine stand (Hamilton et al. 2002). Instead of accumulating in wood, the extra C in the sweetgum stand was allocated to fine-root production, which was significantly stimulated (see Fig. 18.3). The allocation to short-turnover pools (fine roots and leaves) instead of longer-lived pools (wood) has significant implications for the C sequestration potential of this stand (Norby et al. 2002). The C that is allocated to fine roots enters the soil as organic C and therefore creates the potential for sequestration in long-lived soil organic matter pools. Analyses of the pine stand, however, discount the potential for significantly increased C sequestration in the soil (Schlesinger and Lichter 2001).

The vision of the future forest that the sweetgum experiment presents is one in which C cycles through faster but does not accumulate in the trees to a significant extent. This is not the same vision that emerged from previous experiments in field chambers; the difference may be attributable to the shift in allocation that occurs in association with the transition from exponential to linear growth. What are the consequences of faster C cycling? The evidence is likely to be much more subtle compared to changes in tree growth. Increased fine root production or turnover could alter soil microbial populations and processes, which in turn could alter nutrient cycling through the ecosystem. Faster C cycling could alter the interactions between plants, insects, and disease. The many interactions and feedbacks shown in Fig. 18.1 have not been resolved and probably never will be, but it is clear that the effect of atmospheric [CO_2] on future forests cannot be considered only in terms of tree growth.

18.3.4 Water Use

The focus of most of the research on forest responses to [CO_2] concerns the C budget: that is, the global C cycle, stand-level C budgets, or the prospect for additional C sequestration. Global change encompasses other values besides carbon, and there is increasing recognition (e.g., the Third Assessment Report of the Intergovernmental Panel on Climate Change) of the importance of considering other issues of consequence to ecological or human systems, including interactions with insects and disease, invasive species, resource economics, aesthetics, and so forth (Gitay et al. 2001). Water is an especially critical consideration. Forests are important mediators of water quality and supply, and water availability is a critical regulator of ecological processes.

There is a strong expectation that elevated atmospheric [CO_2] will have important consequences for forest water use and, in turn, for ecosystem-scale processes that depend on soil water availability (Wullschleger, Tschaplinski, and Norby 2002). The basis of this expectation is the response of stomatal conductance to elevated CO_2, which has been shown to decrease in numerous studies (Morison 1985; Field, Jackson, and Mooney 1995; Medlyn et al. 2001). Although the leaf-level responses of stomatal conductance to [CO_2] are important, they are by themselves insufficient to draw conclusions about processes operating at longer temporal and larger spatial scales. It is the canopy-scale integration of these effects that will ultimately address higher-order questions about forest water use and impacts of potential water conservation on ecosystem-scale processes.

FACE experiments in forest stands present a unique opportunity to explore feedbacks and interactions associated with stomatal and canopy conductance and their role in dictating the response of whole-plant water use to elevated [CO_2]. Measurements made on upper-canopy leaves from sweetgum trees exposed to ambient and elevated [CO_2] showed that stomatal conductance is almost always lower in leaves measured at elevated [CO_2] (Fig. 18.4A). Treatment-induced reductions ranged from 14% to 40% and, for the data set presented here, significant effects of elevated [CO_2] on stomatal conductance were observed on 7 out of 11 days (Wullschleger et al. 2002). Although these CO_2-mediated effects were strong for leaves of the upper canopy, there were no significant effects of CO_2 enrichment on stomatal conductance for leaves in the middle- to lower-canopy positions. As a result, there were only modest reductions in canopy-averaged conductance due to elevated [CO_2] as calculated from sap-flow data (Wullschleger et al. 2002). Reductions in canopy conductance due to elevated [CO_2] averaged 14% over the season, and only minor effects of elevated [CO_2] were observed on whole-stand transpiration (Fig. 18.4B). Stand transpiration tended to be less for trees in the elevated compared to the ambient [CO_2] treatment and across the entire growing season averaged 2.8 and 3.1 mm d^{-1} in the two treatments, respectively (Wullschleger and Norby 2001). Stand transpiration on a monthly basis indicated few differences in water use between treatments, although a few significant differences were observed early in the season (Fig. 18.4C). Seasonal estimates of canopy transpiration were 540 mm and 484 mm for stands measured at ambient and elevated CO_2 concentrations, respectively: a difference of just 10% (Wullschleger and Norby 2001). There were no effects of elevated CO_2 on stand transpiration in the loblolly pine FACE experiment, where stomatal conductance of the dominant pines declined with CO_2 enrichment only during severe drought (Schäfer et al. 2002). Indirect effects of litter production on soil moisture were more important to the water budget than were direct effects of [CO_2] on transpiration.

The results of the sweetgum FACE experiment show that large reductions in stomatal conductance at elevated [CO_2] do not necessarily translate to equivalent reductions in rates of canopy transpiration; that is, the response is dampened as the scale increases (Sellers et al. 1996; Raupach 1998; Wilson, Carlson, and Bunce 1999). While FACE experiments can demonstrate some of these scale

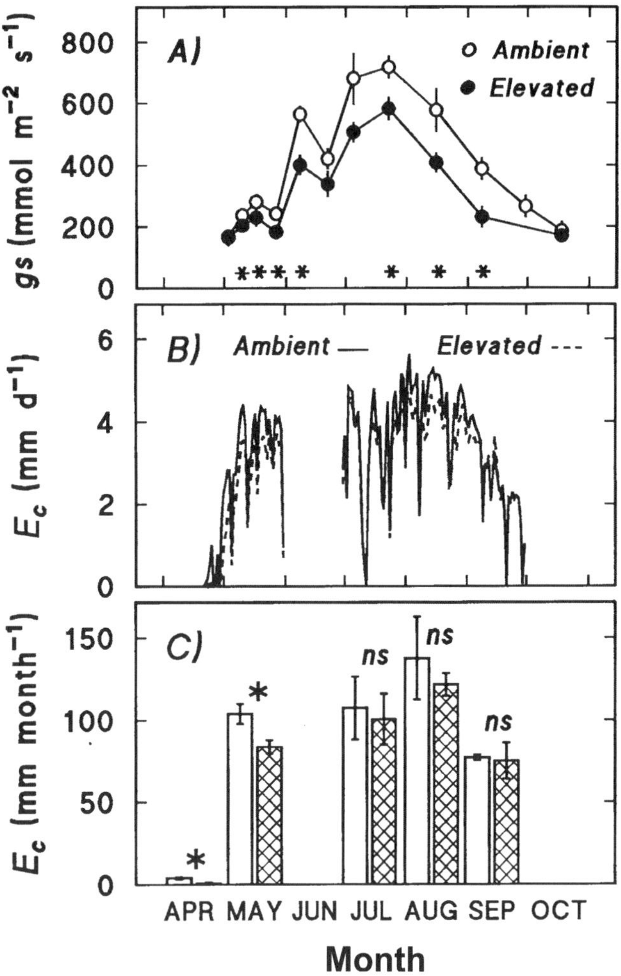

Figure 18.4. Seasonal patterns of: **(A)** stomatal conductance, **(B)** daily stand transpiration, and **(C)** monthly rates of stand transpiration for sweetgum trees measured in both ambient and elevated CO_2 concentration. The open bars in **(C)** are for the ambient [CO_2] treatments, whereas the crosshatched bars are for the elevated [CO_2] treatments. Asterisks indicate significant differences between [CO_2] treatments. ns designates no significant differences. Figures are from Wullschleger and Norby (2001) and Wullschleger et al. (2002); copyright, New Phytologist Trust.

dependencies, other important feedbacks operate at much larger scales. Atmospheric transport processes exert increasingly more control on water vapor exchange as the scale increases from leaf to canopy. The degree to which the canopy is decoupled from the atmosphere (Jarvis and McNaughton 1986) can be expected to increase in canopies that are more extensive than a 25 m diameter FACE plot, and this will further diminish the CO_2 effect on evapotranspiration (Wilson, Carlson, and Bunce 1999). Significant effects of elevated [CO_2] on stand transpiration were difficult to detect from one day to the next because of the dependence of the CO_2 responses on climate and soil resources (Wullschleger et al. 2002). Day-to-day variability in radiation, vapor pressure deficit, and longer-term variation in soil water availability masked what otherwise might have been a much larger CO_2-induced response given more favorable conditions (Gunderson et al. 2002; Wullschleger et al. 2002). Similarly, changes in litter production confounded direct effects of [CO_2] on soil moisture in the loblolly pine FACE experiment (Schäfer et al. 2002). Such conclusions are not usually intuitively obvious, and they complicate, confound, and compromise our ability to measure important biological impacts in CO_2-enrichment studies (see also Ellsworth et al. 1995; Senock et al. 1996; Kellomäki and Wang 1998). Hence, it is not yet at all clear how water balance will be affected by the higher [CO_2] in the forests of the future.

18.4 Modeling the Responses of Future Forests

Experiments have revealed a great deal of information and insight about how trees and forests will respond to the increases in [CO_2] and, to a lesser extent, to the warmer climate that will attend the rising [CO_2]. The experiments, however, cannot create a future reality, even if we knew the exact combination of atmospheric and climatic variables to try to reproduce. Only a small fraction of the tree species and forest environments on Earth have been studied. More experiments with different species, in different environments, and with different combinations of interacting variables will certainly expand our understanding. Larger-scale experiments in intact forests will provide new insights about important forest processes that have not been incorporated in smaller-scale experiments. Nevertheless, predictions of the state of the future forest in a higher-CO_2 world must come from models. A wide variety of models have been employed: biochemical and biophysical, whole-tree, gap-phase dynamics, forest biogeochemistry, and global vegetation models. All of these are informed to some extent by experimental results, and continued model improvement will come with a closer communication between experiments and model building. Here we will discuss how observations and experimental data were used with models to address the important social, economic, and political need for an assessment of the future state of ecosystems in response to rising atmospheric [CO_2] and the climatic change that is expected to accompany it.

18.4.1 National Assessment: Objectives and Approach

The National Assessment of the Potential Consequences of Climate Variability and Change synthesized, evaluated, and reported on what was known about the potential consequences of climatic variability and change for the United States in the twenty-first century (National Assessment Synthesis Team 2001). This first National Assessment brought together both stakeholders and scientific experts to begin a national process of research, analysis, and dialogue about the potential changes in climate, their impacts, and what could be done to adapt to an uncertain and changing climate. Because of limitations on present knowledge, the assessment sought to identify the highest priority uncertainties about which more must be known to understand climate impacts, vulnerabilities, and society's ability to adapt. Regional teams sought to identify key climatic vulnerabilities in the context of other global changes. Similarly, sector teams examined these vulnerabilities in national sectors of water, agriculture, human health, forests, and coastal areas and marine resources. The objectives of Forest Sector assessment were to synthesize what was presently known about the potential consequences of climate variability and change on U.S. forests in consideration of other environmental, social, and economic changes (Joyce et al. 2001).

Modeling analyses were an integral part of the Forest Sector analyses of productivity, natural disturbances, biodiversity, and timber markets. Two climate scenarios were used to provide a quantitative description of plausible future climates for the next 100 years (National Assessment Synthesis Team 2001). Both scenarios suggest that the U.S. climate is going to get warmer. The Canadian model scenario suggests a drier Southeast in the twenty-first century while the Hadley scenario suggests a wetter one. Forest productivity was examined using the biogeochemical models within the VEMAP study (Pan et al. 1998; Aber et al. 2001). These models include the effect of elevated [CO_2] as it interacts with nutrients and water on ecosystem dynamics. Biodiversity responses to climatic change were modeled at the level of the species, communities, and biomes (Hansen et al. 2001). These analyses assumed that the environment-organism relationship remained unaltered in the future, and that natural disturbances regimes were unchanged. Two statistical models describe the potential distribution of individual tree species based on present-day environmental factors, without a CO_2 effect. The biome-level model describes changes in vegetation structure based on light, energy, and water limitations, including the effect of CO_2. Land-use changes or other global changes, such as air quality, were not included in these ecological analyses. The socioeconomic analysis was based on a dynamic optimization model that incorporated changes in forest productivity, shifts between forest and agriculture land, and timber prices to examine climate change impacts on the forestry sector.

18.4.2 National Assessment: Conclusions

Aber et al. (2001) concluded that, under climatic change, forest productivity would likely increase with the fertilizing effect of atmospheric CO_2, but that those increases will be strongly tempered by local conditions, such as moisture stress and nutrient availability. Across a wide range of climate scenarios in several biogeochemical models, modest warming resulted in increased C storage (often greater than 10%) in most forest ecosystems in the conterminous United States (Fig. 18.5). However, under some warmer scenarios, forests, notably in the Southeast and the Northwest, experienced drought-induced losses of C, often greater than 10%. Elevated [CO_2] in the models compensated for negative effects

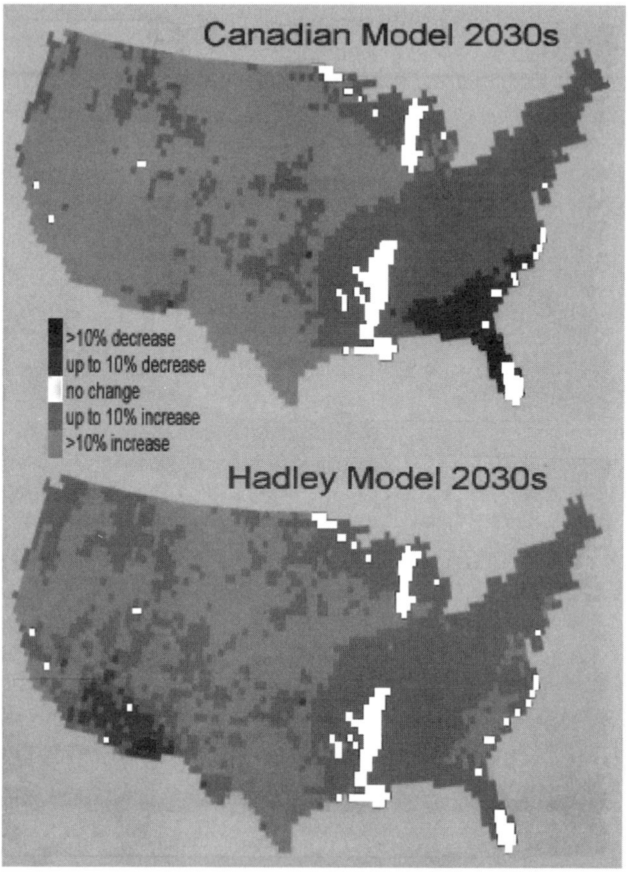

Figure 18.5. Projections of relative changes in vegetation carbon between 1990 and the 2030s for two climate scenarios. Under the Canadian model scenario, vegetation carbon losses of up to 20% are projected in some forested areas of the Southeast in response to warming and drying of the region by the 2030s. A carbon loss by forests is treated as an indication that they are in decline. Under the same scenario, vegetation carbon increases of up to 20% are projected in the forested areas in the West that receive substantial increases in precipitation. Output is from the TEM (Terrestrial Ecosystems Model) as part of the VEMAP II (Vegetation Ecosystem Modeling and Analysis Project) study.

of climatic change and enhanced the positive effects, but the impact of the changed climate (temperature and precipitation) on vegetation C was greater than that of elevated [CO_2]. When the natural disturbance of fire was included, changes in climate alone increased fire frequency (Dale et al. 2001). But, the occasional years of favorable growth conditions (elevated CO_2 and favorable climate) resulted in increased fuel loadings in the western United States and further increased fire frequency and intensity. An important lesson here is that the net effect of multiple factors at the ecosystem scale can be quite different from that implied by an analysis of CO_2 or fire effects alone.

The changes in C storage result from assumptions within the models that describe the influence of favorable growth conditions, including elevated [CO_2], on C accumulation. The biogeochemical models (Biome-BGC, CENTURY, TEM) variously assume that with increased [CO_2], production (or potential production) increases, transpiration or canopy conductance decrease resulting in increased soil moisture, and leaf [N] decreases resulting in decreased litter decomposition (Pan et al. 1998, Aber et al. 2001). There is some experimental evidence to support all of these assumptions, yet as we have discussed above, many of the responses have not proven to be robust or are moderated as the scale of observation or length of exposure increases (e.g., lower transpiration, lower foliar [N], reduced litter quality and decomposition). Just as the results differed across these biogeochemical models because of the differing assumptions within each model, changing the assumptions within each model about the internal dynamics of CO_2 could also result in different C dynamics.

Additionally, these model analyses did not include the effects of other components of global change, such as nitrogen deposition and ground-level ozone concentrations. Interactions between nitrogen, ozone, and elevated CO_2 concentrations can result in increases or decreases in forest productivity depending upon the relative atmospheric concentrations (e.g., Percy et al. 2002). These issues lead to the recognition that model results cannot be taken as an exact representation of future forests, but rather should be seen as providing guidance on the range of possible responses and future vulnerabilities, as well as highlighting the importance of continual dialogue between experimentalists and modelers to ensure the best representation of field experimental results into models.

As discussed above, carbon is not the only endpoint of interest. The effects of climatic change and elevated [CO_2] on species dynamics and biodiversity were evaluated jointly with ecophysiological processes at the biome scale. Across the different modeling approaches used to examine biodiversity, common patterns in the response of forests occurred (Hansen et al. 2001). Oak-hickory and oak-pine in eastern North America and ponderosa pine and arid-tolerant hardwoods in the West expand in area. Suitable habitats projected to greatly decrease in area or disappear from the coterminous United States included: alpine ecosystems, subalpine spruce-fir forests, aspen, maple-beech-birch forests, sagebrush, and loblolly-shortleaf pine ecosystems. These models used only the current organism-environment relationship and did not include population factors, dispersal ability, or disturbance to their environment. In this analysis, most species habitats were projected to move 100 to 530 km north in 100 years, nearly twice the estimated dispersal rates during the Holecene (10–45 km/century). The pace of land-use and climatic change is likely to be rapid relative to the adaptability of species, leading to rapid shifts in species, ranges, and extinctions (Hansen et al. 2001). Here, the influences of elevated [CO_2] may be small in comparison to other global changes.

The effects of climatic change on forest productivity will have socioeconomic implications, especially in the timber and recreation markets (Irland et al. 2001). Analyses using the results of the biogeochemical models described above generated forest productivity gains for North America that increased timber inventories over the next 100 years (Irland et al. 2001). The increased wood supply reverberates through the local economy but has minimal impact at the national scale. At a finer scale, climatic change results in product differences (hardwood vs. softwood), regional differences (increased timber output in the South more than in the North), and welfare differences (consumers benefit but timber producer profits decline). These analyses suggested that timber producers could possibly adjust and adapt to climatic change under the relatively benign scenarios used here if new technologies and markets are recognized in a timely manner.

18.5 Summary

The increase in the atmospheric CO_2 concentration over the past century has been an unprecedented change in the global environment. The increase will continue for decades to come, as there is no sign that the human activities causing the change will abate anytime soon. The National Assessment in the United States was based on the observations from experiments and models that increased atmospheric [CO_2], and the associated climatic variability and change, have potential consequences for future forests, including their productivity, carbon storage, composition, fire regimes, and timber markets. Forest response to atmospheric and climatic change must be analyzed in the context of multiple, interacting environmental variables, interactions, and feedbacks, as well as economic and social factors, such as national and international economic activity and markets, population growth, and societal perceptions about these changes in forests. There is a need to enhance the communication between experimental results and models and to develop a comprehensive system to make the enormous amounts of data and information more accessible and useful to public and private decision makers. Rapid changes in forests in response to [CO_2] and climatic change could challenge current management strategies, and these changes will co-occur with such human activities as multiple uses of forests, agricultural and urban encroachment, and the stresses of air pollution. Understanding how to minimize the vulnerability of ecosystems and human society to atmospheric and climatic change and climate variability will require a broader interpretation of those global change factors and novel approaches that integrate the dynamics of environmental, economic, and social systems in response to change.

References

Aber, J., R.P. Neilson, S. McNulty, J.M. Lenihan, D. Bachelet, and R.J. Drapek. 2001. Forest processes and global environmental change: Predicting the effects of individual and multiple stressors. *BioScience* 51:735–51.

Amthor, J.S. 2000. Direct effect of elevated CO_2 on nocturnal in situ leaf respiration in nine temperate deciduous tree species is small. *Tree Physiology* 20:139–44.

Bazzaz, F.A. 1990. Response of natural ecosystems to the rising global CO_2 levels. *Annual Review of Ecology and Systematics* 21:167–96.

Caspersen, J.P., S.W. Pacala, J.C. Jenkins, G.C. Hurtt, P.R. Moorcroft, and R.A. Birdsey. 2000. Contributions of land–use history to carbon accumulation in US forests. *Science* 290:1148–51.

Ceulemans, R., and M. Mousseau. 1994. Tansley Review no. 71: Effects of elevated atmospheric CO_2 on woody plants. *New Phytologist* 127:425–46.

Cotrufo, M.F., P. Ineson, and A.P. Rowland. 1994. Decomposition of tree leaf litters grown under elevated CO_2: Effect of litter quality. *Plant and Soil* 163:121–30.

Curtis, P.S., and X. Wang. 1998. A meta-analysis of elevated CO_2 effects on woody plant mass, form, and physiology. *Oecologia* 113:299–313.

Dale, V.H., L.A. Joyce, S. McNulty, R.P. Neilson, M.P. Ayres, M.D. Flannigan, P.J. Hanson, L.C. Irland, A.E. Lugo, C.J. Peterson, D. Simberloff, F.J. Swanson, B.J. Stocks, and B.M. Wooton. 2001. Climate change and forest disturbances. *BioScience* 51:723–34.

D'Arrigo, R.D., and G.C. Jacoby. 1993. Tree growth–climate relationships at the northern boreal forest tree line of North–America: Evaluation of potential response to increasing carbon dioxide. *Global Biogeochemical Cycles* 7:525–35.

DeLucia, E.H., J.G. Hamilton, S.L. Naidu, R.B. Thomas, J.A. Andrews, A. Finzi, M. Lavine, R. Matamala, J.E. Mohan, G.R. Hendrey, and W.H. Schlesinger. 1999. Net primary production of a forest ecosystem with experimental CO_2 enrichment. *Science* 284:1177–79.

DeLucia, E.H., K. George, and J.G. Hamilton. 2002. Radiation-use efficiency of a forest exposed to elevated concentrations of atmospheric carbon dioxide. *Tree Physiology* 22:1003–1010.

Eamus, D., and P.G. Jarvis. 1989. The direct effects of increase in the global atmospheric CO_2 concentration on natural and commercial temperate trees and forests. *Advances in Ecological Research* 19:1–55.

Ellsworth, D.S., R. Oren, C. Huang, N. Phillips, and G.R. Hendrey. 1995. Leaf and canopy response to elevated CO_2 in a pine forest under free-air CO_2 enrichment. *Oecologia* 104:139–46.

Field, C.B., F.S. Chapin, P.A. Matson, and H.A. Mooney. 1992. Responses of terrestrial ecosystems to the changing atmosphere: A resource-based approach. *Annual Review of Ecology and Systematics* 23:201–235.

Field, C.B., R.B. Jackson, and H.A. Mooney. 1995. Stomatal responses to increased CO_2: Implications from the plant to global scale. *Plant, Cell, and Environment* 18:1214–25.

Gitay, H., S. Brown, W. Easterling, and B. Jallow. 2001. Ecosystems and their goods and services. In *Climate change 2001: Impacts, adaptation, and vulnerability,* ed. J.J. McCarthy, O.F. Canziani, N.A. Leary, D.J. Dokken, and K.S. White, 235–342. Cambridge: Cambridge University Press.

Graumlich, L.J. 1991. Sub–alpine tree growth, climate, and increasing CO_2: an assessment of recent growth trends. *Ecology* 72:1–11.

Graumlich, L.J., L.B. Brubaker, and C.C. Grier. 1989. Long–term trends in forest net primary productivity: Cascade Mountains, Washington. *Ecology* 70:405–10.

Graybill, D.A., and S.B. Idso. 1993. Detecting the aerial fertilization effect of atmospheric CO_2 enrichment in tree-ring chronologies. *Global Biogeochemical Cycles* 7: 81–95.

Gunderson, C.A., J.D. Sholtis, S.D. Wullschleger, D.T. Tissue, P.J. Hanson, and R.J. Norby. 2002. Environmental and stomatal control of photosynthetic enhancement in the canopy of a sweetgum (*Liquidambar styraciflua* L.) plantation during 3 years of CO_2 enrichment. *Plant, Cell, and Environment* 25:379–94.

Hamilton, J.G., E.H. DeLucia, K. George, S.L. Naidu, A.C. Finzi, and W.H. Schlesinger. 2002. Forest carbon balance under elevated CO_2. *Oecologia* 131:250–60.

Hansen, A.J., R.P. Neilson, V.H. Dale, C.H. Flather, L.R. Iverson, D.J. Currie, S. Shäfer, R. Cook, and P.J. Bartlein. 2001. Global change in forests: responses of species, communities, and biomes. *BioScience* 51:765–79.

Hättenschwiler, S., F. Miglietta, A. Raschi, and C. Körner. 1997. Thirty years of *in situ* tree growth under elevated CO_2: A model for future forest responses? *Global Change Biology* 3:463–71.

Hendrey, G.R., D.S. Ellsworth, K.F. Lewin, and J. Nagy. 1999. A free-air enrichment system for exposing tall forest vegetation to elevated atmospheric CO_2. *Global Change Biology* 5:293–309.

Houghton, R.A., J.L. Hackler, and K.T. Lawrence. 1999. The US carbon budget: Contributions from land-use change. *Science* 285:574–78.

Intergovernmental Panel on Climate Change (1992): Climate Change 1992: The Supplementary Report to the IPCC Scientific Assessment (Eds. Houghton, J.T., Callander, B.A. & Varney, S.K.). Cambridge University Press, Cambridge. U.K.

Irland, L.C., D. Adams, R. Alig, C.J. Betz, C. Chen, M. Hutchins, B.A. McCarl, K. Skog, and B.L. Sohngen. 2001. Assessing socioeconomic impacts of climate change on US forests, wood-product markets, and forest recreation. *BioScience* 51:753–64.

Jacoby, G.C., and R.D. D'Arrigo. 1997. Tree rings, carbon dioxide, and climatic change. *Proceedings of the National Academy of Sciences of the United States of America* 94: 8350–53.

Jarvis, P.G., and K.G. McNaughton. 1986. Stomatal control of transpiration: Scaling up from leaf to region. *Advances in Ecological Research* 15:1–49.

Joos, F., I.C. Prentice, and J.I. House. 2002. Growth enhancement due to global atmospheric change as predicted by terrestrial ecosystem models: consistent with US forest inventory data. *Global Change Biology* 8:299–303.

Joyce, L.A., J. Aber, S. McNulty, V. Dale, A. Hansen, L. Irland, R. Neilson, and K. Skog. 2001. Potential consequences of climate variability and change for the forests of the United States. In *Climate change impacts on the United States: The potential consequences of climate variability and change,* ed. Foundation, National Assessment Synthesis Team. Cambridge: Cambridge University Press.

Karnosky, D.F., B. Gielen, R. Ceulemans, W.H. Schlesinger, R.J. Norby, E. Oksanen, R. Matyssek, and G.R. Hendrey, 2001. FACE systems for studying the impacts of greenhouse gases on forest ecosystems. In *The impact of carbon dioxide and other greenhouse gases on forest ecosystems*, ed. D.F. Karnosky, R. Ceulemans, G.E. Scarascia-Mugnozza, and J.L. Innes, 297–324. Wallingford, U.K.: CABI Publishing.

Kellomäki, S., and K.Y. Wang. 1998. Sap flow in Scots pine growing under conditions of year-round carbon dioxide enrichment and temperature elevation. *Plant, Cell, and Environment* 21:969–81.

Kienast, F., and R.J. Luxmoore. 1988. Tree-ring analysis and conifer growth-responses to increased atmospheric CO_2 levels. *Oecologia* 76:487–95.

Kramer, P.J. 1981. Carbon dioxide concentration, photosynthesis, and dry matter production. *BioScience* 31:29–33.

LaMarche, V.C., D.A. Graybill, H.C. Fritts, and M.R. Rose. 1984. Increasing atmospheric carbon-dioxide: Tree-ring evidence for growth enhancement in natural vegetation. *Science* 225:1019–21.

Lee, H. S.J., and P.G. Jarvis. 1995. Trees differ from crops and each other in their responses to increases in CO_2 concentration. *Journal of Biogeography* 22:323–30.

Lindroth, R.L., S. Roth, E.L. Kruger, J.C. Volin, and P.A. Koss. 1997. CO_2-mediated changes in aspen chemistry: Effects on gypsy moth performance and susceptibility to virus. *Global Change Biology* 3:279–89.

Luo, Y., B. Medlyn, D. Hui, D. Ellsworth, J. Reynolds, and G. Katul. 2001. Gross primary production in the Duke Forest: Modeling synthesis of the free-air CO_2 enrichment experiment and eddy-covariance measurements. *Ecological Applications* 11:239–52.

Medlyn, B.E., C.V.M. Barton, M.S.J. Broadmeadow, R. Ceulemans, P. De Angelis, M. Forstreuter, M. Freeman, S.B. Jackson, S. Kellomaki, E. Laitat, A. Rey, P. Roberntz, B.D. Sigurdsson, J. Strassemeyer, K. Wang, P.S. Curtis, and P.G. Jarvis. 2001. Stomatal conductance of forest species after long-term exposure to elevated CO_2 concentration: A synthesis. *New Phytologist* 149:247–64.

Mielnick, P.C., W.A. Dugas, H.B. Johnson, H.W. Polley, and J. Sanabria. 2001. Net grassland carbon flux over a subambient to superambient CO_2 gradient. *Global Change Biology* 7:747–54.

Morison, J.I.L. 1985. Sensitivity of stomata and water use efficiency to high CO_2. *Plant, Cell, and Environment* 8:467–74.

National Assessment Synthesis Team. 2001. *Climate Change Impacts on the United States: The Potential Consequences of Climate Variability and Change.* Foundation. Cambridge: Cambridge University Press.

Norby, R.J. 1996. Forest canopy productivity index. *Nature* 381:564.

Norby, R.J., M.F. Cotrufo, P. Ineson, E.G. O'Neill, and J.G. Canadell. 2001a. Elevated CO_2, litter chemistry, and decomposition: A synthesis. *Oecologia* 127:153–65.

Norby, R.J., P.J. Hanson, E.G. O'Neill, T.J. Tschaplinski, J.F. Weltzin, R.T. Hansen, W. Cheng, S.D. Wullschleger, C.A. Gunderson, N.T. Edwards, and D.W. Johnson. 2002. Net primary productivity of a CO_2-enriched deciduous forest and the implications for carbon storage. *Ecological Applications* 12:1261–66.

Norby, R.J., E.G. O'Neill, and R.J. Luxmoore. 1986. Effects of atmospheric CO_2 enrichment on the growth and mineral nutrition of *Quercus alba* seedlings in nutrient-poor soil. *Plant Physiology* 82:83–89.

Norby, R.J., D.E. Todd, J. Fults, and D.W. Johnson. 2001b. Allometric determination of tree growth in a CO_2-enriched sweetgum stand. *New Phytologist* 150:477–87.

Norby, R.J., S.D. Wullschleger, C.A. Gunderson, D.W. Johnson, and R. Ceulemans. 1999. Tree responses to rising CO_2: implications for the future forest. *Plant, Cell, and Environment* 22:683–714.

Oren, R., D.S. Ellsworth, K.H. Johnsen, N. Phillips, B.E. Ewers, C. Maier, K.V.R. Schäfer, H. McCarthy, G. Hendrey, S.G. McNulty, and G.G. Katul. 2001. Soil fertility limits carbon sequestration by forest ecosystems in a CO_2-enriched atmosphere. *Nature* 411:469–72.

Pan, Y.D., J.M. Melillo, A.D. McGuire, D.W. Kicklighter, L.F. Pitelka, K. Hibbard, L.I. Pierce, S.W. Running, D.S. Ojima, W.J. Parton, and D.S. Schimel. 1998. Modeled responses of terrestrial ecosystems to elevated atmospheric CO_2: A comparison of simulations by the biogeochemistry models of the Vegetation/Ecosystem Modeling and Analysis Project (VEMAP). *Oecologia* 114:389–404.

Percy, K.E., C.S. Awmack, R.L. Lindroth, M.E. Kubiske, B.J. Kopper, J.G. Isebrands, K.S. Pregitzer, G.R. Hendrey, R.E. Dickson, D.R. Zak, E. Oksanen, J. Sober, R. Harrington, and D.F. Karnosky. 2002. Altered performance of forest pests under atmospheres enriched by CO_2 and O_3. *Nature* 420:403–407.

Raupach, M.R. 1998. Influences of local feedbacks on land-air exchanges of energy and carbon. *Global Change Biology* 4:477–94.

Rogers, A., and D.S. Ellsworth. 2002. Photosynthetic acclimation of *Pinus taeda* (loblolly pine) to long term growth in elevated pCO_2 (FACE). *Plant, Cell, and Environment* 25:851–58.

Schäfer, K.V.R., R. Oren, C.T. Lai, and G.G. Katul. 2002. Hydrologic balance in an intact temperate forest ecosystem under ambient and elevated atmospheric CO_2 concentration. *Global Change Biology* 8:895–911.

Schimel, D.S. 1995. Terrestrial ecosystems and the carbon cycle. *Global Change Biology* 1:77–91.

Schlesinger, W.H., and J. Lichter. 2001. Limited carbon storage in soil and litter of experimental plots under elevated atmospheric CO_2. *Nature* 411:466–69.

Sellers, P.J., L. Bounoua, G.J. Collatz, D.A. Randall, D.A. Dazlich, S.O. Los, J.A. Berry, I. Fung, C.J. Tucker, C.B. Field, and T.G. Jensen. 1996. Comparison of radiative and physiological effects of doubled atmospheric CO_2 on climate. *Science* 271:1402–1406.

Senock, R.S., J.M. Ham, T.M. Loughin, B.A. Kimball, D.J. Hunsaker, P.J. Pinter, G.W. Wall, R.L. Garcia, and R.L. LaMorte. 1996. Sap flow in wheat under free-air CO_2 enrichment. *Plant, Cell, and Environment* 19:147–58.

Stitt, M. 1991. Rising CO_2 levels and their potential significance for carbon flow in photosynthetic cells. *Plant, Cell, and Environment* 14:741–62.

Strain, B.R., and F.A. Bazzaz. 1983. Terrestrial plant communities. In *CO_2 and plants,* ed. E.R. Lemon, 177–222. Boulder, Colorado: Westview Press.

Tognetti, R., P. Cherubini, and J.L. Innes. 2000. Comparative stem-growth rates of Mediterranean trees under background and naturally enhanced ambient CO_2 concentrations. *New Phytologist* 146:59–74.

Tolley, L.C., and B.R. Strain. 1984. Effects of CO_2 enrichment and water stress on growth of *Liquidambar styraciflua* and *Pinus taeda* seedlings. *Canadian Journal of Botany* 62:2135–39.

West, D.C., T.W. Doyle, M.L. Tharp, J.J. Beauchamp, W.J. Platt, and D.J. Downing. 1993. Recent growth increases in old–growth longleaf pine. *Canadian Journal of Forest Research* 23:846–53.

Wilson, K.B., T.N. Carlson, and J.A. Bunce. 1999. Feedback significantly influences the simulated effect of CO_2 on seasonal evapotranspiration from two agricultural species. *Global Change Biology* 5:903–17.

Wullschleger, S.D., C.A. Gunderson, P.J. Hanson, K.B. Wilson, and R.J. Norby. 2002. Sensitivity of stomatal and canopy conductance to elevated CO_2 concentration: Interacting variables and perspectives of scale. *New Phytologist* 153:485–96.

Wullschleger, S.D., and R.J. Norby. 2001. Sap velocity and canopy transpiration in a sweetgum stand exposed to free-air CO_2 enrichment (FACE). *New Phytologist* 150: 489–98.

Wullschleger, S.D., T.J. Tschaplinski, and R.J. Norby. 2002. Plant water relations at elevated CO_2: Implications for water-limited environments. *Plant, Cell, and Environment* 25:39–331.

Zak, D.R., K.S. Pregitzer, P.S. Curtis, and W.E. Holmes. 2000. Atmospheric CO_2 and the composition and function of soil microbial communities. *Ecological Applications* 10:47–59.

19. Modern and Future Semi-Arid and Arid Ecosystems

M. Rebecca Shaw, Travis E. Huxman, and Christopher P. Lund

19.1 Introduction

Arid and semi-arid ecosystems are diverse biotic assemblages with wide variations in topographic, climatic, edaphic, geologic, and biological conditions. These ecosystems include dry and humid grasslands in all latitudes and altitudes, scrub, tropical and subtropical savannas, dry forest, and coastal ecosystems. Such regions comprise up to 40% of the world's land surface and by some estimates account for up to 30% of global terrestrial net primary productivity. Semi-arid grasslands cover approximately 33 million km^2 (25% of the land surface) on all continents except Antarctica. Savanna ecosystems cover an estimated 25 million km^2 in Africa, South America, Asia, and Australia. With several exceptions in the Mediterranean ecosystem, these arid and semi-arid regions are not species rich, but endemism tends to be very high. Functional diversity, however, may be rather high (Smith, Monson, and Anderson 1997), as many species have evolved strategies for growth and survival under extreme environmental conditions, including high and low temperature, low and episodic precipitation, and periodic protracted drought. Annual rainfall averages up to 350 mm and 700 mm for arid and semi-arid systems, respectively. It has been suggested that these systems are likely to be among the most responsive to changes in atmospheric CO_2 (Strain and Bazzaz 1983; Smith et al. 2000).

Semi-arid and arid ecosystems should respond favorably to rising CO_2 since plant function is primarily limited by water availability (Mooney et al. 1991).

This projection derives from the often-measured ability of elevated CO_2 to reduce water stress in many plants (Strain 1992). Additionally, the effects of elevated CO_2 on photosynthesis and stomatal conductance are such that the photosynthesis/transpiration ratio in arid systems should be relatively higher than in more mesic systems (Knapp et al. 1996), directly impacting carbon gain and whole-system water loss and likely enhancing the plant community productivity (Melillo et al. 1993). Rising CO_2 may also affect vegetation composition, promoting plant survival through drought by reducing water stress in adult plants and by establishing seedlings (Smith, Monson, and Anderson 1997). The most significant long-term response to elevated CO_2 may be the manifestation of interactions with additional global change factors that could significantly influence vegetation change and aggregated shifts in other biogeochemical cycles important for ecosystem function. In this regard, how nitrogen and water availability co-vary and influence a CO_2 response in arid ecosystems may be critical to understand (Smith, Monson, and Anderson 1997; Ehleringer 2001).

19.2 Effects of Atmospheric CO_2 Across Multiple Scales of Organization

19.2.1 Physiological Responses of Plants from Arid and Semi-Arid Ecosystems

19.2.1.1 Photosynthetic Gas Exchange and Down-Regulation

A rise in CO_2 often increases net rates of photosynthesis in many C_3 plants Under a doubling of CO_2, assimilation rates may increase by as much as 75% depending on the environmental conditions and plant growth form. Warmer temperatures can additionally stimulate photosynthesis responses (Berry and Björkman 1980; Sage and Sharkey 1987). C_3 photosynthesis is a CO_2-unsaturated biochemical reaction, and higher rates of net photosynthesis occur under elevated CO_2 as a result of increased substrate availability at the site of carboxylation and reduced competitive inhibition by photorespiration activity, increasing the efficiency of ribulose-1,5-biphosphate carboxylase/oxygenase, or, Rubisco (Bowes 1991, 1993). This mechanism does not result in increased photosynthetic performance in C_4 plants because C_4 plants already achieve high CO_2 concentrations around the chloroplasts (Ghannoum et al. 2000).

Long-term exposure to elevated CO_2 can result in a range of responses, including the maintenance of increased rates of photosynthesis (Arp 1991; Körner and Miglietta 1994; Hamerlynck et al. 2000a,b; Huxman and Smith 2001), the down-regulation of photosynthesis (DeLucia, Sasek, and Strain 1985; Tissue and Oechel 1987; Sage, Sharkey, and Seemann 1989; Oechel et al. 1995), or non-significant responses (Jackson et al. 1995; Huxman and Smith 2001). In the case where increased photosynthesis under elevated CO_2 occurs initially, concomitant increases in leaf carbohydrate concentrations can result in a negative feedback on photosynthetic capacity (Tissue, Thomas, and Strain 1993). Referred to as photosynthetic down-regulation, this may represent a homeostatic adjustment that aligns whole-plant processes and carbon gain rates to within some bounds important for coordinated function. In semi-arid systems, photosynthetic down-regulation does not appear to be a major limitation to plant carbon gain, as it has been observed for well-watered plants only under conditions that were not favorable for whole-plant growth (Huxman et al. 1998c; Naumburg et al. 2003). Indeed, many plants in semi-arid ecosystems show increases in net photosynthetic rates and carbon gain under elevated CO_2 even in the presence of down-regulation activity (Cure and Acock 1986; Oechel et al. 1995; Curtis and Wang 1998). This is most likely as a result of plant carbon-sink strength remaining strong when resources are available for plant growth and when stomata are open. The result is that during periods with high resource availability, elevated CO_2 results in greater leaf-level photosynthesis in many arid plants (Hamerlynck et al. 2000a; Huxman and Smith 2001; Hamerlynck et al. 2002; Naumburg et al. 2003).

19.2.1.2 Plant Water Use, Stomatal Conductance, and Transpiration

Stomatal conductance reflects a balance that plants maintain in order to maximize carbon gain while minimizing water loss (Cowan and Farquhar 1977). This is true also for plants growing in semi-arid environments that may have been selected to maximize water-use efficiency in the context of other limiting resources (Bloom, Chapin, and Mooney 1985). Increases in atmospheric CO_2 concentrations cause stomatal closure (Linsbauer 1917), which results in decreased transpiration (Ketellapper 1963). The role of CO_2 in regulating stomatal conductance (g_l) has been well studied, with C_3 and C_4 herbaceous species showing, on average, a 40% decrease in conductance in response to a doubling of ambient CO_2 concentration (Morison 1985). In a tall-grass prairie, stomatal conductance in the dominant C_4 species decreased by 21% in a dry year and by 59% in a wet year (Knapp, Hamerlynck, and Owensby 1993). In Mediterranean grassland, elevated CO_2 led to a 45% decrease in stomatal conductance in mid-season. This decrease in stomatal conductance under elevated CO_2 can result from either decreases in stomatal aperture or decreases in stomatal density (Clifford et al. 1995; Webb et al. 1996; Beerling, McElwain, and Osborne 1998). Continuous, long-term exposure to elevated CO_2 does not appear to dampen this response as has been shown for plants growing near natural CO_2 springs where for generations plants have been exposed to an elevated CO_2 environment (Betterini, Vaccari, and Miglietta 1998). Thus, rising atmospheric CO_2 allows the potential for plants in arid ecosystems to meet the growth demand for carbon substrate with less water lost through transpiration.

Decreases in leaf-level conductance in response to exposure to elevated CO_2 have been observed in numerous studies in semi-arid ecosystems (Jackson et al. 1994; Oechel et al. 1995; Bremer, Ham, and Owensby 1996; Nijs et al. 1997; Huxman and Smith 2001). This response to elevated CO_2 allows plants to increase water use efficiency (WUE), which is the mass of carbon fixed per mass of water transpired. Herbaceous plants often show the largest decreases in stomatal conductance and increases in WUE, while woody species tend to show smaller or nonsignificant responses (Curtis and Wang 1998; Saxe and Heath 1998; Ellsworth 1999). In the most arid ecosystems, the decreases in stomatal conductances by plants have been seen primarily during periods of high resource availability when photosynthetic and transpiration fluxes are greatest (Pataki et al. 2000; Nowak et al. 2001).

While it has been frequently reported that elevated CO_2 can compensate for the expected decrease in carbon gain elicited by water deficits (Tyree and Alexander 1993), the mechanisms responsible extend beyond simple increases in leaf-level water-use efficiency (Chaves and Pereira 1992; Tyree and Alexander 1993) and include a range of functionally co-related traits (Table 19.1). Increases in leaf-level water-use efficiency under elevated CO_2 conditions are a partial result of photosynthetic stimulation, which leads to greater carbon gain per unit water loss (Tyree and Alexander 1993).

Changes in root characteristics also play an important role in increased leaf-level water-use efficiency under elevated CO_2 at the whole-plant under water-limiting conditions when exposed to elevated CO_2 may be related to selective carbon allocation toward roots (i.e., higher root/shoot ratio), which may be beneficial in the survival of water-limited plants through enhanced balance between water uptake and canopy transpiration (Bazzaz 1990; Chaves and Pereira 1992). Exposure of plants to elevated CO_2 and water limitation has been shown to alter the capacities of roots to transport water through their xylem conduits. In soy-

Table 19.1. The interaction between elevated CO_2 and drought on several inter-related physiological processes and plant organs for sunflower (*Helianthus annuus*) grown in glasshouse conditions (data from Huxman 1999). It should be noted that any single character change associated with growth at elevated CO_2 does not completely predict the "drought-alleviation" in biomass production exhibited by this species.

	Treatment			
	Ambient CO_2		Elevated CO_2	
Character	well-watered	droughted	well-watered	droughted
Root/Shoot biomass ratio $g\ g^{-1}$	0.45+.0.5	0.48+0.02	0.42+.0.4	0.41+0.03
Root/Shoot area ratio $cm^2\ cm^{-2}$	1.91+0.35	2.01+0.12	2.38+0.30	3.15+.024
Assimilation rate $\mu mol\ m^{-2}s^{-1}$	10.7+0.79	9.57+2.8	19.4+2.27	10.76+1.93
Stomatal conductance $mmol\ m^{-2}s^{-1}$	130+14	103+28	174+28	48+27
Water-use efficiency $\mu mol\ CO_2\ mmol^{-1}\ H_2O$	4.43+0.44	4.52+066	7.06+1.18	9.84+0.20
Root hydraulic conductivity $m\ s^{-1}\ MPa^{-1}\ X\ 10^{-8}$	7.27+1.3	3.54+.054	4.25+1.27	4.67+0.30

bean, root hydraulic conductance decreased 26% at elevated CO_2 (Bunce 1996) and in sunflower whole root system hydraulic conductivity decreased by nearly 50% (Huxman, Smith, and Neuman 1999) compared to ambient-CO_2-grown plants. We are still in our infancy in understanding the influences of elevated CO_2 on different physiological processes, plant organs, and their inter-relationships. Yet from the information available today there is a clear suggestion that multiple mechanisms exist to affect continued plant growth under water-limited systems.

19.2.2 Community Responses of Water Limited Ecosystems to Rising Atmospheric CO_2

Changes in individual plant performance should impact competitive hierarchies and the representation of different species in terrestrial communities at elevated CO_2 (Bazzaz 1990). This assumption is supported by the strong, species-specific effects of elevated CO_2 seen in nearly all ecosystems studied to date. Elevated CO_2 is expected to have a differential impact on C_3 versus C_4 plants as was discussed by Ehleringer in Chapter 10. Studies have shown that higher concentrations of CO_2 might favor plants that fix carbon via the C_3 photosynthesis over C_4 photosynthetic pathways, altering the current pattern of C_4 dominance in many semi-arid ecosystems (Ehleringer, Cerling, and Helliker 1997). Additionally, there is the potential that elevated CO_2 will foster non-native species invasions, especially in conditions where plant growth rates are CO_2 dependent (Dukes and Field 2000). In recent studies, growth rate differences at elevated CO_2 between native and non-native species have been detected that are consistent with the expectation that native species will be out-competed by non-native species in a higher CO_2 environment (Poorter and Navas 2003). However, many of these ideas have been based on the assumption that changes in total biomass production scale directly to changes in fecundity. How seed production and future-offspring growth potential are influenced by elevated CO_2 is not well known (Huxman et al. 1998a; Huxman and Smith 2001; Jablonski, Wang, and Curtis 2002). Will elevated CO_2 alter species composition in plant communities, and, if so, what is the potential for changes in ecosystem functioning? For Ward's discussion of some of the possible fecundity adjustments that can take place as plants adjust to a different CO_2 regime, see Chapter 11.

19.2.2.1 Functional Type Response in Xeric Ecosystems

Existing plant distributions are likely to be impacted by temperature changes, an indirect effect of elevated CO_2. One common life form that is likely to be impacted is the succulent species, the distribution of which tends to be sensitive to thermal regimes. Freezing conditions, even for a brief period, tend to cause extensive mortality among some succulent species. Based on the expected shifts in temperature conditions, the distribution of many succulent species, such as *Yucca brevifolia*, is expected to expand considerably (Dole, Loik, and Sloan 2003).

Arid and semi-arid ecosystems are floristically rich assemblages in contrast to other temperate-region ecosystem types (Beatley 1974; Cody 1986; Smith, Monson, and Anderson 1997). While arid lands themselves do not necessarily have high species diversity per unit area, they represent landscapes diverse in plant functional types. This pattern is in contrast to many other ecosystems, such as temperate grasslands and forests, which have high biological diversity with relatively low functional diversity (Tilman et al. 1997). Applying a "functional type" approach to the question of how vegetation assemblages may respond to rising CO_2 seems applicable. Two approaches can be employed when placing plants into functional groups and attempting to understand how different physiological and life history properties relate to distribution. First, characteristics that are likely to be influenced by climate change can be used to form the functional groups. Examples might include photosynthetic pathway or growth potential (Poorter and Navas 2003). Second, we might focus on those characteristics that are important for current species distribution, which may better help us understand current constraints on distribution patterns. In fact, combining these approaches may provide significant power in the prediction of future plant distributions.

For many ecosystems in the arid southwestern North America, there are significant patterns between climate and the distribution of general life history strategy (e.g., perennial versus annual). It has been suggested that the distribution of perennial plants is related to seedling survival in the abiotic environment near the soil surface (Smith and Nowak 1990). The high rates of seedling mortality are mostly the result of adverse environmental conditions (Fenner 1985; Franklin et al. 1992). How elevated CO_2 influences seedling survival is generally unknown, but the available evidence suggests that elevated CO_2 can offset plant responses to environmental stress during early life stages (Huxman et al. 1998a; Housman et al. 2003) and can increase the likelihood of the establishment of long-lived perennials.

For annual plants and some fast-growing perennials, the ability to produce a substantial seed crop is critical for long-term persistence (Mulroy and Rundel 1977). Annual plants have often been categorized as "stress avoiders" as a result of their life history strategy of being present only in the form of seeds during periods of extended stress. Evaluations of the reproductive characteristics of annual plants in arid ecosystems suggest that seed production in these groups may be affected by growth at elevated CO_2 (Smith et al. 2000). Size of seed crops of desert plants is often a linear function of plant size (Ehleringer 1985), and elevated CO_2 typically increases both plant growth rate and plant size after a period of exposure (Poorter and Navas 2003). Plant size and nutrient status are commonly influenced by plant growth under elevated CO_2 (Bazzaz, Coleman, and Morse 1990; Bazzaz 1997), and it would not be unexpected that a change in reproductive characteristics, such as shifts in seed quality, would also occur in the future (Huxman et al. 1998a).

19.2.2.2 C_3 Versus C_4 Photosynthetic Pathway Dominance

Photosynthetic rates and light-use (quantum yield) efficiencies and growth of C_3 species increase with elevated CO_2 (Carlson and Bazzaz 1980; Cure and Acock 1986; Wand et al. 1999). While not all C_4 species show a significant, positive photosynthetic response, growth in many C_4 species is indirectly stimulated by elevated CO_2 as a consequence of new leaf area production and changes in whole-plant water status (Woodward, Thompson, and McKee 1991; Polley, Johnson, and Derner 2002). Across a range of taxa, C_4 plants do not appear to increase growth rate at elevated CO_2 to the same extent as C_3 plants (Poorter, Roumet, and Campbell 1996; Poorter and Navas 2003). As discussed in Chapter 10, quantum yield and photorespiration rates in C_3 leaves are sensitive both to changes in atmospheric CO_2 and to changes in temperature. Rising CO_2 coupled with moderate increases in temperatures are thus expected to provide a relatively greater enhancement for C_3 plants as compared to C_4 plants (Long 1991). In a future high CO_2 world, C_3 species may not exhibit the energetic tradeoff associated with increasing temperature that has allowed for the proliferation of C_4 species in the low CO_2 world that plants have experienced for the last several million years (Hatch 1992). Thus, in conditions typical of many arid and semi-arid ecosystems, the relative competitive ability of C_4 plants might be diminished (see Chapters 9 and 10).

Leaf-level photosynthesis measurements in mixed C_3/C_4 communities often show a greater proportional response of photosynthesis or stomatal conductance by C_3 species as compared to C_4 species at elevated CO_2 (Owensby et al. 1993; Owensby et al. 1999; Morgan et al. 2001; Polley, Johnson, and Derner 2002). However, fractional productivity or representation in the community does not always reflect a similar pattern. This is most likely due to the effects that elevated CO_2 has on whole-system water budgets (Owensby et al. 1993; Polley, Johnson, and Derner 2002) and the energetic implications of increases in whole-system productivity. In a tallgrass prairie, production by C_4 grasses increased considerably at elevated CO_2, while production from C_3 species did not change (Owensby et al. 1993; Owensby et al. 1999). In the shortgrass steppe, proportional production of C_3 and C_4 species did not change at elevated CO_2 because of indirect feedbacks associated with the availability of soil water (Morgan et al. 2001). When constrained to comparisons within the grass morphological form, C_3 and C_4 species have similar growth responsiveness to elevated CO_2 (Wand et al. 1999) but show variation in the form of that response (e.g., increases in vegetative vs. clonal growth). The C_3 species that have increased in abundance at elevated CO_2 in CO_2-exposure experiments involving mixed communities have been the herbaceous and woody C_3 life forms rather than grasses, which could be a simple reflection of grasses being shaded and out-competed by taller life forms. This may be the most consistent trend in the C_3 versus C_4 literature and suggests that when systems are comprised of plants where there is covariation between life form and photosynthetic pathway, shifts in the abundance of C_3

and C_4 species may be expected. Interactions with other global change factors may be especially important for understanding the relative performance and abundance of C_4 and C_3 plants in arid regions at elevated CO_2. Relaxation of the cold season temperature constraint by general atmospheric warming may extend the C_4 growing season. Greater primary production at elevated CO_2 combined with increased growing season precipitation may alter fire disturbance characteristics in arid land, differentially affecting C_3 and C_4 species (Sage 1996). Nitrogen deposition may result in a homogeneous landscape, favoring grassland physiognomies over shrublands (Reynolds, Hilbert, and Kemp 1993). The frequency and intensity of episodic drought may alter the relative performance of C_3 and C_4 species in nonintuitive manners (Ward et al. 1999). Additional traits, such as seed quality, may differentially affect species that "escape" drought in a different life history stage (Huxman et al. 1998a). Thus, the issue of C_3 versus C_4 responsiveness in water-limited ecosystems is quite complex and requires an understanding of interactions among factors. For these functional types, the ecological outcome of elevated CO_2 may be strongly influenced by temperature, precipitation, and disturbance scenarios in arid lands. In this regard, there is not likely to be a single expected outcome from these multifactor interactions.

19.2.2.3 Woody Plant Encroachment, Invasive Species, and Rising CO_2

How elevated CO_2 may influence shifts in plant community composition through changes in woody-plant encroachment and non-native species introductions may substantially alter the structure/function properties of semi-arid and arid ecosystems. Invasion by non-native species has the ability to shift patterns of nutrient availability (Evans et al. 2001) and disturbance (D'Antonio and Vitousek 1992) such that further dominance by the non-native is facilitated. The representation of woody versus nonwoody plants in the flora of water-limited regions is important for local, regional, and global biogeochemical and hydrological processes (Reynolds, Virginia, and Schlesinger 1997; Jackson et al. 2002). As a result, water-limited regions are classified based on total cover and functional diversity, which is related primarily to the presence of woody plant species (Smith, Monson, and Anderson 1997). In contracted desert regions, perennial cover is sparse, low in–stature, and restricted to washes and runon areas. True deserts are dominated by shrubs and dwarf shrubs with $<10\%$ total perennial cover. Semidesert shrublands (steppes) have approximately 10% to 30% perennial cover. Shrublands dominated by large shrubs (0.5–2 m high) have approximately 30% perennial cover. Perennial grasses dominate grasslands with fairly continuous cover. The relative change in the total abundance of woody species has, therefore, a large impact on the properties of the material and energy exchange patterns in the whole system.

Rising CO_2 may affect several plant and system characteristics that would favor increased woody plant species encroachment or the maintenance and increased vigor of currently established woody plant species. Rising CO_2 may differentially affect C_3 woody plant seedlings as compared to C_4 grasses, facilitating shrub establishment (Polley, Johnson, and Tischler 2003). Reduced transpiration rates from grasses may provide greater water recharge to depth and may facilitate coexistence of the two growth forms (Polley et al. 1997). Additionally, increased carbon balance during and following disturbances such as fire may facilitate woody plant dominance in a number of systems. Bond and Midgely (2000) suggest that woody plant seedlings will be able to reach heights critical for escaping the detrimental impacts of periodic fire, typical of many semi-arid savannas, much quicker as a result of increased growth rates and carbon balance associated with rising CO_2 (Fig. 19.1).

A similar argument for growth-rate induced changes in species composition has been made for non-native species success with rising CO_2 (Smith, Strain, and Sharkey 1987; Huxman, Hamerlynck, and Smith 1999; Ziska 2003). Rising CO_2 may allow fast-growing plant species to accumulate biomass at proportionally greater rates than slower-growing plant species (Poorter and Navas 2003). This would allow these fast-growing species to reach important life history stages (such as reproduction) quickly, increasing their competitive abilities and probabilities of success in novel habitats on a model that is conceptually similar to the shrub encroachment pattern (Fig. 19.1). Such life history shifts, as a result of changes in growth rate, are likely critical for understanding non-native species dynamics with global change (Dukes and Mooney 1999; Dukes and Field 2000). Empirical data for both changes in woody-plant encroachment and non-native species success are lacking but critical for understanding the structure of semi-

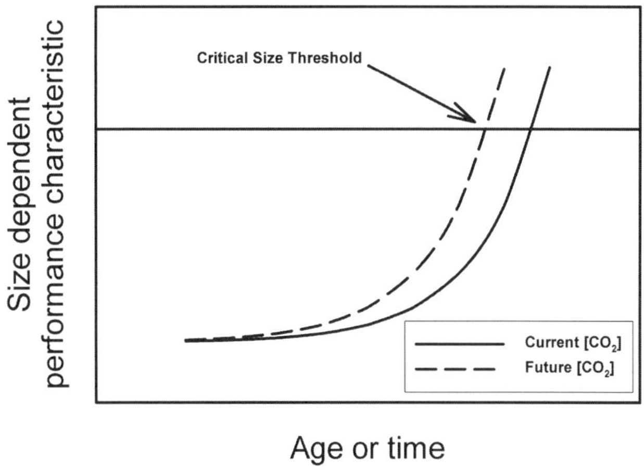

Figure 19.1. The relationship between size-related growth characteristics and age or time for two different conditions; ambient (solid line) and elevated CO_2 (dashed line). The horizontal line represents a threshold for important life history events or escape from the destructive nature of disturbance (adapted from Bond and Midgely 2000), which occurs at an earlier time or age in elevated CO_2 grown plants.

arid and arid ecosystems in a future high-CO_2 world. How interactions between woody species and non-native invasions (typically grasses) may play out for native species composition, productivity, and disturbance regimes is unknown but is likely to have impacts on ecological processes in arid lands.

Across a gradient of water limitation, disturbance by fire has differential probabilities associated with fuel load (flammable biomass), which is directly proportional to total production and the fraction of shrub and tree to grass life forms upon the landscape (Trabaud, Christensen, and Gill 1993). Woody plant encroachment and non-native species invasion in semi-arid and arid ecosystems may increase the probability and recurrence of fire (Torn and Fried 1992; Mayeux, Johnson, and Polley 1994). Shifts in species composition toward more fast-growing grass life forms may result in lower ignition requirements yet greater water balance in whole-plants and ecosystems may increase ignition heat requirements (Sage 1996). The ability for plants to quickly return significant biomass to the landscape following a disturbance event may reduce the time between events. In water-limited ecosystems where fire has a large impact on ecosystem structure and function, changes in fire frequency and intensity are very important. However, for more xeric ecosystems, where fire has not been present in recent history, it could now significantly alter the vegetative characteristics of the landscape such as species composition. How fire may alter the responsiveness of water-limited ecosystems to rising CO_2 and other global change factors is unknown, but an important implication of shifts in species composition (Smith and Huxman 2001).

19.2.3 Ecosystem Responses to Rising Atmospheric CO_2

19.2.3.1 Ecosystem Carbon and Nitrogen Cycling

Numerous papers describe plant or ecosystem responses to elevated CO_2 (Körner 2000), but the evidence of generally stimulatory responses from studies with individual plants (Kimball 1983) is mixed at the ecosystem scale, including examples of substantial net primary production increases as well as modest responses or no change (Mooney et al. 1999). In semi-arid ecosystems, aboveground biomass generally increases under elevated CO_2, but this response is highly site-dependent and variable within systems across years (see Mooney et al. 1999 for a detailed review). Belowground biomass appears to increase in conjunction with aboveground biomass, but data are limited and highly variable (Canadell et al. 1996). The response of grassland net primary production to elevated CO_2 is greatest when water is limiting (Volk, Niklaus, and Körner 2000), suggesting that indirect increases in water availability are more important than are the direct impact on rates of carbon fixation. Additionally, at the extreme end of xeric conditions, dry years and wet years have the opposite response, where CO_2 enhancements in production are seen only during extremely wet years (Smith et al. 2000).

Focusing on net primary production provides information only about the flow of carbon in and out of plants. For a comprehensive ecosystem-level understanding of carbon cycling dynamics in response to elevated CO_2, an evaluation of net ecosystem production, or net primary production minus carbon losses, is necessary. Net ecosystem production incorporates autotrophic activity along with heterotrophic use of carbon energy sources that represents an actual measure of total carbon flux from an ecosystem. This can be estimated from measurements of net ecosystem exchange of CO_2 by large-scale material and energy flux measurements. These summed for a year provide a more robust estimate of net ecosystem production. A number of studies have attempted to estimate ecosystem C uptake under elevated CO_2 using net ecosystem CO_2 exchange measurements (Nie et al. 1992; Oechel et al. 1994; Fredeen and Field 1995; Ham et al. 1995; Kimball et al. 1995; Drake et al. 1996; Stocker, Leadley, and Körner 1997; Diemer and Körner 1998; Fredeen, Koch, and Field 1998). Such studies frequently report increases in net ecosystem C uptake under elevated CO_2. However, very few studies have been conducted in arid or semi-arid ecosystems, and few have conducted measurements over a period of time that would be required to understand lag and interactive effects in whole-ecosystem variables. Furthermore, the majority of these studies have measured net ecosystem exchange using open dynamic chambers (but see Oechel et al. 1994), a technique subject to pressure anomalies that can lead to substantial overestimates of ecosystem C uptake (Lund et al. 1999).

In arid and semi-arid ecosystems, the dynamics of carbon and nitrogen cycling are closely regulated and coupled to inputs of precipitation (Schlesinger 1997). It may be expected that elevated CO_2 influences patterns of nitrogen cycling, and this effect occurs through linkages to changes in ecosystem carbon status. Elevated CO_2 could decrease nitrogen availability through production of litter with high C:N ratios and increases in labile carbon in the soil via root exudation. Each of these result in the immobilization of nitrogen (Hungate et al. 1999). Some experiments have documented the reduction in tissue N concentration as plants increase production without increased N demand, but data on this from water-limited regions is lacking. There is little evidence for decreases in decomposition rates, and therefore in N availability, as a result of changes in litter quality with elevated CO_2. Considering the differential environmental control over decomposition in arid lands, any change in nitrogen cycling that involves decomposition as a mechanism may be speculative.

A more plausible effect that elevated CO_2 may have on nitrogen cycling would occur through changes in soil moisture resulting from reduced evapotranspiration. Increases in soil moisture could stimulate microbial activity and lead to increase nitrogen mineralization when carbon sources are not limiting (Hungate et al. 1997). Elevated CO_2 may also stimulate N mineralization in the following ways: through increased in-root exudation of highly labile C compounds combined with this soil water mechanism (Hungate et al. 1997), or through increased carbon supply to nitrogen-fixing bacterial symbionts. However, large shifts in the magnitude of nitrogen available in arid-land ecosystems may not be expected from rising CO_2, as linkages between production and nitrogen cycling may maintain the systems within certain bounds defined by the relationship between evapotranspiration and ecosystem leaf area (Hungate et al. 1999). Only with very large shifts in biomass on a landscape or shifts in system water balance should large effects on nitrogen cycling be expected. This could occur if it were mediated through shifts in the composition of vegetation specifically involving a change in dominant life form that affected the magnitude and characteristics of evapotranspiration.

19.2.3.2 Ecosystem Water Budgets

Soil Moisture and Water Availability. In arid and semi-arid plant communities, decreases in stomatal conductance can facilitate a higher efficiency of water use at the ecosystem level and increases in the availability of soil water (Fig. 19.2, Field, Jackson, and Mooney 1995). Many studies conducted in water-limited systems have shown that, especially in dry years, elevated CO_2 decreases ecosystem evaportranspiration so that water-use efficiency is enhanced (Fredeen and Field 1995; Ham et al. 1995; Field et al. 1997; Grunzweig and Körner 2001). In arid and semi-arid systems, this indirect impact of elevated CO_2 can enhance whole-system productivity, depending on the timing and magnitude of the availability of the additional soil water resource (Lockwood 1999). The degree of system aridity may also be important, as increased growth and leaf area production may offset decreases in stomatal conductance at the whole-plant scale (Pataki et al. 2000).

If decreased leaf-level transpiration does lead to decreased evapotranspiration

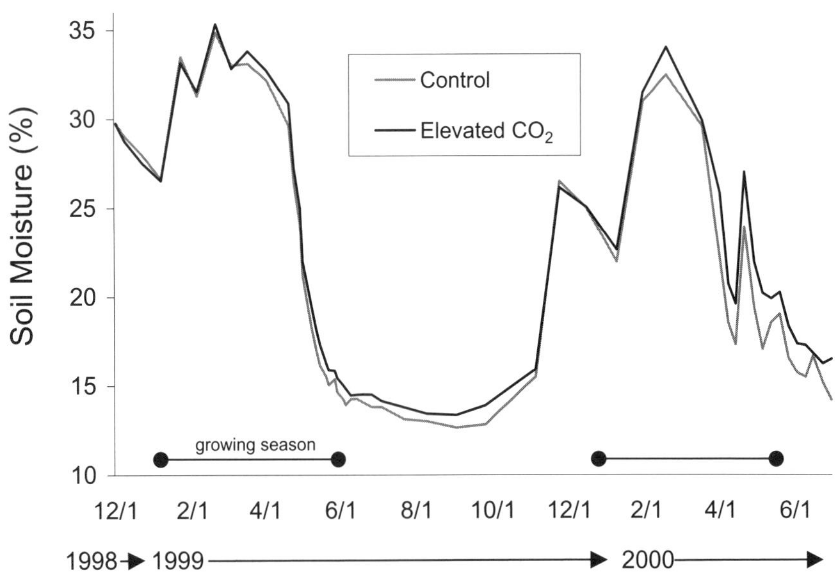

Figure 19.2. Soil moisture increases in late growing season in response to elevated CO_2.

under elevated CO_2, the effects on ecosystem primary production could be profound in water-limited ecosystems. A number of studies in semi-arid ecosystems have reported an increase in soil water content and leaf-water potential under elevated CO_2 (Jackson et al. 1994; Bremer, Ham, and Owensby 1996; Field et al. 1997; Niklaus, Spinnler, and Körner 1998; Owensby et al. 1999). Soil water content has been shown to remain high under elevated CO_2 conditions relative to control plots by as much as 18% in the tallgrass prairie (Owensby et al. 1999) and by 22% in a calcareous grassland (Niklaus, Spinnler, and Körner 1998). A similar pattern has also been reported for shortgrass steppe ecosystems (Morgan et al. 2001), but in the driest ecosystem, no significant differences in soil water content between elevated CO_2 and control plots has been detected in a Mohave Desert ecosystem (S. Smith, unpublished data).

Runoff and Deep Drainage. In many ecosystems, the effects on persistent residual soil moisture also have potential consequences for runoff, deep drainage, and nutrient leaching. Studies addressing the effects of elevated CO_2 on groundwater recharge and subsurface flow (collectively called drainage), runoff, and stream flow have largely been limited to modeling analyses (Aston 1984; Idso and Brazel 1984; Hatton et al. 1992; Jackson et al. 1998) and container experiments (Casella, Soussana, and Loiseau 1996). Modeling analyses predict an increase in both drainage and streamflow, but none of these studies have been adequately constrained by experimental data. In one of the few studies to make direct measurements, Casella, Soussana, and Loiseau (1996) found a 9% increase in drainage for perennial ryegrass swards grown in containers under elevated CO_2. In a modeling study constrained by experimental data, Lund (2001) also showed that elevated CO_2 increased annual drainage by 7% in a Mediterranean grassland. In general, increases in drainage will be greatest when large differences in evapotranspiration coincide with heavy precipitation to create a large pool of stored soil water that is not immediately lost to the atmosphere (Huxman et al. in review).

In semi-arid ecosystems where the leaf to air vapor pressure gradient is large, decreases in stomatal conductance under elevated CO_2 are associated with decreases in leaf-level transpiration (Pataki et al. 2000). Decreases in leaf-level transpiration can influence ecosystem-level evapotranspiration but only if canopy leaf area index does not increase proportionally. Several studies of semi-arid systems have examined the influence of elevated atmospheric concentrations of CO_2 on canopy transpiration and soil evaporation and have shown that decreased leaf-level transpiration can result in a 10% to 25% decrease in evapotranspiration (Nie et al. 1992; Ham et al. 1995; Stocker, Leadley, and Körner 1997). The relative reduction in evapotranspiration under elevated CO_2 is often less than the relative decrease in leaf-level transpiration for three reasons. First, soil evaporation and canopy interception of precipitation are not under stomatal control; yet these two factors are a significant proportion of ET in semi-arid systems (Huxman et al. in review). Second, evaporation from the soil may increase with an increase in soil water content as a result of decreased transpiration and increased soil hydraulic conductance. Third, a decrease in total leaf-level transpiration can be offset by a CO_2-driven increase in overall leaf area index, or, LAI (Gifford 1988; Nie et al. 1992; Schapendonk et al. 1997).

19.3 Plant and Ecosystem Responses to CO_2 in the Context of Multiple Global Change Factors

At this stage in our understanding, most of the experimental research on plant and ecosystem responses to global change in semi-arid and arid ecosystems have addressed responses to single global change factors, with relatively few studies exploring responses to two or more interacting treatments (Oechel et al. 1995; Oren et al. 2001). Yet with the information available from more mesic ecosystems, we can speculate on how future arid and semi-arid ecosystems might respond to those global factors that change indirectly as a consequence of a change in atmospheric CO_2.

19.3.1 Temperature

Concurrent with the predicted increase in CO_2, changes in the mean temperature of Earth's atmosphere of 1.5°C to 4°C are distinctly possible (Watson et al. 1990). This change in heat load on the planet may consist of both increases in mean growth temperatures and changes in the number and frequency of high temperature events (Wagner 1996)—both of which are important for understanding the overall impacts of climate change (Woodward and Williams 1987; Peters and Lovejoy 1992; Loik and Harte 1996; Roden and Ball 1996). While plant responses to a change in mean growing season temperature are much better understood, responses to high temperature extreme events that could interact with changes in CO_2 concentrations are less well known (Roden and Ball 1996; Sage 1996).

Rising CO_2 concentrations affect the processing of light energy through photochemical membranes (Sharkey 1985; Sage 1994), resulting in either a greater load of electron transport (Long 1991) or a decreased photochemical efficiency, leading to an increased susceptibility of photoinhibition (Roden and Ball 1996). In addition, stomatal conductance can decrease in elevated CO_2 (Knapp et al. 1996), potentially reducing the ability of the plant to reduce heat load (Sage 1996). Each of these factors suggest that while plants may perform greater at slightly higher growth temperature and elevated CO_2, the reduced ability to dissipate heat through nonphotochemical mechanisms may be exacerbated by the interaction between elevated CO_2 and an extreme temperature event. Results of studies evaluating these patterns are inconclusive but are important to continue, considering that species distributions in arid regions are limited by the ability of seedlings to survive the abiotic environment near the soil surface (Fenner 1985; Smith and Nowak 1990; Franklin et al. 1992). What little work has been done suggests that plants with high growth rates, or plants that experience temperature extremes concurrent with their growth season, may actually experience an increase in ability to tolerate high temperatures when exposed to elevated CO_2 (Huxman et al. 1998b; Taub, Seemann, and Coleman 2000). However, there is considerable variation in this response and the response to low temperature extremes, which suggests the potential for complex shifts in species boundaries in arid regions (Hamerlynck et al. 2000a; Loik et al. 2000; Dole, Loik, and Sloan 2003).

19.3.2 Precipitation

Anthropogenic emissions of greenhouse gases are expected to substantially influence hydrologic cycling, altering global and regional precipitation regimes (Houghton et al. 2001). We might see these alterations either through changes in the seasonality of precipitation (winter vs. summer) or in total precipitation amounts. Shifts in precipitation amounts in arid-land ecosystems have the potential to far exceed the impact of rising [CO_2] and temperature on terrestrial ecosystems, since primary productivity in arid and semi-arid regions is often a linear function of total precipitation. General Circulation Models (GCMs) predict mean increases in global precipitation of up to 7% during this century (Houghton et al. 2001). Precipitation intensity and the frequency of extreme events are predicted to increase throughout the globe, raising the potential for episodic events that might have long-term consequences to ecosystems (Easterling et al. 2000). Yet at this point, the scenarios for specific geographic regions remain ambiguous, with unresolved discrepancies between the outputs of different models (Weltzin et al. 2003).

For arid and semi-arid regions, the availability of water more than any other factor dominates recruitment, growth and reproduction, nutrient cycling, and net ecosystem productivity (Smith et al. 1997; Neilson and Drapek 1998; Weltzin and McPherson 2000). For example, predicted increases in summer precipitation might contribute to a substantial "greening" across wide areas of the arid Southwest, primarily by increasing the density and relative production of C_4 grasses (Neilson and Drapek 1998). It was initially suggested that under relatively high water availability, plants from arid and semi-arid regions would show a photosynthetic down-regulation and thereby limit the potential CO_2 stimulatory effect (Oechel et al. 1995; Huxman et al. 1998c). This is apparently not the case in the arid Mohave Desert ecosystem, where without water availability, elevated CO_2 has no effect because of the absence of plant activity (Smith et al. 2000). As a result, an enhancement in year-to-year variability in production, beyond the extreme values already seen, could result from the coupling of CO_2 effects on semi-arid and arid ecosystems and enhanced variability in year-to-year rainfall.

19.3.3 Interactions among Multiple Climate Change Factors

A few ecosystem studies address effects of elevated CO_2 in combination with other global changes at the ecosystem scale. Oechel et al. (1994) found that warming could extend the time over which elevated CO_2 stimulated net primary production in arctic tundra. Oren et al. (2001) observed a parallel result with N addition in a pine plantation. Reich et al. (2001) report increased sensitivity of net primary production to elevated CO_2 with added nitrogen but only when species diversity is high. Several studies from agricultural ecosystems also have shown that enhanced nitrogen availability can lead to larger increases in net primary production in response to elevated CO_2 (Idso and Idso 1994). In each of these studies involving nitrogen-limited ecosystems, the sensitivity of net primary production to elevated CO_2 increased with the addition of nitrogen or, in the tundra example, with a treatment that indirectly increased nitrogen availability.

Several modeling studies have addressed ecosystem responses to multifactor global changes (Schimel et al. 2000; McGuire et al. 2001), but the theoretical foundation for predicting ecosystem responses to simultaneous changes in multiple factors, including elevated concentration of CO_2 and temperature and alterations in the timing and rates of precipitation, N deposition, and species movements, is incomplete. For some processes (e.g., photosynthesis), well-tested mechanistic models support the simulation and interpretation of multifactor responses (Farquhar, von Caemmerer, and Berry 1980). For many others, however, including biomass allocation, the timing of seasonal activity, the competitive balance between C_3 and C_4 plants, and other species changes, the empirical data are too sparse to support credible models or to allow comprehensive hypothesis tests.

Both empirical and modeling studies highlight potential contrasts between responses to single global changes versus multiple, interacting global changes. Stimulation of plant growth by elevated CO_2, for example, may be strongest when water is limiting (Owensby et al. 1999), when nutrients are abundant (Kimball and Mauney 1993), or when plant species diversity is high (Reich et al. 2001). Ecosystem responses to future global changes depend strongly on the interactions. A more plausible explanation of ecosystem responses to realistic global changes depends on a thorough understanding of responses to multiple, simultaneous changes.

The results from the Jasper Ridge Global Change Experiment, which manipulated atmospheric CO_2, temperature, precipitation, and N deposition in a full-factorial manipulation in Mediterranean annual grassland, suggest a fundamentally different kind of ecosystem response than would be expected from the combination of responses to a single manipulated factor. Over the first two years of the manipulations and across all of the treatment combinations (Fig. 19.3), elevated CO_2 had no significant effect on net primary production. In the third year, 2001, the mean net primary production for all treatment combinations with elevated CO_2 was 988 ± 52 g m^{-2} versus 1089 ± 54 g m^{-2} for all treatment combinations with ambient CO_2 (Shaw et al. 2002). Indeed, in all treatment combinations involving elevated CO_2, elevated CO_2 suppressed the allocation of carbon to the roots, decreasing the positive effects of increased

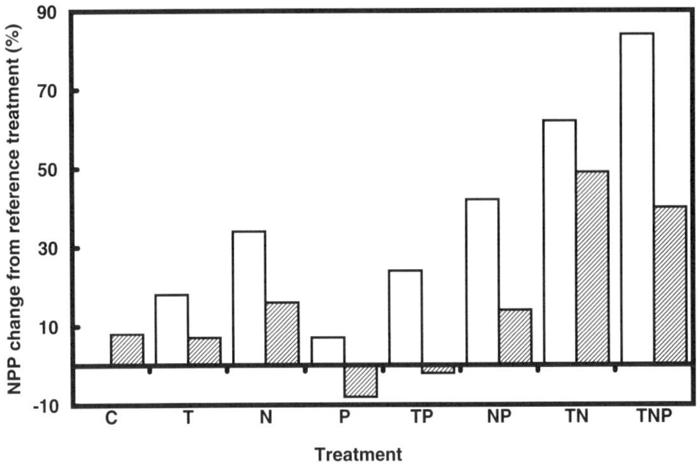

Figure 19.3. Percent changes in net primary production for each treatment, relative to corresponding reference. For each treatment combination, the reference is defined as all ambient CO_2 plots in which the variable(s) that defines each treatment combination is also at ambient (from Shaw et. al 2002). amb, ambient; C, elevated CO_2; N, N deposition; T, increased temperature; P, increased precipitation. For treatments T, N, and P, n=24 plots. For treatment TP, NP, and TN, n=12. For treatments C and TNP, n=6. For example, the open bar for the increased temperature pair (T) is calculated using all treatments with increased temperature but no elevated CO_2 (n=24). The gray bar in the same pair (T) is calculated using all treatments with increased temperature and elevated CO_2 (n=24).

temperature, precipitation, and N deposition on net primary production (Shaw et al. 2002).

19.4 Summary

We have a good understanding of ecological importance of CO_2 effects on growth, survival, and reproduction in plants from water-limited regions. We know that increases in water use-efficiency occur as a result of multiple character changes in many plants and that in some systems, these changes in water use can impact whole-system water balance. We additionally know that during periods that are not limited by other important resources, elevated CO_2 can lead to greater productivity on a landscape for water-limited systems. However, despite considerable effort being directed toward understanding plant and ecosystem responses to elevated CO_2 in semi-arid and arid ecosystems, we have limited knowledge of the expected pattern beyond the same range of biomes with experimental systems. This stands in contrast to the importance of these biome types in their rank cover of the globe. The fraction of the human population that is directly affected by processes such as desertification, decreased productivity of rangelands, and water yield of semi-arid landscapes is substantial, and greater attention to these systems is certainly warranted. Additionally, there is little understanding of how important modifiers of the CO_2 response may shape these future landscapes (Smith, Jordon, and Hamerlynck 1999). For example, in the southwestern United States shifts in the relative frequency of winter-versus-summer rainfall have been predicted, with either greater summer or winter precipitation possible. It has also been suggested that the magnitude of rainfall may shift along with interannual variability. Each of these climatic factors has been said to be critically important in shaping the structure and function of the arid regions, perhaps to a greater degree than rising CO_2.

Several concepts emerge relative to how elevated CO_2 will impact structure, composition, and process in water-limited systems. The covariation in life form with photosynthetic pathway may result in shifts in the relative abundances of C_3 versus C_4 taxa. Within semi-arid and arid ecosystems, the responses to elevated CO_2 may be a function of the magnitude and interannual variablility of precipitation (Fig 19.4). In semi-arid ecosystems, CO_2 effects in relatively wet years can result in limited overall impacts on the ecosystem. However, in extremely arid systems, the greatest effect occurs in wetter years (Smith et al. 2000). As a result, increasing CO_2 may have two general effects across a range of water-limited ecosystems: (1) *reducing* year-to-year variation in production for relatively mesic systems by disproportionately increasing production in dry rather than wet years, and (2) *increasing* year-to-year variation in more xeric systems, where production

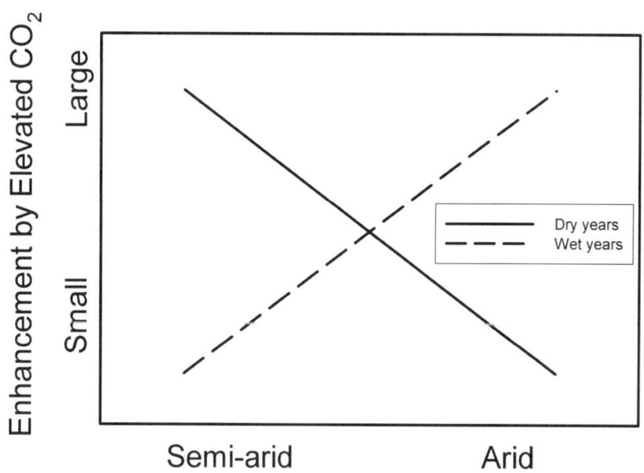

Figure 19.4. The conceptualized effects of elevated CO_2 on plant and ecosystem function across a gradient of water availability for semi-arid and arid ecosystem types. The dashed line represents relatively wet years (annual hydrologic year precipitation), and the solid line is for relatively dry years.

is greatly enhanced in wet years and not affected in dry years. That is, the characteristics of water limitation may diminish in relatively mesic ecosystems and become more pronounced in relatively xeric ecosystems. It is possible that elevated CO_2 may make some water-limited regions more arid rather than less.

References

Arp, W.J. 1991. Effects of source-sink relations on photosynthetic acclimation to elevated CO_2. *Plant, Cell, and Environment* 14:869–76.

Aston, A.R. 1984. The effect of doubling atmospheric CO_2 on streamflow: A simulation. *Journal of Hydrology* 67:273–280.

Bazzaz, F.A. 1990. The response of natural ecosystems to the rising global CO_2 levels. *Annual Review of Ecology and Systematics* 21:167–96.

———. 1997. Allocation of resources in plants: State of the science and critical questions. In *Plant resource allocation*, ed. F.A. Bazzaz and J. Grace, 1–37. San Diego: Academic Press.

Bazzaz, F.A., J.S. Coleman, and S.R. Morse. 1990. Growth responses of seven major cooccuring tree species of the northeastern United States to elevated CO_2. *Canadian Journal of Forest Research* 19:1479–84.

Beatley, J.C. 1974. Phenological events and their environmental triggers in Mojave-desert ecosystems. *Ecology* 55:856–63.

Beerling, D.J., J.C. McElwain, and C.P. Osborne. 1998. Stomatal responses of the 'living fossil' *Ginkgo biloba* L. to changes in atmospheric CO_2 concentrations. *Journal of Experimental Botany* 49:1603–1607.

Berry, J., and O. Björkman. 1980. Photosynthetic response and adaptation to temperature in higher plants. *Annual Review of Plant Physiology* 31:491–543.

Betterini, I., F.P. Vaccari, and F. Miglietta. 1998. Elevated CO_2 concetrations and stomatal density: observations from 17 plant species growing in a CO_2 spring in Central Italy. *Global Change Biology* 4:17–22.

Bloom, A.J., F.S. Chapin, and H.A. Mooney. 1985. Resource limitation in plants: an economic analogy. *Annual Review of Ecology and Systematics* 16:363–92.

Bond, W.J. and G.F. Midgley. 2000. A proposed CO2-controlled mechanism of woody plant invasion in grasslands and savannas. *Global Change Biology* 6:865–70.

Bowes, G. 1991. Growth at elevated CO_2: photosynthetic responses mediated through RUBISCO: Commissioned review. *Plant, Cell, and Environment* 14:795–806.

———. 1993. Facing the inevitable: plants and increasing atmospheric CO_2. *Annual Review of Plant Physiology and Molecular Biology* 44:309–22.

Bremer, D.J., J.M. Ham, and C.E. Owensby. 1996. Effect of elevated atmospheric carbon dioxide and opten top chambers on transpiration in a tallgrass prairie. *Journal of Environmental Quality* 25:691–701.

Bunce, J.A. 1996. Short Communication: Growth at elevated carbon dioxide concentration reduces hydraulic conductance in alfalfa and soybean. *Global Change Biology* 2: 155–58.

Canadell, J., R.B. Jackson, J.R. Ehleringer, H.A. Mooney, O.E. Sala, and E.-D. Schulze, 1996. Maximum rooting depth of vegetation types at the global scale. *Oecologia* 108: 583–94.

Carlson, R.W., and F.A. Bazzaz. 1980. The effects of elevated CO_2 concentrations on growth, photosynthesis, transpiration, and water use efficiency of plants. In *Environmental and climatic impact of coal utilization*, ed. J. Singh and A. Deepak, 609–23. New York: Academic Press.

Casella, E., J.F. Soussana, and P. Loiseau. 1996. Long-term effects of CO_2 enrichment and temperature increase on a temperate grass sward: 1. Productivity and water use. *Plant and Soil* 182:83–99.

Chaves, M., and J. Pereira. 1992. Water-stress, CO_2 and climate change. *Journal of Experimental Botany* 43:1131–39.
Clifford, S.C., C.R. Black, J.A. Roberts, I.M. Stronach, P.R. Singleton-Jones, A.D. Mohamed, and S.N. Azamali 1995. The effect of elevated atmospheric CO_2 and drought on stomatal frequency in groundnut (*Arachis hypogaea* L.). *Journal of Experimental Botany* 46:847–52.
Cody, M.L. 1986. Spacing patterns in Mojave Desert plant communities: Near-neighbor analyses. *Journal of Arid Environments* 11:199–217.
Cowan, I.R., and G.D. Farquhar. 1977. Stomatal function in relation to leaf metabolism and environment. *Symposium for the Society of Experimental Biology* 31:471–505.
Cure, J.D., and B. Acock. 1986. Crop responses to CO_2 doubling: A literature survey. *Agricultural and Forest Meteorology* 38:127–45.
Curtis, P.A., and X. Wang. 1998. A meta-analysis of elevated CO_2 effects on woody plant mass, form, and physiology. *Oecologia* 113:299–313.
D'Antonio, C.M., and P.M. Vitousek. 1992. Biological invasions by exotic grasses, the grass/fire cycle, and global change. *Annual Review of Ecology and Systematics* 23: 63–87.
DeLucia, E.H., T.W. Sasek, and B.R. Strain. 1985. Photosynthetic inhibition after long-term exposure to elevated levels of atmospheric carbon dioxide. *Photosynthesis Research* 7:175–84.
Diemer, M., and C. Körner 1998. Transient enhancement of carbon uptake in an alpine grassland ecosystem under elevated CO_2. *Arctic and Alpine Research* 30: 381–87.
Dole, K., M. Loik, and L. Sloan. 2003. The relative importance of climate change and the physiological effects of CO_2 on freezing tolerance for the future distribution of Yucca brevifolia. *Global and Planetary Change* 36:137–46.
Drake, B.G., G. Peresta, E. Beugeling, and R. Matamala. 1996. Long-term elevated CO_2 exposure in a Chesapeake Bay wetland: Ecosystem gas exchange, primary production, and tissue nitrogen. In *Carbon dioxide and terrestrial ecosystems*, ed. G.W. Koch and H.A. Mooney, 197–214. San Diego: Academic Press.
Dukes, J., and C. Field. 2000. Diverse mechanisms for CO_2 effects on grassland litter decomposition. *Global Change Biology* 6:145–54.
Dukes, J.S., and H.A. Mooney. 1999. Does global change increase the success of biological invaders? *Trends in Ecology and Evolution* 14:135–39.
Easterling, D.R., G.A. Meehl, C. Parmesan, S.A. Changnon, T.R. Karl, and L.O. Mearns. 2000. Climate extremes: Observations, modeling, and impacts. *Science* 289:2068–74.
Ehleringer, J.R. 1985. Annuals and perennials of warm deserts. In *Physiological ecology of North American plant communities*, ed. B.F. Chabot and H.A. Mooney. New York: Chapman and Hall.
———. 2001. Productivity of Deserts. In *Terrestrial global productivity*, ed. J. Roy, B. Saugier, and H.A. Mooney. New York: Academic Press.
Ehleringer, J.R., T.E. Cerling, and B.R. Helliker. 1997. C_4 photosynthesis, atmospheric CO_2 and climate. *Oecologia* 112:285–99.
Ellsworth, D.S. 1999. CO_2 enrichment in a maturing pine forest: Are CO_2 exchange and water status in the canopy affected? *Plant, Cell, and Environment* 22:461–72.
Evans, R., R. Rimer, L. Sperry, and J. Belnap. 2001. Exotic plant invasion alters nitrogen dynamics in an arid grassland. *Ecological Applications* 11:1301–10.
Farquhar, G.D., S. von Caemmerer, and J.A. Berry. 1980. A biochemical model of photosynthetic CO_2 assimilation in leaves of C_3 species. *Planta* 149:78–90.
Fenner, M. 1985. *Seed ecology*. New York: Chapman and Hall.
Field, C.B., R.B. Jackson, and H.A. Mooney. 1995. Stomatal responses to increased CO_2: Implications form the plant to the global scale. *Plant, Cell, and Environment* 18:1214–26.
Field, C.B., C.P. Lund, N.R. Chiariello, and B.E. Mortimer. 1997. CO_2 effects on the

water budget of grassland microcosm communities. *Global Change Biology* 3:197–206.

Franklin, J.F., F. Swanson, M. Harmon, D. Perry, T. Spies, V. Dale, A. McKee, W. Ferrel, J.E. Means, S.V. Greogory, J. Lattin, T.D. Schowalter, and D. Larson. 1992. Effects of global climate change on forests in northwestern North America. In *Global warming and biological diversity*, ed. R.L. Peters and T.E. Lovejoy. New Haven: Yale University Press.

Fredeen, A.L., and C.B. Field. 1995. Contrasting leaf and 'ecosystem' CO_2 and H_2O exchange in *Avena fatua* monoculture: Growth at ambient and elevated CO_2. *Photosynthesis Research* 43:263–71.

Fredeen, A.L., G.W. Koch, and C.B. Field. 1998. Influence of fertilization and atmospheric CO_2 enrichment on ecosystem CO_2 and H_2O exchanges in single- and multiple-species grassland microcosms. *Environmental and Experimental Botany* 40: 147–57.

Ghannoum, O., S. Von Caemmerer, L. Ziska, and J. Conroy. 2000. The growth response of C_4 plants to rising atmospheric CO_2 partial pressure: A reassessment. *Plant, Cell, and Environment* 23:931–42.

Gifford, R.M. 1988. Direct effects of higher carbon dioxide concentrations on vegetation. *Greenhouse: Planning for climate change,* ed. G.I. Pearman, 506–19. Melbourne: CSIRO Publications.

Grunzweig, J.M., and C. Körner. 2001. Growth, water and nitrogen relations in grassland model ecosystems of the semi-arid Negev of Israel exposed to elevated CO_2. *Oecologia* 128:251–62.

Ham, J.M., C.E. Owensby, P.I. Coyne, and D.J. Bremer. 1995. Fluxes of CO_2 and water vapor from a prairie ecosystem exposed to ambient and elevated atmospheric CO_2. *Agricultural and Forest Meteorology* 77:73–93.

Hamerlynck, E., T. Huxman, T. Charlet, and S. Smith. 2002. Effects of elevated CO_2 (FACE) on the functional ecology of the drought-deciduous Mojave Desert shrub, *Lycium andersonii. Environmental and Experimental Botany* 48:93–106.

Hamerlynck, E.P., T.E. Huxman, M.E. Loik, and S.D. Smith. 2000a. Effects of extreme high temperature, drought and elevated CO_2 on photosynthesis of the Mojave Desert evergreen shrub, *Larrea tridentata. Plant Ecology* 148:183–93.

Hamerlynck, E.P., T.E. Huxman, R.S. Nowak, S. Redar, M.E. Loik, D.N. Jordan, S.F. Zitzer, J.S. Coleman, J.R. Seeman, and S.D. Smith. 2000b. Photosythetic responses of *Larrea tridentata* to a step-increase in atmospheric CO_2 at the Nevada Desert FACE facility. *Journal of Arid Environments* 44:425–36.

Hatch, M.D. 1992. C_4 Photosynthesis: An unlikely process full of surprises. *Plant and Cell Physiology* 33:333–342.

Hatton, T.J., J. Walker, W.R. Dawes, and F.X. Dunin. 1992. Simulations of hydroecological responses to elevated CO_2 at the catchment scale. *Australian Journal of Botany* 40:679–696.

Houghton, J.T., Y. Ding, D.J. Griggs, M. Noguer, P.J. van der Linden, and D. Xiaosu, eds. 2001. *IPCC 2001: The Scientific Basis.* Contribution of Working Group I to the Third Assessment Report of the Intergovernmental Panel on Climate Change. Cambridge: Cambridge University Press.

Housman, D.C., S.F. Zitzer, T.E. Huxman, and S.D. Smith. 2003. Functional ecology of shrub seedlings after a natural recruitment event at the Nevada Desert FACE Facility. *Global Change Biology* 9:718–28.

Hungate, B.A., P. Dijkstra, D.W. Johnson, C.R. Hinkle, and B.G. Drake. 1999. Elevated CO_2 increases nitrogen fixation and decreases soil nitrogen mineralization in Florida scrub oak. *Global Change Biology* 5:781–89.

Hungate, B.A., C.P. Lund, H.L. Pearson, and F.S. Chapin III. 1997. Elevated CO_2 and nutrient addition alter soil N cycling and trace gas fluxes with early season wet-up in a California annual grassland. *Biogeochemistry* 37:89–109.

Huxman, K.A. 1999. The importance of root function in the CO_2 response to drought. *Master of Science*. Las Vegas: University of Nevada.
Huxman, K.A., S.D. Smith, and D.S. Neuman. 1999. Root hydraulic conductivity of *Larrea tridentata* and *Helianthus annuus* under elevated CO_2. *Plant, Cell, and Environment* 22:325–30.
Huxman, T., E. Hamerlynck, D. Jordan, K. Salsman, and S. Smith. 1998a. The effects of parental CO_2 environment on seed quality and subsequent seedling performance in *Bromus rubens*. *Oecologia* 114:202–208.
Huxman, T., E. Hamerlynck, M. Loik, and S. Smith. 1998b. Gas exchange and chlorophyll fluorescence responses of three south-western Yucca species to elevated CO_2 and high temperature. *Plant, Cell, and Environment* 21:1275–83.
Huxman, T., E. Hamerlynck, and S. Smith. 1999. Reproductive allocation and seed production in *Bromus madritensis* ssp *rubens* at elevated atmospheric CO_2. *Functional Ecology* 13:769–77.
Huxman, T.E., E.P. Hamerlynck, B.D. Moore, S.D. Smith, D.N. Jordan, S.F. Zitzer, R.S. Nowak, J.S. Coleman, and J.R. Seemann. 1998c. Photosynthetic down-regulation in *Larrea tridentata* exposed to elevated atmospheric CO_2: Interaction with grought under glasshouse and field (FACE) exposure. *Plant, Cell, and Environment* 21:1153–61.
Huxman, T.E., and S.D. Smith. 2001. Photosynthesis in an invasive grass and native forb at elevated CO_2 during an El Niño year in the Mojave Desert. *Oecologia* 128:193–201.
Huxman, T.E., B.P. Wilcox, R.L. Scott, K. Snyder, K. Hultine, E. Small, D. Breshears, W. Pockman, and Jackson. Ecohydrological implications of woody plant encroachment. *Ecology.* In review.
Idso, S.B., and A.J. Brazel. 1984. Rising atmospheric carbon dioxide may increase streamflow. *Nature* 312:51–53.
Idso, K.E., and S.B. Idso. 1994. Plant responses to atmospheric CO_2 enrichment in the face of environmental constraints: A review of the past 10 years' research. *Agricultural and Forest Meteorology* 69:153–202.
Jablonski, L., X. Wang, and P. Curtis. 2002. Plant reproduction under elevated CO_2 conditions: A meta-analysis of reports on 79 crop and wild species. *New Phytologist* 156: 9–26.
Jackson, R., O. Sala, J. Paruelo, and H. Mooney. 1998. Ecosystem water fluxes for two grasslands in elevated CO_2: a modeling analysis. *Oecologia* 113:537–46.
Jackson, R.B., J.L. Banner, E.G. Jobbagy, W.T. Pockman, and D.H. Wall. 2002. Ecosystem carbon loss with woody plant invasion of grasslands. *Nature* 418:623–26.
Jackson, R.B., Y. Luo, Z.G. Cardon, O.E. Sala, C.B. Field, and H.A. Mooney. 1995. Photosynthesis, growth, and density for the dominant species in a CO_2-enriched grassland. *Journal of Biogeography* 22:1225–29.
Jackson, R.B., O.E. Sala, C.B. Field, and H.A. Mooney 1994. CO_2 alters water use, carbon gain, and yield in a natural grassland. *Oecologia* 98:257–62.
Ketellapper, H. 1963. Stomatal Physiology. *Annual Review of Plant Physiology and Plant Molecular Biology* 14:249–270.
Kimball, B.A. 1983. Carbon dioxide and agricultural yield: An assemblage and analysis of 430 prior observations. *Agronomy Journal* 75:779–88.
Kimball, B.A., and J.R. Mauney. 1993. Response of cotton to varying carbon dioxide, irrigation, and nitrogen: Yield and growth. *Agronomy Journal* 85:706–12.
Kimball, B.A., P.J.J. Pinter, R.L. Garcia, R.L. Lamorte, G.W. Wall, D.J. Hunsaker, G. Wechsung, F. Wechsung, and T. Kartschall. 1995. Productivity and water use of wheat under free-air CO_2 enrichment. *Global Change Biology* 1:429–43.
Knapp, A.K., E.P. Hamerlynck, J.M. Ham, and C.E. Owensby 1996. Responses in stomatal conductance to elevated CO_2 in 12 grassland species that differ in growth form. *Vegetatio* 125:31–41.
Knapp, A.K., E.P. Hamerlynck, and C.E. Owensby. 1993. Photosynthetic and water re-

lations responses to elevated CO_2 in the C_4 grass *Andropogon gerardii. International Journal of Plant Sciences* 154:459–66.

Körner, C. 2000. Biosphere responses to CO_2 enrichment. *Ecological Applications* 10: 1590–1619.

Körner, C., and F. Miglietta. 1994. Long term effects of naturally elevated CO_2 on Mediterranean grassland and forest trees. *Oecologia* 99:343–51.

Linsbauer, K. 1917. Beitrage zur Kenntnis der Spaltoffnungsbewegung. *Flora* 9:100–143.

Lockwood, J.G. 1999. Is potential evapotranspiration and its relationship with actual evapotranspiration sensitive to elevated atmospheric CO_2 levels? *Climatic Change* 41: 193–212.

Loik, M., T. Huxman, E. Hamerlynck, and S. Smith. 2000. Low temperature tolerance and cold acclimation for seedlings of three Mojave Desert Yucca species exposed to elevated CO_2. *Journal of Arid Environments* 46:43–56.

Loik, M.E., and J. Harte. 1996. High-temperature tolerance of *Artemisia tridentata* and *Potentilla gracilis* under a climate change manipulations. *Oecologia* 108:224–31.

Long, S.P. 1991. Modification of the response of photosynthetic productivity to rising temperature by atmospheric CO_2 concentrations: Has its importance been underestimated? Opinion. *Plant, Cell, and Environment* 14:729–40.

Lund, C.P. 2001. *Ecosystem carbon and water budgets under elevated atmospheric carbon dioxide concentration in two California grasslands.* Stanford: Stanford University.

Lund, C.P., W.J. Riley, L.L. Pierce, and C.B. Field. 1999. The effects of chamber pressurization on soil-surface CO_2 flux and implications for NEE measurements under elevated CO_2. *Global Change Biology* 5:269–82.

Mayeux, H.S., H.B. Johnson, and H.W. Polley. 1994. Potential interactions between global change and intermountain annual grasslands. In *Ecology and management of annual rangelands,* ed. S.B. Monsen and S.G. Kitchen. United States Forest Service General Technical Report INT-GTR-313. Ogden, Utah: Intermountain Research Station.

McGuire, A.D., S. Sitch, J.S. Clein, R. Dargaville, G. Esser, J. Foley, M. Heimann, F. Joos, J. Kaplan, D.W. Kicklighter, and R.A. Meier. 2001. Carbon balance of the terrestrial biosphere in the twentieth century: Analyses of CO_2, climate and land use effects with four process-based ecosystem models. *Global Biogeochemical Cycles* 15: 183–206.

Melillo, J.M., D.W. Kicklighter, A.D. McGuire, B. Moore, III, C.J. Vorosmarty, and A.L. Grace. 1993. Global climate change and terrestrial net primary production. *Nature* 363:234–40.

Mooney, H.A., J. Canadell, F.S. Chapin, III, J. Ehleringer, C. Körner, R. McMurtrie, W.J. Parton, L. Pitelka, and E.-D. Schulze. 1999. Ecosystem physiology responses to global change. In *The terrestrial biosphere and global change: Implications for natural and managed ecosystems,* B.H. Walker, W.L. Steffen, J. Canadell, and J.S.I. Ingram, 141–89. Cambridge: Cambridge University Press.

Mooney, H.A., B.G. Drake, R.J. Luxmoore, W.C. Oechel, and L.F. Pitelka. 1991. Predicting ecosystem responses to elevated CO_2 concentrations. *Bioscience* 41:96–104.

Morgan, J.A., D.R. LeCain, A.R. Mosier, and D.G. Milchunas. 2001. Elevated CO_2 enhances water relations and productivity and affects gas-exchange in C_3 and C_4 grasses of the Colorado shortgrass steppe. *Global Change Biology* 7:451–66.

Morison, J.I.L. 1985. Sensitivity of stomata and water use efficiency to high CO_2. *Plant, Cell, and Environment* 8:467–74.

Mulroy, T.W., and P.W. Rundel. 1977. Annual plants: Adaptations to desert environments. *Bioscience* 27:109–14.

Naumburg, E., D. Housman, T. Huxman, T. Charlet, M. Loik, and S. Smith. 2003. Photosynthetic responses of Mojave Desert shrubs to free air CO_2 enrichment are greatest during wet years. *Global Change Biology* 9:276–85.

Neilson, R.P., and R.J. Drapek. 1998. Potentially complex biosphere responses to global warming. *Global Change Biology* 4:505–22.
Nie, D., H. He, G. Mo, M.B. Kirkham, and E.T. Kanemasu 1992. Canopy photosynthesis and evapotranspiration of rangeland plants under doubled carbon dioxide in closed-top chambers. *Agricultural and Forest Meteorology* 61:205–17.
Nijs, I., R. Ferris, H. Blum, G. Hendry, and I. Impens. 1997. Stomatal regulation in a changing climate: a field study using Free Air Temperature Increase (FATI) and Free Air CO_2 Enrichment (FACE). *Plant Cell and Environment* 20:1041–50.
Niklaus, P.A., D. Spinnler, and C. Körner. 1998. Soil moisture dynamics of calcareous grassland under elevated CO_2. *Oecologia* 117:201–208.
Nowak, R., L. DeFalco, C. Wilcox, D. Jordan, J. Coleman, J. Seemann, and S. Smith. 2001. Leaf conductance decreased under free-air CO_2 enrichment (FACE) for three perennials in the Nevada desert. *New Phytologist* 150:449–58.
Oechel, W.C., S. Cowles, N. Grulke, S.J. Hastings, B. Lawrence, T. Prudhomme, G. Riechers, B. Strain, D. Tissue, and G. Vourlitis. 1994. Transient nature of CO_2 fertilization in Arctic tundra. *Nature* 371:500–503.
Oechel, W.C., S.J. Hastings, G.L. Vourlitis, M.A. Jenkins, and C.L. Hinkson. 1995. Direct effects of elevated CO_2 in Chaparral and Mediterranean-Type ecosystems. In *Global Change and Mediterranean-Type Ecosystems,* ed. J.L. Moreno and W.C. Oechel, 58–75. New York: Springer-Verlag.
Oren, R., D.S. Ellsworth, K.H. Johnsen, N. Phillips, B.E. Ewers, C. Maier, K.V.R. Schafer, H. McCarthy, G. Hendrey, S.G. McNulty, and G.G. Katul. 2001. Soil fertility limits carbon sequestration by forest ecosystems in a CO_2-enriched atmosphere. *Nature* 411:469–72.
Owensby, C.E., P.I. Coyne, J.M. Ham, L.M. Auen, and A.K. Knapp. 1993. Biomass production in a tallgrass prairie ecosystem exposed to ambient and elevated CO_2. *Ecological Applications* 3:644–53.
Owensby, C.E., J.M. Ham, A.K. Knapp, and L.M. Auen. 1999. Biomass production and species composition change in a tallgrass prairie ecosystem after long-term exposure to elevated atmospheric CO_2. *Global Change Biology* 5:497–506.
Pataki, D., T. Huxman, D. Jordan, S. Zitzer, J. Coleman, S. Smith, R. Nowak, and J. Seemann. 2000. Water use of two Mojave Desert shrubs under elevated CO_2. *Global Change Biology* 6:889–97.
Peters, R.L., and T.E. Lovejoy, eds. 1992. *Global warming and biological diversity.* New Haven: Yale University Press.
Polley, H., H. Johnson, and J. Derner. 2002. Soil- and plant-water dynamics in a C_3/C_4 grassland exposed to a subambient to superambient CO_2 gradient. *Global Change Biology* 8:1118–29.
Polley, H., H. Johnson, and C. Tischler. 2003. Woody invasion of grasslands: Evidence that CO_2 enrichment indirectly promotes establishment of *Prosopis glandulosa*. *Plant Ecology* 164:85–94.
Polley, H., H. Mayeux, H. Johnson, and C. Tischler. 1997. Viewpoint: Atmospheric CO_2, soil water, and shrub/grass ratios on rangelands. *Journal of Range Management* 50: 278–84.
Poorter, H., M. and Navas. 2003. Plant growth and competition at elevated CO_2: On winners, losers and functional groups. *New Phytologist* 157:175–98.
Poorter, H., C. Roumet, and B.D. Campbell. 1996. Interspecific variation in the growth response of plants to elevated CO_2: A search for functional types. In *Carbon, dioxide, populations, and communities,* ed. C. Körner and F.A. Bazzaz, 375–412. San Diego: Academic Press.
Reich, P.B., J. Knop, D. Tilman, J. Craine, D. Ellsworth, M. Tjoelker, T. Lee, D. Wedin, S. Naeem, D. Bahauddin, G. Hendrey, S. Jose, K. Wrage, J. Goth, and W. Bengston. 2001. Plant diversity enhances ecosystem responses to elevated CO_2 and nitrogen deposition. *Nature* 410:809–12.

Reynolds, J.F., D.W. Hilbert, and P.R. Kemp 1993. Scaling ecophysiology from the plant to the ecosystem: A conceptual framework. In *Scaling physiological processes: Leaf to globe,* ed. J.R. Ehleringer and C.B. Field, 127–40. San Diego: Academic Press.
Reynolds, J.F., R.A. Viginia, and W.H. Schlesinger. 1997. Defining functional types for models of desertification. In *Plant functional types: Their relevance to ecosystem properties and global change,* ed. T.M. Smith, H.H. Shugart, and F.I. Woodward III. Cambridge: Cambridge University Press.
Roden, J.S., and M.C. Ball. 1996. The effect of elevated [CO_2] on growth and photosynthesis of two Eucalyptus species exposed to high temperatures and water deficits. *Plant Physiology* 111:909–19.
Sage, R.F. 1994. Acclimation of photosynthesis to increasing atmoshperic CO_2: The gas enchange perspective. *Photosynthesis Research* 39:351–68.
Sage, R.F., and T.D. Sharkey. 1987. The effect of temperature on the occurrence of O_2 and CO_2 insensitive photosynthesis in field grown plants. *Plant Physiology* 84:658–64.
Sage, R.F., T.D. Sharkey, and J.R. Seeman. 1989. Acclimation of photosynthesis to elevated CO_2 in five C_3 species. *Plant Physiology* 89:590–96.
Sage, R.W. 1996. Modication of fire disturbance by elevated CO_2. In *Carbon, dioxide, populations, and communities,* ed. C. Körner and F.A. Bazzaz, 231–49. New York: Academic Press.
Saxe, H.D.S.E., and J. Heath. 1998. Tree and forest functioning in an enriched CO_2 atmosphere. *New Phytologist* 139:395–436.
Schapendonk, A.H.C.M., P. Dijkstra, J. Groenwald, C.S. Pot, and S.C. van de Geijn. 1997. Carbon balance and water use efficiency of frequently cut *Lolium perenne* L. swards at elevated carbon dioxide. *Global Change Biology* 3:207–17.
Schimel, D., J. Melillo, H. Tian, A.D. McGuire, D. Kicklighter, T. Kittel, N. Rosenbloom, S. Running, P. Thornton, D. Ojima, W. Parton, R. Kelly, M. Sykes, R. Neilson, and B. Rizzo. 2000. Contribution of increasing CO_2 and climate to carbon storage by ecosystems in the United States. *Science* 287:2004–2006.
Schlesinger, W.H. 1997. *Biogeochemistry: An analysis of global change.* New York: Academic Press.
Sharkey, T.D. 1985. Photosynthesis in intact leaves of C_3 plants: Physics, physiology, and rate limitations. *Botanical Review* 51:53–105.
Shaw, M.R., E.S. Zavaleta, N.R. Chiariello, E.E. Cleland, H.A. Mooney, and C.B. Field. 2002. Grassland responses to global environmental changes suppressed by elevated CO_2. *Science* 298:1987–90.
Smith, S.D., and T.E. Huxman. 2001. Elevated atmospheric CO_2 and deserts: Will increasing CO_2 alter deserts and desertification processes? *Arid Lands Newsletter,* no. 49.
Smith, S.D., T.E. Huxman, S.F. Zitzer, T.N. Charlet, D.C. Housman, J.S. Coleman, L.K. Fenstermaker, J.R. Seemann, and R.S. Nowak. 2000. Elevated CO_2 increases productivity and invasive species success in an arid ecosystem. *Nature* 408:79–82.
Smith, S.D., D.N. Jordan, and E.P. Hamerlynck. 1999. Effects of elevated CO_2 and temperature stress on ecosystem processes. In *Carbon dioxide and environmental stress,* ed. Y. Luo and H.A. Mooney. New York: Academic Press.
Smith, S.D., R.K. Monson, and J.E. Anderson. 1997. *Physiological ecology of North American desert plants.* Berlin: Springer-Verlag.
Smith, S.D., and R.S. Nowak. 1990. Physiological ecology of plants in the Intermountain lowlands. In *Plant biology of the basin and range,* ed. C.B. Osmond, L.F. Pitelka, and G. Hidy, 181–241. Berlin: Springer-Verlag.
Smith, S.D., B.R. Strain, and T.D. Sharkey. 1987. Effects of CO_2 enrichment on four Great Basin grasses. *Functional Ecology* 1:139–43.
Stocker, R., P.W. Leadley, and C. Körner. 1997. Carbon and water fluxes in a calcareous grassland under elevated CO_2. Functional Ecology 11:222–31.

Strain, B.R. 1992. Field measurements of CO_2 enhancement and change in natural vegetation. *Water, Air, and Soil Pollution* 64:45–60.
Strain, B.R., and F.A. Bazzaz. 1983. Terrestrial plant communities. In *CO_2 and plants: The response of plants to rising levels of carbon dioxide,* ed. E. Lemon, 177–222. Washington, D.C.: American Association for the Advancement of Science.
Stulen, I., and J. Denhertog. 1993. Root-growth and functioning under atmospheric CO_2 enrichment. *Vegetatio* 104:99–115.
Taub, D.R., J.R. Seemann, and J.S. Coleman. 2000. Growth in elevated CO_2 protects photosynthesis against high-temperature damage. *Plant, Cell, and Environment* 23: 649–56.
Tilman, D., J. Knops, D. Wedin, P. Reich, M. Richie, and E. Siemann. 1997. The influence of functional diversity and composition on ecosystem processes. *Science* 277: 1300–1302.
Tissue, D.T., and W.C. Oechel. 1987. Response of *Eriophorum vaginatum* to elevated CO_2 and temperature in the Alaskan tussock tundra. *Ecology* 68:401–10.
Tissue, D.T., R.B. Thomas, and B.R. Strain. 1993. Long-term effects of elevated CO_2 and nutrients on photosynthesis and Rubisco in loblolly pine seedlings. *Plant Cell and Environment* 16:859–65.
Torn, M.S., and J.S. Fried. 1992. Predicting the impacts of global warming on wildland fire. *Climatic Change* 21:257–74.
Trabaud, L.V., N.L. Christensen, and A.M. Gill. 1993. Historical biogeography of fire in temperate and Mediterranean ecosystems. In *Fire in the environment,* ed. P.J. Crutzen and J.G. Goldammer. Chichester, England: Wiley.
Tyree, M., and J. Alexander. 1993. Plant water relations and the effects of elevated CO_2: A review and suggestion for future research. *Vegetatio* 104:47–62.
Volk, M., P.A. Niklaus, and C. Körner. 2000. Soil moisture effects determine CO_2 responses of grassland species. *Oecologia* 125:380–88.
Wagner, D. 1996. Scenarios of extreme temperature events. *Climate Change* 33:385–407.
Wand, S., G. Midgley, M. Jones, and P. Curtis. 1999. Responses of wild C_4 and C_3 grass (Poaceae) species to elevated atmospheric CO_2 concentration: A meta-analytic test of current theories and perceptions. *Global Change Biology* 5:723–41.
Ward, J.K., D.T. Tissue, B.R. Thomas, and B.R. Strain. 1999. Comparative responses of model C_3 and C_4 plants to drought in low and elevated CO_2. *Global Change Biology* 5:857–67.
Watson, R.T., H. Rhodhe, H. Oeschger, and U. Siegenthaler. 1990. Greenhouse gases and aerosols. In *Climate change: The IPCC scientific assessment,* ed. J.T. Houghton, G.J. Jenkins, and J.J. Ephraums, 1–40. Cambridge: Cambridge University Press.
Webb, A.A.R., M.R. McAinsh, T.A. Mansfield, and A.M. Hetherington. 1996. Carbon induces increases in guard cell cytosolic free calcium. *The Plant Journal* 9:297–304.
Weltzin, J., and G. McPherson. 2000. Implications of precipitation redistribution for shifts in temperate savanna ecotones. *Ecology* 81:1902–13.
Weltzin, J.F., M.E. Loik, S. Schwinning, D.G. Williams, P. Fay, B. Haddad, J. Harte, T.E. Huxman, A.K. Knapp, G. Lin, W.T. Pockman, M.R. Shaw, E. Small, M.D. Smith, S.D. Smith, D.T. Tissue, and J.C. Zak. In Press. Assessing the response of terrestrial ecosystems to potential changes in precipitation. *Bioscience.*
Woodward, F.I., G.B. Thompson, and I.F. McKee. 1991. The effects of elevated concentrations of carbon dioxide on individual plants, populations, communities, and ecosystems. *Annals of Botany* 67:23–38.
Woodward, F.I., and B.G. Williams. 1987. Climate and plant distribution at global and local scales. *Vegetatio* 69:189–97.
Ziska, L. 2003. Evaluation of the growth response of six invasive species to past, present and future atmospheric carbon dioxide. *Journal of Experimental Botany* 54:395–404.

20. Effects of CO_2 on Plants at Different Timescales

Belinda E. Medlyn and Ross E. McMurtrie

20.1 Introduction

How do plants respond to changes in atmospheric CO_2 concentration? The answer depends strongly on the timescale of interest to the questioner. Changes in atmospheric CO_2 concentration ($[CO_2]$) have direct and immediate effects on plant physiology, but as time progresses, a series of plant and ecosystem feedbacks act to modify this direct response. Different feedbacks have different time constants and hence become important on different timescales (Fig. 20.1). For example, changes in growth patterns can occur over months to years, changes in nutrient availability due to soil feedbacks may occur over decades, while adaptation may alter plant responses over centuries or more. Evidence for the effects of changes in atmospheric $[CO_2]$ on plants is available on many different timescales, from short-term experiments to longer-term models to paleobotanical studies. Evidence from one timescale can have bearing on responses at another timescale, but to interpret this evidence correctly, it is important to understand how plant responses to atmospheric $[CO_2]$ change over time.

The aim of this paper is to review our current knowledge of CO_2 effects on plants on different timescales and to examine how responses at different times relate to each other. We consider four general timescales. The first, minutes to hours, is the timescale of direct physiological responses. The second, months to years, is the experimental timescale, or, the duration of most experiments. The third, decades, we denote the human timescale, since it represents the timescale

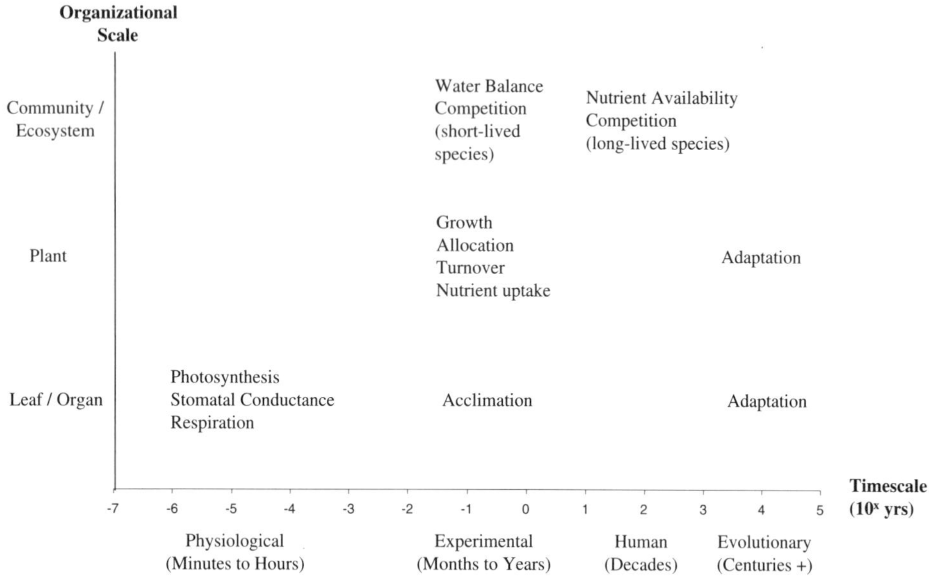

Figure 20.1. A summary of the timescales on which different processes become important in determining plant response to changes in atmospheric CO_2 concentration, and the organizational scales at which these processes occur. The four general timescales considered in this chapter are indicated on the x-axis.

of most concern to human society. The fourth and final timescale, centuries and beyond, may be regarded as the evolutionary timescale. We discuss how evidence from all four timescales may be used to predict plant responses to elevated [CO_2] on the human timescale. Our primary focus is on responses of woody perennials, which grow to maturity on the human timescale.

20.2 The Physiological Timescale

Changes in atmospheric [CO_2] directly affect several plant physiological processes. The most significant effect is on photosynthetic rate, a response that is well documented and mechanistically understood. Changes in [CO_2] can also affect stomatal conductance, although the magnitude of the response varies between species and the mechanism by which it occurs has yet to be identified. There is also some evidence that changes in [CO_2] have a direct effect on plant respiration, though whether this response is real or artifactual is actively debated.

20.2.1 Photosynthesis

The direct effect of changes in atmospheric [CO_2] on photosynthetic rate in C_3 plants has been studied for well over a century (Hall and Rao 1999) and is now well enough understood to be modeled based on a knowledge of leaf biochemistry (Farquhar, Von Caemmerer, and Berry 1980). The response is largely determined by the activity of the enzyme catalysing CO_2 fixation, ribulose bisphosphate carboxylase/oxygenase, or, Rubisco. In C_3 plants that have been genetically engineered to have a low Rubisco concentration, the CO_2 response of photosynthesis follows typical enzyme kinetics (Von Caemmerer et al. 1994). In wild-type C_3 plants, the CO_2 response follows a similar pattern unless another factor—electron transport or triose-phosphate utilization rate—is limiting to photosynthesis. Where electron transport is the limiting factor, the CO_2 response is much reduced, but is still not zero, owing to the suppression of photorespiration, which is the catalyzation of oxygen rather than CO_2. The triose-phosphate utilization rate is related to the rate of export of photosynthate and hence limits photosynthesis only under conditions of very high photosynthetic rate; in this case, there is no response to $[CO_2]$ (Sharkey 1985). A typical response of light-saturated C_3 photosynthesis to $[CO_2]$ is shown in Fig. 20.2A.

The photosynthetic response to $[CO_2]$ in C_4 plants is quite different. C_4 plants differ from C_3 plants in that CO_2 is initially fixed in the mesophyll by PEP carboxylase, which is more efficient than Rubisco. The product is transported to gas-impermeable bundle sheath cells, where it undergoes decarboxylation and refixation by Rubisco in a high-CO_2 environment, thus avoiding photorespiration. These properties of C_4 plants result in more efficient fixation of CO_2 at low atmospheric $[CO_2]$, as shown in Fig. 20.2A. The CO_2-concentrating mechanism of C_4 plants is energetically costly, however, so although the mechanism gives an advantage to C_4 plants at low atmospheric $[CO_2]$, it becomes a disadvantage at higher $[CO_2]$—a point to which we return below.

20.2.2 Stomatal Conductance

There is also clear evidence that stomata in both C_3 and C_4 plants respond to changes in $[CO_2]$, although we have much less mechanistic understanding of this response (Morison 1987, 1998). To be precise, plants do not respond to the $[CO_2]$ at the leaf surface, C_s, but rather the intercellular concentration, C_i (Mott 1988). Stomata tend to respond to C_i in such a way as to maintain a constant ratio of C_i:C_s, except under conditions of changing vapor-pressure deficit or soil-water availability (Wong, Cowan, and Farquhar 1979). This tendency results in a generally nonlinear response of stomatal conductance, g_s, to C_s (Fig. 20.2B). The magnitude of the response is variable between species: for example, stomata of coniferous trees tend to be less sensitive to changes in $[CO_2]$ than those of deciduous trees (Medlyn et al. 2001a).

Stomatal closure with increasing $[CO_2]$ can be said to "make sense" from an optimization argument that runs as follows: Plants both gain CO_2 and lose water through their stomata. Cowan and Farquhar (1977) argue that the optimal stomatal opening is that for which the total amount of CO_2 taken up per unit water lost is maximized. If stomata do not shut with increasing atmospheric $[CO_2]$, CO_2 gain will be increased while water loss will remain steady. However, water

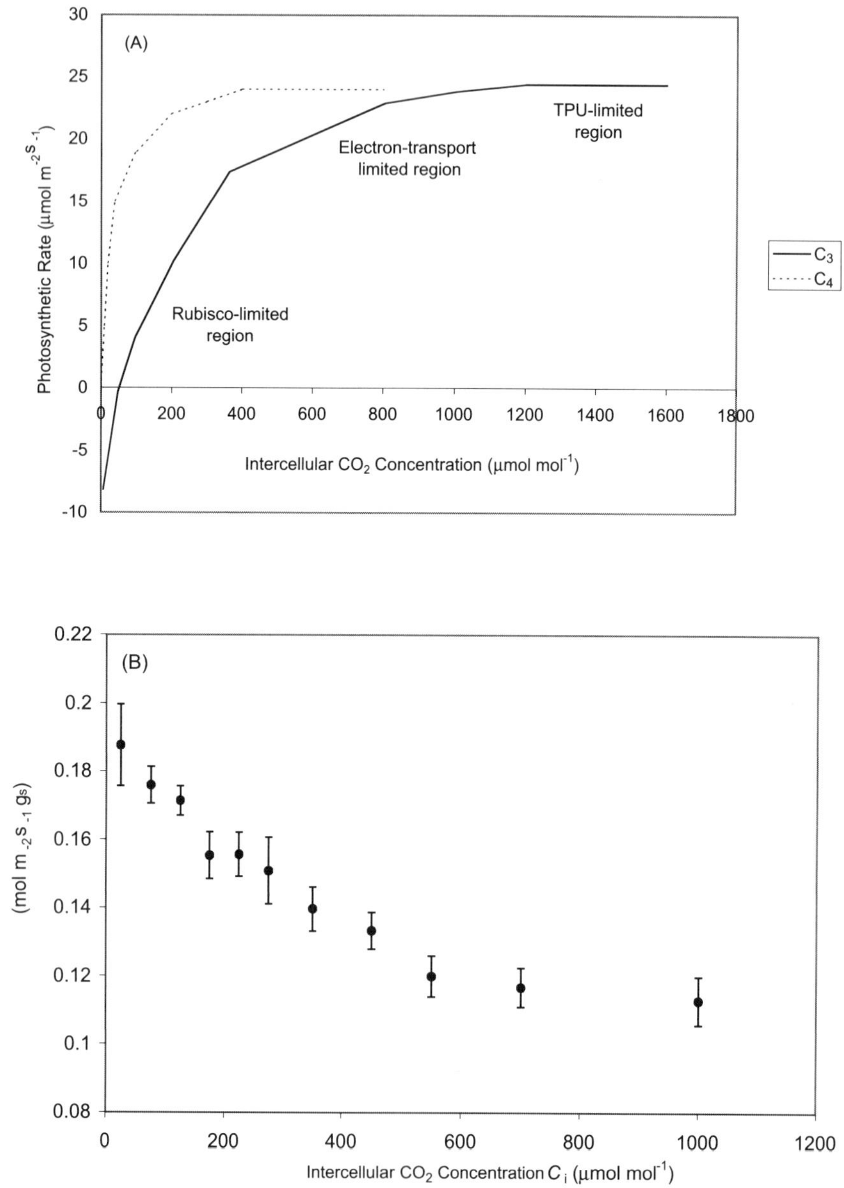

Figure 20.2. **(A)** Typical direct responses of photosynthesis to intercellular CO_2 concentration. Photosynthetic responses are shown for C_3 and C_4 plants (after Collatz, Ribas-Carbo, and Berry 1992). For C_3 plants, photosynthesis is limited by Rubisco activity, electron-transport rate, or triose-phosphate utilization rate (TPU). **(B)** Typical direct response of stomatal conductance to intercellular CO_2 concentration (Barton and Jarvis 1999).

loss is more sensitive to changes in stomatal conductance than is CO_2 gain. Therefore, a decrease in stomatal conductance would reduce water loss more than it would reduce photosynthesis, so CO_2 gain per unit water lost would be increased. Despite this understanding in terms of optimal plant function, the physical or biochemical mechanism that causes stomatal closure under increased [CO_2] remains unknown.

The amount of carbon or biomass gained per unit water transpired is generally termed the plant water-use efficiency. Whether stomata close or not, water-use efficiency generally increases with increasing [CO_2], owing to the significant increase in photosynthetic rate (Morison 1993; Wullschleger, Tschaplinski, and Norby 2002).

20.2.3 Respiration

Our understanding of the direct effects of [CO_2] on plant dark respiration is also poor. Several studies have shown that an increase in [CO_2] produces an immediate and reversible reduction in dark respiration (see Drake et al. 1999 for a review). However, the response observed in some of these studies has been shown to be an artifact of leaks in the system used to measure gas exchange (Amthor 2000). Several recent studies, in which care was taken to avoid or correct for leaks, have shown no or nonsignificant effects of [CO_2] on leaf or fine root dark respiration (Tjoelker et al. 1999; Amthor 2000; Amthor et al. 2001; Hamilton, Thomas, and DeLucia 2001; Burton and Pregitzer 2002). Although it has clearly been demonstrated that [CO_2] can directly affect several of the enzymes involved in mitochondrial respiration (Gonzalez-Meler et al. 1996), these enzymes have little control over respiration rates and hence are unlikely to cause a significant change in respiration (Gonzalez-Meler and Siedow 1999).

It has been found that altering [CO_2] at nighttime alone can have an effect on plant growth (Bunce 1995; Reuveni, Gale, and Zeroni 1997; Griffin, Sims, and Seemann 1999), suggesting that the direct effects of [CO_2] are not limited to photosynthesis and stomatal opening, since these processes take place only in the light. Some aspects of these experiments hint at other direct effects of [CO_2]: for example, Griffin, Sims, and Seemann (1999) found that nighttime [CO_2] affected biomass accumulation in soybean but only during the reproductive phase, not the vegetative phase. This intriguing result is hard to explain only in terms of effects on respiration. However, it is generally accepted that the major direct effects of [CO_2] are those on photosynthesis, stomatal opening, and possibly respiration. Below, we focus on these processes.

20.3 The Experimental Timescale

If the CO_2 concentration of the air surrounding a plant is altered for any length of time, the plant's carbon economy will be changed significantly. The CO_2 effect on the rate of photosynthesis leads to a large change in carbon uptake, shifting the balance between sources and sinks of carbon, and between the availability of carbon and other nutrients. These shifts in resource availability set in motion a cascading series of feedback effects.

There is a wealth of experimental data examining these feedbacks, from hundreds of studies carried out over the past two decades. The experiments have varied in species and age of plant material; exposure facility (controlled environment, open-top chamber, free-air CO_2 enrichment, or, FACE, among others); duration, ranging from several days to several years; CO_2 levels; and interactions with other factors, including nutrient and water availability, temperature, light, and ozone levels. Even where experiments are comparable, results tend to be highly variable. Fig. 20.3, for example, shows that reported responses to doubled $[CO_2]$ of aboveground woody biomass of trees grown in field-based chamber experiments range from a slight decrease to a fourfold increase. Drawing general conclusions from this literature is a challenging task. One analysis technique that is proving powerful is meta-analysis, a statistical method of estimating mean responses across a number of similar experiments (Curtis and Wang 1998; Gurevitch, Curtis, and Jones 2001). However, variation about the mean response is also important, as it indicates that different feedbacks may operate in different situations (Norby et al. 1999; Zak et al. 2000). For this outline, we draw on several excellent recent reviews that use alternative approaches to examine the experimental database.

20.3.1 Growth and Allocation

The major feedback on the experimental timescale is through changes in plant growth and carbon allocation patterns. Consider a doubling in $[CO_2]$ from 350 to 700 $\mu mol\ mol^{-1}$—a common experimental treatment. Such an increase in $[CO_2]$ will lead to an increase in photosynthetic C uptake. What happens to this additional C—whether it goes to leaves, stems, roots, reproductive organs, or soil—has major implications for plant and ecosystem functioning, which may be summarized as follows: Increased C in leaves will generally mean an increase in leaf area, which would increase canopy or whole-plant photosynthesis further (a positive feedback) but also increase both canopy transpiration, potentially countering any gains in water availability from the reduction in stomatal conductance, and plant nutrient demand. Stems represent a long-term storage of carbon, so increased C in stems leads to enhanced C sequestration. Increased C in root systems may increase uptake of water and nutrients, giving positive feedbacks through improved plant water status and nutrition. If additional C were allocated to reproductive organs, the plant's fitness could increase. All these outcomes would tend to enhance total biomass. Other outcomes can be envisaged that would give no net increase in plant biomass, such as an increase in turnover rates, which would see most of the additional C going to litter and soil. This outcome would have a strong effect on soil processes. Alternatively, if plants do not have strong sinks for C, there may be a down-regulation of photosynthesis, such that C uptake is not increased.

With this plethora of potential feedbacks, it is easy to understand why experimental results are diverse (Fig. 20.3). Different feedbacks will dominate depending on experimental conditions. For example, several reviews report that both leaf mass and leaf area increase under elevated [CO_2] (Ceulemans and Mousseau 1994; Curtis and Wang 1998; Norby et al. 1999), leading to feedbacks on both photosynthesis and transpiration. However, most of the experiments covered by these reviews were carried out on young plants growing individually. The response is different in experiments where the plant canopy is closed, and competition for light occurs. In microcosm experiments, a sustained increase in leaf area has only been observed in high-nutrient conditions (Körner 1996). In some open-top chamber experiments where the canopy closes over the course of the experiment, an initial increase in leaf area is often observed to diminish (Tissue, Thomas, and Strain 1997; Jach, Laureysens, and Ceulemans 2000). In FACE experiments in closed canopies, leaf area is generally not observed to increase (Gielen et al. 2001; Norby et al. 2001a). Finally, trees growing near CO_2 springs have a similar leaf area to those growing in ambient conditions (Hättenschwiler et al. 1997). Taken as a whole, this evidence suggests that although [CO_2] stimulates leaf area in plants grown individually, the leaf area of plants with closed canopies may not be greatly changed under elevated [CO_2].

The importance of the feedback to photosynthesis through leaf area can be seen in the responses of aboveground woody biomass to elevated [CO_2]. Above-

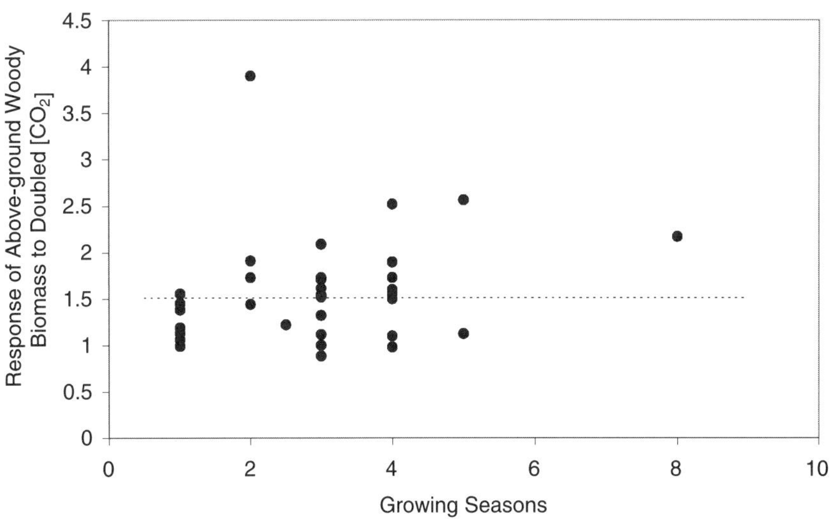

Figure 20.3. Ratio of aboveground woody dry matter of trees grown in elevated [CO_2] (ca. 700 μmol mol^{-1}) to that of trees grown in ambient [CO_2] (ca. 350μmol mol^{-1}) versus length of exposure to [CO_2] treatment. Data from harvests of trees grown in field-based chamber experiments (from Norby et al. 1999) and the ECOCRAFT database (Medlyn and Jarvis 1999). The dotted line shows the mean response.

ground woody biomass is generally found to increase (Ceulemans and Mousseau 1994; Curtis and Wang 1998; Saxe, Ellsworth, and Heath 1998; Norby et al. 1999), but there is tremendous variability across experiments even in comparable experiments (see Fig. 20.3). One possible cause for this variability is feedbacks arising from allocation patterns during the exponential growth phase. Where additional carbon is allocated to leaf area during this phase, there is a large positive feedback on carbon uptake, whereas the reverse may be true if additional carbon is allocated to roots (Medlyn et al. 2001b). Again, it is questionable how relevant such responses are for natural systems, where LAI is unaffected by [CO_2]. Norby (1996) suggested the use of the canopy productivity index, CPI, which is stem production per unit peak LAI. Norby et al. (1999) showed that CPI increased, on average, by 26% with a doubling of [CO_2], and this increase was much less variable than the increase in aboveground woody biomass. Since the rate of production is normalized by the LAI, it is argued that this increase should also be applicable in closed-canopy systems. Initial increases in stem biomass in two forest FACE experiments also were of this order, but, interestingly, the CO_2 stimulation of stem production in both these experiments has diminished over time (Norby et al. 2001a; Oren et al. 2001). Proposed mechanisms for the observed decline in the stimulation of stem production include nutrient limitation (Oren et al. 2001) and changes in allocation patterns (Norby et al. 2001a).

The effect of changes in [CO_2] on allocation to reproductive structures is also quite variable. Although fruit or seed yield is generally enhanced by elevated [CO_2] in high-yielding agricultural crops (Kimball 1983), the overall response in natural ecosystems is much less clear-cut (Bazzaz 1990; Reynolds et al. 1996), with both positive and negative responses being observed. It is also unclear to what extent an increase in allocation to reproductive structures affects species' dispersal and recruitment patterns. For example, allocation to cones and seeds in a 19-year-old loblolly pine forest increased greatly after three years of fumigation with CO_2 (LaDeau and Clark 2001). Seeds were taken from this forest and germinated in nutrient-deficient soils in greenhouse chambers to examine whether the increased seed weight and lipid content of high-CO_2 grown plants led to more successful establishment (Hussain, Kubiske, and Conner 2001). Although germination success was enhanced, down-regulation of photosynthesis was observed in all seedlings, indicating that increased reserves in high-CO_2 grown seeds did not necessarily improve seedling survival.

Turning to root biomass, it has often been predicted that allocation of carbon to root biomass should increase with increasing [CO_2], in order to bring nutrient uptake into balance with carbon uptake (Norby, Pastor, and Melillo 1986; Eamus and Jarvis 1989; Medlyn and Dewar 1996). As with stem and foliar biomass, root biomass is generally increased with elevated [CO_2] (Curtis and Wang 1998). The important question for nutrient uptake, however, is not whether total root biomass increases, but whether root biomass as a fraction of total biomass is altered. This question is notoriously difficult to answer because of inherent ontogenetic drift in allocation patterns (Poorter and Nagel 2000). To overcome this problem, Poorter and Nagel (2000) used an allometric analysis, and showed that, on average, plants grown in elevated [CO_2] do not change their allocation patterns: that is, root and leaf biomass as fractions of total biomass were unaltered by growth in elevated [CO_2].

However, the analysis of Poorter and Nagel (2000) refers to the standing biomass of plants, which overlooks possible changes in turnover rate with elevated [CO_2]. In several experiments where the carbon balance has been calculated, the extra carbon taken up by plants in elevated [CO_2] cannot completely be accounted for by simply measuring biomass (Luo et al. 1997; Wang, Rey, and Jarvis 1998; Cheng et al. 2000). One hypothesis for this "missing sink" of carbon is increased belowground flows, including fine root turnover, root exudation, and carbon flow to mycorrhizae. Thus, although the root mass fraction may be unchanged, the plants may still be partitioning more carbon to processes that are below–ground. An additional flow of carbon to the soil may not directly increase nutrient uptake, as might an increase in fine root biomass, but it could indirectly enhance nutrient uptake by stimulating decomposition and mycorrhizal activity.

Processes of nutrient uptake have been most closely studied for nitrogen (N). It is generally found that, on the experimental timescale, nitrogen uptake by plants is stimulated by growth in elevated [CO_2] but not to the same extent as biomass accumulation. In consequence, average tissue nitrogen concentration declines (Cotrufo, Ineson, and Scott 1998). The decline in tissue [N] varies between functional types, with C_3 plants having a larger reduction (-16%) than C_4 plants or nitrogen-fixing species (-7%) (Cotrufo, Ineson, and Scott 1998). This reduction in tissue [N] has important implications for both plant physiological processes and soil feedbacks, as discussed below.

20.3.2 Physiological Acclimation

The direct effects of [CO_2] on the processes of photosynthesis, stomatal opening, and respiration may be modified by prolonged growth in elevated [CO_2]. Such adjustments are often referred to as acclimation, although this term implies a response that is advantageous to the plant, which may not necessarily be the case.

For photosynthesis, the direct short-term stimulation by elevated [CO_2] may be attenuated by longer-term exposure to high [CO_2]. Meta-analysis suggests that, for field-grown plants, this attenuation is significant but relatively small (Medlyn et al. 1999). There are two major hypotheses to explain the attenuation of the CO_2 response of photosynthesis. The first is related to the decline in tissue nitrogen concentration noted above. Maximum photosynthetic rates are generally strongly related to leaf N (Field and Mooney 1986), so a decline in leaf N content may act as a feedback to the CO_2 stimulation of photosynthesis. The second hypothesis is based on the observation that photosynthesis may be limited by low sink demand for C (Stitt 1991; Koch 1996). Elevated CO_2 levels are likely to result in high C availability relative to demand. Carbon may thus accumulate in leaves as starch grains, which can impede photosynthetic ability. There is evidence that either or both of these feedbacks may occur in elevated [CO_2] (Medlyn et al. 1999).

Acclimation of stomatal conductance to growth in elevated [CO_2] has not been as widely studied as acclimation of photosynthesis. Available evidence suggests, however, that there is little acclimation of stomatal conductance, although the short-term response to changes in [CO_2] may be altered slightly (Medlyn et al. 2001a). Woodward and Kelly (1995) suggested that stomatal density, in addition to stomatal aperture, may be affected by changes in growth [CO_2], but a recent review showed no consistent response in experimental studies performed to date (Royer 2001). Since there is only weak acclimation of photosynthesis, and little or no acclimation of stomatal conductance, plant water-use efficiency on the experimental timescale is almost always significantly increased with increasing [CO_2], as it is on the immediate timescale (Drake, Gonzalez-Meler, and Long 1997) Dark respiration, on the other hand, is fairly consistently modified by longer-term changes in [CO_2]. In general, respiration rates are reduced in elevated [CO_2] (Drake, Gonzalez-Meler, and Long 1997; Curtis and Wang 1998). This suppression of respiration is thought to be related to reduced tissue nitrogen and protein concentrations (Drake, Gonzalez-Meler, and Long 1997).

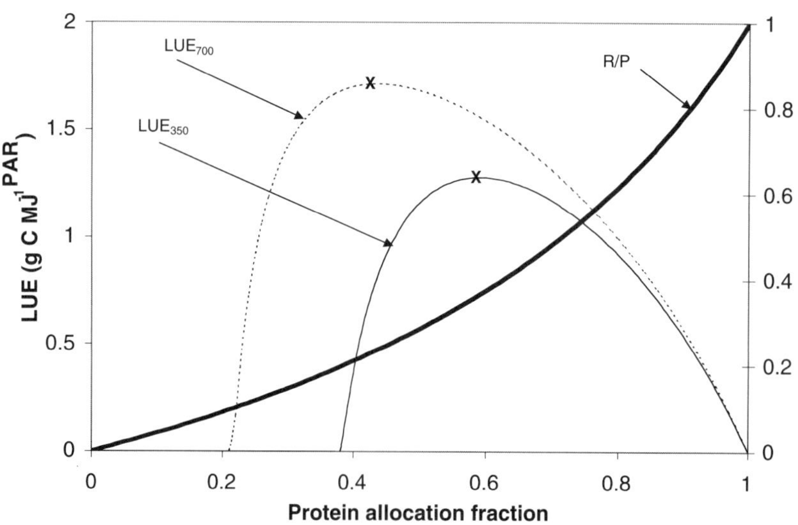

Figure 20.4. Theoretical model of leaf acclimation to [CO_2] (after Dewar, Medlyn, and McMurtrie 1998). Solid thin and dotted lines show the relationship between steady-state light-use efficiency (LUE) and leaf protein allocation (a_p) for atmospheric CO_2 levels of 350 and 700 μmol mol^{-1}, respectively. The optimal value of a_p, marked by X, decreases from 0.59 to 0.44 when atmospheric [CO_2] is increased from 350 to 700 μmol mol^{-1}. The solid thick line shows the dependence of the ratio of leaf respiration to gross photosynthesis (R:P) on leaf protein allocation. At optimal a_p R:P is 0.37 and 0.24 at [CO_2] of 350 and 700 μmol mol^{-1}, respectively.

A theoretical perspective on physiological acclimation to [CO_2] was presented by Dewar, Medlyn, and McMurtrie (1998), who modeled the dynamics of leaf protein and nonstructural carbon pools following a step increase in [CO_2]. At their model's steady state there is an optimal leaf-protein content that maximises plant light-use efficiency. The optimum (Fig. 20.4) reflects a trade-off between photosynthesis and leaf respiration. At elevated [CO_2] the point of optimal balance between photosynthesis and respiration is predicted to shift to lower protein contents.

Thus, the commonly observed shift to lower protein content (i.e., reduced leaf [N]) at high [CO_2] may be beneficial to plant productivity, which is a view counter to the usual interpretation that a shift to lower leaf [N] represents a down-regulation of photosynthesis. Concurrent with the predicted decline in leaf-protein content at elevated [CO_2], this model predicts a reduction in the ratio of leaf respiration to photosynthesis, providing a theoretical explanation for the observed acclimation of respiration rates discussed above.

20.3.3 Community Composition

On the basis of differences in photosynthetic response to [CO_2] (see Fig. 20.2A), it is commonly thought that C_3 plants will gain a competitive advantage over C_4 plants with increasing atmospheric [CO_2]. However, recent experimental evidence shows that this conclusion is not so clear-cut. Studies of the response of C_4 grasses to increasing [CO_2] show that these plants do respond with a significant increase in total biomass, which is only marginally less than that of C_3 grasses (Wand et al. 1999). It is suggested that this response arises through both a reduction in stomatal conductance and an increase in photosynthetic rate (Ghannoum et al. 2000). It appears that C_4 photosynthesis is not always CO_2-saturated at current [CO_2] levels, particularly under environmental conditions occurring in the field (Ghannoum et al. 2000).

It is therefore not entirely surprising that, when grown together in competition at high [CO_2], C_3 plants do not always out-compete C_4 plants (Reynolds 1996). One oft-cited counter-example is an open-top chamber experiment in tallgrass prairie (Owensby et al. 1993; Owensby et al. 1999), where it was found that C_3 forbs and sedges increased their biomass, C_4 plants changed little, and the biomass of other C_3 plants decreased, after eight years' exposure to elevated [CO_2]. Similarly, an open-top chamber experiment in shortgrass steppe found no difference in the response to elevated [CO_2] of C_3 and C_4 plants after two years. Such results have renewed research interest in a question that many scientists thought was settled, and we may expect to see new information on this topic in the near future.

20.4 The Human Timescale

We have denoted the timescale of decades as the human timescale, since it is the timescale of most relevance to human society. Unfortunately, it is also the timescale on which there is the least evidence for effects of altered [CO_2], because there is no real parallel for the changes in atmospheric [CO_2] currently occurring, namely, a 1 to 2 μmol mol^{-1} increase per annum. Experiments with elevated [CO_2] differ from this scenario in several ways. First, they are shorter in duration, so some important feedbacks with longer timescales will not be observed. Second, experiments must use a step change in [CO_2] rather than a gradual increase. The suddenness of the step change sets in motion many transient effects, which must be carefully analyzed for their relevance to a situation of gradual change (Luo and Reynolds 1999). Third, the current increase in atmospheric [CO_2] coincides with a number of other environmental changes, such as changes in temperature and precipitation and increased nitrogen deposition, which will modify the response to [CO_2]. Experiments are rarely able to examine all these interacting factors in concert. Turning to longer timescales, the evidence from paleobiology is not directly relevant either, since previous changes in atmospheric [CO_2] have been much slower than the current anthropogenically forced increase. Thus, rather than relying on observation, the chief method used to analyze plant CO_2 responses on the human timescale is theoretical modeling. Mathematical models incorporating our best understanding of plant ecosystem processes can be used to simulate plant responses over decades or centuries (Thornley and Cannell 1996).

20.4.1 Models and Assumptions

The major feedbacks that modify plant responses on human timescales are related to soil processes. These feedbacks act slowly because of the slow rate of turnover of soil organic matter, which comes into equilibrium over several decades. Models used to study these questions therefore need to include some representation of plant-soil interactions. Such models range from relatively simple conceptual models (Comins and McMurtrie 1993; Rastetter, Ågren, and Shaver 1997), to models of intermediate complexity (Parton et al. 1995; Ryan et al. 1996a; Thornley and Cannell 1996; Kirschbaum 1999; Ollinger et al. 2002), through to highly detailed simulations of particular systems (Kellomäki and Väisänen 1997; Thornley and Cannell 1997; Grant and Nalder 2000).

If our understanding of ecosystem processes were perfect, constructing and running such models would be relatively easy, and we could expect that all models would give consistent outcomes. However, over the past decade there has been considerable debate among the researchers who implement and interpret the models about the extent to which soil processes either reduce or amplify direct CO_2 effects. Early publications questioned whether nutrient-limited ecosystems would respond to rising [CO_2] (Melillo et al. 1991). Subsequently, the Inter-governmental Panel on Climate Change (IPCC) concluded from modeling studies that plant growth responses to high [CO_2] should be reduced by approximately half when soil nutrient and water limitations are taken into account (Houghton et al. 1995). Another IPCC report concluded that the CO_2 effect should be diminished under nutrient limitation but enhanced under water limitation (Kirschbaum et al. 1996). The latest IPCC report predicts a large (80%) increase in global NPP due to an anticipated doubling of [CO_2] over the next century (Cramer et al. 2001). This prediction is still uncertain, however, because 4 of the 6 models upon which it was based did not simulate nutrient cycling feedbacks. The debate stems from our limited understanding of soil processes, which makes it necessary to make a number of assumptions when building a model. These assumptions can strongly affect model outcomes (Kirschbaum et al. 1994; Ryan et al. 1996b). Key assumptions where there is still considerable uncertainty include (1) the effects of elevated [CO_2] on nutrient cycling processes and (2) the basic mechanisms underlying carbon and nutrient flows in the soil. We consider each of these sets of assumptions in turn.

Depending on how elevated [CO_2] affects nutrient cycling processes, there are several potential alternative pathways by which soil feedbacks could modify plant responses. Three chief pathways can be identified: the "litter quality" feedback, the "litter quantity" feedback, and the "stimulation of N mineralisation" feedback (Berntson and Bazzaz 1996; McMurtrie et al. 2000). In the litter quality feedback, reduced N content of live plant tissue leads to reduced N content of plant litter, which retards litter decomposition, slowing down N release from litter and thus limiting N availability for uptake (Melillo et al. 1991; Norby et al. 2001b). In the litter quantity feedback, the increased flow of litter to the soil tends to increase soil C content, causing increased N immobilization and reduced plant N availability (Diaz et al. 1993). In the feedback called stimulation of N mineralisation, the enhanced flow of C to the soil, whether as litter, root exudates, or transfer to mycorrhizae, stimulates microbial activity and thus N mineralization and N fixation rates, leading to enhanced N availability for plants (Zak et al. 1993). The first two pathways result in a negative feedback, while the third yields a positive feedback.

McMurtrie et al. (2000) encoded each of these alternative assumptions in their plant-soil ecosystem model G'DAY and found that the predicted plant response to [CO_2] over several decades varied according to the assumptions made. Assumptions about litter quality at high [CO_2] had little impact on the simulated CO_2 response. The litter quantity feedback, however, caused a large reduction in the CO_2 response of net primary productivity (NPP). Following a step increase of [CO_2] from 350 to 700 μmol mol^{-1}, simulations showed an initial increase in NPP of 25%, similar to the average increase observed in experiments. This increase was reduced to just 1% to 2% after several decades and remained this low for centuries, because of increasing N limitations to plant production. In contrast, in the simulation where N mineralization was stimulated through increased belowground carbon allocation, the stimulation of NPP was reduced to zero after several decades but rose to 180% after approximately 200 years. In this simulation, the increased belowground carbon allocation caused significant nitrogen immobilization, and it took several decades before the stimulation in nitrogen mineralization was sufficiently large to overcome the consequent nutrient limitation.

The second set of assumptions with the potential to modify model outcomes significantly are those that describe the basic mechanisms of carbon and nutrient fluxes in the soil. One assumption of particular importance is the C:N ratio of soil organic matter (McMurtrie and Comins 1996; McMurtrie, Medlyn, and Dewar 2001). If this ratio is assumed to be fixed, increased C flows to the soil must be accompanied by an increase in N immobilization, limiting N availability for plants. However, if this ratio is variable, soil C storage may be increased without a concomitant reduction in N mineralization. Simulations by McMurtrie et al. (2000) show that if the soil C:N ratio is assumed to be variable rather than fixed, the increase of 1% to 2% in NPP after decades of elevated [CO_2] rises to become a 44% increase.

These gaps in our knowledge will take years to resolve. In the interim we must use models to make best guesses regarding several pressing environmental issues. One vitally important issue that has been addressed by several modellers is the future of the global terrestrial carbon sink. Contrasting possibilities that have been suggested are: that the sink will decline over the next few decades due to soil C loss through temperature stimulation of soil respiration and saturation of the CO_2 response (Pearce 1999; Walker, Steffen, and Langridge 1999); or that the sink will be sustained because growth increases due to CO_2 fertilization and temperature stimulation of nutrient availability will override soil C losses (Cramer et al. 2001). Here we present our best guess for the response of the carbon sink of a stand of Norway spruce at Flakaliden, northern Sweden, to a gradual increase of [CO_2] and to gradual increases of both [CO_2] and temperature (McMurtrie, Medlyn, and Dewar 2001; Pepper et al. submitted). The Flakaliden site is N-limited, but not water-limited. Simulations were initiated by running G'DAY to equilibrium under the current climate at Flakaliden, and then imposing gradual increases in [CO_2] (+3 μmol mol^{-1} year^{-1}) and temperature (+0.03°C year^{-1}), which approximate the IPCC "business as usual" climate change scenario IS92a (Houghton et al. 1995). For these simulations, it was assumed under elevated [CO_2] that: (a) litter [N] is reduced (litter quality feedback); (b) litter quantity is increased and causes N immobilization (litter quantity feedback); (c) there is no stimulation of N mineralization feedback; and (d) soil C:N ratios are flexible. Model outputs shown in Fig. 20.5 are net primary production (NPP), and the C sink (= NPP minus heterotrophic soil respiration).

Under increasing [CO_2], NPP increases steadily but tends to decelerate. The C sink reaches a peak after 20 years followed by a gradual decline. The C sink declines for several reasons, including saturation of the photosynthetic CO_2 response, and N immobilization in accumulating soil organic matter. The simulation with gradual increases of both [CO_2] and temperature predicts a sustained positive C sink over the twenty-first century. The explanation for this sustained C sink is that warming stimulates soil decomposition, which causes a release of soil N that enhances NPP. Unlike the high CO_2 simulation, there is little tendency for the C sink to decline over time.

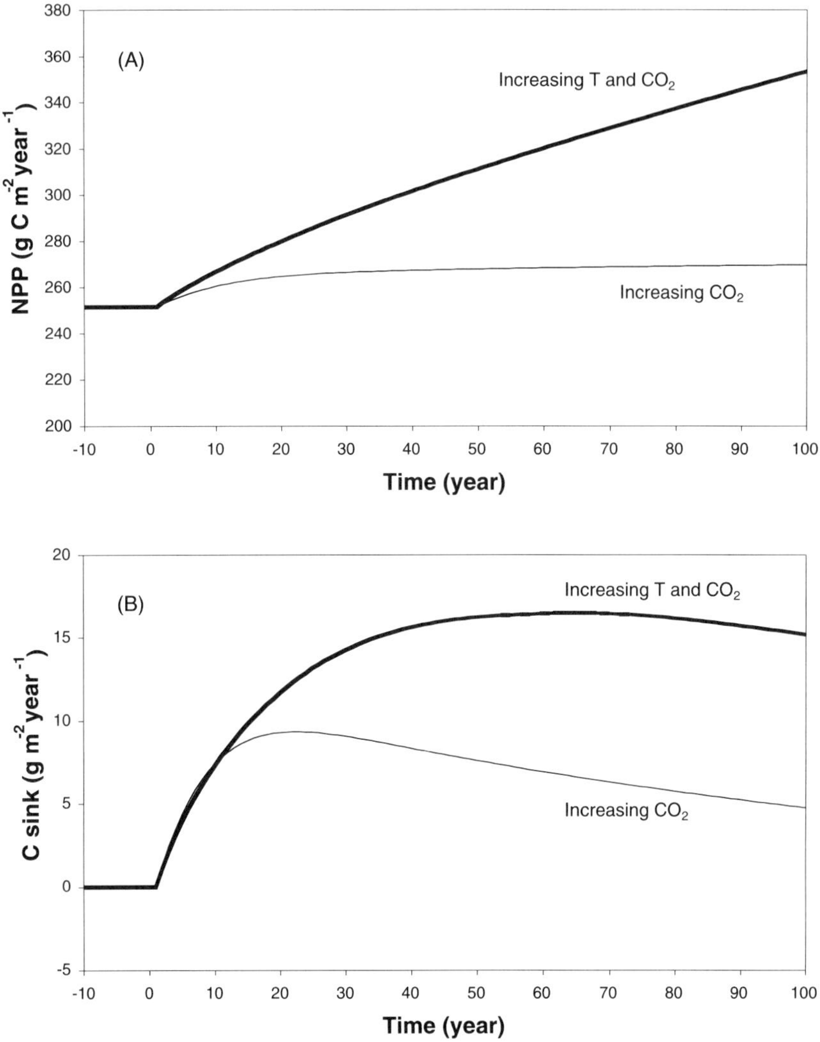

Figure 20.5. Simulated responses of G'DAY to gradual increases in [CO_2] of +3 μmol mol^{-1} $year^{-1}$ alone (thin line), and of both [CO_2] and temperature (+0.03°C $year^{-1}$) (thick line) **(A)** net primary production (NPP); **(B)** the carbon sink. Each simulation was initiated by running G'DAY to equilibrium at current T and [CO_2] (after McMurtrie, Medlyn, and Dewar 2001).

20.4.2 Experimental Testing of Model Assumptions

Our confidence in model results should increase in the future because our understanding of the uncertain processes is improving through experimental work. In particular, elevated CO_2 experiments, although they cannot directly reveal plant responses on the human timescale, can give much useful information to improve models of this timescale. The litter quality feedback is a case in point. Careful review of short-term experimental data shows that, although plant tissue N does decline in elevated [CO_2], there is no significant decline in plant litter N, and decomposition rates of litter grown in elevated [CO_2] are not different from those grown in ambient [CO_2] (Norby et al. 2001b). Hence, the hypothesized litter quality feedback is unlikely to occur. Experimental data relevant to the litter quantity and stimulation of N mineralization feedbacks have also been carefully reviewed (Zak et al. 2000), but here the outcome is not so clear. Soil respiration is consistently increased by growth in elevated [CO_2], indicating an increased flow of C belowground. However, this increased flow can lead to either increased or reduced N mineralization, with no consistent response across experiments. Zak et al. (2000) hypothesize that both effects can and do occur in different experimental situations. Aspects of fine-root biology that we do not as yet understand may determine which effect will occur in a particular situation. This is an area that we must study carefully in the future.

Improving our understanding of C and N dynamics in soil is also a key priority for experimentation. Here, N addition experiments with isotopic tracers can provide useful insights. Nadelhoffer et al. (1999a) reviewed a number of such experiments and found that only about 20% of the N added to soils found its way into plants, implying considerable sequestration of N in soil and thus a variable soil C:N ratio. These results have been somewhat controversial (Jenkinson, Goulding, and Powlson 1999; Nadelhoffer et al. 1999b), and experimental work is continuing to try to resolve this question.

Despite the uncertainty, a major conclusion from the modeling work is that soil feedbacks certainly have the potential to significantly modify plant CO_2 responses observed in short-term experiments. There is some evidence, from CO_2 springs, to support this conclusion. Newton et al. (2001) carried out a transplant experiment in an area close to a CO_2 spring and a similar area with normal CO_2 levels. Plants grown in high atmospheric [CO_2] but ambient CO_2 soil had much less biomass accumulation than plants grown in high atmospheric [CO_2] and high CO_2 soil. Newton et al. (2001) concluded that long-term responses to elevated [CO_2] are highly likely to differ from transient responses, qualitatively confirming model predictions.

20.5 The Evolutionary Timescale

Since atmospheric [CO_2] has varied dramatically in the geologic past, paleobotany provides us with another promising source of information about CO_2

effects on plants. However, as with other sources of information, there are several caveats that must be observed when interpreting these data. Here we consider in particular the caveats arising from timescale: Are the processes of relevance to shorter timescales operating on the very long timescale? Are additional processes operating on the very long timescale?

It has been strongly argued by Gifford (1992, 1994) that soil nitrogen availability is unlikely to limit plant production on century-plus timescales, but rather is likely to track the carbon cycle. He argues that with increased carbon availability, there would be small annual increases in nitrogen fixation that, over a long enough period, would be sufficient to alleviate any limitation. This argument is supported by model simulations (McMurtrie et al. 2000; Pepper et al. submitted). Hence, it is possible that soil nutrient feedbacks of concern on the human timescale do not operate on the very long timescale.

On the very long timescale, however, we must take into account evolutionary processes. To demonstrate how evolutionary changes can modify plant responses, Hovenden and Schimanski (2000) conducted an experiment with three genotypes of southern beech, *Nothofagus cunninghamii*. Plants were exposed to ambient or subambient levels of [CO_2]. The mean stomatal density across the three genotypes was significantly reduced by growth in subambient [CO_2]. However, there was considerable variation between genotypes (-20% to $+7\%$), and the reduction in stomatal density was linearly related to the reduction in plant biomass. If biomass accumulation is taken to indicate fitness, the genotype with increased stomatal density at low [CO_2] could thus be the most successful, so that over several generations, growth in low [CO_2] would cause an increase in stomatal density rather than the decrease observed in the first generation.

20.5.1 Potential Evolutionary Changes

It has been questioned whether changes in atmospheric [CO_2] would, indeed, drive evolutionary responses in plants. It is generally thought that the selective pressure of [CO_2] is weak because it is nonlethal and acts on many characters (Thomas and Jasienski 1996). Experimental evidence does suggest, however, that there is some evolutionary response. For example, Bunce (2001) examined whether today's plants are adapted to the current atmospheric [CO_2]. He studied the way in which various aspects of growth of several herbaceous annuals responded to CO_2 concentrations from 90 μmol mol^{-1} below ambient to 90 μmol mol^{-1} above ambient. Bunce found that, in general, the efficiency of use of CO_2 increased up to ambient [CO_2] but then decreased, and he drew the conclusion that there was some adaptation to ambient CO_2 levels. Using a different approach, Ward et al. (2000) studied the effectiveness of atmospheric [CO_2] as an evolutionary agent by conducting a selection experiment over five generations using *Arabidopsis thaliana*. Selection for seed number at low [CO_2] resulted in plants with increased biomass accumulation compared to controls, while selection at high [CO_2] resulted in plants with earlier initiation of flowering. Taken together, these results suggest that there is an evolutionary response of plants to changes in [CO_2].

An important question is which plant traits might be acted upon by evolutionary pressure. Likely traits can be identified either from theoretical arguments or by analogy with adaptive responses to other environmental factors. One candidate trait is the allocation of resources within the leaf. It has often been observed that leaf nitrogen is allocated in such a way that light and dark reactions are commonly co-limiting to photosynthesis at ambient [CO_2], resulting in an efficient use of nitrogen (Evans 1989). At elevated [CO_2], the photosynthetic rate would be optimized by a shift in nitrogen away from CO_2-fixation processes and toward light-capture processes (Medlyn 1996). Such a shift is rarely seen in first-generation plants exposed to elevated [CO_2] (Medlyn et al. 1999), but it seems a possible adaptive response—one that would serve to increase plant nitrogen-use efficiency in elevated [CO_2]. Similarly, Dewar, Medlyn, and McMurtrie (1998) showed, using a theoretical argument, that light-use efficiency could be optimized for different CO_2 levels by a shift in the amount of photosynthate allocated to leaf protein, suggesting another potential adaptive response (see Fig. 20.4). A second major candidate trait identified from optimization arguments is stomatal density. As described above, at high levels of [CO_2], a reduction in stomatal conductance acts to optimize water-use efficiency (Cowan and Farquhar 1977). A reduction in stomatal density would have a similar effect.

Other potential traits can be identified by comparison to the adaptive response of plants to other environmental factors. Adaptive responses to low [CO_2], for example, may be similar to adaptations to other resource limitations and thus may include enhancement of carbon uptake through increased leaf area and increased investment in photosynthetic enzymes, increased allocation of carbon to storage, and limitation of growth rate to avoid resource depletion during stressful periods (Sage and Cowling 1999). In contrast, adaptive responses to high [CO_2] may be similar to responses to competitive interactions, such as selection for early germination and rapid early growth (Thomas and Jasienski 1996).

20.5.2 Evidence from Paleobotany

Any effects of [CO_2] on plants observed in the paleobotanical record are, as discussed above, subject to myriad potential genetic modifications. Bearing this in mind, we now turn to the evidence available from the paleobotanical record. This evidence, broadly speaking, relates to the following characters: foliar chemistry, stomatal density, water-use efficiency, annual stem growth, and community composition.

Trends in foliar chemistry have been examined in herbarium specimens dating back to AD 1750 (Peñuelas and Matamala 1990; Peñuelas and Estiarte 1997). From these studies it appears that leaf N concentrations have declined significantly over the past 250 years. Concurrent measurements of leaf $\delta_{15}N$ also show a declining trend, suggesting that plants may have reduced N losses and increased N fixation and mineralization over this period (Peñuelas and Estiarte 1997). These results suggest that nitrogen limitation of plant response to [CO_2] is both a long-term and a short-term response. Unfortunately, longer records of foliar chemistry, for example from lake sediments, are more difficult to interpret because of potential decomposition (Hedges and Weliky 1989; Rundgren, Loader, and Beerling 2000). More information on nitrogen cycling in the past would be of great interest.

With respect to stomatal density, fossil sequences show a strong negative relationship between stomatal density and atmospheric [CO_2] (Royer 2001). Indeed, this relationship is so strong that stomatal density is commonly used as a proxy for atmospheric [CO_2] (Royer, Berner, and Beerling 2001). This relationship is much stronger in fossil plants than it is in experimental plants exposed to changes in [CO_2] (Royer 2001). Experimental plants had particularly weak responses to increases in [CO_2] above the current ambient level, which is the maximum level experienced over the past 3 million years. The fact that fossil plants show a relationship with [CO_2] above the current ambient level strongly suggests that there is a strong genetic component to this response (Royer 2001).

Plant water-use efficiency in the past can be estimated from the leaf stable carbon isotope composition, $\delta^{13}C$, of fossil and herbarium samples. Herbarium data show a decreasing trend in $\delta^{13}C$ since AD 1750, implying that water-use efficiency has increased with increasing [CO_2] over this period (Peñuelas and Azcon-Bieto 1992). Interpretation of isotopic composition of fossil samples is more complicated; environmental factors, including temperature and humidity, have varied greatly over this timescale, and these factors also have an important impact on water-use efficiency. This problem was tackled by Beerling and Woodward (1997) who employed a mechanistic model of leaf gas exchange to estimate water-use efficiency for [CO_2] and temperature conditions over the past 400 million years and to compare model predictions with fossil data. The model is based on observations of direct responses of photosynthesis and stomatal conductance to changes in [CO_2] and temperature. The model reproduced well the isotopic record of the fossil data, indicating that the response of plant water-use efficiency to changes in [CO_2] and temperature on the geologic timescale is likely to be similar to the response on the immediate and experimental timescales.

The sensitivity of stem growth to changes in atmospheric [CO_2] can be examined from tree-ring widths. Several studies have examined trends in tree-ring widths over the past century, but the evidence from these is inconclusive thus far (Jacoby and D'Arrigo 1997). Although some studies suggest a positive effect of increasing [CO_2] on tree-ring width over the past century, there are also a considerable number of studies that find no effect (Knapp, Soule, and Grissino-Mayer 2001). Interpretation of tree-ring records is difficult because of the large number of factors simultaneously affecting stem growth, including climatic change, nitrogen deposition, and improved methods of forest management (Lebourgeois et al. 2000; Knapp, Soule, and Grissino-Mayer 2001). On a longer timescale these difficulties become insurmountable, and so there is a complete absence of data on CO_2 effects on stem growth rates in the geologic past. There have been efforts to reconstruct net primary productivity of forests on very long timescales using $\delta^{13}C$ data (Beerling 1997). This approach relies on the use of models to scale up from isotopic discrimination data. Such models assume that the parameters and functioning of leaf gas exchange and canopy structure are the same as they are today, despite evidence that these parameters may acclimate or adapt in response to changes in [CO_2]. These models can therefore give us only an idea of what plant productivity might have been, rather than providing evidence that could be used to test our theories of CO_2 effects on plant production.

A final insight from the paleobotanical record is the relative abundances of C_3 and C_4 plants in the geologic past. Phylogenetic studies indicate that the C_4 photosynthetic pathway is a relatively recent innovation and evolved independently several times (Ehleringer, Cerling, and Helliker 1997). Comparison of the timing of the appearance of C_4 plants with environmental conditions shows that this pathway is likely to have evolved in response to low atmospheric [CO_2] (Sage 2001). However, it is debatable whether atmospheric [CO_2] has been the major control on relative C_4 abundance since its evolution. Studies of the current distribution of C_4 plants indicate that it tends to be favored by warm, arid conditions (Ehleringer, Cerling, and Helliker 1997), and recent evidence from sediment cores suggests that climatic conditions, rather than [CO_2], is the major control on the relative abundance of C_4 species (Huang et al. 2001). This finding supports experimental data showing that C_3 plants do not always out-compete C_4 plants at elevated CO_2 levels and suggests that environmental conditions need to be taken into account when attempting to predict the future relative abundance of the two pathways.

The paleobotanical record thus reinforces several of our current ideas about the effects of [CO_2] on plants, including effects on stomatal density, water-use efficiency, and community composition. However, some of the key questions regarding CO_2 effects on plant carbon and nutrient cycling remain unanswered.

20.6 Summary

The response of plants to atmospheric CO_2 concentration depends strongly on the timescale over which the response is measured, because the response is determined by different plant and ecosystem-level feedbacks on different timescales. The main goal of this review was to integrate our current understanding of CO_2 effects on plants, particularly woody perennials, across timescales ranging from the immediate to the immense. A step change in [CO_2] causes changes in photosynthesis, stomatal conductance, and possibly respiration rates within minutes. On the timescale of most ecosystem experiments, that is, months to several years, these processes may re-adjust, while plant growth rate and carbon partitioning will also be affected. On the timescale of decades, soil feedbacks through nutrient cycling may further modify the response. Responses on this timescale are of particular interest to human society, because the timescale of decades is comparable to that of the current increase in atmospheric [CO_2]. Because experiments are logistically difficult on this timescale, these responses are generally analyzed using plant-soil models. On even longer timescales, of centuries to millennia, processes such as adaptation and competition may dominate plant responses.

Our understanding of plant CO_2 responses on different timescales has improved dramatically over the past decade. However, several key questions still hinder attempts to predict the response to the increase in [CO_2] expected over the next 50 years. At the experimental timescale, observed responses of plant growth, allocation, and nitrogen mineralization to increases in [CO_2] are highly variable, and thus difficult to interpret, let alone extrapolate to the human timescale, where additional soil-feedback processes must be taken into account. Paleobotanical data can shed some light, but their usefulness is limited by the availability of relevant data and the difficulty of their interpretation.

How do plants respond to changes in atmospheric CO_2 concentration? The question is simple to ask but is hard to answer without more patient, detailed, and thoughtful research on all timescales.

References

Amthor, J. 2000. Direct effect of elevated CO_2 on nocturnal *in situ* leaf respiration in nine temperate deciduous tree species is small. *Tree Physiology* 20:139–44.

Amthor, J., G. Koch, J. Willms, and D. Layzell. 2001. Leaf O_2 uptake in the dark is independent of coincident CO_2 partial pressure. *Journal of Experimental Botany* 52: 2235–38.

Barton, C.V.M., and P.G. Jarvis. 1999. Growth response of branches of *Picea sitchensis* to four years exposure to elevated atmospheric carbon dioxide concentration. *New Phytologist* 144:233–43.

Bazzaz, F. 1990. The response of natural ecosystems to the rising global CO_2 levels. *Annual Review of Ecology and Systematics* 21:167–96.

Beerling, D. 1997. The net primary productivity and water use of forests in the geological past. *Advances in Botanical Research* 26:193–227.

Beerling, D., and F. Woodward. 1997. Changes in land plant function over the Phanerozoic: Reconstructions based on the fossil record. *Botanical Journal of the Linnean Society* 124:137–53.

Berntson, G.M., and F.A. Bazzaz. 1996. Belowground positive and negative feedbacks on CO_2 growth enhancement. *Plant and Soil* 187:119–31.

Bunce, J. 1995. Effects of elevated carbon dioxide concentrations in the dark on the growth of soybean seedlings. *Annals of Botany* 75:365–68.

———. 2001. Are annual plants adapted to the current atmospheric concentration of carbon dioxide? *International Journal of Plant Science* 162:1261–66.

Burton, A., and K.S. Pregitzer. 2002. Measurement carbon dioxide concentration does not affect root respiration of nine tree species in the field. *Tree Physiology* 22:67–72.

Ceulemans, R., and M. Mousseau. 1994. Effects of elevated atmospheric CO_2 on woody plants. *New Phytologist* 127:425–46.

Cheng, W.X., D.A. Sims, Y.Q. Luo, D.W. Johnson, J.T. Ball, and J.S. Coleman. 2000. Carbon budgeting in plant-soil mesocosms under elevated CO_2: Locally missing carbon? *Global Change Biology* 6:99–109.

Collatz, G., M. Ribas-Carbo, and J.A. Berry. 1992. Coupled photosynthesis-stomatal conductance model for leaves of C_3 plants. *Australian Journal of Plant Physiology* 19:519–38.

Comins, H.N., and R.E. McMurtrie. 1993. Long-term response of nutrient-limited forests to CO_2 enrichment; equilibrium behavior of plant-soil models. *Ecological Applications* 3:666–81.
Cotrufo, M., P. Ineson, and A. Scott. 1998. Elevated CO_2 reduces the nitrogen concentration of plant tissues. *Global Change Biology* 4:43–54.
Cowan, I.R., and G.D. Farquhar. 1977. Stomatal function in relation to leaf metabolism and environment. In *Integration of activity in the higher plant,* ed. D.H. Jennings, 471–505. Cambridge: Cambridge University Press.
Cramer, W., A. Bondeau, F.I. Woodward, I.C. Prentice, R.A. Betts, V. Brovkin, P.M. Cox, V.A. Fisher, J.A. Foley, A.D. Friend, C.J. Kucharik, M.R. Lomas, N. Ramankutty, S. Sitch, B. Smith, A. White, and C. Young-Molling. 2001. Global response of terrestrial ecosystem structure and function to CO_2 and climate change: Results from six dynamic global vegetation models. *Global Change Biology* 7:357–73.
Curtis, P.S., and X.Z. Wang. 1998. A meta-analysis of elevated CO_2 effects on woody plant mass, form, and physiology. *Oecologia* 113:299–313.
Dewar, R.C., B.E. Medlyn, and R.E. McMurtrie. 1998. A mechanistic analysis of light and carbon use efficiencies. *Plant Cell and Environment* 21:573–88.
Diaz, S., J. Grime, J. Harris, and E. McPherson. 1993. Evidence of a feedback mechanism limiting plant response to elevated carbon dioxide. *Nature* 364:616–17.
Drake, B., J. Azcon-Bieto, J. Berry, J. Bunce, P. Dijkstra, J. Farrar, R. Gifford, M. Gonzalez-Meler, G. Koch, H. Lambers, J. Siedow, and S. Wullschleger. 1999. Does elevated atmospheric CO_2 concentration inhibit mitochondrial respiration in green plants? *Plant Cell and Environment* 22:649–57.
Drake, B., M. Gonzalez-Meler, and S. Long. 1997. More efficient plants: A consequence of rising atmospheric CO_2? *Annual Review of Plant Physiology and Plant Molecular Biology* 48:609–39.
Eamus, D., and P.G. Jarvis. 1989. The direct effects of increase in the global atmospheric CO_2 concentration on natural and commercial temperate trees and forests. *Advances in Ecological Research* 19:1–55.
Ehleringer, J., T. Cerling, and B. Helliker. 1997. C_4 photosynthesis, atmospheric CO_2 and climate. *Oecologia* 112:285–99.
Evans, J.R. 1989. Photosynthesis and nitrogen relationships in leaves of C_3 plants. *Oecologia* 78:9–19.
Farquhar, G.D., S. Von Caemmerer, and J.A. Berry. 1980. A biochemical model of photosynthetic carbon dioxide assimilation in leaves of 3-carbon pathway species. *Planta* 149:78–90.
Field, C., and H. Mooney. 1986. The photosynthesis-nitrogen relationship in wild plants. In *On the economy of plant form and function,* ed. T. Givnish, 25–55. Cambridge: Cambridge University Press.
Ghannoum, O., S. Von Caemmerer, L. Ziska, and J. Conroy. 2000. The growth response of C_4 plants to rising atmospheric CO_2 partial pressure: a reassessment. *Plant Cell and Environment* 23:931–42.
Gielen, B., C. Calfapietra, M. Sabatti, and R. Ceulemans. 2001. Leaf area dynamics in a closed poplar plantation under free-air carbon dioxide enrichment. *Tree Physiology* 21:1245–55.
Gifford, R.M. 1992. Interaction of carbon dioxide with growth-limiting environmental factors in vegetation productivity: Implications for the global carbon cycle. *Advances in Bioclimatology* 1:24–58.
———. 1994. The global carbon cycle: A viewpoint on the missing sink. *Australian Journal of Plant Physiology* 21:1–15.
Gonzalez-Meler, M., M. Ribas-Carbo, J. Siedow, and B. Drake. 1996. Direct inhibition of plant mitochondrial respiration by elevated CO_2. *Plant Physiology* 112: 1349–55.
Gonzalez-Meler, M., and J. Siedow. 1999. Direct inhibition of mitochondrial respiratory

enzymes by elevated CO_2: Does it matter at the tissue or whole-plant level? *Tree Physiology* 19:253–59.
Grant, R.F., and I.A. Nalder. 2000. Climate change effects on net carbon exchange of a boreal aspen-hazelnut forest: Estimates from the ecosystem model ecosys. *Global Change Biology* 6:183–200.
Griffin, K., D. Sims, and J. Seemann. 1999. Altered night-time CO_2 concentration affects the growth, physiology and biochemistry of soybean. *Plant Cell and Environment* 22: 91–99.
Gurevitch, J., P. Curtis, and M. Jones. 2001. Meta-analysis in ecology. *Advances in Ecological Research* 32:199–247.
Hall, D.O., and K.K. Rao. 1999. *Photosynthesis.* Cambridge: Cambridge University Press.
Hamilton, J., R. Thomas, and E. DeLucia. 2001. Direct and indirect effects of elevated CO_2 on leaf respiration in a forest ecosystem. *Plant Cell and Environment* 24:975–82.
Hättenschwiler, S., F. Miglietta, A. Raschi, and C. Körner. 1997. Thirty years of in situ tree growth under elevated CO_2: A model for future forest responses? *Global Change Biology* 3:463–71.
Hedges, J., and K. Weliky. 1989. Diagenesis of conifer needles in a coastal marine environment. *Geochimica et Geophysica Acta* 53:2659–2673.
Houghton, J.T., L.G. Meira Filho, J. Bruce, H. Lee, B.A. Callander, E. Haites, N. Harris, and K. Maskell. 1995. *Climate change 1994: Radiative forcing of climate change and an evaluation of the IPCC IS92 emission scenarios.* Intergovernmental Panel on Climate Change. Cambridge: Cambridge University Press.
Hovenden, M., and L. Schimanski. 2000. Genotypic differences in growth and stomatal morphology of Southern Beech, *Nothofagus cunninghamii*, exposed to depleted CO_2 concentrations. *Australian Journal of Plant Physiology* 27:281–87.
Huang, Y., F. Street-Perrott, S. Metcalfe, M. Brenner, M. Moreland, and K. Freeman. 2001. Climate change as the dominant control on glacial-interglacial variations in C_3 and C_4 plant abundance. *Science* 293:1647–51.
Hussain, M., M. Kubiske, and K. Conner. 2001. Germination of CO_2-enriched *Pinus taeda* L. seeds and subsequent seedling growth responses to CO_2 enrichment. *Functional Ecology* 15:344–50.
Jach, M.E., I. Laureysens, and R. Ceulemans. 2000. Above- and below-ground production of young Scots pine (*Pinus sylvestris* L.) trees after three years of growth in the field under elevated CO_2. *Annals of Botany* 85:789–98.
Jacoby, G., and R. D'Arrigo. 1997. Tree rings, carbon dioxide, and climatic change. *Proceedings of the National Academy of Sciences of the United States of America* 94: 8350–53.
Jenkinson, D., K. Goulding, and D. Powlson. 1999. Nitrogen deposition and carbon sequestration. *Nature* 400:629.
Kellomäki, S., and H. Väisänen. 1997. Modelling the dynamics of the forest ecosystem for climate change studies in the boreal conditions. *Ecological Modelling* 97:121–40.
Kimball, B. 1983. Carbon dioxide and agricultural yield: An assemblage and analysis of 430 prior observations. *Agronomy Journal* 75:779–88.
Kirschbaum, M.U.F. 1999. Modelling forest growth and carbon storage in response to increasing CO_2 and temperature. *Tellus* 51B:871–88.
Kirschbaum, M.U.F., P. Bullock, J.R. Evans, K. Goulding, P.G. Jarvis, I.R. Noble, M. Rounsevell, and T.D. Sharkey. 1996. Ecophysiological, ecological and soil processes in terrestrial ecosystems: A primer on general concepts and relationships. In *Climate change 1995 impacts, adaptations, and mitigation of climate change: Scientific-technical analyses, IPCC Second Assessment Report,* ed. R.T. Watson, M.C. Zzinyowera, and R.H. Moss, 57–74. Cambridge and New York: Cambridge University Press.

Kirschbaum, M.U.F., D.A. King, H.N. Comins, R.E. McMurtrie, B.E. Medlyn, S. Pongracic, D. Murty, H. Keith, R.J. Raison, P.K. Khanna, and D.W. Sheriff. 1994. Modeling forest response to increasing CO_2 concentration under nutrient-limited conditions. *Plant Cell and Environment* 17:1081–99.

Knapp, P.A., P.T. Soule, and H.D. Grissino-Mayer. 2001. Detecting potential regional effects of increased atmospheric CO_2 on growth rates of western juniper. *Global Change Biology* 7:903–17.

Koch, K. 1996. Carbohydrate-modulated gene expression in plants. *Annual Review of Plant Physiology and Plant Molecular Biology* 47:509–40.

Körner, C. 1996. The response of complex multispecies systems to elevated CO_2. In *Global change and terrestrial ecosystems,* ed. B. Walker, and W. Steffen, 20–42. Cambridge: Cambridge University Press.

LaDeau, S., and J. Clark. 2001. Rising CO_2 levels and the fecundity of forest trees. *Science* 292:95–98.

Lebourgeois, F., M. Becker, R. Chevalier, J.I. Dupouey, and J.M. Gilbert. 2000. Height and radial growth trends of Corsican pine in western France. *Canadian Journal of Forest Research* 30:712–24.

Luo, Y., J.-L. Chen, J. Reynolds, C. Field, and H. Mooney. 1997. Disproportional increases in photosynthesis and plant biomass in a Californian grassland exposed to elevated CO_2: A simulation analysis. *Functional Ecology* 11:696–704.

Luo, Y.Q., and J.F. Reynolds. 1999. Validity of extrapolating field CO_2 experiments to predict carbon sequestration in natural ecosystems. *Ecology* 80:1568–83.

McMurtrie, R.E., and H.N. Comins. 1996. The temporal response of forest ecosystems to doubled atmospheric CO_2 concentration. *Global Change Biology* 2:49–57.

McMurtrie, R.E., R.C. Dewar, B.E. Medlyn, and M.P. Jeffreys. 2000. Effects of elevated [CO_2] on forest growth and carbon storage: A modelling analysis of the consequences of changes in litter quality/quantity and root exudation. *Plant and Soil* 224:135–52.

McMurtrie, R.E., B.E. Medlyn, and R.C. Dewar. 2001. Increased understanding of nutrient immobilization in soil organic matter is critical for predicting the carbon sink strength of forest ecosystems over the next 100 years. *Tree Physiology* 21:831–39.

Medlyn, B.E. 1996. The optimal allocation of nitrogen within the C_3 photosynthetic system at elevated CO_2. *Australian Journal of Plant Physiology* 23:593–603.

Medlyn, B.E., F.W. Badeck, D.G.G. De Pury, C.V.M. Barton, M. Broadmeadow, R. Ceulemans, P. De Angelis, M. Forstreuter, M.E. Jach, S. Kellomäki, E. Laitat, M. Marek, S. Philippot, A. Rey, J. Strassemeyer, K. Laitinen, R. Liozon, B. Portier, P. Roberntz, K. Wang, and P.G. Jarvis. 1999. Effects of elevated [CO_2] on photosynthesis in European forest species: A meta-analysis of model parameters. *Plant Cell and Environment* 22:1475–95.

Medlyn, B.E., C.V.M. Barton, M.S.J. Broadmeadow, R. Ceulemans, P. De Angelis, M. Forstreuter, M. Freeman, S.B. Jackson, S. Kellomäki, E. Laitat, A. Rey, P. Roberntz, B.D. Sigurdsson, J. Strassemeyer, K. Wang, P.S. Curtis, and P.G. Jarvis. 2001a. Stomatal conductance of forest species after long-term exposure to elevated CO_2 concentration: A synthesis. *New Phytologist* 149:247–64.

Medlyn, B.E., and R.C. Dewar. 1996. A model of the long-term response of carbon allocation and productivity of forests to increased CO_2 concentration and nitrogen deposition. *Global Change Biology* 2:367–76.

Medlyn, B.E., and P.G. Jarvis. 1999. Design and use of a database of model parameters from elevated [CO_2] experiments. *Ecological Modelling* 124:69–83.

Medlyn, B.E., A. Rey, C.V.M. Barton, and M. Forstreuter. 2001b. Above-ground growth responses of forest trees to elevated CO_2. In *The impact of carbon dioxide and other greenhouse gases on forest ecosystems,* ed. D. Karnosky, G. Scarascia-Mugnozza, R. Ceulemans, and J. Innes, 127–46. Wallingford, U.K.: CABI Publishing.

Melillo, J., T. Callaghan, F. Woodward, E. Salati, and S. Sinha. 1991. Effects on ecosystems. In *Climate change: The IPCC scientific assessment,* ed. J. Houghton, G. Jenkins, and J. Ephraums, 282–310. Cambridge: Cambridge University Press.

Morison, J.I.L. 1987. Intercellular CO_2 concentration and stomatal response to CO_2. In *Stomatal function,* ed. E. Zeiger, I.R. Cowan, and G.D. Farquhar, 229–51. Stanford: Stanford University Press.

———. 1993. Responses of plants to CO_2 under water limited conditions. *Vegetatio* 104/105:193–209.

———. 1998. Stomatal response to increased CO_2 concentration. *Journal of Experimental Botany* 49:443–52.

Mott, K.A. 1988. Do stomata respond to carbon dioxide concentrations other than intercellular? *Plant Physiology* 86:200–203.

Nadelhoffer, K.J., B. Emmett, P. Gundersen, O. Kjonaas, C. Koopmans, P. Schleppi, A. Tietema, and R. Wright. 1999a. Nitrogen deposition makes a minor contribution to carbon sequestration in temperate forests. *Nature* 398:145–48.

Nadelhoffer, K.J., B. Emmett, P. Gundersen, C. Koopmans, P. Schleppi, A. Tietema, and R. Wright. 1999b. Nitrogen deposition and carbon sequestration—Reply. *Nature* 400: 630.

Newton, P.C.D., H. Clark, G. Edwards, and D.J. Ross. 2001. Experimental confirmation of ecosystem model predictions comparing transient and equilibrium plant responses to elevated atmospheric CO_2. *Ecology Letters* 4:344–47.

Norby, R. 1996. Forest canopy productivity index. *Nature* 381:564.

Norby, R., J. Pastor, and J. Melillo. 1986. Carbon-nitrogen interactions in CO_2-enriched white oak: Physiological and long-term perspectives. *Tree Physiology* 2:233–41.

Norby, R., D. Todd, J. Fults, and D. Johnson. 2001a. Allometric determination of tree growth in a CO_2-enriched sweetgum stand. *New Phytologist* 150:477–87.

Norby, R., S. Wullschleger, C. Gunderson, D. Johnson, and R. Ceulemans. 1999. Tree responses to rising CO_2 in field experiments: Implications for the future forest. *Plant Cell and Environment* 22:683–714.

Norby, R.J., M.F. Cotrufo, P. Ineson, E.G. O'Neill, and J. Canadell. 2001b. Elevated CO_2, litter chemistry, and decomposition: A synthesis. *Oecologia* 127:153–65.

Ollinger, S.V., J.D. Aber, P.B. Reich, and R.J. Freuder. 2002. Interactive effects of nitrogen deposition, tropospheric ozone, elevated CO_2 and land use history on the carbon dynamics of northern hardwood forests. *Global Change Biology* 8:545–62.

Oren, R., D. Ellsworth, K. Johnsen, N. Phillips, B. Ewers, C. Maier, K. Schafer, H. McCarthy, G. Hendrey, S. McNulty, and G. Katul. 2001. Soil fertility limits carbon sequestration by forest ecosystems in a CO_2-enriched atmosphere. *Nature* 411:469–72.

Owensby, C., P. Coyne, J. Ham, L. Auen, and A. Knapp. 1993. Biomass production in a tallgrass prairie ecosystem exposed to ambient and elevated levels of CO_2. *Ecological Applications* 3:644–53.

Owensby, C., J. Ham, A. Knapp, and L. Auen. 1999. Biomass production and species composition change in a tallgrass prairie ecosystem after long-term exposure to elevated atmospheric CO_2. *Global Change Biology* 5:497–506.

Parton, W.J., J. Scurlock, D.S. Ojima, D.S. Schimel, and D.O. Hall. 1995. Impact of climate-change on grassland production and soil carbon worldwide. *Global Change Biology* 1:13–22.

Pearce, F. 1999. That sinking feeling. *New Scientist* 164:20–21.

Peñuelas, J., and J. Azcon-Bieto. 1992. Changes in leaf $\delta^{13}C$ of herbarium plant species during the last 3 centuries of CO_2 increase. *Plant Cell and Environment* 15: 485–89.

Peñuelas, J., and M. Estiarte. 1997. Trends in plant carbon concentration and plant demand for N throughout this century. *Oecologia* 109:69–73.

Peñuelas, J., and R. Matamala. 1990. Changes in N and S leaf content, stomatal density and specific leaf area of 14 plant species during the last three centuries of CO_2 increase. *Journal of Experimental Botany* 41:1119–24.

Pepper, D.A., S.J. Del Grosso, R.E. McMurtrie, and W.J. Parton. [submitted] Simulated carbon sink response of shortgrass steppe, tallgrass prairie and forest ecosystems to rising [CO_2], temperature and nitrogen input. *Global Biogeochemical Cycles.*

Poorter, H., and O. Nagel. 2000. The role of biomass allocation in the growth response of plants to different levels of light, CO_2, nutrients and water: A. quantitative review. *Australian Journal of Plant Physiology* 27:595–607.

Rastetter, E.B., G. Ågren, and G. Shaver. 1997. Responses of N-limited ecosystems to increased CO_2: A balanced-nutrition, coupled-element-cycles model. *Ecological Applications* 7:444–60.

Reuveni, J., J. Gale, and M. Zeroni. 1997. Differentiating day from night effects of high ambient [CO_2] on the gas exchange and growth of *Xanthium strumarium* L. exposed to salinity stress. *Annals of Botany* 79:191–96.

Reynolds, H. 1996. Effects of elevated CO_2 on plants grown in competition. In *Carbon dioxide, populations, and communities,* ed. C. Körner and F. Bazzaz, 273–86. San Diego: Academic Press.

Reynolds, J., P. Kemp, B. Acock, J.-L. Chen, and D. Moorhead. 1996. Progress, limitations, and challenges in modeling the effects of elevated CO_2 on plants and ecosystems. In *Carbon dioxide and terrestrial ecosystems,* ed. G. Koch and H. Mooney, 347–80. San Diego: Academic Press.

Royer, D. 2001. Stomatal density and stomatal index as indicators of paleoatmospheric CO_2 concentration. *Review of Palaeobotany and Palynology* 114:1–28.

Royer, D., R. Berner, and D. Beerling. 2001. Phanerozoic atmospheric CO_2 change: Evaluating geochemical and paleobiological approaches. *Earth Science Reviews* 54: 349–92.

Rundgren, M., N. Loader, and D. Beerling. 2000. Variations in the carbon isotope composition of late-Holocene plant macrofossils: A comparison of whole-leaf and cellulose trends. *Holocene* 10:149–54.

Ryan, M.G., E.R. Hunt, R. McMurtrie, G. Ågren, J. Aber, A.D. Friend, E.B. Rastetter, W. Pulliam, R.J. Raison, and S. Linder. 1996a. Comparing models of ecosystem function for temperate conifer forests. I. Model description and validation. In *Global change: Effects on coniferous forests and grasslands,* ed. A. Breymeyer, D.O. Hall, J. Melillo, and G. Ågren, 313–62. Chichester, U.K.: John Wiley.

Ryan, M.G., R. McMurtrie, G. Ågren, E.R. Hunt, J. Aber, A.D. Friend, E.B. Rastetter, and W. Pulliam. 1996b. Comparing models of ecosystem function for temperate conifer forests. II. Simulations of the effect of climate change. In *Global change: Effects on coniferous forests and grasslands,* ed. A. Breymeyer, D.O. Hall, J. Melillo, and G. Ågren, 363–87. Chichester, U.K.: John Wiley.

Sage, R.F. 2001. Environmental and evolutionary preconditions for the origin and diversification of the C_4 photosynthetic syndrome. *Plant Biology* 3:202–13.

Sage, R.F., and S. Cowling. 1999. Implications of stress in low CO_2 atmospheres of the past: are today's plants too conservative for a high CO_2 world? In *Carbon dioxide and environmental stress,* ed. Y. Luo, and H.A. Mooney, 289–308. San Diego: Academic Press.

Saxe, H., D. Ellsworth, and J. Heath. 1998. Tree and forest functioning in an enriched CO_2 atmosphere. *New Phytologist* 139:395–436.

Sharkey, T. 1985. Photosynthesis in intact leaves of C_3 plants: Physics, physiology, and rate limitations. *Botanical Review* 51:53–105.

Stitt, M. 1991. Rising CO_2 levels and their potential significance for carbon flow in photosynthetic cells. *Plant Cell and Environment* 14:741–62.

Thomas, S.C., and M. Jasienski. 1996. Genetic variability and the nature of microevolutionary responses to elevated CO_2. In *Carbon dioxide, populations, and communities,* ed. C. Körner, and F.A. Bazzaz, 3–12. San Diego: Academic Press.

Thornley, J.M.H., and M.G.R. Cannell. 1996. Temperate forest responses to carbon dioxide, temperature and nitrogen: A model analysis. *Plant Cell and Environment* 19: 1331–48.
———. 1997. Temperate grassland responses to climate change: An analysis using the Hurley pasture model. *Annals of Botany* 80:205–21.
Tissue, D., R. Thomas, and B. Strain. 1997. Atmospheric CO_2 enrichment increases growth and photosynthesis of *Pinus taeda*: A 4 year experiment in the field. *Plant Cell and Environment* 20:1123–34.
Tjoelker, M., J. Oleksyn, T. Lee, and P. Reich. 1999. Direct inhibition of leaf dark respiration by elevated CO_2 is minor in 12 grassland species. *New Phytologist* 150: 419–24.
von Caemmerer, S., J.R. Evans, G.S. Hudson, and T.J. Andrews. 1994. The kinetics of ribulose-1,5-bisphosphate carboxylase/oxygenase in vivo inferred from measurements of photosynthesis in leaves of transgenic tobacco. *Planta* 195:88–97.
Walker, B.H., W.L. Steffen, and J. Langridge. 1999. Interactive and integrated effects of global change on terrestrial ecosystems. In *The terrestrial biosphere and global change. Implications for natural and managed ecosystems,* ed. B.H. Walker, W.L. Steffen, J. Canadell, and J.S.I. Ingram, 329–75. Cambridge: Cambridge University Press.
Wand, S., G. Midgeley, M. Jones, and P.S. Curtis. 1999. Responses of wild C_4 and C_3 grass (*Poaceae*) species to elevated atmospheric CO_2 concentration: A meta-analytic test of current theories and perceptions. *Global Change Biology* 5:723–41.
Wang, Y.-P., A. Rey, and P.G. Jarvis. 1998. Carbon balance of young birch trees grown in ambient and elevated atmospheric CO_2 concentrations. *Global Change Biology* 4: 797–807.
Ward, J., J. Antonovics, R.B. Thomas, and B.R. Strain. 2000. Is atmospheric CO_2 a selective agent on model C_3 annuals? *Oecologia* 123:330–41.
Wong, S.C., I.R. Cowan, and G.D. Farquhar. 1979. Stomatal conductance correlates with photosynthetic capacity. *Nature* 282:424–26.
Woodward, F., and C. Kelly. 1995. The influence of CO_2 concentration on stomatal density. *New Phytologist* 131:311–27.
Wullschleger, S.D., T.J. Tschaplinski, and R.J. Norby. 2002. Plant water relations at elevated CO_2: Implications for water-limited environments. *Plant Cell and Environment* 25:319–31.
Zak, D., K. Pregitzer, J. King, and W. Holmes. 2000. Elevated atmospheric CO_2, fine roots and the response of soil microorganisms: A review and hypothesis. *New Phytologist* 147:201–22.
Zak, D.R., K.S. Pregitzer, P.S. Curtis, J.A. Teeri, R. Fogel, and D. Randlett. 1993. Elevated atmospheric CO_2 and feedback between carbon and nitrogen cycles. *Plant and Soil* 151:105–17.

21. Herbivory in a World of Elevated CO_2

Richard L. Lindroth and M. Denise Dearing

21.1 Introduction

All carbon fixed into plant biomass via photosynthesis is eventually consumed, either by herbivores (living tissues) or by saprophytes and detritivores (litter). Thus, the processes of herbivory and decomposition play pivotal roles in the cycling and storage of carbon in the biosphere (Fig. 21.1).

For several reasons, herbivory is uniquely important among ecological processes. First, it governs the flow of energy and nutrients to all higher trophic levels, thus strongly influencing the organization and dynamics of biological communities. Second, an extraordinary number of taxa are involved in the process; approximately half of known species consist of plants and their associated herbivores (Strong et al. 1984). Third, herbivory has contributed markedly as an agent of natural selection to the evolutionary radiation and diversification of both plants and animals (Spencer 1988; Sues 2000; Herrera and Pellmyr 2002). Interactions between herbivores and plants have been studied from a multitude of perspectives, ranging from the molecular basis of interactions to ecosystem level consequences and from evolutionary theory to agricultural application. A synthesis of these studies is beyond the scope of this chapter. Readers desiring additional information are encouraged to see recent reviews (Simmonds 1998; Barrett and Willis 2001; Herrera and Pellmyr 2002; Strauss et al. 2002).

Atmospheric CO_2 influences herbivory via a complex array of direct, indirect, and interactive processes (Fig. 21.2). For example, changing levels of atmo-

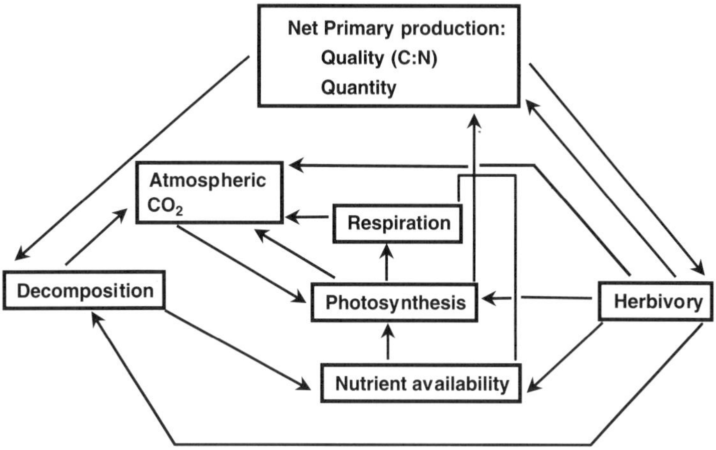

Figure 21.1. Processes affecting carbon cycling and storage in terrestrial ecosystems. Herbivory directly and indirectly influences the amounts and the rate of production of plant biomass.

spheric CO_2 are likely to alter plant chemical composition, which may alter host selection and population dynamics of herbivores. These changes, in turn, may effect shifts in nutrient cycling and community composition, thereby feeding back to influence concentrations of atmospheric CO_2.

Concentrations of atmospheric CO_2 have exhibited striking variation over the history of life on Earth (Chapters 1, 2, 4, and 5). In this chapter, we describe how such variation has likely served as a driver of, and constraint on, the evolution of plants, herbivores, and their interactions. We focus on the direct effects of CO_2 rather than effects mediated by climate change. Bale et al. (2002) recently reviewed the effects of climate change on plant-insect interactions.

21.2 Determinants of Herbivore Food Selection

A host of factors determines which plants are, and are not, selected as food by herbivores. These can be broadly categorized as factors affecting either the *quality* or *quantity* (e.g., availability) of food. While ultimately the diet of herbivores is determined by the interplay of both quality and quantity, for simplicity we initially discuss these factors independently.

21.2.1 Plant Quality

The three primary determinants of plant quality are nutrients, chemical defenses, and physical defenses (Schultz 1988; Lindroth 1989; Augner 1995). Herbivores derive all their nutrient requirements directly from plant tissues, or from sym-

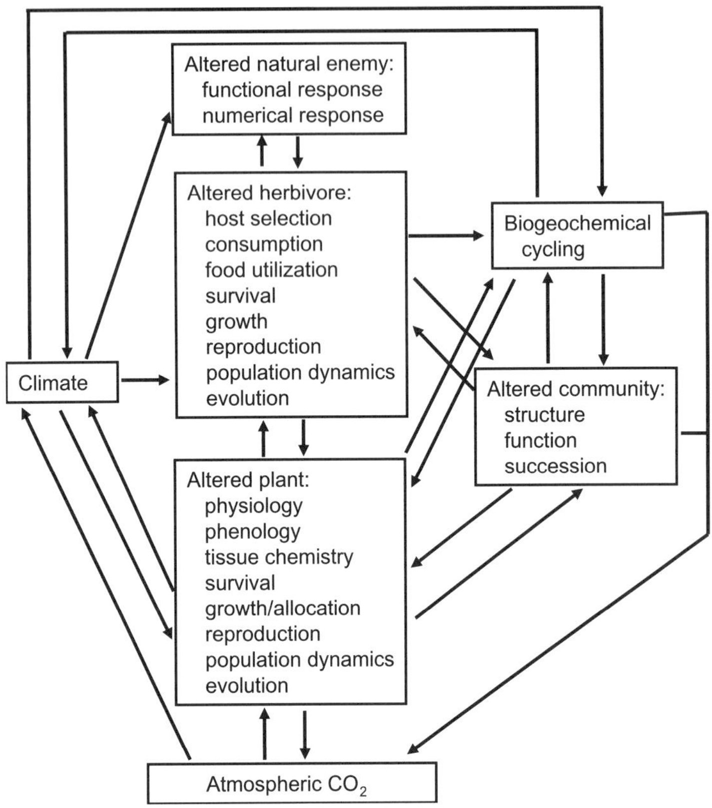

Figure 21.2. A complex of interacting factors determines herbivore responses to atmospheric CO_2 concentrations, with feed-forward and feedback effects culminating in changes in ecosystem structure and function (adapted from Lindroth 1996a).

biotic gut microflora (e.g., rumen microbes) that digest plant tissues. Thus, plant chemical composition with respect to essential nutrients, such as protein, carbohydrates, lipids, vitamins, and minerals, can determine a plant's suitability as food for herbivores. Of the various nutrients, protein (measured as total nitrogen) is generally regarded as the most limiting to herbivore growth and reproductive performance (Mattson 1980). Indeed, exceptionally low nutritive quality in some plant tissues has been postulated to evolve as a form of defense against herbivores (Augner 1995).

Plant chemical defenses encompass a vast number of secondary metabolites, including carbon-based compounds, such as phenolics and terpenoids, and nitrogen-based compounds, such as alkaloids and nonprotein amino acids. Generally recognized examples of secondary compounds include condensed tannins, which are phenolics present in beverages produced from red grapes and cranberries, and caffeine, which is an alkaloid present in coffee and chocolate. Although nearly all plant species produce multiple forms of secondary metabolites, individual plant taxa are often characterized by particular secondary metabolites (the basis for the discipline of chemotaxonomy). These compounds protect plants by causing toxicosis or disrupting digestive processes in herbivores.

Finally, a variety of physical defenses influence the quality of plant tissues as foods for herbivores. The "barbed wire syndrome" refers to the possession of various bristles, spines, thorns, and trichomes, which ostensibly evolved as defense against herbivores (Levin 1973; Grant 1984). Internal constituents can also provide a form of physical defense. The best known of these is silica, which accelerates tooth wear and reduces the digestibility of plant tissues (McNaughton et al. 1985). Fiber is another physical defense that is ubiquitous in photosynthetic tissues and can be present in high concentrations (up to 50% dry weight). Most animals lack the enzymes required to digest fiber constituents, such as cellulose. However, many vertebrate herbivores have circumvented this problem through symbiotic relationships with bacteria and protists capable of digesting fiber. Microbially aided digestion (fermentation) generates byproducts that herbivores use and rely on as a significant energy source (Robbins 1993).

These determinants of plant quality vary both among and within species, and over time. Variation in nutrient concentrations, chemical defenses, and physical defenses can be attributed to a number of factors, including inter- and intraspecific genetic variation; availability of critical resources (e.g., light, mineral nutrients); prior exposure to herbivory (e.g., induced defenses); and developmental stage (Denno and McClure 1983; Karban and Baldwin 1997). Genetic variation, the magnitude of which differs among different habitats (gene by environment interactions), provides the basis for differential selection and evolution of these plant traits.

Of course, herbivores have evolved an array of counter-adaptations such that no plant is completely resistant to all potential consumers. Herbivores have detoxification systems consisting of scores of enzymes capable of transforming secondary compounds from plants into inactive compounds that are then excreted from the animal's body (Scheline 1978; Lindroth 1991). In addition, many herbivores produce specialized salivary proteins that bind to dietary tannins and reduce the potential toxicity of these compounds (Hagerman and Robbins 1993; Dearing 1997a). Some herbivores behaviorally manipulate levels of secondary compounds by storing plants until toxins decay, or mix and match plant foods to minimize the dose of any one toxin (Dearing 1997b; Dearing and Cork 1999). These examples represent a few of the sundry strategies employed by herbivores to deal with plant secondary compounds.

21.2.2 Plant Availability

The absolute availability of plant material is largely a function of plant productivity, which is governed by levels of soil nutrients, light, water, temperature, and CO_2 (Bloom, Chapin, and Mooney 1985). Although plant species differ with respect to optimal levels of resources required for growth, in general productivity is greatest when none of these factors is limiting or in excess.

Plant availability can exhibit tremendous temporal variation in strongly seasonal environments and thus can have consequences for herbivores in such habitats. For example, the Great Plains in North America have abundant vegetation in the spring and summer, but far less plant material is available during the winter. Herbivores that use such ephemeral resources have a number of strategies for coping with temporal reductions in plant availability. Examples include bison that migrate to other habitats, voles that remain but feed on lower quality vegetation, and insects that overwinter in a dormant state.

Plant availability coupled with plant quality ultimately shape the diets and diet breadth of herbivores. When a choice is available, herbivores typically select plants of the highest quality: that is, low in secondary compounds and fiber and high in nutrients. Not surprisingly, high-quality forage is rare in nature (Demment and Van Soest 1985). The limited quantity of high-quality forage constrains specialists on this resource to small body sizes. Most herbivores larger than a kilogram are forced to feed on a variety of plant species that vary in quality and quantity (Demment and Van Soest 1985).

21.3 Effects of Elevated CO_2 on Plant Quality

Over the past several decades, the factors responsible for the evolution and diversification of plant chemical defenses have been the subject of considerable conjecture and debate (Bryant, Chapin, and Klein 1983; Coley, Bryant, and Chapin 1985; Herms and Mattson 1992; Hamilton et al. 2001). Central to many theories of plant defense is the notion that resource availability, particularly the balance of carbon to nutrient (e.g., nitrogen) availability, plays a major role in defining the types and amounts of chemical defenses accumulated by plants. Bryant, Chapin, and Klein (1983) introduced the "carbon-nutrient balance" hypothesis to explain intraspecific chemical variation on the basis of physiological responses of plants to shifts in the relative availability of carbon and nutrients. Coley, Bryant, and Chapin (1985) incorporated resource availability into explanations of interspecific variation in plant defense over evolutionary timescales in the "resource availability" hypothesis.

According to these resource-based theories of plant defense, the amount and type of chemical defense expressed by plants is typically determined by two variables interacting over evolutionary time: inherent plant growth rate and resource availability, both in absolute and in relative terms. Thus, when nutrient resources are limited, in absolute or relative amounts, plants with inherently slow growth rates will be favored over those with fast growth rates. In turn, slow growth rates favor production of carbon-based defenses; although many of these compounds (e.g., tannins) have high costs of initial investment, they are relatively stable (low turnover rates). Production of nitrogen-based defenses (e.g., alkaloids) would be cost-prohibitive, due to limited nitrogen availability or higher turnover rates, or both, for such compounds. Current examples of these types of plants would include many gymnosperms (e.g., spruce and fir trees), which tend to be slow growing, adapted to nutrient-deplete environments, and defended by large concentrations of such carbon-based secondary metabolites as tannins and terpenoids. Alternatively, when nutrient resources are abundant, plants with rapid growth rates will be favored. The relative cost of nitrogen-based defenses is reduced, so plants expressing these forms of defense are more common. Current examples include the brassicas (containing glucosinolates) and solanaceous plants (containing alkaloids).

21.3.1 Geologic History

We propose that fluctuating levels of atmospheric CO_2 mediated the evolution and diversification of plant chemical defenses during the history of life on Earth. CO_2 concentrations were extraordinarily high during most of the Paleozoic and Cenozoic Eras, as much as four times higher than present levels. Plant defense theory suggests that during periods of high CO_2 concentrations, plant growth rates may have been restricted primarily by nutrient (principally nitrogen) deficiencies (e.g., Oren et al. 2001). Allocation of nitrogen resources to N-based secondary metabolites would have exacted an exceptionally high cost in terms of plant growth or reproduction. Thus, the evolution of N-based defense systems would have been constrained to a greater extent than that of C-based systems. During the late Cretaceous period, however, coincident with the diversification of angiosperms, atmospheric CO_2 levels declined precipitously over a period of approximately 55 million years. At the beginning of the Tertiary, atmospheric CO_2 levels were only 35% to 40% of those of the previous 100 million years (Chapters 2, 4, and 5). CO_2 levels continued to decline during the Quaternary, but to a lesser extent and at a much more gradual rate. Thus at the onset of the Tertiary, the availability of nitrogen relative to that of carbon likely increased. We suggest that this shift may have facilitated the evolution and extraordinary diversification of nitrogen-based defenses in angiosperms (Table 21.1).

The evolution of nitrogen-based compounds vis-à-vis changes in atmospheric CO_2 levels is presented as a simple graphical model in Fig. 21.3. Under high levels of atmospheric CO_2, as occurred prior to the Cretaceous, the construction cost of C-based compounds would have been reduced relative to the cost under current conditions. As atmospheric CO_2 levels declined during the Cretaceous, the cost of C-based compounds would have increased. We suggest that this decline reduced the differential in cost between C-based and N-based defenses, such that N-based secondary metabolites increased in abundance, frequency, and diversity in rapidly growing plant species. We extend this idea to the extremely low levels of atmospheric CO_2 that occurred during the glacial periods of the Quaternary. During those times, atmospheric CO_2 concentrations were so low (<200 ppm) that plant growth rates were constrained by CO_2 availability (Chapter 11). These conditions would have accentuated the cost of C-based defenses,

Table 21.1. Evolution appearance of classes of defensive chemicals in plants, adapted from Swain (1978)

Class of Compound	Psilotum (400)*	Horsetails (370)	Ferns (320)	Gymnosperms (200)	Angiosperms (70)
Phenolics:					
Simple	+	+	+	+	+
Tannins	–	(+)	++	++	++
Terpenoids:					
Mono-	–	–	–	++	++
Sequi- and Di-	–	–	+	++	++
Tri-	–	–	+	++	++
N-Containing:					
Nonprotein amino acids	–	–	+	(+)	+
Alkaloids	–	–	–	(+)	++
Cyanogenic glycosides	–	–	(+)	(+)	+
Glucosinolates	–	–	–	–	+

• Age of dominance (million years ago)

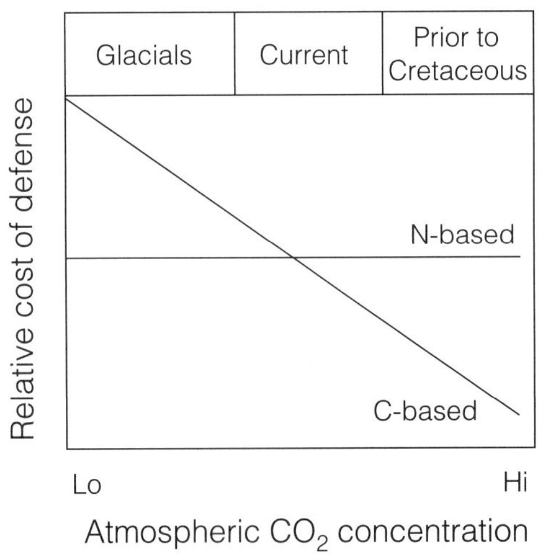

Figure 21.3. A model of how variation in concentrations of atmospheric CO_2 over geologic timescales likely influenced the evolution of chemical defense traits in plants. The relative costs of defense for compounds based on carbon (C) and nitrogen (N) are presented at varying atmospheric CO_2 concentrations. As carbon becomes less abundant during the Cretaceous, the cost of C-based compounds relative to N-based compounds increases such that plants should invest in N-based compounds. As carbon becomes even less abundant during the interglacials of the Quaternary, plants should increasingly invest in N-based compounds.

such that N-based defenses would become more prevalent during periods of low CO_2 concentrations. Thus, over geologic time scales, even recent geologic time such as the last 150 ky, optimal plant defense traits may have been strongly influenced by CO_2 concentrations.

The primary goal of this simplistic model is to illustrate that the relative cost of C-based defenses correlates negatively with CO_2 levels, whereas the relative costs of N-based defenses remain constant over variable CO_2 levels. A limitation of this model is that it does not take into account other factors that also influence allocations to plant defense. For example, life history traits such as leaf lifetime may interact with CO_2 levels to ultimately determine the cost of a particular defense. Furthermore, the model presents only the cost side of the equation. "Expensive" defensives may be selected for if the benefits in terms of future reproduction outweigh the costs. Thus is it feasible for individual plants to produce both C-based and N-based compounds when these other factors are incorporated.

21.3.2 Current and Future Trends

The concentrations of atmospheric CO_2 predicted to occur in the next one hundred years are relatively low compared to levels prior to the Cenozoic. However, the rates of change in concentrations documented over recent decades and predicted for the near future are much greater than experienced previously on Earth. Thus, biochemical and physiological adjustments of plants to high CO_2 conditions will occur without the benefit of significant periods of time for evolutionary adaptation. This consequence is especially relevant for long-lived species, such as most trees.

A growing body of literature describes the effects of enriched CO_2 on foliar chemistry. In general, levels of nitrogen and minerals decline, whereas those of C-based storage and defensive compounds (e.g., starch and tannins, respectively) increase (Lincoln, Fajer, and Johnson 1993; Le Thiec et al. 1995; McGuire, Miller, and Joyce 1995; Watt et al. 1995; Lindroth 1996b; Peñuelas et al. 1996; Poorter et al. 1997; Bezemer and Jones 1998). Overall, changes in chemical composition are consistent with carbon-nutrient balance theory (Koricheva et al. 1998), although many exceptions (especially for terpenoid compounds) exist. Considerable variability in response occurs, however, among species, among genotypes within species, and in relation to the availability of other plant resources such as nutrients and water (Lindroth, Kinney, and Platz 1993; Roth, McDonald, and Lindroth 1997; Koricheva et al. 1998; Lindroth, Roth, and Nordheim 2001).

A study by Lindroth, Roth, and Nordheim (2001) illustrates that a variety of interacting genetic and environmental factors determines the impact of high CO_2 concentrations on plant quality. Six aspen genotypes were grown under ambient and elevated (700 ppm) concentrations of CO_2, in low- or high-nutrient soil. Foliar nitrogen levels declined by an average of 16% under high CO_2, and the magnitude of decline did not vary among genotypes or in relation to nutrient availability. Starch concentrations doubled under high CO_2, and responses were again consistent across genotypes and nutrient conditions. Levels of the phenolic glycoside tremulacin were highly variable among genotypes. Moreover, genotypes varied in their response to CO_2, with most exhibiting increases, but others exhibiting decreases or no response, to CO_2. Levels of condensed tannins showed the most complex responses to enriched CO_2, with marked increases in concentrations in some, but not all, genotypes under low nutrient conditions, but no change in levels under high nutrient conditions. The complexity of these types of interactions, both within and between species, confounds the development of general predictions about the effects of CO_2 on plant quality.

Genetic variation in plant chemical response to elevated CO_2 provides the raw material necessary for evolutionary adaptations to occur. Interestingly, in the previously described study with aspen, phenotypic variance in chemical attributes was greater under enriched, rather than ambient, CO_2 concentrations (Lindroth, Roth, and Nordheim 2001). Thus, selection differentials may shift under future CO_2 conditions, affecting both the rate and direction of plant evolutionary responses.

Differences between C_3 and C_4 plants with respect to chemical responses to enriched atmospheric CO_2 are of particular interest. Under low levels of atmospheric CO_2, C_4 plants are less carbon-limited than are C_3 plants. Thus we would predict that the magnitude of chemical change would be less for C_4 plants than for C_3 plants as the world becomes increasingly CO_2-enriched. A meta-analysis of 62 papers comparing responses of C_4 and C_3 grasses to elevated CO_2 suggests this is the case (Wand et al. 1999). Foliar concentrations of nonstructural carbohydrates increased, whereas those of nitrogen decreased, in C_3 species. Levels of carbohydrates and nitrogen in C_4 species, however, were not significantly altered. Insufficient research has been conducted to determine whether C_3 and C_4 plants differ in terms of allelochemical production under high CO_2. Lack of research in this area may simply reflect the fact that most C_4 plants are monocotyledonous and are thus considered to be generally depauperate (relative to dicots) in terms of secondary metabolite production.

21.4 Effects of Elevated CO_2 on Plant Availability

The effects of elevated CO_2 on plant availability can be profound. Since approx 1990 more than 100 studies have been conducted on the effects of elevated CO_2 on plant productivity. These studies have evaluated representatives of the major plant functional groups, such as trees, shrubs, herbs, and grasses. Moreover, they included species that use the various photosynthetic pathways (C_3, C_4, CAM). In addition, they employed a wide range of experimental conditions, from growth chambers to in situ environments as afforded by Free Air CO_2 Enrichment (FACE) technology. A common finding has emerged from these various studies despite the large differences in the species and methods used. In general, elevation of CO_2 concentrations by approx 2x ambient increases plant productivity by approx 25% to 50% (Wand et al. 1999; Morgan et al. 2001; Curtis and Wang 1998).

21.4.1 Geologic History

The enormous changes in CO_2 concentration over geologic time presumably had huge effects on the productivity, and thus availability, of plants in the past. Over the past 125 million years, CO_2 concentrations have generally declined, although some oscillations, such as during the Pleistocene, have occurred. The responses of extant plants to high and low CO_2 concentrations suggest that plant productivity could have been much greater during the Cretaceous than the Tertiary. Many extant species of plants do not exhibit photosynthetic saturation until 1000 ppm CO_2 (Chapter 11). Thus, productivity of plants during the Cretaceous, when CO_2 levels were at least 1000 ppm, could have exceeded that of current plants by 50%.

Data on the effects of CO_2 on plant availability over geologic time are scarce. However, changing levels of CO_2 may have affected at least three well-documented changes in plant availability: the evolution of angiosperms in the Cretaceous, the expansion of the grasslands in the Miocene, and the changes in plant abundance and distribution during the Pleistocene. Because these changes have been deduced in part from the fossil remains of herbivores, these changes are discussed in Section 5.1 below.

21.4.2 Current and Future Trends

Plants with different photosynthetic pathways (C_3 vs. C_4) are predicted to have disparate responses to increases in CO_2 concentrations. Plants that use the C_4 photosynthetic pathway have been hypothesized to be relatively insensitive to elevated CO_2 concentrations in terms of productivity compared to that of C_3 species (Henderson et al 1994). This difference is due to the ability of C_4 species to concentrate CO_2 inside the leaf to levels that saturate photosynthetic enzymes, even at very low ambient CO_2 concentrations. In contrast, levels of CO_2 within leaves of C_3 species roughly reflect ambient concentrations, which are below the level of enzymatic saturation (ca. 1000 ppm). Thus, productivity of C_3 species may be positively correlated with CO_2 concentration whereas productivity of C_4 species may be constant regardless of CO_2 concentration. For this reason, C_3 species have been predicted to have a greater productivity response to increasing atmospheric CO_2 than C_4 species (Bazzaz 1990; Bowes 1993; Ehleringer and Monson 1993). Moreover, this difference in performance was predicted to give C_3 plants a competitive advantage over C_4 plants, such that areas currently dominated by C_4 species would become increasingly dominated by C_3 species under CO_2 conditions predicted for the near future.

Unexpectedly, several recent studies have demonstrated the opposite from that predicted based on photosynthetic pathways. C_4 plants exhibit greater productivity under elevated CO_2 compared to current levels and the response is comparable or even superior to that of C_3 species. Wand et al. (1999) conducted a meta-analysis of the results of elevated CO_2 studies on C_3 and C_4 grasses in the Poaceae. Total biomass of C_4 species increased by 33% whereas total biomass of C_3 species increased by 44%. The CO_2 response in C_3 species was due to an increase in tiller biomass whereas C_4 species increased leaf area. In a long-term study of species in tallgrass prairie, C_4 species showed no change in percent abundance after 8 years of elevated CO_2 whereas C_3 species declined significantly (Owensby et al. 1999). However, this response appears to be highly dependent on the ecosystem. Morgan et al. (2001) found no change in the relative abundance of C_3- C_4 species in the shortgrass prairie with increasing CO_2.

21.5 Responses of Herbivores to CO_2-Mediated Changes in Vegetation

As expressed by Thompson (1999): ". . . interactions between species are as evolutionarily malleable as the species themselves and have played a central role in the diversification and organization of life." Such is certainly true for herbivory as a means for the acquisition of essential nutrients by animals. Herbivory evolved frequently and independently in numerous phylogenetic lineages over the past 300 million years of life on Earth (Weis and Berenbaum 1989; Sues 2000). Indeed, the advent and proliferation of herbivory as a dominant feeding form was fundamentally important in the evolution of life, contributing to explosive speciation (driven by coevolutionary interactions) as well as to the establishment of expansive, grazing-dominated ecosystems. We suggest that atmospheric CO_2, through its effects on plant quality and availability, acted as an abiotic driver of interactions between plants and herbivores, with consequent effects on the emergence, composition, and dynamics of entire ecosystems.

21.5.1 Geologic History

The fossil record provides little in the way of evidence of trophic interactions between plants and animals, so inferences about such interactions must be drawn from data such as biochemistry (Swain 1978), morphology (Sues 2000), and parallel cladograms (Farrell, Mitter, and Futuyma 1992). The ideas we present here are thus speculative but nonetheless suggest a key role for atmospheric CO_2 in mediating the evolution of plant-herbivore associations.

Shifting concentrations of atmospheric CO_2 over geologic timescales likely impacted herbivorous animals via changes in plant chemical composition. Indeed, Swain (1978) postulated that the extinction of dinosaurs may have been caused by the appearance and proliferation of alkaloids as flowering plants became dominant on Earth! Given, however, that various lineages of dinosaurs exhibited divergent and specialized morphological and digestive adaptations for accommodating plant diets, we find it difficult to believe that the reptiles were incapable of evolving effective metabolic adaptations for dealing with alkaloids in their diets. For example, extant herbivorous reptiles discern between foods with and without alkaloids (Schall 1990).

Did dinosaurs and angiosperms coevolve, and were changes mediated to any extent by CO_2? Recent work suggests that major groups of herbivorous dinosaurs were affected little by the initial radiation of angiosperms; how they responded to shifts in the distribution of flowering plants by the late Cretaceous remains speculative (Weishampel and Jianu 2000). However, during approx 25 million years of the Cretaceous, a distinct increase in diversity occurred in dinosaurs with feeding morphologies apparently suited to high fiber diets (Weishampel and Jianu 2000). The increase was in part due to the appearance of the ceratopsians with novel parrotlike bills, as exemplified by *Triceratops*. The emergence of these unique feeding morphologies is suggestive of an increase in abundance of fiber-rich plants. It is possible that declining CO_2 levels during the Cretaceous resulted in more fibrous foliage as fiber content is inversely correlated with CO_2 levels (Runion et al. 1999).

A stronger argument can be made for the role of atmospheric CO_2 as a forcing factor in the coevolution of plant-insect associations. The explosive radiation of flowering plants and herbivorous insects is attributed, at least in part, to the reciprocal nature of their coevolutionary relationships (Ehrlich and Raven 1964; Farrell, Mitter, and Futuyma 1992). Indeed, evolutionary interactions between plants and insects are considered responsible, directly or indirectly, for much of the diversity of terrestrial life on Earth (Farrell, Mitter, and Futuyma 1992). We suggest that low atmospheric concentrations of CO_2 in the Cretaceous afforded plants with an entirely new form of armament against herbivorous insects: N-based allelochemicals. Concomitant with a reduction in the relative "cost" of N-based defenses, entirely new biosynthetic pathways evolved for the production of novel alkaloids, glucosinolates, cyanogenic glycosides, and so forth. The proliferation of various forms of these defenses was likely driven, at least in part, by coevolutionary interactions between plants and specialist insect feeders.

Changes in atmospheric CO_2 concentrations over geologic timescales also likely influenced herbivorous animals through changes in plant availability. One of the best-documented changes in plant availability occurred during the Miocene, when C_4 grasses expanded on a global scale (Cerling et al. 1997). The forcing factor behind the C_4 grass expansion is thought to be the interplay of low CO_2 and increased temperature (Ehleringer, Cerling, and Helliker 1997; Pagini, Freeman, and Arthur 1999). Major changes in the vertebrate herbivore fauna occurred concomitant to the grassland expansion. In general, species diversity of woodland herbivores declined while herbivores suited for savannas increased (Cerling et al. 1997). A guild of grazing vertebrates appeared for the first time in history (MacFadden 2000). The evolution of grazers during the Miocene is exemplified by the radiation of the Bovidae (cow family), a group defined by its specializations for grassland feeding (Van Soest 1994).

Changes in overall plant productivity during the Miocene may also have affected communities of mammalian herbivores. CO_2 concentrations during the

Miocene were low enough (180–320 ppm; Pagani, Freeman, and Arthur 1999) to potentially cause a significant decrease in plant productivity, particularly in C_3 species. Herbivores, such as browsers that fed almost exclusively on C_3 vegetation, would have been especially susceptible to such changes. Janis, Damuth, and Theodor (2000) proposed that the decrease in species diversity of ungulate browsers during the Miocene was due to reduced plant productivity mediated by declining CO_2 levels. Furthermore, the decline in the diversity of grazing mammals, particularly horses, at the end of the Miocene, may have been the result of declining plant productivity (MacFadden 2000).

Low CO_2 concentrations have also been implicated in the extinctions of the Pleistocene megafauna. Gutherie (1984) proposed that environmental change during the last glacial period caused a decrease in plant availability. Moreover, they suggested that the predominant plant defenses (alkaloids, cyanide) selected for under this climatic regime would have been more toxic to the megafauna (mammals with primarily simple-stomach digestion, e.g., mammoths) than to ruminants (e.g., caribou). They base this notion on the premise that the microflora of ruminants detoxify compounds such as alkaloids whereas the absence of a forestomach with microbes would have subjected the megafauna to poisoning. Differential detoxification of plant toxins by monogastrics and ruminants is still unsubstantiated. However, our model of plant defenses based on resource availability is consistent with Gutherie's predicton that alkaloids and cyanide (both N-based defenses) should have been more common during the glacial period.

Cowling (1999) presented an alternative interpretation for megafaunal extinctions. She suggested that levels of plant defense are positively correlated with concentrations of atmospheric CO_2. Her model predicts that plants are most nutritious under low CO_2. She suggested that as CO_2 levels rose during the last interglacial, concentrations of secondary compounds in plants also increased. The increased toxin concentrations in the diets of herbivores may have given humans an advantage in overhunting megafauna. A potential problem with her model of plant defense, however, is that she predicts both N-based and C-based defenses to have responded similarly to changes in CO_2.

21.5.2 Current and Future Trends

Levels of atmospheric CO_2 predicted for the next one hundred years are unlikely to have detectable *direct* effects on herbivore fitness parameters, such as survival, growth, and reproduction (e.g., Agrell, McDonald, and Lindroth 2000; Kopper, Welson, and Lindroth unpublished data). In cases in which CO_2 serves as a sensory cue, however, elevated levels may disrupt herbivore behavior. For example, the pyralid moth *Cactoblastis cactorum* uses CO_2 gradients around potential host plants as stimuli for oviposition. For this moth, the rates of oviposition decline when host plants are exposed to high levels of CO_2 (Stange 1997).

In contrast to the minimal direct effects of CO_2 on herbivore performance, effects mediated through changes in food quality are anticipated to be significant. Numerous studies have evaluated such impacts for the performance of insect herbivores, as reviewed most recently by Bezemer and Jones (1998), Coviella and Trumble (1999), and Whittaker (1999). For leaf-chewing insects, high CO_2 environments generally lead to increased consumption rates (likely compensatory feeding responses due to dilute foliar nitrogen), prolonged development times, reduced efficiencies of conversion of food into insect biomass, and minor reductions or no changes in herbivore growth and survival. Long-term impacts at the population level are poorly understood for leaf-chewing insects; the few studies conducted to date indicate that the effects of CO_2-altered plant quality may or may not increase over multiple insect generations (Brooks and Whittaker 1998; Kopper 2001). Sap-feeding (phloem or xylem) insects tend to show no response or improved performance under high CO_2 (reviewed by Bezemer and Jones 1998; Whittaker 1999). These insects are especially amenable to population-level studies, which have revealed positive, negative, and no detectable effects of enriched CO_2 (Bezemer and Jones 1998; Whittaker 1999). An important caveat to this brief summary of insect performance under high CO_2 concentrations is that responses are highly specific with respect to both plant and insect species, as well as to environmental conditions.

Although the majority of herbivorous insects are specialists, many of the most important insect pests and virtually all of the herbivorous mammals are generalists, feeding on plants from a broad range of taxonomic affiliations. This generalist feeding habit, coupled with differential responses of plant species to enriched CO_2, suggests that host preferences of herbivores may shift in the future. Because herbivory is an important determinant of community dynamics (e.g., plant succession and competition), shifts in host preferences could lead to changes in plant community composition (see Fig. 21.2). Such responses have rarely been investigated, although Peters et al. (2000) found that herbivorous slugs shift preferences from nonlegumes to legumes under high CO_2. For grazing mammals in particular, differential responses of C_3 and C_4 plants to enriched CO_2 are likely to elicit shifts in food preferences. In addition to influencing herbivore fitness via bottom-up (plant-related) processes, high levels of atmospheric CO_2 may also affect herbivores via top-down (natural enemy) processes (see Fig. 21.2). The few studies conducted to date indicate that organisms at the third trophic level are unlikely to be negatively affected by CO_2-mediated changes in plant quality as expressed in terms of altered prey quality (Roth and Lindroth 1995; Lindroth et al. 1997; Holton et al. 2001). In other words, higher trophic levels appear to be buffered from the changes in plant chemical composition caused by high CO_2. Moreover, enriched CO_2 may cause insect herbivores to have increased susceptibility to natural enemies due to prolonged development times, or due to shifts in the feeding efficiency of the enemies themselves. Stiling et al. (1999) documented increased parasitization of leaf miners, and Awmack, Woodcock, and Harrington (1997) reported reductions in alarm pheromone responses of aphids in high CO_2 environments.

Perhaps the most marked effects of enriched atmospheric CO_2 on herbivores will be mediated through climate change, particularly warming (see Fig. 21.2). Warmer temperatures affect the physiology, growth, development, phenology, and spatial distribution of herbivores (especially so for heterotherms) as well as those of the plants on which they feed. Such indirect effects of CO_2 are beyond the scope of this chapter and have been covered in other recent reviews (Hughes 2000; Bale et al. 2002).

21.6 Summary

Atmospheric CO_2 has likely served as a significant driver of, and constraint on, the evolution of plants, herbivores, and their interactions. We propose that CO_2 has had a major impact on the nutritional quality and quantity of plants over evolutionary time. Low levels of CO_2 may have provided opportunity for evolution of novel N-based plant compounds, which in turn provided the raw material for an explosive diversification of interacting plant and animal species. In addition, changes in plant abundance and community structure mediated by alterations in atmospheric CO_2 likely had significant impacts on herbivore communities. Although levels of CO_2 predicted to occur in the next century are lower than concentrations witnessed in the past, the rate of change may preclude concomitant evolution of some plant-herbivore assemblages.

Acknowledgments. We thank P. Coley for invaluable comments on plant defense theory and N. Lindroth for preparation of the figures. The National Science Foundation (Ecology Program) and USDA (National Research Initiatives Competitive Grants Program) supported the research of R.R. Lindroth described in this chapter. The National Science Foundation (IBN 007–9865) and the Packard Foundation supported the research of M.D. Dearing.

References

Agrell, J., E.P. McDonald, and R.L. Lindroth. 2000. Effects of CO_2 and light on tree phytochemistry and insect performance. *Oikos* 88:259–72.

Augner, M. 1995. Low nutritive quality as a plant defense: Effects of herbivore-mediated interactions. *Evolutionary Ecology* 9:605–16.

Awmack, C.S., C.M. Woodcock, and R. Harrington. 1997. Climate change may increase vulnerability of aphids to natural enemies. *Ecological Entomology* 22:366–68.

Bale, J.S., G.J. Masters, I.D. Hodkinson, C.S. Awmack, T.M. Bezemer, V.K. Brown, J. Butterfield, A. Buse, J.C. Coulson, J. Farrar, J.E.G. Good, R. Harrington, S.E. Hartley, T.H. Jones, R.L. Lindroth, M.C. Press, I. Symrnioudis, A.D. Watt, and J.B. Whittaker. 2002. Herbivory in global climate change research: Direct effects of rising temperature on insect herbivores. *Global Change Biology* 8:1–16.

Barrett, P.M., and K.J. Willis. 2001. Did dinosaurs invent flowers? Dinosaur-angiosperm coevolution revisited. *Biological Reviews* 76:411–47.

Bazzaz, F.A. 1990. The response of natural ecosystems to the rising global CO_2 levels. *Annual Review of Ecology and Systematics* 21:167–96.

Bezemer, T.M., and T.H. Jones. 1998. Plant-insect herbivore interactions in elevated atmospheric CO_2: Quantitative analyses and guild effects. *Oikos* 82:212–22.

Bloom, A.J., F.S. Chapin III, and H.A. Mooney 1985. Resource limitation in plants: An economic analogy. *Annual Review of Ecology and Systematics* 16:363–92.

Bowes, G. 1993. Facing the inevitable: Plants and increasing atmospheric CO_2. *Annual Review of Plant Physiology and Plant Molecular Biology* 44:309–32.

Brooks, G.L., and J.B. Whittaker. 1998. Responses of multiple generations of *Gastrophysa viridula*, feeding on *Rumex obtusifolius*, to elevated CO_2. *Global Change Biology* 4:63–75.

Bryant, J.P., F.S. Chapin, and D.R. Klein. 1983. Carbon/nutrient balance of boreal plants in relation to vertebrate herbivory. *Oikos* 40:357–68.

Cerling, T.E., J.M. Harris, B.J. MacFadden, M.G. Leakey, J. Quade, V. Eisenmann, and J.R. Ehleringer. 1997. Global vegetation change through the Miocene/Pliocene boundary. *Nature* 389:153–58.

Coley, P.D., J.P. Bryant, and F.S. Chapin. 1985. Resource availability and plant antiherbivore defense. *Science* 230:895–99.

Coviella, C., and J.T. Trumble. 1999. Effects of elevated atmospheric CO_2 on insect-plant interactions. *Conservation Biology* 13:700–712.

Cowling, S.A. 2001. Plant carbon balance, evolutionary innovation and extinction in land plants. *Global Change Biology* 7:231–39.

Curtis, P.S., and X. Wang. 1998. A meta-analysis of elevated CO_2 effects on woody plant mass, form, and physiology. *Oecologia* 113:299–313.

Dearing, M.D. 1997a. The effects of *Acomastylis rossii* tannins on a mammalian herbivore, the North American pika, *Ochotona princeps*. *Oecologia* 109:122–31.

Dearing, M.D. 1997b. The manipulation of plant secondary compounds by a food-hoarding herbivore, the North American pika, *Ochotona princeps*. *Ecology* 78:774–81.

Dearing, M.D., and S. Cork. 1999. The role of detoxification of plant secondary compounds on diet breadth in mammalian herbivores. *Journal of Chemical Ecology* 25: 1205–20.

Demment, M.W., and P.J. Van Soest. 1985. A nutritional explanation for body-size patterns of ruminant and nonruminant herbivores. *The American Naturalist* 125:641–72.

Denno, R.F., and M.S. McClure, eds. 1983. *Variable plants and herbivores in natural and managed systems*. New York: Academic Press.

Ehleringer, J.R., T.E. Cerling, and B.R. Helliker. 1997. C_4 photosynthesis, atmospheric CO_2, and climate. *Oecologia* 112:285–99.

Ehleringer, J.R., and R.K. Monson. 1993. Evolutionary and ecological aspects of photosynthetic pathway variation. *Annual Review of Ecology and Systematics* 24:411–39.

Ehrlich, P.R., and P.H. Raven. 1964. Butterflies and plants: A study in coevolution. *Evolution* 18:586–608.

Farrell, B.D., C. Mitter, and D.J. Futuyma. 1992. Diversification at the insect-plant interface. *Bioscience* 42:34–42.

Grant, S. 1984. *Beauty and the beast: The coevolution of plants and animals*. New York: Charles Scribner's Sons.

Gutherie, R.D. 1984. Mosaics, allelochemicals, and nutrients: An ecological theory of late-Pleistocene megafaunal extinctions. In *Quarternary extinctions: A prehsitoric revolution,* ed. P.S. Martin, R.G. Klein, 259–98. Tucson: University of Arizona Press.

Hagerman, A.E., and C.T. Robbins. 1993. Specificity of tannin-binding salivary proteins relative to diet selection by mammals. *Canadian Journal of Zoology* 71:628–33.

Hamilton, J.G., A.R. Zangerl, E.H. DeLucia, and M.R. Berenbaum. 2001. The carbon-nutrient balance hypothesis: Its rise and fall. *Ecology Letters* 4:86–95.

Henderson, S., P. Hattersley, S. von Caemmerer, and C.B. Osmond. 1994. Are C_4 pathway plants threatened by global climatic change? In Ecophysiology of photosynthesis, ed. E.D. Schulze and M.M. Caldwell, 529–49. Heidelberg: Springer-Verlag..

Herms, D.A., and W.J. Mattson. 1992. The dilemma of plants: To grow or defend. *Quarterly Review of Biology* 67:283–335.

Herrera, C.M., O. Pellmyr, eds. 2002. *Plant-Animal Interactions: An evolutionary approach.* Malden, Mass.: Blackwell Science.

Holton, M.K., R.L. Lindroth, and E.V. Nordheim. 2003. Foliar quality influences tree-herbivore-parasitoid interactions: effects of elevated CO_2, O_3, and genotype. Oecologia 137:233-244.

Hughes, L. 2000. Biological consequences of global warming: is the signal already apparent? *Trends in Ecology and Evolution* 15:56–61.

Janis, C., M.J. Damuth, and J.M. Theodor. 2000. Miocene ungulates and terrestrial primary productivity: Where have all the browsers gone? *Proceedings of the National Academy of Science* 97:7899–7904.

Karban, R., and I.T. Baldwin. 1997. *Induced responses to herbivory.* Chicago: University of Chicago Press.

Kopper, B.J. 2001. Consequences of elevated carbon dioxide and ozone for interactions between deciduous trees and lepidopteran folivores. Ph.D. thesis. Madison, Wisc.: University of Wisconsin.

Kopper, B.J., A.J. Weldon, and R.L. Lindroth. [submitted] Direct effects of elevated CO_2 and O_3 on caterpillar performance. *Ecological Entomology.*

Koricheva, J., S. Larsson, E. Haukioja, and M. Keinänen. 1998. Regulation of woody plant secondary metabolism by resource availability: Hypothesis testing by means of meta-analysis. *Oikos* 83:212–26.

Le Thiec, D., M. Dixon, P. Loosveldt, and J.P. Garrec. 1995. Seasonal and annual variations of phosphorus, calcium, potassium, and manganese contents in different cross-sections of *Picea abies* (L.) Karst. needles and *Quercus rubra* L. leaves exposed to elevated CO_2. *Trends in Ecology and Evolution* 10:55–62.

Levin, D.A. 1973. The role of trichomes in plant defense. *Quarterly Review of Biology* 48:3–15.

Lincoln, D.E., E.D. Fajer, and R.H. Johnson. 1993. Plant-insect herbivore interactions in elevated CO_2 environments. *Trends in Ecology and Evolution* 8:64–68.

Lindroth, R.L. 1989. Mammalian herbivore–plant interactions. In *Plant-animal interactions,* ed. W.G. Abrahamson, 163–206. New York: McGraw-Hill.

———. 1991. Differential toxicity of allelochemicals to insects: Roles of enzymatic detoxication systems. In *Insect-plant interactions,* vol. 3, ed. E.A. Bernays, 1–33. Boca Raton, Fla.: CRC Press.

———. 1996a. Consequences of elevated atmospheric CO_2 for forest insects. In *Carbon dioxide, populations, and communities,* ed. C. Körner and F.A. Bazzaz, 347–61. San Diego: Academic Press.

———. 1996b. CO_2-mediated changes in tree chemistry and tree-Lepidoptera interactions. In *Carbon dioxide and terrestrial ecosystems,* ed. G.W. Koch and H.A. Mooney, 105–20. San Diego: Academic Press.

Lindroth, R.L., K.K. Kinney, and C.L. Platz. 1993. Responses of deciduous trees to elevated atmospheric CO_2: Productivity, phytochemistry, and insect performance. *Ecology* 74:763–77.

Lindroth, R.L., S. Roth, E.L. Kruger, J.C. Volin, and P.A. Koss. 1997. CO_2-mediated changes in aspen chemistry: Effects on gypsy moth performance and susceptibility to virus. *Global Change Biology* 3:279–89.

Lindroth, R.L., S. Roth, and E.V. Nordheim. 2001. Genotypic variation in response of quaking aspen (*Populus tremuloides*) to atmospheric CO_2 enrichment. *Oecologia* 126: 371–79.

MacFadden, B.J. 2000. Origin and evolution of the grazing guild in Cenozoic New World terrestrial mammals. In *Evolution of herbivory in terrestrial vertebrates: Perspectives from the fossil record,* ed. H.-D. Sues, 223–44. Cambridge: Cambridge University Press.

Mattson, W.J. 1980. Herbivory in relation to plant nitrogen content. *Annual Review of Ecology and Systematics* 11:119–61.

McGuire, A.D., J.M. Melillo, and L.A. Joyce. 1995. The role of nitrogen in the response of forest net primary production to elevated atmospheric carbon dioxide. *Annual Review of Ecology and Systematics* 26:473–503.

McNaughton, S.J., J.L. Tarrants, M.M. McNaughton, and R.H. Davis. 1985. Silica as a defense against herbivory and a growth promoter in African grasses. *Ecology* 66:528–35.

Morgan, J.A., D.R. Lecain, A.R. Mosier, and D.G. Milchunas. 2001. Elevated CO_2 enhances water relations and productivity and affects gas exchange in C_3 and C_4 grasses of the Colorado shortgrass steppe. *Global Change Biology* 7:451–56.

Oren, R., D.S. Ellsworth, K.H. Johnsen, N. Phillips, B.E. Ewers, C. Maier, K.V.R. Schäfer, H. McCarthy, G. Hendrey, S.G. McNulty, and G.G. Katul. 2001. Soil fertility limits carbon sequestration by forest ecosystems in a CO_2-enriched atmosphere. *Nature* 411:469–72.

Owensby, C.E., J.M. Ham, A.K. Knapp, and L.M. Auen. 1999. Biomass production and species composition change in a tallgrass prairie ecosystem after long-term exposure to elevated atmospheric CO_2. *Global Change Biology* 5:497–506.

Pagani, M., K.H. Freeman, and M.A. Arthur. 1999. Late Miocene atmospheric CO_2 concentrations and the expansion of C_4 grasses. *Science* 285:876–79.

Peñuelas, J., M. Estiarte, B.A. Kimball, S.B. Idso, P.J. Pinter, G.W. Wall, R.L. Garcia, D.J. Hansaker, R.L. LaMorte, and D.L. Hendrix. 1996. Variety of responses of plant phenolic concentration to CO_2 enrichment. *Journal of Experimental Botany* 47:1463–67.

Peters, H.A., B. Baur, F. Bazzaz, and C. Körner. 2000. Consumption rates and food preferences of slugs in a calcareous grassland under current and future CO_2 conditions. *Oecologia* 125:72–81.

Poorter, H., Y. Van Berkel, R. Baxter, J. Den Hertog, P. Dijkstra, R.M. Gifford, K.L. Griffin, C. Roumet, J. Roy, and S.C. Wong. 1997. The effect of elevated CO_2 on the chemical composition and construction costs of leaves of 27 C_3 species. *Plant, Cell and* Environment 20:472–82.

Robbins, C.T. 1993. *Wildlife feeding and nutrition,* 2d ed. New York: Academic Press.

Roth, S., E.P. McDonald, and R.L. Lindroth. 1997. Atmospheric CO_2 and soil water availability: Consequences for tree-insect interactions. *Canadian Journal of Forest Research* 27:1281–90.

Roth, S.K., and R.L. Lindroth. 1995. Elevated atmospheric CO_2: Effects on phytochemistry, insect performance and insect-parasitoid interactions. *Global Change Biology* 1: 173–82.

Runion, G.B., J.A. Entry, S.A. Prior, R.J. Mitchell, H.H. Rogers. 1999. Tissue chemistry and carbon allocation in seedlings of *Pinus palustris* subjected to elevated atmospheric CO_2 and water stress. *Tree Physiology* 19:329–35.

Schall, J.J. 1990. Aversion of whiptail lizards cnemidophorus to a model alkaloid. *Herpetologica* 46:34–38.

Scheline, R.R. 1978. *Mammalian metabolism of plant Xenobiotics.* New York: Academic Press.

Schultz, J.C. 1988. Many factors influence the evolution of herbivore diets, but plant chemistry is central. *Ecology* 69:896–97.

Simmonds, M.S.J. 1998. Chemoecology: The legacy left by Tony Swain. *Phytochemistry* 49:1183–90.

Spencer, K.C., ed. 1988. *Chemical mediation of coevolution.* New York: Academic Press.

Stange, G. 1997. Effects of changes in atmospheric carbon dioxide on the location of hosts by the moth, *Cactoblastis cactorum. Oecologia* 110:539–45.

Stiling, P., A.M. Rossi, B. Hungate, P. Dijkstra, C.R. Hinkle, W.M. Knott, and B. Drake. 1999. Decreased leaf-miner abundance in elevated CO_2: Reduced leaf quality and increased parasitoid attack. *Ecological Applications* 9:240–44.

Strauss, S.Y., J.A. Rudgers, J.A. Lau, R.E. Irwin. 2002. Direct and ecological costs of resistance to herbivory. *Trends in Ecology and Evolution* 17:278–85.
Strong, D.R., J.H. Lawton, and R. Southwood. 1984. *Insects on plants.* Cambridge, Mass.: Harvard University Press.
Sues, H.-D., ed. 2000. Evolution of herbivory in terrestrial vertebrates: Perspectives from the fossil record. Cambridge: Cambridge University Press.
Swain, T. 1978. Plant-animal coevolution: A synoptic view of the Paleozoic and Mesozoic. *Annual Proceedings of the Phytochemical Society of Europe* 15:3–19.
Thompson, J.N. 1999. The evolution of species interactions. *Science* 284:2116–18.
van Soest, P.J. 1994. *Nutritional ecology of the ruminant.* Ithaca: Cornell University Press.
Wand, S.J.E., G.F. Midgley, M.H. Jones, and P.S. Curtis. 1999. Responses of wild C_4 and C_3 grass (Poaceae) species to elevated atmospheric CO_2 concentration: A meta-analytic test of current theories and perceptions. *Global Change Biology* 5:723–41.
Watt, A.D., J.B. Whittaker, M. Docherty, G. Brooks, E. Lindsay, and D.T. Salt. 1995. The impact of elevated atmospheric CO_2 on insect herbivores. In *Insects in a changing environment,* ed. R. Harrington and N.E. Stork, 197–217. New York: Academic Press.
Weis, A.E., and M.R. Berenbaum. 1989. Herbivorous insects and green plants. In *Plant-animal interactions,* ed. W.G. Abrahamson, 123–62. New York: McGraw-Hill.
Weishampel, D.B., and C.-M. Jianu. 2000. Plant-eaters and ghost lineages: Dinosauran herbivory revisited. In *Evolution of herbivory in terrestrial vertebrates: Perspectives from the fossil record,* ed. H.-D. Sues, 123–43. Cambridge: Cambridge University Press.
Whittaker, J.B. 1999. Impacts and responses at population level of herbivorous insects to elevated CO_2. *European Journal of Entomology* 96:149–56.

22. Borehole Temperatures and Climate Change: A Global Perspective

Robert N. Harris and David S. Chapman

22.1 Introduction

The rise in atmospheric concentrations of CO_2 and other radiatively active gases since the industrial revolution has significantly modified the atmosphere and the terrestrial environment. One expected result of increasing concentrations of such greenhouse gases is surface warming. Over the past 150 years or so, the period for which instrumental records are widely available, global mean surface air temperatures (SAT) have increased 0.6°C (Jones et al. 1999). However, to assess the full spectrum of temperature change and climate variability over the past 500 to 1000 years, it is necessary to turn to proxy indicators of surface temperature change. Proxy records of temperature change includes dendroclimatic measurements (tree-ring width and density), layer thicknesses of laminated sediment cores; accumulation of annually resolved ice cores; and coral isotopes. Global networks of these multiproxy data have proven invaluable at assessing global and hemispheric patterns of climate in past centuries (Mann, Bradley, and Hughes 1998, 1999; Jones et al. 1998; Crowley and Lowery 2000). Overall, these data indicate that the observed increase in SAT during the past century is unprecedented during the past 1000 years (Mann, Bradley, and Hughes 1998). While reconstructions of temperature change from conventional proxy indicators offer high temporal resolution of relative climatic changes, uncertainties in the calibration procedures, the long-term mean temperature, and the ability to resolve low-frequency components remain problematic.

One method of temperature reconstruction that is a direct measure of changing surface temperature and is not currently incorporated into proxy networks comprises temperature-depth profiles measured in boreholes drilled to depths of 200 m or deeper. These temperature records are a direct measure of changing thermal conditions at the ground surface (Lachenbruch and Marshall 1986; Pollack and Chapman 1993; Harris and Chapman 2001), thus bypassing the temperature calibration step inherent with proxy records of climate change. Further, as we will show, these temperature profiles have information in common with surface air temperature (SAT) records (Harris and Chapman 1997, 1998), supporting the contention that these records can be fruitfully employed to investigate climate change. Additionally, because thermal diffusion in the Earth filters out short-period surface temperature fluctuations but retains long-period temperature information, borehole temperatures hold the promise of improving temperature reconstructions based on high-resolution multi-century paleoclimatic data.

22.2 The Geothermal Method of Temperature Reconstruction

In the absence of a changing surface temperature, the temperature structure of the shallow continental crust increases with depth reflecting the outward flow of heat from Earth's hot interior to the cooler surface. The local thermal gradient is a function of both the heat flow and the thermal properties of the crust. Temporal changes in continental heat flow occur on geologic timescales of millions of years and therefore within the context of contemporary climate change can be considered constant. A direct consequence of diffusive heat transfer is that variations in surface ground temperature (SGT) propagate into the subsurface and are manifested as departures from the background temperature field. The linear combination of these two signals, background heat flow and a time-varying surface temperature, permit us to express the subsurface temperature regime as

$$T(z,t) = T_s + \frac{q}{k}z + T(z,t), \qquad (22.1)$$

where T is temperature, z is depth, t is time, T_s is the long-term surface temperature, q is the background heat flow, k is the thermal conductivity, and $T'(z,t)$ represents the transient diffusion of the SGT history into the ground. The first two terms in Equation 22.1 represent the background thermal regime, and the last term represents the transient component due to the SGT history. This last term is governed by the differential equation

$$\frac{\partial T'(z,t)}{\partial t} = \alpha \frac{\partial^2 T'(z,t)}{\partial z^2}, \qquad (22.2)$$

where α is thermal diffusivity. Equation 22.2 says that a changing SGT produces curvature in the subsurface temperature field; the link between time and depth is governed by thermal diffusivity (typically 10^{-6} m^2 s^{-1} for the subsurface). Because SGT variations give rise to curvature in the subsurface temperature field and the background thermal field is often linear (given uniform rock properties), it is possible to extract and interpret the record of SGT variations in the past. In contrast to traditional proxy methods in which the link between the measured data and temperature requires calibration, temperature-depth profiles bear a direct relationship to SGT through the simple physics of heat conduction.

In the forward problem, changes in temperature at Earth's surface propagate slowly downward such that excursions in surface temperature 10, 100, and 500 years ago produce temperature anomalies to depths of approximately 35, 110, and 250 m, respectively. The past 500 years of surface temperature change and an estimate of the average value prior to that time are therefore potentially recoverable from temperature-depth profiles in boreholes 300 m deep.

In the inverse problem of estimating a SGT history from temperature-depth measurements, the crux is to isolate the climatic signal from the background thermal regime and sources of noise. An ability to identify noise in temperature-depth profiles is particularly important because the amplitude of a thermal wave generated at the surface decreases with depth, thereby decreasing the signal-to-noise ratio. Additionally, because we are solving the backward heat equation, as we look farther into the past, noise has the potential to become exponentially amplified in the solution. These two consequences of thermal diffusion mean that extending surface temperature information farther into the past becomes more hazardous both because the signal becomes more difficult to identify and because the penalty for misidentifying noise becomes larger. Two strategies have been developed to suppress noise in geothermal climatic studies. The first is through inverse techniques that limit noise through, for example, the use of a priori constraints (i.e., the cutoff of small eigenvalues in singular value decomposition (Beltrami and Mareschal 1992; Harris and Chapman 1995) or a priori estimates of noise in the case of function space inversions (Shen and Beck 1991; Shen et al. 1992; Shen et al. 1995). The second approach, adopted in this study, is to suppress random noise by using a large ensemble of borehole temperature logs (Pollack, Shen, and Huang 1996). Here the idea is that a coherent signal becomes amplified, whereas noise, which is assumed random, is diminished.

22.3 Geothermal Data

Huang and Pollack (1998) recently compiled a global database of borehole temperature profiles for climatic reconstruction. Thermal profiles included in this data set passed a careful quality assessment, which included having: (1) a depth extent of at least 200 m, (2) a measurement interval of 20 m or less, and (3) a smooth variation of temperature with depth free of evidence of advective perturbations or permafrost zones. Over 600 boreholes passed this screening. The database includes details for each thermal profile, including borehole location (Fig. 22.1), thermal gradient, thermal conductivity, and heat flow for each site.

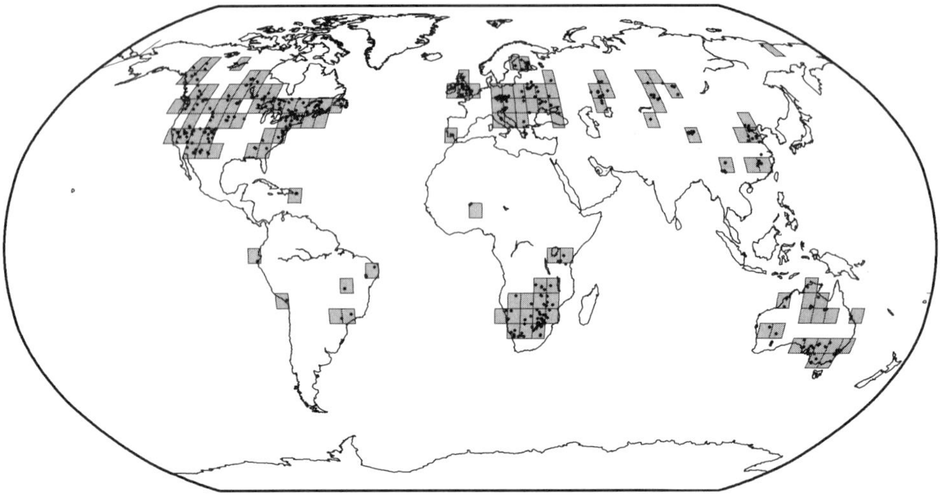

Figure 22.1. Global borehole temperature database for climatic reconstruction (Haung and Pollack 1998). Individual borehole locations are shown by dots. The mean surface air temperature time series are constructed from the shaded 5° × 5° grid cells.

While all continents (with the notable exception of Antarctica) are represented in the database, the data are not geographically uniform (Fig. 22.2). The data fall within two latitudinal bands with the majority of the data coming from the Northern Hemisphere between 30° and 60°N. In the Southern Hemisphere, the majority of the data falls between 20° and 40° S. South America is only sparsely represented.

We have processed these data to facilitate hemispheric and global reconstructions of SGT change following the procedure given in Harris and Chapman (2001). By removing the background thermal regime it is possible to isolate borehole temperature anomalies that may be attributed to climate change. This process is illustrated with a specific borehole (US-VA6-60) logged in 1964 (Fig. 22.3).

Individual temperature depth measurements are shown as closed circles and indicate the two features discussed above. The general increase in temperature with depth, representing background continental heat flow, is most prominent in the lower part of the borehole where the temperature profile is nearly linear. The upper part of the temperature profile, in contrast, displays conspicuous curvature consistent with a changing ground surface temperature condition. Positive curvature, as shown here, with anomalous temperature increasing toward the surface indicates recent and ongoing surface warming (Equation 22.2). We estimate the background thermal gradient and surface temperature intercept using the deeper, linear part of the temperature profile. A fitting interval of 160 m to the bottom (based on a trial-and-error approach) represents a trade-off between using as

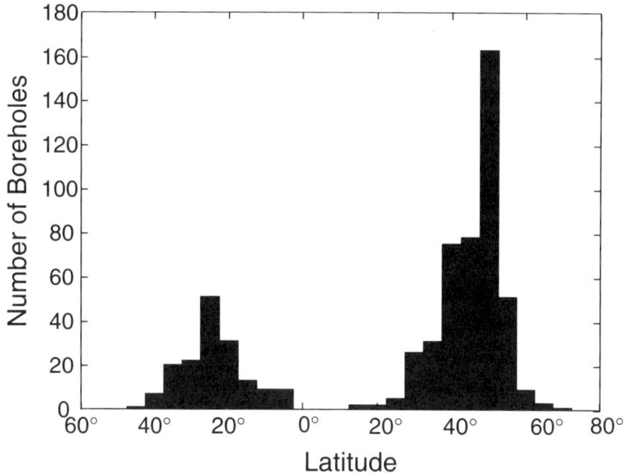

Figure 22.2. Histogram of borehole sites as a function of latitude. There are a total of 599 boreholes that satisfy the quality criteria, 73% of which are in the Northern Hemisphere.

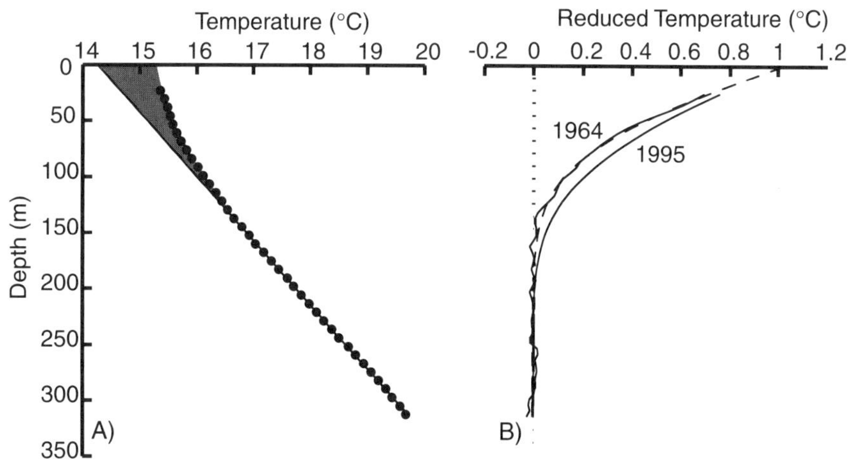

Figure 22.3. Processing borehole temperature-depth data to isolate climate-change transients. **(A)** Example log is borehole US-VA6-60 measured in 1964 showing individual temperature measurements (*dots*) and background thermal regime (*solid line*). The shaded area represents ground warming resulting from surface ground temperature variations to 1964. **(B)** Reduced temperature profile (*solid line labeled 1964*) constructed by subtracting the background thermal regime. *Dashed line:* the best fitting step function used to estimate the mean surface ground temperature. *Solid line labeled 1995:* the forward continued reduced-temperature profile assuming a constant surface temperature between 1964 and 1995.

much data as possible to obtain a robust fit, and biasing the fit to the background thermal regimes by using perturbed values in the shallow subsurface. It is important to recognize that temperatures below 160 m are still affected by SGT variations in the past and our background thermal state represented by T_s and q is not truly steady state. The solid line shows our estimate of the background thermal regime and indicates that the data below 160 m are fit very well. Removing the background thermal regime is a necessary precondition to averaging temperature profiles because it removes differences in the profiles due to deep variations in heat flow, thermal conductivity, latitude, and elevation. Curvature at or below 160 m remains unaffected by removing a linear trend. The shaded region in the upper profile indicates the anomalous temperatures or curvature that we interpret as variations in SGT (see Fig. 22.3A) and are relative to the long-term surface temperature.

To highlight this anomaly, we plot reduced temperatures, defined as the difference between the observed temperature data and the background thermal field (see Fig. 22.3B). The expanded scale produces a better display of the anomaly as well as instrumental/geologic noise inherently present in the data. The dotted vertical line corresponds to zero residual temperature. The average noise level in this profile below 160 m depth is 0.011°C.

Temperature-depth logs in this database were measured over a time span of 37 years (Fig. 22.4). Because Earth is a frequency-dependent low-pass filter, as SGT variations propagate into the subsurface, high frequencies are filtered out relative to low frequencies. As a result the SGT signal contained in a temperature log measured in 1958 has a different frequency content than one logged

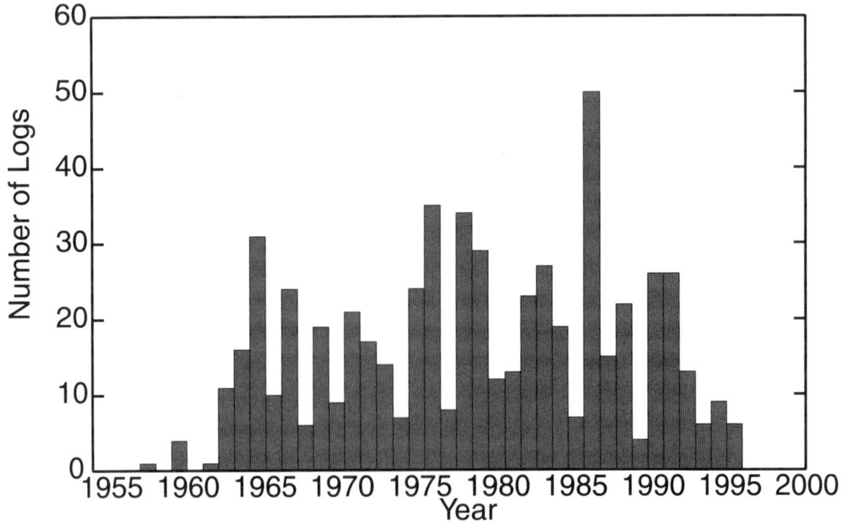

Figure 22.4. Histogram of boreholes as a function of the year of temperature logging.

in 1995, and it is not appropriate simply to average reduced temperature profiles measured at different times. Consequently, we have adjusted all of the profiles to a common datum by forward continuing all logs to a standard year of 1995.

To forward continue the temperature logs we use the following formulation (Carslaw and Jaegar 1959, p. 102):

$$\begin{aligned} \frac{\partial T}{\partial t} &= \alpha \frac{\partial^2 T}{\partial z^2} \; (0 < z < l), \\ T &= T_1(t) \text{ when } z = 0, \\ T &= T_2(t) \text{ when } z = l \\ \text{and} \quad T &= f(z) \text{ when } t = 0, \end{aligned} \qquad (22.3)$$

where $T_1(t)$ and $T_2(t)$ describe temperature as a function of time at the ends of the profiles ($z = 0$ and $z = l$). In this formulation the initial condition, $f(z)$, is simply the reduced temperature profile. We assume that the bottom of the reduced temperature profile is unaffected by the transient over the time span of the forward continuation and set $T_2(t, z = l) = 0$.

There are a number of choices for $T_1(t)$. The best choice for $T_1(t)$ is the actual SGT variation at the location of the borehole; since this is not known, however, we make the conservative choice of assuming that over the period of the forward continuation the surface temperature remains constant. Since a temperature measurement is not made at Earth's surface, we must estimate it. For each temperature profile we estimate the mean recent surface temperature, at the time the profile was logged, by fitting the reduced temperature profile in terms of a step function. The dashed line (see Fig. 22.3B) results from the best fitting step function to the reduced temperature profile and indicates a surface temperature of 1°C above the surface temperature intercept (i.e., reduced temperature of zero). This surface temperature is held constant over the time span of the forward continuation, in this case from 1964 to 1995, and the reduced temperature profile is diffused forward in time. The resulting reduced temperature profile (marked 1995) shows that the high-frequency noise is attenuated, the anomalous reduced temperatures extend a little deeper in the subsurface consistent with time moving forward, and the magnitude of the reduced temperature below the surface is somewhat greater, as dictated by Equation 22.3. All reduced temperature profiles are forward continued in this manner.

22.4 Observations

Reduced-temperature profiles for the Northern Hemisphere, Southern Hemisphere, and Global are shown in Fig. 22.5. The anomalous temperatures simply represent the response of the subsurface to recent temperature change from some long-term mean value, and the curvature in the profile is caused by variations in temperature at Earth's surface. These profiles indicate modest variability with

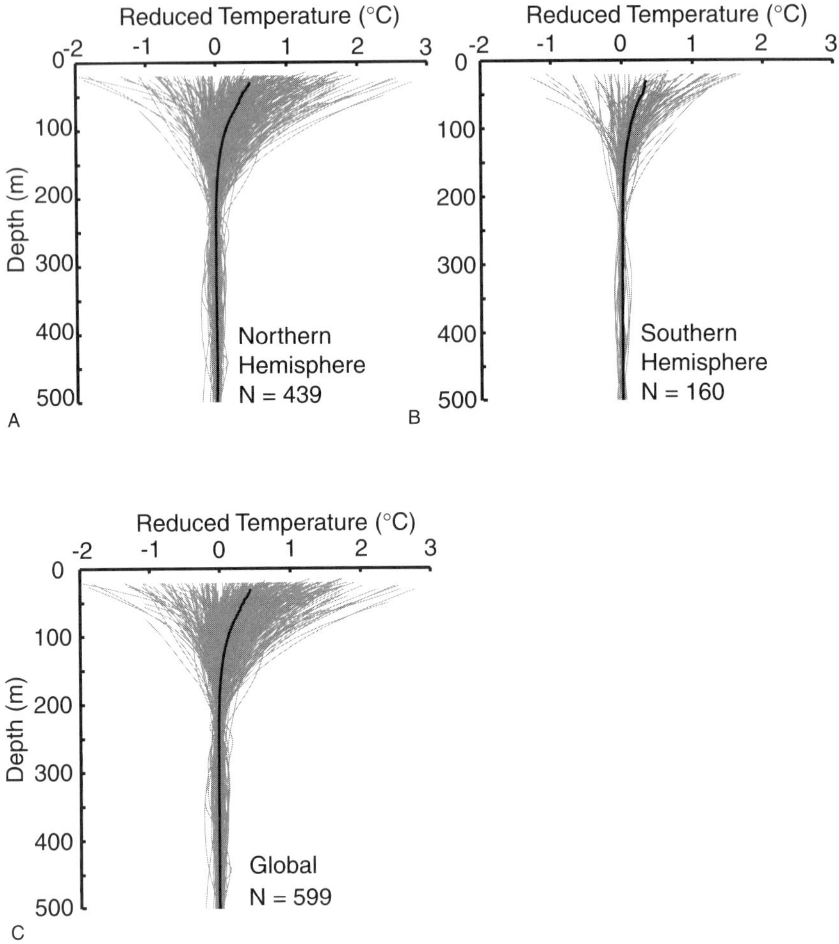

Figure 22.5. Reduced-temperature profiles and average: **(A)** Northern Hemisphere, **(B)** Southern Hemisphere, **(C)** Global. Individual temperature profiles are shown forward continued to 1995. *Bold line:* average reduced-temperature profile.

maximum reduced temperature anomalies at 30 m between $-2°$ and 3°C. The variation observed among the reduced-temperature profiles is not unexpected and presumably reflects local site conditions and spatial climatic variability. This variation illustrates the importance of using many profiles to obtain robust averages (Pollack, Shen, and Huang 1996).

Regional and global average reduced-temperature profiles can now be extracted from these data, but one must ensure that the average is not unduly affected by a concentration of sites in particular regions or continents. Each temperature profile is inversely weighted by the number of profiles falling in its

5° × 5° cell (see Fig. 22.1). The average reduced-temperature profiles are shown in bold (see Fig. 22.5). The amplitudes of the average reduced-temperature profiles at 30 m are 0.49°, 0.33°, and 0.45°C, for the Northern Hemisphere, Southern Hemisphere, and Global, respectively. In each case, anomalies extends to approximately 190 m. A thermal length calculation based on the onset time of a single step function indicates that these reduced-temperature profiles are sensitive to at least the past 300 years of SGT change.

We obtain a quantitative estimate of the amount and timing of SGT warming these profiles represent by inverting the average reduced-temperature profiles for a linear change in surface temperature characterized in terms of the amplitude of change, ΔT, and the onset time τ. The transient temperature-depth profile resulting from a linear (ramp) change in temperature can be given by (Carslaw and Jaegar 1959)

$$\Delta T(z) = 4\Delta T i^2 erfc\left(\frac{z}{\sqrt{4\alpha\tau}}\right), \qquad (22.4)$$

where i^2erfc is the second integral of the complementary error function. The best fitting ramp parameters are listed in Table 22.1, and the RMS value is calculated between the average reduced temperature profile and the model profile.

As a check on these results, we also performed a joint inversion for each set of temperature profiles (see Table 22.1). The ramp amplitudes between the two methods are similar with the 95% confidence intervals overlapping. The joint inversion however suggests earlier onset times. Chisholm and Chapman (1992) investigated parameter resolution using a grid search technique and found that the ramp amplitude was much better resolved than the onset time. The good agreement between the two inversion techniques suggests that while this is a nonlinear problem it is only mildly so. In the remainder of the discussion we use the average reduced-temperature profiles.

Table 22.1. Ground surface temperature changes obtained from borehole temperatures using a ramp model

Data Set	ΔT (95% C.I.) °C	τ (95% C.I.) Year	RMS[a] °C
Solution based on average reduced-temperature profile			
Northern Hemisphere	0.88 (0.02)	1842 (7)	0.006
Southern Hemisphere	0.76 (0.02)	1872 (10)	0.007
Global	0.82 (0.01)	1847 (6)	0.006
Solution based on joint inversion of reduced-temperature profiles			
Northern Hemisphere	0.86 (0.02)	1791 (11)	0.017
Southern Hemisphere	0.79 (0.04)	1846 (18)	0.017
Global	0.85 (0.02)	1801 (9)	0.021

[a] RMS between model and average reduced-temperature profile

22.5 Comparisons Between Borehole Temperatures and Meteorological Records

One of the most important indicators of climate change is the surface air temperature (SAT) record. Annual global mean surface temperature has increased by 0.6°C from 1856 to 1999 (Jones et al. 1999). SAT warming magnitude is slightly less than that given by the simple ramp inversion approach employed above. However, as indicated in Table 22.1, the estimated time span of warming is longer than the meteorological record of climate change. This qualitative first-order comparison suggests that ground and surface temperatures are responding to similar forcings.

What is the quantitative relationship between SAT records and reduced-temperature profiles? How much information do these signals share? Before answering these questions it is important to recognize differences between SAT records and reduced-temperature profiles (Harris and Chapman 1998). There are several fundamental differences between these signals. The first difference is that SAT temperatures are normally measured at a height of 2 m above the ground, while borehole temperatures are measured below the ground surface. The interface between the land and atmosphere is a complicated and dynamic boundary. The second difference in these two signals stems from differences in heat transfer governing these systems. Air temperatures respond rapidly to convective heat transfer, whereas subsurface rock temperatures respond more slowly by conductive heat transfer. As a result the response of these two systems is very different. A third difference concerns the different frequency content of the two signals. SATs represent discrete measurements, either at specific times of the day or the average of daily minimum and maximum temperatures. The record of annual temperatures is calculated from averaging daily and monthly temperatures. In contrast, curvature in borehole temperatures represents the continuous sensing of surface ground temperature and the integration of SGT through time. Earth is a frequency-dependent low pass filter. Thus while the frequency content of SAT records maintains fidelity with time in the past, ground temperature records do not.

A fourth difference between the signals of SAT records and reduced-temperature profiles concerns the temporal content of the signals. Air temperature measurements have been made only at a few observing sites back to 1750. Comprehensive SAT coverage is restricted primarily to this century. Because Earth's subsurface has a relatively low value of thermal diffusivity, the upper 300 m of Earth contains a thermal memory of surface temperature events over at least the past 500 years. Obviously, the time span these signals represent is quite different. To compound matters, because of the diffusive nature of the heat conduction there is not a simple correlation between time in the past and depth. As temperature variations at Earth's surface propagate downward, they spread, intermingling with other surface temperature events. Each temperature-depth point represents a running average in which the averaging window increases as we look back in time. Thus it is not straightforward to separate surface temperature events that precede SAT records from those that are coincident in time.

These factors (the different temporal and frequency content) lead to important differences between the natures of these signals, such that simply overlaying SGT solutions on SAT time series may not be appropriate. One appropriate method for comparing these data is to filter SAT records in the same way that Earth filters SGT signals. This formulation has the advantage that the heat equation is solved in the forward sense and is a more stable approach than solving the backward heat equation. In contrast to most inversion techniques employed in other geothermal climate reconstruction studies, this formulation has the advantage of very few free parameters.

We compute a transient temperature-depth profile $T_r(z)$ by expressing SAT time series as a sequence of N steps of amplitude ΔT_i and time prior to the reduced-temperature profile τ_i (Carslaw and Jaegar 1959; Beck 1977) as follows:

$$T_r(z) = \sum_{i=1}^{N} \Delta T_i \, erfc\left(\frac{Z}{\sqrt{4\alpha\tau_i}}\right) \tag{22.5}$$

where *erfc* is the complementary error function. Because the SAT data have a limited time duration and the reduced-temperature profile indicates sensitivity to the past 500 years, we parameterize the time before the start of the SAT record in terms of the simplest initial temperature distribution possible, a constant temperature termed the pre-observational mean, or, POM (Chapman, Chisholm, and Harris 1992; Chisholm and Chapman 1992; Harris and Chapman 1995). The POM represents a weighted average of surface temperatures prior to the onset of the SAT record at τ_l (Harris and Chapman 1998). Just as reduced-temperature profiles are defined relative to the background surface temperature, T_s, changes in SATs can be defined relative to a POM. The POM is determined by finding the temperature that minimizes the misfit between the reduced-temperature profile and the SAT data, and in the comparison between the reduced-temperature profile and synthetic profile, this is the only free parameter. Fig. 22.6 illustrates this technique for the Northern Hemisphere.

Departures from the 1961-to-1990 annual-mean SAT record (Jones et al. 1999) are shown in Fig. 22.6A; this time series is used as a forcing function at Earth's surface to compute a synthetic transient temperature-depth profile. For illustration purposes, three possible choices of POM, each separated by 0.5°C, are shown and quoted relative to the 1961-to-1990 mean SAT value. A POM of −0.2°C represents minimal long-term warming, a POM of −0.7°C represents a century of warming from a baseline temperature representative of temperatures at the beginning of the century, and a POM of −1.2°C is illustrative of extreme warming. By coupling these particular values of POM with the SAT data shown, and using this combination as the forcing function, it is possible to calculate transient-temperature profiles (see solid lines, Fig. 22.6B).

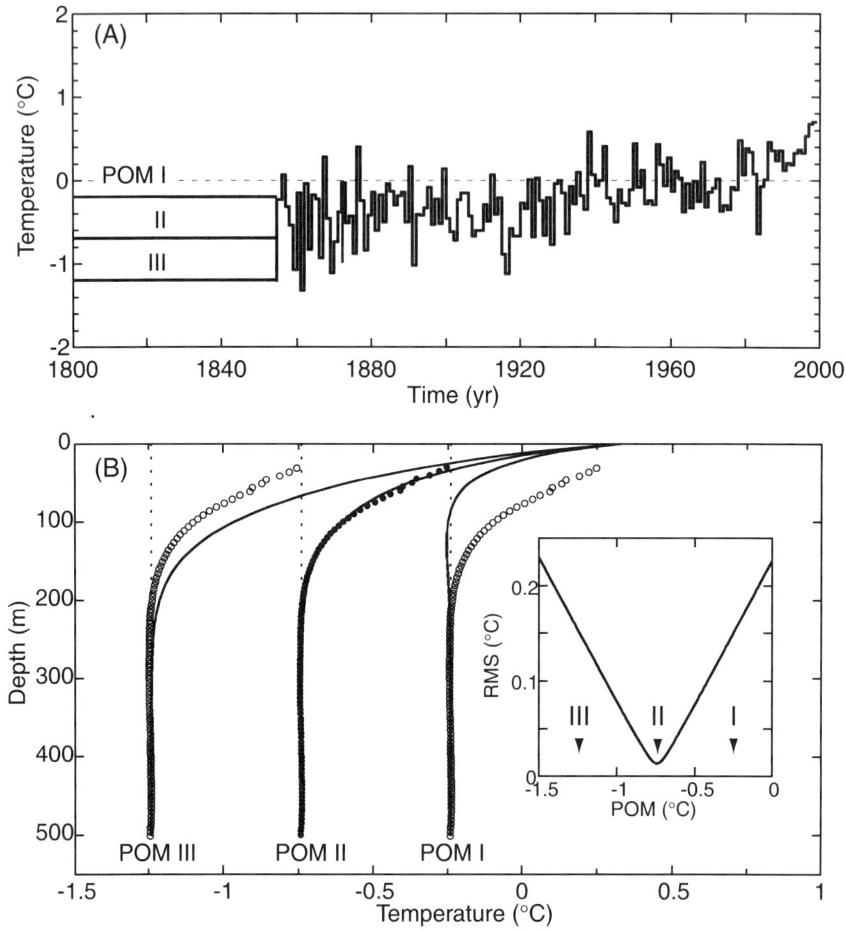

Figure 22.6. Determining the pre-observational mean (POM). **(A)** Mean annual departures of Northern Hemisphere SAT data for the period 1856 to 1995 for 5° × 5° grid cells where boreholes are present. *Horizontal bold lines (POM I, II, and III):* three different choices of POM. The SAT data coupled with a particular choice of POM are used as a forcing function to compute synthetic transient temperature-depth profiles. **(B)** Average reduced-temperature profile for the Northern Hemisphere (*circles*). Three synthetic transient temperature-depth profiles constructed using the SAT data coupled with a corresponding choice of POM (*solid lines*). Temperature profiles are plotted relative to POM (*dashed lines*). Inset shows RMS misfit as a function of the POM and illustrates the best fit for POM II.

For the minimal warming scenario (POM I), the SAT record in the early 1900s is cooler on average than the POM. This is manifested in the model profile as a slight negative anomaly in the depth range 200 to 100 m. Recent warming produces positive anomalies in the shallow subsurface. Note the attenuation of the high-frequency SAT fluctuations; air temperature variations of almost 1.5°C in the annual means are damped to a few tenths in the subsurface temperature variations. In contrast, for the most extreme warming scenario (POM III), the SAT record is nearly always warmer than this long-term average, and the model profile exhibits positive departures throughout its entire depth range. An intermediate initial value (POM II) produces an intermediate model profile.

The model curves computed from the POM-SAT model can be compared to the average reduced-temperature profile for the Northern Hemisphere (see Fig. 22.6B). Among the simulations shown, POM II clearly matches the Northern Hemisphere reduced-temperature profile (solid circles) most closely; this comparison definitely eliminates the surface temperature histories of POM I and POM III.

The inset to Fig. 22.6B shows the sensitivity of the POM. A sharp trough in the RMS misfit indicates that the POM is a robust temperature estimate. Furthermore, the excellent correspondence between the temperature profile and the model in both amplitude and depth produced with just one free parameter (i.e., the POM) supports our contention that subsurface borehole temperatures can be gainfully combined with meteorological data in climate change studies.

Fig. 22.7 shows the comparison between hemispheric and global reduced-temperature profiles and their respective SAT records. The reduced-temperature and POM-SAT models correlate well in both amplitude and depth (Fig. 22.7B). The inset panels show the RMS misfit as a function of POM values for each comparison. The RMS misfits indicate that model fits are in excellent agreement with the reduced-temperature profile having RMS misfits of 9, 12, and 12 mK for the Northern Hemisphere, Southern Hemisphere, and Global, respectively (Table 22.2). POMs relative to the 1961-to-1990 mean SAT are plotted in Fig. 22.7A, and they represent the mean temperature prior to the advent of recorded SAT data. In each case the POM is significantly below the 1961-to-1990 mean SAT.

This analysis demonstrates that the average reduced-temperature profiles computed from the SAT time series are quite sensitive to the choice of POM. To demonstrate that these models are also sensitive to the SAT records, we calculate the best fitting step function (Equation 22.5, $N = 1$) to the reduced-temperature profiles. The step model corresponds in time with SAT time series and represents the null SAT change information hypothesis (Harris and Gosnold 1999). If the step model fits the reduced-temperature profile better than the POM-SAT model then we can infer that the SAT data are not adding information but, in fact, are degrading model performance. If the comparison is not statistically different then we can infer that our model is insensitive to the SAT data. If, however, the

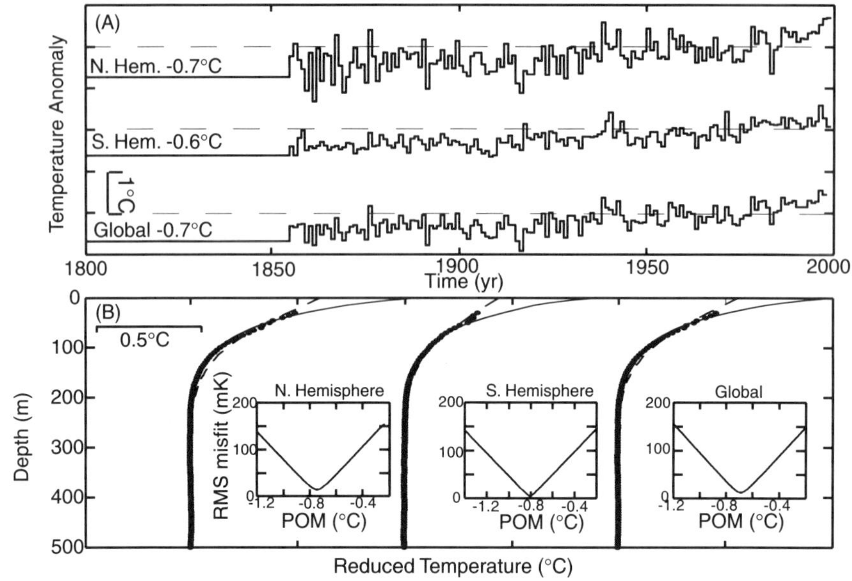

Figure 22.7. Comparison between average meteorological data and average reduced-temperature profiles. **(A)** SAT records plotted as a departure from the 1961-to-1990 mean (*dashed lines*) and best fitting POMs for Northern Hemisphere, Southern Hemisphere, and Global. Records are offset to avoid overlap. **(B)** Average reduced-temperature profiles (*circles*), synthetic temperature profiles constructed from the POM-SAT models (*solid lines*). Profiles are offset to avoid overlap. Insets show RMS misfit as a function of POM.

POM-SAT model fit is statistically better than the step function fit to the reduced-temperature profile then we can infer that these signals have common information. This conclusion would support our contention that reduced-temperature profiles and SAT time series can be combined in meaningful ways. It is important to note that each of these comparisons depends on only one free parameter: the amplitude of the step change.

We constrain the step to occur at the time of the first annual mean. Thus, we are not claiming that the POM-SAT models fit better than an arbitrary step model or some other model. In fact, because these models are not unique, there are models that fit the reduced-temperature profile better than the POM-SAT models (the two free parameter ramp models, for example), although these other models will have more free parameters. The step model merely serves as a point of reference for evaluating the POM-SAT model fit. One potential pitfall in using a step function in this manner is that the SAT record may not contain significant low-frequency trends and therefore approximate a step function. Results of these comparisons are shown in Table 22.2 and indicate that in all cases the POM-SAT models fit the reduced-temperature profiles better than do the step models. The F-ratio test, formulated as a simple ratio of variances (Bevington 1969), is

Table 22.2. Ground surface temperature changes obtained from borehole temperatures using a POM-SAT Model

Data Set	POM[a] °C	RMS[b] °C	Step °C	RMS[b] °C
Northern Hemisphere	−0.7	0.009	0.58	0.032
Southern Hemisphere	−0.6	0.012	0.44	0.032
Global	−0.7	0.012	0.54	0.028

[a] POM is determined relative to 1961-to-1990 SAT average
[b] RMS between model and average reduced-temperature profile

used to measure the improvement in fit between the POM-SAT model and the step model. In this example the ratio of variances exceeds the F statistic at the 99% confidence level, so we conclude that the POM-SAT model provides a superior fit to the data than does the step model alone.

22.6 Results of SGT and SAT Comparisons

The ramp function fits and the POM-SAT models suggest that twentieth-century warming is a real and significant departure from preindustrial times (see Table 22.1, 22.2). These models (supported by the ramp inversion) indicate that a portion of the warming comes before the advent of widespread SAT data. In the POM-SAT model this warming is represented by the step from the POM to the first annual mean SAT. A number of functions could be used to make this join, but for now we leave it simply as a step. The POMs are relative to the 1961-to-1990 mean annual SAT temperatures and when coupled with recent warming observed during the 1990s suggest an overall warming of 1.12°, 0.82°, and 0.98°C for Northern Hemisphere, Southern Hemisphere, and Global, respectively. These magnitudes are significantly larger than observed using the SAT data alone.

The observation that warming trends derived from borehole temperature profiles are often (but not always) greater than SAT warming trends has led to a number of hypothesis. Gosnold, Todhunter, and Schmidt (1997) found good agreement between warming trends derived from SAT and SGT models in South Dakota, Nebraska, and Texas, but they also found that SGT indicated significantly greater warming than did SAT data in Southern Manitoba and North Dakota. This discrepancy was explained in terms of significant seasonal ground freezing. According to this model, latent heat released during freezing of soil moisture buffers soil temperature at 0°C until all of the moisture has frozen. The larger magnitude of SGT warming rates relative to SAT trends is consistent with a presumed secular increase in soil moisture corresponding with an observed increase in precipitation during the past 50 years (Gosnold, Todhunter, and Schmidt 1997). However, spring thawing should soak up heat as the frozen water melts, producing no net gain or loss in thermal energy. Harris and Gosnold (1999), using techniques similar to those presented here, showed the discrepancy could be equally well explained in terms of an onset time of warming prior to the recorded SAT data. Skinner and Majorowicz (1999), using a larger data set than that of Gosnold, Todhunter, and Schmidt (1997), also concluded that warming trends derived from SGT models indicated significantly greater warming than did SAT data for northwestern North America. They hypothesized that land cover changes alter the radiation budget and energy balance and concluded that the discrepancy between SGT solutions and SAT data may be a step change in ground temperature associated with land-cover and land-use change. However, the comparisons between the SGT solutions and SAT data were based on linear fits over the period of overlap. Further, radiation balance models suggest that conversion of land from forest to crops in the eastern and midwestern United States produces a net cooling due to enhanced evapotranspiration and increased winter albedo (Bonan 1999). Our preferred hypothesis is that warming began before the advent of widespread SAT records. Thus the full amplitude of warming may be captured in borehole temperature profiles but obviously not captured by SAT data.

Clearly, the processes governing the energy transfer between the ground and atmosphere are complex (Geiger, Aron, and Todhunter 1995) and imperfectly understood. Studies investigating the relationship between air and ground temperatures are mostly oriented toward agriculture (De Vries 1975; Vladimirovich 1985; Sharrat et al. 1992). With the recognition that the ground may contain a record of past surface temperature change, several studies have investigated the relationship between meteorological variables and borehole temperatures within the context of climate change. Baker and Ruschy (1993) found that warming trends, based on a climatological observatory in Minnesota, between air and ground (at a depth of 12.7 m) temperatures compare favorably. Putnam and Chapman (1996) document the energy flux over a year at a coupled borehole/meteorological site in northwestern Utah. They found that ground and air temperatures track each other with time varying seasonal offsets and that heat conduction is the primary mode of heat transport in the subsurface at this location. More research is needed, however, to understand details of energy exchange across the land-air interface.

22.7 Comparison to Other Multicentury Records

Climate reconstructions from borehole estimates and proxy reconstructions of climate change both offer the opportunity to examine surface temperature variations over longer timescales than are available from SAT records. Because most of the borehole profiles and most of the proxy time series are from the Northern Hemisphere, we focus on the Northern Hemisphere. Huang, Pollack, and Shen (2000) using the same borehole data set as that presented here parameterized their solution in terms of five century-long ramps. Their solution indicates a temperature increase over the past 5 centuries of about 1°C (Fig. 22.8A). The

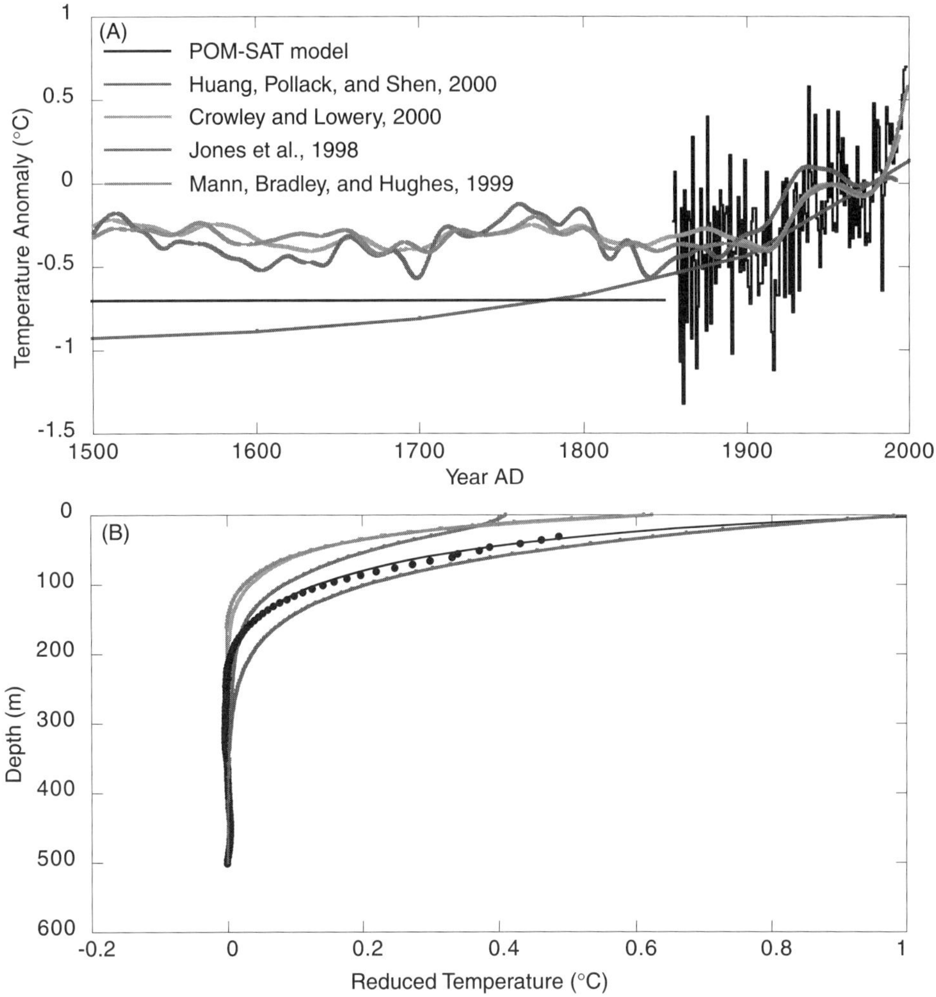

Figure 22.8. Northern Hemisphere climate reconstructions and their geothermal consequences. **(A)** Mean surface air temperatures from 1856 to 1999 (Jones et al. 2000) with the pre-observational mean temperature (POM) prior to 1856. Climatic models are plotted relative to the 1961-to-1990 mean temperature. **(B)** Comparison of Northern Hemisphere reduced-temperature profile and synthetic transients computed from proxy reconstructions. Each synthetic profile is shown relative to the 1500-to-1700 mean temperature, which was used to initialize the models.

overall magnitude of warming determined by these two methods is in excellent agreement with our results. These ramps corroborate the unusual warming of the twentieth century.

Northern Hemisphere surface temperature proxy reconstructions (Jones et al. 1998; Mann, Bradley, and Hughes 1999; Crowley and Lowery 2000) have much in common (see Fig. 22.8A). These records show cooler temperatures in the seventeenth century, warmer temperatures in the eighteenth century, cooler temperatures in the nineteenth century (the so called little ice age), and rapid warming in the twentieth century. A simple measure of warming since preindustrial time can be computed by taking the difference between the AD 1500 to −AD 1700 mean temperature and the 1961-to-1990 mean temperature. Using this measure, the proxy records indicate warming of 0.3° to 0.4°C, about half the warming captured by the borehole temperature profiles (Table 22.3).

At this point it is difficult to know the significance of this measure of warming. It may simply be due to the different geographic distribution of the boreholes and the proxy records, an unrecognized process in the heat exchange across the land-air interface, the calibration procedures in transforming the proxy record to temperature, or some combination of these factors. Jones, Briffa, and Osborn (2003) in an analysis of long-term seasonal temperature trends in Europe, China, Japan, and Korea find that century-scale warming is dominated by winter warming with summer temperatures showing little change. It is possible then that the proxies (e.g., tree growth) are responsive to summer temperatures while borehole temperatures integrate summer and winter temperatures yielding greater warming trends that are more representative of overall warming.

We note again, however, that due to the different frequency content of borehole temperature profiles and SAT or proxy reconstructions, it can be misleading to compare the reconstructions directly. A better comparison is illustrated by using the proxy temperature records as a forcing function at the surface of Earth (see Fig. 22.8B). Synthetic temperature profiles for each proxy reconstruction are calculated relative to respective AD 1500 to AD 1700 mean temperatures. The ground response to hemispheric proxy reconstructions indicates significantly less ground warming than the Northern Hemisphere reduced-temperature profile indicates. Thus the borehole temperature records suggest about twice the cli-

Table 22.3. Estimated warming magnitudes

Time Series	ΔT,°C[a]
Huang et al. (2000)	0.9
POM-SAT model	0.7
Jones et al. (1998)	0.4
Mann, Bradley, and Hughes (1999)	0.3
Crowley and Lowery (2000)	0.3

[a] ΔT = 1960-to-1991 mean temperature—1500-to-1700 mean temperature

matic sensitivity to greenhouse gases than do the proxy records of surface temperature change.

22.8 Summary

We have shown that the average reduced-temperature profiles for the Northern Hemisphere, Southern Hemisphere, and Global are sensitive to, and compare well with, their respective SAT records. Our analysis indicates 0.7 ± 0.1°C of ground warming between preindustrial time and the 1961-to-1990 interval. Meteorologic data show another 0.4°C of most recent warming. Thus the total surface warming from preindustrial times to the end of the twentieth century may be as much as 1.1°C, although it will take some time to see if the extreme warming in the 1990s is sustained. This study supports the hypothesis that borehole temperature profiles contain a valuable signal for measuring the magnitude and timing of global warming since preindustrial time. The POM-SAT approach taken here indicates that aggregates of borehole temperature profiles are sensitive to both the long-term mean surface temperature and hemispheric SAT records.

This approach allows a quantitative comparison between the signals of ground warming and proxy reconstructions of climate change, at appropriate frequencies and temporal scales; such a comparison is necessary if we are to understand and learn from the similarities and differences. These tests have assumed a constant relationship (offset) between proxy surface temperatures and surface ground temperatures. Potential candidates for a time-varying offset relationship include: (1) an imperfect or changing calibration between proxy records and air temperatures at long periods (Ellsaesser et al. 1986), (2) changing patterns of snow cover (Groisman, Karl, and Knight 1994), and (3) changing patterns of land-use and land-cover (Lewis 1998; Skinner and Majorowicz 1999). Furthermore, borehole temperature profiles respond to the continuous variation of surface ground temperature throughout the year, whereas various proxies are weighted toward a seasonal temperature, such as growing season for tree-ring proxies and snow season for ice cores.

References

Baker, D.G., and D.L. Ruschy. 1993. The recent warming in eastern Minnesota shown by ground temperatures. *Geophysical Research Letters* 20:371–74.

Beck, A.E. 1977. Climatically perturbed temperature gradients and their effect on regional and continental heat flow means. *Tectonophysics* 41:17–39.

Beltrami, H., and J.-C. Mareschal. 1992. Ground temperature histories for central and eastern Canada from geothermal measurements; Little Ice Age signature. *Geophysical Research Letters* 19:689–92.

Bevington, P.R. 1969. *Data reduction and error analysis for the physical sciences.* New York: McGraw-Hill.

Bonan, G.B. 1999. Frost followed the plow: impacts of deforestation on the climate of the United States. *Ecological Applications* 9:1305–15.

Carslaw, H.S., and J.C. Jaegar. 1959. *Conduction of heat in solids.* Oxford: Oxford University Press.

Chapman, D.S., T.J. Chisholm, and R.N. Harris. 1992. Combining borehole temperature and meteorologic data to constrain past climate change. *Global and Planetary Change* 6:269–81.

Chisholm, T.J., and D.S. Chapman. 1992. Climate change inferred from analysis of borehole temperatures; an example from western Utah. *Journal of Geophysical Research* 97:14,155–14,175.

Crowley, T.J., and T.S. Lowery. 2000. How warm was the Medieval Warm Period? *Ambio* 29:51–54.

De Vries, D.A. 1975. Heat transfer in soils. In *Heat and mass transfer in the biosphere, vol. I: Transfer processes in plant environments,* ed. D.A. De Vries and N.H. Afgan. New York: Wiley.

Ellsaesser, H.W., M.C. MacCraken, J.J. Walton, and S.L. Grotch. 1986. Global climatic trends as revealed by the recorded data. *Reviews of Geophysics* 24:745–92.

Geiger, R., R.H. Aron, and P.E. Todhunter. 1995. *The climate near the ground.* Wiesbaden: Vieweg.

Gosnold, W.D., P.E. Todhunter, and W. Schmidt. 1997. The borehole temperature record of climate warming in the mid-continent of North America. *Global and Planetary Change* 15:33–45.

Groisman, P.Y., T.R. Karl, and R.W. Knight. 1994. Observed impact of snow cover on the heat balance and the rise of continental spring temperatures. *Science* 263:198–200.

Harris, R.N., and D.S. Chapman. 1995. Climate change on the Colorado plateau of eastern Utah inferred from borehole temperatures. *Journal of Geophysical Research* 100:6367–81.

———. 1997. Borehole temperature and a baseline for 20th-century global warming estimates. *Science* 275:1618–21.

———. 1998. Geothermics and climate change; 2, Joint analysis of borehole temperature and meteorological data. *Journal of Geophysical Research* 103:7371–83.

———. 2001. Mid-latitude (30°–60°N) climatic warming inferred by combining borehole temperatures with surface air temperatures. *Geophysical Research Letters* 28:747–50.

Harris, R.N., and W.D. Gosnold. 1999. Comparisons of borehole temperature-depth profiles and surface air temperatures in the northern plains of the USA. *Geophysical Journal International* 138:541–48.

Huang, S., and H.N. Pollack. 1998. Global borehole temperature database for climate reconstruction, Contribution Series 1998-044 International Geosphere-Biosphere Program Past Global Changes/World Data Center A for Paleoclimatology data, National Oceanographic and Atmospheric Administration/National Geophysical Data Center, Paleoclimatology Program, Boulder, Colo.

Huang, S., H.N. Pollack, and P.-Y. Shen. 2000. Temperature trends over the past five centuries reconstructed from borehole temperatures. *Nature* 403:756–58.

Jones, P.D., K.R. Briffa, T.P. Barnett, and S.F.B. Tett. 1998. High-resolution palaeoclimatic records for the last millennium; interpretation, integration and comparison with general circulation model control-run temperatures. *The Holocene* 8:455–71.

Jones, P.D., K.R. Briffa, and T.J. Osborn. 2003. Changes in the seasonal cycle: Implications for paleoclimatology? *Geophysical Research Abstracts* (European Geophysical Society) 5:01100.

Jones, P.D., M. New, D.E. Parker, S. Martin, and I.G. Rigor. 1999. Surface air temperature and its changes over the past 150 years. *Reviews of Geophysics* 37:173–99.

Lachenbruch, A.H., and B.V. Marshall. 1986. Changing climate: Geothermal evidence from permafrost in the Alaskan Arctic. *Science* 234:689–696.

Lewis, T.J. 1998. The effect of deforestation on ground surface temperature. *Global and Planetary Change* 18:1–13.

Mann, M.E., R.S. Bradley, and M.K. Hughes. 1998. Global-scale temperature patterns and climate forcing over the past six centuries. *Nature* 392:779–87.

———. 1999. Northern Hemisphere temperatures during the past millennium: inferences, uncertainties, and limitations. *Geophysical Research Letters* 26:759–62.

Pollack, H.N., and D.S. Chapman. 1993. Underground records of changing climate. *Scientific American* 268:44–50.

Pollack, H.N., P.Y. Shen, and S. Huang. 1996. Inference of ground surface temperature history from subsurface temperature data: Interpreting ensembles of borehole logs. *Pure and Applied Geophysics* 147:537–50.

Putnam, S.N., and D.S. Chapman. 1996. A geothermal climate change observatory: First year results from Emigrant Pass in Northwest Utah. *Journal of Geophysical Research* 101:21,877–21,890.

Sharrat, B.S., D.G. Baker, D.B. Wall, R.H. Skaggs, and D.L. Ruschy. 1992. Snow depth requirements for near steady-state soil temperatures. *Agriculture and Forest Meteorology* 57:243–51.

Shen, P.Y., and A.E. Beck. 1991. Least squares inversion of borehole temperature measurements in functional space. *Journal of Geophysical Research* 96:19,965–19,979.

Shen, P.Y., H.N. Pollack, S. Huang, and K. Wang. 1995. Effects of subsurface heterogeneity on the inference of climate change from borehole temperature data; model studies and field examples from Canada. *Journal of Geophysical Research* 100:6383–6396.

Shen, P.Y., K. Wang, H. Beltrami, and J.C. Mareschal. 1992. A comparative study of inverse methods for estimating climatic history from borehole temperature data. *Global and Planetary Change* 6:113–27.

Skinner, W.R., and J.A. Majorowicz. 1999. Regional climatic warming and associated twentieth century land-cover changes in north-western North America. *Climatic Research* 12:39–52.

Vladimirovich, N.S. 1985. *Heat and mass transfer in the plant-soil-air system.* Rotterdam: Balkema.

Index

D

T

Ecological Studies

Volumes published since 1998

Volume 134
Oxidant Air Pollution Impacts in the Montane Forests of Southern California: A Case Study of the San Bernardino Mountains (1999)
P.R. Miller and J.R. McBride (Eds.)

Volume 135
Predation in Vertebrate Communities: The Bialowieza Primeval Forest as a Case Study (1998)
B. Jedrzejewska and W. Jedrzejewski

Volume 136
Landscape Disturbance and Biodiversity in Mediterranean-Type Ecosystems (1998)
P.W. Rundel, G. Montenegro, and F.M. Jaksic (Eds.)

Volume 137
Ecology of Mediterranean Evergreen Oak Forests (1999)
F. Roda, J. Retana, C.A. Gracia, and J. Bellot (Eds.)

Volume 138
Fire, Climate Change, and Carbon Cycling in the Boreal Forest (2000)
E.S. Kasischke and B.J. Stocks (Eds.)

Volume 139
Responses of Northern U.S. Forests to Environmental Change (2000)
R.A. Mickler, R.A. Birdsey, and J. Hom (Eds.)

Volume 140
Rainforest Ecosystems of East Kalimantan: El Nino, Drought, Fire and Human Impacts (2000)
E. Guhardja, et al. (Eds.)

Volume 141
Activity Patterns in Small Mammals: An Ecological Approach (2000)
S. Halle and N.-C. Stenseth (Eds.)

Volume 142
Carbon and Nitrogen Cycling in European Forest Ecosystems (2000)
E.-D. Schulze (Ed.)

Volume 143
Global Climate Change and Human Impacts on Forest Ecosystems: Postglacial Development, Present Situation and Future Trends in Central Europe (2001)
P. Joachim and U. Bernhard

Volume 144
Coastal Marine Ecosystems of Latin America (2001)
U. Seeliger and B. Kjerfve (Eds.)

Volume 145
Ecology and Evolution of the Freshwater Mussels Unionoida (2001)
B. Gerhard and K. Wächtler (Eds.)

Volume 146
Inselbergs: Biotic Diversity of Isolated Rock Outcrops in Tropical and Temperate Regions (2001)
S. Porembski and W. Barthlott (Eds.)

Volume 147
Ecosystem Approaches to Landscape Management in Central Europe (2001)
J.D. Tenhunen, R. Lenz, and R. Hantschel (Eds.)

Volume 148
A Systems Analysis of the Baltic Sea (2001)
F.V. Wulff, L.A. Rahm, and P. Larsson (Eds.)

Volume 149
Banded Vegetation Patterning in Arid and Semiarid Environments: Ecological Processes and Consequences for Management (2001)
D.J. Tongway, C. Valentin, and J. Seghieri (Eds.)

Volume 150
Biological Soil Crusts: Structure, Function and Management (2001)
J. Belnap and O.L. Lange (Eds.)

Volume 151
Ecological Comparisons of Sedimentary Shores (2001)
K. Reise (Ed.)

Volume 152
Future Scenarios of Global Biodiversity (2001) F.S. Chapin, III, O.E. Sala, and E. Huber-Sannwald (Eds.)

Volume 153
UV Radiation and Arctic Ecosystems (2002)
D.O. Hessen (Eds.)

Volume 154
Geoecology of Antartic Ice-Free Costal (2002)
L. Beyer and M. Bolter (Eds.)

Volume 155
Conserving Biodiversity in East African Forests: A Study of the Eastern Arc Mountains (2002)
W.D. Newmark (Eds.)

Volume 156
Urban Air Pollution and Forests: Resources at Risk in the Mexico City Air Basin (2002)
M.E. Fenn, L.I. de Bauer, and
T. Hernández-Tejeda (Eds.)

Volume 157
Mycorrhizal Ecology (2002, 2003)
M.G.A van der Heijden and I.R. Sanders

Volume 158
Diversity and Interaction in a Temperate Forest Community: Ogawa Forest Reserve of Japan (2002)
T. Nakashizuka and Y. Matsumoto

Volume 159
Big-Leaf Mahogany: Genetics, Ecology, and Management (2003)
A.E. Lugo, J.C. Figueroa Colón, and
M. Alayon (Eds.)

Volume 160
Fire and Climatic Change in Temperate Ecosystems of the Western Americas (2003)
T.T. Veblen, W.L. Baker, G. Montenegro, and T.W. Swetnam (Eds.)

Volume 161
Comptition and Coexistence (2003)
U. Sommer and B. Worm (Eds.)

Volume 162
How Landscapes Change (2003)
G.A. Bradshaw and P.A. Marquet (Eds.)

Volume 163
Fluxes of Carbon, Water and Energy of European Forests (2003)
R. Valentini (Eds.)

Volume 164
Herbivory of Leaf-Cutting Ants (2003)
R. Wirth, H. Herz, R.J. Ryel, W. Beyschlag, And B. Hölldobler (Eds.)

Volume 165
Population Viability in Plants (2003)
C.A. Brigham and M.W. Schwartz (Eds.)

Volume 166
North American Temperate Deciduous Forest Response to Changing Precipitation Regimes (2003)
P. Hanson and S.D. Wullschleger (Eds.)

Volume 167
Alpine Biodiversity in Europe (2003)
L. Nagy, G. Grabherr, C. Körner, and
D.B.A. Thompson (Eds.)

Volume 168
Root Ecology (2003)
H. de Kroon and E.J.W Visser (Eds.)

Volume 169
Fire in Tropical Savannas: The Kapalga Experiment (2003) A.N. Andersen, G.D Cook, and
R.J. Williams (Eds.)

Volume 170
Molecular Ecotoxicology of Plants (2004)
H. Sandermann (Eds.)

Volume 171
Coastal Dunes: Ecology and Conservation (2004)
M.L. Martínez and N. Psuty (Eds.)

Volume 172
Biogeochemistry of Forested Catchments in a Changing Environment: A German Case Study (2004)
E. Matzner (Eds.)

Volume 173
Insects and Ecosystem Function (2004)
W.W. Weisser and E. Siemann (Eds.)

Volume 174
Pollination Ecology and the Rain Forest: Sarawak Studies (2005)
D. Roubik, S. Sakai, and A.A. Hamid (Eds.)

Volume 175
Antartic Ecosystems: Enviromental Contamination, Climate Change, and Human Impact (2005)
R. Bargagli (Eds.)

Volume 176
Forest Diversity and Function: Temperate and Boreal Systems (2005)
M. Scherer-Lorenzen, C. Körner, and
E.-D. Schulze (Eds.)

Volume 177
A History of Atmospheric CO_2 and Its Effects on Plants, Animals, and Ecosystems (2005)
J.R. Ehleringer, T.E. Cerling, and M.D. Dearing (Eds.)

Volume 178
Photosynthetic Adaptation: Chloroplast to Landscape (2004)
W.K. Smith, T.C. Voqelmann, and Christa Critchley (Eds.)